ANNUAL REVIEW OF NUCLEAR AND PARTICLE SCIENCE

EDITORIAL COMMITTEE (1984)

GORDON A. BAYM
G. F. BERTSCH
GERALD T. GARVEY
HARRY E. GOVE
J. D. JACKSON
LEE G. PONDROM
ROY F. SCHWITTERS

Responsible for the organization of Volume 34
(Editorial Committee, 1982)

CHARLES BALTAY
GORDON A. BAYM
JAMES D. BJORKEN
GERALD T. GARVEY
HARRY E. GOVE
ERNEST M. HENLEY
J. D. JACKSON
ROY F. SCHWITTERS
GEORGE BRANDENBURG (Guest)
LEE GRODZINS (Guest)

Production Editor   MARGOT PLATT

ANNUAL REVIEW OF NUCLEAR AND PARTICLE SCIENCE

Volume 34, 1984

J. D. JACKSON, *Editor*
University of California, Berkeley

HARRY E. GOVE, *Associate Editor*
University of Rochester

ROY F. SCHWITTERS, *Associate Editor*
Harvard University

ANNUAL REVIEWS INC.　　4139 EL CAMINO WAY　　PALO ALTO, CALIFORNIA 94306 USA

ANNUAL REVIEWS INC.
Palo Alto, California, USA

COPYRIGHT © 1984 BY ANNUAL REVIEWS INC., PALO ALTO, CALIFORNIA, USA. ALL RIGHTS RESERVED. The appearance of the code at the bottom of the first page of an article in this serial indicates the copyright owner's consent that copies of the article may be made for personal or internal use, or for the personal or internal use of specific clients. This consent is given on the condition, however, that the copier pay the stated per-copy fee of $2.00 per article through the Copyright Clearance Center, Inc. (21 Congress Street, Salem, MA 01970) for copying beyond that permitted by Sections 107 or 108 of the US Copyright Law. The per-copy fee of $2.00 per article also applies to the copying, under the stated conditions, of articles published in any *Annual Review* series before January 1, 1978. Individual readers, and nonprofit libraries acting for them, are permitted to make a single copy of an article without charge for use in research or teaching. This consent does not extend to other kinds of copying, such as copying for general distribution, for advertising or promotional purposes, for creating new collective works, or for resale. For such uses, written permission is required. Write to Permissions Dept., Annual Reviews Inc., 4139 El Camino Way, Palo Alto, CA 94306 USA.

International Standard Serial Number: 0163-8998
International Standard Book Number: 0-8243-1534-0
Library of Congress Catalog Card Number: 53-995

Annual Review and publication titles are registered trademarks of Annual Reviews Inc.

Annual Reviews Inc. and the Editors of its publications assume no responsibility for the statements expressed by the contributors to this *Review*.

TYPESET BY AUP TYPESETTERS (GLASGOW) LTD., SCOTLAND
PRINTED AND BOUND IN THE UNITED STATES OF AMERICA

PREFACE

The contents of the present volume reflect some of the shifts of emphasis that are occurring among the fields of astrophysics, nuclear physics, and elementary particle physics. In the past five or ten years, particle physics has made great strides in the unification of the fundamental forces of nature. The discovery at CERN in 1983 of the W and Z vector bosons of the weak interactions, with roughly the masses expected by the unified electroweak theory, has given theorists faith that their speculations on still greater unifications may not be idle. Grand unified theories that unite the leptons and the quarks almost always have two consequences. One is the instability of the nucleon against decay into leptons and mesons (failure of conservation of baryon number) and the other is the existence of very massive gauge monopoles (generalizations of Dirac's magnetic monopole, possessing magnetic charge, but also color charges of the strong interactions). Evidently nucleons are quite stable, but the search for their instability is a necessity. The chapter by Perkins describes authoritatively the situation up to the end of 1983. Gauge monopoles and how they emerge from the theories are the subject of Preskill's clear, if somewhat mathematical, review, which also touches on the astrophysical limits on their abundance.

The great advances in high-energy physics are cause for satisfaction, but expose further perplexities and unanswered questions that need higher energies for their resolution. During the past year, in the US and also in Western Europe, there have been discussions and planning for a next big step in accelerator-colliders, even as a 1-TeV superconducting accelerator begins operation at Fermilab and a large electron-positron ring (LEP) begins to be constructed at CERN. For proton-proton or antiproton-proton collisions, such a new facility would exploit the technology of superconducting magnet systems, the subject of the chapter by Palmer & Tollestrup. The fundamental physical principles of particle accelerators are discussed by Lawson & Tigner. As high-energy-physics experimentation moves more and more toward collider facilities, questions of vertex location and precise charged-particle tracking become crucial. The chapter on high-resolution electronic particle detectors by Charpak & Sauli is therefore a welcome addition.

The quest for unification, with its consideration of ultra-high energies, has led inevitably to consideration of the earliest moments in "big bang" cosmology, where such energies presumably existed. Much fruitful interplay now occurs among astrophysicists and particle theorists over the consequences of unification on cosmology and, conversely, on limits set on the theories by the evolution of the universe to its present, observed state. Neutrinos, their interactions and possible masses, the subject of the chapter by Boehm & Vogel, have an important bearing on cosmology and the matter density of the universe in addition to their inherent interest in the submicroscopic world. Although not primarily concerned with the highest energies and the earliest times of the universe, Truran's subject, nucleosynthesis, is one with particle physics and cosmology—the elemental and isotopic abundances are governed, admittedly in a complicated way, by the fundamental properties and interactions of particles.

In nuclear physics changes have been occurring, too. Nuclei are no longer viewed as groupings of neutrons and protons. Other degrees of freedom are considered; pions and nucleonic isobars are excited within nuclei in collisions. The dispersive (virtual) effects of such excitations can cause modifications of basic properties inside nuclei. Exchange currents and the quenching of g_A are two such effects. The explanation of these and other phenomena is the subject of the chapter on pion interactions within nuclei by Rho. A related chapter by Friar, Gibson & Payne is on the three-nucleon system.

At still higher energies of collision between nuclei, it is appropriate to discuss the phenomena in terms of quarks, not nucleons, although traditional approaches (hydrodynamic flow, shock waves, nucleonic cascade) are also used. In the past year or two, part of the nuclear physics community, and the high-energy community, too, has become excited over the possibility of creating a phase transition within nuclear matter, from a phase of largely nucleons to one of almost-free quarks and gluons, the building blocks of quantum chromodynamics. Thus there is much discussion of future relativistic heavy-ion colliders. Nagamiya, Randrup & Symons address the question of what has been learned so far about the collisions of relativistic heavy ions, largely at the Bevalac, and what the future might hold.

The remaining chapters are on important topics in nuclear physics, particle physics, and applications. Moretto & Wozniak discuss the role of rotational degrees of freedom in heavy-ion collisions at low and moderate energies. Gaillard & Sauvage review the subject of hyperon beta decays, with prominence given to the impressive recent work at CERN. The analysis of materials via nuclear reaction techniques is the subject of the review by Amsel & Lanford.

The present volume is the seventh under the current editorship. Looking at the scope and quality of the chapters herein, we consider ourselves fortunate to have colleagues who are both expert and responsible, who share with us the perception that serious reviews are more than ever necessary. We thank them for their efforts. May you, the reader, benefit and enjoy.

J. D. JACKSON, EDITOR

CONTENTS

PROTON DECAY EXPERIMENTS, *D. H. Perkins*	1
NUCLEOSYNTHESIS, *James W. Truran*	53
THE PHYSICS OF PARTICLE ACCELERATORS, *J. D. Lawson and M. Tigner*	99
LOW-ENERGY NEUTRINO PHYSICS AND NEUTRINO MASS, *F. Boehm and P. Vogel*	125
NUCLEAR COLLISIONS AT HIGH ENERGIES, *S. Nagamiya, J. Randrup, and T. J. M. Symons*	155
THE ROLE OF ROTATIONAL DEGREES OF FREEDOM IN HEAVY-ION COLLISIONS, *L. G. Moretto and G. J. Wozniak*	189
SUPERCONDUCTING MAGNET TECHNOLOGY FOR ACCELERATORS, *R. Palmer and A. V. Tollestrup*	247
HIGH-RESOLUTION ELECTRONIC PARTICLE DETECTORS, *G. Charpak and F. Sauli*	285
HYPERON BETA DECAYS, *Jean-Marc Gaillard and Gilles Sauvage*	351
RECENT PROGRESS IN UNDERSTANDING TRINUCLEON PROPERTIES, *J. L. Friar, B. F. Gibson, and G. L. Payne*	403
NUCLEAR REACTION TECHNIQUES IN MATERIALS ANALYSIS, *G. Amsel and W. A. Lanford*	435
MAGNETIC MONOPOLES, *John Preskill*	461
PION INTERACTIONS WITHIN NUCLEI, *Mannque Rho*	531

INDEXES
  Cumulative Index of Contributing Authors, Volumes 24–34   583
  Cumulative Index of Chapter Titles, Volumes 24–34   585

ANNUAL REVIEWS INC. is a nonprofit scientific publisher established to promote the advancement of the sciences. Beginning in 1932 with the *Annual Review of Biochemistry*, the Company has pursued as its principal function the publication of high quality, reasonably priced *Annual Review* volumes. The volumes are organized by Editors and Editorial Committees who invite qualified authors to contribute critical articles reviewing significant developments within each major discipline. The Editor-in-Chief invites those interested in serving as future Editorial Committee members to communicate directly with him. Annual Reviews Inc. is administered by a Board of Directors, whose members serve without compensation.

1984 Board of Directors, Annual Reviews Inc.

Dr. J. Murray Luck, Founder and Director Emeritus of Annual Reviews Inc.
  Professor Emeritus of Chemistry, Stanford University
Dr. Joshua Lederberg, President of Annual Reviews Inc.
  President, The Rockefeller University
Dr. James E. Howell, Vice President of Annual Reviews Inc.
  Professor of Economics, Stanford University
Dr. William O. Baker, *Retired Chairman of the Board, Bell Laboratories*
Dr. Winslow R. Briggs, *Director, Carnegie Institution of Washington, Stanford*
Dr. Sidney, D. Drell, *Deputy Director, Stanford Linear Accelerator Center*
Dr. Eugene Garfield, *President, Institute for Scientific Information*
Dr. Conyers Herring, *Professor of Applied Physics, Stanford University*
Mr. William Kaufmann, *President, William Kaufmann, Inc.*
Dr. D. E. Koshland, Jr., *Professor of Biochemistry, University of California, Berkeley*
Dr. Gardner Lindzey, *Director, Center for Advanced Study in the Behavioral Sciences, Stanford*
Dr. William D. McElroy, *Professor of Biology, University of California, San Diego*
Dr. William F. Miller, *President, SRI International*
Dr. Esmond E. Snell, *Professor of Microbiology and Chemistry, University of Texas, Austin*
Dr. Harriet A. Zuckerman, *Professor of Sociology, Columbia University*

Management of Annual Reviews Inc.
  John S. McNeil, Publisher and Secretary-Treasurer
  William Kaufmann, Editor-in-Chief
  Mickey G. Hamilton, Promotion Manager
  Donald S. Svedeman, Business Manager
  Richard L. Burke, Production Manager

ANNUAL REVIEWS OF
Anthropology
Astronomy and Astrophysics
Biochemistry
Biophysics and Bioengineering
Cell Biology
Earth and Planetary Sciences
Ecology and Systematics
Energy
Entomology
Fluid Mechanics
Genetics
Immunology
Materials Science
Medicine
Microbiology
Neuroscience
Nuclear and Particle Science
Nutrition
Pharmacology and Toxicology
Physical Chemistry
Physiology
Phytopathology
Plant Physiology
Psychology
Public Health
Sociology

SPECIAL PUBLICATIONS

Annual Reviews Reprints:
  Cell Membranes, 1975–1977
  Cell Membranes, 1978–1980
  Immunology, 1977–1979

Excitement and Fascination of Science, Vols. 1 and 2

History of Entomology

Intelligence and Affectivity, by Jean Piaget

Telescopes for the 1980s

A detachable order form/envelope is bound into the back of this volume.

SOME RELATED ARTICLES IN OTHER *ANNUAL REVIEWS*

From the *Annual Review of Astronomy and Astrophysics,* Volume 22 (1984):

An Astronomical Life, Jesse L. Greenstein
Alternatives to the Big Bang, G. F. R. Ellis
The Origin of Ultra-High-Energy Cosmic Rays, A. M. Hillas
Neutron Stars in Interacting Binary Systems, Paul C. Joss and Saul A. Rappaport

From the *Annual Review of Earth and Planetary Sciences,* Volume 12 (1984):

Applications of Accelerator Mass Spectrometry, Louis Brown
Cooling Histories from $^{40}Ar/^{39}Ar$ Age Spectra: Implications for Precambrian Plate Tectonics, Derek York

From the *Annual Review of Energy,* Volume 9 (1984):

Progress and Directions in Magnetic Fusion Energy, K. I. Thomassen

From the *Annual Review of Materials Science,* Volume 14 (1984):

Tailoring of Piezoelectric Ceramics, W. Heywang and H. Thomann
Brittle Fracture and Toughening Mechanisms in Ceramics, S. M. Wiederhorn

From the *Annual Review of Physical Chemistry,* Volume 35 (1984):

When Polymer Science Looked Easy, Walter H. Stockmayer and Bruno H. Zimm
Human Effects on the Global Atmosphere, Harold S. Johnston
Quantum Ergodicity and Spectral Chaos, E. B. Stechel and E. J. Heller

PROTON DECAY EXPERIMENTS

D. H. Perkins

CERN, European Organization for Nuclear Research, Geneva, Switzerland; and Department of Nuclear Physics, University of Oxford, England

CONTENTS

1. INTRODUCTION 2
   1.1 Summary and Overview 2
   1.2 History of Baryon Conservation 3
2. PROBLEMS FOR BARYON CONSERVATION 5
   2.1 Limits on Long-Range Field Coupled to Baryon Number 6
   2.2 CP Violation in the Early Universe 9
3. THEORETICAL PREDICTIONS ON PROTON DECAY 11
   3.1 Proton Decay in SU(5) Version of GUTs 11
   3.2 Proton Decay in Supersymmetric GUTs 14
   3.3 Consequences for Proton Decay Experiments 15
4. EARLY RESULTS ON NUCLEON STABILITY 15
   4.1 Decay-Mode Independent Methods (Geochemical and Radiochemical) 15
   4.2 Summary of Early Searches for Nucleon Decay by Direct Methods 16
5. BACKGROUND IN PROTON DECAY EXPERIMENTS 18
   5.1 Atmospheric Neutrino Fluxes 19
   5.2 Results of Flux Calculations 22
   5.3 Latitude Dependence, Up-Down Ratios, and Solar Modulation 24
   5.4 Event Rates 26
   5.5 Topology of Neutrino Background 26
   5.6 Neutron Background 29
6. PROTON DECAY EXPERIMENTS—RESULTS 30
   6.1 Water Čerenkov Detectors 30
   6.2 The IMB Experiment 32
   6.3 The HPW Experiment 38
   6.4 The Kamiokande Experiment 38
   6.5 Tracking Calorimeter Detectors 39
   6.6 The Soudan I Detector 40
   6.7 The KGF Calorimeter 40
   6.8 The NUSEX Calorimeter 43
   6.9 The Frejus Experiment 46
7. LIMITS ON MONOPOLE CATALYSIS OF PROTON DECAY 46
8. COMPARISON OF RESULTS, FUTURE POSSIBILITIES, AND CONCLUSIONS ... 48
   8.1 Present Results 48
   8.2 Future Experiments 49
   8.3 Conclusions 50

1. INTRODUCTION

1.1 Summary and Overview

At the time of writing this review, there is neither convincing evidence for believing that protons (and bound neutrons in nuclei) undergo decay, nor any good reasons for believing that they can live forever. The situation is a dramatic one. Some of the limits set to the lifetime for particular decay modes are reaching perilously close to the limit of experimental feasibility. Nevertheless, there are circumstantial grounds, based on the absence of any detectable long-range field coupled to the baryon number and on the observed matter-antimatter asymmetry of the Universe, for believing that protons cannot be completely stable. The grand unified theories of the fundamental interactions (GUTs) predict that protons will indeed decay, at a very small but quite possibly measurable rate, determined by the mass M_X of the postulated gauge bosons of the grand unifying symmetry.

One particular form of GUT, incorporating minimal SU(5) as the gauge group, with $M_X \approx 10^{15}$ GeV, makes clear predictions for the lifetime in terms of virtual X-boson exchange, and indicates that the mode $p \to e^+ \pi^0$ should dominate. The postulated decay of a proton, containing u and d quarks, to a positron and nonstrange meson results from putting together the lightest quarks (u, d) and lightest leptons (e, v_e) into an SU(5) multiplet. The present experimental limit for $\tau(p \to e^+ \pi^0) > 1.5 \times 10^{32}$ yr is, however, in clear conflict with the SU(5) prediction of $\tau(p \to e^+ \pi^0) < 10^{31}$ yr. Other forms of GUT make less definite lifetime predictions but emphasize other decay modes. Some incorporate supersymmetry in which bosons like the Higgs (H) have supersymmetric fermion partners ($\tilde{H}$). They suggest that protons decay via fermion ($\tilde{H}, \tilde{W}$) exchanges, and, because of the symmetries involved, heavier quarks and leptons should be frequent decay products, for example $p \to \bar{v}_\tau K^+$ or $\mu^+ K^0$. Since the only reasonably accurate prediction, that of SU(5), seems excluded, the question of baryon stability, lifetime, and decay modes has become an almost purely empirical matter.

A few major experiments, carried out deep underground to reduce cosmic-ray background, have now reported preliminary results. Apart from excluding $p \to e^+ \pi^0$ at the 10^{32}-yr level, they have found odd events that are apparently difficult to understand in terms of background and even look tantalizingly like examples of nucleon decay in modes consistent with supersymmetric grand unified theories (SUSYGUTs). The mention of lifetimes of 10^{32} yr implies that the detectors must have sensitive masses of 100s to 1000s of tons, i.e. containing 10^{32}–10^{33} protons, so as to have any chance of finding a signal in a reasonable running time. It is, however, very difficult to instrument a 1000-ton detector with high resolution, and in fact designs have to be a compromise among size, precision, and cost. This means that proton decay events will not be individually unique, but must be distinguished on a statistical basis against the dominant background, that due to interactions of atmospheric neutrinos. The relative rates—baryon decay and neutrino interactions—are equal for a lifetime of order 10^{31} yr. In trying to push lifetime limits to 10^{32} or even 10^{33} yr or, one hopes, to detect a signal, an understanding of the neutrino background problem is crucially important and considerable space in this review is devoted to it.

Two principle techniques have been used so far in proton decay searches. Water Čerenkov detectors exploit the fact that Čerenkov light is emitted by relativistic particles in water, and can be detected by photomultipliers placed in or at the surface of the water volume. They were originally oriented to the search for the SU(5)-favored decay mode $p \rightarrow e^+ \pi^0$, in which most of the energy appears in the form of relativistic electrons in electromagnetic showers. Large volumes (10^4 m^3) can be used because pure water is transparent over much of the Čerenkov spectrum and, for a surface array, the phototube cost rises only as the 2/3 power of the water mass.

The second approach uses tracking calorimeters, consisting of a matrix of iron sheets (the target medium) separated by layers of counters (proportional, flash or streamer tubes, or drift chambers) familiar in high-energy accelerator experiments. They record the tracks of any charged decay products. The cost per unit mass is 10 times that of the water detectors, and they have disadvantages because of nuclear absorption effects in the iron. On the other hand, decay or interaction vertices can be located with typically 100 times the precision of the water detectors, they can be built in modular form and tested in accelerator beams, and, most importantly, they are sensitive to a wider range of possible decay modes and should be much better at finding the unexpected than are the water Čerenkov devices.

This review begins by tracing some of the early history of baryon decay, followed by a discussion of the formal problems associated with complete baryon stability (absence of a long-range field coupled to baryon number, baryon asymmetry of the Universe). The third section deals with theoretical lifetime predictions, and this is followed by an account of the early experimental limits on nucleon lifetime. The problems of cosmic-ray background are then discussed, before entering the main description of the recent experiments and their results.

1.2 History of Baryon Conservation

The apparent stability of matter was first given a concrete formulation in terms of a conservation law in 1929 by Weyl (1). At that time, the two pairs of components of the Dirac wavefunction were interpreted as the electron and the proton, and the conservation of both was associated by Weyl with a "two-fold gauge invariance." In 1938, Stueckelberg (2) postulated a conservation law for "heavy charge"—what we now call baryon number—expressing the fact that transformations of heavy particles (nucleons) to light particles (electrons and neutrinos) were not observed. In 1949 (also 1952) Wigner (3) independently restated the conservation law of baryons, drawing an explicit parallel between conservation of electric charge and that of baryon number, associated with stability of the electron and proton respectively. As he said, "It will be assumed ... that the two conservation laws have similar causes and that these have similar consequences."

The experimental limits on nucleon stability were, even in the 1950s, known to be formidably long. For example, M. Goldhaber pointed out that the very existence of advanced life forms on Earth implied $\tau_p > 10^{16}$ yr. Goldhaber also quoted a limit of 10^{21} yr from the nonexistence of spontaneous Th^{232} fission induced by nucleon decay (4), a limit that was to increase to 10^{23} yr by 1958.

Despite these increasingly stringent limits, the consequences of complete stability of the nucleon and the possibility of nucleon decay were also starting to be discussed more than 20 years ago. Following the extension of the local gauge principle to non-Abelian fields by Yang & Mills (5) in 1954, Lee & Yang (6) pointed out in 1955 that a consequence of a local gauge symmetry associated with an absolute conservation law for baryons would be the existence of a long-range field coupled to baryon number. As is discussed in more detail below, there is no evidence for such a field. Here it is just worth recalling that, a quarter of a century earlier, Weyl had also stressed gauge invariance in this context.

In 1959, Yamaguchi (7) in a little-quoted paper noted the possibility of a "superweak" interaction responsible for nucleon decay. He stated that "the stronger the interaction, the greater the number of symmetry properties it possesses" and, by the inverse argument, he envisaged that the new superweak interaction would be characterized by CP and possibly CPT violation. Nucleon decay to leptons was assumed to proceed through 4-fermion-type interactions, $p \to 2e^+ + e^-$ or the K-capture $e^- + p \to e^+ + e^-$, and obeyed the selection rule $\Delta B = \Delta L$. This paper is interesting not simply as (apparently) the very first detailed theoretical discussion of nucleon decay via a new interaction on a new energy scale: it stimulated the first deep underground experiment to search for nucleon decay, using Čerenkov counters, in 1960 (8).

More compelling arguments for nucleon instability were brought forward by Sakharov (9) in 1966—in a paper well-known today but also little noticed at the time. He was the first to point to the importance of a CP-violating mechanism in accounting for the observed cosmological baryon-antibaryon asymmetry, if it is produced in the initial stages of the Hot Big

Bang model of the Universe. Sakharov emphasized the inevitability of proton decay with such a mechanism, and estimated the lifetime as 10^{50} yr or more, using a model in which the mass scale of the bosons—what he called maximons—mediating the B-violating interactions was taken as the Planck mass.

From 1973 onward, proton instability has been postulated from a different viewpoint, as a consequence of the grand unification of the strong and electroweak interactions [Pati and Salam (10), Georgi and Glashow (11), Georgi, Quinn & Weinberg (12), and many others, as discussed below]. Grand unification schemes (GUTs) incorporate leptons and quarks into the same multiplets of the "leptoquark symmetry" so that decay of quarks to leptons is, at some level, almost unavoidable.

Complete baryon stability is not excluded, but requires a peculiar combination of both local and global gauge symmetries (81). The energy scale determining the nucleon lifetime is set by the unification mass, where the strong and electroweak couplings g_s, g, and g' are equal within Clebsch-Gordan coefficients and mixing angles. These theoretical speculations over the years, and particularly the last decade, have further stimulated experimentalists to search seriously with large-scale detectors for evidence of proton decay.

2. PROBLEMS FOR BARYON CONSERVATION

As mentioned above, there are two possible difficulties of a formal nature, backed by circumstantial experimental evidence, that might arise if baryon conservation were absolute. First, we expect such conservation to be associated with a long-range field coupled to baryon number, which so far is unobserved. Second, in the framework of the Big Bang model of the Universe, the mechanism postulated for generating a baryon-antibaryon asymmetry is also expected to lead to baryon decay.

At this stage, it is worth reiterating the point made by Goldhaber & Sulak (13) that charge conservation would allow proton decay only if the numerical values of the electric charges of a baryon (proton) and a charged lepton (e^+, μ^+) were identical. The limits on the electron-proton charge difference ΔQ and the neutron charge ΔQ_N are both $< 10^{-19} |e|$, as obtained by Hughes and his collaborators (14, 15) by measuring the deflection of (CsI) molecular beams in an electric field. Indirect methods (16–18) give similar limits. Thus the equivalence of the charges of the electron and proton has been checked with high precision. Their exact equality is of course assumed, by construction, in all grand unified models incorporating leptons and baryons in multiplets, and in which the charge operator is one of the generators of the symmetry.

2.1 Limits on Long-Range Field Coupled to Baryon Number

Over the last 50 years or so, a succession of experiments of ever-increasing precision has been carried out to test the Einstein Principle of Equivalence. Some of these provide, as a by-product, stringent limits on the strength of any new long-range field coupling to baryon number.

The experiments in question essentially compare the inertial and gravitational masses of bodies using the principle illustrated in Figure 1. A body A at the Earth's surface at latitude λ is subject to two forces; the gravitational force F_G along the line AB toward the Earth's center 0, and a centrifugal force F_I along AC arising from the Earth's rotation, where

$$F_G = K \frac{M_G M_E}{r^2} \qquad \qquad 1.$$

$$F_I = M_I \omega^2 r \cos \lambda. \qquad \qquad 2.$$

Here, K is the Newtonian constant, r is the Earth's radius, ω is the angular velocity of rotation relative to distant stars, M_E is the gravitational mass of the Earth, and M_G and M_I are respectively the gravitational and inertial masses of body A. If a body is suspended by a string, the string lies along the resultant AD of F_I and F_G, the angle θ to the local vertical depending on the ratio $R = M_I/M_G$. According to the Principle of Equivalence, R should be the same for all bodies of whatever material, and for convenience we then define units so that $M = M_I = M_G$ is simply referred to as the mass of the

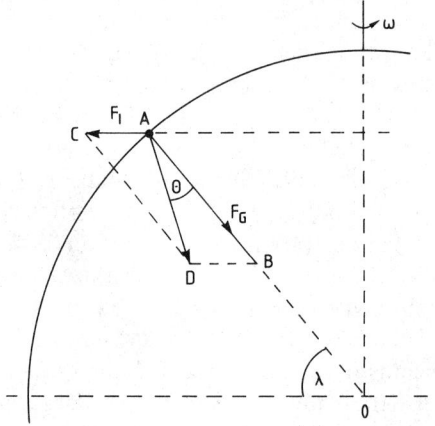

Figure 1 The force on a body A at latitude λ at sea level is the resultant, AD, of the gravitational force, F_G, proportional to the gravitational mass, m_G, and the centrifugal force, F_I, proportional to the inertial mass, m_I.

body. This hypothesis has been checked by suspending two bodies of equal masses but different materials from either end of a horizontal beam, itself hanging from a torsion fiber. If R is different for different materials, so also will be the value of θ and the result is a net couple that changes sign upon rotating the entire apparatus through 180°. In 1922, Eötvös et al (19) obtained a null result, which can be expressed as a limit on

$$\frac{\Delta R}{R} = \frac{2(R_1 - R_2)}{(R_1 + R_2)}, \qquad 3.$$

where R_1, R_2 refer to the two materials.

If we *assume* the validity of the Equivalence Principle, then the limit $\Delta R/R$ determines the maximum strength of any long-range field coupling to baryon number. In Figure 1, such a field would produce an extra force along AB:

$$F_B = K_B M_P^2 \frac{BB_E}{r^2}, \qquad 4.$$

where B and B_E are the baryon numbers of the body and of the Earth; M_P is the proton mass; K_B is the coupling of the new field analogous to K. It is assumed that the static potential is central and has the form $1/r$. For a tensor (vector) field, F_B will be parallel (antiparallel) to F_G. For the same inertial mass, two bodies will not have the same baryon number because of the dependence of nuclear binding energy on mass number A, the neutron-proton mass difference, and variations of the neutron-proton ratio. The atomic mass can be written (neglecting electronic binding) as

$$M = AM_N\left[1 - \frac{Z}{A}\frac{(M_N - M_H)}{M_N} - \frac{W}{AM_N}\right], \qquad 5.$$

where W is the total nuclear binding energy in MeV, M_N and M_H are the masses of the neutron and the hydrogen atom. Then the fractional mass difference of two different elements containing the same number of nucleons will be

$$\frac{\Delta M}{M} = \left[8.3\Delta\left(\frac{Z}{A}\right) + 10.6\Delta\left(\frac{W}{A}\right)\right] \times 10^{-4}. \qquad 6.$$

In the pairs of elements measured, the first term is only about 10% of the second and can be neglected. For the *same mass* of the two elements, the fractional difference in baryon number will also be

$$\frac{\Delta B}{B} \approx 10^{-3}\Delta\left(\frac{W}{A}\right)$$

with a corresponding limit

$$\frac{K_B}{K} \leq \frac{(\Delta R/R)}{(\Delta B/B)}. \qquad 7.$$

Table 1 shows results from some experiments since 1922. The Dicke et al (20) and Braginsky & Panov (21) experiments achieved high sensitivity by utilizing the gravitational field of the Sun, rather than the Earth, and searching for a 12-hr oscillation period of the torsion system, as the magnitude of the centripetal force due to the Earth's rotation changes sign relative to the Sun's gravitational field.

The result of these experiments is the limit

$$K_B < 10^{-9} K. \qquad 8.$$

This limit of course proves nothing directly about the stability of the nucleon. It shows that, if any such long-range field coupled to baryon number exists, the coupling is extremely weak. Since it is hoped to incorporate the various fundamental interactions into a unified theory, a new interaction with such very weak coupling would clearly make such a task much more difficult. The alternative is to postulate a unified theory with both global and local symmetries, such that a suitable linear combination of the global and local charges can correspond to an absolutely conserved baryon number (81). This is arbitrary and rather ugly and involves extra particles with new quantum numbers, and is generally

Table 1 Experiments leading to limits on a new interaction coupled to baryon number

Authors	Materials	$\Delta B/B^a$	90% CL upper limit on $\Delta R/R^b$	K_B/K^c
Eötvös, Pekar & Fekete 1922 (19)	Pt, Cu Cu, CNO	8×10^{-4} -10×10^{-4}	5×10^{-9} 3×10^{-9}	6×10^{-6} 3×10^{-6}
Renner 1935 (22)	Pt, Cu Bi, Cu	8×10^{-4} 9×10^{-4}	2×10^{-9}	2×10^{-6}
Dicke 1962 Roll, Krotkov & Dicke 1964 (20)	Pb, Cu Au, Al	9×10^{-4} 4×10^{-4}	$2. \times 10^{-10}$ 3×10^{-11}	2×10^{-7} 7×10^{-8}
Braginsky & Panov 1972 (21)	Pt, Al	4×10^{-4}	1×10^{-12}	2×10^{-9}

[a] $\Delta B/B$ = fractional difference in baryon number per unit mass.
[b] $\Delta R/R$ = fractional difference in ratio of inertial to gravitational mass.
[c] K_B/K = ratio of baryon coupling to gravitational coupling.

considered to be much less attractive than the unification models in which unstable baryons occur "naturally."

2.2 CP Violation in the Early Universe

The Big Bang theory is at present the most credible model of the evolution of the Universe and has had two notable successes: it is able to account for the 3-K background radiation, as the cooled remnant of the primordial radiation emitted with the Big Bang; and it correctly predicts the observed mass ratio (≈ 0.25) of helium to hydrogen, in terms of nucleosynthesis in the first 100 s after the Big Bang. However, a major problem is to account for the currently observed ratio of matter density (baryons) to radiation (photons), $n_B/n_\gamma \approx 10^{-9}$, and the preponderance of matter over antimatter [$> 10^3 : 1$ in our Galaxy (45)]. These questions were discussed at length in the recent review of this subject by Kolb & Turner (23) and we only outline a few salient points.

In the initial stage of the Big Bang ($t < 10^{-6}$ s, temperature $T > 1$ GeV), nucleons and antinucleons should have been generated as abundantly as photons and leptons. As the Universe cooled, the creation of baryon-antibaryon pairs by radiation would no longer compensate annihilation of pairs to photons, and the resulting density would fall as

$$n_B/n_\gamma \approx (M/T)^{3/2} \exp(-M/T), \qquad 9.$$

where M is the nucleon mass, and $n_{\bar{B}} = n_B$ by symmetry. The annihilation rate per nucleon is given by $R = n_B \sigma v$, where σ and v are the annihilation cross section and the relative velocity of nucleon and antinucleon. Thus R falls exponentially as T decreases and must eventually decrease below the expansion rate H, varying as T^2. When $R < H$ (in fact for $T < 22$ MeV) nucleons can no longer find antinucleons to annihilate with, and the $B\bar{B}$ residue becomes "frozen out" with a value, from Equation 9, of $n_B/n_\gamma \approx 10^{-18}$ and $n_B = n_{\bar{B}}$ (assuming no initial asymmetry). Thus the model predicts a catastrophically smaller proportion of baryons than is observed and is unable to account for the preponderance of matter over antimatter. An alternative is to postulate a baryon number $n_B/n_\gamma \approx 10^{-9}$ as an initial condition, but this seems extremely artificial and unlikely.

On the other hand, cosmological models incorporating grand unification, while at present unable to predict unambiguously the observed n_B/n_γ ratio, do offer a more reasonable and viable scenario, in terms of an initially baryon-symmetric Universe. They appeal to CP violation to generate the eventual asymmetry. Such CP violation is already known to occur in at least one physical situation, that of the K^0-$\bar{K}^0$ system.

The first models of a dynamically-generated baryon excess were described by Sakharov (9) and Kuz'min (24), who listed three necessary conditions:

1. A baryon number nonconserving process.
2. *CP* and *C* violation (so that the rates of production of quarks and antiquarks are unequal).
3. The baryons and antibaryons should be out of thermal equilibrium. In thermal equilibrium, the baryon density can depend only on the temperature and particle mass, which is the same for particle and antiparticle, by *CPT*. Hence an "arrow of time" is also necessary, provided by a nonequilibrium expansion.

The generation of baryon number in the context of GUTs was considered by Yoshimura (25), Weinberg (26), Toussaint et al (27), and others. The heavy gauge bosons X, X̄ and the Higgs H, H̄ of the GUT are supposed to be created initially (at temperatures $T \approx \mathcal{M}_{Planck}$) in equilibrium with the other fundamental fermions and bosons. The *equilibrium* abundance of X bosons when T falls below M_X will be given by Equation 9 with $M = M_X$. This abundance will not be achieved unless the X bosons can stay in equilibrium, and the most important process for decreasing the density n_X is decay (X, X̄ annihilation is too slow). The X decay rate $\Gamma \propto M_X$, on dimensional grounds, while the expansion rate $H \propto T^2$. For values $T \approx M_X$, whether Γ is large or small compared with H depends on the X mass and couplings for decay processes. If $\Gamma \gg H$, then X bosons remain in thermal equilibrium and no baryon asymmetry can develop, with *CP*-violating decays exactly counterbalanced by the inverse decay process. If $\Gamma \ll H$, however, the abundance n_X becomes larger than the equilibrium value and an asymmetry is possible. This condition requires a minimum value for M_X, which calculations show is $M_X > 10^{16}$ GeV (28–30). This condition is not satisfied by the gauge bosons X, Y of the SU(5) model, where $M_X \approx 10^{15}$ GeV. On the other hand, because of their different couplings, the corresponding condition for GUT Higgs bosons is the weaker requirement $M_H > 10^{13}$ GeV, and this is very likely to be fulfilled, since one expects $M_H \approx M_X$.

In the minimal SU(5) model, the predicted baryon asymmetry turns out to be much too small, resulting in $n_B/n_\gamma \lesssim 10^{-16}$. For nonminimal SU(5), with more complex Higgs structure, it seems possible to account for the observed asymmetry, although the result depends sensitively on the degree of *CP* violation in the model. In particular, since several *CP*-violating phases are involved, it is not possible to relate the baryon asymmetry directly, in either magnitude or sign, to the observed *CP* violation parameters in the neutral kaon system. However, it can at least be claimed that GUTs offer the possibility of accounting numerically, and in a nonartificial way, for the present huge preponderance of matter over antimatter in terms of tiny asymmetries in the early stages of the Big Bang.

In the sense that grand unified theories so far seem to offer the only natural way to account for the baryon asymmetry of the Universe, the existence of that asymmetry is tangible, if circumstantial, evidence in support of grand unification and, therefore, of the existence of baryon instability at some level.

3. THEORETICAL PREDICTIONS ON PROTON DECAY

3.1 *Proton Decay in SU(5) Version of GUTs*

A thorough discussion of proton decay in GUTs was given in the review by Langacker (32), and we discuss here only a few salient points.

Although it is possible to construct unified theories in which baryons are absolutely stable—even if they have undesirable features— nonconservation of B is a feature of most GUT models. Estimates of lifetime depend primarily on the type of operators involved in the decay mechanism, and on the mass scale at which it is expected that ΔB interactions will be commonplace. For example, Sakharov (9) had taken the scale as the Planck mass $\mathcal{M}_{\text{Planck}} = 2 \times 10^{19}$ GeV and obtained $\tau_p > 10^{50}$ yr. In GUTs the mass scale is the energy at which the electroweak and strong couplings merge. In the minimal SU(5) version (11), quarks and leptons are placed in SU(5) multiplets and the scale of the grand unified coupling is $M_X \approx 10^{15}$ GeV, where M_X is the mass of the X and Y "leptoquark" bosons, of charges 4/3 and 1/3, mediating quark-lepton and quark-antiquark transitions. Some diagrams—those involving the dominant "dimension 6" operators—for proton decay in this model are shown in Figure 2. Figure 2*a* shows the two-quark fusion graph corresponding to the processes

$$p \to e^+ u\bar{u} \quad \text{or} \quad n \to e^+ d\bar{u}$$
$$\to \bar{v}_e d u \qquad\qquad \to \bar{v}_e d\bar{d} \qquad\qquad 10.$$

where the $Q\bar{Q}$ pairs can form $\pi, \rho, \omega, \ldots$ mesons. Figure 2*b* shows the process of three-quark annihilation to a lepton preceded by meson emission, giving the same result as Equation 10 and of comparable amplitude. Other graphs contribute but may be neglected compared with the ones shown. Since the X propagator introduces a factor M_X^{-4} in the decay rate, dimensional arguments suggest that

$$\tau_p = \frac{A M_X^4}{\alpha_g^2 M_p^5}, \qquad\qquad 11.$$

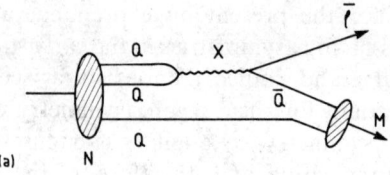

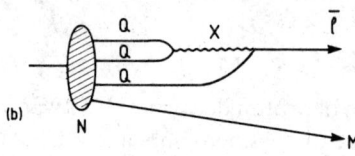

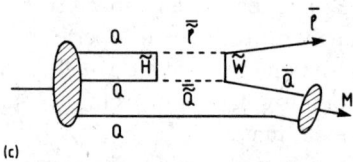

Figure 2 Diagrams describing proton decay, p → lepton + meson. Panels (*a*) and (*b*) show, respectively, decay by two-quark fusion, and that via three-quark fusion preceded by meson emission, in the SU(5) model. These are the dominant diagrams involving X-boson exchange, and have comparable amplitudes. Panel (*c*) shows decay via exchange of supersymmetric fermions H̃ and W̃, in which heavy mesons and leptons (i.e. $\bar{\nu}_\tau K$, μK) are the dominant decay products.

where $\alpha_g \approx 1/40$ is the grand unified coupling, and A contains details of hadronic matrix elements and is of order unity. For $A = 1$ and $M_X = 10^{15}$ GeV, $\tau_p = 6 \times 10^{31}$ yr. In principle, decays can also occur via super-heavy scalar (Higgs) exchange. However, because the coupling to light fermions is weaker for H than for X and Y, the contribution from Higgs exchange is small, unless $M_H \ll M_X$, which appears unnatural. In SU(5), the predominance of e^+ and $\bar{\nu}_e$ leptons in the decay products is a consequence of putting the light leptons and the light quarks in the same family, and this is to some extent justified by the limited success of relations between quark and lepton masses.

There are a few selection rules for nucleon decay that are generally applicable. Weinberg (26) showed that a necessary result of baryon nonconservation in leading order is the conservation of $(B-L)$, so that only ℓ^+ and $\bar{\nu}_\ell$ should appear among the decay products. The observation of negative electrons or muons, for example, would be quite disastrous for the GUT schemes. Within the context of vector exchange [as in SU(5) or SO(10)] there are two fundamental couplings, whose ratio can be denoted by r. For $\Delta S = 0$ processes it can be shown from isospin arguments (31) that, for example,

$$\Gamma(p \to e^+ \pi^0) = \left(\frac{1+r^2}{2}\right)\Gamma(p \to \pi^+ \bar{v}) = \frac{1}{2}\Gamma(n \to \pi^- e^+)$$
$$= (1+r^2)\Gamma(n \to \pi^0 \bar{v})$$

$$\Gamma(p \to e^+ + X_{NS}) = (1+r^2)\Gamma(n \to \bar{v} + X_{NS}) \geq \left(\frac{1+r^2}{2}\right)\Gamma(p \to \bar{v} + X_{NS}) \quad 12.$$

and so on, where X_{NS} is any nonstrange hadronic state. In the SU(5) model, $r = 2$, while in SO(10) $r \approx 0$.

Estimates of SU(5) branching ratios are given in Table 2, after the review by Langacker (32). We do not list here the numerous theoretical estimates of the lifetime. A dominant factor is the unification mass, M_X, which in terms of the QCD scale parameter $\Lambda_{\overline{ms}}$(GeV) has the value

$$M_X \approx 2.4 \times 10^{14}[\Lambda_{\overline{ms}}/0.16] \text{ GeV.} \quad 13.$$

In his 1981 review Langacker, using this value of M_X, quotes a "best" value of the lifetime as

$$\tau_p = 3.2 \times 10^{29 \pm 1.3}[\Lambda_{\overline{ms}}/0.16]^4 \text{ yr.} \quad 14.$$

The usually quoted value is $\Lambda_{\overline{ms}} = 0.16$ GeV, and this is considered reliable within a factor of two. The main changes to τ_p calculated recently in SU(5) have come from variations in A in Equation 11. This quantity involves specification of matrix elements relating the fundamental process involving

Table 2 Nucleon decay branching ratios predicted in SU(5). Average values from review by Langacker (32)

Mode	BR (%)	Mode	BR (%)
$p \to e^+\pi^0$	30	$n \to e^+\pi^-$	54
$e^+\rho^0$	14	$e^+\rho^-$	23
$e^+\eta$	4	$\bar{v}_e\eta$	~1
$e^+\omega$	30	$\bar{v}_e\omega$	7
$\bar{v}_e\pi$	11	$\bar{v}_e\pi^0$	7
$\bar{v}_e\rho$	4	$\bar{v}_e\rho^0$	4
$\mu_e^+K^0$	7	$\bar{v}_\mu K^0$	4
$\bar{v}_\mu K^+$	<1		

quark decay via X exchange to the final state hadrons. The value in Equation 14 comes from nonrelativistic SU(6) or bag models. Berezinsky et al (33) determine A from QCD sum rules and find a value

$$\tau_p/B(p \to e^+\pi^0) = 1.1 \times 10^{29 \pm 1.3}[\Lambda_{\overline{ms}}/0.16]^4 \text{ yr} \qquad 15a.$$

with $B = 0.5$. More recently Brodsky et al (34) found another way of calculating A, relating it to nucleon form factors and $\psi \to p\bar{p}$ decay rates. These calculations give even shorter lifetime estimates, and they obtain

$$\tau_p/B(p \to e^+\pi^0) = 1.7 \times 10^{28 \pm 1.3}[\Lambda_{\overline{ms}}/0.16]^4 \text{ yr} \qquad 15b.$$

that is a factor of 20 shorter than the value in Equation 14. As we discuss below, the IMB collaboration's observed 90% confidence limit (CL), $\tau_p/B(p \to e^+\pi^0) > 1.5 \times 10^{32}$ yr, places the prediction in Equation 14 in serious trouble and is in complete disagreement with that in Equations 15a or 15b. The SU(5) prediction can be increased by invoking more complex structure than for minimal SU(5). We have already seen that SU(5) also fails to predict the correct cosmological baryon/photon density ratio. Within the restrictions imposed by the need to obtain the correct value of $\sin^2\theta_w$, there is considerable freedom also in other forms of GUT [E(6), O(10), ...] to obtain larger values of τ_p.

3.2 Proton Decay in Supersymmetric GUTs

In supersymmetry, each fermion (boson) is duplicated by a boson (fermion) partner. These extra particles slow down the logarithmic q dependence of the running coupling constants so that the unification mass grows to $M_X \gtrsim 10^{17}$ GeV, and the predicted proton lifetime may become very long. However, as pointed out by Weinberg (35) and Sakai & Yanagida (36), in this case "dimension 5" operators could contribute, corresponding to the supersymmetric Higgs ($\tilde{H}$) exchange as in Figure 2c. Since $\tilde{H}$ is a fermion, the lifetime varies only as $M_{\tilde{H}}^2$. Further, the mass $M_{\tilde{H}}$ can be very different from M_X or M_H, so that a lifetime similar to that in minimal SU(5) is easily possible. Indeed, extra symmetries are invoked to suppress some of the operators and avoid short lifetimes, with the result that decays to quarks and leptons in other generations than (u, d, e, v_e) are strongly favored, for example

$$p \to \bar{v}_\tau K^+ \qquad n \to \bar{v}_\tau K^0$$

with
$\qquad\qquad\qquad\qquad\qquad\qquad\qquad\qquad\qquad\qquad\qquad\qquad$ 16.

$$p \to \mu^+ K^0 \qquad n \to \bar{v}_\mu \pi$$

less probable (37–39). Decay $N \to e\pi, \mu\pi$ is suppressed in these models.

3.3 Consequences for Proton Decay Experiments

The SU(5) model predicts that two-body decay modes into positron and meson will dominate. Since π^0, ρ, η, ω also give high-energy γ's among the decay products, double electron-photon showers with a "back-to-back" configuration will be prominent, and it was on this basis that the large water Čerenkov detectors were built, since

1. They are very good detectors of electron-photon showers separated in space by large angles, as in two-body decay.
2. In the most favored SU(5) decay modes, nearly all the energy appears in relativistic charged particles, and the total Čerenkov signal is a simple and fairly precise indicator of the total energy in the decay.
3. The momenta of the individual decay products can be measured, since Čerenkov light indicates the direction of motion of the particle producing it and this, together with the energy measurement, is an essential kinematic constraint in combatting the neutrino background.

In GUTs incorporating supersymmetry, on the other hand, the lepton is frequently a neutrino, and strange hadrons are favored. Topologically, such events are much more difficult to identify. For example, the decay n → $\bar{\nu}K^0$, K^0 → $2\pi^0$ → 4γ provides an adequate Čerenkov pulse of well-defined energy (if we neglect Fermi motion) but there is no "back-to-back" configuration, because the K^0 has low velocity, and the Čerenkov light is almost isotropic in direction. Generally speaking, multiprong decays, involving charged particles with a decay sequence, as in p → $\mu^+ K^0$, K^0 → $\pi^+\pi^-$, or p → $\bar{\nu}K^+$, K^+ → $\mu^+ + \nu$, should be more easily distinguished by means of fine-grained tracking calorimeters with good (≤ 1 cm) vertex resolution.

4. EARLY RESULTS ON NUCLEON STABILITY

4.1 Decay-Mode Independent Methods (Geochemical and Radiochemical)

The first useful limits on nucleon lifetime were obtained using nuclear methods, the principle being that, when a nucleon is removed from a nucleus, it may be left in an excited state, undergoing radioactive decay or fission, which can be detected radiochemically. Alternatively, a rare isotope may be formed as a result of decay of a common nuclide, and can be detected by geochemical techniques. Both methods have been discussed by Rosen (46).

Table 3 summarizes the results obtained. The Goldhaber (4) and Flerov

Table 3 Geochemical and radiochemical limits on nucleon lifetime

Authors	Experiment	Depth (mwe)	τ_{min} (yr)
Reines, Cowan & Goldhaber (4)	Th^{232} fission		10^{21}
Flerov et al (40)	Th^{232} fission		10^{23}
Evans & Steinberg (41)	$Te^{130} \to Xe^{129}$	~400	1.6×10^{25}
Bennett (44)	mica spallation	10,000	2×10^{27}
Fireman (47)	$K^{39} \to Ar^{37}$	4,400	2×10^{26}

(40) limits from Th^{232} have already been mentioned. Evans & Steinberg (41) used the measured Xe^{129} abundance (42, 43) in telluride ores to set a limit on nucleon decay in Te^{130}.

The most precise geochemical limit is that of Bennett (44), who used etched mica samples from a mine of depth 10,000 meters of water equivalent (mwe). He measured short nuclear spallation tracks that could have followed nuclear absorption of hadrons from nucleon decay. From the rate of such etch-pits he found $\tau > 2 \times 10^{27}$ yr.

Fireman (47) used a 1.7-ton sample of potassium acetate to measure the rate of production of radioactive Ar^{37} from nucleon decay, followed by nucleon emission in K^{39}. In this technique, locating the sample at great depth is an essential and crucial feature in reducing the background.

4.2 Summary of Early Searches for Nucleon Decay by Direct Methods

In this section, we describe briefly the results of early searches (48) for nucleon decay by direct detection of decay products. The detectors were not primarily designed for this purpose, and in particular could not fully contain the charged particles, electrons, or γ rays from such decays inside the detector volume. The list of experiments is given in Table 4.

Reines, Cowan & Goldhaber (4) recorded charged particles of kinetic energy above 100 MeV in 300 liters of liquid scintillator under 100 ft of rock, setting a limit $\tau > 10^{22}$ yr determined by crossing cosmic-ray background. In 1958, Reines, Cowan & Kruse (49) improved this limit to $\tau > 4 \times 10^{23}$ yr by using a 170-kg target of heavy water containing $CdCl_2$ and requiring a delayed pulse (within 20 μs of the initial pulse) from moderation and capture of the neutron produced in the reaction $d \to n +$ (proton decay products).

The first deep underground experiment was undertaken by the CERN group of Backenstoss et al (8) in 1960 in the Lötschberg railway tunnel (2400 mwe). A 50-liter liquid Čerenkov counter recorded upward-travelling relativistic charged particles (for example, from proton decay in the surrounding rock). Giamati & Reines (50) and Kropp & Reines (51)

Table 4 Early limits on nucleon lifetime by direct methods

Authors	Experiment	Decay mode	Depth (mwe)	τ_{min} (yr)
Reines, Cowan & Goldhaber 1954 (4)	300-liter liquid scintillator	all ($E_{ch} > 100$ MeV)	200	10^{22}
Reines, Cowan & Kruse 1958 (49)	As above, with delayed neutron pulse	all	200	4×10^{23}
Backenstoss et al 1960 (8)	50-liter liquid Čerenkov, upward relativistic secondary	at least one secondary of >250 MeV	2400	3×10^{26}
Giamati & Reines 1962 (50)	200-liter liquid scintillator	all	1760	6×10^{27}
Kropp & Reines 1965 (51)				$\sim 10^{28}$
Gurr et al 1967 (52)	scintillator hodoscope	all	8000	2×10^{28}
Reines & Crouch 1974 (53)	scintillator hodoscope + μ decay	muon	8000	3×10^{29} –3×10^{30}
Bergamesco & Picchi 1974 (54)	500-liter liquid scintillator	all	4270	1.3×10^{29}
Learned, Reines & Soni 1979 (55)	liquid scintillator	muon	8000	10^{30}
Cherry et al 1981 (56)	150-ton H$_2$O Čerenkov + μ decay	muon	4400	1.5×10^{30}

carried out experiments in 1962–1965 at the Fairport Harbor Salt Mine, Ohio (1760 mwe), using 200 liters of liquid scintillator plus a Čerenkov anticoincidence shield to veto crossing cosmic rays.

Gurr, Kropp, Reines & Meyer (52) and Reines & Crouch (53) made deep underground observations in a South African gold mine (8000 mwe), using a large (20 ton) scintillator hodoscope, intended originally to record secondary muons from neutrino reactions in the surrounding rock. On the basis of the muon rate at wide zenith angle (thus excluding atmospheric muons), Gurr et al found $\tau > 2 \times 10^{28}$ yr, while Reines & Crouch based their estimate on muons stopping and decaying in the scintillator. Five events were observed, which could be accounted for in terms of neutrino origin, but if ascribed to nucleon decay, gave $\tau > 2 \times 10^{30}$ yr.

In the same year (1974) Bergamesco & Picchi (54) operated a 500-liter liquid scintillator detector plus anticoincidence shield in the Mont Blanc tunnel (4270 mwe). On the basis of the rate of secondaries with pulse heights exceeding 10 MeV, both originating and ending in the scintillator (i.e. with range $< 130 \text{ g cm}^{-2}$), they found $\tau > 1.3 \times 10^{29}$ yr. This seems to have been the first experiment actually requiring that a charged particle should be totally confined in the detector, but of course it would reject two-body decays in which charged decay products could obtain up to 500 MeV energy.

The last two experiments in the list are those of Learned et al (55) and Cherry et al (56). Learned et al (1979) re-evaluated the Reines-Crouch data, while Cherry et al (1981) operated 150 tons of water Čerenkov detectors (in several tanks of 2 m^3 volume) in the Homestake Mine (4200 mwe). They observed three decays of upward or horizontal muons contained in the tanks and involving a prompt pulse of 50–600 MeV equivalent energy. Taking account of the surrounding veto counter efficiency, they concluded $\tau > 1.5 \times 10^{30}$ yr.

All the limits we have listed depend on assumptions about decay modes, and in this sense are *partial* lifetime limits. The requirement of a muon decay, for example, assumes pions or muons among the decay products. Such limits are irrelevant if the dominant decays are $p \to e^+ + 2\nu$ or $n \to e^+ + e^- + \nu$, for example. Nevertheless, since most theories of nucleon decay involve at least one meson secondary, with at least a 25% probability of escaping nuclear absorption and of producing a muon via the decay chain, it is true to say that these early experiments set limits on the nucleon lifetime of order 10^{30} yr.

5. BACKGROUND IN PROTON DECAY EXPERIMENTS

There are several potential sources of background in proton decay experiments. In calorimeter experiments in which decay products are to be tracked through a large array of gas counters (which might total a million in a one-kiloton array), radioactivity is not negligible; it produces a singles rate of about 1 Hz in a counter of 1 cm^2 cross section and 1 m in length. Thus a basic requirement of the trigger is that counters from several contiguous or nearly contiguous layers must fire simultaneously in order to be accepted as an "event," so that a threshold energy release of perhaps 100 MeV in the calorimeter is involved.

Crossing cosmic-ray muons cannot be confused with decay events, but they are a nuisance that can be kept down to reasonable levels by trigger requirements, by use of anticoincidence shields, and by going deep underground (see Figure 5 for the muon flux as a function of depth). The muons also generate neutral hadrons in the surrounding rock and they are a potential source of background, but not, fortunately, an important one, as discussed in Section 5.5 below.

The dominant background to proton decay experiments is that due to atmospheric neutrinos, which are distributed nearly isotropically in space angle and are the ultimate limiting factor determining the sensitivity to long proton lifetimes. It has become abundantly clear from the experiments now under way that a proper understanding of this background is at least 90% of the battle to discover a proton decay signal.

5.1 *Atmospheric Neutrino Fluxes*

There have been several calculations of atmospheric neutrino fluxes over the past 20 years, with two broadly different approaches to the problem. The procedure in the earlier calculations was to deduce an empirical production spectrum of pions (and kaons) in the atmosphere, on the basis of the muon fluxes measured at sea level or at high altitudes. The original object of these calculations was to compare expected and observed neutrino event rates deep underground and to obtain information about neutrino cross sections above the then available energy range of accelerator beams, i.e. $E_\nu \gg 1$ GeV. They made power-law approximations to production spectra that are not necessarily relevant in the calculation of low-energy neutrino fluxes ($E_\nu < 1$ GeV). It is precisely these, however, that are important for nucleon decay background.

More recently, calculations have started from the measured primary spectrum of protons and α particles, and have used a Monte Carlo cascade program to propagate these through the atmosphere, employing accelerator data on pion and kaon production to derive absolute muon and neutrino fluxes at sea level or elsewhere. These calculations have the advantage that they provide neutrino spectra at low as well as high energy, and that geomagnetic and solar modulations of the primary spectrum are easily incorporated. The precision of the predicted neutrino fluxes is, however, limited, because the pion and kaon yield data from accelerators exist for only a limited range of incident energy, and of secondary angle and momentum, so that interpolations are necessary. In all the flux calculations, normalization is made to the sea-level muon spectrum, measured using magnetic spectrometers (57, 58).

A few general remarks can be made before discussing the detailed results. The primary cosmic-ray protons generate secondary mesons with interaction length $\lambda \approx 100$ g cm^{-2}, small in comparison with the total atmospheric depth $X \approx 1000$ g cm^{-2}. The pions and kaons can either be absorbed by nuclear interaction, with absorption length λ, or decay in flight. Integrating over the source distribution, the fraction of mesons of energy E that have decayed by sea level is (59)

$$P_{\pi,k}(\text{decay}) = \frac{q \sec \theta}{(1+q \sec \theta)}, \qquad 17.$$

where

$$q = H/\gamma \beta c \tau \approx A/E(\text{GeV}). \qquad 18.$$

Here, $\gamma = E/mc^2$ and τ are the Lorentz factor and proper lifetime of the meson, θ is the zenith angle, and $H = RT/Mg \approx 6.5$ km is the scale height of the (isothermal) atmosphere above 10 km. At large zenith angles, the flux is enhanced by the sec θ factor, expressing the larger decay probability for inclined particles. The values of A for pions and kaons are:

	$\pi^\pm$	$K^\pm$	K_L^0
A(GeV)	117	866	208

19.

For values of $E \ll A$, which is the energy region relevant to neutrino background to proton decay, we see that essentially all parents decay before interacting ($P \to 1$). Thus, if the primary spectrum follows a power law and we assume Feynman scaling, the muons and neutrinos will follow the same power law. (For $E \gg A$, however, it is clear that the μ and ν spectra will have an index one unit steeper than the primaries.)

Muons also contribute to the neutrino fluxes (ν_e and ν_μ). If we crudely assume all muons are generated in a thin layer of atmosphere at depth x_0 (or $x_0/\sec \theta$ for inclined primaries) the proportion that decay by sea level (depth X g cm^{-2}), neglecting ionization loss in the atmosphere, will be

$$P_\mu(\text{decay}) = 1 - \exp[-q \sec \theta \ln (X \sec \theta/x_0)], \qquad 20.$$

where, for muons, $q = 1.03/E(\text{GeV})$. With $x_0 \approx 100$ g cm^{-2} and $X \approx 1000$ g cm^{-2}, one finds that for $E_\mu < 1$ GeV nearly all muons decay before sea level whereas at higher energy ($E_\mu \cos \theta > 10$ GeV) the decay probability varies as E_μ^{-1}. The sea-level muon spectrum is affected by ionization loss in the atmosphere, which is of order 2 sec θ GeV. Thus nearly all muons recorded at sea level originate from pions with energy above 3 GeV, whereas the bulk of neutrinos of low energy, especially $E_v < 0.5$ GeV, come from decay of pions of energy below 3 GeV. The sea-level muon spectrum is therefore a good monitor for high-energy neutrino fluxes, but is not relevant at low E_v. For this reason, muon flux measurements at high altitudes are important (but, unfortunately, very meager). The important sources of neutrinos and antineutrinos, the branching ratios, and average fractional energies they receive from the parents in the relativistic limit are as shown in Table 5.

Regarding contributions to the neutrino fluxes, it is found that the rapidly falling primary spectrum combines with the rapid increase of neutrino yield with proton energy E_0 (61) to give a peak in the proton energy contributing to the neutrino flux for $E_v \lesssim 1$ GeV, in the region $E_0 \approx 10$ GeV. The relevant K^+/π^+ and K^-/π^- ratios are of order 0.07 and 0.03 respectively, considerably below the value 0.20 for $K^\pm/\pi^\pm$ used in early flux calculations (60). We also note that despite the larger energy fractions involved, the lower decay branching ratios and production cross sections of kaons as compared with pions means the v_μ flux contribution from kaons is almost an order of magnitude smaller than for pions, at least for $E_v \lesssim 1$ GeV. Muons and pions therefore emerge as the dominant and roughly equal sources of both v_e and v_μ, until one reaches energies $E_v > 5$ GeV, when the $1/E$ decay probability for muons cuts down their contribution.

For electron neutrinos, the flux is totally dominated by μ decay. When branching ratios and energy fractions are taken into account, the v_e contribution from K decay is negligible in comparison. The v_e flux from μ decay is slightly less than the v_μ flux from the same source, because of the smaller energy fraction in the decay.

In summary, v_e fluxes in the 1-GeV energy region are dominated by muon decay, and v_μ fluxes have their main (and roughly equal) contributions from both pion and muon decay. The v_e flux is about one half the v_μ

Table 5 Sources of atmospheric neutrinos

Neutrino flavor	Source	Branching fraction	Mean energy $\bar{E}_v$
$v_\mu, \bar{v}_\mu$	$\pi^+ \to \mu^+ + v_\mu$	1	$0.21\, E_\pi$
	$K^+ \to \mu^+ + v_\mu$	0.64	$0.48\, E_K$
	$K^0 \to \pi^- + \mu^+ + v_\mu$	0.13	$0.30\, E_K$
	$\mu^- \to e^- + \bar{v}_e + v_\mu$	1	$0.35\, E_\mu\, (= 0.28\, E_\pi)$
$v_e, \bar{v}_e$	$K^+ \to \pi^0 + e^+ + v_e$	0.05	$0.30\, E_K$
	$K^0 \to \pi^- + e^+ + v_e$	0.20	$0.30\, E_K$
	$\mu^+ \to e^+ + v_e + \bar{v}_\mu$	1	$0.30\, E_\mu\, (= 0.24\, E_\pi)$

flux, for vertical incidence. At large zenith angles, the sec θ factor enhances the muon-produced neutrino flux. At $E_\nu \approx 1$ GeV, the horizontal fluxes of ν_e and ν_μ from muon decay are about twice those at vertical incidence. On the other hand, Equations 17–19 tell us that the horizontal and vertical fluxes of ν_μ from pion (or kaon) decay are nearly equal.

The relative fluxes of neutrinos and antineutrinos are of great interest, since both neutral and charged current cross sections are so different for the two, and this affects total rates. The sea-level μ^+/μ^- ratio has been extensively measured, is practically energy-independent, and has a value $R = 1.25$–1.30 (58, 62). This means the same value for the effective π^+/π^- ratio as well as $(\nu_\mu/\bar{\nu}_\mu)_\pi$ from pion decay. From muon decay, however, $(\nu_\mu/\bar{\nu}_\mu)_\mu \approx 1/R = 0.8$. In the GeV region, pion and muon contributions to the ν_μ flux are about equal, so the overall $\nu_\mu/\bar{\nu}_\mu$ flux ratio should be close to unity. On the other hand, electron neutrinos originate almost entirely from muon decay, so that the ratio $(\nu_e/\bar{\nu}_e)_\mu = R \approx 1.25$.

5.2 Results of Flux Calculations

Early estimates of atmospheric neutrino fluxes from pion decays were made in 1961 by Markov & Zheleznyk (63), to be followed by calculations by Zatsepin & Kuz'min (64), who included the contribution from muon decay. They assumed a power-law energy spectrum for the pions and included ionization loss of muons in the atmosphere. In 1965, Osborne et al (60) made similar calculations but included also the various decay chains of kaons. Their results are included in Figures 3 and 4, showing the vertical $(\nu_\mu + \bar{\nu}_\mu)$ spectrum, the ratio $(\nu_e + \bar{\nu}_e)/(\nu_\mu + \bar{\nu}_\mu)$, and the ratio of vertical to horizontal fluxes. All the above calculations applied to $E_\nu > 1$ GeV.

Tam & Young (65) extended the Osborne et al flux calculations down to $E_\nu = 0.2$ GeV. As explained above, high-altitude muon fluxes are required to provide information on the low-energy part of the pion source spectrum, and on the daughter neutrinos. Tam & Young employed the pion spectrum deduced by Olbert (66) using the technique of Ascoli (69), and based on the altitude dependence of low-energy (~ 300 MeV/c) muons measured in aircraft flights by Conversi (67) and Sands (68). It is not clear if Tam & Young included effects of kaon production or not. More recently (1980), fresh calculations of neutrino flux were made by Volkova (70), using the most recent accelerator data for K/π ratios, and presenting the zenith angle and energy dependence of the flux in empirical analytical forms. The results of Tam & Young and of Volkova are included in Figures 3 and 4.

An alternative approach to the problem is to start from the primary flux and employ a Monte Carlo program to trace the hadronic cascade through the atmosphere to determine the source spectrum of neutrinos (and muons). This is the approach taken by Gaisser et al (61). As input they used primary

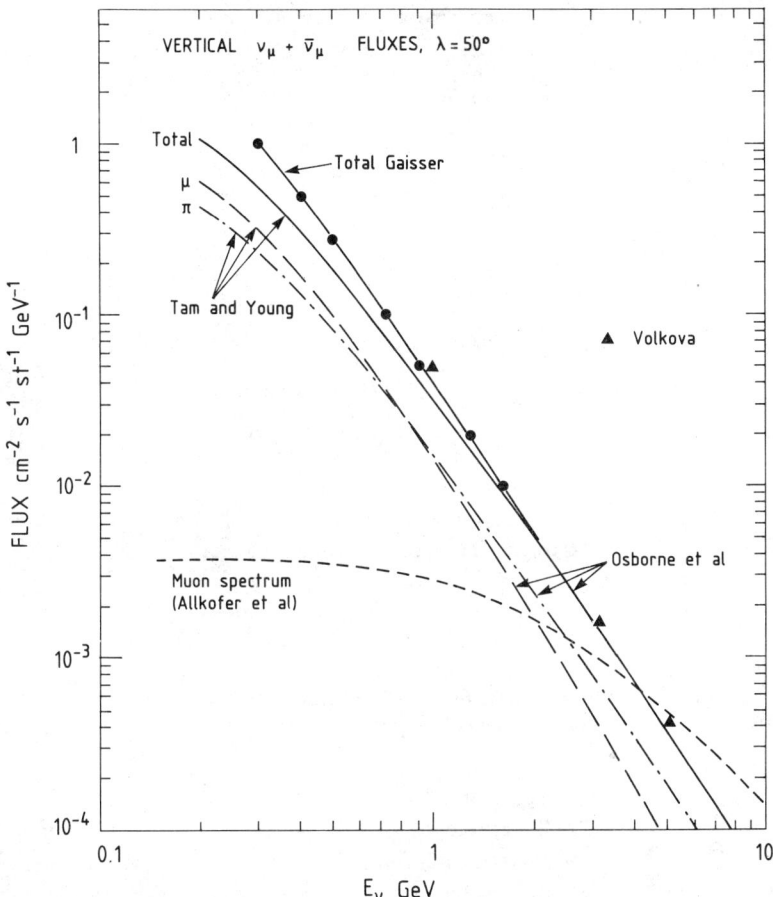

Figure 3 Vertical fluxes versus energy of muon neutrinos and antineutrinos at $\lambda = 50°$ magnetic latitude, calculated by various authors, as detailed in the text. The contributions from muon (μ) and pion (π) decay are shown separately. The sea-level muon flux is shown for comparison.

flux data on protons and heavier nuclei (71) and accelerator yield data on pion and kaon production as a function of energy and angle, for proton–light nucleus collisions (71a). Their calculated fluxes extend down to $E_\nu = 0.3$ GeV and are included in Figures 3 and 4.

Comparing the vertical ($\nu_\mu + \bar{\nu}_\mu$) fluxes at $\lambda = 50°$ in Figure 3, one can see that for $E_\nu > 2$ GeV the Osborne et al, Volkova, and Gaisser et al results are in good agreement. This is not surprising, since they are normalized to the sea-level muon spectrum and the only differences can be in the K/π ratio. For $E_\nu < 2$ GeV, substantial discrepancies are apparent. The Volkova flux

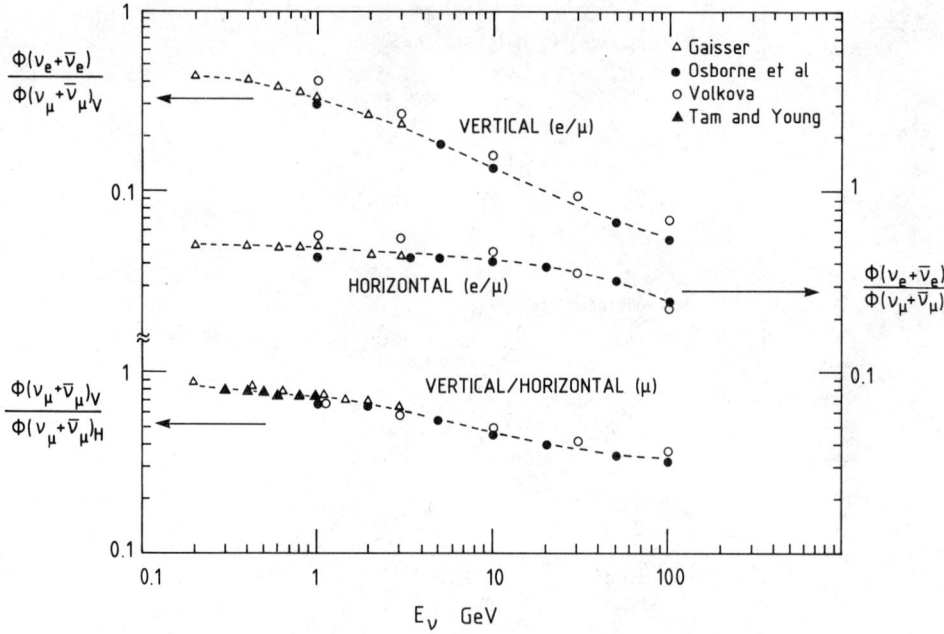

Figure 4 The flux ratios versus energy of electron to muon neutrinos and antineutrinos, in the vertical (*top curve*) and horizontal (*middle curve*) directions. The lower curve shows the vertical/horizontal flux ratio for muon neutrinos. For references, see text.

at $E_\nu = 1$ GeV is 60% larger than that of Osborne et al, despite their higher K/π ratio. The Gaisser et al fluxes are 30% larger than those of Osborne et al at 1 GeV, and a factor of two larger than those of Tam & Young at 0.3 GeV. These discrepancies are probably a fair reflection of the present uncertainties in absolute neutrino flux in the low-energy region.

The *relative* intensities of ν_e and ν_μ flux and of horizontal and vertical fluxes are shown in Figure 4, and for such ratios there is good agreement among the various calculations.

5.3 Latitude Dependence, Up-Down Ratios, and Solar Modulation

As first noted by Tam & Young, neutrino intensities vary appreciably with geomagnetic latitude, on account of the momentum cut-off in the primary proton spectrum, for the allowed trajectories that can penetrate to the Earth's atmosphere. This cut-off is given only approximately by the Störmer formula

$$p_c = 59 \cos^4 \lambda / [1 + (1 - \cos^3 \lambda \sin \theta \sin \phi)^{1/2}]^2 \text{ GeV}/c, \qquad 21.$$

where λ is the magnetic latitude, θ the zenith angle and ϕ the azimuth measured clockwise from magnetic north. For example, protons from the eastern (western) horizons at $\lambda = 0°$ have $p_c = 59$ GeV/c and 10 GeV/c respectively. The effective cut-offs, properly calculated, are somewhat larger (72). Note that there is an east-west as well as a latitude effect in the flux.

The magnitude of the latitude effect has been computed by Tam & Young (65), by Gaisser et al (61), and by Dar (73). Tam & Young used the pion spectrum deduced by Olbert (66) as a function of latitude, which in turn was based on the Conversi (67) muon fluxes. Broadly, the factor of two reduction in low-energy muon flux between $\lambda = 50°$ and $\lambda = 0°$, is mirrored by the variation in neutrino flux for $E_\nu \approx 0.5$ GeV. Table 6 shows the various calculations of latitude effect for downward-travelling neutrinos. The calculations of Dar (73) show a very large latitude dependence (a factor of ten reduction in flux at $E_\nu = 0.2$ GeV, between $\lambda = 50°$ and $0°$) and they are, presumably, in error.

The geomagnetic cut-off also affects the up-down flux ratio. At high latitudes, the downward flux is hardly affected while the upward flux, integrated over equatorial as well as high latitudes, is smaller. At the equator, the reverse is the case. Broadly, the downward neutrino flux decreases toward low latitudes, while the upward flux is practically latitude independent. Table 6 shows calculated values of the up-down ratio from the work of Gaisser et al (61). Finally, Table 8 quotes neutrino event rates for both solar maximum and solar minimum, with differences of the order of

Table 6 Dependence of $(\nu_\mu + \bar{\nu}_\mu)$ flux ϕ on latitude
(a) Latitude ratio $\phi(\lambda = 0°)/\phi(\lambda = 50°)$ for downward neutrinos

Authors	E_ν (GeV)	0.2	0.5	1.0	2.0
Tam & Young (65)		0.58	0.62	0.79	0.84
Gaisser et al (61)		—	0.58	($E > 0.6$)	—
Dar (73)		0.13	0.20	0.31	0.70

(b) Up/down flux ratio [Gaisser et al (61); $E_{\nu_\mu} > 0.6$ GeV, $E_{\nu_e} > 0.4$ GeV]

	Up-down rates per kton yr	
Latitude	solar maximum	solar minimum
52° (IMB)	51.4/59.7 = 0.86	59/73 = 0.81
27° (Kamioka)	46.3/43.4 = 1.07	53/46 = 1.15
2° (KGF)	44.3/33.0 = 1.34	50/35 = 1.43

10–20%. It should be emphasized that, in all these calculations, neutrino oscillation effects have been neglected.

5.4 Event Rates

The expected neutrino event rates at $\lambda \approx 50°$ as estimated by Battistoni et al (74) on the basis of the Tam-Young fluxes and measured accelerator cross sections for the elastic and various inelastic channels are listed in Table 7. Table 8 includes these results and those of Krishnaswamy et al (83), also based on the Tam-Young fluxes. Although total rates are compatible, there are severe discrepancies between elastic and inelastic rates and in the neutral current (NC) channel. The NC rate quoted by Battistoni is binned in visible (hadronic) energy; that quoted by Krishnaswamy et al is surely overestimated. The Gaisser & Stanev rates (84) are for somewhat different energy cuts but seem to be in reasonable agreement with the other two calculations, considering that they employed a completely independent flux calculation.

5.5 Topology of Neutrino Background

Neutrino interactions form the principle background in proton decay experiments. They have, apart from the rates, energy spectra, up-down ratios, and latitude dependences discussed above, quite characteristic topologies that can be used to differentiate them from possible proton decays. These topologies are described in more detail under discussion of experimental results, but we outline the important points here.

Charged current neutrino interactions should have the property that summing over all the secondaries yields $|\sum \mathbf{p}_s| = \sum E_s = E_\nu$, with an uncertainty due to Fermi motion of the nucleon target in the parent nucleus ($p_f \approx 100\text{--}200$ MeV/c). Of course, additional energy and momentum are lost in the form of nuclear excitation, missing neutrals, etc. Neutral current interactions should also exhibit a strong correlation between total visible energy and resultant momentum, with (within the Fermi motion uncertainty) $|\sum \mathbf{p}_s| \geq \sum E_s = E_\nu - E'_\nu$, where E_ν, E'_ν are the energies of the incident and outgoing neutrinos. On the other hand, nucleon decays should be characterized by $\sum E_s \approx M_p c^2$ and $|\sum \mathbf{p}_s| \approx 0$, if all secondaries are measured and again with an uncertainty in momentum of order p_f. A two-prong nucleon decay has therefore a "back-to-back" topology, while a two-prong neutrino interaction usually has the two prongs correlated in the "forward" direction.

Battistoni et al (97) exposed a module, similar to that in the calorimeter they employed to search for proton decay, to an accelerator beam from the CERN PS. The accelerator and atmospheric neutrino energy spectra were closely similar. About 400 events were recorded, with the plane of the plates

Table 7 Neutrino event rates, per kton yr, $\lambda = 50°$ [Battistoni et al (97)]

E_ν (or E_{vis}) (GeV)	ν_μ		$\bar{\nu}_\mu$		$\nu_\mu + \bar{\nu}_\mu$ total CC	$\nu_e + \bar{\nu}_e$ total CC[a]	NC[a]	Total
	elastic	inelastic	elastic	inelastic				
0.3–0.4	6.8	1.2	2.5	—	10.5	5.3	1.0	16.8
0.4–0.6	9.6	3.6	3.7	1.3	18.2	9.1	1.8	29.1
0.6–0.8	6.3	3.7	2.3	1.3	13.6	6.3	1.2	21.1
0.8–1.0	4.2	2.0	1.3	0.9	9.4	4.9	1.3	15.6
1–2	7.9	10.3	3.2	2.6	24.0	11.4	3.7	39.1
2–5	4.1	14.1	2.2	3.2	23.6	11.4	0.9	35.9
								157

[a] CC = charged current; NC = neutral current.

Table 8 Comparison of predicted event rates per kton year, $\lambda = 50°$, $E_v = 0.3$–5 GeV

	$(v_\mu + \bar{v}_\mu)$			$(v_e + \bar{v}_e)$		
	Elastic	Inelastic	Total	Total	NC[a]	Total
Battistoni et al (97)	54.1	44.2	98.3	48.4	10.0	157
Krishnaswamy et al (85)	68.4	27.8	96.2	41.6	25.8	164
Gaisser et al (61)						
Solar maximum			67.8	43.3		111 (CC)[a]
Solar minimum						132 (CC)[a]

[a] CC = charged current; NC = neutral current.

at 90° and 45° to the neutrino beam. Disregarding differences at the two angles, they obtained measurements of background in a few specific channels, applying of course to the particular granularity and steel thickness (1 cm) employed. They considered the modes:

(a) $p \to \mu^+ \pi^0$ and $n \to \mu^+ \pi^-$,

(b) $n \to \bar{v} K^0$, $K^0 \to \pi^+ \pi^-$,

(c) $p \to \mu^+ K^0$, $K^0 \to \pi^+ \pi^-$.

From a Monte Carlo calculation they estimated that in 50% of genuine decays of type (a) the lepton and pion would have opening angle $\theta > 120°$ (in the remainder, the pion is scattered or absorbed inside the iron nucleus). Among the 400 neutrino events, seven were two-prong events compatible with $\mu \pi^\pm$, energy 940 ± 210 MeV and $\theta > 120°$, while only one simulated $p \to \mu \pi^0$. Thus, 1.7% of neutrino events simulate $n \to \mu^+ \pi^-$ and about 0.3%, $p \to \mu^+ \pi^0$.

Battistoni et al tried to fit two-prong neutrino events to the $n \to \bar{v} K^0$ hypothesis (b), and found nine events out of 400 doing so, a 2.5% background. Finally, hypothesis (c) $p \to \mu^+ K^0$ will produce three-prong events. In the accelerator data, only two events had three prongs, total energy below 1.2 GeV, and with angles between pions and muon in the correct region to mimic $p \to \mu^+ K^0$ decay. So, this background is at the level of 0.5% of the total neutrino events.

Although the number of neutrino events examined by Battistoni et al was not very large, their work is unique in the sense that they can directly compare, in similar modules, underground events with neutrino events. All other experiments have had to rely on Monte Carlo calculations to estimate background, or use accelerator neutrino data from a completely different detector.

5.6 Neutron Background

The cosmic-ray muon flux is shown as a function of depth in Figure 5, after Menon (75). At the depths of nucleon decay experiments, the mean muon energy is 200–400 GeV. Such muons can generate nucleon cascades in inelastic scattering in the rock, and isolated neutrons from them may enter the detector and conceivably simulate nucleon decay events.

Upper limits (76) on neutron rates have been deduced from stopping muon rates (87, 88). At great depth, these muons result mostly from decay of

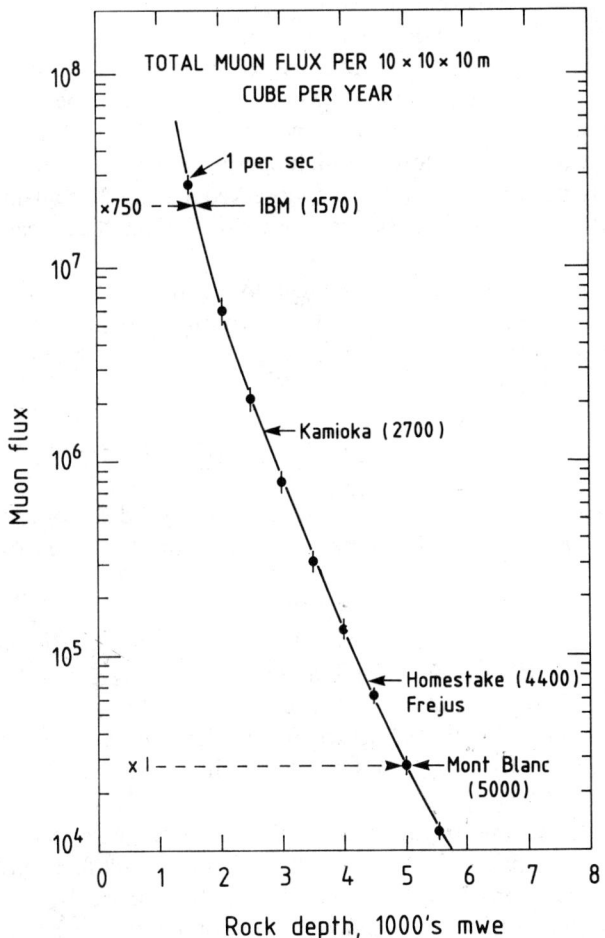

Figure 5 The underground muon flux as a function of depth in kilometers of water equivalent, after Menon (75).

pions and thus reflect hadron production by energetic muons. The actual number of isolated neutrons has been estimated by Grant (77) from a Monte Carlo cascade program. Grant considered neutrons of kinetic energy $T > 0.7$ GeV; actually, neutrons of at least 3 GeV are required to produce pions with high probability, and to mimic nucleon decay, so the Grant estimates should be reduced by a factor ≥ 30. We deduce that, for a $10 \times 10 \times 10$ m cubic detector at $h = 1600$ mwe depth (that of the IMB experiment), there are $\ll 10$ isolated neutron interactions of $E_n > 3$ GeV per year. At $h = 5000$ mwe (Mont Blanc tunnel) this would be reduced to $\ll 1$ event per year. For a detector of unit density (i.e. of 1 kiloton), the neutrino rate would be ~ 150 events per year.

These calculations indicate that neutron background is $\ll 10\%$ of neutrino background at all depths exceeding 1600 mwe. Because of the short interaction length, neutrons can in any case be vetoed by a fiducial volume cut. The practical situation is that, in the NUSEX (Mont Blanc) experiment discussed below, no high-energy (≥ 1 GeV) neutral hadrons accompanying crossing muons were observed among 10 neutrino interactions. Thus *isolated* neutron background, calculated to be at least an order of magnitude smaller than that accompanied by muons through the detector, must be less than a few percent of neutrino background.

In summary, neutron background in proton decay experiments is at a low level relative to neutrino background and may be disregarded in comparison with it.

6. PROTON DECAY EXPERIMENTS—RESULTS

As indicated in the introduction, two main types of detector are in use to study proton decay: water Čerenkov detectors and tracking calorimeters.

6.1 *Water Čerenkov Detectors*

The water Čerenkov method relies on the fact that Čerenkov light is emitted along the surface of a cone of maximum half-angle $41° (= \cos^{-1} 1/n)$ about the trajectory of a relativistic charged particle of $\beta > 1/n = 0.75$. It is therefore suitable for recording secondary electrons and photons, using photomultipliers to sample the Čerenkov light, and indeed such detectors were proposed largely because $p \rightarrow e^+\pi^0$ was the favored decay mode in SU(5). The Čerenkov cone from a track in the water will intersect the water surface in an elliptical ring, as shown in Figures 6 and 8. The ring width depends on the track length, and the relative times of firing of the photomultipliers allow the direction in space of the track to be determined. The identification of multitrack events depends on reconstruction of several Čerenkov cones. For two-track "back-to-back" events, as might be obtained in two-body proton decay, the cones are well separated in space and should be easily identified, whereas for multitrack events the pattern can be complicated and difficult to analyze.

Among the advantages of the Čerenkov technique are the following:

1. The water medium responds uniformly, and track directions are unambiguous.
2. Muon decays are detectable as delayed (2 μs) pulses.
3. Very large volumes of water can be employed. For very pure water, the absorption length in the blue region is as large as 40 m. Because of the large dimensions, proton decay events are easily contained inside the volume (radiation length in water $X_0 = 40$ cm).
4. A fraction of the nucleons (11%) are free protons.
5. Nuclear absorption of hadronic decay products in a parent oxygen

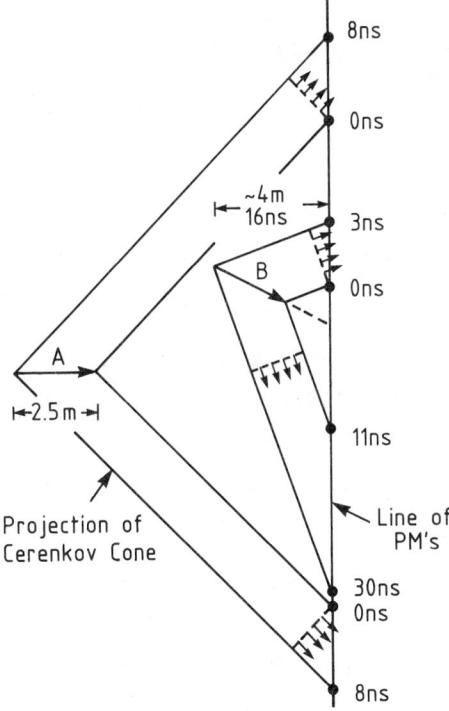

Figure 6 Principle of the water Čerenkov detector. Light is emitted on the surface of a cone of maximum half-angle 41° with axis along the track, *A* or *B*. The light hits the water surface in an elliptical ring (see also Figure 8). Directions and relative distances of tracks are determined by the time of firing of photomultipliers placed at the water surface.

nucleus can take place, but this is a smaller effect than in calorimeters employing an iron medium.

Table 9 gives a list of water Čerenkov detectors now in operation, their characteristics, and the expected photoelectron yields for different decay modes.

6.2 The IMB Experiment

The IMB experiment has been mounted in the Morton Salt Mine, Cleveland, Ohio (78, 79 and J. van der Velde, private communication). The volume of water has a total mass (fiducial mass) of 7000 (3300) tons. It is almost cubical in shape and the 2048 5-inch photomultipliers are mounted over the surface on a 1-m grid (see Figure 7). Crossing cosmic-ray muons provide the absolute energy calibration, with an error of $\pm 15\%$. The relative timing and pulse heights of the phototubes are calibrated using a pulsed light source of variable intensity, placed inside the water. Examples of the Čerenkov signals from a crossing and stopping muon are given in Figure 8. The Čerenkov light from a muon decay is detected as a delayed pulse, with efficiency $60 \pm 10\%$.

The trigger for event selection requires more than 12 photomultiplier tubes (PMTs) to fire within 50 ns, or that more than 3 PMTs in any 2 of 32 groups of 64 fire within 150 ns. Any PMT signal within 7.5 μs of the initial trigger is also recorded. The time resolution was 11 ns (FWHM) resulting in

Table 9 Water Čerenkov detectors

Experiment	IMB	HPW	Kamiokande
Location	Cleveland, Ohio	Park City, Utah	Kamioka, Japan
Depth (mwe)	1570	1500	2700
Shape and dimensions (m)	rectangular 23 × 18 × 17	cylindrical 7.3 × 12 (ϕ)	cylindrical 16 × 15.6 (ϕ)
Mass in tons (fiducial mass)	7000 (3300)	780 (560)	3000 (880)
Number of PMTs % surface	2048 5" 2%	704 5" volume	1000 20" 20%
Expected photoelectron yield for:			
$p \to e^+ \pi^0$	170	600	3560
$\to \mu^+ K^0$ $\hookrightarrow 2\pi^0$	140	—	2670
$\to \mu^+ \pi^0$	—	410	2670

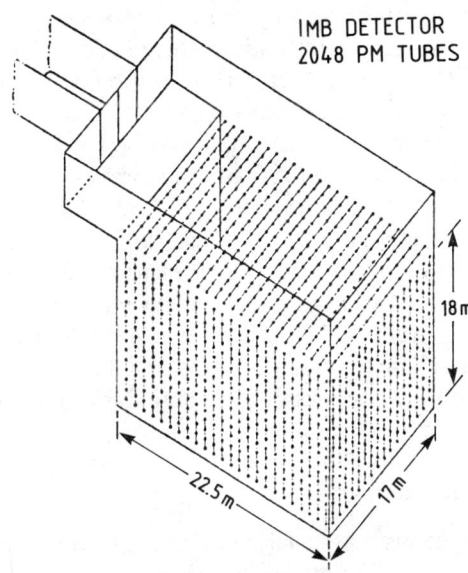

Figure 7 Sketch of the surface array of photomultipliers in the IMB experiment.

a "vertex" resolution (as judged by the pulsed point light source) for single tracks of ± 1 m and for $p \to e^+ \pi^0$ of ± 0.6 m. Two-track events are recognizable if each track fires >40 PMTs and the opening angle exceeds $100°$.

The trigger rate is 2.3×10^5 per day, due mostly to crossing muons.

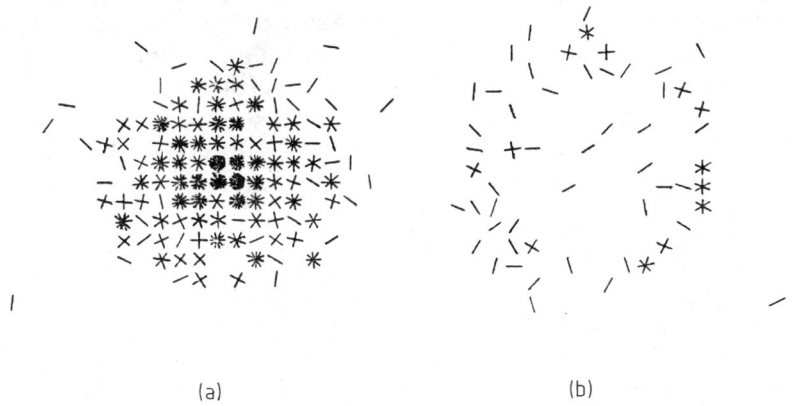

Figure 8 Photomultiplier pulses recorded for (*a*) a muon crossing the IMB detector, and (*b*) a muon stopping in the water, leaving an open Čerenkov ring. The number of crossing lines at each point indicates the PMT pulse height.

These events are subject to an energy selection (40–300 PMTs) corresponding to an energy loss for a relativistic particle of 250–1700 MeV, plus the requirement that the reconstructed vertex position be more than 2 m from the detector surface (for short tracks from neutrino reactions or proton decay, it is a good approximation to treat the light as from a point source). This selection reduces the event rate by a factor 1000, to 230 per day, and these are analyzed by three independent methods. They are mostly due to short entering and stopping muon tracks and corner-clipping muons. Based on the points on the surface where the earliest tubes fired, and using a fitting procedure, all the muons can be rejected and one is left with "contained" events inside the fiducial volume, at the rate of 3 per day with >40 PMTs firing. These candidates are examined visually using a graphics system, and about 1 per day is retained. Monte Carlo simulations show that 75% of neutrino interactions and 90% of proton decays should be retained by the filtering procedure. The selected events are due primarily to single-track neutrino interactions.

NEUTRINO INTERACTIONS After a live time of 132 days (or 4×10^{32} "proton years" of exposure), 112 neutrino events were observed (78). They are uniformly distributed through the detector volume, and allowing for the detection efficiency, correspond to a rate of 125 ± 12 events per kton yr, in reasonable agreement with expectations (see Tables 11, 7, and 8). Figure 9 shows the estimated true energy spectrum of neutrino events, after adding 230 MeV to the "Čerenkov energy" to allow for the muon mass and the fact that, for part of the range, the muon is below Čerenkov threshold. The dashed curve shows the expected distribution.

The up-down ratio in the events is 0.88 ± 0.17, consistent with the prediction $U/D = 0.86$ in Table 6. It is, however, to be emphasized that the typical angle of emission of the muon relative to the neutrino at energies below 1 GeV is large and therefore, the U/D ratio of the muons will be closer to unity than for the neutrinos. The observed fraction of events with muon decays is 25%. This is consistent with the $v_\mu/(v_e + v_\mu)$ flux ratios (Table 7), the muon detection efficiency (60%), and the higher sensitivity of the Čerenkov detector to electrons than to muons.

THE DECAY MODES $p \to e^+ \pi^0$, $p \to \mu^+ \pi^0$ All except 3 of the 112 events described above fitted the "one-track hypothesis"; that is, they are single- or multitrack events events in which the Čerenkov light is mainly concentrated in one hemisphere, with fewer than 40 PMTs firing in the backward hemisphere. The three events are "wide-angle two-track events"; that is, with a back-to-back configuration of Čerenkov cones, and are potential candidates for decay $p \to e^+ \pi^0$. (Although $\pi^0 \to 2$ photons, the value of E_π/m_π is 3.55 so that the average angle $\bar{\varepsilon}$ between the photons is

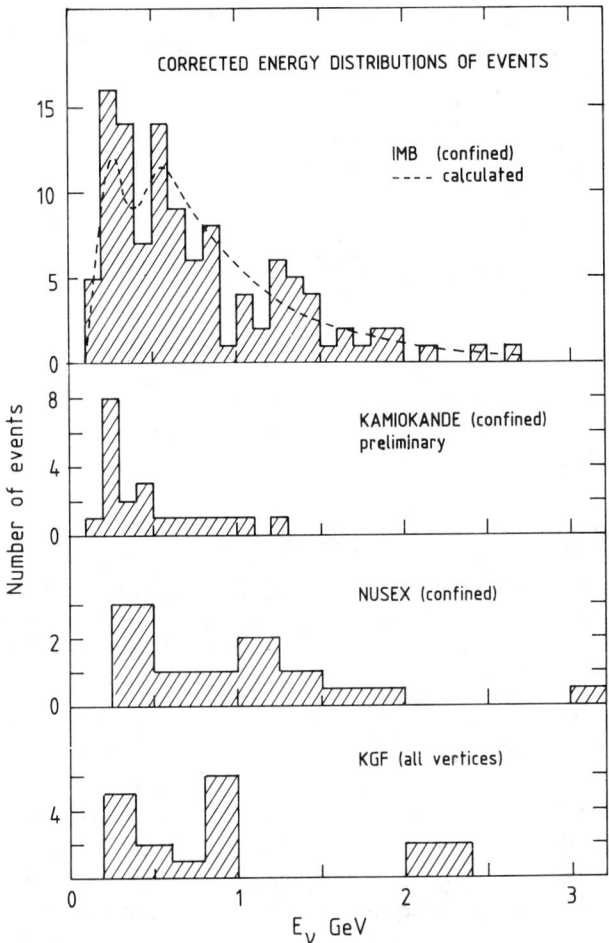

Figure 9 Energy spectra of (neutrino) events as measured in proton decay detectors. For the IMB and Kamiokande experiments, the equivalent Čerenkov energy is corrected for the fact that muon secondaries are below Čerenkov threshold at low energy. The dashed curve is the expected distribution for the IMB experiment.

smaller than the Čerenkov angle of 41°. This fact, together with the disparity between photon energies for $\varepsilon > \bar{\varepsilon}$, means the π^0 signal will appear as a single cone.) The three events have opening angles between the cones of $\theta = 115\text{–}135°$. They are, however, not acceptable as $p \to e^+\pi^0$ decays because (*a*) two have a muon decay signal, (*b*) one event triggers 340 PMTs and has total energy 1.7 GeV, and (*c*) for this event $\theta = 115°$, which is outside the range expected for decay of a free or bound proton, taking

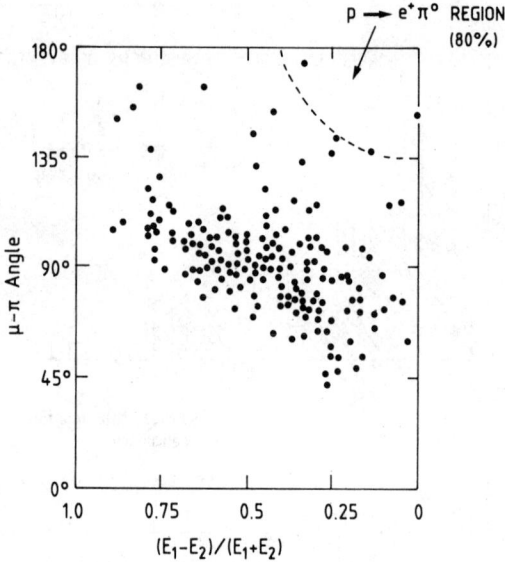

Figure 10 Plot of μ-π angle versus the energy ratio $(E_1 - E_2)/(E_1 + E_2)$ for simulated neutrino interactions, $\nu + N \to \mu + \pi + N$, in water. E_1 and E_2 are the energies of the particles, where $E_1 > E_2$. For proton decay, $p \to e^+ \pi^0$ in water, 80% of e and π energies, and angle should lie within the circular quadrant [after Foster (98)].

account of Fermi motion in the parent nucleus. It is expected that $\theta > 140°$ in the majority of genuine decays.

The three "two-track" events with $\theta > 100°$ are attributed to inelastic neutrino reactions of the form $\nu N \to \mu\pi N$ or $\nu N \to e\pi N$. The characteristics have been estimated from the configuration of events in the Gargamelle heavy liquid chamber exposed to a neutrino beam at the CERN PS [Deden et al (80)]. The opening angle between μ and π is plotted in Figure 10 against the energy ratio $(E_1 - E_2)/(E_1 + E_2)$ where E_1 is the energy of the higher energy particle. The $p \to e^+\pi^0$ events should occur in the circular quadrant, while the points are simulated neutrino data, equivalent to what would be obtained from the atmospheric neutrino flux in a 1.75-year run. We note that 12 events have $\theta > 115°$–135° so that in a 132-day exposure, two or three neutrino events would be expected.

On the basis that no events attributable to $p \to e^+\pi^0$ are observed in a more extended 250-day run, Bionta et al (79), Foster (98), and J. van der Velde (private communication) place the following 90% CL limit

$$\tau/B(p \to e^+\pi^0) > Nf\ \varepsilon_n \varepsilon T/2.3 = 1.5 \times 10^{32} \text{ yr}, \qquad 22.$$

where $N = 2 \times 10^{33}$ is the number of nucleons, $f = 10/18$ is the proton fraction, $\varepsilon_n = 0.68$ is the computed π^0 nuclear survival probability, $\varepsilon = 0.9$ is the detection efficiency, and T the time. The IMB group also quote a similar limit for

$$\tau/B(p \to \mu^+\pi^0) > 1.5 \times 10^{32} \text{ yr.} \qquad 23.$$

THE DECAY MODES $p \to \mu^+ K^0$, $n \to \nu K^0$ The first method employed to search for the decay $p \to \mu^+ K^0$ is based on the decay $K_s^0 \to 2\pi^0$, giving four γ's with a nearly isotropic distribution in angle (the Lorentz factor of the K^0 is only 1.2). The Čerenkov rings from four isotropically distributed showers are not easily separable. Bionta et al calculate a quantity called the isotropy, I, for the event, which is the vector sum of unit vectors from the vertex to each PMT, divided by the number of PMT hits. For a single track, $I \approx \cos 42° = 0.7$, while for an isotropic light source, or a "back-to-back" two-track event such as $p \to e^+\pi^0$, $I \approx 0$. Figure 11 shows the scatter plot for 109 events with Čerenkov energy $500 < E_c < 850$ MeV. Of $p \to \mu^+ K^0$ events, 90% are expected to fall within the dashed curve, and three events are found there. One is excluded on the basis of too high an energy on one track, while the vertex of a second can be adjusted so that all the light is contained in one hemisphere. The one remaining candidate yields a 90% CL limit of

$$\tau/B(p \to \mu^+ K^0) > 1.8 \times 10^{31} \text{ yr} \qquad \text{(for } K^0 \to 2\pi^0) \qquad 24.$$

after inserting the appropriate branching ratio for $K^0 \to 2\pi^0$ including a

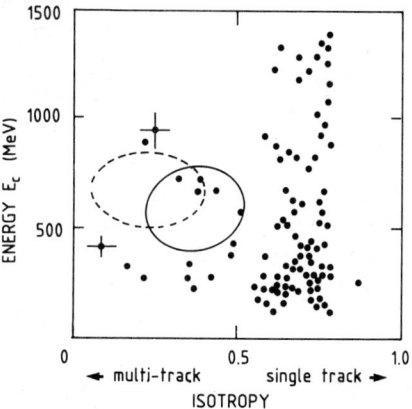

Figure 11 Čerenkov energy, E_c, from IMB events plotted against isotropy, I, defined as the sum of the unit vectors from the vertex to the hit PMTs, divided by the number of hit PMTs. For single tracks, $I \approx$ cosine (Čerenkov angle) ≈ 0.7, while for decays such as $p \to e^+\pi^0$, $p \to \nu K^0$, or $p \to \mu^+ K^0$, one expects $I < 0.5$. The dashed and full curves indicate the allowed region for decays $p \to \mu^+ K^0$, $K^0 \to 2\pi^0$ and $n \to \nu K^0$, $K^0 \to 2\pi^0$, respectively.

contribution from K_L^0 interactions. A search has also been made for the decay $p \to \mu^+ K^0$ via the mode $K_s^0 \to \pi^+\pi^-$, with the two subsequent decays $\pi^\pm \to \mu^\pm \to e^\pm$. From two events within the required pulse height range, each with two muon decays, one is rejected because it again contains a single track of too high an energy, and the remaining candidate gives

$$\tau/B(p \to \mu^+ K^0) > 1.3 \times 10^{31} \text{ yr } (K^0 \to \pi^+\pi^-). \qquad 25.$$

The results in Equations 24 and 25 can be combined to give

$$\tau/B(p \to \mu^+ K^0) > 2.6 \times 10^{31} \text{ yr}. \qquad 26.$$

The above methods have also been applied to the decay $n \to \nu K^0$, $K^0 \to 2\pi^0$. The 90% contour for such events in Figure 11 is shown by the full curve. Of the five events in this region, only three survive after rejecting one event with a muon decay and one with too much energy in one track, leading to the limit

$$\tau/B(n \to \nu K^0) > 0.8 \times 10^{31} \text{ yr}. \qquad 27.$$

Other limits (79, 98; J. van der Velde, private communication) obtained by the IMB group are given in Table 13. In summary, this experiment has given by far the most stringent limits on lifetime for a variety of decay modes of the nucleon. It is clear that, with increased statistics, the limit for the decay $p \to \mu^+\pi^0$ or $e^+\pi^0$ can be pushed proportionately further, possibly by another order of magnitude. On the other hand, the multiprong decays such as $p \to \mu^+ K^0$, or those without a clear back-to-back signature such as $n \to \nu K^0$, are more difficult to differentiate from background, and the results (Equations 24–27) can most likely not be improved by a large factor.

6.3 *The HPW Experiment*

The water detector employed by the HPW Collaboration (82) in a mine at Park City, Utah, is in the form of a cylindrical tank of diameter 12 m and depth 7.3 m. The 700 5-inch PMTs are mounted on a 1-m grid through the water volume. The inside of the tank is covered with an aluminum reflector to increase light output and a veto shield of proportional wire chambers surrounds the tank. Operational since early 1983, the equipment has been checked by measuring muon angular distributions and the stopping muon lifetime. Pattern recognition in the volume array, with reflection of light by the walls, is clearly more difficult than for the IMB surface array. No proton lifetime limits have yet been reported.

6.4 *The Kamiokande Experiment*

Constructed by groups from Tokyo, KEK, Niigata, and Tsukuba (99 and M. Koshiba, private communication), this detector is a cylindrical water tank of diameter 15.6 m and height 16 m, with (fiducial) mass of 3000 (1000) tons. Twenty-inch diameter photomultipliers are mounted on a surface grid inset 1.5 m from the tank walls. The photocathodes cover 20% of the surface area, that is about 10 times that of the IMB detector.

On the basis of 98 days of running, 29 events, presumably of neutrino origin, have been observed in the fiducial volume, corresponding to a rate of 123 ± 23 per kton yr, in good agreement with expectations (see Table 11). The energy spectrum of a sample of single-track neutrino reactions is included in Figure 9. Limits have been set on the decay modes discussed for the IMB experiment (see Table 13). They are about one order of magnitude smaller, because of the smaller fiducial mass and running time to date. In principle, this device is sensitive to decay modes which would be below the trigger threshold of the IMB experiment. Furthermore, with a tenfold larger fractional area covered by photocathode, it should be able to identify multitrack events and make more rigorous cuts on background, for example on the basis of coplanarity.

6.5 Tracking Calorimeter Detectors

The water Čerenkov detectors described above possess a sensitivity to nucleon decay, in terms at least of the mass of active medium, that is hard to match by other methods. Nevertheless, although they have given by far the most stringent limits on two-body decay modes such as $p \rightarrow e^+ \pi^0$ and $\mu^+ \pi^0$, they have some disadvantages when dealing with events of higher multiplicity or with nonrelativistic secondaries, and their ability to discover unexpected decay modes is limited. Tracking calorimeters were developed in parallel with the water Čerenkovs, because they depended on well-proven techniques at a time when the true potentiality of the water Čerenkov method was unknown (see Table 10). The advantages of an iron calorimeter instrumented with layers of gas counters are the following:

1. Because of the high density and high mass number of the medium, proton decay products can be confined within a volume of dimensions one order of magnitude smaller than for water detectors. Even for arrays of order 100 tons only, most of the mass can be useful for containing events.
2. Vertices can be reconstructed with a precision of order 1 cm, compared with 1 m, and this enormously aids pattern recognition.
3. Tracks can be followed in detail only limited by the sampling frequency, and particle direction inferred from scattering (or, if it is measured, ionization), or existence of muon decays.
4. The detector can be easily built in modular form, is transportable, and can be tested with accelerator beams.

Table 10 Tracking calorimeter detectors

Experiment	KGF	NUSEX	Frejus
Location	Kolar Gold Fields, southern India	Mont Blanc tunnel	Frejus tunnel
Depth (mwe)	7600	5000	4400
Dimensions (m)	6 × 4 × 4	3.5 × 3.5 × 3.5	6 × 6 × 13
Mass (tons)	140	150	1000
Steel thickness (cm)	1.2	1	0.3
Counters, dimensions (cm)	10 × 10 proportional counters	1 × 1 resistive streamer tubes	0.5 × 0.5 flash tubes
Year operational	1980	1982	1984

6.6 The Soudan I Detector

The Soudan I calorimeter consists of an array of 3456 proportional tubes, each 4.5 cm in diameter, embedded in an iron/concrete matrix. The dimensions are 2.9 × 2.9 × 1.9 m and the total mass is 31.5 tons. It was built and is operated by a Minnesota-ANL Collaboration [Bartelt et al (92)]. It is located at a depth of 1800 mwe at the Soudan mine in northern Minnesota. In 0.38 yr of operation, one contained event has been reported. It is interpreted as a neutrino reaction and the expected rate (Table 7) is consistent with this. The experiment places a lower limit on the proton lifetime $\tau > 10^{30}$ yr.

6.7 The KGF Calorimeter

The first iron calorimeter specifically designed to search for nucleon decay was built by the Tata-Osaka-Tokyo Collaboration [Krishnaswamy et al (85, 86)], and installed since 1980 in a mine at the Kolar Gold Fields (KGF) in southern India. At a depth of 7600 mwe, it is by far the deepest detector in operation. The detector consists of a rectangular box of dimensions 6 × 4 × 4 m consisting of 1.2-cm thick iron plates separated by 34 horizontal layers of proportional counters, each counter being 10 × 10 cm in cross section. Alternate layers of counters are mounted at right angles to give three-dimensional track coordinates. The total mass is 140 tons and the mean density is 1.5. Because of the low density, many events originating in the detector are unconfined. The basic trigger is a five-fold coincidence of pulses among any successive 11 layers.

After 2.2 yr of operation (as of July 1983) a total of 17 events have been observed with a vertex inside the detector. This tally includes five single-track events. Since the detector has no directional information, elastic neutrino events, consisting of a single exiting muon, are indistinguishable from muons entering and stopping in the detector. The number of muons generated by neutrinos outside the detector, and stopping in it, must be equal to the number of muons generated inside and leaving. Using this principle, and up/down equality of the fluxes, the five single tracks entering or leaving the bottom represent a fair estimate for the total of one-prong neutrino events. Making an allowance for trigger efficiency, this provides a gross event rate of 72 ± 17 per kton yr, compatible with the charged current neutrino rate at $\lambda = 0°$ magnetic, of 107 per kton yr calculated by Gaisser & Stanev (84). The number of totally confined events is 7 (the detector is 26 radiation lengths deep, and v_e, $\bar{v}_e$ interactions are expected to form 30% of the total neutrino rate).

In the list of events published by the KGF Collaboration, three are claimed to be nucleon decay candidates with all tracks fully contained in the detector—and another three as partially confined nucleon decay candidates. Figure 12 shows two confined candidates. Notice that because of the large (10 × 10-cm cross section) gas counters spatial resolution is poor and the number of hits per track per stereo view is very small (2–3 on average). This means that track and vertex reconstruction is extremely difficult. Event 587, containing three or four prongs is interpreted as $p \to e^+ \pi^0$ because of the scatter of hits, implying electromagnetic origin. Event 877 is interpreted as a two-prong event with a back-to-back configuration; on the basis of the criteria of Battistoni et al described previously, it is also unlikely as a neutrino interaction. This and the third event are interpreted as $p \to \nu \pi^+$ or νK^+, and $p \to \mu^+ K^0$ or $n \to e^+ \pi^-$ respectively. If the reconstruction of these events is accepted, they are unlikely to be of neutrino origin.

What are the KGF events? Are they nucleon decays, as the authors claim, or simply background? I believe the answer does not lie in just trying to reconstruct tracks in these events and showing that they have unacceptable topologies as neutrino interactions; the very reconstruction of tracks and vertices must be highly subjective and therefore suspect. For example, secondary hadrons undergo Coulomb and nuclear scattering (and absorption) in iron, at a rate critically dependent on particle momentum. There is no way one can take this into account with a coarse-grain detector or on an event-by-event basis, because the information level is too low. Rather one must expose the detector to an accelerator beam, to observe a sample of neutrino reactions directly for comparison, and on a statistical basis; or, failing that, generate neutrino events by a Monte Carlo program using as input actual data (e.g. from bubble chamber accelerator neutrino events), and propagating such Monte Carlo events through the detector. Unless some additional information of this kind is fed in, it seems impossible to reach definite conclusions about the KGF events.

In making these criticisms, it should, however, be emphasized that the KGF detector was, by a long way, the first in the field, and that if the proton lifetime had been a factor 100 shorter than the present limit, this group would have discovered proton decay, just from the gross event rate. Their great contribution in the field was to be there first with a large detector that

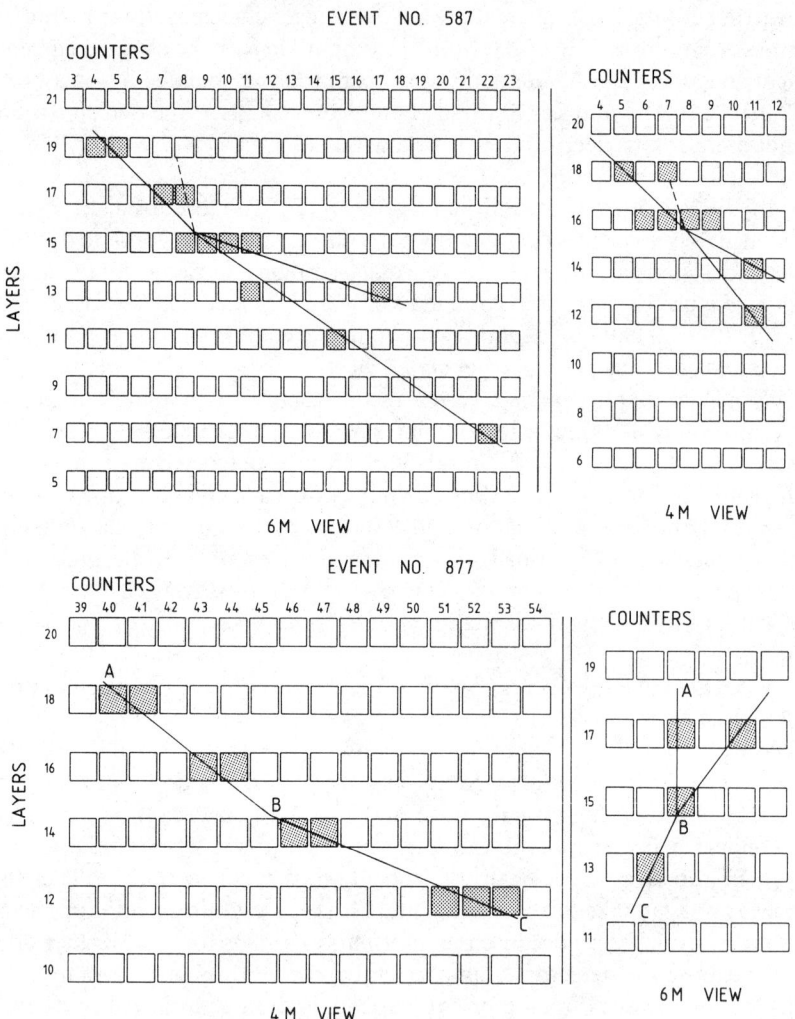

Figure 12 Events in the KGF detector, showing, in the two orthogonal views, the proportional tubes that fire. Event 587 is ascribed to the decay $p \to e^+ \pi^0$ (on the basis of energy and scatter of hit tubes indicating electromagnetic origin). Event 877 is ascribed to the decay $n \to e^+ \pi^-$, or $p \to \mu^+ K^0$, $K^0 \to \pi^+ \pi^-$.

Table 11 Event rates observed

Experiment	KGF (101)	Kamioka (99)[a]	IMB (79)	NUSEX (91)
Latitude	2°	27°	50°	50°
Number of (neutrino) events	~17[b]	29	112	10
Kiloton yr (fiducial volume)	0.22	0.27	1.20	0.113
Rate per kton yr (corrected for efficiency)	77 ± 19	123 ± 23	125 ± 12	118 ± 37[c]
Prediction (61) solar minimum (see Table 6b)	85	99	132	132

[a] Also M. Koshiba, private communication.
[b] Total vertex rate.
[c] Assumes 75% containment efficiency.

could contain decay events, and to stimulate the building of larger and more finely grained detectors.

6.8 *The NUSEX Calorimeter*

The NUSEX calorimeter consists of a 3.5 × 3.5 × 3.5-m cube of mass 150 tons built from 136 layers of 1-cm thick iron plates separated by layers of 1 × 1-cm plastic streamer tubes. The experiment is being carried out by a collaboration of Frascati, Milano, Torino, and CERN (89–91) in the Mont Blanc tunnel, at 5000 mwe depth. Each one of the 4300 tubes can provide a three-dimensional track coordinate via read-out from anode wires and cathode pick-up strips orthogonal to the anodes (Figure 13). The mean density is 3.6 g cm^{-3} and the radiation length $X_0 = 4.5$ cm. The event

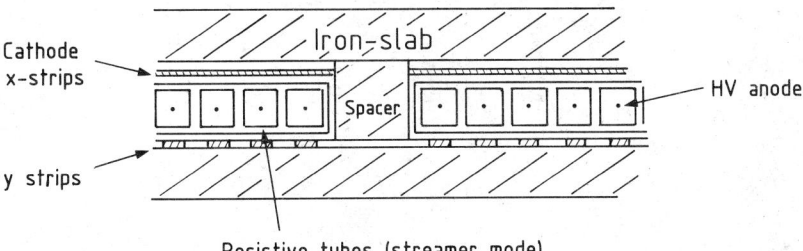

Figure 13 Configuration of steel plates and resistive plastic streamer tubes in the NUSEX detector. Readout of the streamer coordinates is by pick-up on orthogonal (x, y) cathode strips.

trigger requires a coincidence of any four adjacent planes of counters, or of three adjacent planes plus two other adjacent planes elsewhere in the array.

NEUTRINO EVENTS After a live time of 0.87 yr (130 ton yr total, 113 ton yr fiducial) the collaboration have observed 10 contained events (97). The gross event rate of 118 ± 37 per kton yr is compatible with that expected from neutrino background (see Table 11). The identified neutrino events (Figure 14) include six elastic and up to four inelastic events and two due to $v_e, \bar{v}_e$. In bins of visible energy, the sample of 10 events is consistent with rates expected in Table 8, with five events below 1 GeV and five above (see also Figure 9).

PROTON DECAY CANDIDATE Among the 10 events in the NUSEX calorimeter is one stated to be incompatible with a neutrino interpretation

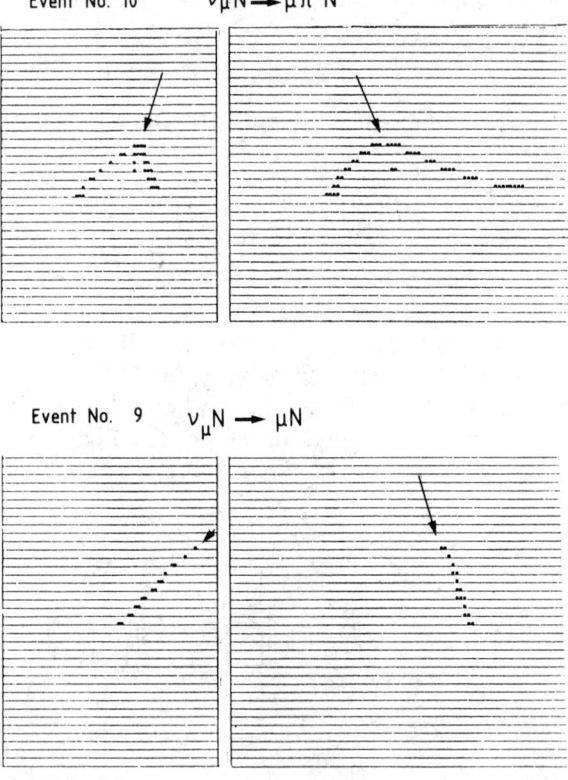

Figure 14 Examples of interactions of atmospheric neutrinos in the NUSEX detector. Only a small section of the detector is displayed in each case. The x and y coordinate "views" are shown, side by side.

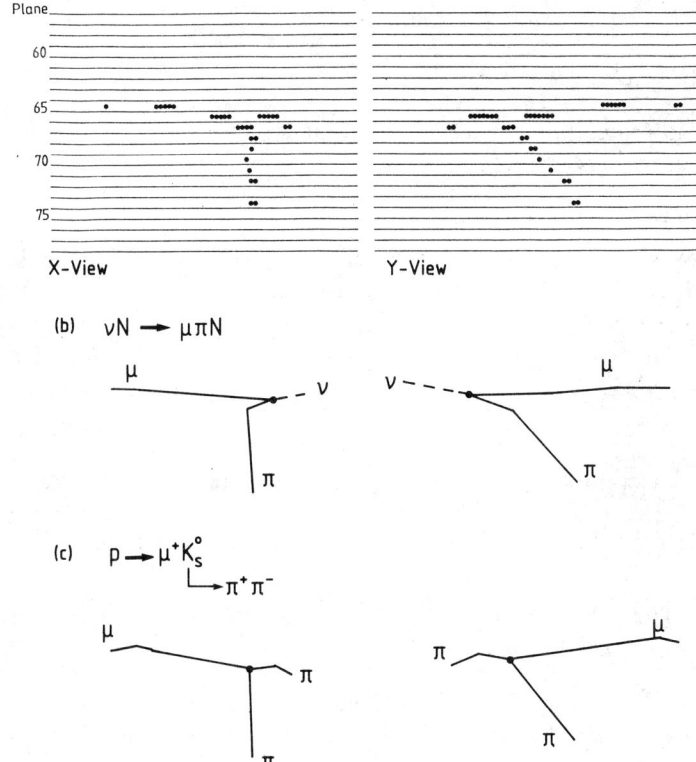

Figure 15 (*a*) The orthogonal views of a NUSEX event that is a candidate for proton decay. (*b*) Interpretation of event in terms of a neutrino interaction $\nu_\mu N \to \mu\pi N$, with large-angle scattering of the pion. (*c*) Interpretation according to the decay $p \to \mu^+ K_s^0$, $K_s^0 \to \pi^+\pi^-$.

(90). It is shown in Figure 15*a*. If the event were to be ascribed to a reaction of the form $\nu N \to \mu\pi N$, then the pion would have to undergo a large-angle scatter as shown (Figure 15*b*), and the total energy and momentum would then be $E_{vis} = 0.84 \pm 0.20$ GeV, $p_{vis} = 0.81 \pm 0.20$ GeV/c. The probability that a neutrino interaction could have such a topology is found directly from the accelerator data in the same calorimeter, discussed in Section 5.4, to be 2/400. Hence the expected number of neutrino events of this type is 0.05. An alternative interpretation (Figure 15*c*) is of the decay $p \to \mu^+ K^0$, $K^0 \to \pi^+\pi^-$. In that case, the visible energy and momentum are $E_{vis} = 1.0 \pm 0.2$ GeV and $p_{vis} = 0.4 \pm 0.2$ GeV/c. A finite momentum is permitted for proton decay when allowance is made for nuclear Fermi motion. If it is so interpreted, then, after correcting for efficiency and probability of observing K^0 if it decays in the K_L^0 mode instead of $K_s^0 \to \pi^+\pi^-, \pi^0\pi^0$, the NUSEX group find a 90% CL limit of

$$\tau/B(p \to \mu^+ K^0) \geqq 0.9 \times 10^{31} \text{ yr.} \qquad 28.$$

Other interpretations in terms of alternative decay modes ($p \to 3\mu$, $p \to K^0\pi\nu$) are also possible for this event.

6.9 The Frejus Experiment

The device located in the Frejus tunnel near Modane (4400 mwe) consists of an array of 3-mm steel plates separated by layers of 0.5×0.5-cm polypropylene flash chambers. There are 500 flash tube planes (a total of 10^6 tubes), and 124 planes of Geiger tubes for triggering, with 10.5 cm between trigger planes. The experiment is being carried out by a collaboration of Aachen, Orsay, Palaiseau, Saclay, and Wuppertal, and the full detector mass will be 1 kton (103).

The detector has three times finer granularity than the NUSEX detector and its expected energy resolution for electrons and photons is 15%, and for pions, 12–20%. The full device should be operating in mid 1984.

7. LIMITS ON MONOPOLE CATALYSIS OF PROTON DECAY

Several of the proton decay experiments have set limits on fluxes of magnetic monopoles. Emphasis is on the proposed superheavy GUT monopoles of mass $\sim 10^{16}$ GeV, which were supposedly created in the early stages of the Universe and which are likely to have present velocities in the range $10^{-4} < \beta < 10^{-2}$. These could certainly penetrate deeply into the Earth and be recorded in proton decay detectors.

Flux limits have been set by the Soudan I Collaboration (93), the NUSEX Collaboration (91), and the KGF Collaboration (101) by assuming that monopoles produce sufficient ionization to leave tracks and by utilizing a trigger that ensures that the particle velocity is in the above range. No events with the required linearity in traversal time through successive layers of detector have been observed in any of these experiments. The corresponding flux limits, of order $F_m < 10^{-13}$ cm^{-2} sr^{-1} s^{-1}, are given in Table 12.

Rubakov (94) and Callan (95) have argued that GUT monopoles M may catalyze nucleon decay via a strong interaction process of the type

$$M + p \to M + e^+ + \text{mesons} \qquad 29.$$

with a cross section

$$\sigma_c = \sigma_0/\beta_r,$$

where $\sigma_0 \approx O(1 \text{ GeV})^{-2} \approx 0.1$ mb and the catalysis cross section σ_c is expected to follow the $1/\beta_r$ dependence typical of an exothermic capture reaction. Here, β_r is the relative velocity of monopole and proton. In a nucleus, for slow monopoles, $\beta_r \approx 0.1$, the velocity of Fermi motion. If σ_c is large enough, so that the interaction mean free path $\lambda (= 1/N\sigma_c$, where N is the nucleon density) is comparable with or smaller than the detector dimensions, multiple proton decays should be observed. The question of whether such decays can be recognized will depend on the monopole velocity, β, and the sensitive time or dead time of the detector, following the first proton decay event.

Limits have been set on such a process by the IMB Collaboration (96) and by the NUSEX Collaboration (91). Errede et al (96) observe no multiple events inside an 8-μs time slot (corresponding to $\beta > 10^{-4}$), in a 100-day run. The corresponding limit on flux of monopoles is $F_m < 7 \times 10^{-15}$ cm^{-2} sr^{-1} s^{-1} for large values of $\sigma_c \geq 10$ mb, and for $10^{-4} < \beta < 10^{-1}$, where the efficiency for detecting multiple events would have been essentially 100%. For smaller values of σ_c the limit is less stringent (see Table 12). Battistoni et al (91) also observe no double interactions, but in

Table 12 Monopole flux limits in proton decay detectors

Experiment	Range of β	F_m (cm^{-2} sr^{-1} s^{-1}) (90% CL upper limit)
Time of flight		
Bartelt et al (93)	10^{-3}–10^{-2}	4×10^{-13}
Battistoni et al (91)	10^{-4}–10^{-2}	4×10^{-12}
Krishnaswamy et al (101)	10^{-3}–10^{-2}	1×10^{-13}
Monopole-induced proton decay		
Absence of double events:		
Battistoni et al (91)	10^{-4}–10^{-1} $\sigma_c \geq 5$ mb	2×10^{-14}
Errede et al (96)	10^{-4}–10^{-1} $\sigma_c \geq 10$ mb	7×10^{-15}
Single-candidate rate:		
Battistoni et al	$\sigma_c = 1$ mb	10^{-13}
	$= 0.1$ mb	3×10^{-13}
Errede et al	$= 1$ mb	2×10^{-12}
	$= 0.1$ mb	10^{-11}

this case, they would have to occur either within a 5-μs time slot or after a recovery interval of 20 ms. They find for $\sigma_c \geq 5$ mb, $F_m < 2 \times 10^{-14}$ cm^{-2} sr^{-1} s^{-1}.

If the catalysis cross section $\sigma_c \ll 1$ mb, multiple proton decays would be infrequent in any of these detectors, since λ is larger than the detector size ($\sigma_c = 1$ mb corresponds to 16 m of water). Nevertheless, a flux limit can be set from the frequency of single interactions. Errede et al (96) argue that, because of the "brick-wall" kinematics between a nucleon and massive monopole, a monopole-catalyzed nucleon decay should not have zero net momentum for the decay products, and can indeed look like any neutrino reaction of energy below 2 GeV. They therefore conservatively assign all 66 neutrino events in 100 days run time to monopole-induced proton decays. The flux limit in this case is $F_m < 2 \times 10^{-12}$ cm^{-2} sr^{-1} s^{-1} for $\sigma_c = 1$ mb, rising to $F_m < 10^{-11}$ for $\sigma_c = 0.1$ mb. Battistoni et al use their single proton decay candidate to set a limit on monopole flux of $F_m < 10^{-13}$ for $\sigma_c = 1$ mb, rising to three times this limit for $\sigma_c = 0.1$ mb.

In summary, no experiments have direct evidence for GUT monopoles in the form of tracks due to slow particles, and limit the flux to the level $F_m < 10^{-12}$–10^{-13} cm^{-2} sr^{-1} s^{-1}, for velocities in the range $\beta = 10^{-4}$–10^{-2}. These limits depend on assumptions about monopole velocity and ionization in gases that are still the subject of debate. There is also no evidence for proton decay catalyzed by monopoles. The flux limits also depend on the monopole velocities and of course on the catalysis cross sections, but are in the region $F_m < 10^{-14}$–10^{-12} cm^{-2} sr^{-1} s^{-1}. None of these limits yet reaches inside the Parker bound (104) of $F_m < 10^{-15}$ cm^{-2} sr^{-1} s^{-1}, which is the maximum flux that could be tolerated without destroying the galactic magnetic field, of order 3 μG.

8. COMPARISON OF RESULTS, FUTURE POSSIBILITIES, AND CONCLUSIONS

8.1 Present Results

All experiments described find events attributable to neutrino interactions. If we assign *all* events in this way, the total rates observed at different locations (Table 11) are in agreement with expectations based on calculated fluxes. Further, the energy spectrum of events (Figure 9), the relative numbers due to upward and downward neutrinos, and the relative numbers attributed to ν_μ and ν_e, respectively, are all in accord with predictions. In particular, there are no peaks in energy spectra or anomalies in prong distributions that might suggest another major source of events.

Table 13 summarizes the proton lifetime limits from different detectors and for various decay modes. Regarding the decay mode p $\rightarrow$ e$^+\pi^0$, IMB set

Table 13 Recent limits on proton lifetime[a]

Experiment (Reference)	KGF (101)	NUSEX (91)	KAMIOKA (99)[b]	IMB (79)
$p \to e^+\pi^0$	~2 (1)	1.5	2.0	15
$\to \mu^+\pi^0$	~2 (1)	1.0	2.0	15
$\to e^+\rho^0$	—	—	2.5	—
$\to \mu^+ K^0$	~2 (1)	0.6 (1)	0.8 (1)	2.6 (1)
$\to \mu^+\eta^0$	—	—	0.4 (1)	5.0
$\to e^+ K^0$	—	—	0.9	3.1
$\to \nu K^+$	~2 (1)	0.2	0.5 (2)	—
$\to \nu\pi^+$	—	0.2 (≤ 3)	0.3 (2)	—
$n \to e^+\pi^-$	~2 (1)	1.5	1.2	—
$\to \nu\pi^0$	—	0.7	0.6	—
$\to \nu K^0$	—	0.5	0.3	0.8 (3)

[a] In units of 10^{31} yr × branching ratio. 90% CL lower limits, unless ~ indicates signal claimed. Number of candidates, if any, given in parentheses.
[b] Also M. Koshiba, private communication.

a limit $\tau/B(p \to e^+\pi^0) > 1.5 \times 10^{32}$ yr, while the KGF Collaboration claim one event as a serious candidate, with negligible background. The corresponding lifetime is $\sim 2 \times 10^{31}$ yr. The joint probability that these two experiments are statistically compatible has a maximum value of 2.2%, and, in view of the remarks in a previous section, the KGF claim has to be rejected. For the decay mode $p \to \mu^+ K^0$, the limits are less severe. On the basis of one possible candidate each, IMB give a limit $\tau/B(p \to \mu^+ K^0) > 2.6 \times 10^{31}$ yr while NUSEX quote $> 0.6 \times 10^{31}$ yr, with neutrino background at the 7% level. These two results are compatible, with a joint probability of 15% [for $\tau/B(p \to \mu^+ K^0) \approx 7 \times 10^{31}$ yr].

8.2 Future Experiments

For the decay mode $p \to e^+\pi^0$, the water detectors have little background and, in time, the existing IMB experiment should push the limit to around 10^{33} yr. On the other hand, there does not appear to be much prospect of pushing the limits on the modes $p \to \mu^+ K^0$, $n \to \bar{\nu} K^0$ by more than a factor of three or so, because of the branching ratio $K^0 \to 2\pi^0$ and the absence of clean kinematic handles, as in $p \to e^+\pi^0$. For these decay modes, large (kiloton) tracking calorimeters are probably more suitable. The new Frejus calorimeter should be able to attain a limit for $N \to \mu K, \bar{\nu} K$ of 10^{32} yr or beyond.

Our discussions concerned results from existing detectors. New projects are also proposed, including Soudan II, a 1-kiloton calorimeter instrumented with drift chambers, to be located in northern Minnesota and intended to provide information on particle direction from track ionization; a multikiloton calorimeter equipped with flash chambers, with track direction from timing, to be located in the Gran Sasso tunnel; and various "second generation" projects, for example using liquid argon as the calorimeter medium.

8.3 Conclusions

At present, there is simply no evidence that protons decay. The IMB limit on $p \to e^+ \pi^0$ definitely excludes the minimal SU(5) version of GUTs, but a signal in other modes, such as μK^0 or νK^0, is by no means excluded at a level that should be detectable in the future (10^{32} yr or even longer) and compatible with other versions of GUT. I may quote here from my assessment at the Paris Conference (July 1982), since what I said then (102) is still true today: "Nucleon decay, if it is ever discovered, will have to be based on unimpeachable evidence from several independent experiments using different techniques. We are a long, long way from such a goal." The fact that there are physicists actively engaged on the search is mute testimony to the baryon asymmetry of the Universe, and by inference, that protons do decay at some level. It is important to find such a process, not least because its existence would be a unique test of our ideas of grand unification, and the observed decay modes would give us an important insight to the nature of the mechanisms and gauge groups involved.

Literature Cited

1. Weyl, H. 1929. *Z. Phys.* 56:330
2. Stueckelberg, E. C. G. 1938. *Helv. Phys. Acta* 11:225–99
3. Wigner, E. P. 1949. In *Proc. Am. Philos. Soc.* 93:521; 1952. *Proc. Natl. Acad. Sci.* 38:449
4. Reines, F., Cowan, C. L., Goldhaber, M. 1954. *Phys. Rev.* 96:1157
5. Yang, C. N., Mills, R. L. 1954. *Phys. Rev.* 96:191
6. Lee, T. D., Yang, C. N. 1955. *Phys. Rev.* 98:1501
7. Yamaguchi, Y. 1959. *Proc. Theor. Phys.* 22:373
8. Backenstoss, G. K., et al. 1960. *Nuovo Cimento* 16:749
9. Sakharov, A. 1967. *JETP Lett.* 5:24
10. Pati, J. C., Salam, A. 1973. *Phys. Rev. Lett.* 31:661
11. Georgi, H., Glashow, S. 1974. *Phys. Rev. Lett.* 32:438
12. Georgi, H., Quinn, H. R., Weinberg, S. 1974. *Phys. Rev. Lett.* 33:451
13. Goldhaber, M., Sulak, L. R. 1981. *Comments Nucl. Part. Phys.* 10:215
14. Hughes, V. W. 1949. *Phys. Rev.* 76:474; 1957. 105:170
15. Zorn, J. C., Chamberlain, G. E., Hughes, V. W. 1963. *Phys. Rev.* 129:2566
16. Piccard, A., Kessler, E. 1925. *Arch. Sci. Phys. Nat.* 7:340
17. Hillas, A. M., Crawshaw, T. E. 1959. *Nature* 184:892
18. King, J. G. 1960. *Phys. Rev. Lett.* 5:562
19. Eötvös, R. V., Pekar, D., Fekete, F. 1922. *Ann. Phys. (Leipzig)* 68:11
20. Dicke, R. H. 1962. In *Proc. Int. Sch. Phys. "Enrico Fermi"*, ed. G. Polyani. New York: Academic; Roll, P. G., Krotkov, R., Dicke, R. H. 1964. *Ann. Phys.* 26:442
21. Braginsky, V. B., Panov, V. I. 1972. *Sov. Phys. JETP* 34:463
22. Renner, J. 1935. *Mat. Term. Ert.* 53:542
23. Kolb, E. W., Turner, M. S. 1983. *Ann. Rev. Nucl. Part. Sci.* 33:645
24. Kuz'min, V. A. 1970. *JETP Lett.* 12:228
25. Yoshimura, M. 1978. *Phys. Rev. Lett.* 41:281

26. Weinberg, S. 1979. *Phys. Rev. Lett.* 42: 850; 43: 1566
27. Toussaint, D., Treiman, S. B., Wilczek, F., Zee, A. 1979. *Phys. Rev.* D19: 1036
28. Yoshimura, M. 1979. *Phys. Lett.* 88B: 294
29. Kolb, E. W., Wolfram, S. 1980. *Phys. Lett.* 91B: 217
30. Fry, J. N., Olive, K. A., Turner, M. S. 1980. *Phys. Rev.* D22: 2953, 2977
31. Wilczek, F., Zee, A. 1979. *Phys. Rev. Lett.* 43: 1571
32. Langacker, P. 1981. *Phys. Rep.* 72: 186
33. Berezinsky, V. S., Ioffe, B. L., Kogan, Ya. I. 1981. *Phys. Lett.* 105B: 33
34. Brodsky, S. T., Ellis, J., Haglin, J. S., Sachrajda, C. T. 1983. *SLAC-PUB 3141*
35. Weinberg, S. 1982. *Phys. Rev.* D26: 287
36. Sakai, N., Yanagida, T. 1982. *Nucl. Phys.* B197: 533
37. Dimopoulos, S., Raby, S., Wilczek, F. 1982. *Phys. Lett.* 112B: 133
38. Ellis, J., Nanopoulos, D. V., Rudaz, S. 1982. *Nucl. Phys.* B202: 43
39. Ellis, J., Hagelin, J., Nanopoulos, D. V., Tamvakis, K. 1983. *Phys. Lett.* 124B: 484
40. Flerov, G. N., Kolchkov, D. S., Skobkin, V. S., Terent'ev, V. V. 1958. *Sov. Phys. Dokl.* 3: 79
41. Evans, J. C., Steinberg, R. I. 1977. *Science* 197: 989
42. Kirsten, T., Müller, H. W. 1969. *Earth Planet. Sci. Lett.* 6: 271
43. Hennecke, E. W., Manuel, O. K., Sabu, D. D. 1975. *Phys. Rev.* C11: 1378
44. Bennett, C. L. 1981. In *Proc. 2nd Workshop on Grand Unification, Ann Arbor*. Cambridge, Mass: Birkhäuser Boston
45. Buffington, A., Schindler, S. M. 1981. *Astrophys. J.* 247: L105
46. Rosen, S. P. 1975. *Phys. Rev. Lett.* 34: 774
47. Fireman, E. L. 1978. *Proc. Neutrino '77, Baksan Valley*, p. 53 (publ. in *Nauka*, Moscow)
48. Goldhaber, M., Langacker, P., Slansky, R. 1980. *Science* 210: 851
49. Reines, F., Cowan, C. L., Kruse, H. W. 1957. *Phys. Rev.* 109: 609
50. Giamati, C. C., Reines, F. 1962. *Phys. Rev.* 126: 2178
51. Kropp, W. R., Reines, F. 1965. *Phys. Rev.* 137B: 740
52. Gurr, H. S., Kropp, W. R., Reines, F., Meyer, B. 1967. *Phys. Rev.* 158: 1321
53. Reines, F., Crouch, M. F. 1974. *Phys. Rev. Lett.* 32: 493
54. Bergamesco, L., Picchi, P. 1974. *Lett. Nuovo Cimento* 11: 636
55. Learned, J., Reines, F., Soni, A. 1979. *Phys. Rev. Lett.* 43: 907
56. Cherry, M. L., et al. 1981. *Phys. Rev. Lett.* 47: 1507
57. Allkofer, O. C., Carstensen, K., Dau, W. D. 1971. *Phys. Lett.* 36B: 425
58. Appleton, I. C., Hogue, M. T., Rastin, B. C. 1971. *Nucl. Phys.* B26: 365
59. Duthie, J., Fowler, P. H., Kaddoura, A., Perkins, D. H., Pinkau, K. 1962. *Nuovo Cimento* 24: 122
60. Osborne, J. L., Said, S. S., Wolfendale, A. W. 1965. *Proc. Phys. Soc.* 86: 93
61. Gaisser, T. K., Stanev, T., Bludman, S. A., Lee, H. 1983. In *Proc. 4th Workshop on Grand Unification, Univ. Penn. Phys. Rev. Lett.* 51: 223
62. Liland, A. 1979. In *Proc. 16th Int. Conf. Cosmic Rays, Kyoto 1979*, Vol. 13, p. 353. Tokyo: Univ. Tokyo Press
63. Markov, M. A., Zheleznyk, I. M. 1960. In *Proc. Int. Conf. High Energy Phys., Rochester*. New York: Univ. Rochester Press/Interscience
64. Zatsepin, G. T., Kuz'min, V. A. 1962. *JETP* 14: 1294
65. Tam, A. C., Young, E. C. M. 1970. In *Proc. 11th Int. Conf. Cosmic Rays, Budapest; Acta Phys. Acad. Sci. Hung.* 29: S4, 307
66. Olbert, S. 1954. *Phys. Rev.* 96: 1400
67. Conversi, M. 1950. *Phys. Rev.* 79: 749
68. Sands, M. 1950. *Phys. Rev.* 77: 180
69. Ascoli, G. 1950. *Phys. Rev.* 79: 812
70. Volkova, L. V. 1980. *Sov. J. Nucl. Phys.* 31: 784
71. Ormes, J., Freier, P. 1978. *Astrophys. J.* 222: 471
71a. Eichten, T., et al. 1972. *Nucl. Phys.* B44: s333
72. Cooke, D. J. 1983. *Phys. Rev. Lett.* 51: 320
73. Dar, A. 1983. *Phys. Rev. Lett.* 51: 227
74. Battistoni, G., et al. 1983. *CERN/EP 83-92*. Submitted to *Nucl. Instrum. Methods*
75. Menon, M. G. K. 1977. In *Proc. Neutrino Conf., Aachen, 1976*. Braunschweig: Vieweg
76. Perkins, D. H. 1979. *CERN/EP/DP/mk, September*
77. Grant, A. L. 1979. *CERN/EF/ALG/ed, December*
78. Bionta, R. M., et al. 1983. *Phys. Rev. Lett.* 51: 27
79. Bionta, R. M., et al. 1983. In *Proc. Int. Conf. High Energy Phys., Brighton*
80. Deden, H., et al. 1975. *Nucl. Phys.* B85: 269
81. Gell-Mann, M., Ramond, P., Slansky, R. 1978. *Rev. Mod. Phys.* 50: 721
82. Gaidos, J. A., et al. 1982. In *Proc. 1982 Summer Workshop on Proton Decay Experiments*. Argonne National Lab, Ill.

83. Krishnaswamy, M. R., et al. 1982. *Pramana* 19:525
84. Gaisser, T. K., Stanev, T. 1983. In *Proc. Conf. on Low Energy Tests of Conservation Laws, Blacksburg, VA. Bartol BA 83-37*
85. Krishnaswamy, M. R., et al. 1982. *Proc. Int. Coll. Baryon Nonconservation, Bombay, Pramana Suppl.* p. 115
86. Krishnaswamy, M. R., et al. 1981. *Phys. Lett.* 106B:339; 1982 *Phys. Lett.* 115B:349
87. Cassiday, G. L., et al. 1973. *Phys. Rev.* D7:2023
88. Grupen, G., et al. 1972. *Nuovo Cimento* 10B:144
89. Battistoni, G., et al. 1982. *Nucl. Instrum. Methods* 202:459
90. Battistoni, G., et al. 1982. *Phys. Lett.* 118B:461
91. Battistoni, G., et al. 1983. *CERN/EP 83-147*
92. Bartelt, J., et al. 1983. *Phys. Rev. Lett.* 50:651
93. Bartelt, J., et al. 1983. *Phys. Rev. Lett.* 50:655
94. Rubakov, V. A. 1981. *JETP Lett.* 33:644; 1982. *Nucl. Phys.* B203:311
95. Callan, C. G. 1983. *Nucl. Phys.* B212:391; 1982. *Phys. Rev.* D26:2058
96. Errede, E., et al. 1983. *Phys. Rev. Lett.* 51:245
97. Battistoni, G., et al. 1983. *CERN/EP 83-92*
98. Foster, G. W. 1983. PhD thesis, Harvard Univ.
99. Arisake, K., et al. 1982. *LICEPP preprint, (UTLICEPP 82-04)*. Univ. Tokyo
100. Morse, R. (J. Blandino et al.) 1982. In *Proc. 3rd Workshop on Grand Unification, Chapel Hill, NC*. Cambridge, Mass: Birkhäuser Boston
101. Krishnaswamy, M. R., et al. 1983. In *Proc. 4th Workshop on Grand Unification, Univ. Penn.* Cambridge, Mass: Birkhäuser Boston
102. Perkins, D. H. 1982. *Oxford preprint NP 60-82*; review by Sulak L. 1982. In *Proc. Int. Conf. on High Energy Paris*
103. Barloutaud, R., et al. 1982. See Ref. 85, p. 143
104. Turner, M. S., Parker, E. N., Bogdan, T. J. 1982. *Phys. Rev.* D26:1296

NUCLEOSYNTHESIS

James W. Truran

Department of Astronomy, University of Illinois, Urbana, Illinois 61801

CONTENTS

1. INTRODUCTION 53
2. ABUNDANCES 55
   2.1 Solar System Abundances 55
   2.2 Cosmic Ray Abundances 57
   2.3 Abundances in Stars and Galaxies 58
   2.4 Isotopic Abundances in Meteorites 60
3. EARLY HISTORICAL DEVELOPMENT 60
4. STELLAR EVOLUTION AND NUCLEOSYNTHESIS 62
   4.1 Hydrogen Burning 63
   4.2 Helium Burning 64
   4.3 The α-Process: Carbon and Oxygen Burning 66
   4.4 The Equilibrium Process: Silicon Burning 67
   4.5 Neutron Capture Synthesis: s-Process and r-Process 68
   4.6 p-Process Nucleosynthesis 72
   4.7 The x-Process: Synthesis of the Light Elements 73
5. NUCLEAR PARAMETERS 74
   5.1 Thermonuclear Reaction Rates 75
   5.2 Weak Interaction Rates 76
6. ACTIVE AREAS OF RESEARCH IN NUCLEOSYNTHESIS 78
   6.1 Cosmological Nucleosynthesis 79
   6.2 s-Process Synthesis in Red Giants 81
   6.3 r-Process Nucleosynthesis in Supernovae 84
   6.4 Explosive Nucleosynthesis in Supernovae 87
   6.5 Hydrogen Burning at High Temperatures 90
7. CONCLUDING REMARKS 91

1. INTRODUCTION

Nucleosynthesis theory has advanced at a significant pace over the past quarter century. Factors contributing to these rapid developments include progress in nuclear physics and an improved knowledge of abundances in astrophysical environments. Refined experimental techniques in nuclear physics have made possible the determination of the very low cross sections for reactions involving light nuclei at low energies, which are characteristic of thermonuclear processes in stellar environments (fewer than one in every 10^{30} collisions of two protons in the Sun's core leads to the formation of a deuterium nucleus). Theoretical investigations in nuclear physics have provided methods for estimating both the thermonuclear reaction and the weak interaction properties of intermediate and heavy nuclei. Improved determinations of the distributions of element abundances in diverse astronomical objects provide clues to and impose increasingly stringent boundary conditions upon the character of the nuclear transformations by which nuclei heavier than hydrogen and helium are synthesized in stellar and supernova environments. Coupled with the advances in computer technology, which now allow extremely detailed calculations to be carried out of the nuclear abundance evolution of the matter processed through stars and supernovae, these factors have stimulated significant research efforts in nucleosynthesis.

Progress in nucleosynthesis theory is chronicled in a series of review articles and conference proceedings. Significant early developments in nuclear astrophysics were reviewed by Cameron (1) and by Burbidge (2). There also exist a number of recent and more specialized reviews of explosive nucleosynthesis (3, 4), of mechanisms of neutron-capture synthesis (5–7), of the origin of the light elements (8, 9), and of the related subjects of stellar evolution (10–12), supernovae (13, 14), nucleocosmochronology (15), and the chemical evolution of the galaxy (16, 17). Excellent discussions of various aspects of nucleosynthesis theory may also be found in a number of books and conference proceedings (18–21). Finally, a very extensive and excellent survey of the subject of the origin and abundances of the elements has been presented by Trimble (22).

The aim of this article is to provide a review of the present status of research in nucleosynthesis, with emphasis on problems of nucleosynthesis in stellar and supernova environments. We begin with a survey of abundance patterns in the Sun, our galaxy and other galaxies, calling attention to the manner in which such data serve to guide and to restrict nucleosynthesis theories. A brief overview of early historical developments is then followed by a survey of stellar evolution and nucleosynthesis, which provides an introduction to and a definition of specific nucleosynthesis mechanisms (23, 24), as they are now understood. Experimental and theoretical results from research in nuclear physics, which provide the empirical foundations of nucleosynthesis theories, are then briefly summarized. A more detailed discussion of specific mechanisms of heavy element synthesis, emphasizing the character of their operation in astrophysical environments, is then presented.

2. ABUNDANCES

The major constraints upon theories of nucleosynthesis are those provided by abundance data. Historically, it was the important early data on abundances for solar system matter compiled by Goldschmidt (25), Brown (26), and Suess & Urey (27) that provided the basis upon which the defining nucleosynthesis studies of Burbidge et al (23) and Cameron (24) were built. We now have available a significant body of abundance data involving both galactic and extragalactic sources. Cosmic rays and supernova ejecta are assumed to represent direct probes of nucleosynthesis environments, while abundance distributions in other galaxies provide measures of the integrated effects of galactic nucleosynthesis. Evidence for the existence of the extinct radioactivities ^{26}Al$(\tau_{1/2} = 7 \times 10^5$ y) and ^{107}Pd$(\tau_{1/2} = 7 \times 10^6$ y) in primitive solar system matter attests to the fact that the solar system received a last minute contamination from a nucleosynthetic event. As such abundance data generally provide important clues regarding the characteristics and sites of nucleosynthesis processes, it is appropriate to provide an overview of the essential features. In particular, we review abundance patterns in solar system matter, in cosmic rays, in the gas and stars in the Milky Way galaxy and other galaxies, and in meteorites.

2.1 Solar System Abundances

The major boundary conditions within which we form theories of nucleosynthesis are those shown in Figure 1—the solar system abundances. These are commonly taken to define the "cosmic" abundances. Important recent compilations of solar system abundances of the elements include those by Anders & Ebihara (28) and by Cameron (29), from which this figure is taken. The primary source of abundance information in both compilations was the class of C1 carbonaceous chondritic meteorites (30). This was necessarily supplemented, for the volatile elements, by solar abundance data derived from spectral analyses of the Sun's atmosphere. It may prove useful at this stage to scan the important abundance features and to keep these in mind in our subsequent discussions.

We note first that hydrogen and helium are far and away the most abundant elements in solar system matter. Together they comprise 98% of the mass while the entirety of the heavy elements represents only 2% of the mass. The abundance of helium relative to hydrogen is a critical parameter with respect to cosmological models, and it is therefore interesting to note that the helium abundance differs in these two abundance compilations. Anders & Ebihara adopt the solar wind value for the He/H ratio of 0.07, yielding a helium mass fraction 0.24, while Cameron adopts a compromise

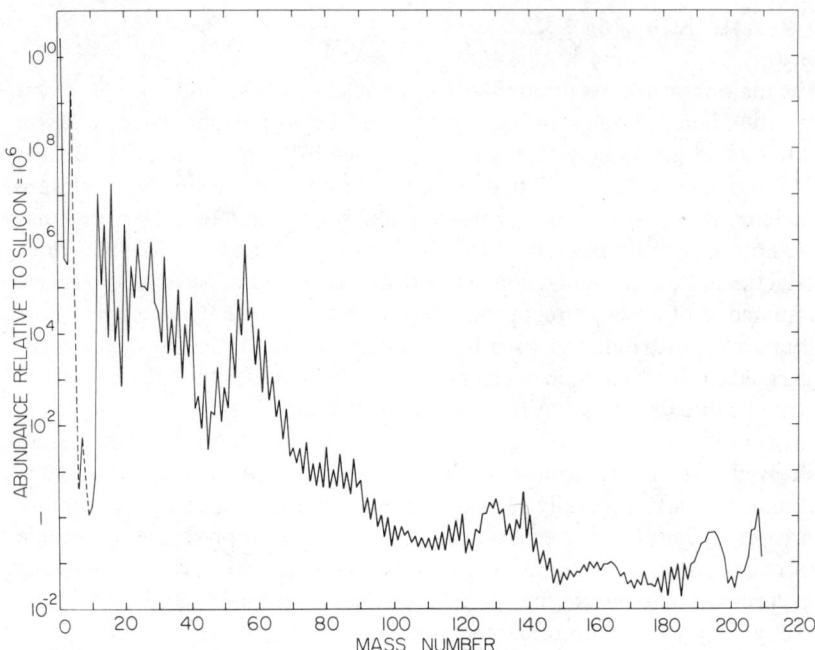

Figure 1 The solar system abundances as compiled by Cameron (29) are plotted as a function of mass number on a scale with silicon = 10^6.

between the solar cosmic ray value and the solar abundance given by Ross & Aller (31), which yields 0.21. The lower value is inconsistent with the standard model of big bang nucleosynthesis.

The distribution of heavy-element abundances shows many features that reflect the characteristics of the nuclear processes by which these nuclei are synthesized. The extremely low concentrations of lithium, beryllium, and boron reflect the fact that they are not formed in abundance at the temperatures prevailing in the interiors of stars. With the exception of ^{7}Li, for which both stellar production and cosmological production are possible, their formation is now understood as resulting from the interactions of cosmic rays with the more abundant constituents of the interstellar gas in our galaxy. The large abundances of ^{12}C and ^{16}O are interpreted as the products of stellar helium burning. Subsequent processing of this matter in the cores of massive stars and under explosive burning conditions in supernovae forms the bulk of the isotopes of elements from neon to nickel, including those in the distinctive equilibrium abundance peak centered on ^{56}Fe. The formation of most nuclei past mass 60 is generally attributed to neutron-capture processes. We shall see that the abundance features evident at mass numbers $A \approx 88$, 130, 138, 195, and 208 may be very naturally interpreted in terms of neutron-shell structure and its implied enhanced nuclear stability.

2.2 Cosmic Ray Abundances

Abundance determinations for matter outside the solar system can also provide critical constraints upon the nature of nucleosynthesis sources. Cosmic rays are particularly interesting in this regard since they are often viewed as providing direct probes of nucleosynthesis environments. Supernovae and their remnants have historically been considered to be perhaps the most likely source of cosmic rays in our galaxy. The energy requirements for filling the galaxy with cosmic rays are easily met and the general heavy-element enrichment of cosmic rays identified in earlier studies (32) suggested their origin at a site of nucleosynthesis. Recent observations indicate that the situation may be more complicated, with several distinct sources contributing to the cosmic rays.

Our knowledge of cosmic ray abundances has advanced considerably over the past decade as a result of advances in space technology and instrumentation. Reviews of our current knowledge of the elemental and isotopic characteristics of the galactic cosmic ray source abundances include those by Mewaldt (33) and Simpson (34). It is, of course, well known that arriving cosmic rays exhibit abundance features attributable to effects of propagation (e.g. substantial overabundances of the light elements lithium, beryllium, and boron and a reduced odd-even effect in Z). These features arise from spallation reactions that occur as the cosmic rays traverse the interstellar medium in our galaxy. Detailed quantitative studies of this question allow one to infer the distribution of path lengths to which the cosmic rays have been exposed, and thereby to correct for the effects of propagation and determine the composition of the cosmic rays at their sources.

A comparison of the inferred cosmic ray source composition with that of solar system matter reveals several features relevant to nucleosynthesis theories. Perhaps the most significant finding is that the two abundance patterns show a general overall similarity. While the cosmic ray data reveal a correlation of cosmic ray abundance with first ionization potential (elements of high first ionization potential are significantly depleted), the fact that a similar behavior is exhibited by solar cosmic rays indicates that this is not a source abundance effect. The existence of such an overall similarity suggests that the cosmic rays might have their origin in the acceleration of galactic matter of average composition rather than at specific stellar or supernova sites.

The general consistency of the cosmic ray and solar system abundance patterns for the elements heavier than hydrogen does not, however, extend to all levels of detail. There exist several clear and interesting exceptions. The cosmic ray source abundance ratio for C/O is roughly a factor of 1.5 larger than the solar system value of approximately 0.6. Nitrogen appears to be depleted relative to oxygen in the cosmic ray source by a factor of ~ 2. These data argue against models in which cosmic rays are assumed to arise from the acceleration of average interstellar matter and instead favor a stellar or supernova origin. Interesting deviations from solar system isotopic patterns were also recently identified. The ratio $^{22}Ne/^{20}Ne$ was found to be high by a factor of $\sim 3-6$ with respect to $^{22}Ne/^{20}Ne$ ratios characteristic of solar system matter. This again suggests the possibility that singular contributions from diverse stellar and supernova environments, which do not represent average galactic matter, may significantly influence cosmic ray abundances. Finally, the neutron-rich isotopes of both magnesium and silicon in the cosmic rays have been found to be enriched, relative to solar, by 60–80%. Nucleosynthesis calculations (35) indicate that such variations in isotope ratios—a gradual enrichment over the course of our galaxy's history—might naturally arise as a consequence of the nature of the explosive nucleosynthesis processes by which they are formed; the typical interstellar material in our galaxy today might then be expected to be characterized by such isotopic enrichments for magnesium and silicon. Determinations of isotopic abundances from observations of interstellar molecules can in principle serve to confirm this picture (36).

2.3 *Abundances in Stars and Galaxies*

A detailed survey of the enormous body of available abundance data in our galaxy and other galaxies is clearly beyond the scope of this article. There exist a number of recent review articles to which the reader may refer for more extensive discussion. Abundances in stars in our galaxy were reviewed by Pagel (37) and Baschek (38). Abundances in interstellar matter were discussed by Spitzer & Jenkins (39), Wannier (36), York (40), and Peimbert (41). Abundances in stellar populations in our galaxy and other galaxies are surveyed by Pagel, Edmunds & Kellermann (42), Mould (43), and van den Bergh (44). Abundances in supernova ejecta and remnants are reviewed by Oke & Searle (45) and by Trimble (13, 14). Other useful reviews include those on the chemical compositions of globular clusters (46, 47) and on gamma-ray astronomy (48).

There exist several broad underlying features of the abundance patterns of these seemingly diverse sources, and these provide important clues regarding nucleosynthesis processes and histories. Most significant of these is the observation that the local pattern of abundances is generally representative. The gross abundance features throughout our galaxy, in other galaxies, and even apparently in quasars are generally similar to those of solar system matter, testifying to the fact that the underlying stellar systems share the same nucleosynthesis processes. Variations in the total metal concentrations by mass, Z, for stellar systems and populations are, with some exceptions, roughly correlated with the extent to which star formation has proceeded and the available gas has been exhausted. Radial variations in Z through the disks of our galaxy and other spiral galaxies similarly reflect the relative efficiency of star formation and perhaps the effects of Z redistribution by gas flows as well.

Observations of individual stars and stellar remnants also provide critical input. Extreme metal-deficient stars in the halo of our galaxy, characterized by metal concentrations of order 10^{-3} that of solar system matter, exhibit detailed abundance patterns different from solar patterns. It is generally argued that these anomalous patterns reflect the preferential contributions from very massive stars, which evolve rapidly on the timescale of halo evolution and can contaminate the interstellar gas at the very earliest stages of galactic evolution (49). At the opposite extreme, evolved stars and stellar remnants can provide rather direct evidence of the consequences of nuclear processes in stars. The discovery by Merrill (50) of the presence of the element technetium in the atmospheres of red giant stars, for example, provided an absolutely critical clue. Technetium has no stable isotopes and, because its longest-lived isotope has a half-life considerably less than the lifetime of the stars in which it was observed, the conclusion that element synthesis had taken place recently in those very stars was unescapable. Red giant stars also reveal other evidence for internal nuclear processing (51): the high abundance of carbon in "carbon stars" is believed to result from helium burning in the interior, assuming that such processed matter has been carried to the surface by convection. Furthermore, observed carbon isotope ratios approaching $^{12}C/^{13}C \approx 4$ in some red giants reflect the fact that the envelope has been contaminated by material processed through carbon-nitrogen-oxygen (CNO) cycle hydrogen burning: detailed numerical analyses of these nuclear burning sequences indeed predict a ratio $^{12}C/^{13}C \approx 4$, which represents a ratio of experimentally determined thermonuclear reaction rates for the $^{12}C(p,\gamma)^{13}N$ and $^{13}C(p,\gamma)^{14}N$ reactions.

The terminal stages of stellar evolution find stars of masses less than approximately $4M_\odot$ ejecting shells of matter to form planetary nebulae and leaving behind white dwarf remnants. More massive stars undergo supernova explosions and leave behind neutron star or black hole remnants or perhaps, in some instances, no condensed remnant at all. Observations both of planetary nebulae (52) and of supernova ejecta (53–57) reveal enrichments of some heavy-element abundances relative to solar abundances, which confirm that significant nuclear processing of this material has indeed occurred.

The abundance of ^{4}He, so critical to considerations of the nature of the early universe, deserves some further attention. We noted previously the differences in the helium mass fractions in solar system matter adopted by Anders & Ebihara and by Cameron: 0.24 and 0.21, respectively. The lower value is obtained using the rather more uncertain value for the Sun (31). In contrast, somewhat higher ^{4}He abundances are inferred from observations of galactic and extragalactic sources. Pagel (58) reviewed the situation and estimated the primordial helium content resulting from cosmological nucleosynthesis to be $Y_p = 0.23 \pm 0.01$, using the more accurate observations of emission nebulae in irregular galaxies and compact extragalactic HII regions.

2.4 Isotopic Abundances in Meteorites

We have seen that abundance compilations for solar system matter rely heavily on meteoritic data (30). While this is essential to considerations of elemental abundances, since terrestrial matter has experienced much chemical fractionation, it has until recently been assumed that solar system matter is isotopically homogeneous. Recent studies of isotopic abundance patterns in meteorites have shown that this is not the case. Nonsolar isotopic abundance patterns for a number of elements (O, Ne, Mg, Si, Ca, Kr, Sr, Xe, Ba, Nd, and Sm) have been identified in samples of meteoritic material. These abundance anomalies and their possible interpretations are discussed by several authors (59–64). Additionally, scrutiny of the isotopic patterns reveals that short-lived radioactive nuclei (^{26}Al, ^{107}Pd, ^{129}I, and ^{244}Pu) were present in the early solar system. Of particular importance are the very short-lived extinct radionuclides ^{26}Al($\tau_{1/2} = 7.2 \times 10^5$ y) and ^{107}Pd($\tau_{1/2} = 6.5 \times 10^6$ y). The presence of these nuclei in primitive solar system matter at the levels inferred from the observations (65–67) is viewed by some as suggesting that solar system matter may have received a last minute contamination from a nearby supernova (68, 69) or nova (70) event.

3. EARLY HISTORICAL DEVELOPMENT

Nucleosynthesis is, for very obvious reasons, a twentieth century science. The occurrence in nature of natural radioactive elements, which was recognized only early in this century, testifies to the fact that at least some of the elements are of finite age. The development of spectroscopy in the late nineteenth century first allowed information to be gathered concerning the compositions of the Sun and other stars. Early geological evidence that many earth rocks had ages in excess of a billion years provided a critical challenge, as it was recognized that the release of gravitational energy alone would suffice to maintain the Sun's luminosity for only some tens of millions of years. A knowledge of nuclear physics was clearly required. Advances in nuclear physics and in spectroscopy have subsequently provided the tools and the boundary conditions, respectively, which are essential to investigations of stellar evolution and nucleosynthesis.

The earliest spectral studies of the Sun and stars were able only to establish that the elements of which they and the Earth are composed were qualitatively the same. It was therefore quite reasonable to assume a universe of uniform chemical composition and to search within the framework of cosmology for a set of physical conditions that would give rise to the present abundance distribution. The details of these early theoretical efforts were summarized in a review article by Alpher & Herman (71). As some aspects of these early attempts at theories of universal synthesis have survived, it is instructive to examine their essential features.

One broad class of early theory was governed by considerations of nuclear statistical equilibrium. It was assumed that at an early stage in the universe the matter was gathered together in a high-temperature, high-density state, ensuring that nuclear transformations would take place very rapidly and lead to the establishment of a true statistical equilibrium distribution. A major problem with such theories arose from the fact that, for a specified temperature and density, the equilibrium abundance patterns were found to be extremely sharply peaked in a narrow range of mass number. It quickly became clear that the universal synthesis of the elements could not have occurred in this manner. It was recognized, however, that the distinctive iron abundance peak centered on ^{56}Fe could be quite reasonably fit by such an equilibrium distribution (72). Current theories attribute the iron-peak abundances observed in nature to a modified equilibrium distribution, which is realized in the expansion and cooling of matter from high temperatures and densities that accompanies its ejection in supernova events. The formation of mass 56 in situ as ^{56}Ni, as we shall discuss, holds important implications for the nature of supernova light curves.

Alpher, Bethe & Gamow (73) advanced an important alternative model for the universal synthesis of the elements tied to a cosmological "big bang." They recognized that free neutrons would constitute a substantial component of matter at high densities and temperatures. During the early stages of expansion of this neutron-rich matter following the big bang, the interaction of these neutrons with protons would, in their model, subsequently lead via successive neutron captures to the production of all of the heavier elements. The correlation of the positions of the heavy-element abundance peaks observed in nature with regions in which the neutron-capture probabilities were small, due to shell closures, was encouragingly compatible with this model. We have since learned, however, that universal big bang nucleosynthesis cannot have been responsible for the formation of most of the nuclear species heavier than helium. Revised analyses of the weak interaction processes proceeding at early epochs now indicate that the neutron-to-proton ratio characterizing the onset of the nucleosynthesis era is of order 1/7, which leads to the production of a ^{4}He mass fraction of order 25% but only trace amounts of heavy elements. The absence of stable nuclei with mass numbers five and eight inhibits the production of nuclei heavier than helium under cosmological conditions. However, the crucial role played by successive neutron captures in the formation of the heavy elements has been incorporated into current theories.

Even as these studies proceeded, advances in nuclear physics and discoveries in observational astronomy were paving the way toward a more comprehensive understanding of stellar evolution and the mechanisms of origin of the elements. The pivotal role played by thermonuclear reactions in providing an energy source sufficient to account for stellar lifetimes of billions of years was established in the late 1930s when Bethe [74], von Weizsäcker [75], and Bethe & Critchfield [76] first identified the reaction sequences by which the fusion of four protons to helium ("hydrogen burning") can occur. It remained, however, to be realized that nuclear processes in stellar interiors might be responsible for the formation of many of the heavier elements observed in nature. The recognition that heavy-element synthesis is an ongoing process in stellar interiors followed Merrill's discovery [50] of the presence of technetium in red giant stars. That the integrated effects of such stellar processing over the course of galactic evolution might account for the heavy-element abundances in solar system matter was further suggested by the discovery that there exist abundance differences between certain broad classes of stars in our galaxy. Stars in the halo of our galaxy that are assumed, for dynamic reasons, to constitute an earlier stellar generation are observed to have significantly lower metal contents—as low as 10^{-3} that of the Sun. This emphasized the likely importance of mechanisms of stellar nucleosynthesis and set the stage for the defining work on this subject in the late 1950s.

4. STELLAR EVOLUTION AND NUCLEOSYNTHESIS

Stellar evolution is characterized by a sequence of alternate stages of gravitational contraction to higher temperatures and densities and the thermonuclear burning of the available fuel at these temperatures. Since these nuclear burning stages give rise to the production of successively heavier nuclei, it is clear that questions of stellar energy generation and nucleosynthesis are tied closely together. The various nuclear processes occurring in stars and supernovae responsible for nuclear energy generation and for the formation of the heavier elements observed in nature were first defined in the now classic papers by Burbidge et al (23) and by Cameron (24). The eight distinct processes thus defined are identified briefly below, with emphasis on those aspects relevant to current research in nucleosynthesis.

4.1 Hydrogen Burning

The energy release accompanying the conversion of hydrogen to helium in a core comprising roughly 10% of its mass powers a star for 90% of its active burning lifetime. The burning temperature is dependent upon the stellar mass and ranges from 10 to 50 million degrees Kelvin. Two nuclear reaction sequences contribute to this burning (Table 1): the proton-proton chains and the CNO cycles, in which these heavier nuclei serve as catalysts in the conversion of hydrogen to helium. The proton-proton reactions dominate the burning in stars of mass less than or comparable to that of the Sun, while the CNO cycles become important in more massive stars, which typically burn hydrogen at higher temperatures. Intense interest particularly in the proton-proton burning reactions has been stimulated over the past two decades by the ongoing solar neutrino detection experiment of Davis (77), the situation regarding which has most recently been reviewed by Bahcall (78).

We are concerned here rather with mechanisms of nucleosynthesis. In this regard, the fraction of the ^{4}He produced in hydrogen burning in stellar cores that is ultimately returned in this form to the interstellar gas is quite

Table 1 Hydrogen-burning reaction sequences

	Proton-proton chains	Carbon-nitrogen-oxygen cycles
	^{1}H + ^{1}H → ^{2}D + e^+ + ν	^{12}C + ^{1}H → ^{13}N + γ
PP I	^{2}D + ^{1}H → ^{3}He + γ	^{13}N → ^{13}C + e^+ + ν
	^{3}He + ^{3}He → ^{4}He + 2^{1}H	^{13}C + ^{1}H → ^{14}N + γ
	or	^{14}N + ^{1}H → ^{15}O + γ
	^{3}He + ^{4}He → ^{7}Be + γ	^{15}O → ^{15}N + e^+ + ν
PP II	^{7}Be + e^- → ^{7}Li + ν	^{15}N + ^{1}H → ^{12}C + ^{4}He
	^{7}Be + ^{1}H → ^{8}Be → 2^{4}He	or
	or	^{15}N + ^{1}H → ^{16}O + γ
	^{7}Be + ^{1}H → ^{8}B + γ	^{16}O + ^{1}H → ^{17}F + γ
PP III	^{8}B → ^{8}Be + e^+ + ν	^{17}F → ^{17}O + e^+ + ν
	^{8}Be → 2^{4}He	^{17}O + ^{1}H → ^{14}N + ^{4}He

small. The bulk of the ^{4}He observed in the gas and stars in our galaxy and other galaxies is believed rather to have originated in the cosmological big bang. This is usually assumed to be true as well for ^{2}D, since deuterium is destroyed so rapidly at the temperatures characteristic of stellar interiors (e.g. in the core of the Sun, ^{2}D/^{1}H $\approx 10^{-18}$). The quantities of other nuclei formed in hydrogen burning may, however, represent important nucleosynthesis contributions. Stellar evolution models (79–81) indicate that ^{3}He is produced in low mass stars in sufficient concentrations to make an important contribution to the interstellar gas over the history of the galaxy (82–84); significant production of ^{3}He also accompanies the cosmological big bang.

Under most circumstances, the surviving abundance of ^{7}Li is expected to be small since the ^{7}Li+^{1}H reaction proceeds rapidly. Observations confirm, however, that substantial ^{7}Li production can occur, at least as a transient phase, in the evolution of red giants (85). Mass loss from such lithium-rich stars might also contribute to the ^{7}Li enrichment of galactic matter (82, 86). Cameron & Fowler (87) proposed that high ^{7}Li concentrations in red giant envelopes could be realized if nuclei of mass 7, in the form of ^{7}Be (which is not as readily destroyed by proton interactions), were convectively mixed outward from the hotter shell-burning regions at the base of the envelope. Unfortunately, detailed predictions of the surface enhancement of ^{7}Li are rendered unreliable by uncertainties in theories of convective transport (88–90). We note that the uncertainties associated with quantitative estimates of stellar production of both ^{3}He and ^{7}Li restrict the use of these nuclei as probes of cosmological nucleosynthesis conditions (91).

One extremely important consequence of hydrogen burning via the CNO cycles is the synthesis of ^{14}N. It is generally agreed that virtually all of the ^{14}N in galactic matter originated in this environment. Calculations of CNO cycle hydrogen burning (92) reveal that quite generally ^{14}N will be the most abundant nucleus under steady-state burning conditions, as a result of the relatively low rate of the ^{14}N(p,γ)^{15}O reaction. Most of the initial carbon and oxygen present in the hydrogen-burning region will thereby be converted to ^{14}N. This is an important result, since the formation of ^{12}C and ^{16}O in helium burning is not accompanied by the production of an amount of ^{14}N consistent with solar system abundances. The isotopic abundance patterns resulting from these reaction sequences are also found to be distinctly unlike those of solar system matter.

4.2 Helium Burning

The exhaustion of hydrogen in a stellar core is followed by a stage of gravitational contraction to higher temperatures and densities, which continues until nuclear transformations can again provide a sufficient energy source (or, for stars less massive than approximately 0.5 $M_\odot$, until electron degeneracy pressure becomes dominant and the star becomes a helium white dwarf). At temperatures approaching 2×10^8 K and associated densities of 10^2–10^4 g cm^{-3}, thermonuclear burning of helium (Table 2) is initiated in the core. The star is now a red giant.

The stellar helium-burning phase of evolution is dominated by the "triple alpha" reaction $3\,^4\text{He} \rightarrow\,^{12}\text{C}$, first discussed by Salpeter (93) and Hoyle et al (94), and by the ensuing alpha-capture reaction $^{12}\text{C}(\alpha,\gamma)^{16}\text{O}$. Significant buildup past ^{16}O is impeded by the relatively slow effective rate of the $^{16}\text{O}(\alpha,\gamma)^{20}\text{Ne}$ reaction. It is generally believed that the ^{12}C and ^{16}O in galactic matter had their origin in stellar helium-burning phases.

The subsequent evolution of a star is critically dependent upon the relative concentrations of ^{12}C and ^{16}O in the core following the exhaustion of helium and these abundances, in turn, are sensitive to the relative rates of these two reactions. While the rate of the $3\,^4\text{He} \rightarrow\,^{12}\text{C}$ reaction is assumed to be known to acceptable accuracy (95–97), there is some uncertainty associated with the rate of carbon destruction by $^{12}\text{C}(\alpha,\gamma)^{16}\text{O}$ (98, 99). Under typical astrophysical helium-burning conditions, the capture of an alpha particle by ^{12}C proceeds primarily in the tail of the 7.12-MeV excited state of ^{16}O. Experimental and theoretical determinations of the reduced alpha width of this state span a considerable range, within which limits the abundances predicted to result from helium burning may vary dramatically. Currently accepted values are consistent with the production of roughly equal amounts of ^{12}C and ^{16}O in stellar helium burning, with

Table 2 Helium and heavy-ion reactions

Helium-burning reactions	Carbon- and oxygen-burning reactions
$3\,^4\text{He} \rightarrow\,^{12}\text{C}+\gamma$	$^{12}\text{C}+^{12}\text{C} \rightarrow\,^{20}\text{Ne}+^4\text{He}$
$^{12}\text{C}+^4\text{He} \rightarrow\,^{16}\text{O}+\gamma$	$^{12}\text{C}+^{12}\text{C} \rightarrow\,^{23}\text{Na}+^1\text{H}$
$^{16}\text{O}+^4\text{He} \rightarrow\,^{20}\text{Ne}+\gamma$	$^{12}\text{C}+^{12}\text{C} \rightarrow\,^{23}\text{Mg}+n$
$^{20}\text{Ne}+^4\text{He} \rightarrow\,^{24}\text{Mg}+\gamma$	
also	
	$^{16}\text{O}+^{16}\text{O} \rightarrow\,^{28}\text{Si}+^4\text{He}$
$^{14}\text{N}+^4\text{He} \rightarrow\,^{18}\text{F}+\gamma$	$^{16}\text{O}+^{16}\text{O} \rightarrow\,^{31}\text{P}+^1\text{H}$
$^{18}\text{F} \rightarrow\,^{18}\text{O}+e^++\nu$	$^{16}\text{O}+^{16}\text{O} \rightarrow\,^{31}\text{S}+n$
$^{18}\text{O}+^4\text{He} \rightarrow\,^{22}\text{Ne}+\gamma$	
and	
$^{18}\text{O}+^4\text{He} \rightarrow\,^{21}\text{Ne}+n$	
$^{22}\text{Ne}+^4\text{He} \rightarrow\,^{25}\text{Mg}+n$	

variations attributable to dependences of the relative rates upon the prevailing temperatures and densities. The net production, integrated over contributions from massive stars (100, 101) and red giants (84, 102), is then compatible with a solar system value of C/O of roughly 0.6.

Another sequence of nuclear transformations accompanying helium burning (Table 2), though not significant with respect to nuclear energy generation, plays a critical role in the generation of a neutron source for heavy-element synthesis. We recall that hydrogen burning acts to convert most preexisting carbon, nitrogen, and oxygen nuclei to ^{14}N as a consequence of the CNO cycles. Cameron (103) noted that this ^{14}N is readily converted into ^{18}O and subsequently ^{22}Ne during helium burning by the reactions ^{14}N$(\alpha,\gamma)^{18}$F$(e^+\nu)^{18}$O$(\alpha,\gamma)^{22}$Ne. An important neutron source is then provided by the ^{22}Ne$(\alpha,n)^{25}$Mg reaction. As we discuss in a subsequent section, recent investigations (90, 104–106) indicate that the operation of this ^{22}Ne$(\alpha,n)^{25}$Mg reaction in providing a neutron source in red giants leads to the production of many of the elements more massive than iron in essentially solar proportions.

4.3 The α-Process: Carbon and Oxygen Burning

Early calculations indicated that substantial concentrations of ^{20}Ne as well as ^{12}C and ^{16}O might be formed in helium burning. Burbidge et al (23) envisioned that, following helium exhaustion, the release of alpha particles by photodisintegration reactions like ^{20}Ne$(\gamma,\alpha)^{16}$O occurring at temperatures $T \gtrsim 10^9$ K would provide a source of alpha particles that would then be recaptured by remaining ^{20}Ne nuclei to form heavier elements by the sequence of reactions ^{20}Ne$(\alpha,\gamma)^{24}$Mg, ^{24}Mg$(\alpha,\gamma)^{28}$Si, ^{28}Si$(\alpha,\gamma)^{32}$S, etc through nuclei in the mass range $A = 40$–48. We now know rather that, before this can take place, heavy-ion thermonuclear reactions (24) involving the interaction of carbon and oxygen nuclei with themselves—carbon and oxygen burning—will begin to contribute. Note that these burning phases can proceed only in more massive stars: for stars of core mass less than the Chandrasekhar (107) limit of approximately 1.4 $M_\odot$, or total mass less than approximately 8–10 $M_\odot$, electron degeneracy pressure becomes dominant prior to the onset of carbon burning and the star evolves to a carbon-oxygen white dwarf configuration.

The dominant reactions associated with carbon and oxygen burning are shown in Table 2. The experimental situation regarding these reactions has been reviewed by Fowler, Caughlan & Zimmerman (95, 96). Both the ^{12}C+^{12}C and the ^{16}O+^{16}O interactions are characterized by roughly comparable yields in the proton and alpha-particle channels and a relatively weak endothermic neutron branching. Contributions from ^{12}C+^{16}O as well can be important under some conditions. The carbon-burning phase of evolution of massive stars proceeds at temperatures of approximately 8×10^8 K, while oxygen burning requires temperatures approaching 2×10^9 K; densities of 10^5–10^6 g cm^{-3} are characteristic in both instances. The ^{12}C + ^{12}C and ^{16}O + ^{16}O reactions alone provide a reasonable estimate of the nuclear energy generation associated with these burning epochs. Detailed calculations of carbon-burning (108) and of oxygen-burning (109) nucleosynthesis also explore the consequences of the increasing numbers of thermonuclear reactions involving protons, alpha particles, and neutrons associated with these heavy-ion burning stages.

The carbon- and oxygen-burning phases of evolution of massive presupernova stars have been investigated by Arnett (110–112), Weaver, Zimmerman & Woosley (101), and Woosley, Axelrod & Weaver (113). The existence of a brief, intermediate "neon-burning" phase, which mimics the classical α-process of Burbidge et al (23), has also been established. Here, the reaction ^{20}Ne$(\gamma,\alpha)^{16}$O releases alpha particles, which are subsequently captured by the abundant nuclei present in the gas at the end of carbon burning: ^{16}O, ^{20}Ne, and ^{24}Mg. The oxygen-burning phase then follows. The contributions of these burning stages in massive stars to the abundances of elements in the interstellar gas and stars in our galaxy can only be determined by continuing the evolution through the supernova phase (100, 114).

4.4 *The Equilibrium Process: Silicon Burning*

The nuclei in the iron abundance peak observed in nature (Figure 1) were proposed by Burbidge et al (23) and Cameron (24) to have been formed in an equilibrium process (72). It was assumed that a true nuclear statistical equilibrium, favoring these nuclei characterized by high binding energies per nucleon, would be achieved at the extreme temperatures and densities existing in massive stellar cores during the final stages of presupernova evolution. The physical conditions providing the best overall fit to the abundances of the iron group nuclei were identified. It rapidly became apparent, however, that no single equilibrium configuration could account for every detail of the abundance pattern in the iron-peak region. Fowler & Hoyle (115) found that the best fit to the iron isotope ratios was obtained for a rather neutron-rich equilibrium centered on ^{56}Fe and defined by $T = 3.8 \times 10^9$ K, $\rho = 3.1 \times 10^6$ g cm^{-3} and $\bar{Z}/\bar{N} \approx 0.87$ ($\bar{Z}$ and $\bar{N}$ are the average numbers of protons and neutrons, respectively, per nucleus). For these conditions, however, the predicted abundances of ^{50}Cr, ^{54}Fe, and particularly ^{58}Ni failed to account for their solar system ratios relative to iron. A neutron-poor equilibrium configuration ($\bar{Z}/\bar{N} \approx 0.99$) dominated by ^{56}Ni provides a very attractive alternative, and one that fits rather well the relative abundances of the more proton-rich stable isotopes of even-Z

nuclei (^{48}T, ^{49}Ti, ^{50}Cr, ^{52}Cr, ^{53}Cr, ^{54}Fe, ^{56}Fe, ^{57}Fe, ^{58}Ni), and the odd-Z nuclei ^{51}V, ^{55}Mn, and ^{59}Co. Such distributions are expected to be the natural consequence of burning at higher temperatures: on the correspondingly shorter timescales, weak interactions cannot provide a significant change in the ratio $\bar{Z}/\bar{N} \approx 1$ characterizing the core matter. Discussions of such equilibrium fits to solar system abundances have been provided by a number of authors (116–119).

The primary products of hydrostatic carbon, neon, and oxygen burning are the self-conjugate nuclei ^{20}Ne, ^{24}Mg, ^{28}Si, and ^{32}S, with lesser amounts of ^{36}Ar and ^{40}Ca. This implies a value of $\bar{Z}/\bar{N}$ close to unity for the core matter, which will ultimately ensure that the equilibrium configuration in situ will be dominated by ^{56}Ni. As core contraction increases the central temperature, readjustments due to photodisintegration and capture reactions first favor an increased concentration of ^{28}Si. Subsequently, at temperatures above $\sim 3 \times 10^9$ K, the burning of ^{28}Si proceeds by means of a complex sequence of nuclear transformations. Photodisintegrations of ^{28}Si and lighter nuclei release protons, neutrons, and alpha particles that can be recaptured on the remaining ^{28}Si and heavier nuclei, gradually building up toward the iron region. Detailed numerical studies and analyses of the thermonuclear reaction sequences involved in silicon burning and the accompanying approach to nuclear statistical equilibrium have been performed by several authors (120–126). The influence of expansion and cooling on the abundance patterns (freeze-out) was first considered by Truran, Arnett & Cameron (127). This study called attention to the fact that realistic calculations must include the effects of evolution in the appropriate stellar or supernova environment. Recent studies of hydrostatic silicon burning have been carried out in the context of models of the evolution of massive stars (101, 128). Calculations of silicon burning in supernova environments are reviewed in a later section.

4.5 Neutron Capture Synthesis: s-Process and r-Process

The formation of most of the nuclear species of mass $A \gtrsim 60$ occurs in nature as a consequence of neutron-capture processes. The abundance features in the heavy-element region (Figure 1), which are correlated with neutron-shell closures at neutron numbers $N = 50$, 82, and 126, strongly support this view. Scrutiny of these abundance patterns indeed reveals signatures of two distinct neutron fluxes. This has led historically to the definition of two nucleosynthesis processes that are identified with quite different astrophysical environments.

The distinction is made here largely on the basis of relative lifetimes for neutron captures (τ_n) and electron decays (τ_β). The condition that $\tau_n > \tau_\beta$, where τ_β is a characteristic lifetime for beta-unstable nuclei near the valley of beta stability, ensures that as captures proceed the neutron-capture path will itself remain close to the valley of beta stability. This defines the astrophysical s-process of neutron capture. The s-process capture flow path through a representative portion of the heavy-element region is illustrated in Figure 2. Considerations of beta decay timescales both constrain the flow to the vicinity of the valley of beta stability and ensure that no production of nuclei past $A = 209$ results from this nucleosynthesis mechanism. When $\tau_n < \tau_\beta$, it follows that successive neutron captures will proceed into the neutron-rich regions off the beta-stable valley. Following exhaustion of the neutron flux, the capture products approach the position of the valley of beta stability by beta decay; such nuclei in this figure as ^{176}Yb, ^{186}W, and ^{187}Re, which lie off the s-process capture path, can be produced in this manner. This r-process neutron-capture mechanism is expected to operate in an environment characterized by a more violent epoch of generation of neutrons.

The early work by Cameron (14, 129) and Burbidge et al (23) established the basic characteristics of these two processes and the timescales for their operation. The neutron flux associated with the r-process is sufficiently high that the most neutron-rich isotopes along the r-process capture path have beta decay lifetimes typically $\lesssim 1$ s. In contrast, lower limits on the lifetimes

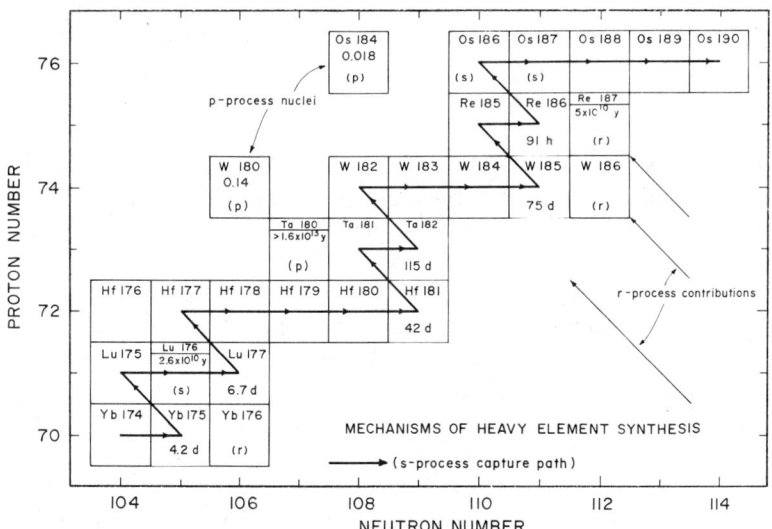

Figure 2 A typical portion of the heavy-element region illustrating the characteristics of the three processes of heavy-element synthesis. The s-process path, which is constrained to the valley of beta stability, is indicated. Proton-rich isotopes (p) formed by the p-process and neutron-rich isotopes (r) formed by the r-process are also identified.

of critical nuclei participating in the s-process range from ~10–100 years. Subsequent analyses of these neutron capture mechanisms (130, 131) took advantage of the observed constancy of the product of neutron-capture cross section and s-process abundance between closed shells and the precipitous falls at the positions of the shell closures to provide a more solid quantitative means of disentangling the r-process and s-process contributions. Current estimates of the r-process and s-process components in solar system matter are provided by Cameron (29, 132). The problem of distinguishing s-process and r-process patterns in other stars is complicated by the facts that most heavy elements receive contributions from both sources and that there exist only a few elements, like europium, that are formed primarily by the r-process.

The modes of operation of these two neutron-capture processes are fundamentally the same. Both demand an appropriate source of free neutrons and an abundance of seed nuclei on which they are to be captured. Iron peak nuclei provide s-process seeds while neutrons are provided by reactions like $^{13}C(\alpha,n)^{16}O$ and $^{22}Ne(\alpha,n)^{25}Mg$, which arise during the normal course of evolution of red giants. Successive neutron captures interspersed by electron decays then define flow patterns progressing steadily toward higher mass nuclei. The s-process forms nuclei beyond the iron peak all the way to lead and bismuth while the timescale of the r-process allows the buildup to proceed through the actinide region. The conditions required for the r-process include a high ratio of free neutrons to heavy seed nuclei, suggestive of a supernova environment. For the large neutron fluxes characteristic of the r-process, the closed neutron shells at neutron numbers $N = 82$ and 126 are encountered in the neutron-rich regions at lower proton numbers, and the subsequent decays following neutron exhaustion form the abundance features (Figure 1) at $A \approx 130$ and 195 (approximately 10 masses lower than the corresponding shell features at $A \approx 138$ and 208 attributed to the s-process).

The identification of red giant stars as an active site of s-process nucleosynthesis seems firm. The presence of technetium and of enhanced abundances of strontium, yttrium, zirconium, barium, and lanthanum in some classes of red giants generally supports the view that heavy elements are synthesized in the interiors of these stars and subsequently brought to the surface by convection. A working model for effecting both neutron-capture synthesis and mixing provided the challenge. As studies proceeded, the complexity of the s-process mechanism began to emerge. Clayton et al (130) determined that no *single* neutron exposure could account for the distribution of s-process elements in solar system matter. Seeger, Fowler & Clayton (131) demonstrated that an exponential distribution of exposures does provide a reasonable fit. Questions concerning the solution of the s-process network equations, the role of beta decays, and the timescale of the s-process have been addressed by a number of authors (133–139). None of these studies unambiguously identified the stellar interior environment. Peters (140) performed detailed calculations of s-process synthesis in massive stellar cores near the end of the helium-burning phase and concluded that the maximum neutron exposures that could be achieved were less than that characterizing solar system matter; subsequent studies have confirmed this result (141, 142). Recent studies of the more promising s-process environment provided by the convective helium-burning shells of thermally pulsing red giant stars are surveyed in a later section.

Interest in the astrophysical r-process event has been maintained by the continued demand for accurate predictions of the important cosmochronological production ratios $^{232}Th/^{238}U$, $^{235}U/^{238}U$, $^{244}Pu/^{238}U$, and $^{129}I/^{127}I$ (15, 143). This ensured that r-process nucleosynthesis remained an active area for research. The fact that nuclei from the valley of beta stability through to the neutron drip line can in principle participate in the r-process buildup of heavy and possibly superheavy nuclei, the properties of which must be estimated from nuclear systematics, reflects the complexity of this problem. In an early study of the r-process mechanism by Seeger, Fowler & Clayton (131), a constant temperature and neutron flux were assumed. These, together with a knowledge of the neutron capture, neutron photodisintegration, and beta decay rates, allow one to define a steady flow path on the neutron-rich side of the valley of beta stability. Competition between neutron capture and photodisintegration reactions, for the specified conditions, determines the distribution of nuclei along each isotope chain while the beta-decay lifetimes of the more neutron-rich isotopes dictate the rate of buildup to higher Z. Cameron, Delano & Truran (144) have argued rather that a proper dynamical study of r-process synthesis must be based on the time-dependent conditions expected to obtain in the appropriate astrophysical environment. For such conditions, a more accurate and realistic treatment of the complex sequences of nuclear transformations may be demanded (145, 146).

The site of r-process nucleosynthesis remains a mystery. The required high neutron densities and corresponding short timescales are strongly suggestive of a violent event associated perhaps with a supernova environment. A number of possible models have been explored, with varying degrees of success. These range from considerations of the expansion of neutron-rich matter from the innermost layers ejected in supernova explosions to studies of conditions resulting from the traversal of supernova shocks through the helium and carbon shells. A more thorough discussion of progress in our understanding of the r-process is presented in a later section.

4.6 p-Process Nucleosynthesis

A number of stable nuclei lying on the proton-rich side of the valley of beta stability are shielded from production by neutron-capture mechanisms by the existence of stable nuclei of lower proton number on their respective isobars. Examples of such "bypassed" nuclei are ^{180}Ta, ^{180}W, and ^{184}Os, shown in Figure 2. Burbidge et al (23) and Cameron (24) attributed the formation of these nuclei to a combination of (p,γ), (p,n), and (γ,n) thermonuclear reactions proceeding on preexisting heavy nuclei in stellar interiors at temperatures $T \approx 1-2 \times 10^9$ K and densities of protons of roughly 100 g cm^{-3}: this is the "p-process." The fact that the abundance level of the p-process nuclei parallels that of nuclei formed by neutron capture, but it is systematically lower by a factor of 100 to 1000, is compatible with the view that preexisting s-process and r-process nuclei serve as seeds. The passage of a supernova shock wave through the hydrogen-rich outer envelope of a star was believed to provide an appropriate environment.

Several alternative mechanisms for the production of the p-process nuclei have been proposed and examined. The relative scarcity of the p-process nuclei compared to their neighbors led to the suggestion that they might be formed by spallation reactions (147–149). The possible role of weak interaction processes—positron decay and excited-state beta decays ("photobeta" reactions)—has also been considered (133, 150–153). In both instances it was determined that, while significant enhancements might result for a number of specific cases, these mechanisms cannot account generally for the observed abundances of the p-process nuclei.

The thermonuclear reaction model advanced by Burbidge et al (23) and Cameron (24) for p-process nucleosynthesis has been explored in greater detail in subsequent studies (147, 154–161). Arnould (160) considered in particular the formation of the p-process nuclei during the course of hydrostatic oxygen burning in stellar interiors. It is not clear that these nuclei can subsequently survive the process of supernova ejection into the interstellar medium without suffering significant reprocessing. For this reason, models in which the formation of the p-process nuclei is tied to supernova events seem more promising. Two such models have recently been examined in detail. Both envision the p-process as occurring in supernovae with preexisting (primordial) abundances of s-process and r-process nuclei serving as seeds. Audouze & Truran (159) demonstrated that there exists a restricted range of post-shock temperature-density conditions in hydrogen-rich supernova envelope matter ($T \approx 2 \times 10^9$ K; $\rho \approx 10^4$ g cm^{-3}) for which the enhanced production of most of the designated p-process nuclei occurs in relative concentrations consistent with solar abundances. The dominant role under these conditions is played by (p,γ) and (p,n) reactions, although (γ,n) reactions become increasingly important as one proceeds to higher masses. Woosley & Howard (161) explored higher temperature conditions and determined that a series of photodisintegration reactions, both (γ,n) and (γ,p), operating on a distribution of heavy seeds for a continuum of temperatures in the range 2–3 $\times$ 10^9 K can also produce the bypassed nuclei. Their model avoids the requirement of a large proton density to drive (p,γ) and (p,n) reactions, but is sensitive to the temperature structure and history of the processed matter. Further calculations performed in the context of realistic hydrodynamic models of supernova events are required to establish which of these p-process models is appropriate.

4.7 The x-Process: Synthesis of the Light Elements

The light elements lithium, beryllium, and boron, together with deuterium and ^{3}He, presented a special challenge to early theories (8, 162). It was recognized that they are quite readily destroyed by proton-initiated thermonuclear reactions in the interiors of stars at temperatures below those required for hydrogen burning, but an appropriate environment for their formation was not immediately identified. The possibility that spallation reactions might contribute was first examined in the context of models involving reactions in stellar surfaces (163–166). Such theories have been criticized on very general grounds (167), on the basis of the fact that the enormous energy requirements are inconsistent with a stellar origin for galactic level concentrations of these isotopes.

The important role played by galactic cosmic rays in the formation of the light elements was first discussed by Reeves, Fowler & Hoyle (168) and subsequently confirmed by more detailed calculations (169–171). These studies indicated that substantial concentrations of lithium, beryllium, and boron, but not of deuterium and ^{3}He, can result from the interaction of cosmic ray protons and alpha particles with the more abundant heavy constituents of the interstellar gas (^{4}He, ^{12}C, ^{14}N, ^{16}O, and ^{20}Ne) over the lifetime of the galaxy. Calculations of the rates of formation of the light elements were performed in the context of a diffusion model for the propagation of galactic cosmic rays. It was concluded, specifically, that the concentrations of ^{6}Li, ^{9}Be, ^{10}B, and ^{11}B thus formed are quantitatively in agreement with the abundance levels of the isotopes characterizing primordial solar system matter. It is now generally accepted that the abundances of these four isotopes in our galaxy are attributable to these galactic cosmic ray interactions.

The predicted contribution from this mechanism to the abundance of ^{7}Li, however, represents only a small fraction of the observed abundance. A variety of possible alternative sites for ^{7}Li production have been considered. In addition to the red giant source of ^{7}Li discussed previously, possible sites for the production of ^{7}Li include supermassive stars (172–174), shock waves in supernova envelopes (175, 176), nova explosions (177) and the cosmological big bang (8, 172, 178–180). Although the cosmological conditions compatible with ^{7}Li production are nearly coincident with those for deuterium, detailed calculations indicate that the abundance level of ^{7}Li resulting from this source is low by approximately a factor of ten. This result is very nicely compatible with the low levels of lithium recently determined for unevolved halo stars and old disk stars (181, 182). A galactic origin for ^{7}Li is then demanded. The identification of red giants and perhaps novae as the major contributors to ^{7}Li in our galaxy would seem to be the best choice.

5. NUCLEAR PARAMETERS

Studies of stellar evolution and nucleosynthesis require a substantial variety of nuclear parameters. The extended phases of hydrogen and helium burning that comprise the major fraction of the active burning lifetimes of stars are dominated by reactions involving a relatively small number of nuclear species. Experimental determinations of the cross sections for many of these reactions are now available. However, it clearly is not feasible to obtain cross-section information for each and every nuclear reaction that may be involved in the later stages of nuclear energy generation in stellar interiors or in nucleosynthesis processes associated with stars and supernovae. Calculations of nucleosynthesis accompanying carbon, oxygen, neon, and silicon burning in stars and supernovae demand the use of complex networks of nuclear transformations involving tens to hundreds of nuclear species and literally hundreds to thousands of individual nuclear reaction and weak interaction processes. For purposes of illustration, in Figure 3 is shown the nuclear reaction network used in early numerical studies of carbon, oxygen, and silicon burning (35, 120, 121). In addition to the triple-alpha, $^{12}C+^{12}C$, and $^{16}O+^{16}O$ reactions, all reactions involving neutrons, protons, alpha particles, and photons were included as well as critical electron decay, positron decay, and electron-capture reactions. Studies of the neutron-capture processes responsible for the production of the bulk of the nuclear species more massive than iron similarly involve the use of vast networks of nuclear reactions.

The mechanisms of heavy-element nucleosynthesis thus fall naturally into two broad categories, dependent upon whether the nuclear transformations are dominated by charged particle reactions or by neutron-capture reactions. Of particular importance to such studies are thermonuclear reaction rates for both neutron and charged-particle reactions and rates of electron decay, positron decay, and electron capture.

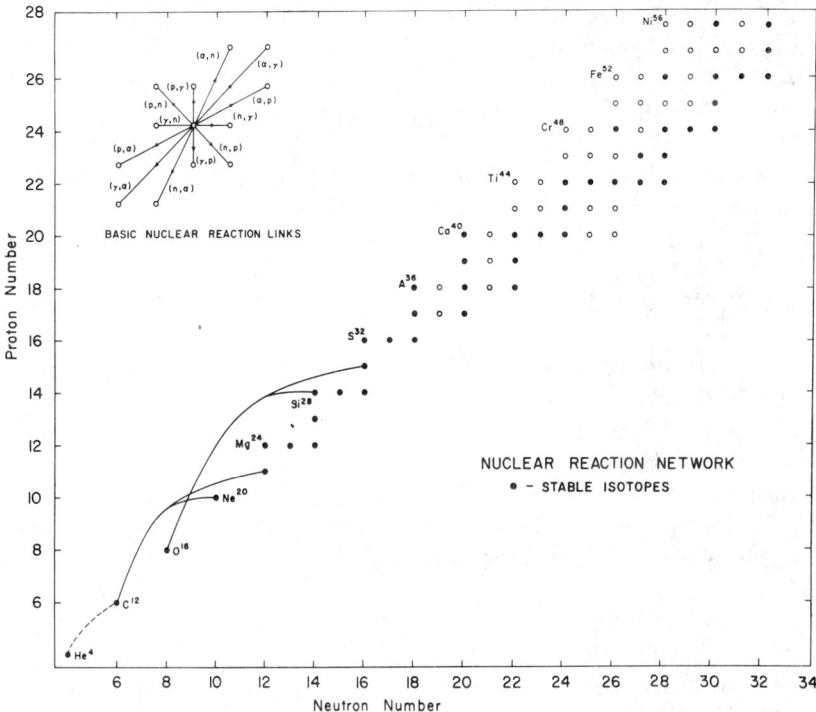

Figure 3 A representative nuclear reaction network used in investigations of carbon-, oxygen-, and silicon-burning nucleosynthesis. Note that in addition to the $3\,^4\text{He} \to {}^{12}\text{C}$, ${}^{12}\text{C} + {}^{12}\text{C}$, and ${}^{16}\text{O} + {}^{16}\text{O}$ reactions, all single-particle reactions involving neutrons, protons, alpha particles, and photons were considered.

5.1 Thermonuclear Reaction Rates

Thermonuclear reaction rates appropriate to stellar and supernova environments are computed as averages over Maxwellian distributions. The effective energies at which charged-particle reactions proceed represent a compromise of the low thermal energies ($kT = 1.3$ keV at the Sun's central temperature of 15×10^6 K) and the significantly higher Coulomb barrier energies. The critical ingredient to be specified here, from either experiment or theory, is of course the appropriate nuclear reaction cross section as a function of energy in the energy region of interest. One requires, specifically, a knowledge of charged-particle cross sections at extremely low energies where the cross sections are themselves correspondingly low.

The empirical foundations of theories of stellar nuclear energy generation and nucleosynthesis are provided by experimentally determined cross sections and reaction rates for a number of critical nuclear reactions. In particular, experimental information is available regarding most of the reactions involved in the hydrogen-burning sequences—both the proton-proton chains and the CNO cycles—and for the reactions $3\,^4\text{He} \rightarrow {}^{12}\text{C}$, ${}^{12}\text{C}(\alpha,\gamma){}^{16}\text{O}$, and ${}^{16}\text{O}(\alpha,\gamma){}^{20}\text{Ne}$, which define the helium-burning phase. This is true as well for the ${}^{12}\text{C}+{}^{12}\text{C}$ and ${}^{16}\text{O}+{}^{16}\text{O}$ reactions, which dominate the phases of carbon and oxygen burning in massive stars. Reviews of the experimental situation regarding charged-particle reactions have been provided by Reeves (183) and Barnes (184) and an important series of compilations of experimental thermonuclear reaction rate parameters by Fowler, Caughlan & Zimmerman (95, 96), and Harris et al (97). Experimental cross section data is also available for an increasing number of the neutron-capture reactions that participate in the buildup of heavy nuclei (185–188).

There are literally hundreds of nuclear reactions involving the interactions of neutrons, protons, and alpha particles with heavier nuclei that can participate in the later stages of nuclear burning in stellar interiors and in the synthesis of heavy nuclei in supernovae. Since it is clearly impractical to obtain direct experimental cross sections in every case, it is necessary to develop some model of nuclear reactions to the point where reliable theoretical predictions become possible. Substantial success has been achieved in calculations of thermonuclear reaction rates in the context of the statistical model. Here one adopts a Hauser-Feshbach (189) or equivalent expression for the energy-averaged reaction cross section. Crucial input to such calculations includes a knowledge of, or reliable theoretical estimates of, both nuclear level densities and the nuclear reaction widths for radiative decay and particle emission.

Theoretical calculations of thermonuclear reaction rates appropriate to temperatures in the range one to ten billion degrees Kelvin have been performed by a number of researchers. Early studies (120, 190–192) were largely restricted to considerations of nuclei in their ground states: it was assumed that only the ground states of the target and residual nuclei contribute to the widths in the nucleon and alpha-particle channels. Arnould (193) called attention to the fact that excited-state interactions can make considerable contributions and recent studies (194–196) include such effects. These papers provide the most complete compilations of data on neutron and charged-particle thermonuclear reaction rates for conditions compatible with nuclear burning in the late stages of evolution of massive stars and thermonuclear processing of matter ejected in supernovae.

5.2 Weak Interaction Rates

Weak interaction processes serve two particularly important purposes with regard to nucleosynthesis: (*a*) they produce neutrinos and antineutrinos, the loss of which results in cooling and contraction of stellar and presupernova cores and accelerates these phases of evolution (197); and (b) they typically act to increase the total neutron-to-proton ratios characterizing the core matter and thereby dictate important features of the abundance distributions resulting from thermonuclear burning. The significance of both of these consequences is such that accurate weak interaction rates represent critical input to calculations of stellar evolution, supernova hydrodynamics, and nucleosynthesis. As with nuclear reaction rates, it is not possible here to obtain experimental data for all of the thousands of electron decay, positron decay, and electron-capture reactions that can proceed in stellar environments. It is once again necessary to provide theoretical determinations of the rates of a very large number of reactions.

The situation for weak interactions is further complicated by the effects of the extreme temperature and density conditions characterizing astrophysical environments. Weak decay rates can be strongly enhanced because excited states are thermally populated (133); indeed, nuclei that are stable in their ground states under terrestrial conditions may undergo substantial decay via thermally populated excited states. Pauli principle inhibition of phase space can become a critical factor in electron degenerate matter at high densities (198, 199). It should be recognized that the magnitudes of these effects can be substantial. For example, the half-life of ^{99}Tc, a nucleus critical to considerations of s-process neutron-capture synthesis in the envelopes of red giant stars, is reduced from its terrestrial value of 2.13×10^5 years to ~ 0.5 year as a result of thermal population effects at the temperatures believed to characterize the red giant envelope environment.

Calculations of weak interaction rates appropriate to astrophysical conditions require a knowledge of the properties of nuclear ground and excited states that can contribute. Nuclear matrix elements for these transitions must be taken from experiment or otherwise assigned on the basis of averaged strengths for transitions of different orders of forbiddenness. This latter approach necessarily introduces uncertainties into the calculations. Rates for a number of weak decays involving iron group nuclei were determined by Fowler & Hoyle (115) on the basis of then existing experimental data. Hansen (200, 201) and later Mazurek, Truran & Cameron (202) supplemented the available experimental information with an integration over a density of nuclear excited states (203) of appropriately averaged transition strengths. Iben (204) calculated weak interaction rates directly from the existing excited-state data for a limited number of intermediate mass nuclei ($20 \lesssim A \lesssim 60$). Electron and positron decay and electron-capture rates for a large number of heavier nuclei in the vicinity of the valley of beta stability that participate in the s-process reaction sequences have also been similarly calculated from existing excited-state data by a number of authors (136, 137, 205, 206). Calculations of beta decay rates for many heavy nuclei have been performed based both upon the statistical treatment of the gross theory of beta decay (207–210) and on the Klapdor treatment of the Gamow-Teller resonance (211–213).

The most complete and up-to-date calculations and tables of stellar electron and positron emission rates and continuum electron- and positron-capture rates are those provided in a recent series of papers by Fuller, Fowler & Newman (214–216). As in the earlier studies by Hansen (200, 201) and Mazurek et al (202), experimental nuclear level and matrix element information is used where available and is supplemented by theoretical estimates of transition strengths and nuclear level densities. They find that the contributions to the rates from discrete states, for which experimental information is generally available and Fermi selection rules apply, are dominant at temperatures and densities typical of the hydrostatic phases of presupernova evolution. In contrast, at the more extreme conditions of temperature and density characteristic of the supernova collapse phase, the nuclear rates were found to be dominated by the Fermi and the Gamow-Teller collective resonance contributions. The effects of neutron-shell closure blocking of electron capture on extremely neutron-rich nuclei were also included (217). Tables of rates of electron and positron decay and continuum electron and positron capture and associated neutrino energy loss rates, for a range of temperature density conditions, are presented by Fuller, Fowler & Newman (216) for several nuclei in the mass range $21 \leqslant A \leqslant 60$.

6. ACTIVE AREAS OF RESEARCH IN NUCLEOSYNTHESIS

In this section, we focus on developments in active areas of research in nucleosynthesis. We begin with a brief review of nucleosynthesis associated with the cosmological big bang. We then address our attention to the processes responsible for the formation of the great majority of the heavy elements: the s-process and r-process neutron-capture mechanisms and the charged-particle-dominated thermonuclear burning processes that produce the elements from neon through the iron-peak region. We note that the emphasis in current research is on the establishment of the manner of operation of these nucleosynthesis processes in the context of realistic stellar or supernova models. We deal specifically with s-process nucleosynthesis in red giant envelopes, with models for r-process synthesis in supernovae and with explosive charged-particle nucleosynthesis in supernovae. The pervasive role of supernovae in heavy-element synthesis is suggested by several factors: the extreme temperature and density conditions therein achieved allow increasingly complex nuclear processes to occur; evolution on a hydrodynamic timescale is compatible with inferred nuclear timescales; and the nuclear transmutations occur as a concomitant of the expansion and cooling associated with mass ejection, thereby ensuring that the abundance patterns thus formed reach the interstellar medium without further alteration. It will become clear how these factors act to dictate the characteristics of the mechanisms of heavy-element synthesis.

6.1 Cosmological Nucleosynthesis

Interest in the subject of big bang or "primordial" nucleosynthesis was stimulated when Penzias & Wilson (218) discovered the 2.7-K blackbody microwave background radiation. Its identification as the relic of a cosmological big bang implies that the earliest moments of the universe were characterized by extreme temperatures and densities. Nuclear reactions are expected to accompany the subsequent expansion and cooling. Commencing with Peebles (219, 220) and Wagoner, Fowler & Hoyle (172), the predictions of big bang nucleosynthesis have been examined and reviewed by a number of authors (8, 91, 162, 178–180, 221–223).

Nucleosynthesis occurs as a natural consequence of the expansion and cooling of the universe. At temperatures above 10^{11} K, the concentrations of neutrons and protons are maintained in thermodynamic equilibrium by the rapidity of the weak interaction processes

$$p + e^- \rightleftarrows n + \nu_e$$

$$n + e^+ \rightleftarrows p + \bar{\nu}_e.$$

The relative abundances of free neutrons and protons at the onset of the nucleosynthesis era are dictated by the freezing out of these reactions upon expansion and cooling to a temperature $\gtrsim 10^{10}$ K, which yields a ratio of neutrons to protons of approximately 1/7. The critical temperature for nucleosynthesis is about 10^9 K: production of heavy nuclei is inhibited at higher temperatures by nuclear photodisintegrations and at lower temperatures by the inability of charged particles to penetrate their mutual Coulomb barriers. Deuterium formation first proceeds by $n + p \rightarrow D + \gamma$. Deuterium interaction with neutrons and protons and with itself, and subsequently with the intermediate reaction products tritium and ^{3}He, then leads rapidly to the production of ^{4}He. Most of the initial neutrons are ultimately utilized in the production of ^{4}He, which yields a helium mass fraction of ~ 0.23–0.25.

One rather firm conclusion follows from these nucleosynthesis studies: for a current universal temperature of 2.7 K and a current universal density compatible with observational limits, no significant synthesis of nuclei past ^{4}He is found to occur. The triple-alpha reaction ($3\,^4$He → ^{12}C), which serves successfully to bridge the instability gaps at masses $A = 5$ and 8 in stellar interiors, and other possible three-particle reactions as well are simply ineffective at the very low densities existing in the expanding fireball at the appropriate burning temperature $\sim 10^9$ K.

Significant concentrations of ^{4}He and other light elements—^{2}D, ^{3}He, and ^{7}Li—can be formed under appropriate burning conditions in the big bang. ^{4}He production occurs over a range of densities for the standard model, and the uniformity of the ^{4}He concentrations characterizing diverse astronomical objects (58) is strongly supportive of a cosmological interpretation of its origin. Although the greater fractions of the deuterium and ^{3}He that are formed are quite rapidly burned to form ^{4}He, some amounts can survive. The surviving concentrations of these elements will be extremely sensitive to the competition between the rates of the nuclear reactions by which they are destroyed and the expansion rate of the universe. The predicted ^{2}D and ^{3}He concentrations thus decrease with increasing matter density. ^{7}Li is an exception to the rule that nuclei past ^{4}He cannot be formed in this environment, its production being attributable to the interactions of ^{4}He with both ^{3}H and ^{3}He.

The abundances of ^{2}D, ^{3}He, and ^{7}Li in the galaxy today reflect both contributions and losses due to stellar processing. ^{2}D is readily destroyed in stellar interiors and there is no identified galactic origin for its production, so its abundance in the interstellar medium today (224, 225) of D/H ≈ 2×10^{-5} is presumed to represent a lower limit on the primordial value. Numerical models of galactic evolution (16, 82) suggest a reduction in deuterium of a factor of ~ 2–3. In contrast, there exist galactic sources for the synthesis of ^{3}He and ^{7}Li, as described in our previous discussions of stellar hydrogen burning and the synthesis of the light elements, for which corrections must be made. A primordial solar system ratio ^{3}He/^{4}He ≈ 1.4×10^{-4} may be inferred from planetary data (226). The low lithium abundance ^{7}Li/H = 1.12×10^{-10} recently determined by Spite & Spite (181, 182) for unevolved halo stars and old disk stars in our galaxy presumably provides a more realistic appraisal of the primordial concentration. Given such considerations as these and the resulting best estimates of the primordial concentrations of H, ^{2}D, ^{3}He, ^{4}He, and ^{7}Li in galactic matter, it is possible to identify a particularly interesting universe ($T_0 = 2.7$ K; $\rho_0 \approx 1$–4×10^{-31} g cm^{-3}) for which the production of these elements will occur. More generally, if one assumes the validity of the standard big bang model, the observed abundances of the light elements that can be formed in primordial nucleosynthesis can be used to impose constraints upon both cosmology and particle physics (91, 180, 227–234); of particular recent interest have been the possible constraints imposed on the number of two-component neutrino species.

6.2 s-Process Synthesis in Red Giants

The past decade has seen significant progress toward an understanding of the stellar environments in which s-process nucleosynthesis can proceed (for review, see 6, 7, 12). This recent work is built upon the operation of two neutron sources long ago identified by Cameron (103, 129): the $^{13}C(\alpha,n)^{16}O$ and $^{22}Ne(\alpha,n)^{25}Mg$ reactions. The ^{13}C and ^{22}Ne source elements both are formed in the normal course of stellar evolution: ^{13}C follows from the reactions $^{12}C(p,\gamma)^{13}N(e^+v)^{13}C$ and ^{22}Ne is produced from ^{14}N, the main product of CNO cycle hydrogen burning by the reaction sequence $^{14}N(\alpha,\gamma)^{18}F(e^+v)^{18}O(\alpha,v)^{22}Ne$. The seed nuclei for the s-process operating in red giant environments are provided by the primordial concentrations of iron-peak nuclei in the stellar envelope.

Schwarzschild & Härm (235) first identified the environment provided by low and intermediate mass stars possessing both hydrogen- and helium-burning shells as a promising site for neutron-capture synthesis. They determined that thermal pulses associated with the helium shells in such stars during the asymptotic giant phase of evolution trigger the growth of a convective shell encompassing most of the matter within the ^4He and ^{12}C region, between the inert carbon-oxygen core and the hydrogen envelope. Sanders (236) demonstrated that only a very small admixture of protons into the helium- and carbon-rich shell, such as might occur if the convective region were to extend outward beyond the hydrogen-helium interface, could provide a substantial flux of neutrons by means of the reactions $^{12}C(p,\gamma)^{13}N(e^+v)^{13}C(\alpha,n)^{16}O$. Building extensively on this model, Ulrich (237) was able to show that a succession of thermal pulses (with successive convective regions partially overlapping one another in mass) naturally provides an exponential distribution of neutron exposures, comparable to that determined by Seeger, Fowler & Clayton (131) to be necessary to fit the solar system distribution of s-process abundances. In Ulrich's (237) notation, the relative abundance of nuclei that have received a neutron exposure τ after experiencing a large number of shell flashes is given by $\rho(\tau) = \Lambda \exp(-\Lambda\tau)$, where Λ is an "exposure" parameter that may be calculated from the detailed characteristics of the thermal pulse. The main difficulty with this model arises from the fact that calculations of thermal pulses for a wide variety of stellar masses predict that the required extension of the convective shell through the hydrogen-helium interface will not occur: Iben (238) has shown that such an extension is inhibited by the presence of an entropy barrier. Additionally, the influx of protons into the helium- and carbon-burning shell (239), should it occur, must be restricted to a very narrow range of possibilities (240) if it is to provide a ratio of neutrons to seed nuclei that is compatible with the formation of the s-process nuclei in solar system proportions.

In contrast, the neutron release associated with the operation of the ^{22}Ne(α,n)^{25}Mg reaction as a neutron source during thermal pulses is not subject to these difficulties (90, 104, 238, 241, 326, 327). On this model, the outward progression of the hydrogen-burning shell during the interpulse phase leaves in its wake a substantial concentration of ^{14}N, which is readily transformed into ^{22}Ne by ^{14}N(α,γ)^{18}F($e^+\nu$)^{18}O(α,γ)^{22}Ne when subsequently mixed downward into the helium shell during the ensuing thermal pulse. As has been emphasized by Iben (90, 104), this neutron source thus arises as a natural consequence of the evolution of intermediate mass stars along the asymptotic giant branch. Truran & Iben (105) demonstrated that the operation of the ^{22}Ne(α,n)^{25}Mg neutron source in thermally pulsing stars will produce the s-process abundances in the mass range $70 < A < 204$ in solar system proportions. Indeed, a particular strength of this model arises from the fact that it is constrained to consistency with solar system abundances by a combination of nuclear physics and stellar evolutionary considerations. Iben (90, 104) and Truran & Iben (105) demonstrated that the exposure parameter Λ is equal, to first order, to the mean neutron-capture cross section of the light-element progeny of ^{22}Ne (specifically, ^{25}Mg and its neutron-capture progeny). Numerically, this gives $\Lambda \approx 4$–6, compatible with the production of a distribution of s-process nuclei in agreement with that of solar system material (131, 237). Iben & Truran (84) also demonstrated that sufficient matter is processed in this manner in red giant envelopes to explain the abundance level of the s-process nuclei in the galaxy.

Some specific questions concerning the operation of this ^{22}Ne source s-process model remain. Despain (242) has argued that the ^{22}Ne source cannot be responsible for the production of the s-process elements, since flashes occurring at sufficiently high core masses to ensure neutron production via ^{22}Ne(α,n)^{25}Mg invariably liberate the neutrons too rapidly: the average flux is too high to be consistent with the branches at beta-unstable nuclei like ^{79}Se and ^{85}Rb, which set the timescale of the s-process. Detailed calculations performed by Cosner, Iben & Truran (106, 243) indicate that his conclusion is incorrect, since the time dependence of the neutron flux must be considered. While the neutron density at the peak of the shell flash may indeed be much higher than is generally assumed to be appropriate for s-process nucleosynthesis, final relative abundances are determined rather by the much lower neutron density that characterizes the terminal stages of the flash—when the last few neutrons are captured. The situation is further improved when one takes into account the suggestion of Ward (244) that the isomeric levels of several critical nuclei do not become thermalized on the timescale of a thermal pulse. Scalo & Miller (245) challenge the ^{22}Ne model on quite different grounds: they concluded, on the basis of observations of technetium abundances as a function of luminosity in red giants, that envelope mixing operated only at degenerate core masses below the critical value ~ 0.9 $M_\odot$ demanded for the ^{22}Ne(α,n)^{25}Mg reaction to proceed in thermal pulses. Cosner, Despain & Truran (246) demonstrated, however, that the concentration of ^{99}Tc that may be expected to characterize the s-processed matter in such stars at the termination of a thermal pulse is an extremely sensitive function of the temperature and timescale conditions. The ^{99}Tc beta decay rate (206) reaches values as high as roughly 10^5 times its terrestrial value, because of contributions from thermally populated excited states at the prevailing temperatures.

A critical constraint on the operation of the ^{22}Ne(α,n)^{25}Mg reaction as a neutron source is that the temperature at the base of the convective burning shell exceed 300 million degrees Kelvin: only at higher temperatures will the timescale for neutron liberation be less than that of the thermal pulse. This implies (84) that significant s-process abundance enhancements can be realized in red giant atmospheres only for relatively massive stars, $M \gtrsim 3$ $M_\odot$, of high luminosities, $L \gtrsim 10^4$ $L_\odot$. Observational surveys (245, 247) of the peculiar red giants that show s-process abundance enhancements reveal, however, that many of these stars have masses and luminosities too low for the ^{22}Ne reaction to operate. An alternative neutron-capture environment appears to be demanded. It has generally been assumed that the neutron source appropriate to these peculiar red giants must be provided by the ^{13}C(α,n)^{16}O reaction, in spite of the problems encountered in establishing the nature of the mixing mechanisms (88, 239). In realistic astrophysical environments, a significant ^{13}C neutron source can be achieved only when the admixture of protons into a carbon-rich region proceeds at a controlled rate yielding ^{12}C(p,γ)^{13}N($e^+\nu$)^{13}C. If too many protons are introduced, the reaction ^{13}C(p,γ)^{14}N acts both to reduce the absolute ^{13}C concentration and to reduce its effectiveness: the ^{14}N(n,p)^{14}C reaction has a very large cross section and ^{14}N competes favorably with heavy elements for the neutrons subsequently released by ^{13}C(α,n)^{16}O. A promising model for such mixing was recently discussed by Iben & Renzini (248, 249). They demonstrate that a restricted mixing can occur by semiconvection following thermal pulses in asymptotic giant branch stars of small core mass. This would allow for the realization of both carbon and s-process enrichments of the envelopes in low-luminosity red giant stars.

Moreover, the reaction product, ^{16}O, has a very low neutron cross section and therefore does not provide competition for neutrons as does ^{25}Mg, the product of ^{22}Ne$(\alpha,n)^{25}$Mg. This allows in principle for the generation of very large neutron exposures and heavy-element distributions quite unlike those of solar system matter (250, 251).

6.3 r-Process Nucleosynthesis in Supernovae

Proposed models for r-process nucleosynthesis may be conveniently grouped into two broad classes: those involving the expansion and cooling of neutron-rich matter and those concerned with the passage of supernova shock waves through the carbon and helium shells. The former class involves such diverse models as the expansion of neutronized matter from the vicinity of the mass cut in supernova explosions leaving neutron star remnants (144, 252–259); neutronized jets resulting from the collapse of a rotating, magnetized star (260, 261); and neutronized jets realized in the collision of a black hole and a neutron star (262). This work has been nicely reviewed in far greater detail by Hillebrandt (5), while a very useful general discussion of the conditions required for the r-process is given by Norman & Schramm (263). Studies of r-process synthesis occurring after the passage of a supernova shock wave through the outer layers of the stellar core have dealt specifically with the helium shell (212, 264–270) and the carbon shell (271). These more recent studies have been motivated in part by the many new identifications of stable isotope anomalies in meteorites and the suggestion of their possible relation to neutron-capture properties. The general characteristics of these two classes of studied r-process environments are reviewed below.

The sequence of events by which heavy nuclei form in the expansion and cooling of highly neutronized matter is well defined. We consider here specifically the case of mass ejected in supernovae. The initial conditions predicted by early hydrodynamic studies of supernova explosions (260, 272–274) for the innermost ejected layers were temperatures $T \approx 10^{10}$–10^{11} K, densities $\rho \approx 10^{11}$ g cm^{-3}, and expansion on a hydrodynamic timescale. One can expect matter characterized by total neutron-to-proton ratios ranging from ~ 1.5 to ~ 8, established by weak interaction processes at temperatures $T > 10^{10}$ K and densities $\rho > 10^{10}$ g cm^{-3}. The final n/p ratio is quite sensitive to the expansion timescale (255). When the temperature falls below 10^{10} K, thermonuclear reactions forming heavier nuclei proceed rapidly: free neutrons and protons first interact to form deuterium and then ^{4}He; the $3\,^4$He $\rightarrow$ ^{12}C reaction and other three-body reactions that can operate at high densities serve to bridge the mass instability gaps at $A = 5$ and 8, and the production of heavier nuclei then follows. Charged-particle and neutron reactions both contribute to the buildup into the iron region and beyond at temperatures down to approximately $4\text{--}5 \times 10^9$ K (252, 254, 275). These reaction sequences proceed on timescales short compared to hydrodynamic expansion timescales ($\tau_{\text{HYD}} \approx 10^{-2}$ s at $\rho = 10^9$ K), which ensures that the abundances of heavy nuclei are maintained in nuclear statistical equilibrium. The extremely neutron-rich equilibrium distributions achieved under these conditions, dominated by exotic neutron-rich species like ^{78}Ni and characterized by high ratios of free neutrons to heavy nuclei (276), provide an effective starting point for r-process synthesis. Charged-particle reactions are frozen out as the temperature falls below $\sim 3\text{--}4 \times 10^9$ K. One is then left with an extremely neutronized heavy-element seed distribution and a high ratio of free neutrons to seed nuclei.

The subsequent nuclear evolution of this matter is dominated by neutron capture and photodisintegration reactions and beta decays. The operation of these reactions on a hydrodynamic timescale defines an r-process. Cameron, Delano & Truran (144), Schramm (257), and Hillebrandt, Takahashi & Kodama (259) demonstrated that the formation of heavy nuclei by neutron captures interspersed by beta decays will occur in the subsequent expansion and cooling of this matter. The gross features of solar system r-process abundance patterns can be reasonably well reproduced in such calculations: the positions of the r-process peaks, their relative sizes, and the absence of substantial odd-even abundance variations. There are, of course, matters of detail on which the calculations differ. A major question here concerns the rates of beta decay. For any model of the r-process to be successful, it is essential that the beta decay lifetimes be compatible both with timescales dictated by the prevailing neutron fluxes and with the imposed stellar or dynamic timescales. Very different estimates of beta decay rates were utilized in these studies. The revised beta decay rates of Klapdor (213) tend to be higher than either the Senbetu (277) rates utilized by Schramm (257) or the gross theory predictions of Takahashi, Yamada & Kondoh (209) used by Hillebrandt, Takahashi & Kodama (259), but they are a factor of a few slower than those calculated by Cameron, Delano & Truran (144) based on the methods described by Hansen (200). A new examination of this r-process model thus seems appropriate. Critical questions concerning the operation of this process in the context of realistic hydrodynamic models of supernova events must be addressed. It is worth noting that a mass of only $\sim 10^{-6}$ $M_\odot$ per supernova event need be processed in this manner to provide galactic requirements of r-process nuclei (82).

The mode of operation of the r-process in the environment defined by the shock heating of the helium or carbon shells of supernovae is similar in many ways to that of s-process synthesis. The seed heavy nuclei are here again provided by preexisting, and presumably primordial, concentrations of iron-peak and heavier nuclei. The most effective neutron sources are again those provided by the $^{13}C(\alpha,n)^{16}O$, $^{22}Ne(\alpha,n)^{25}Mg$, and, to a lesser degree, $^{18}O(\alpha,n)^{21}Ne$ reactions, which are driven by the high post-shock temperatures. The timescale, set by the hydrodynamic response of the matter to the passage of the shock, is of course quite short—though the conditions are not nearly as extreme as those accompanying the ejection of highly neutronized matter.

The possibility that the helium zones of massive stars subjected to supernova shock-wave heating might provide a promising site for the r-process was first considered by Hillebrandt & Thielemann (264) and Truran, Cowan & Cameron (265, 266). Subsequent studies have dealt with both the helium layer (212, 267–269, 278) and the carbon layer (271). It was originally anticipated that the neutrons released from the reaction $^{22}Ne(\alpha,n)^{25}Mg$ on a short timescale and subsequently captured on iron-peak nuclei might be capable of forming the r-process heavy elements in essentially solar system proportions. It has been determined, however, that for abundances of ^{22}Ne (and ^{18}O) of order 2% by mass, compatible with realistic models of the prior evolution of this helium- or carbon-shell matter, the strength of the $^{22}N(\alpha,n)^{25}Mg$ neutron source is inadequate to the task. Both the high temperatures required for rapid neutron release by this reaction and the fact that the immediate progeny of ^{22}Ne, specifically ^{25}Mg, represent strong competitors for the neutrons thus released constrain the numbers of neutrons available for capture on the appropriate timescale. While it now appears that the bulk of the r-process heavy elements cannot have been formed in this manner, it remains possible that such neutron-capture episodes might be responsible for producing some of the diverse anomalous isotopic abundance patterns identified in meteorites.

A number of authors have also examined the possible role of the $^{13}C(\alpha,n)^{16}O$ reaction as a neutron source for the r-process. Under normal (equilibrium) conditions in CNO hydrogen burning, the ^{13}C concentration is quite small, particularly with respect to ^{14}N. This is a critical factor, since ^{14}N serves as an extremely effective competitor for neutrons due to the large cross section for $^{14}N(n,p)^{14}C$. ^{13}C concentrations compatible with the generation of significant neutron fluxes in the helium layers of stars can only be realized in the presence of convective mixing, as outlined in our previous discussion of the s-process. Even so, the production of at least a few percent by mass of ^{13}C necessary to provide the neutron requirements for the r-process is difficult (278). Similar problems are encountered by models in which the $^{13}C(\alpha,n)^{16}O$ reaction is used to provide neutron fluxes under conditions consistent with helium core flashes in low mass stars (270, 279). Nevertheless, the ^{13}C neutron source can in principle provide neutron fluxes capable of explaining heavy-element anomalies in meteorites (249, 280).

6.4 Explosive Nucleosynthesis in Supernovae

The elements from neon to the iron peak are now believed to have been formed collectively in the environment provided by the shock heating and ejection of helium-exhausted core matter in supernova explosions. Successive exoergic stages of burning of hydrogen, helium, carbon, neon, oxygen, and silicon fuels define the presupernova evolution of the cores of stars more massive than 10 $M_\odot$ (100, 101), while the growth of degenerate carbon-oxygen or oxygen-neon-magnesium (328) cores to the Chandrasekhar (107) limiting mass occurs for stars in the approximate mass range 4–10 $M_\odot$ (12). It is evident that a dynamic phase of evolution must follow in both instances: the iron cores of massive stars lack further nuclear fuel and are compelled to contract under gravity while the growth of degenerate cores beyond the Chandrasekhar limiting mass similarly leads to collapse. When the ashes of prior burning epochs are subsequently subjected to high temperatures and densities accompanying supernova ejection, further thermonuclear processing yields elemental and isotropic abundance patterns that mimic very closely those of solar system matter.

Early calculations of explosive nucleosynthesis in supernovae, previously reviewed by Arnett (3) and Truran (4), were necessarily performed separately from hydrodynamic calculations of these events. In these studies, one chose an initial composition consistent with post helium-burning configurations of stellar cores and then explored the range of temperature and density conditions consistent with the expulsion of the matter in a supernova event. The expansion timescale was equated with the hydrodynamic timescale. Calculations of explosive carbon-, oxygen-, and silicon-burning nucleosynthesis performed in this manner proved extremely successful in their predictions of detailed abundance features. The names of these various burning processes are carryovers from the corresponding nuclear burning stages of stellar evolution; the feature distinguishing these explosive processes is the peak shock temperature to which the matter is subjected. We note specifically: for temperatures $T \approx 2 \times 10^9$ K ($\rho \sim 10^5$ g cm^{-3}), explosive carbon burning forms many of the nuclei in the mass range $20 \leq A \lesssim 30$ (281–285); at slightly higher temperatures ($T \approx 3.6 \times 10^9$ K; $\rho \approx 5 \times 10^5$ g cm^{-3}), explosive oxygen burning will reproduce most of the observed features in the mass range $28 \lesssim A \lesssim 44$ (126, 286); for temperatures $T \gtrsim 4.5 \times 10^9$ K ($\rho \gtrsim 10^6$ g cm^{-3}), explosive silicon burning produces most isotopes in the iron-peak region $48 \lesssim A \lesssim 62$ in solar proportions (35, 126, 127). It is anticipated that the peak post-shock temperatures imparted to helium-exhausted core matter ejected in supernova events will necessarily span the interesting range $10^9 \lesssim T \lesssim 5\text{--}7 \times 10^9$ K. Subsequently, it is the nuclear properties of the species formed that dictate the resulting abundance distributions.

The nuclear abundances by mass resulting from explosive silicon burning are illustrated in Figure 4. The calculated abundances represent an equally weighted average of two mass zones, burning from peak temperatures and densities $T = 4.7 \times 10^9$ K, $\rho = 10^6$ g cm^{-3} and $T = 5.9 \times 10^9$ K, $\rho = 2 \times 10^6$ g cm^{-3}, respectively. Note that, with the exception of the neutron-rich species ^{50}Ti, ^{54}Cr, and ^{58}Fe, all isotopes in the mass range $48 \le A \le 62$ are formed in solar system relative proportions, to within a factor of two. While the overall character of the resulting iron abundance peak is dictated by considerations of nuclear statistical equilibrium, the detailed features are sensitive both to the rates of individual nuclear

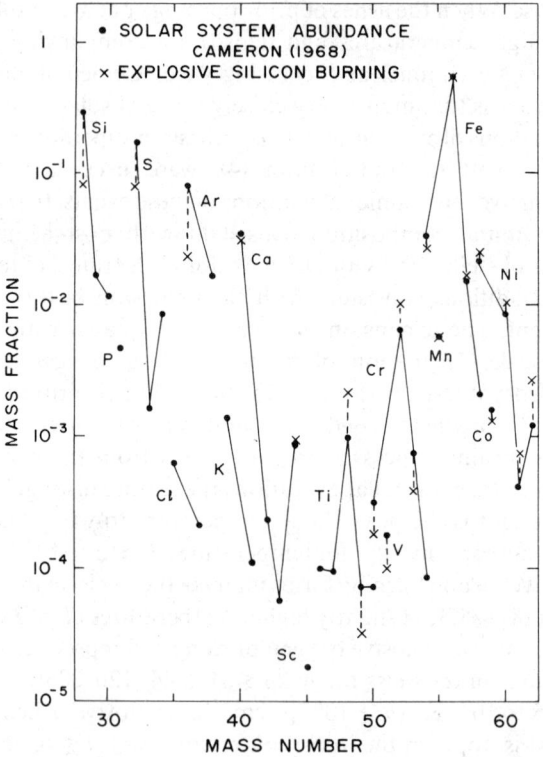

Figure 4 Abundances by mass of nuclei formed under explosive silicon-burning conditions in supernovae are compared to the solar system abundances (29) in this mass region; the two distributions are normalized at ^{56}Fe.

reactions as a function of temperature during the course of expansion and cooling and to the relative numbers of neutrons and protons in the processed stellar matter. We call attention, in particular, to the fact that the most abundant nuclear constituents of the core immediately prior to the supernova event are self-conjugate nuclei (^{4}He, ^{12}C, ^{16}O, ^{20}Ne, etc). Trace amounts of neutron-rich isotopes of these elements and odd-Z elements are also present in the gas, such that the total number of neutrons exceeds that of protons by a factor of order only several parts in a thousand. Thermonuclear processing of such matter in the wake of supernova shocks typically occurs on timescales of the order of seconds or less; this ensures that weak interaction processes cannot act to convert any appreciable fraction of protons into neutrons.

It follows that the final products of these explosive burning episodes must lie along or very near to the $Z = N$ line. This remains true even when the build up proceeds past the last stable alpha-particle nucleus (^{40}Ca): the dominant species in situ include the nuclei ^{44}Ti, ^{48}Cr, ^{52}Fe, ^{56}Ni, and ^{60}Zn. When these nuclei and their immediate neighbors subsequently decay, isotopic abundance patterns compatible with those of solar system matter are realized. We note, for example, that the isotopic composition of chlorine formed in such circumstances is dictated by the nuclear properties of ^{35}Cl and ^{37}Ar: following the decay of the ^{37}Ar formed in situ, the terrestrial ^{35}Cl/^{37}Cl ratio is quite accurately reproduced. This same phenomenon occurs for many other cases: the potassium isotopes ^{39}K and ^{41}K (formed as ^{41}Ca), the chromium isotopes ^{50}Cr, ^{52}Cr, and ^{53}Cr (the last two formed as ^{52}Fe and ^{53}Fe), the iron isotopes ^{54}Fe, ^{56}Fe, and ^{57}Fe (the last two formed as ^{56}Ni and ^{57}Ni), and the nickel isotopes ^{58}Ni, ^{60}Ni, ^{61}Ni, and ^{62}Ni (the last three formed as ^{60}Zn, ^{61}Zn, and ^{62}Zn). Furthermore, the relative abundances of ^{51}V (formed as ^{51}Mn), ^{55}Mn (formed as ^{55}Co), and ^{59}Co (formed as ^{59}Cu) are reproduced very well in these calculations. This general behavior and the fact that the resulting isotopic patterns are so closely in agreement with those of solar system matter strongly support the view that supernovae provide the appropriate nucleosynthesis environment. Moreover, the fact that the formation of mass $A = 56$ occurs in the form of ^{56}Ni holds important implications for the nature of the light curves of supernovae: the subsequent decays of ^{56}Ni($\tau_{1/2} = 6.1$ d) and its daughter ^{56}Co($\tau_{1/2} = 78.5$ d) are capable of providing significant energy output on the appropriate timescales (287–291).

Recent investigations of explosive nucleosynthesis have been carried out in the context of hydrodynamic models of the supernova events in which it is believed to occur. Nucleosynthesis occurring in stars of masses ranging from 10 $M_\odot$ to approximately 100 $M_\odot$ has been calculated by Arnett (100) and by Weaver & Woosley (113, 114, 292). Stars in this mass range are expected to represent the progenitors of supernovae of Type II. While there exist some problems with detailed questions of relative abundances of individual isotopes, the overall picture is quite encouraging in that the bulk of the nuclear species in the mass range $20 \leq A \lesssim 62$ are formed in relative concentrations compatible with solar system abundances. Such stars do tend to overproduce oxygen relative to carbon, which suggests that perhaps the ^{12}C contributions from lower mass stars are important (84). Nucleosynthesis associated with the explosions of degenerate carbon-oxygen and oxygen-neon-magnesium cores has been calculated by Arnett, Truran & Woosley (293) and Bruenn (294) for the carbon detonation case and more recently by Nomoto (295) and Nomoto, Thielemann & Wheeler (296) for carbon deflagration models. Detonation models form almost exclusively iron-peak nuclei, while deflagration models synthesize significant quantities of intermediate mass elements (Si, Ar, S, Ca) as well. Finally, explosive nucleosynthesis accompanying the evolution of stars with initial masses in the range 100 to 300 $M_\odot$ (113, 297, 298) can form significant amounts of heavy elements through approximately calcium. While such massive stars are far too rare to represent the major source of heavy elements in our galaxy, it is believed that their contributions may have been important at early epochs of galactic history.

6.5 Hydrogen Burning at High Temperatures

Hydrogen burning proceeding at high temperatures and under explosive conditions can play a critical role in diverse astrophysical environments, with implications both for nucleosynthesis and for energy generation and dynamic evolution. These environments include supermassive stars (299), supernovae (300), novae (301, 302), and accreting neutron stars (303–307). The defining characteristics of such burning have been discussed by Audouze, Truran & Zimmerman (308) by Wallace & Woosley (309), and most recently by Hanawa, Sugimoto, and Hashimoto (329, 330). At a temperature of approximately 10^8 K, the rates of proton reactions in the CNO cycles become comparable to or faster than the rates of beta decay. The modified CN reaction sequence (see Table 1 for comparison) at high temperature is therefore

$$^{12}C(p,\gamma)^{13}N(p,\gamma)^{14}O(e^+v)^{14}N(p,\gamma)^{15}O(e^+v)^{15}N(p,\alpha)^{12}C.$$

The operation of this "β-limited" CNO cycle is such that the rate of energy generation on a hydrodynamic timescale is restricted by the rates of beta decay of the proton-rich oxygen isotopes, ^{14}O and ^{15}O. For novae, the imposed restrictions on the energetics of the critical early stages of the runaway serve to define the characteristics of novae in outburst (310). At significantly higher temperatures ($T > 5 \times 10^8$ K), such as may be realized in thermonuclear responses to accretion onto neutron stars, Wallace & Woosley (309) demonstrated that leakage out of the hot CNO cycle can occur via ^{15}O(α,p)^{19}Ne(p,γ)^{20}Na. The subsequent buildup can proceed through the iron-peak region accompanied by an energy release far in excess of that available from the β-limited CNO cycles.

These high-temperature hydrogen-burning environments can also contribute to galactic nucleosynthesis. Studies of hot CNO cycle hydrogen burning (308, 311–315) indicate that the less abundant isotopes of carbon, nitrogen, and oxygen, particularly ^{15}N and ^{17}O, may be produced in such environments. Quantitative predictions are rendered uncertain by their sensitivity to the temperature history of the matter. The production of both ^{22}Na and ^{26}Al can also occur in explosive hydrogen burning (316–319). The formation of significant concentrations of mass 22 as ^{22}Na ($\tau_{1/2} = 2.6$ y) provides an attractive explanation for the highly ^{22}Ne-enriched Ne-E anomaly (320–322) identified in meteorites. ^{26}Al is of course important in that the existence of ^{26}Mg excesses in meteoritic material that correlate with Al/Mg ratios (323, 324) confirm the presence of live ^{26}Al($\tau_{1/2} = 7.2 \times 10^5$ y) in the early solar system. The formation of ^{26}Al in explosive hydrogen burning, very likely in nova environments (319), is compatible both with its presence in primordial solar system matter and with observations of diffuse galactic gamma-ray line emission from ^{26}Al (70, 325).

7. CONCLUDING REMARKS

I hope that this review has successfully conveyed some sense of the level of activity that characterizes current research in nucleosynthesis. Researchers over the past quarter century have achieved an impressive record of success: stars and supernovae within our own galaxy have been determined to be the most important nucleosynthesis sites and the diverse nuclear processes that play a role in the synthesis of heavy nuclei have been identified. Significant problems and challenges to nucleosynthesis theorists nevertheless remain. The source(s) of the galactic cosmic rays and the nature of the mechanism of their acceleration remain to be established. The site of r-process neutron-capture synthesis of the heavy elements remains to be identified. A refined determination of the rate of the critical ^{12}C(α,γ)^{16}O reaction is demanded to set the initial core conditions for presupernova models. A definitive statement regarding the operation of the s-process in the atmospheres of red giant stars is not yet possible because of uncertainties associated with the treatment of convective mixing. Advances in high energy physics continue to necessitate reevaluations of the consequences of primordial nucleosynthesis and further detailed numerical studies of the mechanism of mass ejection in supernovae. It seems clear that nucleosynthesis should remain an exciting research area for some years to come.

Acknowledgments

Useful conversations with D. Branch, A. G. W. Cameron, G. M. Fuller, J. S. Gallagher, D. Lambert, G. Steigman, and S. E. Woosley are gratefully acknowledged. I also wish to express my sincere thanks to J. Audouze, D. A. Hunter, and J. S. Gallagher for their more detailed comments on an early version of this paper. Finally, I wish to thank Joanne Klitzing for her patient and competent typing of many versions of this manuscript. The author's continuing research in nucleosynthesis has been supported in part by the National Science Foundation under grants AST 80-18198 and AST 83-14415.

Literature Cited

1. Cameron, A. G. W. 1958. *Ann. Rev. Nucl. Sci.* 8:299
2. Burbidge, G. R. 1963. *Ann. Rev. Nucl. Sci.* 13:507
3. Arnett, W. D. 1973. *Ann. Rev. Astron. Astrophys.* 11:73
4. Truran, J. W. 1973. *Space Sci. Rev.* 15:23
5. Hillebrandt, W. 1978. *Space Sci. Rev.* 21:639
6. Truran, J. W. 1980. *Nukleonika* 25:1463
7. Ulrich, R. K. 1982. See Ref. 21, p. 301
8. Reeves, H. 1974. *Ann. Rev. Astron. Astrophys.* 12:437
9. Audouze, J., Reeves, H. 1982. See Ref. 21, p. 355
10. Iben, I. Jr. 1967. *Ann. Rev. Astron. Astrophys.* 5:571
11. Iben, I. Jr. 1974. *Ann. Rev. Astron. Astrophys.* 12:215
12. Iben, I. Jr., Renzini, A. 1983. *Ann. Rev. Astron. Astrophys.* 21:271
13. Trimble, V. 1982. *Rev. Mod. Phys.* 54:1183
14. Trimble, V. 1983. *Rev. Mod. Phys.* 55:522
15. Schramm, D. N. 1974. *Ann. Rev. Astron. Astrophys.* 12:383
16. Audouze, J., Tinsley, B. M. 1976. *Ann. Rev. Astron. Astrophys.* 14:43
17. Tinsley, B. M. 1980. *Fundamen. Cosmic Phys.* 5:287
18. Arnett, W. D., Hansen, C. J., Truran, J. W., Cameron, A. G. W. 1968. *Nucleosynthesis.* New York: Gordon & Breach
19. Clayton, D. D. 1968. *Principles of Stellar Evolution and Nucleosynthesis.* New York: McGraw Hill
20. Schramm, D. N., Arnett, W. D., eds. 1973. *Explosive Nucleosynthesis.* Austin: Univ. Texas Press
21. Barnes, C. A., Clayton, D. D., Schramm, D. N., eds. 1982. *Essays in Nuclear Astrophysics.* Cambridge: Cambridge Univ. Press
22. Trimble, V. 1975. *Rev. Mod. Phys.* 47:877
23. Burbidge, E. M., Burbidge, G. R., Fowler, W. A., Hoyle, F. 1957. *Rev. Mod. Phys.* 29:547
24. Cameron, A. G. W. 1957. *Atomic Energy of Canada Ltd.*, CRL-41
25. Goldschmidt, V. M. 1937. *Skr. Nor. Vidensk. Akad. Oslo I* No. 4
26. Brown, H. S. 1949. *Rev. Mod. Phys.* 21:625
27. Suess, H. E., Urey, H. C. 1956. *Rev. Mod. Phys.* 28:53
28. Anders, E., Ebihara, M. 1982. *Geochim. Cosmochim. Acta* 46:2363
29. Cameron, A. G. W. 1982. See Ref. 21, p. 23
30. Mason, B. 1979. In *Data of Geochemistry*, ed. M. Fleischer, p. 23. Washington, DC: US GPO
31. Ross, J. E., Aller, L. H. 1976. *Science* 191:1223
32. Shapiro, M. M., Silberberg, R. 1970. *Ann. Rev. Nucl. Sci.* 20:323
33. Mewaldt, R. A. 1983. *Rev. Geophys. Space Phys.* 21:295
34. Simpson, J. A. 1983. *Ann. Rev. Nucl. Part. Sci.* 33:323
35. Truran, J. W., Arnett, W. D. 1971. *Astrophys. Space Sci.* 11:430
36. Wannier, P. G. 1980. *Ann. Rev. Astron. Astrophys.* 18:399
37. Pagel, B. E. J. 1979. In *Les Elements et leur Isotopes dans l'Univers*, ed. A. Boury, N. Grevesse, L. Remy Battiau, p. 261. Liège: Univ. Liège Press
38. Baschek, B. 1979. See Ref. 37, p. 261
39. Spitzer, L. Jr., Jenkins, E. B. 1975. *Ann. Rev. Astron. Astrophys.* 13:133
40. York, D. G. 1982. *Ann. Rev. Astron. Astrophys.* 20:221

41. Peimbert, M. 1975. *Ann. Rev. Astron. Astrophys.* 13:113
42. Pagel, B. E. J., Edmunds, M. G., Kellermann, K. I. 1981. *Ann. Rev. Astron. Astrophys.* 19:77
43. Mould, J. R. 1982. *Ann. Rev. Astron. Astrophys.* 20:91
44. van den Bergh, S. 1975. *Ann. Rev. Astron. Astrophys.* 13:217
45. Oke, J. B., Searle, L. 1974. *Ann. Rev. Astron. Astrophys.* 12:315
46. Kraft, R. P. 1979. *Ann. Rev. Astron. Astrophys.* 17:309
47. Freeman, K. C., Norris, J. 1981. *Ann. Rev. Astron. Astrophys.* 19:319
48. Clayton, D. D. 1982. See Ref. 21, p. 401
49. Truran, J. W. 1984. In *Formation and Evolution of Galaxies and Large Structures in the Universe*, ed. J. Audouze, J. Tran Thanh Van, p. 391. Dordrecht: Reidel
50. Merrill, P. W. 1952. *Science* 115:484
51. Wallerstein, G. 1973. *Ann. Rev. Astron. Astrophys.* 11:115
52. Kaler, J. B. 1983. In *Planetary Nebulae*, ed. D. R. Flower, p. 245. Dordrecht: Reidel
53. Chevalier, R. A., Kirshner, R. P. 1979. *Astrophys. J.* 233:154
54. Branch, D., Falk, S. W., McCall, M. L., Rybski, P., Uomoto, A. K., Wills, B. J. 1981. *Astrophys. J.* 244:780
55. Shull, J. M. 1982. *Astrophys. J.* 262:308
56. Branch, D., Lacy, C. H., McCall, M. L., Southerland, P. G., Uomoto, A., Wheeler, J. C., Willis, B. J. 1983. *Astrophys. J.* 270:123
57. Branch, D. 1984. In *Stellar Nucleosynthesis*, ed. C. Chiosi, A. Renzini. Dordrecht: Reidel
58. Pagel, B. E. J. 1982. *Philos. Trans. R. Soc. London Ser. A* 307:19
59. Clayton, R. N. 1978. *Ann. Rev. Nucl. Part. Sci.* 28:501
60. Podosek, F. A. 1978. *Ann. Rev. Astron. Astrophys.* 16:293
61. Lee, T. 1979. *Rev. Geophys. Space Phys.* 17:1591
62. Begemann, F. 1980. *Rep. Prog. Phys.* 43:1309
63. Wasserburg, G. J., Papanastassiou, D. A., Lee, T. 1980. In *Early Solar System Processes and the Present Solar System*, p. 144. Bologna: Soc. Italiana di Fisica
64. Clayton, D. D. 1982. *Q. J. R. Astron. Soc.* 23:174
65. Lee, T., Papanastassiou, D. A., Wasserburg, G. J. 1976. *Geophys. Res. Letters* 3:109
66. Kelley, W. R., Wasserburg, G. J. 1978. *Geophys. Res. Letters* 5:1079
67. Wasserburg, G. J., Papanastassiou, D. A. 1982. See Ref. 21, p. 77
68. Cameron, A. G. W., Truran, J. W. 1977. *Icarus* 30:447
69. Lattimer, J. M., Schramm, D. N., Grossman, L. 1978. *Astrophys. J.* 219:230
70. Clayton, D. D. 1984. *Astrophys. J.* 280:144
71. Alpher, R. A., Herman, R. C. 1950. *Rev. Mod. Phys.* 22:153
72. Hoyle, R. 1946. *Mon. Not. R. Astron. Soc.* 106:343
73. Alpher, R. A., Bethe, H. A., Gamow, G. 1948. *Phys. Rev.* 73:803
74. Bethe, H. A. 1939. *Phys. Rev.* 55:434
75. von Weizsäcker, C. F. 1938. *Phys. Z.* 39:633
76. Bethe, H. A., Critchfield, C. L. 1938. *Phys. Rev.* 54:248
77. Davis, R. Jr. 1955. *Phys. Rev.* 97:766
78. Bahcall, J. N. 1982. See Ref. 21, p. 242
79. Iben, I. Jr. 1967. *Astrophys. J.* 147:624
80. Iben, I. Jr. 1967. *Astrophys. J.* 177:681
81. Rood, R. T. 1972. *Astrophys. J.* 177:
82. Truran, J. W., Cameron, A. G. W. 1971. *Astrophys. Space Sci.* 14:179
83. Rood, R. T., Steigman, G., Tinsley, B. M. 1976. *Astrophys. J. Lett.* 207:L57
84. Iben, I. Jr., Truran, J. W. 1978. *Astrophys. J.* 220:980
85. Wallerstein, G., Sneden, C. 1982. *Astrophys. J.* 255:577
86. Scalo, J. M. 1976. *Astrophys. J.* 206:795
87. Cameron, A. G. W., Fowler, W. A. 1971. *Astrophys. J.* 164:11
88. Sackman, I.-J., Smith, R. L., Despain, K. H. 1974. *Astrophys. J.* 187:555
89. Iben, I. Jr. 1973. *Astrophys. J.* 185:209
90. Iben, I. Jr. 1975. *Astrophys. J.* 196:525
91. Yang, J., Turner, M. S., Steigman, G., Schramm, D. N., Olive, K. A. 1984. *Astrophys. J.* In press
92. Caughlan, G. R. 1965. *Astrophys. J.* 141:688
93. Salpeter, E. E. 1952. *Astrophys. J.* 115:326
94. Hoyle, F., Dunbar, D. N. F., Wenzel, W. A., Whaling, W. 1953. *Phys. Rev.* 92:1095
95. Fowler, W. A., Caughlan, G. R., Zimmerman, B. A. 1967. *Ann. Rev. Astron. Astrophys.* 5:525
96. Fowler, W. A., Caughlan, G. R., Zimmerman, B. A. 1975. *Ann. Rev. Astron. Astrophys.* 13:69
97. Harris, M. J., Fowler, W. A., Caughlan, G. R., Zimmerman, B. A. 1983. *Ann. Rev. Astron. Astrophys.* 21:165
98. Tombrello, T. A., Koonin, S. E., Flanders, B. A. 1982. See Ref. 21, p. 233
99. Rolfs, C. 1983. *Status of Helium Burning of ^{12}C.* Univ. Münster. Preprint
100. Arnett, W. D. 1978. *Astrophys. J.* 219:1008
101. Weaver, T. A., Zimmerman, G. B.,

101. Woosley, S. E. 1978. *Astrophys. J.* 225: 1021
102. Renzini, A., Voli, M. 1981. *Astron. Astrophys.* 94:175
103. Cameron, A. G. W. 1960. *Astron. J.* 65:485
104. Iben, I. Jr. 1975. *Astrophys. J.* 196:549
105. Truran, J. W., Iben, I. Jr. 1977. *Astrophys. J.* 216:797
106. Cosner, K. R., Iben, I. Jr., Truran, J. W. 1980. *Astrophys. J. Lett.* 238:L91
107. Chandrasekhar, S. 1939. *An Introduction to the Study of Stellar Structure.* New York: Dover
108. Arnett, W. D., Truran, J. W. 1969. *Astrophys. J.* 157:339
109. Woosley, S. E., Arnett, W. D., Clayton, D. D. 1972. *Astrophys. J.* 175:731
110. Arnett, W. D. 1972. *Astrophys. J.* 169:699
111. Arnett, W. D. 1974. *Astrophys. J.* 193:169
112. Arnett, W. D. 1974. *Astrophys. J.* 194:373
113. Woosley, S. E., Axelrod, T. S., Weaver, T. A. 1984. See Ref. 57
114. Weaver, T. A., Woosley, S. E. 1980. *Ann. N. Y. Acad. Sci.* 336:335
115. Fowler, W. A., Hoyle, F. 1964. *Astrophys. J. Suppl.* 9:201
116. Clifford, F. E., Taylor, R. J. 1965. *Mem. R. Astron. Soc.* 69:21
117. Clayton, D. D., Woosley, S. E. 1969. *Astrophys. J.* 157:1381
118. Truran, J. W. 1972. *Astrophys. J.* 177:453
119. Hainebach, K. L., Clayton, D. D., Arnett, W. D., Woosley, S. E. 1974. *Astrophys. J.* 193:157
120. Truran, J. W. 1966. PhD thesis, Yale Univ. (Unpublished)
121. Truran, J. W., Cameron, A. G. W., Gilbert, A. 1966. *Can. J. Phys.* 44:563
122. Bodansky, D., Clayton, D. D. Fowler, W. A. 1968. *Astrophys. J. Suppl.* 16:299
123. Truran, J. W. 1968. *Astrophys. Space Sci.* 2:384
124. Truran, J. W. 1968. *Astrophys. Space Sci.* 2:391
125. Michaud, G., Fowler, W. A. 1972. *Astrophys. J.* 173:157
126. Woosley, S. E., Arnett, W. D., Clayton, D. D. 1973. *Astrophys. J. Suppl.* 26:231
127. Truran, J. W., Arnett, W. D., Cameron, A. G. W. 1967. *Can. J. Phys.* 45:2315
128. Arnett, W. D. 1977. *Astrophys. J. Suppl.* 35:145
129. Cameron, A. G. W. 1955. *Astrophys. J.* 121:144
130. Clayton, D. D., Fowler, W. A., Hull, T. C., Zimmerman, B. A. 1961. *Ann. Phys.* 12:121
131. Seeger, P. A., Fowler, W. A., Clayton, D. D. 1965. *Astrophys. J. Suppl.* 97:121
132. Cameron, A. G. W. 1982. *Astrophys. Space Sci.* 82:123
133. Cameron, A. G. W. 1959. *Astrophys. J.* 130:452
134. Clayton, D. D., Ward, R. A. 1974. *Astrophys. J.* 193:397
135. Blake, J. B., Schramm, D. N. 1975. *Astrophys. J.* 232:831
136. Ward, R. A., Newman, M. J., Clayton, D. D. 1976. *Astrophys. J. Suppl.* 31:33
137. Ward, R. A., Newman, M. J. 1978. *Astrophys. J.* 219:195
138. Ward, R. A., Beer, H. 1981. *Astron. Astrophys.* 103:189
139. Beer, H., Käppeler, F., Wisshak, K., Ward, R. A. 1981. *Astrophys. J. Suppl.* 46:295
140. Peters, J. G. 1968. *Astrophys. J.* 154:225
141. Couch, R. G., Schmiedekamp, A. B., Arnett, W. D. 1974. *Astrophys. J.* 190:95
142. Lamb, S. A., Howard, W. M., Truran, J. W., Iben, I. Jr. 1977. *Astrophys. J.* 217:213
143. Schramm, D. N. 1982. See Ref. 21, p. 325
144. Cameron, A. G. W., Delano, M. D., Truran, J. W. 1970. In *The Properties of Nuclei far from the Region of Beta Stability*, Vol. 2, p. 735. Geneva: CERN
145. Cameron, A. G. W., Cowan, J. J., Klapdor, H. V., Metzinger, J., Oda, T., Truran, J. W. 1983. *Astrophys. Space Sci.* 91:221
146. Cameron, A. G. W., Cowan, J. J., Truran, J. W. 1983. *Astrophys. Space Sci.* 91:235
147. Frank-Kamenetskii, D. A. 1961. *Sov. Astron.* 5:66
148. Audouze, J. 1970. *Astron. Astrophys.* 8:436
149. Hainebach, K. L., Schramm, D. N., Blake, J. B. 1976. *Astrophys. J.* 205:920
150. Reeves, H., Stewart, P. 1965. *Astrophys. J.* 141:1432
151. Agnese, A., LaCamera, M., Wataghin, A. 1969. *Mon. Not. R. Astron. Soc.* 146:58
152. Arnould, M., Brihaye, C. 1969. *Astron. Astrophys.* 1:193
153. Joukoff, A. A. 1969. *Astron. Astrophys.* 1:193
154. Ito, K. 1961. *Prog. Theor. Phys.* 26:990
155. Malkiel, G. S. 1963. *Sov. Astron.* 7:207
156. Macklin, R. L. 1970. *Astrophys. J.* 162:353
157. Truran, J. W., Cameron, A. G. W. 1972. *Astrophys. J.* 171:89
158. Truran, J. W. 1973. See Ref. 20, p. 102
159. Audouze, J., Truran, J. W. 1975. *Astrophys. J.* 202:204
160. Arnould, M. 1976. *Astron. Astrophys.* 46:117

161. Woosley, S. E., Howard, W. M. 1978. *Astrophys. J. Suppl.* 36:285
162. Reeves, H., Audouze, J., Fowler, W. A., Schramm, D. N. 1973. *Astrophys. J.* 179:909
163. Fowler, W. A., Burbidge, G. R., Burbidge, E. M. 1955. *Astrophys. J. Suppl.* 2:167
164. Bernas, R., Gradsztjan, E., Reeves, H., Schatzman, E. 1967. *Ann. Phys.* 44:426
165. Brancazio, P. J., Cameron, A. G. W. 1967. *Can. J. Phys.* 45:3297
166. Reeves, H., Audouze, J. 1968. *Astrophys. Lett.* 1:197
167. Ryter, C., Reeves, H., Gradsztjan, E., Audouze, J. 1970. *Astron. Astrophys.* 8:289
168. Reeves, H., Fowler, W. A., Hoyle, F. 1970. *Nature* 226:727
169. Meneguzzi, M., Audouze, J., Reeves, H. 1971. *Astron. Astrophys.* 15:337
170. Mitler, H. E. 1972. *Astrophys. Space Sci.* 17:186
171. Meneguzzi, M., Reeves, H. 1975. *Astron. Astrophys.* 40:99
172. Wagoner, R. F., Fowler, W. A., Hoyle, F. 1967. *Astrophys. J.* 148:3
173. Nørgaard, H., Arnould, M. 1975. *Astron. Astrophys.* 40:331
174. Nørgaard, H., Fricke, K. J. 1976. *Astron. Astrophys.* 49:337
175. Colgate, S. A. 1973. *Astrophys. J. Lett.* 181:L53
176. Epstein, R. I., Arnett, W. D., Schramm, D. N. 1976. *Astrophys. J. Suppl.* 31:111
177. Starrfield, S., Truran, J. W., Sparks, W. M., Arnould, M. 1978. *Astrophys. J.* 222:600
178. Wagoner, R. V. 1973. *Astrophys. J.* 179:343
179. Schramm, D. N., Wagoner, R. V. 1977. *Ann. Rev. Nucl. Sci.* 27:37
180. Olive, K. A., Schramm, D. N., Steigman, G., Turner, M. S., Yang, J. 1981. *Astrophys. J.* 246:557
181. Spite, F., Spite, M. 1982. *Astron. Astrophys.* 115:257
182. Spite, M., Spite, F. 1982. *Nature* 297:483
183. Reeves, H. 1965. In *Stars and Stellar Systems*, ed. L. H. Aller, D. B. McLaughlin, Vol. VIII, p. 113. Chicago: Univ. Chicago Press
184. Barnes, C. A. 1971. *Adv. Nucl. Phys.* 4:133
185. Macklin, R. L., Gibbons, J. H. 1965. *Rev. Mod. Phys.* 37:166
186. Allen, B. J., Gibbons, J. H., Macklin, R. L. 1971. *Adv. Nucl. Phys.* 4:205
187. Käppeler, F., Beer, H., Wisshak, K., Clayton, D. D., Macklin, R. L., Ward, R. A. 1982. *Astrophys. J.* 257:821
188. Almeida, J., Käppeler, F. 1983. *Astrophys. J.* 265:247
189. Hauser, W., Feshbach, H. 1952. *Phys. Rev.* 87:366
190. Truran, J. W., Hansen, C. J., Gilbert, A., Cameron, A. G. W. 1966. *Can. J. Phys.* 44:151
191. Michaud, G., Fowler, W. A. 1970. *Phys. Rev. C* 2:2041
192. Truran, J. W. 1972. *Astrophys. Space Sci.* 18:308
193. Arnould, M. 1972. *Astron. Astrophys.* 19:82
194. Holmes, J. A., Woosley, S. E., Fowler, W. A., Zimmerman, B. A. 1976. *At. Data Nucl. Data Tables* 18:306
195. Woosley, S. E., Fowler, W. A., Holmes, J. A., Zimmerman, B. A. 1978. *At. Data Nucl. Data Tables* 22:371
196. Thielemann, F.-K. 1980. Thesis, Technischen Hochschule Darmstadt. (Unpublished)
197. Barkat, Z. 1975. *Ann. Rev. Astron. Astrophys.* 13:45
198. Bahcall, J. N. 1961. *Phys. Rev.* 124:495
199. Bahcall, J. N. 1962. *Astrophys. J.* 136:445
200. Hansen, C. J. 1966. PhD thesis, Yale Univ. (Unpublished)
201. Hansen, C. J. 1968. *Astrophys. Space Sci.* 1:499
202. Mazurek, T. J., Truran, J. W., Cameron, A. G. W. 1974. *Astrophys. Space Sci.* 27:261
203. Gilbert, A., Cameron, A. G. W. 1965. *Can. J. Phys.* 43:1446
204. Iben, I. Jr. 1978. *Astrophys. J.* 219:213
205. Newman, M. J. 1973. MS thesis, Rice Univ. (Unpublished)
206. Cosner, K. M., Truran, J. W. 1981. *Astrophys. Space Sci.* 78:85
207. Takahashi, K., Yamada, M. 1969. *Prog. Theor. Phys.* 41:1470
208. Takahashi, K. 1971. *Prog. Theor. Phys.* 45:1466
209. Takahashi, K., Yamada, M., Kondoh, T. 1973. *At. Data Nucl. Data Tables* 12:101
210. Takahashi, K., El Eid, M. F., Hillebrandt, W. 1978. *Astron. Astrophys.* 67:185
211. Klapdor, H. V. 1976. *Phys. Lett.* 65B:35
212. Klapdor, H. V., Oda, T., Metzinger, J., Hillebrandt, W., Thielemann, F.-K. 1981. *Z. Phys. A* 299:213
213. Klapdor, H. V. 1984. *Prog. Part. Nucl. Phys.* In press
214. Fuller, G. M., Fowler, W. A., Newman, M. J. 1980. *Astrophys. J. Suppl.* 42:447
215. Fuller, G. M., Fowler, W. A., Newman, M. J. 1982. *Astrophys. J.* 252:715
216. Fuller, G. M., Fowler, W. A., Newman, M. J. 1982. *Astrophys. J. Suppl.* 48:279
217. Fuller, G. M. 1982. *Astrophys. J.* 252:741

218. Penzias, A. A., Wilson, R. W. 1965. *Astrophys. J.* 142:419
219. Peebles, P. J. E. 1966. *Phys. Rev. Lett.* 16:410
220. Peebles, P. J. E. 1966. *Astrophys. J.* 146:542
221. Greenstein, G. S. 1968. *Astrophys. Space Sci.* 2:155
222. Wagoner, R. V. 1969. *Astrophys. J. Suppl.* 18:247
223. Wagoner, R. V. 1980. In *Physical Cosmology*, ed. R. Balian, J. Audouze, D. N. Schramm, p. 272. Amsterdam: North Holland
224. York, D. G., Rogerson, J. B. Jr. 1976. *Astrophys. J.* 203:378
225. Vidal-Madjar, A., Laurent, C., Gry, C., Bruston, P., Ferlet, R., York, D. G. 1983. *Astron. Astrophys.* 120:58
226. Reynolds, J. H., Frick, V., Neil, J. M., Phinney, D. L. 1978. *Geochim. Cosmochim. Acta* 42:1775
227. Gott, J. R., Gunn, J. F., Schramm, D. N., Tinsley, B. M. 1974. *Astrophys. J.* 194:543
228. Steigman, G. 1979. *Ann. Rev. Nucl. Sci.* 29:313
229. Turner, M. S., Schramm, D. N. 1979. *Phys. Today* 32:42
230. Yang, J., Schramm, D. N., Steigman, G., Rood, R. T. 1979. *Astrophys. J.* 227:697
231. Olive, K. A., Schramm, D. N., Steigman, G. 1981. *Nucl. Phys.* B280:497
232. Szalay, A. S. 1981. *Phys. Lett.* B101:543
233. Kolb, E. W., Scherrer, R. J. 1982. *Phys. Rev.* D25:1481
234. Steigman, G. 1982. See Ref. 21, p. 519
235. Schwarzschild, M., Härm, R. 1967. *Astrophys. J.* 150:961
236. Sanders, R. H. 1967. *Astrophys. J.* 150:971
237. Ulrich, R. K. 1973. See Ref. 20, p. 139
238. Iben, I. Jr. 1976. *Astrophys. J.* 208:165
239. Scalo, J. M., Ulrich, R. K. 1973. *Astrophys. J.* 183:151
240. Truran, J. W. 1973. In *Proc. Red Giant Conf.*, ed. H. R. Johnson, J. P. Mutschlecner, B. F. Peery, p. 394. Bloomington: Indiana Univ. Press
241. Iben, I. Jr. 1977. *Astrophys. J.* 217:788
242. Despain, K. H. 1980. *Astrophys. J. Lett.* 236:L165
243. Cosner, K. R. 1982. PhD thesis, Univ. Ill. (Unpublished)
244. Ward, R. A. 1977. *Astrophys. J.* 216:540
245. Scalo, J. M., Miller, G. E. 1981. *Astrophys. J.* 246:251
246. Cosner, K. R., Despain, K. H., Truran, J. W. 1984. *Astrophys. J.* In press
247. Scalo, J. M., Miller, G. E. 1979. *Astrophys. J.* 233:596
248. Iben, I. Jr., Renzini, A. 1982. *Astrophys. J. Lett.* 259:L79
249. Iben, I. Jr., Renzini, A. 1982. *Astrophys. J. Lett.* 263:L23
250. Danziger, I. J. 1966. *Astrophys. J.* 143:527
251. Sneden, C., Parthasarathy, M. 1983. *Astrophys. J.* 267:757
252. Truran, J. W., Arnett, W. D., Tsuruta, S., Cameron, A. G. W. 1968. *Astrophys. Space Sci.* 1:129
253. Arnett, W. D., Truran, J. W. 1970. *Astrophys. J.* 160:959
254. Delano, M. D., Cameron, A. G. W. 1971. *Astrophys. Space Sci.* 10:203
255. Schramm, D. N., Barkat, Z. 1972. *Astrophys. J.* 173:195
256. Kodama, T., Takahashi, K. 1973. *Phys. Lett.* 43B:167
257. Schramm, D. N. 1973. *Astrophys. J.* 185:293
258. Sato, K. 1974. *Prog. Theor. Phys.* 51:726
259. Hillebrandt, W., Takahashi, K., Kodama, T. 1976. *Astron. Astrophys.* 52:63
260. LeBlanc, J. M., Wilson, J. R. 1970. *Astrophys. J.* 161:541
261. Meier, D. L., Epstein, R. I., Arnett, W. D., Schramm, D. N. 1976. *Astrophys. J.* 204:869
262. Lattimer, J. M., Schramm, D. N. 1974. *Astrophys. J. Lett.* 192:L145
263. Norman, E. B., Schramm, D. N. 1979. *Astrophys. J.* 228:881
264. Hillebrandt, W., Thielemann, F.-K. 1977. *Mitt. Astron. Ges.* 43:24
265. Truran, J. W., Cowan, J. J., Cameron, A. G. W. 1978. *Astrophys. J. Lett.* 222:L63
266. Truran, J. W., Cowan, C. J., Cameron, A. G. W. 1978. In *Proc. Int. Symp. on Superheavy Elements*, ed. M. A. K. Lohai, p. 515. New York: Pergamon
267. Thielemann, F.-K., Arnould, M., Hillebrandt, W. 1979. *Astron. Astrophys.* 74:175
268. Cowan, J. J., Cameron, A. G. W., Truran, J. W. 1980. *Astrophys. J.* 231:1090
269. Blake, J. B., Woosley, S. E., Weaver, T. A., Schramm, D. N. 1981. *Astrophys. J.* 248:315
270. Cowan, J. J., Cameron, A. G. W., Truran, J. W. 1982. *Astrophys. J.* 252:348
271. Lee, T., Schramm, D. N., Wefel, J. P., Blake, J. B. 1979. *Astrophys. J.* 232:854
272. Colgate, S. A., White, R. H. 1966. *Astrophys. J.* 143:626
273. Arnett, W. D. 1967. *Can. J. Phys.* 45:1621
274. Wilson, J. R. 1971. *Astrophys. J.* 163:209
275. Thielemann, F.-K. 1976. Thesis, Technische Hochschule Darmstadt. (Unpublished)
276. Tsuruta, S., Cameron, A. G. W. 1965. *Can. J. Phys.* 43:2056

277. Senbetu, L. 1973. *Phys. Rev.* C7:1254
278. Cowan, J. J., Cameron, A. G. W., Truran, J. W. 1983. *Astrophys. J.* 265:429
279. Cole, P. W., Deupree, R. G. 1981. *Astrophys. J.* 247:607
280. Sandler, D. G., Koonin, S. E., Fowler, W. A. 1982. *Astrophys. J.* 259:908
281. Arnett, W. D. 1969. *Astrophys. J.* 157:1369
282. Hansen, C. J. 1971. *Astrophys. Space Sci.* 14:389
283. Arnett, W. D., Wefel, J. P. 1978. *Astrophys. J. Lett.* 224:L139
284. Truran, J. W., Cameron, A. G. W. 1978. *Astrophys. J.* 219:226
285. Morgan, J. A. 1980. *Astrophys. J.* 238:674
286. Truran, J. W., Arnett, W. D. 1970. *Astrophys. J.* 160:181
287. Colgate, S. A., McKee, C. 1969. *Astrophys. J.* 157:623
288. Arnett, W. D. 1979. *Astrophys. J. Lett.* 230:L37
289. Colgate, S. A., Petschek, A. G., Kriese, J. T. 1980. *Astrophys. J. Lett.* 237:L81
290. Weaver, T. A., Axelrod, T. S., Woosley, S. E. 1980. In *Proc. Texas Workshop on Type I Supernova*, ed. J. C. Wheeler, p. 113. Austin: Univ. Texas
291. Chevalier, R. A. 1981. *Astrophys. J.* 246:267
292. Woosley, S. E., Weaver, T. A. 1982. See Ref. 21, p. 377
293. Arnett, W. D., Truran, J. W., Woosley, S. E. 1971. *Astrophys. J.* 165:87
294. Bruenn, S. W. 1971. *Astrophys. J.* 168:203
295. Nomoto, K. 1984. See Ref. 57
296. Nomoto, K., Thielemann, F.-K., Wheeler, J. C. 1984. *Astrophys. J. Lett.* 279:L23
297. Bond, J. R., Arnett, W. D., Carr, B. J. 1984. *Astrophys. J.* In press
298. Ober, W., El Eid, W., Fricke, K. J. 1983. *Astron. Astrophys.* 119:61
299. Fricke, K. J. 1973. *Astrophys. J.* 183:941
300. Howard, W. M., Arnett, W. D., Clayton, D. D. 1971. *Astrophys. J.* 165:495
301. Starrfield, S., Sparks, W. M., Truran, J. W. 1976. In *Structure and Evolution of Close Binary Systems*, ed. P. Eggleton, S. Mitton, J. Whelan, p. 155. Dordrecht: Reidel
302. Gallagher, J. S., Starrfield, S. 1978. *Ann. Rev. Astron. Astrophys.* 16:171
303. Hansen, C. J., Van Horn, H. M. 1975. *Astrophys. J.* 195:735
304. Taam, R. E., Picklum, R. E. 1978. *Astrophys. J.* 224:210
305. Joss, P. C. 1978. *Astrophys. J. Lett.* 225:L123
306. Starrfield, S., Kenyon, S., Sparks, W. M., Truran, J. W. 1982. *Astrophys. J.* 258:683
307. Wallace, R. K., Woosley, S. E., Weaver, T. A. 1982. *Astrophys. J.* 258:696
308. Audouze, J., Truran, J. W., Zimmerman, B. A. 1973. *Astrophys. J.* 184:493
309. Wallace, R. K., Woosley, S. E. 1981. *Astrophys. J. Suppl.* 45:389
310. Truran, J. W. 1982. See Ref. 21, p. 467
311. Arnould, M., Beelen, W. 1974. *Astron. Astrophys.* 33:215
312. Arnould, M., Nørgaard, H. 1975. *Astron. Astrophys.* 42:55
313. Nørgaard, H. 1977. *Astrophys. J.* 215:200
314. Lazareff, B., Audouze, J., Starrfield, S., Truran, J. W. 1979. *Astrophys. J.* 228:875
315. Rodney, W. S., Rolfs, C. 1982. See Ref. 21, p. 171
316. Arnould, M., Nørgaard, H. 1978. *Astron. Astrophys.* 64:195
317. Arnould, M., Nørgaard, H., Thielemann, F.-K., Hillebrandt, W. 1980. *Astrophys. J.* 237:931
318. Vangioni-Flam, E., Audouze, J., Chièze, J.-P. 1980. *Astron. Astrophys.* 82:234
319. Hillebrandt, W., Thielemann, F.-K. 1982. *Astrophys. J.* 255:617
320. Black, D. C. 1972. *Geochim. Cosmochim. Acta* 36:377
321. Eberhardt, P., Jungck, M. H. A., Meier, F. O., Niederer, F. 1979. *Astrophys. J. Lett.* 234:L169
322. Lewis, R. S., Alaerts, L., Matsuda, J.-I., Anders, E. 1979. *Astrophys. J. Lett.* 234:L165
323. Lee, T., Papanastassiou, D. A., Wasserburg, G. J. 1977. *Astrophys. J. Lett.* 211:L107
324. Lee, T., Papanastassiou, D. A., Wasserburg, G. J. 1977. *Geochim. Cosmochim. Acta* 41:1473
325. Mahoney, W. A., Ling, J. C., Jacobson, A. S., Lingenfelter, R. E. 1982. *Astrophys. J.* 262:742
326. Sugimoto, D., Nomoto, K. 1974. In *Late Stages of Stellar Evolution*, ed. R. J. Taylor, p. 105. Dordrecht: Reidel
327. Sugimoto, D., Nomoto, K. 1975. *Publ. Astron. Soc. Jpn.* 27:197
328. Nomoto, K. 1984. *Astrophys. J.* 277:791
329. Hanawa, T., Sugimoto, D., Hashimoto, M. 1983. *Publ. Astron. Soc. Jpn.* 35:491
330. Hashimoto, M., Hanawa, T., Sugimoto, D. 1983. *Publ. Astron. Soc. Jpn.* 35:1

THE PHYSICS OF PARTICLE ACCELERATORS

J. D. Lawson

Rutherford Appleton Laboratory, Chilton, Oxon, OX11 0QX, United Kingdom

M. Tigner

Cornell University, Ithaca, New York 14853

CONTENTS

1. INTRODUCTION 99
2. CLASSIFICATION OF METHODS OF ACCELERATION 101
3. DISCUSSION OF THE VARIOUS CATEGORIES IN THE TABLE.. 102
   3.1 *Harmonic Fields, Near and Far*.. 102
   3.2 *Integrated Two-Beam Systems*... 106
   3.3 *Nonharmonic Fields in Free Space*... 106
   3.4 *Collective Schemes with Accelerated Particles in Vacuum*............................ 106
   3.5 *Harmonic Collective Schemes with Particles in Beam or Plasma*................... 107
   3.6 *Nonharmonic Collective Schemes with Particles in Beam or Plasma*............. 109
4. CRITERIA FOR NEW ACCELERATORS AND NEW ACCELERATOR CONCEPTS.................. 110
   4.1 *General Aims in Planning New Accelerators*... 110
   4.2 *The Demands of High-Energy Physics*... 110
   4.3 *Some Economic Criteria for High-Energy Machines*..................................... 111
5. APPLICATION OF THE VARIOUS ACCELERATION METHODS AT HIGH ENERGIES 112
   5.1 *Synchrotrons*.. 112
   5.2 *Linear Accelerators*... 117
   5.3 *Status of Undeveloped Ideas*.. 121
6. CONCLUDING REMARKS.. 122

1. INTRODUCTION

The development of particle accelerators during the half century since Cockcroft and Walton's high-voltage generator and Lawrence's original cyclotron has been remarkable. Not only has the maximum particle energy increased by a factor of about 25 per decade, but a wide variety of applications has been found that could hardly have been anticipated by the early pioneers. The spearhead of the development has always been toward the highest possible energies for fundamental research into the basic constituents of matter, but interesting large-scale applications such as accelerator breeding of fissile fuel and heavy-ion-driven inertial confinement fusion may become economic if high currents can be reliably produced at moderate energies. Demand for more modest machines for applications in the fields of nuclear structure research, medicine, chemistry, and other branches of science and industry continues to grow.

Up to the present time, the progress to ever-higher energies has proceeded steadily, as new ideas and more sophisticated techniques have become available. The rate of cost decrease per unit energy is, however, less than the rate of energy increase. For a constant rate of expenditure, likely in the foreseeable future, this leads to fewer and larger machines. Already this is very evident but the present trend cannot continue indefinitely, and it is not clear what will happen after the next round of large machines is completed.

It is timely, therefore, to take a broad look at the whole field, to discern patterns of development in the past, and enquire what might happen in the future. Is the field sufficiently well understood that all we need is good engineering extrapolations from existing technology, or can we discover some new direction that will enable us to continue increasing energy without increasing physical size and cost? This question is often being asked and it deserves careful consideration. If new techniques are to be developed, then experience shows that both time and substantial resources are required; the question, therefore, is urgent.

Many methods of accelerating particles are known (1a–f), though only a limited number of these have been developed into useful devices. At the highest energies, radiofrequency linear accelerators and synchrotrons dominate the scene; after several decades of development, the understanding of how they operate is by now good but hardly complete. Experience with colliding-beam operation in storage rings, for achieving a high center-of-mass energy is being accumulated. At the highest electron energies, synchrotron radiation from the electrons poses a constraint on the orbit curvature, and linear colliders will be needed. Plans for the first of such machines are well under way.

Although synchrotrons and linear accelerators are well understood in the regimes in which they have been operated, our understanding of the whole gamut of conceivable accelerator schemes is sketchy indeed. In this paper we attempt first to classify in a simple physical way accelerating configurations that have been suggested and discuss some of their limitations. Attention is then directed to conventional accelerators, and to new ideas of topical interest, specifically those involving lasers and the two-beam concept. In conclusion, some tentative suggestions are made for profitable activity in the future.

2. CLASSIFICATION OF METHODS OF ACCELERATION

Many schemes have been used for accelerating particles; many others have been suggested but not tried. Different methods are, of course, appropriate to different applications. The emphasis here is on attaining the highest energies, but it must be recalled that some important applications require high current or beam power, where the considerations are somewhat different.

Accelerators have two essential components. The driving field, which is always electric, gives energy to the particles, whereas the guiding field, which is normally magnetic, ensures that they are suitably confined and directed. For existing types of accelerator the constraints, both technical and economic, on the strength and form of these fields are by now reasonably well understood. To achieve higher energies in machines that are not at the same time enormous in size evidently requires a higher guide field in synchrotrons and a higher accelerating field in linear machines. It is in this direction that recent suggestions have been made. It is emphasized below that this is by no means the only requirement if meaningful high-energy experiments are to be carried out.

Accelerating concepts may be classified in several different ways; the one proposed here seems simple and convenient, but is certainly not unique. Only six categories are presented, but these can readily be subdivided. Vertically in Table 1 we distinguish between machines in which the

Table 1 Classification of accelerators

	Accelerated particles in free space		Accelerated particles in a medium
	No free charges in system	Free charges in system	
Harmonic accelerating fields	1.1 Linac Synchrotron Inverse free-electron laser	1.2 Two-beam	1.3 Inverse Čerenkov Beam-wave Laser beat-wave
Accelerating fields not harmonic	2.1 Betatron Induction linac	2.2 Ion drag Wake field	2.3 Ionization front Electron ring

accelerating field at a point varies harmonically and those in which it does not. The latter category contains a number of lower-energy machines, such as the betatron and all electrostatic accelerators, whereas the former contains all existing types of very high-energy machine. Horizontally, we distinguish between systems in which the particles move in free space (as in all existing high-energy machines) and those in which they move in some medium such as a plasma or an intense beam of particles of a different type. The first of the horizontal categories is again divided according to whether the charges that produce the required accelerating and focusing fields are entirely bound within metals or dielectrics, or whether they are free, forming part of a plasma or particle beam.

3. DISCUSSION OF THE VARIOUS CATEGORIES IN THE TABLE

3.1 *Harmonic Fields, Near and Far*

The first category in the table is the most familiar. It can be examined by considering the very fundamental problem of the interaction of a plane wave and a particle, since all fields in regions of free space can be constructed from plane wave solutions, provided that evanescent as well as propagating waves are included.

We consider first systems in which a wave with component of electric field in the direction of propagation travels at the same velocity as the particle. It is well known that infinite plane waves propagating in free space cannot satisfy this condition. First, the wave always moves faster than the particle, and, second, the electric field is perpendicular to the direction of motion (Figure 1a).

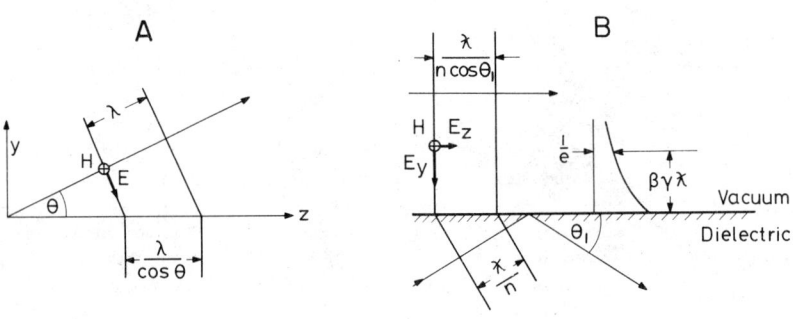

Figure 1 Diagrams to illustrate the essential features of propagating and evanescent plane waves. In the latter, all field components decrease exponentially away from the surface, and the E_z component is $\pi/2$ out of phase with the other components.

Evanescent waves, on the other hand, can provide the type of field required. Single, semi-infinite, evanescent, plane waves are familiar in optics as the fields existing outside a block of dielectric in which total internal reflection is occurring. In the case of an evanescent wave near a surface of discontinuity in refractive index, the angle that the wave vector makes with the surface is complex, since the cosine of the angle of refraction in free space is given as greater than unity by Snell's law. The phase velocity of the wave along the surface is $c/\cos\theta$, which, since $\cos\theta$ exceeds unity, is less than c. We can write the phase velocity as $\beta_w c$ from which it follows that $\sin^2\theta = -1/\beta_w^2 \gamma_w^2$, where $\gamma_w^2 = (1-\beta_w^2)^{-1}$. This type of wave is illustrated in Figure 1b.

Such a wave clearly is synchronous with a particle of energy $\gamma_w m_0 c^2$. Henceforth the subscript w is omitted. It is reasily shown that the fields decay by a factor $1/e$ in a distance from the surface equal to $\beta\gamma\lambdabar$, where $\lambdabar$ is the exciting wavelength divided by 2π, and that for a wave with the appropriate polarization there is a component of electric field in the direction of propagation (parallel to the surface) given by $E_z = E_\perp |\sin\theta| = E_\perp/\beta\gamma$, where $E_\perp$ is the field component perpendicular to the surface. As γ increases, $E_z/E_\perp$ decreases, becoming very small at relativistic energies. This seemingly fundamental disadvantage of using evanescent waves can be overcome in either of two ways. First, if a second surface is placed parallel to the first, then a wave can be propagated between them in which the components of $E_\perp$ and H cancel on the symmetry plane, but the E_z components add, as illustrated in Figure 2a. Away from the central plane $E_\perp$

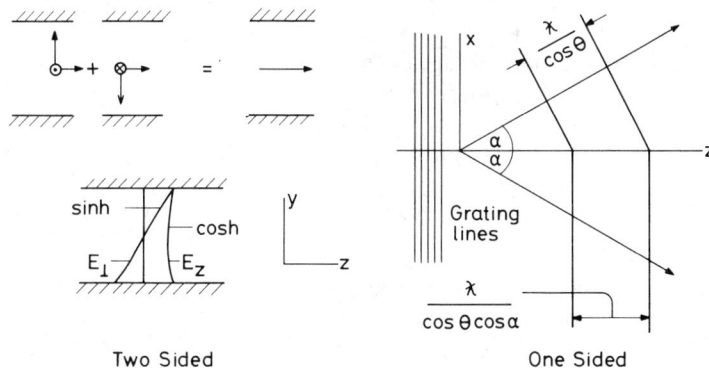

Two Sided   One Sided

Figure 2 Diagrams to show how waves can be produced with phase velocity near c that do not have a vanishingly small component of E_z in the direction of propagation. (A) Two evanescent waves are added in such a way that the components of E_z add, but H and E_y cancel on the median plane. (B) Evanescent waves with phase velocity substantially less than c, so that there is a component of E in the direction of propagation, move at an angle $\pi/2-\alpha$ to the grating lines. This enhances the phase velocity by a factor $\sec\alpha$.

and H increase rapidly, but up to a distance $\pm \lambda$ they are both less than E_z. In conventional linear accelerators a manifold of such evanescent plane waves are combined to produce a system with axial symmetry. Instead of being supported by a dielectric, these evanescent or surface waves can be carried by a corrugated surface exhibiting a reactive surface impedance, or as space harmonics in an appropriate periodic structure. In practical systems other space harmonics (including backward-travelling waves) occur, but the *essential* physical features of the interaction are common to all linear accelerators (linacs) and to synchrotrons. Interaction by means of evanescent waves implies that virtual photons are exchanged between the cavity wall and the accelerated charges. (Particles cannot absorb single real photons, since the conditions for energy and momentum conservation cannot simultaneously be satisfied.)

An alternative configuration, making use of a one-sided system in the form of a grating, has been proposed by Palmer (2) with the aim of utilizing the high fields obtainable from an infrared laser. If we consider a surface wave for which $\sin \theta$ is not small, so that $1/\cos \theta$ is not near unity, then, in a direction parallel to the surface but at an angle α to the wave normal, the phase velocity will be $v/\cos \alpha$. A pair of waves at angles $\pm \alpha$ to the z-direction (taken as perpendicular to the grating lines) combines to produce a wave pattern that travels in the z-direction, but has the form of a standing wave in the direction parallel to the grating. The wavelengths in these two directions are $\lambda/(\cos \theta \cos \alpha)$ and $\lambda/(\cos \theta \sin \alpha)$, respectively, where λ is the free-space wavelength. This is illustrated in Figure 2b. The phase velocity in the z-direction is c when $\cos \theta \cos \alpha = 1$. For example, if $\alpha = 45°$, $\cos \theta = \sqrt{2}$ and the field ratio $E_z/E_\perp = |\sin \theta \cos \alpha/\cos \theta| = \frac{1}{2}$. Such a wave pair can be confined in the transverse direction between a pair of conducting walls perpendicular to the grating spaced by a multiple of $\lambda_0/(2 \cos \theta \sin \alpha)$. Of course, a dielectric sheet backed by a conductor could be used in place of a grating. It does not seem possible to generate a manifold of such waves of the form envisaged by Palmer, in which a finite wave packet on an unbounded grating travels with $v = c$ without dispersion.

Although different in form from existing accelerators, and operating at a different wavelength, the grating (or dielectric sheet) linac is essentially the same class of device. It relies on near fields propagating in the neighborhood of a structure or dielectric with phase velocity less than that of light.

The inverse free-electron laser mechanism is also placed in category 1.1 in the table. Here the particle and wave velocities are different, and there are essentially two waves interacting simultaneously with the particles. One of these may be a static "wiggler" magnet, with finite k but zero frequency. This interaction is examined next, again considering the basic components of plane wave and particle.

We consider first a particle and plane wave moving in the same direction. The wave is characterized by wavelength λ, and the particle by normalized energy γ, where γ is large, so that $\beta \approx 1-(2\gamma^2)^{-1}$. The particle experiences a transverse electric force, almost balanced by a magnetic Lorentz force in the opposite direction, $F = eE(1-\beta) \approx eE/2\gamma^2$. The direction of this force alternates, the particle experiencing one complete cycle in a distance $\Lambda = \lambda/(1-\beta) = 2\gamma^2\lambda$. If now a small transverse deflection of wavelength Λ is induced in the particle orbit, particles at the appropriate phase with respect to the plane wave experience a force that is outward when the particle moves outward, and inward when the particle moves inward. The particle thus gains energy at a rate $eE_\perp v_\perp$. Continuous acceleration is achieved, but this is a second-order process since the electric field and orbit directions are almost perpendicular. This is illustrated in Figure 3.

The transverse deflection can alternatively be provided by a second electromagnetic wave travelling in the direction opposite to the laser beam. The effects of the electric and magnetic fields of this second wave now add (rather than subtract) to produce a transverse deflection of wavelength $\Lambda = (1+\beta)/(1-\beta) = 4\gamma^2\lambda$. Yet another possibility is to use an evanescent wave travelling at an angle to the particle beam.

This type of acceleration makes use of laser light without the severe restrictions arising from proximity to conductors, and from the small transverse dimensions necessary in the grating accelerator described earlier. These advantages create the problem that the accelerating field and particle velocity are almost perpendicular.

Despite these disadvantages, impressive acceleration rates are indicated in a study by Pellegrini and others (3, 3a). An important problem is the design of a practical configuration to provide adequate optical focusing over long distances of the intense laser beam.

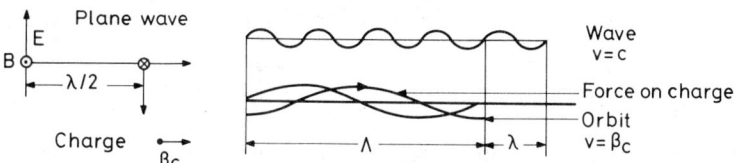

Figure 3 The inverse free-electron laser mechanism. The light wave moves faster than the particle, whose orbit is modulated either by a magnetic wiggler, or by a longer-wavelength wave moving in the opposite direction. The modulation frequency is adjusted in such a way that for particles in the correct phase there is a continuous component of field along the orbit.

3.2 Integrated Two-Beam Systems

Before discussing entry 1.2 in the table, it is of interest to consider what is included in a complete accelerator system. In linacs and synchrotrons the radiofrequency (rf) fields needed to accelerate the particles are derived from power tubes containing a high-current beam, from which energy is extracted by means of some high-impedance resonant circuit or cavity. In a sense, therefore, the whole system consists of a device to transfer energy from a high-current, low-energy beam to one with high energy but low current. Seen in this light the accelerator is a type of transformer. The question now arises, is it possible to have some form of two-beam device, in which coupling is more direct? Can the complication of coupling loops and transmission lines be avoided in a system that might be mechanically simpler and more efficient?

Sessler (4) has suggested a scheme in which two beams run side by side, in separate pipes. The first contains a very high-current electron beam, which alternately loses energy in the form of radiation at the required wavelength in a tapered wiggler using the free-electron laser coupling mechanism, and gains energy by induction from a series of induction linac modules. The radiation, at wavelength of order 1 cm or less, is "piped off" after each wiggler section to feed a disc-loaded linac. Experiments are under way on the amplification of the radiation in the wiggler, but an overall analysis remains to be done. It may be possible eventually to devise a practical scheme with both beams in the same pipe.

3.3 Nonharmonic Fields in Free Space

Several important types of accelerator in which the accelerating fields are not harmonic and the particles move in a vacuum are shown in the table. It is interesting to note that the first of these, the multigap accelerator, now proposed as the initial stage of a high-current accelerator for heavy-ion fusion (5), was first proposed as early as 1924 by Ising (6). At that time the technology to make it work was not available. Today, it appears that none of this class of accelerator will find direct application at the highest energies.

3.4 Collective Schemes with Accelerated Particles in Vacuum

An early example of this class of accelerator, in which free charges are used though the particles to be accelerated are in a vacuum, was suggested by Alfvén & Wernholm in 1952 (7). An intense electron beam is focused through a sharp waist. Near to this waist is an intense radial electric field. If the whole beam is propelled sideways, then positive particles can be dragged along and accelerated. The scheme originally proposed for moving the beam sideways does not look technically very practical, and the interest in this method is now purely historical.

A new method of dragging relativistic particles along has been suggested by Voss & Weiland (8). Termed the wake field accelerator, it is generically related to the two-beam accelerator previously mentioned. In one form the drive beam consists of relativistic rings of electrons preaccelerated in an rf accelerator. These rings are made to propagate through an axisymmetric periodic structure coaxial with the channel for the driven beam. Electromagnetic energy, stripped from the fields of the drive beam by the periodic structure, flows from the periphery of the structure to the axis where the driven beam is to pass. The resulting compression of the broadband wake energy pulses magnifies by ten the associated electric fields, which yields very high longitudinal accelerations for very brief intervals. Study of this scheme is continuing at the DESY laboratory.

3.5 *Harmonic Collective Schemes with Particles in Beam or Plasma*

Three examples of acceleration methods in this category may be cited, in which the medium is respectively a gas, dense particle beam, or plasma. In a gas the phase velocity of electromagnetic radiation is slowed down. If n is the refractive index, continuous interaction can occur between electromagnetic waves travelling at an angle θ to a particle moving with velocity βc when $\beta = 1/n \cos \theta$. If the phase of the particle is such that energy passes from the wave to the particle this may be termed inverse Čerenkov acceleration. Limitations arise from gas breakdown and scattering of the accelerated particles. Scattering is excessive unless the gas density is low, in which case $\cos \theta$ is nearly unity, and the component of electric field in the direction of particle motion is small. This type of acceleration has been analyzed by Fontana & Pantell (9).

The possibility of accelerating particles by the longitudinal electric fields associated with waves on electron beams has received extensive study over a number of years. Early experiments in the USSR are described by Fainberg (10). Sloan & Drummond (11) later suggested using a beam carrying a negative energy wave. As the particles are accelerated, energy is removed from the wave, which therefore grows. Unfortunately, it has proved very difficult to realize this idea in practice. Both experimental experience and general analysis suggest that this concept is not promising for producing high accelerating gradients at particle velocities approaching that of light.

The proposal that so far seems to offer the possibility of the highest accelerating fields is the beat-wave accelerator suggested by Tajima & Dawson (12). The starting point is a neutral plasma rather than a charged electron beam. This enables a much higher initial particle density to be achieved, since the bulk electrostatic force associated with an assembly of particles with net charge is absent. In a neutral plasma purely longitudinal plasma waves (Langmuir waves) can be excited at the electron plasma frequency ω_p, which is equal to $n_0^{1/2} e/(\varepsilon_0 m_e)^{1/2}$. In a cold plasma ω_p is independent of k, so that the phase velocity can assume any value, and the group velocity is zero. Since the ion mass is large, to first approximation the ion background can be assumed stationary, and a small plasma oscillation can be represented by a modulation density $n_1 \sin(\omega_p t - kz)$ in a plasma of electron density n_0. Clearly there is an upper limit to the wave amplitude that can be produced, and this can be shown to be

$$E_z = \alpha m_e c \omega_p / e = \alpha m_e c (n_0/\varepsilon_0 m_e)^{1/2}, \qquad 1.$$

where α is a constant of order, but less than, unity. For a plasma density of 1.6×10^{16} cm^{-3} and $\alpha = \frac{1}{2}$ this corresponds to about 5 GeV m^{-1}.

Rosenbluth & Liu (13) showed that such a wave can be excited by two *transverse* electromagnetic waves with $\omega \gg \omega_p$ propagating in the same direction through the plasma, with frequency difference $\omega_1 - \omega_2 = \omega_p \ll \omega_1$. The superposition of two such travelling waves produces beats with wavelength $2\pi/\Delta k$ where $\Delta k = k_1 - k_2$. In the presence of plasma, a Langmuir wave with frequency ω_p and wavenumber Δk is driven by the "ponderomotive" force proportional to grad$\langle E_0^2 \rangle$, where E_0 is the transverse field.

It is readily verified that the *group* velocity of the transverse electromagnetic plasma waves is equal to the *phase* velocity of the beat wave, $c(1 - \omega_p^2/2\omega^2)$. This velocity is slightly less than that of light, and equal to that of a particle with normalized relativistic mass $\gamma = \omega/\omega_p$. A basic compromise has to be made in choosing ω_p/ω. Large ω_p implies a high accelerating field but a low γ_p, so that phase slip becomes a problem. The theory sketched so far is very much simplified; many factors conspire to limit the value of α and waste laser power by exciting competing processes.

There are two immediate questions to be asked about this mechanism as a candidate for a high-energy accelerator: first, what value of α can be achieved in practice, and, second, how do we deal with the fact that the phase velocity of the wave is less than that of an extremely relativistic particle? The first question requires further research in plasma physics before a decisive answer can be given, and the second seems to imply that the accelerator be split into a number of stages, though Katsouleas & Dawson (14) have pointed out that, if the particle beam is at a small angle to the wave normal, synchronism can be achieved. The small component of electric field at right angles to particle trajectory can be counteracted by a transverse magnetic field.

When an attempt is made to incorporate these ideas into an actual physical system, designed to produce a usable beam of high-energy particles, many severe constraints are encountered. An elegant first analysis of these problems has been made by Ruth & Chao (15), and a somewhat more detailed look by a study group at the Rutherford Appleton Laboratory is reported by Lawson (16). As with other laser schemes, there is concern about power requirements and overall efficiency.

3.6 *Nonharmonic Collective Schemes with Particles in Beam or Plasma*

The final category of the table comprises schemes in which a deep potential well in an electron beam moves forward with a velocity suitable for accelerating positive particles trapped within it. One of the best known and most thoroughly analyzed schemes of this class is the electron ring accelerator proposed originally by Veksler et al (17). Although originally foreseen as a high-energy accelerator, greater understanding of ring stability has revealed limitations to its performance, which suggest that it would be more appropriate as a heavy-ion accelerator at nonrelativistic energies. Experimental work is continuing in the USSR. A detailed survey is given by Olsen & Schumacher (18).

Another important example in this category is the ionization front accelerator. This stems from the unexpected discovery of 5-MeV protons in a 1.3-MeV pulsed electron beam by Graybill & Uglum in 1970 (19) and relies on the intense fields associated with a "virtual cathode." Unneutralized electron beams above a certain limiting current (of the order of 10–50 kA, depending on energy and configuration) cannot be propagated down a hollow metal tube; this is because the space charge sets up retarding fields that are strong enough to reflect the electrons. If, however, there is gas present, this becomes ionized by fast electron collisions. The secondary electrons produced in this process are repelled to the conducting wall, and ions remain to neutralize the space charge of the beam. By this means the virtual cathode region can advance along the pipe, carrying in its intense field ions, which are thereby accelerated. This velocity of advance can be varied by controlling the ionization process, for example by a timed succession of laser pulses that ionize the gas ahead of the beam (20).

This type of scheme has not yet been developed sufficiently for practical application. It would seem to be more appropriate to producing very short but intense bursts of ions at nonrelativistic energies than it is for providing high field acceleration of relativistic particles. In this connection it is interesting to note that the authors of a Department of Energy (DOE) study

(21) conclude that "the acceleration of charged particles to very high energies is an unlikely application of collective accelerators." This study dealt with the schemes mentioned earlier in this review (and many others), with the important exceptions of the beat-wave accelerator and the wake field accelerator discussed in Sections 3.5 and 3.2 above. An extensive account of collective accelerators may be found in the texts by Olson & Schumacher (18) and by Reiser & Rostoker (22).

4. CRITERIA FOR NEW ACCELERATORS AND NEW ACCELERATOR CONCEPTS

4.1 General Aims in Planning New Accelerators

The very diverse range of applications of particle accelerators imposes very different criteria for usefulness. In developing fields, rather than routine applications such as producing medical isotopes, requirements become ever more demanding. Developments for different applications require, with varying emphasis, increases in energy, current, beam quality, and versatility. In some large-scale applications that may be important in the future, such as accelerator breeding of fissile material or inertial confinement fusion by heavy ions, development is envisaged along fairly well perceived lines, and the criteria are to a large extent economic. For high-energy physics, where the objective is to advance our knowledge of the physical world at the very frontier of our understanding, the criteria are different, and it is difficult to express them quantitatively. Already, however, each new accelerator puts a significant strain on available national or regional resources, so that future machines must not only satisfy technical needs, but also offer a significant increase in cost-effectiveness. The problem of what can be done after the next round of large machines is by no means clear. This is a central question for accelerator designers now and in the next decade. Accordingly we concentrate on this problem, and enquire to what extent the requirements to be met in future high-energy machines can be achieved by extrapolation of existing techniques, and to what extent we must rely on new concepts for further progress.

4.2 The Demands of High-Energy Physics

The central emphasis in particle physics has always been the achievement of ever higher projectile energies. High particle energy is not the only requirement, however; the beam current must be sufficient to produce an adequate reaction rate for interesting events, and the beam quality such that meaningful experiments can be done without undue difficulty. At energies of current interest, the ratio of energy to rest mass γ is very large for both protons and electrons. At such high energies the center-of-mass energy obtainable in the colliding-beam mode exceeds that obtainable with a stationary target (of identical particles) by the large factor of $(2\gamma)^{1/2}$. For this reason future accelerators will need to be operable in the colliding-beam mode. What this implies in the way of additional constraints is considered in Section 5.

Before discussing constraints and technical problems, it is of interest to examine the present situation, and aims for the immediate future. Within the next few years accelerators now in operation or under construction will permit extensive study of elementary phenomena in the energy range of order 100 GeV, a rich region encompassing the characteristic mass scale of electroweak unification. The next energy range, which might be reached in the next decade, would be a factor of ten higher, which would imply electrons machines of order 1 TeV and proton energies greater by a factor of 5–10. A further factor of ten in the decade beyond may be taken as a rough guide of what is needed to enable continued steady progress in the field.

Development goals are needed not only for energy, but also for beam current. This requirement is conventionally quantified for colliding-beam operation in terms of the luminosity L, the proportionality constant between reaction rate and cross section. Not only must the luminosity be sufficient, but the time structure of the beam must be such as to allow for the finite resolution time of the detectors.

Current experience with elementary collision energies in the range 10–100 GeV shows that a luminosity of 10^{29} cm^{-2} s^{-1} can be adequate for some discoveries, but factors greater by 100 or so are needed for thorough quantitative work. It is generally agreed that for energies around 1 TeV luminosities up to 10^{32} cm^{-2} s^{-1} or more will be needed to compensate for the decline of the elementary cross section with energy. New ideas for frontline accelerators must be judged on this scale. Further developments of currently used technologies will be able to carry us into the 1-TeV elementary collision range with protons (this implies a proton beam energy of 5–10 TeV), and perhaps with electrons. It is not at all clear that they can carry us significantly beyond that.

4.3 *Some Economic Criteria for High-Energy Machines*

Resource limitations place bounds on both capital and operating costs. Both represent vital considerations, therefore, in planning new accelerators. While detailed computation of cost factors presents complex issues, the rough outlines are easily seen. We have reason to hope that within a decade or so we may be able to build proton synchrotrons at a unit cost of about 10^5 dollars per GeV of beam energy (23). Current projections for electron machine costs are about 20 times higher (24). Experience indicates that the cost of new national or regional facilities for particle physics will probably not exceed a few times 10^9 dollars. The upper bound on acceptable operating costs and electric power usage is expected to be a few times 10^8 dollars per annum and a very few hundred megawatts. Present facilities are close to or already at these limits.

The continuing challenge to new accelerator technologies, then, is to increase available energies by at least an order of magnitude each decade, increase luminosities by two orders of magnitude, and at the same time keep the total cost of each new facility constant. Startling events in physics may reshape this challenge but it is clear that, in any case, dramatic new improvements in accelerator technology will be required to maintain the pace of new discoveries in accelerator-based, high-energy particle physics.

5. APPLICATION OF THE VARIOUS ACCELERATION METHODS AT HIGH ENERGIES

Accelerator design is, in large measure, a problem in systems analysis. The attainment of an economical and efficient system involves battling with a host of interdependent and often conflicting constraints, which arise from the physics of beams and the technology of both accelerator construction and experimental method. Even in the most thoroughly elaborated schemes, much remains that is obscure. Progress in understanding the physics and refining techniques has in recent years helped to increase cost-effectiveness, and we have every expectation that this will continue.

In the following sections an outline of some of the factors determining the performance of the mature synchrotron and linac technologies is given, together with references to more detailed discussions. Finally some general comments on "undeveloped" ideas are made.

5.1 *Synchrotrons*

The synchrotron is perhaps the most highly developed of the near-field accelerators. Most synchrotrons now used for frontier particle physics operate at least part-time in the storage ring–colliding-beam mode (1a, 25, 26), and in future this will certainly be the principal mode of operation. The record for luminosity is about 10^{32} cm^{-2} s^{-1} for proton storage rings (pp) and about 3×10^{31} for electron storage rings (e^+e^-). The record for beam energy in electron storage rings is now just over 21 GeV (e^+e^-) whereas over 700 GeV has been achieved with single beams of protons, and 270 GeV in the p$\bar{\text{p}}$ collision mode. Synchrotron-based storage ring systems now under construction are expected to reach beam energies in excess of 100 GeV for electrons and 1 TeV for protons. The luminosities expected for these accelerators are not likely to surpass those already achieved. No electron storage rings with energies beyond 120 GeV are now planned, but preliminary plans for proton machines of up to 20 TeV in beam energy are being discussed. Because of the crucial importance of high luminosity, the factors that determine its value are discussed next.

5.1.1 LIMITATIONS TO LUMINOSITY In the colliding-beam configuration, oppositely directed beams are focused to a small volume at the desired collision zone. If the beams are bunched with time separation f^{-1} and each bunch contains N particles, the luminosity characteristic of one collision zone is

$$L = N^2 f / A_{\text{eff}} \; [\text{cm}^{-2} \, \text{s}^{-1}], \qquad 2.$$

where A_{eff} is a (suitably projected) effective cross-sectional area of beam overlap. It depends upon the three-dimensional distribution functions, the angle of crossing, the depth of focus at the crossing zone, the relative beam alignment, and the focusing or defocusing effect of one beam upon the other as they interpenetrate. For round gaussian beams colliding perfectly head on, $A_{\text{eff}} = 4\pi\sigma_\perp^2$, where $\sigma_\perp$ is the effective rms beam radius. If the depth of focus is greater than the rms bunch length σ_L and the contribution of the self-fields to the focusing effect is negligible, then the transverse rms width of the beam is given by $\sigma_\perp = (\varepsilon_\perp \beta^*)^{1/2}$, where β^* is the betatron function at the crossing point, a measure of the focusing strength there, and $\varepsilon_\perp$ is the transverse emittance, proportional to the two-dimensional phase-space area of the beams. Similar simple formulae may be written for round or flat, bunched or continuous beams crossing at an angle (1a, b).

While no fundamental limit to the luminosity itself is known, the quantities that enter into it are constrained in many ways. For example, the projected density, N/A, at the crossing zone is limited through the action of the (nonlinear) focusing or defocusing of one beam by the other, the so-called beam-beam effect (27–30). Attempts to exceed this limit result in compensating dilution of density or in sharp decreases in lifetime of the beam. The effect is produced by the collective fields of one bunch deflecting the individual particles in the other. In practice the beam distribution functions tend to be gaussian or quasi-gaussian so that the deflection forces are nonlinear functions of the particle coordinates. It is convenient to characterize the beam-beam forces by a focal length, F, which applies to particles near the axis. Noting that the bunch collective fields E and B aid in deflecting oncoming particles, an elementary calculation gives this focal length for a round gaussian beam, $F = \gamma \sigma_\perp^2 / N r_0$. (This applies strictly only to weak deflections of a particle passing the crossing zone with displacement equal to the transverse rms beam radius.) This beam-dependent nonlinear focusing element acts together with the other forces to which the beam particles are subject in determining the equilibrium distribution functions of the beam particles and their diffusion times to the walls if the perturbed orbits are not stable.

No general procedure for computing the time evolution of the distribution functions under the influence of the beam-beam force is now available, and designers must have recourse to semi-empirical rules. This remains a central and unsolved problem of accelerator physics. In storage ring literature the accepted measure of the strength of the beam-beam effect is the dimensionless ratio of $\beta^*/4\pi$ to F, which gives the spread in particle betatron tunes ΔQ engendered by the beam-beam force acting at the crossing zone

$$\Delta Q = \beta^*/4\pi F = r_0 N \beta^*/4\pi \sigma_\perp^2 \gamma \qquad 3.$$

for round gaussian beams. The betatron tune is the number of oscillations of a particle about its central orbit per revolution in the accelerator (1a, b). The tolerable strength of the beam-beam interaction depends upon whether the beams are bunched or unbunched, whether they cross at an angle or head-on, the particle type and energy, and to a lesser extent on the accelerator configuration. Our rather limited experience with head-on collisions of very high-energy, bunched proton beams indicates that a total time spread on ΔQ for the ring of as much as 0.024 (or 0.004 per crossing) can be sustained (31). In electron machines where beam equilibria are established by the strong radiation damping (1a), values up to about 0.05 per crossing have been measured (32), somewhat independent of the number of crossing points on the ring.

The rules for compounding the beam-beam effect from successive crossings are obscure and subject to controversy and certainly depend upon whether strong synchrotron radiation damping is at work or not. For electrons, the maximum beam-beam limit ΔQ may depend on the radiation damping decrement per crossing, saturating at some ultimate value (33). It is known to be lessened when bunched electron beams cross at an angle (34). For protons the limit is greater for unbunched than for bunched beams but by an uncertain amount. Crossing at an angle is also believed to reduce the limit for bunched proton beams. The limit ΔQ appears also to depend on such factors as the phase oscillation frequency, betatron frequencies, chromaticity, beam energy spread, and betatron phase advance between crossings. Further elucidation of these matters could lead to the achievement of higher luminosity per accelerated particle and thus to needed cost savings. The direct role of the beam-beam limit in luminosity determination is seen by substituting $A_{\text{eff}} = \pi \sigma_\perp^2$ in the expression for L, and expressing $\sigma_\perp^2$

as $\varepsilon_\perp \beta^*$. This yields

$$L = \gamma \Delta Q N f / r_0 \beta^* \qquad \qquad 4.$$

for perfectly aligned, head-on collisions.

In addition to effects directly associated with the collision zone, numerous other factors limiting the beam current in synchrotrons must be taken in account. Well-known criteria exist to ensure that the wall impedance "seen" by the beam is sufficiently low that collective beam instabilities do not occur. A wide variety of types of instability is known, and feedback techniques to control some of these have been developed. While further work is clearly needed, experts believe that by a combination of analytical and numerical simulation methods we may soon be able to compute these limits to beam current for any given accelerator design, including hardware details, from first principles.

A further effect of a different type also needs consideration; the average volume density of a beam can be limited by the dynamic redistribution of energy among its several degrees of freedom through internal Rutherford scattering, an effect now generally referred to as intrabeam scattering (35, 36). Rendered harmless by the strong radiation damping of high-energy electron storage rings, this effect must be reckoned with in high-energy proton rings where it has been observed to enlarge the beam and curtail lifetime (37). The evidence is that this effect can be correctly computed from first principles.

As well as these essentially collective effects, there are, of course, numerous interlinked constraints associated with the complex nonlinear aspects of single-particle dynamics. These play their part in determining the overall design, and hence the luminosity through the factors N, f, and β^*.

In addition to the constraints imposed by accelerator physics, detector technology also dictates the acceptable time structure of the reaction rate. For protons the nondiffractive part of the total cross section is expected to be about 100 mb at 10 TeV of beam energy. Most of these events will have high multiplicity, making it difficult to separate them if more than one occurs per bunch crossing or per detector resolution time. Already at $L/f = 10^{25}$ cm^{-2} s^{-1} we can expect one event per crossing.

If a total luminosity of 10^{33} is needed simply to produce sufficient quantities of rare events, a crossing frequency of 100 MHz will be required to maintain an average of one event per crossing. This permits a detector recovery time of only 10 ns. At the even higher luminosities that may be needed at yet higher energies, substantially new detector technology will be needed. The high cross section for beam-beam bremsstrahlung puts a similar but milder constraint on high-luminosity electron storage rings.

Crossing frequencies of a few hundreds of kHz are employed in the existing rings with highest energy. Boosting this to tens of MHz will be a substantial technical challenge. The use of continuous beams to alleviate the duty factor problem is ruled out in electron storage rings, since the need to replenish continually the synchrotron radiation losses implies tightly bunched beams. This is likely to be true also for proton rings at energies much in excess of 10 TeV.

Luminosities of most, but not all, high-energy storage rings have fallen short of their design goals. The reasons seem to be a complex combination of the constraints mentioned above with factors peculiar to a particular project. It is important to note that advances in accelerator physics have helped measurably in closing the original gap between design and achieved luminosities in all storage rings now operating. While much development of relevant accelerator physics and technology still remains to be accomplished, it seems reasonable to assume that storage ring technology can yield the increasing luminosities needed as their energy is increased.

5.1.2 LIMITATIONS TO ENERGY   The ultimate energy limit for synchrotrons seems to be economic in nature, though continued improvements in technology together with more efficient use of accelerator components made possible by better understanding of the underlying physics have served to drive unit costs down (38). On the other hand, the phenomenon of synchrotron radiation ultimately operates to make the size of a synchrotron scale faster than linearly with the energy, and a term in energy squared appears in the cost equation (39). The electron storage ring is already well into this regime. With the best anticipated technological developments of the next 5 to 10 years, the posited facility cost limit of 10^9 would correspond to an electron storage ring with a center-of-mass energy of a few hundred GeV (40). While synchrotron radiation from protons has so far been negligible, the generation of proton storage rings now on the horizon will have to begin to deal with it. The energy lost per revolution by a charged particle constrained to follow a circular path by a field B is

$$U_s = \frac{4}{3} \pi r_0 c \gamma^3 B \text{ [eV]}. \qquad 5.$$

In terms of the luminosity, the power radiated by two beams, each of current I, can be found by combining this with Equation 3 to give

$$P_b = \frac{8\pi}{3} r_0^2 c e \frac{\beta^*}{\Delta Q} L \gamma^2 B \text{ [W]}. \qquad 6.$$

For an electron of 100 GeV with a 3-km orbit radius, U_s amounts to 3% of its total energy. For a machine with reasonable luminosity, P_b would amount to tens of megawatts. For proton machines the loss is much less since, for the same energy, γ and r_0 are smaller by the ratio m_e/m_p; nevertheless in the TeV range the effect is significant, as the following argument shows.

It seems likely that in future synchrotrons small-aperture superconducting magnets will be essential to keep down both capital and running costs. At 20 TeV of beam energy, and a peak field of 5 Tesla, we find $\beta^* = 2$ m, $\Delta Q = 0.003$, $L = 10^{33}$ cm^{-2} s^{-1}, $P_b = 14$ kW. Unless special means are employed, this power will be deposited at essentially liquid helium temperature. Current refrigeration efficiencies are such that 14 kW deposited at 4 K would require an extra 7 MW of electric power for its removal. Such extra power usage at 20 TeV may be tolerable. At 100 TeV even without increased luminosity or magnetic field $P_b = 350$ kW, corresponding to 175 MW of electric power if absorbed at 4 K. This is clearly impossible, and the use of an absorber maintained at a temperature significantly above 4 K will be required. No satisfactory engineering solution to this problem is known that combines a small aperture with a wall impedance low enough at high frequency to ensure orbit stability. Exacerbating the situation further, it appears that the economies needed to push on to 100 TeV may require yet higher fields. Alloys are known that may make 15-Tesla magnets possible by the problems entailed in their use are totally unexplored. Reconciliation of these conflicting technical and economic constraints will be most difficult.

On present understanding it appears that electron synchrotrons will not be feasible above a few hundred GeV, whereas for proton synchrotrons the technology may be viable at least up to 100 TeV.

5.2 Linear Accelerators

The energy of synchrotrons is ultimately limited by synchrotron radiation; beyond this limit only linear machines are possible. Further, if their length is to be kept within bounds, a high accelerating gradient is called for.

For operations in the colliding-beam mode the requirements are very different. In synchrotrons the beams can be stored for many hours; individual particles spend a long time in the intersection region where they repeatedly meet particles coming in the opposite direction. The total circulating power can be orders of magnitude more than the input power to the whole installation. In linacs, on the other hand, the encounter time of oppositely travelling bunches is short indeed, since each bunch can only be used once. Emphasis must therefore be placed on the achievement of the highest possible repetition rate and interaction probability per particle per bunch collision. This latter requirement implies beams with very high density, and hence low cross-sectional area.

So far the focus of attention in linac colliders has been on electron machines, for which the need is at present most pressing. For proton machines, a comparison with synchrotrons is instructive. The effective energy gradient of a proton synchrotron can be defined quite simply as the ratio of beam energy to the circumference, $\gamma m_0 c^2/2\pi R = eB/2\pi$ (eV m^{-1}). For a 5-Tesla magnet this is 240 MeV m^{-1}. Such magnets can be built for about 7×10^3 dollars per meter. Improvements over the next few years are expected to increase B, perhaps to 10 Tesla or more, while holding or reducing the cost per meter. When synchrotron radiation is negligible, and the cost of the accompanying acceleration system is relatively low, these figures can be used to compare linac achievements (or expectations) with those of synchrotrons. Since existing linacs run at $E \approx 20$ MeV m^{-1} with unit length costs generally in excess of that for the magnets, it follows that substantial advances in the linac art will be needed before they can compete effectively for the acceleration of protons.

Various hybrid schemes in which one attempts to combine the advantages of both synchrotrons and linacs have been suggested (41). Most of them rely on microwave superconductivity and their pursuit will depend on future progress in that area.

5.2.1 LIMITATIONS TO LUMINOSITY AND ENERGY The basic expression for luminosity discussed in Section 5.1, with the appropriate subsidiary conditions, applies as well to linear colliders, but the details of the constraints on the parameters that determine the luminosity are somewhat different. As the first linear collider is just now under construction (42), it is not surprising that our understanding of these constraints is considerably more tentative than it is for storage rings.

The maximum useful areal density at the crossing zone is again likely to be limited by the beam-beam focusing effect. In this context an appropriate measure of the beam-beam strength is the ratio of the rms bunch length to F and is called the disruption parameter, D. Using the expression $F = \gamma \sigma_\perp^2 / N r_0$ quoted in Section 5.1, we find

$$D = N r_0 \sigma_L / \gamma \sigma_\perp^2 \qquad 7.$$

for round gaussian beams. Numerical simulations (43) indicate that for $D \approx 1$, beams of oppositely charged particles pinch during the collision; this decreases their effective area and enhances the luminosity by a factor $H(D)$. The pinch appears to be most effective when $D \approx 2$; for larger values the benefit deteriorates because of instability. As D increases, beam particles are thrown out at increasing angles to the beam and tend to interfere with detection (44). The simulation shows a maximum $H(D)$ of about six. Whether such enhancements will occur in practice remains to be seen.

In this high-density regime, more than 20 times the tolerable limit for electron storage rings, the beam self-magnetic fields at the focus can be enormous. In the proposed Stanford Linear Collider (SLC), $\sigma_L \approx 1$ mm, and $\sigma_\perp \approx 1$ μm. For $D = 1$ the magnetic field at the beam surface is about 200 Tesla, and this gives rise to another important effect. As a result of the strong transverse accelerations suffered by the beam particles during this intense beam-beam focusing, they radiate strongly, a phenomenon referred to as "beam-strahlung." For interpenetrating round gaussian beams an elementary calculation shows that the fractional energy loss suffered by a typical beam particle in a single crossing is (45)

$$\delta = \frac{1}{3\sqrt{3}} \frac{r_0^3 N^2 \gamma}{\bar{\sigma}_\perp^2 \sigma_L}, \qquad 8.$$

where $\bar{\sigma}_\perp$ is the effective average value of the beam area throughout the collision, which will be reduced by the pinch if the colliding beams are of opposite charge. The enhancement for δ is the same as for L, namely $H(D)$. As δ increases, the spray of beam particles and gamma rays emerging from the collision becomes more divergent. In addition, the average energy of the particles will be lowered and their energy spectrum will widen. These undesirable features limit the allowed volume density at the crossing. It has been assumed that δ of a few percent will be tolerable in practice.

Of central technical and economic importance is the beam power required to produce a given luminosity. Combining Equations 2, 3, and 8 we find the power for both beams to be

$$P_b = \frac{5.8 \sigma_L \,[\text{mm}]\, L \,[10^{32}\ \text{cm}^{-2}\ \text{s}^{-1}]}{D\, H(D)} \,[\text{MW}] \qquad 9.$$

per crossing point. If we derive nearly the full predicted gain from the pinch effect, and use $\sigma_L = 1$–2 mm, typical of extant microwave linacs, we get $P_b \approx 5$–10 MW for $L = 10^{33}$ per crossing zone. If the pinch effect is less efficacious, or if we are unable to maintain D while keeping beam-strahlung under control, the power will be correspondingly greater. It is most important that this be kept down, since the input power to the facility will be P_b/η, where η is the net efficiency of the accelerator. Present accelerators operating under constraints needed for small spot size have $\eta \approx 1\%$. Considerable improvement can ultimately be expected from optimization, as can the implementation of schemes now being explored that make use of many bunches. Obtaining good efficiencies requires careful optimization of conflicting requirements, several examples of which are now given. First, high efficiency demands as large an N as possible to extract the maximum possible fraction of the stored energy. On the other hand, the large peak current implied by large N generates larger wakes in the accelerator, and these wakes tend to increase emittance and make the needed tight final focus very difficult to achieve (46). For standard 36-Hz linacs, a value of about 10^{10} particles per bunch seems to be the limit.

Another problem is that use of the highest possible energy gradients, limited only by breakdown, though desirable from many points of view, makes it more difficult to achieve high efficiency. Yet again, for a fixed geometry higher gradient raises stored energy as E^2 whereas the severity of the wake effects is lessened only as E. There is a basic conflict here, too. Preliminary, but as yet inconclusive attempts to understand the scaling and general optimization problem presented by these facts, including the influence of varying drive frequency (bunch length), have been published (47, 48). Major uncertainties include the limits to luminosity enhancement arising from the pinch effect, and to beam size at very high energy where cost-effective designs seem to demand submicrometer- or even Ångstrom-sized foci. Such factors could well result in a cost relation for linear colliders that rises with a power of the energy greater than unity.

5.2.2 THE EXTERNALLY DRIVEN MICROWAVE LINAC   Benefiting from more than 40 years of development, the microwave linac is by far the most developed of the near-field linacs. The high-current, low-voltage drive beams in the power tubes convert dc power to rf power by interacting with specially designed and optimized structures. The most commonly used tubes have been magnetrons and klystrons capable of tens of megawatts peak output power with conversion efficiencies of 25% to more than 60%. Future development of these and some new devices such as gyrotrons and free-electron lasers are expected to result in peak powers exceeding 1 GW (49, 50).

Other stratagems have also been suggested (51). At present the rf power generators are separately packaged and connected through some sort of waveguide to the high-energy accelerating structure. This structure, through resonant build-up of the fundamental accelerating field, acts as a transformer matching the low-voltage, high-current (low-impedance) load to the high-energy beam. Linacs now in operation routinely support gradients of more than 10 MV m^{-1} at about 10-cm wavelength. It is believed that with care this figure might be doubled, and experiments indicate that with specially prepared structures 100 MV m^{-1} might be attained (52, 53). By using shorter wavelengths, yet higher gradients may be expected (53, 54). (At this point we recall that operation of a specific structure at the breakdown limit is in general *less* economical than having a longer machine with lower gradient.)

Microwave superconductivity, offering the promise of high fields at low power, has so far not found useful application in linacs for particle physics.

Recent progress in the field and analyses of its potential indicate that it may yet flourish (54, 55).

Time-tested recipes for the design of economical disc-loaded accelerator structures for relatively long pulse beams are amply described in the literature (56, 57). These are now supplemented by powerful computer codes that can be used for cavity optimization (58). Workers in the field are now engaging the problem of optimizing the microwave linac for linear collider service (59–61). Excellent detailed reviews are available (45, 49, 62).

It appears that it may be possible to meet the necessary luminosity and economic criteria up to perhaps even TeV energies. Before solidly based cost-optimized designs can be produced, however, considerably better understanding of the matters previously discussed must be obtained, and economical rf sources capable of hundreds of MW peak power at the optimum wavelength must be invented.

5.3 *The Status of Undeveloped Ideas*

Compared with decades of effort invested in the study and construction of synchrotrons and linacs, virtually nothing has yet been done on laser accelerators and the two-beam schemes. Although it is agreed that the laser acceleration mechanisms will accelerate particles, there is some scepticism about whether they will ever be able to compete with the performance of conventional machines at the highest energies. While it is not wise to jettison ideas too early, it is important to establish at least in outline what the fundamental constraints are likely to be. Making the most optimistic assumptions compatible with current understanding, we must ask what laser power and what repetition rates and pulse lengths will be needed to give an interesting performance? In the light of present experience with high-power lasers, what are the technical problems and what will be the cost? Until such questions are realistically faced, no meaningful assessment of the status of these ideas can be given.

In order to tackle these questions, some specific designs must be sketched. Indeed, as indicated in Section 3, this has been done, though the schemes described hardly yet look convincing when examined in detail.

The two-beam and wake field ideas are closer to familiar concepts in microwave tube and accelerator technology, and only a careful comparison with more conventional linacs will show what the possibilities are. The identification of a suitable figure of merit for making comparisons would be helpful. In connection with linac studies generally it is important to understand breakdown limitations as a function of frequency, and also to study carefully what developments can be expected in the field of suitable high-frequency generators.

6. CONCLUDING REMARKS

After a general summary of the various concepts that have been considered for particle acceleration, particularly ideas for high-gradient machines using lasers, a more detailed account has been given of the requirements for the next round of high-energy machines, and of the factors that determine their performance.

For the "conventional" machines, which in this context must include the yet untried linear collider, the way ahead involves challenging technical problems of the highest order; to make real progress requires both determination and adequate resources not tied too tightly to existing commitments. Rising to the challenge of higher energies, many contributors have brought forth interesting new accelerator ideas in recent years. Experience of the past has shown that enormous intellectual and material resources were required to bring the classical accelerators now in use to their present highly sophisticated state. From this, one can guess that to bring any of these new ideas to a state where they can surpass present performance will likewise require prodigious efforts. It may be that in the end new developments on one of the old ideas will win out once more. It may be that some combination will emerge victorious, or it may be that neither the old nor the new will be successful in meeting the challenge. What is clear is that there is urgent need for a continuing flow of new ideas and new commitments by physicists who are determined to succeed.

ACKNOWLEDGMENT

The authors gratefully acknowledge a critical reading of the manuscript by M. Pickup.

Literature Cited

1a. Sands, M. 1970. In *Physics with Colliding Beams*, ed. B. Touschek. New York: Academic
1b. Johnsen, K., ed. 1977. *Proc. 1st. Course Int. Sch. Part. Accel.*, 1977, CERN 77–13. Geneva: CERN
1c. Month, M., ed. 1981. *Physics of High Energy Particle Accelerators, AIP Conf. Proc.* 87
1d. Month, M., ed. 1983. *Physics of High Energy Particle Accelerators, AIP Conf. Proc.* 105.
1e. Channell, P. J., ed. 1982. *Laser Acceleration of Particles, AIP Conf. Proc.* 91
1f. Mulvey, J. H., ed. 1982. *The Challenge of Ultra-High Energies, ECFA Report 83/68.* Chilton: Rutherford Appleton Lab.
1g. Cole, F. T., Donaldson, R., eds. 1983. *Proc. 12th Int. Conf. Part. Accel.* Batavia, Ill: Fermilab
1h. Johnsen, K., ed. 1984. *Proc. CERN Sch. Antiprotons for Colliding-Beam Facilities.* Geneva: CERN. To be published
2. Palmer, R. B. 1980. *Part. Accels.* 11:81–90; also see Ref. 1f, pp. 267–85
3. Pellegrini, C. 1982. See Ref. 1f, pp. 138–53
3a. Pellegrini, C., Sprangle, P., Zacowicz, W. 1983. See Ref. 1g, pp. 473–76
4. Sessler, A. M. 1982. See Ref. 1f, pp. 154–59
5. Faltens, A., Keefe, D. 1980. *Proc. Heavy Ion Fusion Workshop, 1979*, ed. W. B. Herrmannsfeldt. *Lawrence Berkeley Lab. Rep. LBL-10301*, pp. 157–81. Berkeley: LBL
6. Ising, G. 1924. *Ark. Mat. Fys.* 18:1

7. Alfvén, H., Wernholm, O. 1952. *Ark. Fys.* 5:175–76
8. Voss, G. A., Weiland, T. 1982. See Ref. 1f, pp. 287–308
9. Fontana, G., Pantell, R. 1983. *J. Appl. Phys.* 54:4285–88
10. Fainberg, Ya. B. 1968. *Sov. Phys. Usp.* 10:750–58
11. Sloan, M. L., Drummond, W. E. 1973. *Phys. Rev. Lett.* 43:267–69
12. Tajima, T., Dawson, J. M. 1973. *Phys. Rev. Lett.* 43:267–70
13. Rosenbluth, M., Liu, C. S. 1972. *Phys. Rev. Lett.* 29:710–15
14. Katsouleas, T., Dawson, J. M. 1983. *Phys. Rev. Lett.* 51:392–95
15. Ruth, R., Chao, A. O. 1982. See Ref. 1e, pp. 94–111
16. Lawson, J. D. 1983. *Rutherford Appleton Lab. Rep. RL83-057.* Chilton: Rutherford Appleton Lab.
17. Veksler, V. I., and 16 others. 1967. *Proc. 6th Int. Conf. on Accel.* Cambridge Electron Accel. Lab., Mass. *CEAL Rep. 2000*, p. 249. Cambridge, Mass: CEAL
18. Olson, C. L., Schumacher, U. 1979. *Collective Ion Acceleration, Springer Tracts in Modern Physics*, Vol. 84. Berlin: Springer-Verlag
19. Graybill, S. E., Uglum, J. R. 1970. *J. Appl. Phys.* 41:236–40
20. Olson, C. L., Frost, C. A., Patterson, E. L., Poukey, J. W. 1983. *IEEE Trans. Nucl. Sci.* NS-30:3189–91
21. Cole, F. T., ed. 1981. *Collective Accelerators, a Study Carried out for the U.S. DOE.* FN-355, 0102.000. Washington, DC: GPO
22. Reiser, M., Rostoker, N., eds. 1978. *Collective Methods of Acceleration.* Chur/New York/London: Harwood Academic
23. Tigner, M., ed. 1983. Report of the 20-TeV Hadron Collider Workshop. Ithaca: Cornell University
24. HEPAP 1983 Report of Subpanel on New Facilities, *DOE/ER-0169*, p. 38. Washington, DC: GPO
25. Keil, E. 1974. *Proc. 9th Int. Conf. Part. Accel.* Stanford, pp. 660–70. Stanford, Calif: Stanford Univ. Press
26. Keil, E. 1983. See Ref. 1g, pp. 98–103.
27. Shonfeld, J. 1981. See Ref. 1c, p. 314 ff.
28. Tennyson, J. 1981. See Ref. 1c, p. 345 ff.
29. Shonfeld, J. 1982. See Ref. 1d, p. 524 ff.
30. Chao, A. O. 1983. Brookhaven Accel. Sch. To be published
31. Evans, L., Gareyte, J. 1983. *IEEE Trans. Nucl. Sci.* NS-30:2347–49
32. Helm, R., and 16 others. 1983. *IEEE Trans. Nucl. Sci.* NS-30:2001–3
33. Keil, E., Talman, R. 1981. *CERN Rep. ISR-TH/81-33.* Geneva: CERN
34. Piwinski, A. 1977. *DESY Rep. 77/18.* Hamburg: DESY
35. Piwinski, A. 1974. See Ref. 25, pp. 405–9
36. Bjorken, J., Mtingwa, S. 1982. *Fermilab Rep. FNAL 82/47-THY.* Batavia, Ill: Fermilab
37. Gareyte, J. 1983. See Ref. 1h
38. Report of the HEPAP Subpanel on Accelerator Research and Development. 1980. *US DOE/ER 0067.* Washington, DC: GPO
39. Richter, B. 1976. *Nucl. Instrum. Methods* 136:47 ff.
40. Tigner, M. 1982. In *Elementary Particle Physics and Future Facilities,* ed. R. Donaldson, p. 299 ff. Batavia, Ill: Fermilab
41. Amaldi, U. 1979. *Int. Symp. Lepton and Photon Interactions,* Fermilab, p. 314 ff. Batavia, Ill: Fermilab
42. Stiening, R. 1983. *IEEE Trans. Nucl. Sci.* NS-30:1976–77
43. Hollebeek, R. 1981. *Nucl. Instrum. Methods* 184:333–48
44. Sah, R. 1980. *Proc. 11th Int. Conf. Part. Accel.,* CERN, pp. 736–40. Basel: Birkhäuser
45. Wiedemann, H. 1981. *SLAC Publ. 2849.* Stanford: SLAC.
46. Brown, K., Servranckz, R. 1980. *Proc. 11th Int. Conf. Part. Accel,* CERN, pp. 656–60. Geneva: CERN
47. Palmer, R. B. 1982. See Ref. 1f, pp. 267–86
48. Tigner, M. 1982. See Ref. 1f, pp. 229–40
49. Wilson, P. 1983. See Ref. 26, pp. 502–7
50. Prosnitz, D. 1983. *IEEE Trans. Nucl. Sci.* NS-30:2754–57
51. Perevedentsev, E. A., Skrinsky, A. N. 1979. *Proc. 2nd ICFA Workshop, Possibilities and Limitations of Accel. and Detectors,* CERN, pp. 61–72. Geneva: CERN
52. Tanabe, E. 1983. *IEEE Trans. Nucl. Sci.* NS-30:3309–12
53. Jameson, R. 1983. See Ref. 26, pp. 497–501
54. Tigner, M. 1982. See Ref. 1d, pp. 229–48
55. Tigner, M. 1983. *IEEE Trans. Nucl. Sci.* NS-30:3309–12
56. Neal, R. 1968. *The Stanford Two Mile Linac.* New York: Benjamin
57. Lapostolle, P., Septier, A., eds. 1970. *Linear Accelerators.* Amsterdam: North-Holland
58. Weiland, T. 1983. *IEEE Trans. Nucl. Sci.* NS-30:2489–91
59. Skrinsky, A. 1983. See Ref. 26, pp. 104–16
60. Balakin, V., Skrinsky, A. 1979. See Ref. 51, pp. 31–43
61. Neal, R. 1983. *SLAC AP-7.* Stanford: SLAC
62. Wilson, P. 1981. See Ref. 1c, p. 450 ff.

LOW-ENERGY NEUTRINO PHYSICS AND NEUTRINO MASS

F. Boehm and P. Vogel

Physics Department, California Institute of Technology, Pasadena, California 91125

CONTENTS

1. INTRODUCTION 125
   1.1 Motivation and Background 126
   1.2 Dirac and Majorana Neutrinos 127
   1.3 Neutrino Mixing and Oscillations 128
   1.4 Neutrino Decay 129
   1.5 Summary of Present Experimental Evidence 131
2. OSCILLATION EXPERIMENTS WITH REACTOR NEUTRINOS 132
   2.1 The Detection Reaction 132
   2.2 The Gösgen Reactor Experiment 133
   2.3 Fission Antineutrino Spectra 137
3. DIRECT AND INDIRECT NEUTRINO MASS MEASUREMENTS 138
   3.1 Electron Antineutrino Mass from Beta Decay Endpoints 138
   3.2 Muon and Tau Neutrino Mass Limits 140
   3.3 Heavy Neutrino Admixing 141
   3.4 Neutrinoless Double Beta Decay 143
4. OUTLOOK 150

1. INTRODUCTION

Among the principal concerns in neutrino physics today are the questions of whether neutrinos are massive and, if so, whether the neutrinos emitted in a weak decay are pure or mixed quantum states. The concept of mixed neutrinos has been with us for more than 20 years, having first been introduced by Maki et al (1) and by Pontecorvo (2) following the demonstration in 1962 that more than one type (flavor) of neutrino existed. After having been dormant for some time, the interest in these issues was reborn in recent years with the advent of grand unified theories, which predict nonvanishing neutrino mass and which can accommodate neutrino mixing in a natural way. Controversial experiments also refueled the excitment (and consternation) of researchers in this endeavor.

The field was reviewed by Bilenky & Pontecorvo (3) in 1978, by Frampton & Vogel (4) in 1982, and by Bullock & Devenish (5) in 1983. Here we focus on recent developments in the phenomenology of low-energy neutrino physics to the extent that it provides information on neutrino mass and mixing. We discuss neutrino decay, experiments on neutrino oscillations, kinematic mass measurements, searches for heavy neutrino admixtures, and studies of neutrinoless double beta decay. Subjects not discussed are accelerator-based (high-energy) experiments, neutrino-electron scattering, and certain other lepton-number-violating processes.

1.1 Motivation and Background

In the standard minimal electroweak theory, neutrinos are purely left-handed and massless and their three distinct lepton numbers (electron, muon, tau) are conserved. Therefore observation of a neutrino mass and of neutrino mixing would signal new physics beyond the minimal standard model.

Numerous theoretical proposals incorporating finite neutrino mass have been presented. At the electroweak level a neutrino mass can be introduced by extension of the minimal model (6–8), but it is often difficult to explain why neutrinos are so much lighter than the other fermions. Grand unified theories, based on various larger symmetry groups, can accommodate small neutrino masses in the range of 10^{-6}–10^{+1} eV (9–11). Scaling of the neutrino mass is often expected (the tau neutrino is the heaviest and the electron neutrino the lightest); the masses are then proportional to the first or second power of the corresponding charged-lepton (or quark) masses.

Neutrino mass in the eV range has dramatic cosmological and astrophysical ramifications. Based on the universally accepted hot big bang model, one predicts, in analogy with the 3-K microwave background, a background for each light ($m_\nu < 1$ MeV) stable neutrino, with the number density of $n_\nu + n_{\bar{\nu}} \approx 110$ cm^{-3}. In that case there are about 10^{10} times as many neutrinos as baryons, and neutrinos heavier than about 1 eV could dominate the total mass of the universe. From observation of the present expansion rate of the universe, one obtains an upper limit for the total average mass density that translates into the condition $\sum m_\nu < 200$ eV (see, for example, 12), where the summation is over all flavors of light stable neutrinos. The only assumption used in deriving this upper limit is that the cosmological constant vanishes (13).

Massive neutrinos could become gravitationally bound to galaxies or galactic clusters. In that case neutrinos of $m \approx 10$ eV would account for the apparent large excess of the dark matter over the luminous matter (14, 15).

Massive neutrinos would also play an important role in the theory of formation and development of inhomogeneities in the universe, leading to superclusters, clusters, and individual galaxies (16).

Several proposals have been made for the detection of the background neutrino sea. Recent analysis (17, 18) shows, however, that all of them lead to immeasurably small effects even when we include the possible local density enhancement of up to 10^5 due to the gravitational binding of the neutrinos.

The "solar neutrino puzzle" is often mentioned in connection with the neutrino oscillation problem. In the experiment by Davis et al (19) based on the $^{37}\text{Cl}(v, e^-)^{37}\text{Ar}$ reaction with the 814-keV threshold only 1/4 to 1/3 of the expected neutrino flux (20) is observed. Maximum oscillations among three neutrino flavors with a wavelength less than the Sun-Earth distance would indeed reduce the flux of the electron neutrinos reaching the earth by the factor of three. Alternative explanations of this puzzle, however, are not exhausted (21).

1.2 Dirac and Majorana Neutrinos

Massive charged leptons, such as electron, muon, or tau, are easily distinguished from their antiparticles. They are described by four-component spinors and there is only one Lorentz-invariant and charge-conserving expression possible for the Lagrangian mass term:

$$L_\text{D} = m_\text{D}\bar{\psi}\psi. \qquad 1.$$

This mass term obviously conserves the lepton number.

The situation is more complicated for neutral fermions, such as neutrinos, because Lorentz invariance alone also allows another mass term

$$L_\text{M} = m_\text{M}(\bar{\psi}^c\psi + \bar{\psi}\psi^c), \qquad 2.$$

where ψ^c is the charge-conjugated spinor. The term L_M changes a neutrino into an antineutrino and thus violates lepton-number conservation. Particles described by L_D are Dirac neutrinos (distinct from their antiparticles); those described by L_M are Majorana neutrinos (identical with their antiparticles). Only two components of the Majorana spinor are independent. A Dirac neutrino is formally a special case of two Majorana neutrinos with identical masses and opposite CP eigenvalues (e.g. 22, 23).

The distinction between the Dirac and Majorana neutrinos becomes important only if $m \neq 0$ or if both left-handed and right-handed currents participate in weak interactions. In particular, the neutrinoless double beta decay, which violates lepton-number conservation, becomes possible for Majorana neutrinos. Most grand unified theories predict massive Majorana neutrinos.

1.3 Neutrino Mixing and Oscillations

In the standard electroweak theory each charged lepton ℓ^- has its left-handed neutrino partner ν_ℓ. The neutrinos ν_ℓ are weak eigenstates, but are not necessarily states with a definite mass. That means that the mass term discussed in Section 1.2 is generally not diagonal in ν_ℓ. One can define a unitary mixing matrix U,

$$\nu_\ell = \sum_{i=1}^{N} U_{\ell i} \nu_i, \qquad 3.$$

where ν_i are states of a definite mass (the mass term is diagonal in ν_i), and $N \geq 3$ is the number of generations (flavors). It is customary to order the ν_i in such a way that $U_{\ell i}$ is as nearly diagonal as possible, and one can then use the approximate terms "electron neutrino mass," etc.

When matrix $U_{\ell i}$ is not exactly diagonal, we are led to the concept of neutrino oscillations. Let ν_ℓ be created by weak charged-current reaction at $t = 0$ with momentum p. The time development of such a state is given by

$$\nu_\ell(t) = \sum_{i=1}^{N} U_{\ell i} \exp\left[-i\left(p + \frac{m_i^2}{2p}\right)t\right], \qquad 4.$$

provided $p \gg m_i$.

The different components of Equation 4 have time-dependent phases leading to typical interference effects. In particular, the probability that one encounters a weak eigenstate ℓ' after time t equals

$$P_t(\ell \to \ell') = \sum_{i=1}^{N}\left[|U_{\ell i}|^2 |U_{\ell' i}|^2 + \sum_{j \neq i} U_{\ell i} U_{\ell j} U^*_{\ell' i} U^*_{\ell' j} \exp\left(-i\frac{m_i^2 - m_j^2}{2p}t\right)\right]; \qquad 5.$$

that is, this probability is an oscillating function of time t, or of distance $L = ct$. Such an effect is called neutrino oscillation and requires nonvanishing, nondegenerate neutrino masses, and at least some nondiagonal matrix elements in U.

Study of oscillations does not furnish the neutrino masses themselves but the quantity $\Delta m^2 = |m_j^2 - m_i^2|$, obtained from the wavelength

$$L_{\text{osc}}(m) = \frac{2.5 \times E \text{ (MeV)}}{|m_i^2 - m_j^2| \text{ (eV)}^2} \qquad 6.$$

associated with each pair of i, j neutrinos with masses m_i, m_j. Mixing coefficients $U_{\ell i}$ are obtained from the oscillation amplitudes. Note that the oscillation pattern depends on L/E, the ratio of the distance to the neutrino energy. In an experiment both the L dependence and the E dependence can be used to explore oscillations.

Neutrino oscillations described above are "flavor" oscillations; the electron, muon, etc numbers are no longer conserved, but their sum (the total lepton number) is still conserved. Neutral-current weak interactions are not affected by the "flavor" oscillations. Observation of such oscillations would mean that at least some neutrinos are massive; no distinction between Dirac and Majorana neutrinos could be made, however.

For Majorana neutrinos, oscillations of the "second" class are possible (24); they also affect the neutral current and violate the total lepton number. In such a case a neutrino beam can produce antileptons (and an antineutrino beam can produce leptons). The probability of such a $\Delta L = 2$ process is, however, suppressed by the helicity factor $(m_\nu/E)^2$ and becomes essentially unobservable (4, 25). Processes with $\Delta L = 2$ also become possible if weak interactions explicitly involve right-handed lepton currents, as discussed in Section 3.5.

1.4 Neutrino Decay

If neutrinos have mass, the heavier ones could decay into the lighter ones. Neutrino decay has never been seen; however, if it were observed it would give information on the masses, and, because at least two neutrino flavors are involved, on neutrino mixing.

The radiative decays

$$\nu_2 \to \nu_1 + \gamma \qquad\qquad 7.$$

$$\nu_2 \to \nu_1 + e^+ + e^- \qquad\qquad 8.$$

are generally considered the most likely candidates (26–28). The photon mode (Equation 7) has its rate suppressed by the factor $(m_\ell/m_W)^4$ (m_ℓ is the charged-lepton mass, m_W is the W mass; these particles appear as intermediate states in the corresponding Feynman graphs), and for the Dirac neutrino one obtains

$$\Gamma_\gamma = \frac{\alpha}{2}\left[\frac{3 G_F}{32\pi^2}\right]^2 m_{\nu_2}^5 \left|\sum_\ell U_{\ell 2} U_{\ell 1}^* \frac{m_\ell^2}{m_W^2}\right|^2. \qquad 9.$$

Substituting the tau mass for m_ℓ we find

$$\Gamma_\gamma \approx (10^{29}\ \text{years})^{-1} \left(\frac{m_{\nu_2}}{30\ \text{eV}}\right)^5 |U_{\tau 1} U_{\tau 2}^*|^2. \qquad 9a.$$

Various proposals to speed up the decay have been discussed (e.g. 26, 29); they involve a very heavy fourth lepton and other assumptions. The decay rate of Majorana neutrinos is more difficult to calculate (29).

In a laboratory experiment one tries to observe the decay of moving, usually highly relativistic, neutrinos. The laboratory decay rate Γ^{Lab} is related to the invariant center-of-mass rate by the time dilatation factor:

$$\Gamma^{\text{Lab}} = \frac{m_v}{E_v} \Gamma^{\text{CM}}.$$

An experiment thus furnishes a value, or an upper limit, of the product $m_v \Gamma_\gamma$. Existing experimental limits are much poorer than the theoretical estimates of Equation 9. For electron neutrinos one obtains (30)

$$m_{v_e} \Gamma_\gamma \leq 3 \times 10^{-3} \text{ eV/s},$$

which is about 10^{35} times greater than Equation 9a for $m_v = 30$ eV and maximum mixing. Nevertheless, the corresponding decay length exceeds the Sun-Earth distance by a factor of 10^5, excluding decay as a possible explanation of the solar neutrino puzzle. For muon neutrinos the corresponding limit is $m_v \Gamma_\gamma < 0.11$ eV/s (31).

The decay into the electron-positron pair (Equation 8) can proceed only when $m_v > 2m_e$. Its rate is much faster (28) than the photon emission mode,

$$\Gamma_{ee} = \frac{G_F^2}{192\pi^2} m_{v_2}^5 |U_{e2} U^*_{e1 = v_e}|^2 h\left(\frac{m_e}{m_{v_2}}\right), \qquad 10.$$

where h is a phase-space factor such that $h(0) = 1$, $h(0.5) = 0$. We know from the neutrino oscillation searches described below that

$$|U_{e2} U_{e1 = v_e}|^2 \leq 0.05$$

and thus

$$\Gamma_{ee} \leq (6 \times 10^5 \text{ s})^{-1} \left(\frac{m_{v_2}}{1 \text{ MeV}}\right)^5.$$

The experimental and theoretical limits for neutrino lifetime are summarized in Figure 1.

Other decay modes, such as $v \to 3v'$ have also been considered but are typically even slower or their description involves additional assumptions.

Astrophysical considerations allow one to exclude certain neutrino mass-lifetime combinations independently of the theoretical decay rates (12, 32). Such considerations use available data on the microwave and diffuse photon backgrounds, supernova energetics and emission, etc. As described for example by Turner (12), one is left with only three allowed "corners" of the neutrino mass-lifetime space: long-lived light neutrinos ($m_v < 200$ eV, $\tau > 10^{22}$ s for the maximum mass), long-lived heavy neutrinos ($m_v > 1$ GeV, $\tau > 10^{24}$ s), and short-lived heavy neutrinos ($m_v > 10$ MeV, $\tau < 10^2$ s for the minimal mass). The neutrino lifetime boundaries based on astrophysical arguments are also shown in Figure 1.

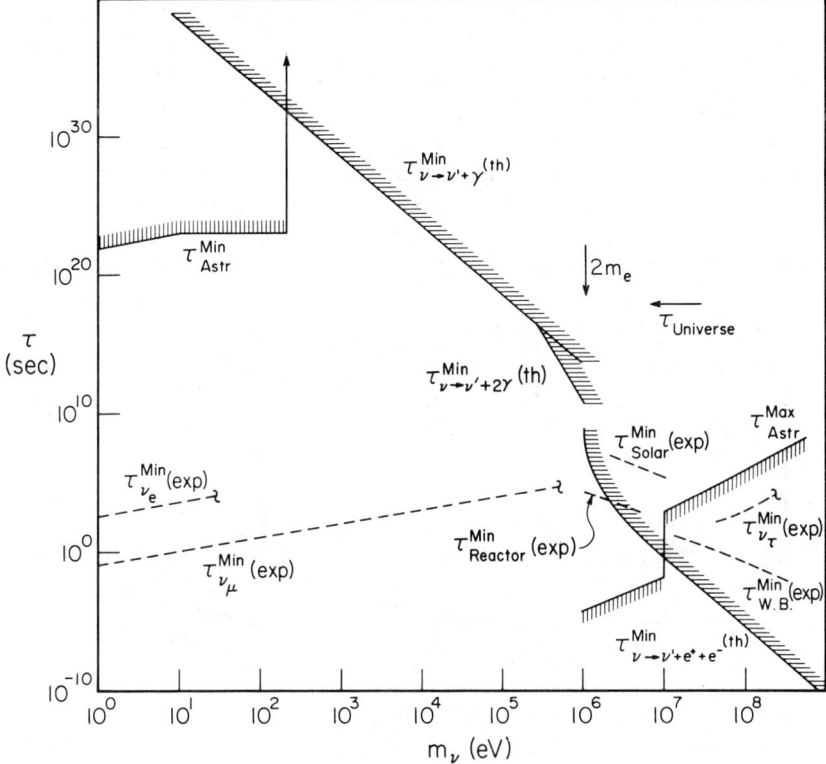

Figure 1 Theoretical and experimental limits of neutrino lifetime: $\tau(\text{th})$ is the lifetime of Equations 9 and 10 for maximum mixing, where one assumes that τ is the heaviest charged lepton; the horizontally shaded regions are allowed. The theoretical lifetime for 2γ decay (96) is also shown. The τ_{Astr} are boundaries based on astrophysical arguments (12); vertically shaded regions are allowed. Experimental lower limits for the dominantly coupled neutrinos extend to the upper limits of the corresponding mass: τ_{ν_e} (30), τ_{ν_μ} (31), τ_{ν_τ} (33). The terms τ_{Solar} (34), τ_{Reactor} (95), and $\tau_{\text{W.B.}}$ (33) are limits for less strongly coupled heavy neutrinos.

1.5 *Summary of Present Experimental Evidence*

To date there is no confirmed evidence that neutrinos have finite mass. A reported deviation in the beta decay endpoint in ^{3}H, if confirmed, may yet indicate a mass in the range of 20–30 eV. Oscillation experiments with low-energy neutrinos from a reactor provide an upper limit for the mass parameter $\Delta m^2 = |m_2^2 - m_1^2| \leq 0.016$ eV^2, if we assume the maximum value of the mixing strength $\sin^2 2\theta$, as well as an upper limit of $\sin^2 2\theta \leq 0.16$ for $\Delta m^2 > 0.05$ eV^2. As to admixtures from heavy (>1 keV) neutrinos, experiments involving two- and three-body decays provide limits of about

10^{-2} to 10^{-3} for the strength of admixture of $m_v = 1$–100 keV neutrinos, and still better limits for neutrinos with $m_v > 100$ keV. If neutrinos are Majorana particles, lepton-number-violating double beta decay may occur. This process has not been observed at the present level of sensitivity, which leads to the conclusion that neutrino mass cannot exceed a value in the range of 5–16 eV, or that right-handed weak currents cannot contribute more than 2–6×10^{-5} with respect to the left-handed current. The smallness (or absence) of neutrino mass is consistent with cosmological bounds confining the heaviest neutrino to masses of less than 200 eV. The solar neutrino puzzle is not inconsistent with the absence of oscillations in laboratory experiments, and may still indicate oscillations with large mixing and small (<0.01 eV^2) mass parameters.

2. OSCILLATION EXPERIMENTS WITH REACTOR NEUTRINOS

2.1 The Detection Reaction

A number of experiments to study neutrino oscillations have been carried out with low- and high-energy neutrinos (5, 35). So far, no evidence for neutrino oscillations has been found. With this in mind, a simple analysis in terms of two neutrino states is therefore appropriate. This description includes two parameters only, a mixing angle θ ($\sin \theta = U_{12} = -U_{21}$; $\cos \theta = U_{11} = U_{22}$) and a mass parameter $\Delta m^2 = |m_2^2 - m_1^2|$.

Low-energy experiments involving electron neutrinos are of the disappearance type since not enough energy is available to create the muon or tau lepton. Thus the probability for a $\bar{v}_e$ having disappeared by undergoing oscillations into another neutrino state (inclusive reaction) is given by the deviation from unity of $P(\ell \to \ell) = 1 - P(\ell \to \ell')$, where $P(\ell \to \ell')$ can be written for our case, following Equation 5, as

$$P(\ell \to \ell') = \frac{\sin^2 2\theta}{2} \left[1 - \cos \frac{2.53 \times \Delta m^2 \text{ (eV)}^2 \times L(m)}{E_v \text{ (MeV)}} \right]. \qquad 11.$$

The disappearance experiments thus test the effect of all oscillation channels (independent of the assumption of only two neutrino states). These experiments are sensitive to a large range of Δm^2; in particular, the lowest Δm^2 is obtained for low E_v values.

Nuclear reactors are powerful sources of electron antineutrinos because the fission fragments are unstable and undergo a series of beta decays. The antineutrinos thus are emitted with energies characteristic of nuclear beta decay. A neutrino detector positioned at varying distances from a reactor is capable of measuring the $\bar{v}_e$ yield as a function of energy and position, thus allowing one to verify both the E and L dependence of oscillations (Equation 11).

To detect the low-energy antineutrinos, the inverse neutron decay (36), $\bar{\nu}_e p \rightarrow e^+ n$, is well suited. This reaction, which can be identified by a time-correlated positron and neutron signature, has a cross section that is a function of the outgoing positron energy, given by

$$\sigma(E_e) = \frac{2\pi^2 \hbar^3}{m_e^5 c^7 f \tau_n} p_e E_e, \qquad 12.$$

where τ_n is the neutron mean life and f is the usual statistical function including the Coulomb correction for $Z = 1$. The outgoing positron and incoming antineutrino energies are related through

$$E_{\bar{\nu}} = E_e + (M_n - M_p). \qquad 13.$$

According to Wilkinson (37), $f = 1.6857$ and the recommended adjusted average neutron lifetime is $\tau_n = 900 \pm 9$ s. However, it should be noted that not all experimental data on τ_n are mutually consistent.

In an experiment at a nuclear reactor one actually measures the positron yield, which is (assuming no oscillations) given by

$$Y(E_e) \sim \sigma(E_e) n [E_{\bar{\nu}} = E_e + (M_n - M_p)], \qquad 14.$$

where $n(E_{\bar{\nu}})$ is the reactor antineutrino flux per unit energy.

With the present good accuracy ($<5\%$) in the experimental yield it is necessary to consider higher order terms in σ, such as neutron recoil corrections, weak magnetism, radiative corrections (bremsstrahlung), and higher order Coulomb terms (38).

Besides the $\bar{\nu}_e p \rightarrow e^+ n$ reaction, the charged-current reaction on the deuteron, $\bar{\nu}_e d \rightarrow e^+ nn$, and the competing neutral-current reaction, $\nu_e d \rightarrow \bar{\nu}_e p n$, have been studied, the latter being insensitive to oscillations. Based on the ratio of the total neutron yields in these reactions Reines et al (39) found indication for neutrino oscillation. In a later paper (40), however, the values were modified and the evidence for oscillations is no longer statistically significant.

2.2 The Gösgen Reactor Experiment

The Caltech-Munich-SIN group has conducted two experiments at the Gösgen reactor in Switzerland; one (41) with the detector at a distance of 38 m from the core and another one (42) at 46 m. Prior to these experiments, a measurement was made at the ILL reactor in Grenoble (France) at a distance of 8.7 m using a similar detector (43). The setup of the Gösgen experiment is sketched in Figure 2. The neutrinos were detected by the reaction $\bar{\nu}_e p \to e^+ n$ using a composite liquid scintillation detector and ^{3}He multiwire proportional chambers. A time-correlated e^+,n event constituted a valid signature.

Pulse shape discrimination in the scintillation counter has proved to be a powerful technique for eliminating correlated neutron background events. These events are caused by cosmic-ray-induced fast neutrons recoiling on protons in the liquid scintillation counter. The recoil gives rise to a scintillation counter trigger, followed, after a thermalization period, by a neutron capture signal in the ^{3}He counter. Neutrons associated with the reactor are entirely absent in these experiments.

About 11,000 neutrino-induced events were recorded in 6–9 months of reactor-on time. Backgrounds for each position were recorded during a one-month reactor-off period. Figure 3 shows the difference spectrum of reactor-on minus reactor-off for both positions, together with a curve representing the expected spectrum for no oscillations. The latter was obtained from the on-line beta spectroscopic measurements at the ILL reactor by Schreckenbach et al (44) studying ^{235}U and ^{239}Pu fission targets. These two isotopes account for about 89% of the total fission energy at the Gösgen reactor. The remaining 11% are due to fission of ^{238}U and ^{241}Pu.

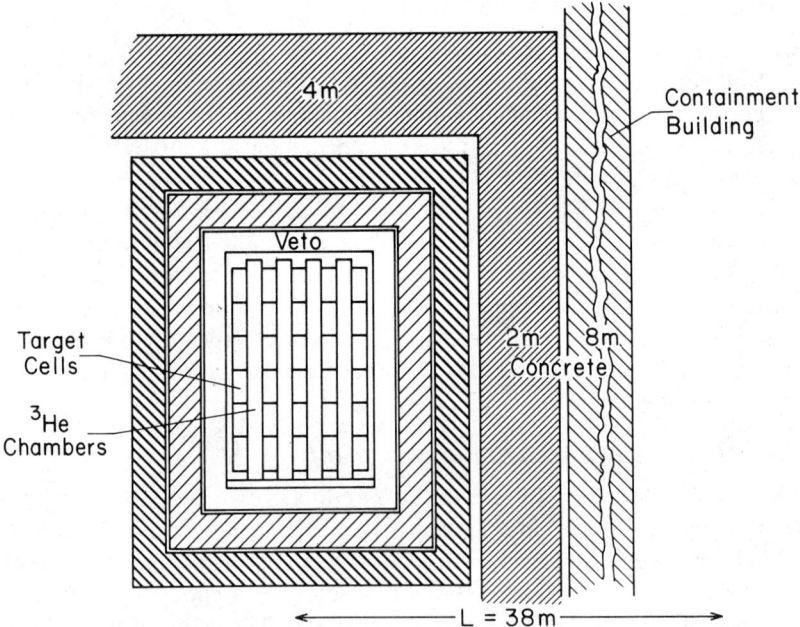

Figure 2 Experimental setup of the neutrino detector at the Gösgen reactor. (The drawing is not to scale.)

The calculations of Vogel et al (45) were used to evaluate the contribution to the antineutrino spectra from ^{238}U and ^{241}Pu. The variation in time of the contributions of each fissioning isotope is well known and was taken into account.

In Figure 4 the observed yields for the 38-m and 46-m positions at Gösgen (41, 42), and the 8.7-m position at ILL (43) are displayed, in units of the no-oscillation yield, as a function of L/E_v. As can be seen in Figures 3 and 4, there is good agreement between experiment and the expectation for no oscillations.

The data at two or three positions can also be analyzed without resorting to the no-oscillation spectrum. The exclusion plots of Figure 5 were obtained by considering the ratios of the data at 8.7, 38, and 46 m for each energy bin and fitting them to calculated ratios for various oscillation parameters. A χ^2 test to all possible values of Δm^2 and $\sin^2 2\theta$ resulted in the 90% confidence limits (CL) displayed in Figure 5. The analysis leading to the dashed curve in Figure 5 is entirely independent of the no-oscillation neutrino spectrum as well as the detector efficiency calibration. It has been concluded that there are no neutrino oscillations with parameters larger than those contained to the right of the curves in Figure 5.

Finally, an important question should be addressed: Where in the Δm^2 vs $\sin^2 2\theta$ plane should one continue to search for oscillations? Unfortunately, there is no guidance whatsoever from theory. As to the

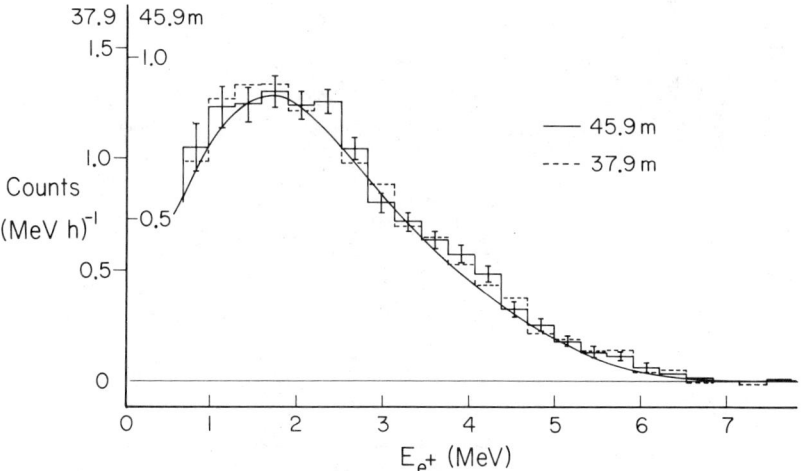

Figure 3 Results of the Gösgen experiments. Positron spectra obtained by subtracting reactor-off from reactor-on spectra for the 38-m and 46-m experiments. The energy bin is 0.305 MeV. The solid curve represents the predicted positron spectrum assuming no neutrino oscillations.

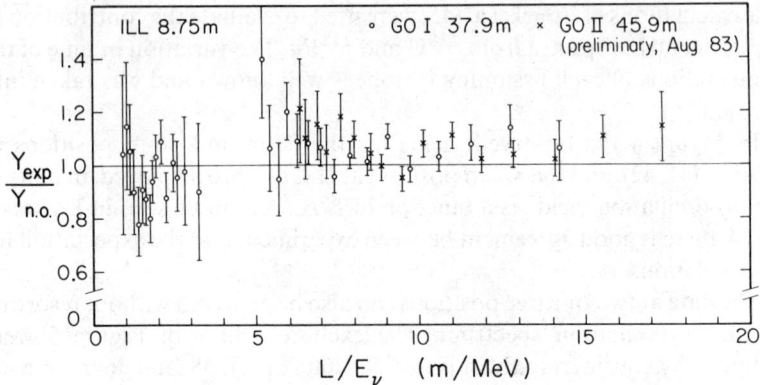

Figure 4 Ratio of experimental to predicted (for no oscillations) positron spectra at 8.7, 38, and 46 m from the reactor core. The errors of the data points shown are statistical (from 46).

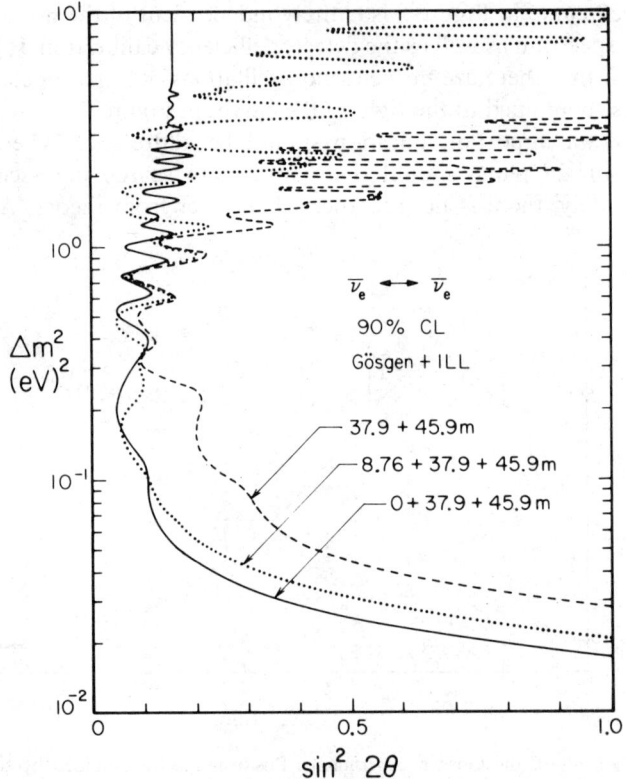

Figure 5 Exclusion plot obtained from the Gösgen and ILL experiment at 90% confidence limit. The curve labelled 37.9+45.9 m refers to the limit obtained by using the 37.9-m and 45.9-m data only, for example. The 0 m label represents the on-line beta spectroscopic measurement (see text).

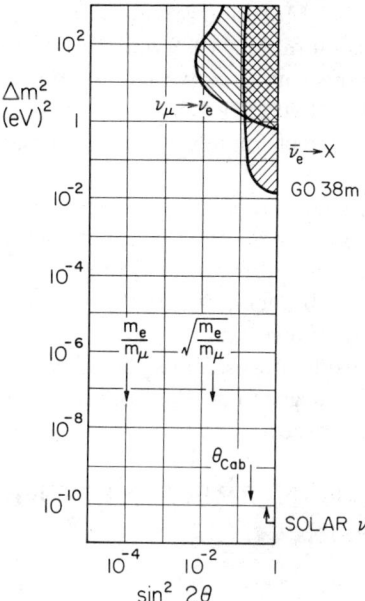

Figure 6 Expanded Δm^2 vs $\sin^2 2\theta$ plane showing current experimental limits and some dimensional guesses.

mixing angle, we can state that present limits are smaller than the Cabibbo angle. Figure 6 shows these limits, together with other possible dimensional guesses (lepton mass ratios). If the solar neutrino experiments are indeed telling us that neutrinos oscillate with large mixing angle, the Δm^2 values must lie between 10^{-2} and 10^{-10} eV^2, a region increasingly difficult to explore.

2.3 Fission Antineutrino Spectra

The above analysis of the reactor oscillation experiments is based on comparison of the positron yields measured at different distances and is therefore independent of the reactor antineutrino flux. Knowledge of this flux is, however, still important because it allows one to correct for the (weak) time dependence of the reactor fuel composition and for the (small) contribution of the ^{238}U and ^{241}Pu fission, and most importantly it makes it possible to determine the maximal allowed mixing angle for large Δm^2 (Figure 5).

Because in each beta decay both e^- and $\bar{\nu}_e$ are emitted with correlated energies, one can deduce the $\bar{\nu}_e$ spectrum from the experimentally determined electron spectrum. For the main reactor fuels, ^{235}U and ^{241}Pu, the electron spectra have been determined (44, 47) very accurately for kinetic energies up to ~ 7 MeV (and somewhat less accurately up to 8 MeV, where the spectrum is more than 1000 times weaker than its maximum).

When the electron spectrum is converted into the antineutrino spectrum, an uncertainty of 4–6% at 90% CL is introduced (47). It is important at this level of accuracy to correct for deviations from the allowed beta decay spectrum shape (38).

The reactor $\bar{\nu}_e$ spectrum can also be obtained by adding spectra of all beta branches of all fission fragments (see, for example, 45). This method, although in principle straightforward, requires knowledge of fission yields, beta decay Q values, and branching ratios. For short-lived fission fragments this information is often uncertain or unavailable and one must resort to nuclear structure considerations. Thus the uncertainties in the resulting spectrum are typically larger than those obtained from conversion of the experimental electron spectra.

3. DIRECT AND INDIRECT NEUTRINO MASS MEASUREMENTS

In this section we review the results of experiments on neutrino mass based on observation of charged particles emitted in weak decays. The momenta and energies of the charged particles, as well as the overall decay rate, are affected by the neutrino mass and mixing. A detailed account of the theory of weak decays with massive mixed neutrinos has been given by Shrock (48).

It is useful to distinguish between dominantly and subdominantly admixed neutrinos. While dominantly admixed neutrinos can be treated as constituting the principal mode, for both massive and massless neutrinos, weakly admixed neutrinos can be regarded as an additional decay channel open only for massive and mixed neutrinos. Neutrinoless double beta decay based on a measurement of the electron spectrum is also included in this section.

3.1 *Electron Antineutrino Mass from Beta Decay Endpoints*

In nuclear beta decay, as in any three-body decay $x \to y + \ell + \bar{\nu}_\ell$, there may be several endpoint energies E^i_{Max} each associated with a neutrino mass m_{ν_i}. They are related to each other by

$$E^i_{\text{Max}} = \frac{M_x^2 + m_\ell^2 - (M_y + m_{\nu_i})^2}{2M_x}. \qquad 15.$$

The corresponding Kurie plot $[N(E)/pEF(Z, E)]^{1/2}$ (or its analog) has kinks at each endpoint and an infinite slope at the maximal endpoint (minimal m_ν). In the neighborhood of each endpoint the neutrino is nonrelativistic; hence deviations of the spectrum from the shape corresponding to massless neutrinos are linear in m_ν.

Here we discuss the dominant mode associated with a single neutrino with mass m_v. The nucleus ^{3}H has the desirable features of a low Q value (to enhance the relative number of decays near the endpoint and to decrease the required relative resolution) and a reasonably short lifetime (to increase specific activity).

Recent results on the ^{3}H beta decay are summarized in Table 1. There is evidence for a nonvanishing electron antineutrino mass from the ITEP experiments (51, 53) but independent confirmation of this important result has yet to come forward.

One of the main problems in the ^{3}H experiments is the effect of the spectrometer resolution and response. In the mentioned experiments the resolution has been comparable or larger than the value or limit of m_v. In the neighborhood of the endpoint the finite spectrometer resolution R causes the slope of the Kurie plot to decrease, whereas the finite neutrino mass m_v causes it to increase. Thus an error ΔR in R results in an assignment of a fictitious neutrino mass related to the true mass m_v by

$$(m_v^{\text{fic}})^2 \approx 2\Delta R \cdot R + m_v^2.$$

An accurate knowledge of the full resolution function, including the effects of electron energy losses in the source, is therefore crucial. It is also necessary to take into account the natural width of the calibration lines in the determination of the spectrometer resolution function, as pointed out by Simpson (54).

Another problem, affecting the interpretation of the results, deals with the fate of the spectator electron originally bound to the ^{3}H atom or molecule. After the sudden change of the nuclear charge this electron does not end up in a single stationary quantum state. In particular, the

Table 1 $\bar{v}_e$ mass determination in ^{3}H decay

Author, Year (Ref.)	Source	Resolution (eV)	$m_{\bar{v}}$(eV), CL	$E_0{}^d$ (eV)
Bergkvist, 1972 (49)	^{3}H in Al	50	<55, 90%	18,623±16[e]
Tretjakov et al, 1976 (50)	valine[a]	45	<33, 90%	18,575±13
Lyubimov et al, 1980 (51)	valine	45	$14 \leq m_{\bar{v}} \leq 46$,[b] 99%	18,577±13
Simpson, 1981 (52)	^{3}H in Si (Li)	220	<60, 95%	18,543±5[f]
Boris et al, 1983 (53)	valine	20	33±1.1,[c] 60%	18,583.2±0.3[e]

[a] 3H-tagged $C_5H_{11}NO_2$.
[b] Limits include different final ^{3}He states. When the neglected intrinsic widths of the calibration lines are taken into account, the lower limit of $m_{\bar{v}} = 14$ eV is reduced to zero (53).
[c] For the theoretical final state of valine as calculated in (55); if a full range of final states is considered, one obtains model-independent lower limits of $m_{\bar{v}} \geq 20$ eV, $E_0 \geq 18,575$ eV at 95% CL (53).
[d] For ^{3}He$^+$ ground state.
[e] Corrected by 13.5 eV to transform from the measured average excitation energy.
[f] Corrected by 24 eV to transform from the measured atomic mass difference.

probability of finding the final ^{3}He ion in its ground state is only 0.6–0.7, depending on the chemical composition of the source, and the average excitation energy of the final state is 13–20 eV. The spectrum of final states can be reliably calculated for the free ^{3}H atom. This spectrum was computed for valine by Kaplan et al (55). Clearly this problem should be considerably alleviated if the resolution could be made smaller than the expected excitation energy of the final atomic or molecular complex.

The background level determines the minimal distance from the endpoint, that is, the maximal electron energy, where data still have statistical significance. There has been a significant reduction, by a factor of about 20, in the background level of the recent ITEP experiment (53) over the results of Lyubimov et al (51).

While the values of m_v in Table 1 are mutually consistent, the endpoints E_0 are not. The value of E_0 has been determined independently by Smith et al (56) by measurement of the ^{3}H-^{3}He atomic mass difference. The resulting value of $E_0 = 18549 \pm 7$ eV is in a significant disagreement with the ITEP result (53). It was recently pointed out (57) that the results of (53) imply that the true endpoint of the electron spectrum (i.e. where the spectrum would end if the resolution were a delta function) is $18580 - 33 \approx 18550$ eV, in agreement with (56) but at the same time indicating vanishing (or very small) neutrino mass.

The previous discussion dealt with the antineutrino mass. Similar studies of positron decays, and thus of the electron neutrino mass, are difficult because electron capture (EC) dominates over positron emission at low decay energies. Information on the electron neutrino mass could be extracted from the study of the endpoint region of inner bremsstrahlung accompanying electron capture (58). The bremsstrahlung arises at low energies mainly from p-capture, and its intensity is enhanced by the resonant process if the energy is not far from some p → s x-ray transition (58, 59). Tests (60) in the EC decay of ^{193}Pt ($Q = 56.6 \pm 0.3$ keV) set an upper limit of $m_{v_e} < 500$ eV for the electron neutrino mass. Electron capture in ^{163}Ho [$Q = 2.58 \pm 0.10$ keV (61), and $T = 4570 \pm 50$ years (62)] has attracted considerable interest as a candidate for further study; other possibilities are ^{158}Tb (63) and ^{157}Tb (64). At the present time the study of bremsstrahlung accompanying EC is considerably less sensitive to neutrino mass than the study of the electron spectrum in ^{3}H decay.

3.2 Muon and Tau Neutrino Mass Limits

In two-body decays, such as the pion decay $\pi \to \mu + v_\mu$, a value or an upper limit for the muon neutrino mass can be determined from kinematics. Again we assume that only one neutrino mass eigenstate is dominantly coupled to the muon.

In the pion rest frame the muon neutrino mass is related to the dominant (maximal) muon momentum by the quadratic dependence

$$m_{\nu_\mu}^2 = m_\pi^2 + m_\mu^2 - 2m_\pi(p_\mu^2 + m_\mu^2)^{1/2}. \qquad 16.$$

Thus in order to obtain a neutrino mass value it is necessary to determine with sufficient accuracy the muon momentum and pion and muon masses.

The study of the pion decay in flight allows one to reduce the absolute value of the neutrino momentum in the laboratory frame and thus increase sensitivity to the neutrino mass; the results are also less dependent on the precise knowledge of the pion mass.

At the present time both methods lead to virtually identical upper limits on the muon neutrino mass. In the pion decay at rest one obtains $m_{\nu_\mu} < 0.52$ MeV (90% CL) using the pion momentum determined by Daum et al (65) and the pion mass of Lu et al (66). A slightly better limit of 0.49 MeV (90% CL) was achieved recently (67). In the pion decay in flight experiment (68) the limit is $m_{\nu_\mu} < 0.50$ MeV (90% CL) and the uncertainty is dominated by the systematic errors in the determination of $p_\mu - p_\pi$ for the forward going muons ($p_\pi \approx 350$ MeV c^{-1}).

In decays having three particles in the final state, one can take advantage of the regime where the neutrino has a small momentum (see the discussion of beta decay above). However, until now only the study of the $K_{\mu 3}$ decay (69) led to a meaningful limit $m_{\nu_\mu} < 0.65$ MeV (90% CL).

Finally, to connect to cosmological and astrophysical considerations discussed in Sections 1.1 and 1.4, we note that the above limits of the muon neutrino mass are in the middle of the "forbidden" region.

The experimental limits on the tau neutrino mass are quite poor. Limits of $m_{\nu_\tau} < 250$ MeV (95% CL) were obtained by Bacino et al (70) and by Blocker et al (71).

3.3 Heavy Neutrino Admixing

We now explore the possibility that one or more heavy neutrinos may be admixed to a light neutrino state. Here we again assume that the weak interaction eigenstates v_ℓ are superpositions of mass eigenstates, v_i (Equation 3). Experimental evidence points to the fact that a state v_ℓ (such as v_e) is predominantly composed of one light neutrino v_i (such as v_1). If heavy neutrinos exist, their admixture must therefore be small. In this case $|U_{\ell,2 \text{ or } 3}| \ll 1$ and a decay proceeding via the heavy neutrinos has a branching ratio proportional to $|U_{\ell i}|^2$.

The two-body decays $K \to \pi v$, $\pi \to \mu v$, or $\pi \to e v$ offer sensitive tests to study these branches (48). In the lepton spectrum each mass eigenstate is expected to manifest itself as a monochromatic peak at some energy below the regular lepton peak associated with the light neutrino, with an intensity proportional to $|U_{\ell i}|^2$. Several spectroscopic experiments aimed at finding these secondary peaks have been carried out, but no evidence for a heavy neutrino decay has been reported so far. The best current limits for the mixing strength $|U_{\ell i}|^2$ from these experiments for admixtures to electron neutrinos ($\ell = e$) and muon neutrinos ($\ell = \mu$) are summarized in Figure 7. References are provided in the caption.

The study of three-body decays also lends itself to the search for heavy neutrino branches (48). As mentioned above, in nuclear beta decay a heavy neutrino would show up as a discontinuity in the electron spectrum, for example. Recent studies by Schreckenbach et al (73) of the electron spectra of ^{64}Cu and by Simpson (72) of the spectrum of ^{3}H provide limits for neutrino mixing in the mass range of 1 keV to 0.5 MeV (see Figure 7).

Limits for $|U_{\ell i}|^2$ from oscillation experiments are also shown in Figure 7. For $\nu_e \to x$ we use $|U_{\ell i}|^2 = (1/4)\sin^2 2\theta$, with $\sin^2 2\theta < 0.16$ (42); existing results from $\nu_\mu \to x$ are not relevant in the mass region shown.

As mentioned in Section 1.4, the mass region between 200 eV and 10 MeV is forbidden for stable or unstable neutrinos based on astrophysical and cosmological arguments.

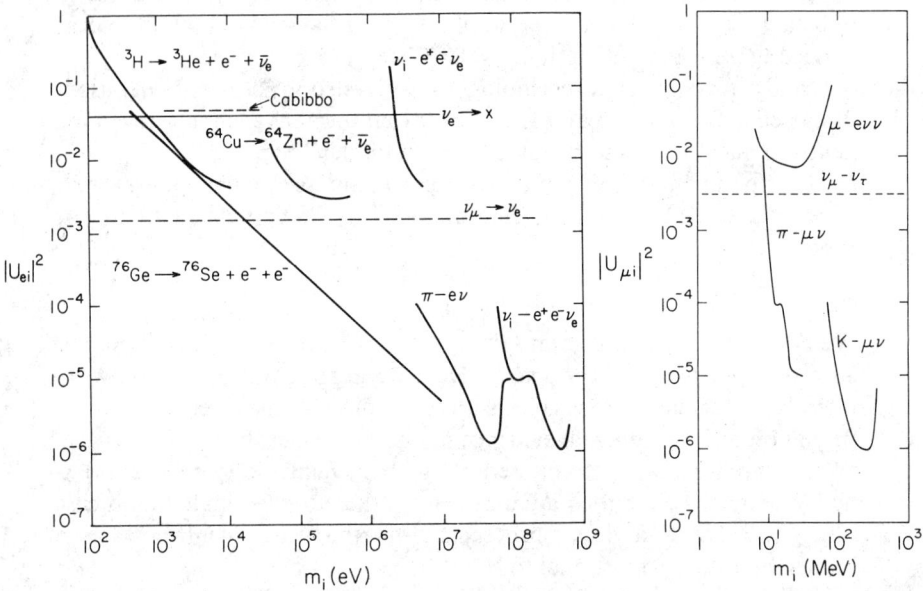

Figure 7 Limits for mixing coefficients $|U_{ei}|^2$ and $|U_{\mu i}|^2$ describing the admixture of heavy neutrinos to the electron neutrino and the muon neutrino, respectively. The regions above the curves are excluded. The curves are based on the following references: ^{3}H → ^{3}He+e$^-$ +$\bar{\nu}$ (72), ^{64}Cu → ^{64}Zn+e$^-$ +$\bar{\nu}$ (73), ^{76}Ge → ^{76}Se+e$^-$ +e$^-$ (74), $\pi \to e\nu$ (75), $\nu \to ee\nu$ (33, 95), $\mu \to e\nu\nu$ (48), $\pi \to \mu\nu$ (77), $K \to \mu\nu$ (78).

3.4 Neutrinoless Double Beta Decay

A sensitive source of information for neutrino mass and right-handed currents is the neutrinoless double beta decay, a semileptonic weak process of second order. Double beta decay proceeds from a nucleus Z to $Z+2$ and should become observable if the first-order process Z to $Z+1$ is energetically forbidden. There are two types of double beta decay: the two-neutrino decay, $Z \to (Z+2)+e_1+\bar{v}_1+e_2+\bar{v}_2$, and the zero-neutrino decay, $Z \to (Z+2)+e_1+e_2$. The former is expected to occur from standard theory; its study is of interest since it might help in estimating the value of the nuclear matrix elements needed to analyze the second process. The zero-neutrino process, if observed, would signal violation of lepton-number conservation, which can be associated with nonzero Majorana neutrino mass or right-handed weak currents (79, 80).

Figure 8 illustrates both the $2v$ and the $0v$ processes. The characteristic signature of the latter is a monochromatic peak in the spectrum of the total electron energy ε_0. The $0v$ decay could proceed by virtual neutrino exchange, as illustrated in the figure. This is a two-nucleon mechanism: a neutron n_1 emits an electron e_1 and a neutrino, and the latter is absorbed by a neutron n_2, which then emits an electron e_2. The process can only proceed

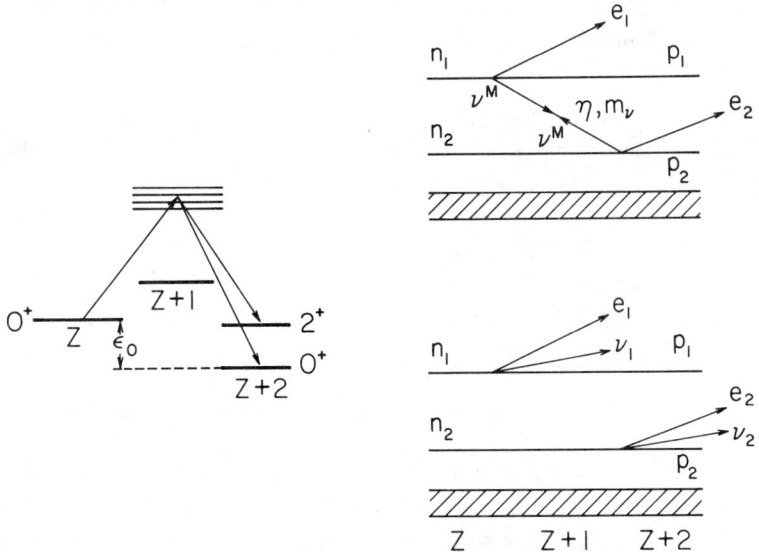

Figure 8 Illustration of double beta decay: (*top right*) neutrinoless, (*bottom right*) two-neutrino.

if the neutrino is a Majorana particle ($v^M \equiv \bar{v}^M$). In addition, in the standard theory angular momentum conservation prevents the $0v$ process from proceeding since only purely right-handed antineutrinos can be emitted, and only purely left-handed neutrinos can be absorbed. Two mechanisms have been identified (79, 80) that violate the perfect helicity of the neutrino and thus allow the $0v$ process to proceed:

1. The charged lepton current in the weak interaction has an explicit right helicity admixture given by

$$j_\lambda \approx e\gamma_\lambda[(1+\gamma_5)+\eta(1-\gamma_5)]v^M. \qquad 17.$$

We assume here that the hadronic current has its standard form. (In the standard theory $\eta = 0$ and $m_v = 0$, hence the subscript M in Equation 17 is not necessary, as explained in Section 1.2.)

2. The neutrino has a mass so that a "wrong" helicity amplitude $\sim m_v/E$ arises even though the interaction is purely "left-handed."

In the discussion below we assume that neutrinoless double beta decay can be described by the above two parameters, the right-handed current parameter η and the Majorana neutrino mass m_v. Other descriptions have been presented in literature, including "quasi-Dirac" neutrinos or the existence of more than one neutrino (81). As to the nuclear aspect, it is assumed that the two-nucleon process sketched above dominates the decay probability and far outweighs the N* mechanism (80). In calculating the rate, the summation over the nuclear intermediate states is carried out in the closure approximation.

TRANSITION PROBABILITIES   The $2v$ process can proceed with massless Dirac or Majorana neutrinos and its rate is given for $0^+ \to 0^+$ transitions by

$$\Gamma^{2v} \approx F_{2v}(\varepsilon_0)C(\varepsilon_0, Z)\left(|M_{GT}|^2 \bigg/ \left\langle \Delta E_N + \frac{\varepsilon_0}{2} + m_e \right\rangle^2\right), \qquad 18.$$

where $F_{2v}(\varepsilon_0)$ is the 4-fermion phase-space factor; $C(\varepsilon_0, Z)$ is the Coulomb function; M_{GT} is the appropriate second-order Gamow-Teller-type matrix element $\langle f \| \sigma^{(1)} \cdot \sigma^{(2)} \tau_\pm^{(1)} \tau_\pm^{(2)} \| i \rangle$ describing the virtual transitions via states in $Z+1$ (see Figure 8); and ΔE_N is the average nuclear energy difference between these states and the initial state. The main uncertainty in this decay rate stems from $|M_{GT}|^2$ and may be as large as a factor of $10^{\pm 2}$. A rough estimate for the $2v$ half-life yields $T_{1/2}^{2v} \approx 10^{22 \pm 2}$ y.

A similar estimate can be made for the $0v$ process. Both right-handed current (RHC) and mass mechanisms have been considered (79). For a transition between ground states ($0^+ \to 0^+$) this rate is given by

$$\Gamma^{0\nu} \approx F_{0\nu}(\varepsilon_0) C(\varepsilon_0, Z) \left| \frac{M_{GT}}{\langle r_{ij} \rangle m_p} \right|^2. \qquad 19.$$

The quantity $F_{0\nu}(\varepsilon_0)$ is the 2-fermion phase-space factor; it contains terms in neutrino mass ($m = m_\nu/m_e$) and in RHC (η) in the following form,

$$F_{0\nu}(\varepsilon_0) = m^2 f_m(\varepsilon_0) + m\eta f_{m\eta}(\varepsilon_0) + \eta^2 f_\eta(\varepsilon_0). \qquad 20.$$

The term f_m has the same energy dependence as in allowed beta decay (s-wave), and f_η has higher powers in energy reflecting the momentum transfer dependence in the RHC process. The quantity | | in Equation 19 represents the Gamow-Teller matrix element divided by an average nucleon separation distance $\langle r_{ij} \rangle$ measured in proton Compton wave length $1/m_p$. The $\langle r_{ij} \rangle$ appears because the virtual neutrino is exchanged between two nucleons within the same nucleus. Here a rough estimate gives $T_{1/2}^{0\nu} \approx 10^{15 \pm 2} \eta^{-2}$ (or m^{-2}) y.

We note that 0ν decay is $\sim 10^7$ times faster than the corresponding 2ν decay if $\eta = 1$ or $m = 1$. Part of this difference can be understood if in Equations 17 and 18 one retains only the leading ε_0 powers and omits all (non-numerical) common factors. This yields $\varepsilon_0^4/\Delta E_N^2$ for $\Gamma^{2\nu}$ and $(m_p/\langle r_{ij} \rangle m_p)^2$ for $\Gamma^{0\nu}$. Using $\langle r_{ij} \rangle m_p \approx \Delta E_N/m_e$ (m_p was in fact introduced into Equation 19 in view of this), one obtains a factor of $\sim 10^4$ in the ratio of the rates. The remaining difference stems from numerical factors in $F_{0\nu}$ and $F_{2\nu}$.

Selection rules could help distinguish between the mass and RHC mechanisms (79, 80). For the mass term, only $0^+ \to 0^+$ transitions are allowed, while for the RHC term, one can have $0^+ \to 0^+, 1^+, 2^+$. (This has to do with the additional momentum transfer dependence of the transition amplitude.) Thus, if a $0^+ \to 2^+$ branch were observed, it would give evidence for right-handed currents.

DISCUSSION OF EXPERIMENTAL EVIDENCE Double beta decay has been studied in several nuclei both by geochemical techniques [extraction of the daughter ($Z+2$) from the parent (Z) in an old ore] and with counters. Geochemical experiments cannot, of course, distinguish between 0ν and 2ν decays. We discuss below the results for some selected cases that are particularly suited for sensitive tests. For a more complete review see, for example, (81).

Te ratio In calculating the rates for double beta decay, the largest uncertainty stems from the nuclear matrix element M_{GT}. The idea of eliminating the matrix elements by comparing two isotopes presumed to have similar nuclear structure is therefore attractive. The ratio of the half-lives is then given by the ratio of the phase-space factors $F_{2\nu}$ and $F_{0\nu}$ for the two isotopes. Such a comparison has been made for ^{128}Te and ^{130}Te. In the 2ν process, one expects $T_{1/2}^{130}(2\nu)/T_{1/2}^{128}(2\nu) = 1.8 \times 10^{-4}$. For the 0ν process, this ratio is always larger and its exact value depends on the parameters m_ν and η. The geochemical work by Kirsten et al (82) gives $T_{1/2}^{130}/T_{1/2}^{128} = (1.0 \pm 1.1) \times 10^{-4}$, in agreement with the 2ν prediction and compatible with the absence of 0ν decay. The upper limits for m_ν and η (ignoring neutrino mixing) are $m_\nu < 5$ eV and $\eta < 2 \times 10^{-5}$, independently of the specifics of the calculation, aside from the assumed equality of the matrix elements. The ratio of Kirsten et al, however, disagrees with another geochemical ratio, obtained by Hennecke et al (83), $T_{1/2}^{130}/T_{1/2}^{128} = (6.3 + 0.2) \times 10^{-4}$. This result could be interpreted as requiring 0ν decay to occur and thus would constitute evidence for lepton-number violation with either $m_\nu \approx 10$ eV or $\eta \approx 5 \times 10^{-5}$. Clearly, no strong case can be made for or against lepton-number nonconservation until this discrepancy is resolved.

^{130}Te The half-life of the decay of ^{130}Te → ^{130}Xe, averaged over the existing geochemical experiments (84) (with large scatter of the data) is 2.6×10^{21} y. Interpreted as 2ν decay, this yields a nuclear matrix element $M_{GT}(2\nu) = 0.24$ (80). Haxton et al (85) calculated this decay and find $M_{GT}(2\nu) = 3.0$ in the same units, i.e. a lifetime of $T_{1/2}^{2\nu} \approx 1.7 \times 10^{19}$ y, about 150 times shorter than the observed one. It should be noted that the geochemical determinations depend critically on the so-called gas retention age of the minerals. Any error in this age would result in an increase in the deduced half-life with respect to the true one.

Notwithstanding the disagreement between the predicted and observed half-lives and setting aside the fact that additional assumptions are needed to relate the nuclear matrix elements for 2ν and 0ν decays, we shall obtain a crude estimate of m_ν based on the hypothesis that *some* fraction of the decay is due to a 0ν process induced by a finite m_ν. Since this fraction is unknown we assume that it is 50%, and furthermore that the ratio of the relevant matrix elements is that calculated by Haxton et al. This set of assumptions yield $m_\nu = 130$ eV; clearly, if the fraction due to 0ν decays is less than half, a correspondingly lower value of m_ν is derived. These arguments are illustrated in Figure 9 and summarized in Table 2.

^{82}Se Geochemical measurements (84) provide an average lifetime value for double beta decay of about 1.5×10^{20} y (with an estimated uncertainty of about 10%). Again, the theoretical prediction (85) disagrees with the geochemical lifetime, as summarized in Table 2 and Figure 10. On the other hand, a cloud chamber experiment (87) gives a 2ν lifetime 15 times shorter than the geochemical one; however, with relatively large uncertainty. The

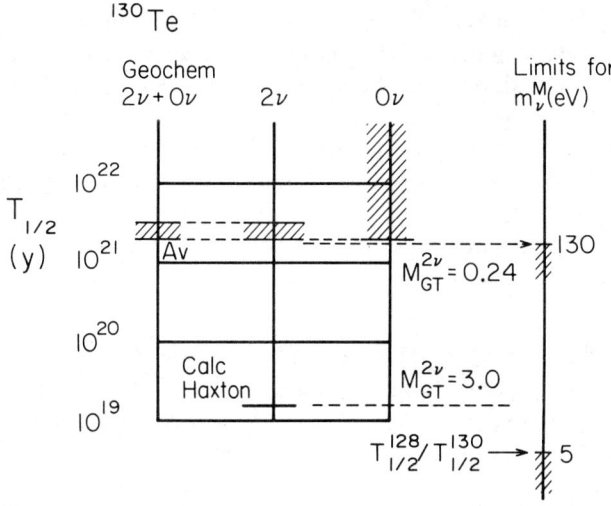

Figure 9 The geochemical and calculated half-lives and inferred neutrino mass limits for ^{130}Te (see text and Table 2 for explanations).

Table 2 Summary of selected double beta decay results

	Experiment		Calculation	
	Geochemistry (Ref.)	Laboratory (Ref.)	Doi et al (80)	Haxton et al (85, 86)
^{76}Ge				
$T_{1/2}(2\nu)$ (y)			2.3×10^{21}	3.7×10^{20}
$T_{1/2}(0\nu)$ (y)		$> 3.7 \times 10^{22}$ (90)	9.4×10^{22a}	
m_ν (eV)			< 16	< 7
^{82}Se				
$T_{1/2}(2\nu)$ (y)	1.5×20^{20} (84)	$(1.0 \pm 0.4) \times 10^{19}$ (87)	1.5×10^{20}	1.7×10^{19}
$T_{1/2}(0\nu)$ (y)		$> 3.1 \times 10^{21}$ (88)	3.2×10^{22a}	
m_ν (eV)			< 33	< 12
^{130}Te				
$T_{1/2}(2\nu)$ (y)	2.6×10^{21c}		2.6×10^{21}	1.7×10^{19}
$T_{1/2}(0\nu)$ (y)			2.5×10^{23a}	
m_ν (eV)			< 130	
$^{130/128}$Te				
$T_{1/2}^{130/128}$	$(1.0 \pm 1.1) \times 10^{-4}$ (82)			
m_ν (eV)[b]			< 5	< 5

[a] Assuming $m_\nu = 10$ eV.
[b] Or $\eta < 2 \times 10^{-5}$.
[c] Average value as quoted in (80).

experimental search for 2ν decay is continuing with an improved apparatus (89) and should help in clarifying the existing discrepancy. It is to be noted that for this transition there exists (88) an experimental limit $T_{1/2}^{0\nu} > 3.1 \times 10^{21}$ y. In analogy with the procedure applied to ^{130}Te, we may assume, using this limit, that $\leq 5\%$ of the geochemical rate is due to 0ν decay. With the same hypotheses as before, and dismissing the cloud chamber result, we obtain then $m_\nu \leq 32$ eV. An alternative approach is to discard the geochemical result, and to rely on theoretical matrix elements, which agree, for the 2ν mode, with the cloud chamber result. This approach yields $m_\nu \leq 12$ eV.

^{76}Ge In the case of ^{76}Ge there now exist several sensitive laboratory results giving tight bounds on $T_{1/2}(0\nu)$. Since there are no geochemical data from which to extract the matrix elements, one has to rely on the calculations by Haxton et al (86), possibly as modified by Doi et al (80). Using only the theoretical matrix elements of (86), and the best current laboratory limit (88) for the $0^+ \rightarrow 0^+$ transition of $T_{1/2}^{0\nu} > 3.7 \times 10^{22}$ y, one obtains $m_\nu < 7$ eV. Recalling the discrepancy, in the case ^{130}Te and ^{82}Se between the geochemical and theoretical 2ν rates, one may, following (80), "scale down" the theoretical matrix element by the factor corresponding to the discrepancy in ^{82}Se (the nuclide closer to ^{76}Ge), and obtain $m_\nu \leq 16$ eV (see Table 2). The half-life limit for the $0^+ \rightarrow 2^+$ branch is 4×10^{21} y (90).

A high-resolution Ge detector is an ideal instrument for obtaining

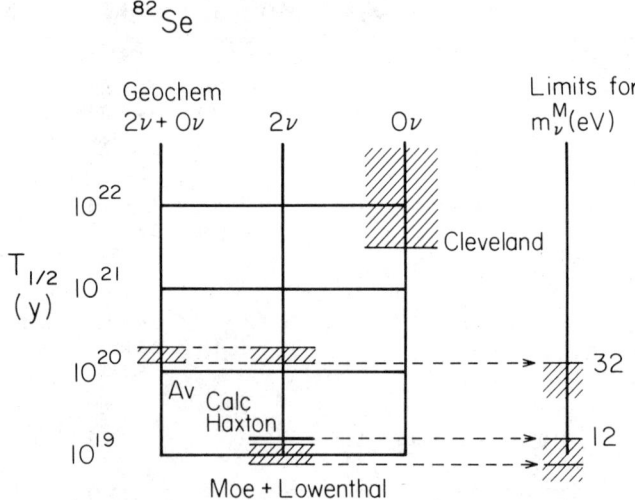

Figure 10 The geochemical and calculated half-lives and inferred neutrino mass limits for ^{82}Se (see text and Table 2 for explanations).

sensitive limits for 0ν decay. ^{76}Ge occurs in germanium with a natural abundance of 7.8%. Fiorini and his group pioneered the Ge experiments and the quoted best current upper limit has been reported by Bellotti et al (90) from an experiment in the Mont Blanc tunnel. Other laboratories (91–93) have also reported results.

Figure 11 depicts the experimental arrangement of the Caltech (93) experiment. The Ge detector is shielded with Cu and Pb and surrounded by a radon tight can. A veto counter serves to reduce cosmic ray background.

The principal limitations for these experiments are detector size and, even more important, detector background. One of the principal sources of the background in the region of the decay energy ε_0 is the Compton contribution of the 2.6-MeV gamma ray accompanying ^{208}Tl decay, a ubiquitous natural contamination. In the Caltech experiment (93) this contamination has been virtually eliminated. Other background components come from cosmic rays. They can be reduced by a veto system, as illustrated in Figure 11. However, high-energy bremsstrahlung and neutrons are not vetoed and to reduce these components one must install the experiment in an underground site, as Bellotti has shown.

Figure 12 illustrates a portion of the spectrum from the Caltech experiment. After 3820 h of running time there is no evidence for a peak at 2.04 MeV. From the number of counts, N, in a 3-keV interval (the detector resolution) and its fluctuation, $\sqrt{N}$, one obtains a 1σ limit for the 0ν lifetime of $T_{1/2}^{0\nu} > 1.9 \times 10^{22}$ y.

As to the future of the ^{76}Ge studies, it is safe to predict that ongoing efforts will stretch the sensitivity for $T_{1/2}^{0\nu}$ to about $T_{1/2}^{0\nu} > 10^{23}$ y, which corresponds to a mass limit of $m_\nu < 10$ eV. To progress substantially below 10 eV, much larger sample sizes will be needed. The largest currently planned Ge experiments envision detectors of about 1000 cm^3, or 3×10^{24}

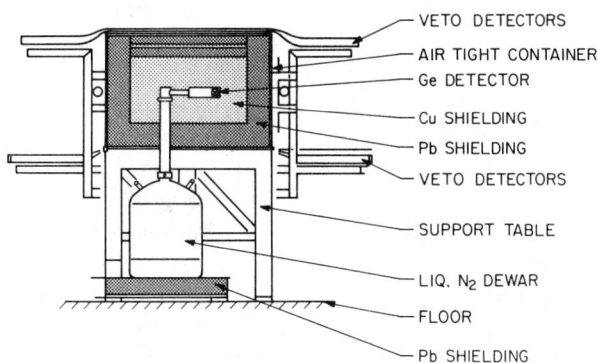

Figure 11 Ge detector setup for the Caltech double beta decay experiment.

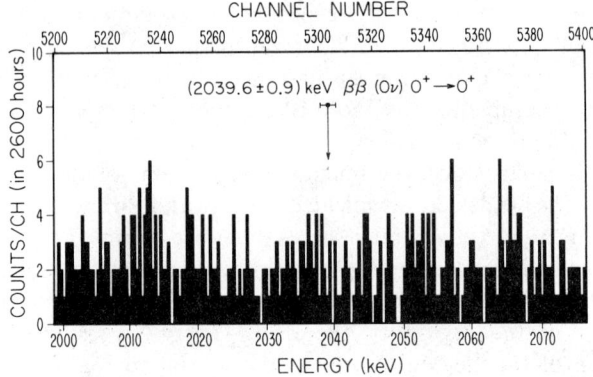

Figure 12 High-resolution Ge spectrum near total decay energy $\varepsilon_0 = 2040$ keV.

atoms. In comparison, a large Xe time projection chamber (TPC) may have in excess of 10^{26} atoms of ^{136}Xe ($\varepsilon_0 = 2.5$ MeV). Both liquid Xe (H. Chen, private communication) and pressured gas Xe (94) TPCs have been proposed. Depending on how well correlated electron tracks with energies up to 2.5 MeV can be identified, these detectors may allow the exploration of neutrino mass down to 1 eV or below.

4. OUTLOOK

Neutrino physics at low energies is capable of providing sensitive tests for neutrino mass and neutrino mixing in a manner complementary to the efforts in high-energy physics.

As to the $\bar{\nu}_e$ mass, the recent much-publicized value of about 30 eV, if confirmed, would provide an important cornerstone for physics, astrophysics, and cosmology. Several independent experiments now underway to study the ^{3}H spectrum with improved resolution are therefore eagerly awaited.

Further progress in improving the sensitivity for inclusive oscillations of reactor neutrinos is expected to be slow, as it can only be accomplished with very much larger detectors. With a detector ten times larger, for example, the sensitivity for Δm^2 for full mixing may be as good as 0.005 eV^2, as compared to the present value of 0.016 eV^2, but the present mixing angle limit of $\sin^2 2\theta < 0.16$, which is based on absolute flux measurements, cannot be improved significantly. On the other hand, progress is expected in the study of exclusive reactions, in particular in their sensitivity to small mixing angles. Work will also continue in the searches of small admixtures of heavy neutrinos.

Double beta decay is one of the most promising topics having a clearly defined program for further study. The discrepancy between the geochemical and calculated lifetimes of the 2ν mode should be resolved, preferably by observing the 2ν mode in the laboratory. That would put the calculation of the nuclear matrix elements on a firmer basis and it would be possible to interpret with greater confidence the experimental lifetimes (or limits) for the 0ν mode in terms of the fundamental parameters of the neutrino Majorana mass or right-handed current. Independently of the uncertainty in the nuclear matrix element, the upper limit on 0ν decay provides a neutrino (Majorana) mass limit of 5–15 eV, which is already below the 30-eV value derived from the ^{3}H experiment. Substantial improvements in sensitivity to 0ν double beta decay are expected with the advent of larger Ge detector arrays and Xe TPCs. These future experiments will probe neutrino mass down to about 1 eV or below.

The solar neutrino puzzle remains unsolved. An independent determination of the solar neutrino flux, in particular of the low-energy neutrinos from the $p+p \rightarrow d+e^+ +\nu_e$ reaction, would help in deciding whether the problem has anything to do with neutrino oscillations.

Last but not least, progress in the predictive power of the underlying particle theory is urgently needed. Guidance as to the expected range of neutrino mass and mixing angles would help reinforce the enthusiasm of experimentalists. It would also help in integrating the problems of neutrino mass and mixing into the broader context of the fundamental properties of the constituents of matter.

ACKNOWLEDGMENT

This review was initiated at the Aspen Center for Physics; its hospitality and stimulating atmosphere are appreciated. Enlightening discussions with V. L. Telegdi are gratefully acknowledged. The work was supported by the US Department of Energy under Contract DEAT-03-81ER40002.

Literature Cited

1. Maki, Z., Nakazawa, M., Sakata, S. 1962. *Prog. Theor. Phys.* 28:870
2. Pontecorvo, B. 1967. *Zh. Eksp. Theor. Fiz.* 53:1717
3. Bilenky, S. M., Pontecorvo, B. 1978. *Phys. Rep.* 41:225
4. Frampton, P. H., Vogel, P. 1982. *Phys. Rep.* 82:339
5. Bullock, F. W., Devenish, R. C. E. 1983. *Rep. Prog. Phys.* 46:1029
6. Cheng, T. P., Li, L. F. 1978. *Phys. Rev. D* 17:2375
7. Mohapatra, R. N., Senjanovic, G. 1980. *Phys. Rev. Lett.* 44:912
8. Wolfenstein, L. 1980. *Nucl. Phys. B* 175:93
9. Gell-Mann, M., Ramond, P., Slansky, R. 1970. In *Supergravity*, ed. P. van Nieuwenhuizen, p. 315. Amsterdam: North-Holland
10. Barbieri, R., Ellis, T., Gaillard, M. K. 1980. *Phys. Lett.* 90B:249
11. Witten, E. 1980. *Phys. Lett.* 91B:81
12. Turner, M. K. 1981. In *Neutrino 81*, ed.

R. J. Cence, E. Ma, A. Roberts. Honolulu: Univ. Hawaii
13. Zeldovich, Ya. B., Sunyaev, R. A. 1980. *Astron. Zh. Pisma* 6:451
14. Tremaine, S., Gunn, J. E. 1979. *Phys. Rev. Lett.* 42:407
15. Schramm, D. N., Steigman, G. 1981. *Astrophys. J.* 243:1
16. Frenk, C. S., White, S. D. M., Davis, M. 1983. *Astrophys. J.* 271:417
17. Langacker, P., Leveille, J. P., Sheiman, J. 1983. *Phys. Rev. D* 27:1228
18. Cabibbo, N., Maiani, L. 1982. *Phys. Lett.* 114B:115
19. Davis, R., Jr., et al. 1983. *Science Underground, AIP Conf. Proc.*, No. 96, p. 2. New York: AIP
20. Bahcall, J. N., et al. 1982. *Rev. Mod. Phys.* 54:767
21. Bahcall, J. N., Davis, R., Jr. 1982. In *Essays in Nuclear Astrophysics*, ed. C. A. Barnes, D. D. Clayton, D. N. Schramm. Cambridge: Cambridge Univ. Press
22. Wolfenstein, L. 1981. *Phys. Lett.* 107B:77
23. Kayser, B., Goldhaber, A. S. 1983. *Phys. Rev. D* 28:2341
24. Barger, V., Langacker, P., Leveille, J. P., Pakvasa, S. 1980. *Phys. Rev. Lett.* 45:692
25. Kobzarev, I. Yu., Martemyanov, B. V., Okun, L. B., Shchepkin, M. G. 1981. *Sov. J. Nucl. Phys.* 32:823
26. DeRujula, A., Glashow, S. L. 1980. *Phys. Rev. Lett.* 45:942
27. Petcov, S. T. 1977. *Sov. J. Nucl. Phys.* 25:340; (E):698
28. Sato, E., Kobayashi, M. 1977. *Prog. Theor. Phys.* 58:1775
29. Pal, P. B., Wolfenstein, L. 1982. *Phys. Rev. D* 25:766
30. Reines, F., Sobel, M. W., Gurr, H. S. 1974. *Phys. Rev. Lett.* 32:180
31. Frank, J. S., et al. 1981. *Phys. Rev. D* 24:2001
32. Cowsik, R. 1980. In *Neutrino Mass*, ed. V. Barger, D. Cline, pp. 50–60. Univ. Wisc.
33. Bergsma, F., et al. 1983. *Phys. Lett.* 128B:361
34. Toussaint, D., Wilczek, F. 1981. *Nature* 289:777
35. Shaevitz, M. 1983. In *Int. Symp. on Lepton and Photon Interactions, Cornell Univ.*, Aug. 1983. Ithaca: Cornell Univ. Press
36. Cowan, C. L., Reines, F. 1957. *Phys. Rev.* 107:528
37. Wilkinson, D. H. 1982. *Nucl. Phys. A* 377:474
38. Vogel, P. 1984. *Phys. Rev. D* 29:1918
39. Reines, F., et al. 1980. *Phys. Rev. Lett.* 45:1307
40. Reines, F. 1983. *Nucl. Phys. A* 396:469
41. Vuilleumier, J.-L., et al. 1982. *Phys. Lett.* 114B:298
42. Gabathuler, K., et al. 1984. *Phys. Lett.* 138B:449
43. Kwon, H., et al. 1981. *Phys. Rev. D* 24:1097
44. Schreckenbach, K., et al. 1981. *Phys. Lett.* 99B:251
45. Vogel, P., et al. 1981. *Phys. Rev. C* 24:1543
46. Boehm, F. 1984. *AIP Conf. Proc.* 112:1. New York: AIP
47. Von Feilitzsch, F., et al. 1982. *Phys. Lett.* 118B:162
48. Shrock, R. E. 1980. *Phys. Lett.* 96B:159; 1981. *Phys. Rev. D* 24:1232, 1275
49. Bergkvist, K. E. 1972. *Nucl. Phys. B* 39:317, 371
50. Tretjakov, E. T., et al. 1976. *Izv. Akad. Nauk. SSSR, Ser. Fiz.* 40:20
51. Lyubimov, V. A., et al. 1980. *Phys. Lett.* 94B:266; 1981. *Soc. Phys. JETP* 54:616
52. Simpson, J. J. 1981. *Phys. Rev. D* 23:649
53. Boris, S., et al. 1983. In *Proc. HEP 83*, ed. J. Guy, C. Costain, pp. 386–89. Oxford: Rutherford Appleton Lab
54. Simpson, J. J. 1983. *Proc. Int. Colloq. Matter Nonconservation*, ed. E. Bellotti, S. Stipcich, p. 279. Frascati, Italy: Serv. Doc. Lab. Frascati
55. Kaplan, I. G., Smutny, V. N., Smelov, G. V. 1982. *Phys. Lett.* 112B:417
56. Smith, L. G., Koets, E., Wapstra, A. H. 1981. *Phys. Lett.* 102B:114
57. Simpson, J. J., Vogel, P. 1984. In *Low Energy Tests of Conservation Laws, AIP Conf. Proc.*, 114:220. New York: AIP
58. De Rujula, A. 1981. *Nucl. Phys. B* 188:414; 1982. *Nucl Phys. A* 379:619
59. Martin, P. C., Glauber, J. 1958. *Phys. Rev.* 109:1307
60. Ravn, H. L., et al. 1983. In *Neutrino Mass and Gauge Structure of Weak Interactions, AIP Conf. Proc.*, No. 99, p. 1. New York: AIP
61. Andersen, J. V., et al. 1982. *Phys. Lett.* 113B:72
62. Baisden, P. A., et al. 1983. *Phys. Rev. C* 28:337
63. Raghavan, R. S. 1983. *Phys. Rev. Lett.* 51:975
64. Beyer, G. J., et al. 1983. *Nucl. Phys. A* 408:87
65. Daum, M., et al. 1979. *Phys. Rev. D* 20:2692
66. Lu, D., et al. 1980. *Phys. Rev. Lett.* 45:1066
67. Abela, R., et al. 1983. *SIN Newsletter* 15:26
68. Anderhub, H. B., et al. 1982. *Phys. Lett.* 114B:76
69. Clark, A. B., et al. 1974. *Phys. Rev. D* 9:533

70. Bacino, W., et al. 1979. *Phys. Rev. Lett.* 42:749
71. Blocker, C. A., et al. 1982. *Phys. Lett.* 109B:119
72. Simpson, J. J. 1981. *Phys. Rev. D* 24:2971
73. Schreckenbach, K., et al. 1983. *Phys. Lett.* 129B:265
74. Simpson, J. J. 1981. *Phys. Lett.* 102B:35
75. Bryman, D. A., et al. 1983. *Phys. Rev. Lett.* 50:1546
76. Dixit, M. S., et al. 1983. *Phys. Rev. D* 27:2216
77. Abela, P. 1981. *Phys. Lett.* 105B:263
78. Hayano, R. S., et al. 1982. *Phys. Rev. Lett.* 49:1305
79. Rosen, S. P. 1982. *Neutrino 81*, 2:76. Honolulu: Univ. Hawaii
80. Doi, M., et al. 1983. *Prog. Theor. Phys.* 69:602
81. Bryman, D., Picciotto, C. 1978. *Rev. Mod. Phys.* 50:11; Zdesenko, Yu. G. 1980. *Sov. J. Part. Nucl.* 11:542
82. Kirsten, T., et al. 1983. *Phys. Rev. Lett.* 50:474
83. Hennecke, E., et al. 1975. *Phys. Rev. C* 11:1378
84. See for example Kirsten, T. 1982. In *Science Underground, AIP Conf. Proc.* 96:396. New York: AIP; Boehm, F. 1982. *AIP Conf. Proc.* 93:321. New York: AIP; and Refs. 80 and 81
85. Haxton, W. C., Stephenson, G. J., Strottman, D. 1982. *Phys. Rev. D* 25:2360
86. Haxton, W. C., Stephenson, G. J., Strottman, D. 1981. *Phys. Rev. Lett.* 47:153
87. Moe, M., Lowenthal, D. 1980. *Phys. Rev. C* 22:2186
88. Cleveland, B., et al. 1975. *Phys. Rev. Lett.* 35:737
89. Moe, M. 1982. *Neutrino 82, Proc. Int. Conf.*, ed. A. Frenkel, L. Jenik. Budapest, Hungary: Balatonfüred
90. Bellotti, E., et al. 1983. See Ref. 53
91. Avignone, F. T., et al. 1983. *Phys. Rev. Lett.* 50:721
92. Leccia, F. 1983. Univ. Bordeaux Preprint
93. Forster, A., et al. 1984. *Phys. Lett. B* 138B:301
94. Forster, A., et al. 1984. *The Time Projection Chamber, AIP Conf. Proc.*, 108:56. New York: AIP
95. Vogel, P. 1984. *Phys. Rev. D.* In press
96. Nieves, J. F. 1983. *Phys. Rev. D* 28:1664

NUCLEAR COLLISIONS AT HIGH ENERGIES[1]

S. Nagamiya

Department of Physics, Faculty of Science, University of Tokyo, Hongo, Bunkyo-ku, Tokyo 113, Japan

J. Randrup and T. J. M. Symons

Nuclear Science Division, Lawrence Berkeley Laboratory, University of California, Berkeley, California 94720, USA

CONTENTS

1. INTRODUCTION 155
2. LIGHT-PARTICLE EMISSION AT LARGE ANGLES 160
   2.1 Energy and Angular Distributions 160
   2.2 Large-Angle Two-Particle Correlations 164
   2.3 Composite Fragments 166
3. SEARCH FOR NEW FORMS OF MATTER 168
   3.1 Efforts to Probe Density and Temperature 168
   3.2 Pion Multiplicity 170
   3.3 Collective Flow 172
   3.4 Cooperative Process 175
4. FRAGMENT EMISSION AT FORWARD ANGLES 177
   4.1 Models of the Fragmentation Process 177
   4.2 Electromagnetic Effects 179
   4.3 Applications of Fragmentation Reactions 180
5. SUMMARY AND FUTURE PERSPECTIVES 182

1. INTRODUCTION

In 1972, the first nuclear beams were accelerated to relativistic energies in synchrotrons at Berkeley and Princeton, ushering in a new field of study for

[1] The US Government has the right to retain a nonexclusive royalty free license in and to any copyright covering this paper.

nuclear physicists. At that time the possibility of producing hot, highly compressed nuclear matter offered great promise for extending our knowledge of the nuclear equation of state away from the equilibrium found in finite nuclei. Ten years later, this remains the principal goal of the field, although progress in this particular direction has been slow. This does not mean, however, that the developing field of high-energy nuclear collisions has moved slowly. Great progress has been made in our understanding of nuclear collisions and results have been found that are intrinsically interesting and will also be invaluable for planning new experiments at higher energies. In this article, we review the progress that has been made in understanding these processes and give some indication of the directions in which the field appears to be heading.

We should emphasize that this is an experimental review. Models will be introduced as needed to aid in interpretation and correlation of the data, but we hardly touch upon the considerable theoretical work concerning potential knowledge that may be gained from high-energy collisions in the future. Thus, the reader will find discussions of hydrodynamics or the abrasion-ablation model but should not expect to learn in detail about pion condensates, Lee-Wick matter, or the quark-gluon plasma. We have restricted our discussion to data from collisions between heavy ($A > 4$) nuclei at beam energies up to a few A GeV. High-energy hadron-nucleus collisions are omitted as are the exciting new experiments at the CERN ISR on collisions of ultra-relativistic light ions. Even within our restricted field, we have had to make difficult choices, and some important topics, such as target fragmentation and the study of anomalons, are not considered.

Before starting this detailed survey, we make some general introductory remarks. In nuclear reactions at beam energies above a few 100 A MeV, the available energy is much greater than the nuclear binding energy, so that the collisions are quite dramatic in comparison to those at low energy. This is illustrated vividly in Figure 1, which displays an emulsion recording of such a collision (H. Heckman, 1983, private communication). The multiplicity of product particles is large and the kinematic domain into which they are emitted is very wide. This is partly because of the larger phase space available at high beam energies and partly because the reaction mechanism is basically different from that of low-energy nuclear collisions. In ordinary nuclear matter the density of nucleons is around $\rho_0 \approx 0.16 \text{ fm}^{-3}$ so that the typical closest-neighbor separation is ≈ 2 fm. This distance is larger than the de Broglie wavelength of a nucleon moving with a kinetic energy above a few hundred MeV. Therefore, at such energies, the projectile nucleons can recognize the individuality of the target nucleons (and vice versa). At the same energies, the mean free path λ of a nucleon moving through normal nuclear matter approaches the "free" value $\lambda_0 = 1/(\rho_0 \sigma_{\text{NN}}) \approx 1.6$ fm (where

σ_{NN} is the nucleon-nucleon interaction cross section, which is fairly constant at high energies). This value is smaller than the nuclear radius $R_A \approx 1.2\ A^{1/3}$ fm $\approx$ 3–7 fm. Furthermore, the differential cross section grows predominantly forward-peaked. Consequently, an incident high-energy nucleon will typically experience a sequence of several collisions with the target nucleons while tending to preserve its forward motion.

As it has turned out, it is possible to understand many of the main features of high-energy nuclear collisions on a very simple conceptual basis in which the colliding nuclei are pictured as two clouds of individual nucleons propagating through each other, with the nucleons suffering sequential hard collisions with those of the other nucleus (see Figure 2). While certainly simplistic, and in many ways inadequate, this picture serves to introduce some concepts that have proven very useful in organizing and discussing the various characteristic features of these processes. In particular, the concept of participants and spectators follows naturally from this picture (2, 3). In a typical collision, such as the one depicted in Figure 2, the outer parts of the two nuclei will miss and the nucleons in them will not experience violent interactions; these parts are denoted the projectile and target spectators, respectively. The remaining two parts interpenetrate and their nucleons suffer several hard collisions; these particles are called the participants. [For the empirical validity of this picture, see (4, 5).]

The state of motion of a particle with mass m, momentum $\mathbf{p} = (\mathbf{p}_\perp, p_\parallel)$

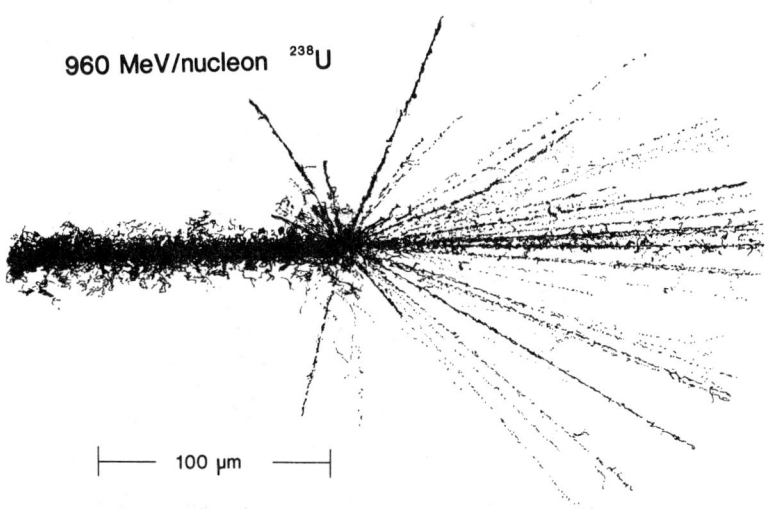

Figure 1 Recording of a collision between a 1-A-GeV U projectile nucleus with a target nucleus in an emulsion (H. Heckman, 1983, private communication).

and energy E can conveniently be characterized by the dimensionless rapidity vector $\mathbf{y} = (\mathbf{p}_\perp/mc, y)$, where $y = \tanh^{-1}(p_\parallel c/E)$ is the ordinary rapidity. (Here $p_\parallel$ denotes the momentum component along the beam and $\mathbf{p}_\perp$ is the transverse momentum.) Since $\mathbf{y}$ is additive under Lorentz boosts along the beam, contour plots of the invariant differential cross section $E\,d\sigma/d\mathbf{p} \approx d\sigma/d\mathbf{y}$ remain undistorted under such transformations. In the nonrelativistic limit $\mathbf{y}$ is simply the velocity $\mathbf{v}$ divided by c.

The spectator matter, which has suffered only little disturbance through the collision, will then emerge with rapidities close to those of the respective initial nuclei, y_P and y_T. This matter will remain rather cold so that fairly large fragments may result. Since the spectators are characterized by the neutron-to-proton ratio associated with their respective primogenitors, the resulting fragments will tend to carry the same ratio and thus ordinarily be excessively neutron-rich for their mass. This special feature of a high-energy nuclear collision can be turned into a very powerful means of producing neutron-rich nuclei far from the stability line. This topic is discussed in Section 4.

The fate of the participant portion of the collision system is more complicated. The nucleons will suffer hard collisions as the two nucleon

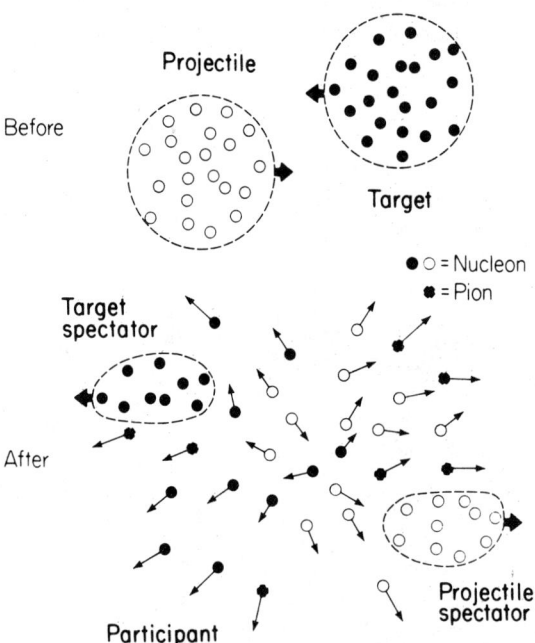

Figure 2 Participant-spectator separation expected in high-energy nucleus-nucleus collisions.

clouds interpenetrate. These collisions may excite some of the nucleons, mostly into Δ resonances, and mesons are produced. Later, as the system disassembles, the longer-range interactions between the emerging particles are important for the formation of the final fragments. Clearly, the first important task is to understand the dynamics of such multiple collisions. Summarizing our basic knowledge of this will occupy Section 2.

There are many systems in nature where microscopic descriptions such as the one we have just described are unnecessarily cumbersome. For example, in studying the behavior of gases, one does not need to follow the fate of individual gas molecules in order to learn how a gas will behave under variation of temperature and pressure; one only needs to know the macroscopic equation of state. Theoretically, various expectations for matter at high density and temperature have been proposed as the highlights of high-energy nuclear collisions; pion condensation (6, 7), abnormal nuclear matter (8, 9), and quark matter (10), as shown in Figure 3. Matter at high density and temperature is a new domain in nuclear physics, since until now we were restricted to the region of densities around ρ_0 and temperatures up to 10–20 MeV. An important caveat, however, is that the system formed must be large enough and live long enough for such a description to be valid. Experimental efforts toward this goal are discussed in Section 3.

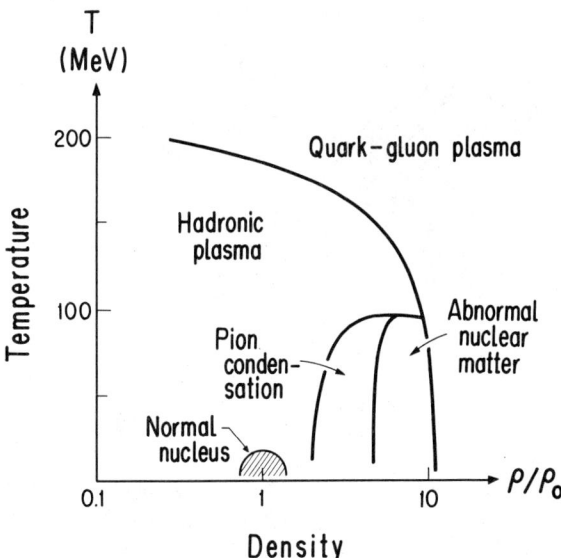

Figure 3 Theoretical expectations of new phases for hot and dense nuclear matter.

Recently, very heavy nuclear beams, such as ^{139}La and ^{238}U, have become available at the Bevalac in Berkeley and the field of high-energy nuclear collisions is growing rapidly. Currently, planning is under way for the construction of accelerators to provide nuclear beams at ultra-relativistic energies. We hope that this review may give a quick impression of the major accomplishments so far and also be helpful in planning for the future.

2. LIGHT-PARTICLE EMISSION AT LARGE ANGLES

2.1 Energy and Angular Distributions

Energy and angular distributions of protons produced in nearly equal-mass collisions at beam energies of 800 A MeV (11) are plotted in Figure 4. The center-of-mass (c.m.) 90° energy distributions are selected, because at this angle the contribution from spectators should be the smallest, as expected from the participant-spectator picture shown in Figure 2. The spectral shapes are nearly identical for all the cases, which implies that the beam

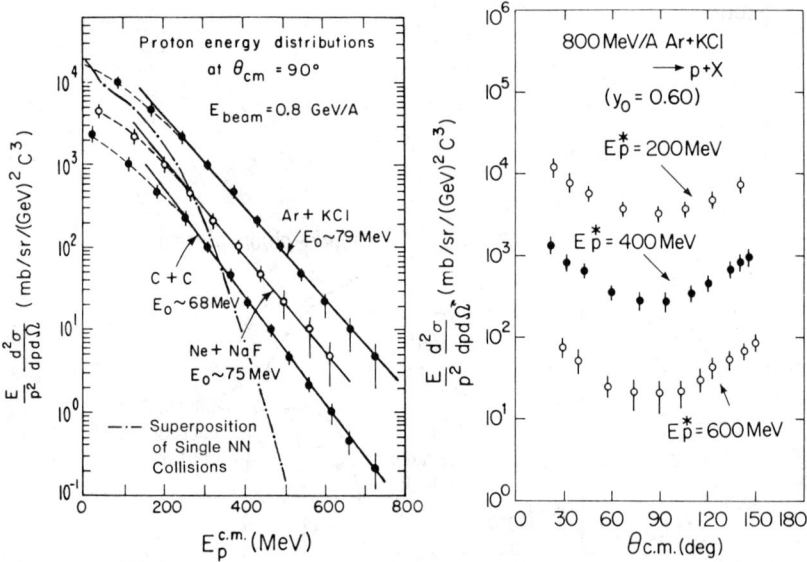

Figure 4 Left: Proton energy spectra at c.m. 90° in collisions of 800-A-MeV C+C, Ne+NaF, and Ar+KCl (11). Dashed curve indicates the calculated result from the single NN collision model with a Gaussian-type Fermi motion included (12). *Right:* Angular distribution of protons in the c.m. frame in 800-A-MeV Ar+KCl collisions for three different proton energies (11).

energy per nucleon, rather than the total beam energy, determines the basic dynamics of proton emission. A "shoulder-arm" type spectrum shape is observed with an approximately exponential form in the high-energy region,

$$E \frac{d\sigma}{d\mathbf{p}} \propto \exp(-E^{c.m.}/E_0). \qquad 1.$$

In particular, a copious production of high-energy protons is observed in the region far beyond the free NN kinematical limit (≈ 182 MeV in this case). Even if a proper Fermi motion is included (12), the production of these high-energy protons cannot be explained by a model in which a nucleon suffers at most a single NN collision. Finally, the angular distribution shows strong forward and backward peakings in the c.m. frame.

Historically, these data were discussed first with a fireball model (13–17) that assumes the total available energy of the participants is completely thermalized. The size of the fireball is given by the geometric clean-cut separation into participants and spectators. While the overall proton yields are fairly well accounted for, this model has various difficulties in explaining the data. First, it predicts a pure exponential shape in the entire kinematic region, which conflicts with the shoulder-arm shape. Second, the predicted angular distribution is isotropic in the c.m. frame of the fireball, in contrast to the observed angular anisotropy shown in Figure 4.

In order to explain the angular anisotropy, two modified thermal models, the firestreak model (18, 19) and the two-fireball model (20), were introduced. The firestreak model assumes that the participant regions are divided into many tubes along the beam direction and thermal equilibrium is assumed in each pair of juxtaposed tubes. The cross section is then given by an incoherent sum of these completely inelastic tube-tube collisions. The two-fireball model introduces the concept of nuclear transparency by assuming that only part of the available energy goes to thermal motion with the rest remaining in translational motion. Both models qualitatively explain the observed large angular anisotropy, but still fail to reproduce the shoulder-arm shape.

In order to make progress on this problem it is useful to consider the pions as well. Pion spectra show an almost pure exponential shape (11), but the observed slope $E_0(\pi)$ is systematically smaller than $E_0(p)$. This difference cannot be explained in the thermal model, which assumes that both protons and pions are emitted from a common source so that they should be characterized by the same temperature.

This puzzle led to a modified thermal model called the thermal explosion model (21). It assumes that the system disassembles in an explosive manner with a radially expanding flow superposed on the chaotic thermal motion. Since heavy, slow particles (nucleons) are affected more by this flow than light, fast ones (pions), the spectral shape for protons deviates substantially from exponential in the low-energy region, and is quite similar to the observed shoulder-arm shape. Furthermore, the model reproduces the observed slope difference between protons and pions for the same reason. However, this model cannot explain the angular anisotropy.

The various characteristic features of the data can also be fairly well understood in the microscopic intranuclear cascade picture outlined in the introduction and illustrated in Figure 2. While several elaborate implementations of this general picture have been made and employed (22–29), the main results can be understood in the simple linear cascade or "rows-on-rows" approximation in which a given nucleon only collides with its juxtaposed partners in the other nucleus (12, 30, 31). In this model high-energy particles result from multiple NN collisions and are in approximate agreement with the thermal results. Furthermore, the low-energy part of the spectrum results in large part from single quasi-elastic NN collisions, which yield a peak at $E_p^{c.m.} \approx E_{Beam}^{c.m.}/A \approx 182$ MeV in the present case, thus producing the shoulder-arm feature. Finally, the angular anisotropy emerges naturally from the different velocities of the various row-row systems, just as in the firestreak model.

A simple model bridging the microscopic rows-on-rows model and the macroscopic firestreak model is the phase-space model (32, 33). It assumes that statistical equilibrium is reached in each row-row system separately. Thus the model can be considered a generalization of the firestreak model to take account of the finite supply of nucleons and energy in each row-row system. In addition to reproducing the spectral shoulder-arm shape and the anisotropy, this model also predicts pions to be colder than nucleons, as a result of the energy expended in producing the pion.

Recently, data for K^+ production with 2.1-A-GeV Ne beams have been reported (34), and are shown in Figure 5. The spectrum shape is again exponential with the inverse slope of $E_0 \approx 142$ MeV, which is larger than E_0 for protons and pions and satisfies the relationship

$$E_0(\pi) < E_0(p) < E_0(K^+). \qquad 2.$$

This is at variance with both the phase-space and thermal explosion models. [The former yields $E_0(K^+) < E_0(\pi) < E_0(p)$ because the threshold energy for K^+ production is much higher than that for π production (35, 36). The latter yields $E_0(\pi) < E_0(K^+) < E_0(p)$ because $m_\pi < m_{K^+} < m_p$.]

A simple argument for the above observation (Equation 2) has been made in terms of the mean free paths of the product particles (37). The energy density of the system reaches a maximum value at a certain time, at which point mesons are expected to be created most copiously. Subsequently the system expands and cools. Since particles with a long mean free path would escape more easily from the system, they would reflect the earlier hot stage of the collision by carrying higher kinetic energies, hence a larger value of E_0. Since the value of the mean free path λ satisfies the relation of $\lambda(\pi) < \lambda(p) < \lambda(K^+)$ in nuclear matter, we expect the above relation (Equation 2).

K^+ production has been calculated within both the microscopic (35, 38–40) and the thermal framework (36, 41–43). In view of the small elementary kaon production cross sections, one would not expect the establishment of chemical equilibrium during the rather short reaction time and indeed the thermal models overpredict the K^+ yields substantially. Contrary to this, the microscopic calculations reproduce the K^+ yield very well, but they tend to give fairly cold kaons. This is at variance with the data, which show rather substantial yields at high energies. A possible mechanism for this feature was suggested to be the elastic scattering of the produced kaons off the surrounding fast-moving nucleons (38). Calculations including the contribution from pion-nucleon interactions to produce K^+ further improve the agreement with the data, although certain deviations still remain (39, 40, 43).

Before finishing this section, it is worthwhile to mention recent data on proton energy distributions at c.m. 90° in 800-A-MeV C+C collisions (44).

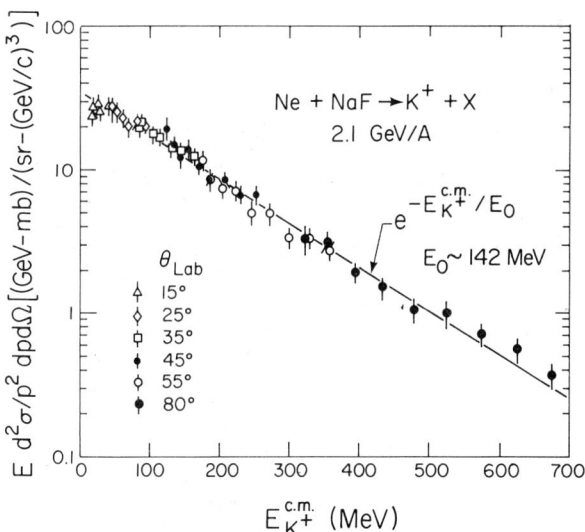

Figure 5 Energy spectra of K^+ in the c.m. frame in 2.1-A-GeV Ne+NaF collisions. The figure is made from the data reported in (34).

Data are shown in Figure 6 covering invariant cross sections over eight orders of magnitude. The data show downward deviation from the exponential shape not only below 200 MeV but also above 700 MeV, most likely because of the limited phase space for emission of high-energy protons. Cascade calculations (27, 28) lead to a steeper slope than the data in the high-energy region, which may suggest that some mechanism other than simple multiple NN collisions is at work.

2.2 Large-Angle Two-Particle Correlations

Measurements of two-particle correlations further clarify the reaction mechanism. The in-plane to out-of-plane coincidence ratio for two protons has been measured in 800-A-MeV C+C collisions (45, 46). Here, the first proton was detected at a fixed angle, $\theta^{\text{Lab}} = 40°$ and the angular distribution of this ratio was measured as a function of the angle of the second proton. As shown in Figure 7, the ratio is larger than unity for C+C and it peaks at $\theta = 40°$, which implies that two protons tend to be emitted at

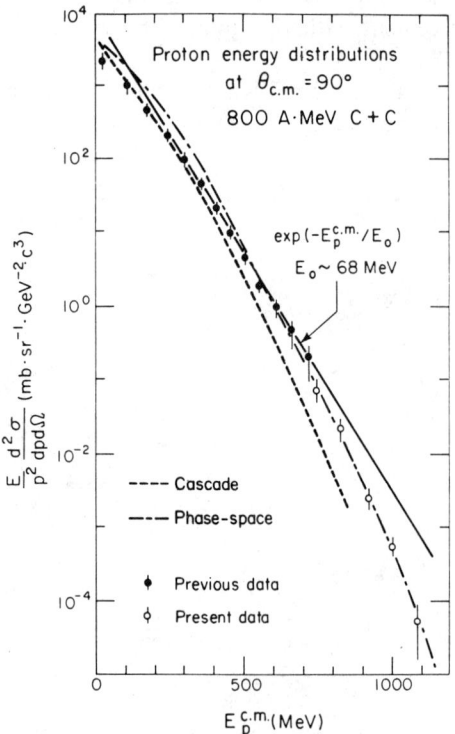

Figure 6 Recent data of proton energy spectra at c.m. 90° in 800-A-MeV C+C collisions (44).

$\theta_1 = \theta_2 = 40°$ on opposite sides, as illustrated in the upper right corner. These kinematics are exactly what we expect from pp quasi-elastic scattering (pp QES), and therefore, the data show the existence of the direct knock-out component in high-energy nuclear collisions. From detailed analysis of the peak height it was estimated that in C + C collisions at $E_{\text{Beam}}^{\text{Lab}}$ = 800 A MeV, about 40% of protons emitted at $E_p^{\text{c.m.}} \approx 182$ MeV experience only a single NN collision.

Roughly speaking, the probability that a nucleon does not suffer an additional collision after the first one is given by $\exp(-R/\lambda)$, where λ and R are, respectively, the mean free path and the radius of the interaction zone. For C + C collisions, the estimated value of R, based on the participant-

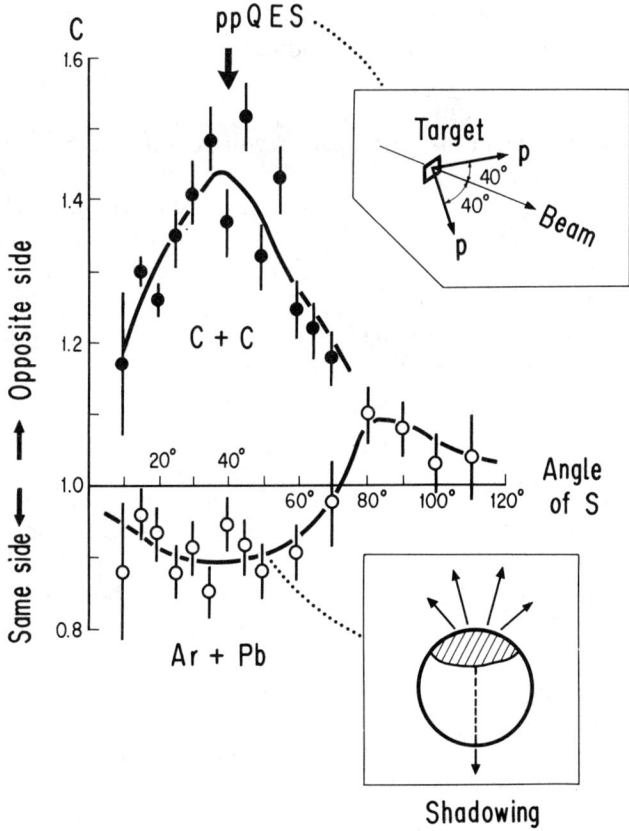

Figure 7 Energy-integrated angular distributions of the in-plane to out-of-plane ratio, C, for two protons emitted in 800-A-MeV C + C and Ar + Pb collisions. Figure made from the data reported in (45, 46, 104).

spectator model, is about 2 fm. On the other hand, recent experiments determined the value of λ to be 2.4 fm at 800 A MeV (47). Therefore, we have $\exp(-R/\lambda) \approx 0.4$, which is in agreement with the data. This number also agrees with several theoretical calculations (26, 31, 32, 48, 49). In addition, this result supports the intuitive expectation, described in Section 1, that a nuclear collision at high energy consists mostly of hard NN collisions, not only single but also multiple, because, even in light-mass systems, more than half the nucleons experience at least two hard NN collisions.

For Ar+Pb collisions the angular distribution of the coincidence ratio shows a completely different pattern from that observed for C+C collisions, as seen from Figure 7. The ratio is smaller than unity at small angles ($\theta < 70°$). This was first ascribed to nuclear shadowing, as illustrated in the lower right corner (45, 46). However, the observation of ratios larger than unity at large angles ($\theta > 80°$) cannot be explained by this mechanism. This puzzle is discussed below in Section 3.3.

2.3 Composite Fragments

Particles emitted from the participant region are mainly single nucleons and, to some extent, single pions. This is expected since the energy transfers in the primary hard NN collisions are large in comparison with typical nucleon separation energies and often also large enough to produce a pion. There is, however, also a significant yield of light composite nuclear fragments, especially d, t, ^{3}He, and α.

The spectral shapes of the light composite fragments are related by a simple power law to the observed proton spectra (11, 50, 51):

$$\left[E\left(\frac{d\sigma}{d\mathbf{p}}\right) \right]_A = C_A \left\{ \left[E\left(\frac{d\sigma}{d\mathbf{p}}\right) \right]_p \right\}^A \qquad 3.$$

for $\mathbf{p}_A = A\mathbf{p}_p$, where A is the mass number of the composite fragment and the normalization constant C_A is adjustable for each fragment species. This feature is illustrated in Figure 8.

A simple interpretation of the above relationship has been made in terms of a coalescence model (50–54), according to which primordial nucleons produced sufficiently close in momentum space may coalesce into a composite fragment because of their final-state interactions. In such a picture, the probability of observing a deuteron with a specified velocity is proportional to the product of the probabilities of having a neutron and a proton with that same velocity. Since the primordial neutron spectra are expected to be nearly proportional to the proton spectra, the above relation (Equation 3) follows. It should be noted, though, that the coalescence model relates the *observed* composite spectra to the *primordial* proton spectra, while the empirical relation is between *observed* spectra only. When the local composite yield is small in comparison with the local proton yield, there is little difference between primordial and observed proton yields. But there are cases where this perturbation condition is not met (55). This fact casts some doubt on the general validity of the coalescence picture.

An alternate framework for discussing the yield of composite fragments is the chemical equilibrium model (16, 17, 56, 57). In this picture, the participant source is considered as a gas of fragments in thermal and chemical equilibrium. The ensuing rate equations determining the equilibrium species composition then lead to relations of the form of Equation 3 between the *observed* spectral shapes. This would seem to favor the chemical equilibrium model. However, the model predictions of the relative yields are too large by nearly an order of magnitude (58). Thus the mechanism of composite fragment formation is still not well understood.

Recently, systematic studies of the d/p ratio as a function of event multiplicity have been performed by the LBL-GSI group (59) at the Bevalac, using a multidetector device called the Plastic Ball/Wall (60). Typical results are shown in Figure 9. As event multiplicity increases, the d/p ratio increases as well. This is understood as due to the finite size of the fragments (59, 61). Accordingly, the ratio approaches a constant at high multiplicities and only in this regime can thermal models be properly applied. However, a quantitative comparison has not yet been made.

Since the d/p ratio can be related to the phase-space density in the source, it has been noted (62) that the specific entropy S can be extracted by the use of the relation $S = 3.95 - \ln(d/p)$. The inherent assumption of isentropic

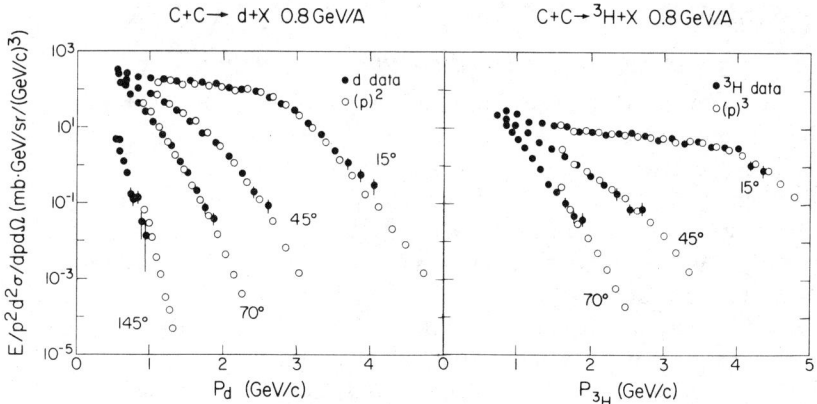

Figure 8 Deuteron and triton spectra in 800-A-MeV C+C collisions, as compared with the second and third power, respectively, of the observed proton spectra (11).

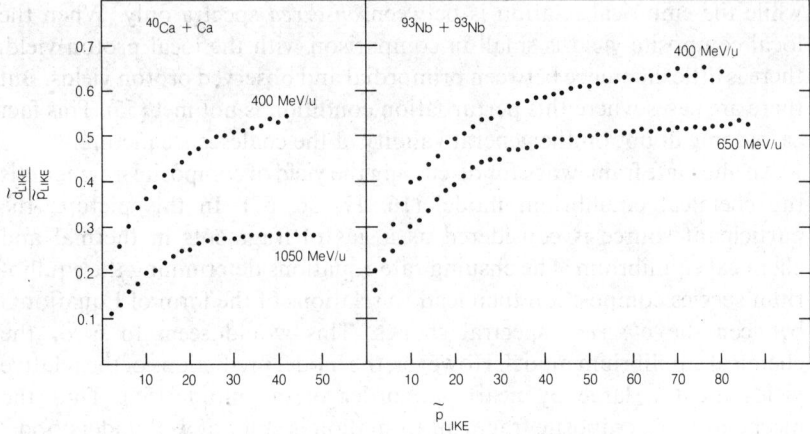

Figure 9 Ratios of d_{Like} plotted as a function of the value of p_{Like}. Data were taken by the Plastic Ball/Wall (58, 59). Here, d_{Like} is defined $N_d + 1.5\,(N_t + N_{^3\text{He}}) + 3N_\alpha$, whereas p_{Like} is defined $N_p + N_d + N_t + 2(N_{^3\text{He}} + N_\alpha)$. Figure taken from (85).

disassembly is supported by various theories (21, 63–65). Thus, a measurement of S would probe the entropy associated with the initial hot stage of the collision. However, the validity of the above simple relation has since been questioned (66) and, furthermore, it has been demonstrated that the observed d/p ratio can be quite different from the primordial value, due to the break-up of unstable fragments (67). Consequently, the question of entropy production is still wide open.

3. SEARCH FOR NEW FORMS OF MATTER

3.1 *Efforts to Probe Density and Temperature*

In order to elucidate the extent to which nuclear matter is compressed or heated during the course of a nuclear collision, we now discuss attempts to deduce the density and temperature experimentally.

Concerning the density, ρ, if we measure both the multiplicity of nucleons emitted from the interaction region, m_N, and its volume, V, then we can extract the density from the relation $\rho = m_N/V$. Since the measurement of m_N is rather straightforward (for charged particles, at least), the question is how to determine the source volume. We recall that stellar radii can be measured by $\gamma\gamma$ interferometry, exploiting the Hanbury-Brown/Twiss effect (68). This idea was applied to high-energy collisions for the purpose of determining the size of the interaction region (69–71) using identical pions rather than photons. Assume that particles are emitted statistically from a source with a certain space-time structure $\rho(\mathbf{r}, t)$. The quantum correlations between two identical particles emitted from the source then produce a structure in the observed two-particle cross section $d^2\sigma(\mathbf{p}_1,\mathbf{p}_2)/d\mathbf{p}_1\,d\mathbf{p}_2$, where $\mathbf{p}_1$ and $\mathbf{p}_2$ are the momenta of the two particles. The scale of this structure reflects the space-time extension of the source via the uncertainty relation. Specifically, if the source $\rho(\mathbf{r},t)$ is assumed to be a Gaussian characterized by the radius R and lifetime τ, then, for identical bosons, the enhancement in the two-particle counting rate relative to the product of the two one-particle counting rates is given by (72, 73)

$$C_2 = \sigma_0 \frac{d^2\sigma(\mathbf{p}_1,\mathbf{p}_2)/d\mathbf{p}_1\,d\mathbf{p}_2}{[d\sigma(\mathbf{p}_1)/d\mathbf{p}_1][d\sigma(\mathbf{p}_2)/d\mathbf{p}_2]}$$
$$= 1 + \exp(-q^2 R^2/2 - \omega^2\tau^2/2), \qquad 4.$$

where $\mathbf{q} = \mathbf{p}_1 - \mathbf{p}_2$ and $\omega = E_1 - E_2$. Thus the structure of the correlation between two identical particles with nearly identical four-momenta can be used to extract information about the space-time structure of the emitting source. Figure 10 shows recent data (74) of $\pi^-\pi^-$ correlations in 1.8-A-GeV Ar+KCl collisions, in which there is a clear enhancement at small q. From these data the source radius R was determined to be 3.2 ± 0.3 fm. Recently, more data have appeared not only for two pions (75) but also for two protons (76, 77).

It is worthwhile to note that Equation 4 holds only if the pions are emitted randomly. If pions were produced coherently (78–81), then the form of C_2 would be substantially distorted (81). For example, in the presence of a

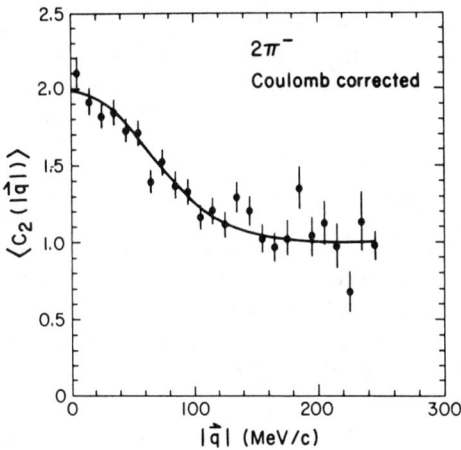

Figure 10 Two-pion interferometry observed in 1.8-A-GeV Ar+KCl collisions. Data are taken from (74).

pion laser (78) the value of C_2 becomes exactly one in the entire range of **q**. Currently, however, no definite signals for such a change in C_2 have been reported.

In order to determine ρ, it is necessary to measure m_N in coincidence with this two-particle detection. Unfortunately, no such measurements have been reported so far. However, using the estimated values of m_N an attempt to extract the value of ρ has been reported (82, 83). For $Ar+KCl$ collisions at 1.8 GeV, the density thus estimated is $\approx 2\rho_0$ when the system was probed by protons and $\approx 0.6\rho_0$ when it was probed by pions. This difference between protons and pions appears reasonable, as mentioned in Section 2.1; pions would tend to probe the coldest and thus the most expanded low-density stage of the collisions. In this regard, future measurements of K^+K^+ interferometry in coincidence with m_N are particularly interesting for probing the density at the initial hot stage of the collisions.

It is not a simple task to define and extract the temperature of the system, as mentioned in Section 2.1. However, the quantity E_0 as defined by Equation 1 is closely related to the average kinetic energy carried by product particles and, thus, indicates how the nuclear matter is heated up at the stage when these product particles are emitted. We therefore may use E_0 as an effective temperature.

With these assumptions, values of density and effective temperature can be determined directly from experiments, albeit with considerable uncertainty. One can then plot these observed values in the plane of ρ and E_0. If we connect the experimental points, we can deduce how the nuclear collision evolves with time in this plane, using the different particles. From such a plot it appears (82, 83) that matter at high density and temperature seems to be created in high-energy nuclear collisions. We wish to emphasize that such measurements in the future should be done under the bias of a fixed value of m_N. This is important for direct extraction of the value of ρ. It is important also for the determination of E_0, since E_0 increases as m_N increases (84, 85).

3.2 Pion Multiplicity

Measurements of pion multiplicity have been performed extensively during the past few years using a streamer chamber (86–88), with which negatively charged tracks, which are mainly from π^-, can be identified easily. Also, the average pion multiplicity, $\langle m_\pi \rangle$, has been measured with a magnetic spectrometer (11) by determining the total integrated pion yield σ_{tot}, which is related to $\langle m_\pi \rangle$ by $\sigma_{tot} = \langle m_\pi \rangle \sigma_0$, where σ_0 is the geometrical cross section. The pion multiplicity increases monotonically with beam energy as shown in Figure 11, which also illustrates various theoretical predictions.

In the past, most theoretical models have succeeded in explaining the absolute proton yield but not the pion yield. This is not surprising since the proton yield is determined mainly by the collision geometry, that is, the nucleon number associated with the participant region, whereas the pion yield directly reflects the collision dynamics. As shown in Figure 12, the thermal model (as well as the phase-space model) overpredicts the pion yield by a factor of 2–3. However, if part of the available energy remains in translational motion due to nuclear transparency or is tied up in macroscopic flow such as a radial explosion, then only the remaining energy is available for pion production, so that the pion yield is reduced (20, 21, 64).

Recent cascade calculations (88) also yield pion multiplicities that are higher than the experimental ones at all the beam energies (Figure 11). A reason for this discrepancy was suggested recently (88). The basic postulate is that part of the available energy is expended as potential energy associated with the compression of the nuclear matter and this part of the energy is thus unavailable for pion production. If this potential energy is equal to E^C, indicated by arrows in Figure 11, then the results of cascade calculations would be consistent with the data. In Figure 12 this energy, E^C, is plotted as a function of the calculated density, ρ, where the latter was extracted from cascade calculations. These "data" points fall on a curve of the expected equation of state of nuclear matter with a compressibility coefficient $K = 240$ MeV. Therefore, it was concluded that measurements

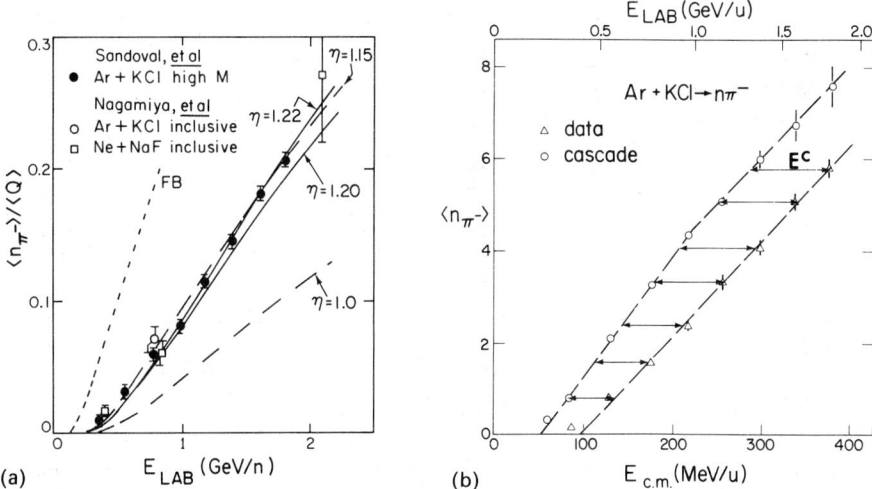

Figure 11 Observed pion multiplicities as compared with the results of (a) thermal (FB), hydrodynamical (with various viscosities), and (b) cascade calculation for central collisions at Ar+KCl. Figures taken from (155) and (88).

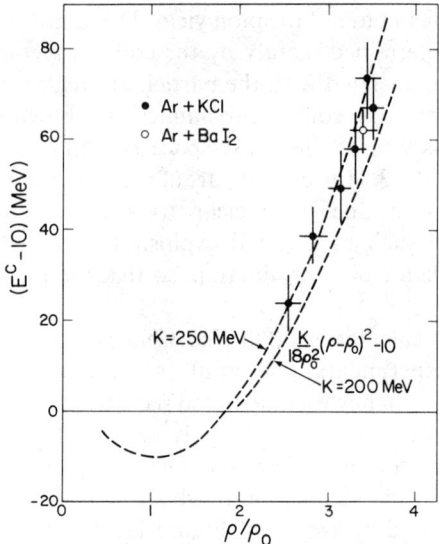

Figure 12 Values of E^c, defined in Figure 11, plotted as a function of the calculated mean baryon density. Dashed curves are based on conventional equations of state. Figure taken from (88).

of pion multiplicity may probe the equation of state of nuclear matter. This is the first attempt at actually measuring the nuclear equation of state in high-energy collisions. It must be emphasized, though, that the method relies on the cascade code being correct in all respects other than the neglect of compressional energy. Since this assumption has been widely criticized (89–91), further investigation is called for.

3.3 Collective Flow

Let us consider an experiment of head-on collisions between U and U. Since the thickness of the U nucleus (≈ 15 fm) is significantly larger than the mean free path of nucleons ($\lambda \approx 2$ fm) in nuclear matter, each nucleon will experience successive NN collisions and, as a result, the local nucleon density should be increased substantially above normal during the collision. On the other hand, the NN potential contains a hard core that counteracts the creation of a local high-density region. Thus, the colliding nucleons may seek to escape from the interaction region into a region in which fewer nucleons exist and thus give rise to a collective flow away from the beam direction (92–100).

The first hint of such collective flow was seen in a broad sideward peak observed in 393-A-MeV Ne+U collisions (101), as shown in Figure 13. In low-multiplicity events the angular distributions are forward peaked, whereas in high-multiplicity events the forward proton emission is highly suppressed. In addition, for low-energy protons ($E^{\text{Lab}} \approx 12$ MeV) a broad peak is observed at $\theta = 70$–$90°$. This broad peak has been attributed to the effect of the collective side splash of nucleons (97). It has been noted (102) that this sideward peak is predicted only with the hydrodynamical model (100) but not with the cascade (23, 25, 29), thermal (19), or thermal-plus-direct (49, 103) models.

A second hint of a side splash is suggested by the two-proton correlations in 800-A-MeV C+Pb and Ar+Pb collisions (46, 104). As described in Section 2.2, the observed value of the in-plane to out-of-plane coincidence ratio for detection of two protons is larger than unity at $\theta > 80°$ in Ar+Pb collisions. From detailed analysis of energy and angular correlations between these two protons (104) it has been found that the ratio is smaller than unity at small angles ($\theta < 70°$) mainly because two high-energy protons tend to be emitted on the same side, whereas it is larger than unity at large angles ($\theta > 80°$) mainly because two protons, one at high energy and the other at low energy, tend to be emitted on opposite sides. These features are what we expect from a fluid dynamical bounce-off (97) (Figure 14) since the projectile matter induces fast-fast correlations on the same side, while the projectile-target matter induces fast-slow correlations on opposite sides.

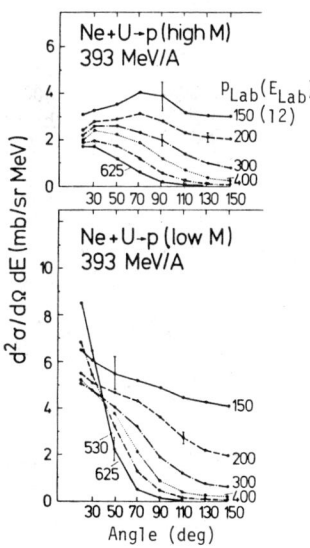

Figure 13 Proton angular distributions for high-multiplicity (*upper*) and low-multiplicity (*lower*) events in 393-A-MeV Ne+U collisions (101).

In spite of the intensive work that has gone into these two measurements, the collective effects are still weak and somewhat speculative. First, in the case of light-mass projectiles the basic assumption involved in the fluid dynamical model, i.e. $\lambda \ll R$, is not well justified even after selection of high-multiplicity events. Second, the fluid dynamical calculations overpredcit both the sideward peak (Figure 13) and the proton-proton correlation.

In the future this topic will be revolutionized by the use of sophisticated 4π detectors such as the Plastic Ball/Wall. Early results from this detector

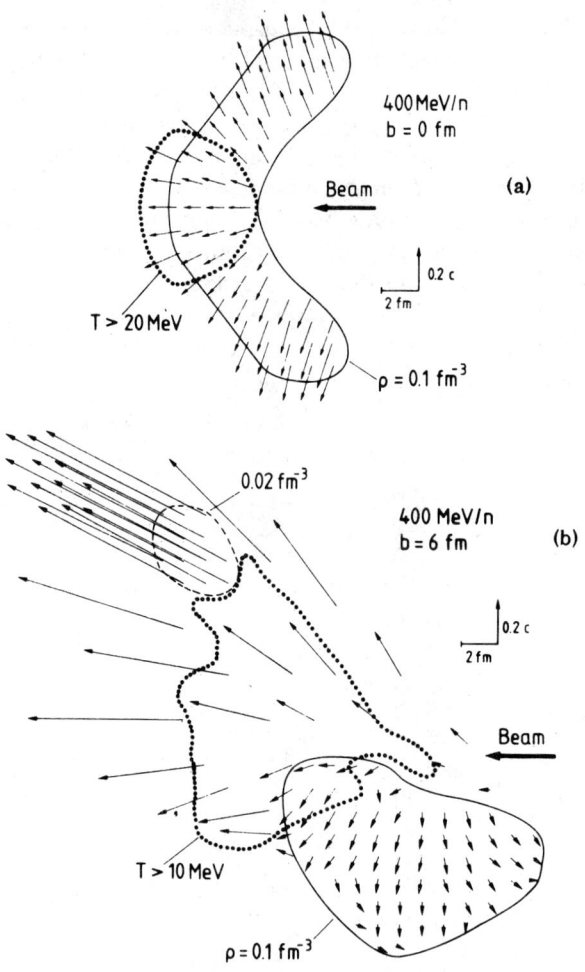

Figure 14 Hydrodynamical side splash (*upper*) and bounce-off (*lower*) predicted by Stöcker et al (97).

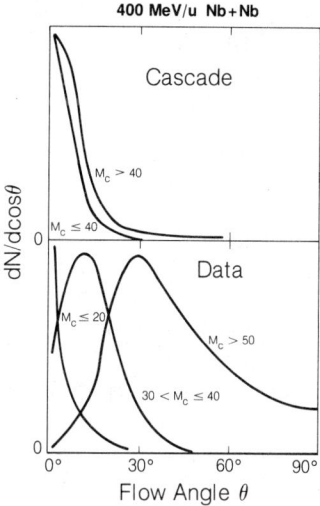

Figure 15 Flow angle distributions for different multiplicity bins in 400-A-MeV Nb+Nb collisions (105). Also shown are the predictions of cascade calculations (156).

are already showing its power for global analysis of complete events. One problem is to find a parameterization of events in which as many as a hundred particles may be detected. An attractive approach is to use the kinetic tensor $F_{ij} = \sum_v p_i(v)p_j(v)/2m(v)$, which approximates the event shape by an ellipsoid whose orientation in space and aspect ratios can be calculated by diagonalizing the tensor. The orientation of the ellipse away from the beam is called the flow angle. Results obtained from Ca+Ca collisions are in reasonable agreement with cascade calculations, but preliminary data from Nb+Nb collisions (105) show clear evidence for flow angles far in excess of the predictions of these calculations (Figure 15). Such behavior is predicted by fluid dynamical calculations (106). Results from heavier systems such as Au+Au are eagerly awaited by theorists and experimentalists alike.

3.4 *Cooperative Process*

If nuclear matter is suddenly compressed along the beam direction as happens in a high-energy nuclear collision, the energy density may increase. This may induce meson production below the threshold energy of free NN collisions. Therefore, pion production below 290 A MeV (107–109) and K^- production below 2.5 A GeV (110) have been studied extensively. Pion spectra at kinetic energies beyond the free NN kinematical limit have also been studied (111; E. Aslanides et al, 1984, private communication). A local high-density region may produce particle emission in the very high-p_T region (44) or in the backward direction (113–115). Such phenomena of particle emission far beyond the kinematical limits associated with free NN scattering may be called cooperative processes, since they require the cooperation of several nucleons.

Let us present an example in which cooperative particle emission is clearly observed. This is an experiment of pion production close to the absolute kinematic limit in 303-A-MeV ^{3}He + ^{6}Li collisions (111; E. Aslanides et al, 1984, private communication). The data cover the region up to the absolute kinematical limit at which all the available energy is converted into a single pion while forming the ground state of the compound nucleus ^{9}C, as shown in Figure 16. In this phenomenon, called pionic fusion, the available energy carried by individual nucleons is concentrated into a small region to create the pion. The mechanism behind this process is not yet understood and further experimental studies are desirable to see if pionic fusion can occur at different beam energies or for much heavier projectile-target combinations.

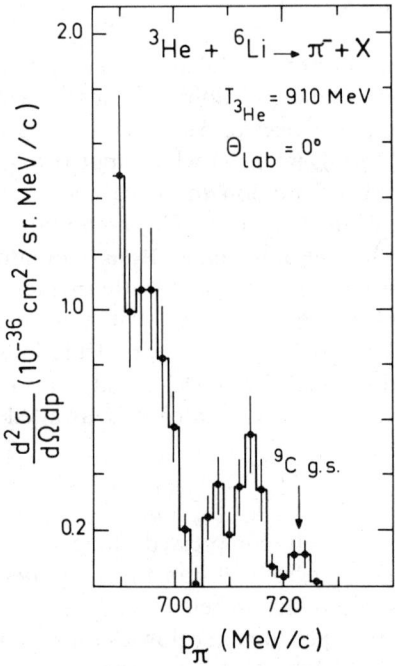

Figure 16 Pion momentum spectrum measured at 0° near the kinematical limit of pion emission in 303-A-MeV ^{3}He + ^{6}Li collisions. Data are provided by E. Aslanides et al (1984, private communication).

4. FRAGMENT EMISSION AT FORWARD ANGLES

In Sections 2 and 3 we focussed our attention on the fate of the participant nucleons. However, some of the earliest experiments using relativistic nuclear beams were concerned with a quite different class of events. These reactions result in the break-up of the projectile or projectile spectator into fragments, one or more of which are moving at close to the original beam velocity.

The inclusive yield of fragments in reactions of this type has been studied in a series of experiments using ^{4}He, ^{12}C, ^{16}O, and ^{18}O beams of 1 and 2 A GeV (116–119). The momentum distributions of the fragments are found to be Gaussian in the projectile frame with widths that are small (typically 200–300 MeV/c) and that do not vary appreciably with bombarding energy. This leads to the introduction of the concept of limiting fragmentation in analogy with high-energy physics. It is also observed that the cross sections for production of individual isotopes can be factored in the following way:

$$\sigma(B, F, T) = \gamma_B^F \gamma_T, \qquad 5.$$

where γ_B^F depends solely on the projectile B and fragment F, and γ_T is a function only of the target T.

4.1 *Models of the Fragmentation Process*

The observation of factorization prompted the development of a statistical model (120, 121) for the reaction process in which it is assumed that the nucleus breaks up as a result of a direct reaction between target and projectile. It was subsequently shown (122) that this model can account nicely for the variation of the measured momentum widths with the mass of the fragment if it is assumed that the fragment is formed by picking the nucleons from a Fermi gas distribution and calculating the dispersion in their momentum. This leads to an expression for the width σ of the form

$$\sigma^2 = \frac{1}{5} p_F^2 \frac{F(B-F)}{B-1} \qquad 6.$$

where B and F are, respectively, the projectile and target masses and p_F is the Fermi momentum in the nucleus and can be measured independently by, for example, electron scattering. The success of this description can be seen in Figure 17. However, it was also shown that the same dependence would be observed if the fragments were to arise from the statistical decay of an excited projectile [a model that has been applied successfully to describe the relative abundance of the different fragments (123)] and does not necessarily reflect the ground-state Fermi momentum. Therefore, the measurement of momentum widths alone may be less sensitive to the reaction mechanism.

Whether the mechanism is direct break-up or excitation followed by decay, the initial interaction between projectile and target has been treated purely phenomenologically. A more complete description of the reaction is provided by the abrasion-ablation model (3, 124–127). In its earliest form (3), the model assumes that the reaction can be divided into two stages. The first stage is a fast one in which the participant nucleons are sheared off (abrasion). It is followed by the second stage: statistical decay of an excited projectile remnant (ablation). This simple picture has been much refined and calculations of the abrasion stage have been made using Glauber's multiple scattering theory (124–127) and using cascade codes (128, 129). These calculations successfully describe ^{12}C and ^{16}O fragmentation (117) and also the fragmentation of ^{40}Ar at 213 A MeV (130).

Despite the success of these relatively simple models, it is probably

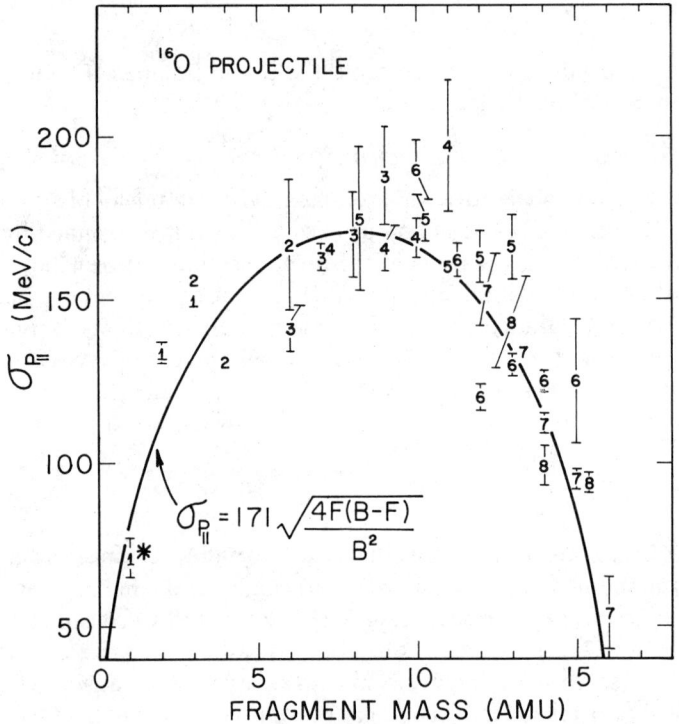

Figure 17 Momentum widths of the isotopes produced in fragmentation of ^{16}O at 2.1 A GeV (117).

premature to infer that fragmentation is well understood. This has been emphasized by several theorists who have attempted to obtain precise agreement between theory and experiment. For example, the measured momentum widths are as much as 30% narrower than would be expected from inelastic electron scattering. At least part of this discrepancy may arise from neglect of local correlations between nucleons (131, 132). A second example is that factorization seems to hold much better than would be expected from the impact parameter dependence predicted by the abrasion-ablation model (133, 134). The process may be more peripheral than can reasonably be predicted using this model (119) although this may be a fortuitous consequence of measuring an inclusive cross section (135). Finally, there are obvious Coulomb effects that become important as the bombarding energy is reduced below 100 A MeV (136–138), which indicates that limiting fragmentation is only valid above this energy.

Small but significant deviations of this kind have led to the introduction of a new, peripheral, model (139) for fragmentation that includes effects of absorption of the fragments by the target. In this model, the separation energy of the fragment becomes the most important physical parameter in determining the momentum width. This model has succeeded in describing simultaneously the detailed structure of the momentum widths, in particular the variation in width of different fragments of the same mass, the relative isotopic yields, and the Coulomb effects. In summary there is a clear need for further experimental information to elucidate the reaction mechanism of projectile fragmentation. The more detailed information that will come from studies using multiparticle spectrometers will obviously be important, but further inclusive experiments using heavy projectiles should not be neglected.

4.2 Electromagnetic Effects

So far, we have assumed that the interaction between projectile and target is a purely nuclear one. This is inadequate in some cases, in particular for those reaction channels in which the corresponding photonuclear cross section is large and when the charge of the target is high. The cross sections for this process can be calculated using the Weizsäcker-Williams theory (140) to obtain the virtual photon spectrum and coupling this to the appropriate photodisintegration cross section. Unfortunately, this process is essentially indistinguishable from a nuclear process leading to the same final state. This problem can, however, be overcome by using the factorization property of fragmentation reactions if the yield is measured from a series of different targets. In general, the electromagnetic process will only contribute strongly to a few channels, in particular those leading to single-particle removal. One can therefore obtain the target factors, γ_T,

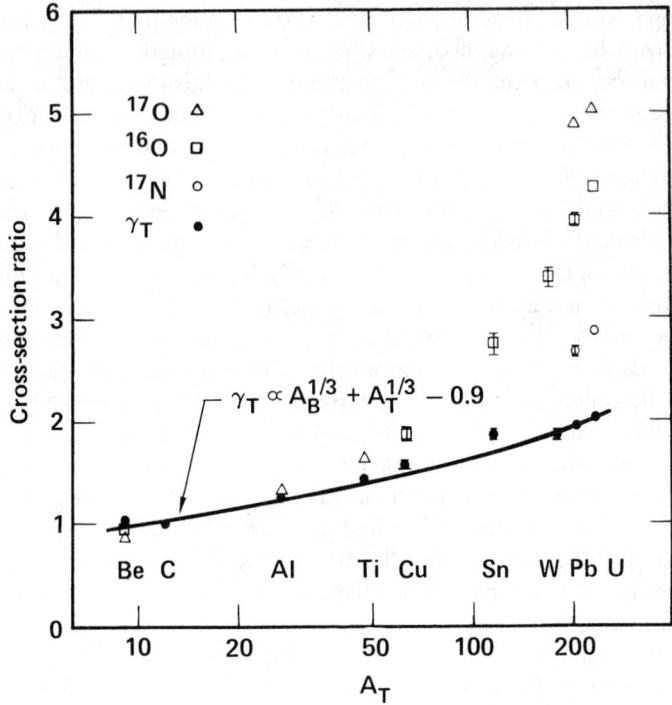

Figure 18 Relative cross sections for isotope production by fragmentation of ^{18}O at 1.8 A GeV showing the enhancement due to electromagnetic dissociation in one- and two-nucleon removal channels (142).

using channels for which electromagnetic dissociation is negligible and thence estimate the nuclear background for the channels of interest and subtract it from the experimentally observed cross section. This procedure has been used with considerable success in studying the fragmentation of ^{16}O, ^{18}O, and ^{56}Fe (141–143) and is illustrated in Figure 18. The absolute cross sections are in good agreement with the Weizsäcker-Williams predictions and with those of more sophisticated models (144) for the virtual photon spectrum.

Another example of electromagnetic effects is seen in pion production at forward angles. Figure 19 shows the π^- yield has a strong peak (and the corresponding π^+ yield has a dip) at the projectile velocity. This feature is understood as resulting from the Coulomb forces of other projectile fragments on the light-mass pion (146).

4.3 Applications of Fragmentation Reactions

The results of fragmentation reactions are having important additional consequences. The most exciting example is the very considerable potential for fragmentation reactions to increase our knowledge of nuclear structure. We have already noted the wide range of nuclear fragments that are produced in these reactions. Indeed almost any combination of the neutrons and protons inside a projectile will be emitted at some level. The important question is whether nuclei far from the valley of stability, such as ^{8}He or ^{34}Mg, will be produced sufficiently often to be measurable in the laboratory. This question has been answered empirically by experiments using high-energy nuclear beams that have uncovered a variety of light, neutron-rich isotopes (147–149). It is important to note that, whereas the cross sections are comparable to those of other techniques, the beam intensities are typically five orders of magnitude less, which demonstrates the very high efficiency of the fragmentation technique. This comes from the excellent collimation of the fragments in the beam direction and the ability to use thick targets.

The persistence of velocity also enables the fragments to be used in

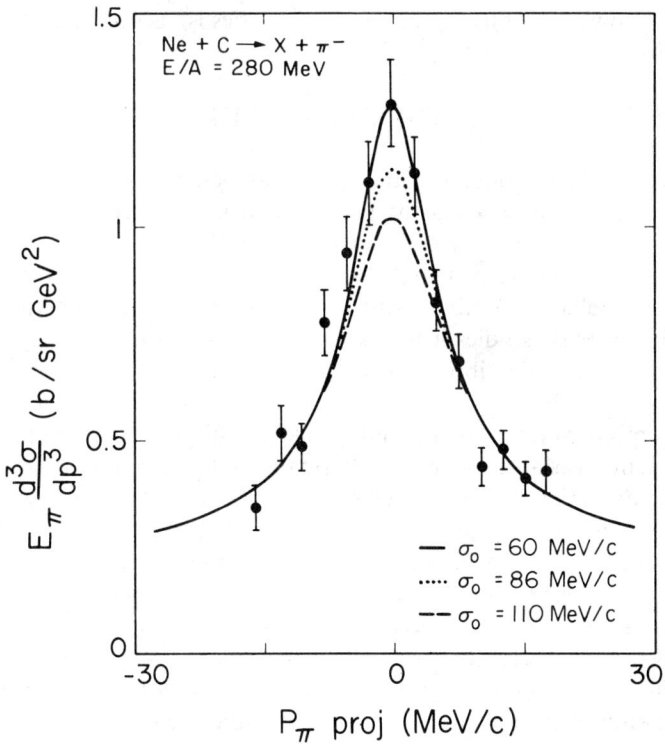

Figure 19 The π^- spectrum at $0°$ for 280 A GeV ^{20}Ne + ^{12}C (145, 146). The theoretical curves (146) are shown for three different choices of fragmentation velocity and dispersion parameter σ_0.

secondary beam experiments with very high efficiency. Many conventional aspects of nuclear structure study can then be extended to unstable nuclei with half-lives as short as a few milliseconds and that are inaccessible to conventional techniques. Some half-life measurements have already been made (150) and several novel experiments are planned, including studies of mirror magnetic moments, electromagnetic dissociation, and reaction cross sections.

A further example of the application of our knowledge of fragmentation reactions is in cosmic ray physics. The species distribution of cosmic rays detected on earth or by satellite may be substantially different from that at their source because of reactions in the interstellar medium. Measured fragmentation cross sections can be used to correct the observed yields.

A final example is in the use of heavy ion beams for medical purposes such as tomography and radiotherapy. In the latter case, one is taking advantage of the fact that the ionization by a heavy ion is greatest at the end of its range, and can thereby reduce the radiation damage to healthy tissue. However, the incident beam can break up before coming to rest and a precise knowledge of fragmentation systematics is needed to calculate the dose distribution correctly.

5. SUMMARY AND FUTURE PERSPECTIVES

As stated in the Introduction, the field of high-energy nuclear collisions is developing rapidly. During the past 5–10 years, a large body of experimental data has been accumulated. In particular, single-particle and two-particle inclusive distributions have been measured for a variety of projectile and target combinations at beam energies from $200\ A$ MeV to $2\ A$ GeV. From these studies it has become clear that a high-energy nuclear collision is a nonequilibrium many-body process including not only the primary NN collisions but also collisions between the product particles, such as pions, and the surrounding nucleons. With the exception of a few experimental variables, such as the d/p ratio and pion multiplicity, it seems that the dominant reaction mechanism can be described quite well using multiple-collision theories with very reasonable theoretical assumptions.

Currently, the following three subjects are being studied intensively. The first one concerns the creation of hot, dense nuclear matter and its properties. As discussed in Section 3.1, it is likely that such nuclear matter is created in high-energy nuclear collisions. We have seen that attempts to study its properties have started from the analysis of d/p ratios (Section 2.3) or pion multiplicity (Section 3.2); the conclusions are still controversial and the subject of lively debate.

The second subject is the search for new dynamic modes that are produced during the course of a nuclear collision. This was discussed in Sections 3.3 and 3.4. Some preliminary results are already available that indicate the presence of nucleonic flow and this is one field where one really can expect to make significant progress in the near future. Sophisticated detectors designed specifically to measure multiparticle final states, such as the Plastic Ball/Wall and HISS (151), are just coming into use, as are the heavy-ion beams needed for these studies. Particle emission beyond the free NN kinematical limit is another indication of cooperative behavior between nucleons. We noted the phenomenon of pionic fusion for light-mass collisions, in which the total available kinetic energy in the c.m. frame is converted to emission of a single pion. We also mentioned subthreshold pion emission and high-p_T particle emission. From these studies it is clear that processes certainly exist that cannot be explained by the superposition of incoherent NN collisions.

The third subject is related to physics of projectile fragments, as discussed in Section 4. Although there are some unknowns in the production mechanism of these fragments, the physics of projectile fragments is a unique feature of high-energy nuclear collisions. Strong Coloumb interactions induce a large photonuclear cross section as well as a large π^-/π^+ ratio. A fascinating aspect of projectile fragments is the production of unstable nuclei far from the stability line and their application to nuclear physics. Since these nuclei are emitted in a very restricted kinematic domain, they can be used as high-quality secondary beams as well.

One of the ultimate goals in the research of high-energy nuclear collisions is to create and study matter that is different from the normal nuclear matter seen in stable nuclei. High-energy nuclear collisions may transform the nucleons into other baryons, such as Δ, N*, Λ, and also create mesons, such as π, ρ, K, so we may expect the production of matter containing these particles. Their quantum numbers differ from those of the nucleons and they may be considered as impurities in the nuclear matter. For example, the behavior of a ρ meson imbedded in nuclear matter is of interest for the understanding of nuclear forces. Another example is the formation of a projectile fragment in which a produced Λ particle is trapped. Such hypernuclei have been studied rather extensively with mesonic probes, but nuclear collisions offer the possibility of forming hypernuclei with multiple strangeness. If the impurity level is high enough, then the matter may change its character. For example, at 700 A MeV, 50% of the NN collisions create Δ's, which might then result in an interesting new form of baryonic matter containing mostly Δ particles (152).

Recently there has been considerable discussion of a more dramatic change in the properties of nuclear matter that may take place in a high-energy nuclear collision. It has been suggested that a phase transition from hadronic matter into a quark-gluon plasma may occur at high density and pressure (153, 154). This transition would involve simultaneously the break-down of quark confinement and the formation of a new type of many-body system and is of fundamental importance to both nuclear and particle physics. At present, the exact form of the transition and the way in which it would become manifest are unclear. However, there does seem to be a consensus that the presently available energies are too low; thus new experimental facilities are currently being planned. We feel confident that the experience already gained from the studies described in this review will prove invaluable for this new venture.

ACKNOWLEDGMENT

We are very grateful to all of our colleagues who have assisted us with material and advice during our preparation of this review.

Literature Cited

1. Deleted in proof
2. Eisenberg, Y. 1954. *Phys. Rev.* 96:1378
3. Bowman, J. D., Swiateki, W. J., Tsang, C. F. 1973. *Lawrence Berkeley Lab. Rep. No. LBL-2908*. Berkeley: LBL
4. Nagamiya, S. 1980. *Nucl. Phys. A* 355:517c
5. Manko, V. I., Nagamiya, S. 1982. *Nucl. Phys. A* 384:475
6. Weise, R., Brown, G. E. 1976. *Phys. Rep.* 27C:1
7. Migdal, A. B. 1978. *Rev. Mod. Phys.* 50:107
8. Lee, T. D., Wick, G. C. 1974. *Phys. Rev. D* 9:2291
9. Lee, T. D. 1976. *Rev. Mod. Phys.* 47:267
10. Jacob, M., Tran Thanh Van, J. 1982. *Phys. Rep.* 88C:321
11. Nagamiya, S., Lemaire, M.-C., Moeller, E., Schnetzer, S., Shapiro, G., et al. 1981. *Phys. Rev. C* 24:971
12. Randrup, J. 1978. *Phys. Lett.* 76B:547
13. Sobel, M., Siemens, P. J., Bondorf, J. P., Bethe, H. A. 1975. *Nucl. Phys. A* 251:502
14. Westfall, G. D., Gosset, J., Johansen, P. J., Poskanzer, A. M., Meyer, W. G., et al. 1976. *Phys. Rev. Lett.* 37:1202
15. Kapusta, J. I. 1977. *Phys. Rev. C* 16:1493
16. Mekjian, A. Z. 1978. *Phys. Rev. C* 17:1051
17. Das Gupta, S., Mekjian, A. Z. 1981. *Phys. Rep.* 72C:131
18. Myers, W. D. 1978. *Nucl. Phys. A* 296:177
19. Gosset, J., Kapusta, J. I., Westfall, G. D. 1978. *Phys. Rev. C* 18:844
20. Das Gupta, S. 1978. *Phys. Rev. Lett.* 41:1450
21. Siemens, P. J., Rasmussen, J. O. 1979. *Phys. Rev. Lett.* 42:844
22. Gudima, K. K., Toneev, V. D. 1978. *Yad. Fiz.* 27:658 (*Sov. J. Nucl. Phys.* 27:351)
23. Stevenson, J. D. 1978. *Phys. Rev. Lett.* 41:1702
24. Gudima, K. K., Iwe, H., Toneev, V. D. 1979. *J. Phys. G* 5:229
25. Yariv, Y., Fraenkel, Z. 1979. *Phys. Rev. C* 20:2227
26. Cugnon, J. 1980. *Phys. Rev. C* 22:1885
27. Cugnon, J., Mizutani, T., Vandermeulen, J. 1981. *Nucl. Phys. A* 352:505
28. Cugnon, J., Kinet, J., Vandermeulen, J. 1982. *Nucl. Phys. A* 379:553
29. Yariv, Y., Fraenkel, Z. 1981. *Phys. Rev. C* 24:488
30. Hüfner, J., Knoll, J. 1977. *Nucl. Phys. A* 290:460
31. Hüfner, J. 1978. *Proc. 4th High-Energy Heavy Ion Summer Study*, LBL-7766, Conf-780766:135. Berkeley: LBL
32. Knoll, J. 1979. *Phys. Rev. C* 20:773
33. Bohrmann, S., Knoll, J. 1981. *Nucl. Phys. A* 356:498
34. Schnetzer, S., Lemaire, M.-C., Lombard, R., Moeller, E., Nagamiya, S., et al. 1982. *Phys. Rev. Lett.* 49:989
35. Randrup, J., Ko, C. M. 1980. *Nucl. Phys. A* 343:519

36. Asai, F. 1981. *Nucl. Phys. A* 356:519
37. Nagamiya, S. 1982. *Phys. Rev. Lett.* 49:1383
38. Randrup, J. 1981. *Phys. Lett.* 99B:9
39. Zwermann, W., Schürmann, B., Dietrich, K., Martschew, E. 1984. *Phys. Lett. B* 134:397
40. Cugnon, J., Lombard, R. 1984. *Phys. Lett. B* 134:392
41. Asai, F., Sato, H., Sano, M. 1981. *Phys. Lett.* 98B:19
42. Ko, C. M. 1981. *Phys. Rev. C* 23:2760
43. Halemane, T. R., Mekjian, A. Z. 1982. *Phys. Rev. C* 25:2398
44. Hamagaki, H., Bai, X. X., Hashimoto, O., Kadota, S., Kimura, K., et al. 1984. *LBL-17276 Preprint*
45. Nagamiya, S., Anderson, L., Brückner, W., Chamberlain, O., Lemaire, M.-C., et al. 1978. *Phys. Lett.* 81B:147
46. Tanihata, I., Lemaire, M.-C., Nagamiya, S., Schnetzer, S. 1980. *Phys. Lett.* 97B:363
47. Tanihata, I., Nagamiya, S., Schnetzer, S., Steiner, H. 1981. *Phys. Lett.* 100B:121
48. Pirner, H. J., Schürmann, B. 1979. *Nucl. Phys. A* 316:461
49. Chemtob, M., Schürmann, B. 1980. *Nucl. Phys. A* 336:508
50. Gutbrod, H. H., Sandoval, A., Johansen, P. J., Poskanzer, A. M., Gosset, J., et al. 1976. *Phys. Rev. Lett.* 37:667
51. Lemaire, M.-C., Nagamiya, S., Schnetzer, S., Steiner, H., Tanihata, I. 1979. *Phys. Lett.* 85B:38
52. Knoll, J., Randrup, J. 1981. *Phys. Lett.* 103B:264
53. Butler, S. F., Pearson, C. A. 1963. *Phys. Rev.* 129:836
54. Schwarzschild, A., Zupančič, C. 1963. *Phys. Rev.* 129:854
55. Sandoval, A., Gutbrod, H. H., Meyer, W. G., Stock, R., Lukner, Ch., et al. 1980. *Phys. Rev. C* 21:1321
56. Bond, R., Johansen, P. J., Koonin, S. E., Garpman, S. 1977. *Phys. Lett.* 71B:43
57. Jennings, B. K., Das Gupta, S., Mobed, N. 1981. *Phys. Rev.* 25:278
58. Bertsch, G. F. 1983. *Nucl. Phys. A* 400:221c
59. Gutbrod, H. H., Löhner, H., Poskanzer, A. M., Penner, T., Riedesel, H., et al. 1983. *Phys. Lett.* 127B:317
60. Gutbrod, H. H., Löhner, H., Poskanzer, A. M., Penner, T., Riedesel, H., et al. 1983. *Nucl. Phys. A* 400:343c
61. Sato, H., Yazaki, K. 1981. *Phys. Lett.* 98B:153
62. Siemens, P. J., Kapusta, J. I. 1979. *Phys. Rev. Lett.* 43:1486, 1690(E)
63. Bertsch, G. F., Cugnon, J. 1981. *Phys. Rev. C* 24:2514
64. Kapusta, J. I., Strottman, D. 1981. *Phys. Rev. C* 23:971, 1282
65. Csernai, L. P., Barz, H. W. 1980. *Z. Phys. A* 296:173
66. Das Gupta, S., Jennings, B. K., Kapusta, J. I. 1982. *Phys. Rev. C* 26:274
67. Fai, G., Randrup, J. 1981. *Nucl. Phys. A* 381:557
68. Hanbury-Brown, R., Twiss, R. Q. 1956. *Nature* 178:1046
69. Goldhaber, G., Goldhaber, S., Lee, W., Pais, A. 1960. *Phys. Rev.* 120:300
70. Cocconi, G. 1974. *Phys. Lett.* 49B:459
71. Kopylov, G. I. 1974. *Phys. Lett.* 50B:572
72. Koonin, S. E. 1977. *Phys. Lett.* 70B:43
73. Yano, F. B., Koonin, S. E. 1978. *Phys. Lett.* 78B:556
74. Zajc, W. A., Bistirlich, J. A., Bossingham, R. R., Bowman, H. R., Clawson, C. W., et al. 1981. *Proc. 5th High-Energy Summer Study*, LBL-12652:350. Berkeley: LBL
75. Beavis, D., Fung, S. Y., Gorn, W., Huie, A., Keane, D., et al. 1983. *Phys. Rev. C* 27:910
76. Zarbakhsh, Z., Sagle, A. L., Brochard, F., Mulera, T. A., Perez-Mendez, V., et al. 1981. *Phys. Rev. Lett.* 36:1268
77. Wieman, H., Gutbrod, H. H., Gustafsson, H. A., Kolb, B., Löhner, H., et al. 1983. *Proc. 6th High-Energy Heavy Ion Summer Study*, LBL-16281:325. Berkeley: LBL
78. Chapline, C. F., Johnson, H. H., Teller, E., Weiss, M. S. 1973. *Phys. Rev. D* 8:4302
79. Deutchmann, P. A., Townsend, L. W. 1980. *Phys. Rev. Lett.* 45:1622
80. Gyulassy, M., Kauffmann, S. K., Wilson, L. W. 1979. *Phys. Rev. C* 20:2267
81. Gyulassy, M. 1982. *Phys. Rev. Lett.* 48:454
82. Nagamiya, S. 1983. *Nucl. Phys. A* 400:399c
83. Nagamiya, S. 1984. *Nucl. Phys.* In press
84. Nagamiya, S., Lemaire, M.-C., Schnetzer, S., Steiner, H., Tanihata, I. 1980. *Phys. Rev. Lett.* 45:602
85. Gutbrod, H. H., Gustafsson, H. A., Kolb, B., Löhner, H., Ludewigt, B., et al. 1982. *Proc. 6th Balaton Topical Conf. on High-Energy Nuclear Phys.*, ed. J. Erö, p. 269. Budapest
86. Fung, S. Y., Gorn, W., Kiernan, G. P., Liu, F. F., Lu, J. J., et al. 1978. *Phys. Rev. Lett.* 40:292
87. Sandoval, A., Stock, R., Stelzer, H. E., Renfordt, R. E., Harris, J. W., et al. 1980. *Phys. Rev. Lett.* 45:874

88. Stock, R., Bock, R., Brockman, R., Harris, J. W., Sandoval, A., et al. 1982. *Phys. Rev. Lett.* 49:1236
89. Malfliet, R., Schürmann, B. 1983. *Phys. Rev. C* 28:1136
90. Bertsch, G. F., Kruse, H., Das Gupta, S. 1983. Mich. State Univ. Preprint
91. Cahay, M., Cugnon, J., Vandermuelen, J. 1983. *Nucl. Phys. A* 411:524
92. Scheid, W., Muller, H., Greiner, W. 1974. *Phys. Rev. Lett.* 32:741
93. Amsden, A. A., Bertsch, G. F., Harlow, F. H., Nix, J. R. 1975. *Phys. Rev. Lett.* 35:905
94. Kitazoe, Y., Matsuoka, K., Sano, M. 1976. *Prog. Theor. Phys.* 56:860
95. Bondorf, J. P., Garpman, S. I. A., Zimanyi, J. 1978. *Nucl. Phys. A* 296:320
96. Nix, J. R. 1979. *Prog. Part. Nucl. Phys.* 2:237
97. Stöcker, H., Maruhn, J. A., Greiner, W. 1980. *Phys. Rev. Lett.* 44:725
98. Tang, H. H. K., Wong, C. Y. 1980. *Phys. Rev. C* 21:1846
99. Stöcker, H., Hofmann, J., Maruhn, J. A., Greiner, W. 1980. *Prog. Part. Nucl. Phys.* 4:133
100. Nix, J. R., Strottman, D. 1981. *Phys. Rev. C* 23:2548
101. Stock, R., Gutbrod, H. H., Meyer, W. G., Poskanzer, A. M., Sandoval, A., et al. 1980. *Phys. Rev. Lett.* 44:1243
102. Stöcker, H., Riedel, C., Yariv, Y., Csernai, L. P., Buchwald, G., et al. 1981. *Phys. Rev. Lett.* 47:1807
103. Schürmann, B., Chemtob, M. 1980. *Z. Phys. A* 294:371
104. Csernai, L. P., Greiner, W., Stöcker, H., Tanihata, I., Nagamiya, S., et al. 1982. *Phys. Rev. C* 25:2482
105. Ritter, H. G., Gustafsson, H. A., Gutbrod, H. H., Kolb, B., Löhner, H., et al. 1983. *Proc. 6th High-Energy Heavy-Ion Study* and *2nd Workshop on Anomalons*, LBL-16281:191. Berkeley: LBL
106. Csernai, L. P., Stöcker, H., Subramanian, P. R., Graebner, G., Rosenhauer, A., et al. 1983. *Phys. Rev. C* 28:2001
107. Benenson, W., Bertsch, G. F., Crawley, G. M., Kashy, E., Nolen, J. A., et al. 1979. *Phys. Rev. Lett.* 43:683; 44:54(E)
108. Johansson, T., Gustafsson, H. A., Jacobsson, B., Kristiansson, P., Norén, B., et al. 1982. *Phys. Rev. Lett.* 48:732
109. Nagamiya, S., Hamagaki, H., Hecking, P., Lombard, R., Miake, Y., et al. 1982. *Phys. Rev. Lett.* 48:1708
110. Shor, A., Ganezer, K., Abachi, S., Carroll, J., Geaga, J., et al. 1982. *Phys. Rev. Lett.* 48:1597
111. Aslanides, E., Fassnacht, P., Hibou, F., Chiavassa, E., Dellacasa, G., et al. 1979. *Phys. Rev. Lett.* 43:1466; 45:1738(E)
112. Deleted in proof
113. Baldin, A. M. 1975. *AIP Conf. Proc.* 26:621
114. Baldin, A. M., Guiordenescu, M., Zubarev, V. N., Moroz, N. S., Povtoreiko, A. A., et al. 1975. *Yad. Fiz.* 20:1201 (*Sov. J. Nucl. Phys.* 20:629)
115. Schroeder, L. S., Chessin, S. A., Geaga, J. V., Grossiord, J. Y., Harris, J. W., et al. 1979. *Phys. Rev. Lett.* 43:1787
116. Anderson, L., Brükner, W., Moeller, E., Nagamiya, S., Nissen-Meyer, S., et al. 1983. *Phys. Rev. C* 28:1224
117. Heckman, H. H., Greiner, D. E., Lindstrom, P. J., Bieser, F. S. 1972. *Phys. Rev. Lett.* 27:926
118. Lindstrom, P. J., Greiner, D. E., Heckman, H. H., Cork, B., Bieser, F. S. 1975. *Lawrence Berkeley Lab. Rep. No. LBL-3650*. Berkeley: LBL
119. Olson, D. L., Berman, B. L., Greiner, D. E., Heckman, H. H., Lindstrom, P. J., et al. 1983. *Phys. Rev. C* 28:1602
120. Feshbach, H., Huang, K. 1973. *Phys. Lett.* 47B:300
121. Bhaduri, K. 1974. *Phys. Lett.* 50B:311
122. Goldhaber, A. S. 1974. *Phys. Lett.* 53B:306
123. Lukyanov, V. K., Titov, A. L. 1975. *Phys. Lett.* 57B:10
124. Hüfner, J., Schaffer, K., Schürmann, B. 1975. *Phys. Rev. C* 12:1888
125. Abul-Magd, A., Hüfner, J., Schürmann, B. 1976. *Phys. Lett.* 60B:327
126. Abul-Magd, A., Hüfner, J. 1976. *Z. Phys. A* 277:379
127. Hüfner, J., Sander, C., Wolschin, G. 1978. *Phys. Lett.* 73B:2891
128. Morrissey, D. J., Oliveira, L. F., Rasmussen, J. O., Seaborg, G. T., Yariv, Y., et al. 1979. *Phys. Rev. Lett.* 43:1139
129. De Oliveira, L. F. 1978. *PhD Thesis, UC Berkeley*; 1978. *Lawrence Berkeley Lab. Rep. LBL-8561*. Berkeley: LBL
130. Viyogi, Y. P., Symons, T. J. M., Doll, P., Greiner, D. E., Heckman, H. H., et al. 1979. *Phys. Rev. Lett.* 42:33
131. Bertsch, G. 1981. *Phys. Rev. Lett.* 46:472
132. Murphy, M. J. 1984. *Phys. Lett.* 135B:25
133. Bleszynski, M., Sander, C. 1979. *Nucl. Phys. A* 326:525
134. Cugnon, J., Sartor, R. 1980. *Phys. Rev. C* 21:2342
135. Stevenson, J. D., Martinis, J., Price, P. B. 1981. *Phys. Rev. Lett.* 47:990
136. Van Bibber, K., Hendrie, D. L., Scott, D. K., Wieman, H. H., Schroeder, L. S., et al. 1979. *Phys. Rev. Lett.* 43:840

137. Mougey, J., Ost, R., Buenerd, M., Cole, A. J., Guet, C., et al. 1981. *Phys. Lett.* 105B:25
138. Wong, C. Y., Van Bibber, K. 1982. *Phys. Rev. C* 25:1460
139. Friedman, W. 1983. *Phys. Rev. C* 27:569
140. Williams, E. J. 1933. *Prog. R. Soc. A* 139:163
141. Heckman, H. H., Lindstrom, P. J. 1976. *Phys. Rev. Lett.* 37:56
142. Olson, D. L., Berman, B. L., Greiner, D. E., Heckman, H. H., Lindstrom, P. J., et al. 1981. *Phys. Rev. C* 24:1529
143. Westfall, G. D., Wilson, L. W., Lindstrom, P. J., Crawford, H. J., Greiner, D. E. 1979. *Phys. Rev. C* 19:1309
144. Jäckle, R. R., Pilkuhn, H. 1975. *Nucl. Phys. A* 247:521
145. Sullivan, J. P., Bistirlich, J. A., Bowman, H. R., Bossingham, R., Buttke, T., et al. 1982. *Phys. Rev. C* 25:1499
146. Radi, H. M. A., Rasmussen, J. O., Sullivan, J. P., Fraenkel, K. K., Hashimoto, O. 1982. *Phys. Rev. C* 25:1518
147. Symons, T. J. M., Viyogi, Y. P., Westfall, G. D., Doll, P., Greiner, D. E., et al. 1979. *Phys. Rev. Lett.* 42:40
148. Westfall, G. D., Symons, T. J. M., Greiner, D. E., Heckman, H. H., Lindstrom, P. J., et al. 1979. *Phys. Rev. Lett.* 43:1859
149. Stevenson, J. D., Price, P. B. 1981. *Phys. Rev. C* 24:2101
150. Murphy, M. J., Symons, T. J. M., Westfall, G. D., Crawford, H. J. 1982. *Phys. Rev. Lett.* 49:455
151. Greiner, D. E. 1983. *Nucl. Phys. A* 400:325c
152. Kondratyuk, L. A., Shapiro, I. S. 1970. *Yad. Fiz.* 12:401 (*Sov. J. Nucl. Phys.* 12:220)
153. Jacob, M., Satz, H., eds. 1982. *Quark Matter Formation and Heavy Ion Collisions.* Singapore: World Scientific
154. Ludlam, T. W., Wagner, H. E., eds. 1984. *Proc. Quark Matter '83.* In *Nucl. Phys. A* 418
155. Stöcker, H. 1981. *Lawrence Berkeley Lab. Rep. LBL-12302.* Berkeley: LBL
156. Gyulassy, M., Fraenkel, K. A., Stöcker, H. 1982. *Phys. Lett.* 110B:185

Ann. Rev. Nucl. Part. Sci. 1984. 34: 189–245

THE ROLE OF ROTATIONAL DEGREES OF FREEDOM IN HEAVY-ION COLLISIONS[1]

L. G. Moretto and G. J. Wozniak

Nuclear Science Division, Lawrence Berkeley Laboratory, University of California, Berkeley, California 94720

CONTENTS

1. INTRODUCTION 190
2. ROLE OF ENTRANCE-CHANNEL ANGULAR MOMENTUM IN EXIT-CHANNEL DISTRIBUTIONS 197
   2.1 Angular Distributions and the Classical Deflection Function 197
   2.2 Width of the Angular Distributions: Quantal or Statistical Fluctuations? 204
   2.3 Angular Distributions as a Function of Mass Asymmetry 207
   2.4 Rigid Rotation and Angular Momentum Fractionation Along the Mass Asymmetry Coordinate 212
3. ANGULAR-MOMENTUM-BEARING MODES AND PARTITION OF ANGULAR MOMENTUM 216
   3.1 The Rotational Degrees of Freedom of the Dinuclear Complex 216
   3.2 Angular Momentum Misalignment 221
   3.3 Angular Distributions of Sequential Fission and of Sequential Light Particle Emission 222
   3.4 Gamma-Ray Angular Distributions 225
   3.5 Experimental Spin Alignment From Gamma-Ray Angular Distributions 228
   3.6 Experimental Data From Sequential Decay 232
4. MORE AMBITIOUS MODELS 237
   4.1 Time-Dependent Hartree-Fock Model 237
   4.2 Coherent Surface Excitation Model 238
   4.3 Transport Models 239
5. CONCLUSION 242

[1] This work was supported by the Director, Office of Energy Research, Division of Nuclear Physics of the Office of High Energy and Nuclear Physics of the US Department of Energy under Contract DE-AC03-76SF00098. The US Government has the right to retain a nonexclusive royalty-free license in and to any copyright covering this paper.

Nec ratio solis simplex et recta patescit
quo pacto aestivis e partibus aegocerotis,
brumalis adeat flexus, atque inde revertens
  cancri se ut vertat metas ad solstitialis,
lunaque mensibus id spatium videatur obire
annua sol in quo consumit tempora cursu.
Non inquam simplex his rebus reddita causast.

Lucretius, *De rerum natura* V 614–620

[And it is not clear nor simple how the sun moves from the hot to the cold regions from solstice to solstice, nor how the moon in a month covers the journey that the sun covers in a year. There is not a simple explanation at hand for this.]

1. INTRODUCTION

Nuclear science, not unlike other human endeavors, has been developed and defined by the tools at its disposal. As late as the early 1970s most accelerators provided a limited range of projectiles, typically protons, deuterons, tritons, ^{3}He, and ^{4}He. The picture of nuclear reactions painted by means of these projectiles was remarkably one-sided, but hardly anyone noticed it. The extreme polarization that ensued can be appreciated in terms of the following modern classification. Two classes of processes were unveiled and studied: On the one hand, the elastic and the nearly elastic reactions involving elementary excitations of the target; on the other, the complete amalgamation of the projectile and the target giving rise to a fully equilibrated intermediate, or compound, nucleus.

The former reactions were instrumental in defining the optical potential and the shell-model picture. Spectroscopy was the vogue, and anything not looking like a peak, and sharp at that, was inexorably discarded as annoying "background." Barely saved from such a "subtraction" of background were some broad structures like the single-particle "strength" and the giant dipole resonance because of special interests and special projectiles, respectively.

The latter reactions, of a more democratic nature, determined the application of statistical mechanics to nuclear physics. The cavalier treatment of matrix elements, spectroscopic factors and the like, made possible the recognition and the treatment of particle evaporation spectra as more or less continuous entities, as well as of statistical branching ratios and excitation functions, but it created a great chasm and an unsympathetic atmosphere between the two areas.

Of special interest in the second class of reactions was the process of fission, which, by involving the coherent motion of many particles at a time, earned the scornful derision of the spectroscopists but sensitized the observers of compound nuclei to collective macroscopic effects. With hindsight, one can argue that the polarization between spectroscopy and equilibrium statistics was more in the mind of the practitioners than in the experimental spectra. The presence of large cross sections in the intermediate inelasticity range was well known (see, for example, 34, 48), but what is now the virgin field of enlightened spectroscopists hunting for resonances was then the wasteland of their conservative forefathers.

The advent of heavy-ion accelerators in the early 1970s made available a range of projectiles and energies leading to reactions in which the complete fusion of the target and projectile was substantially depressed. The room left behind by the failure to form a compound nucleus was filled by reactions with intermediate to complete inelasticities and with a variable amount of mass, charge, spin, and isospin transfer (5, 43, 49, 65, 115). The failure of two nuclei to fuse into a compound nucleus can be understood in terms of the liquid drop model. The surface energy always favors fusion. Coulomb and centrifugal forces oppose it. From a static point of view, it is possible to define the boundary of stability of (liquid drop) nuclei in terms of two parameters,

$$x = \frac{E_{\text{coulomb}}}{2E_{\text{surface}}}, \quad y = \frac{E_{\text{rotational}}}{E_{\text{surface}}},$$

where E_{coulomb}, E_{surface}, and $E_{\text{rotational}}$ are defined for the spherical configuration and rigid rotation. With increasing x and y, the disruptive forces increase and make the nucleus progressively more unstable toward fission. The limits of stability (19, 104), corresponding to the fission barrier going to zero, are shown in Figure 1.

Unfortunately, statics is not sufficient to establish whether or not two colliding nuclei are going to fuse. Dynamics plays a vital role in the fusion process. Early attempts (12, 41, 109) treated the dynamics in a simplified fashion in terms of two spherical nuclei under the action of conservative forces (Coulomb, centrifugal, and nuclear) and dissipative forces (radial and tangential friction). The angular-momentum-dependent potentials increase with decreasing internuclear distance. As the nuclei touch, the nuclear interaction (proximity force) may lead to the appearance of a minimum that eventually disappears (see Figure 2) with increasing orbital angular momentum. Fusion criteria were then adopted on the basis of the presence of a minimum in the relevant potential energy curve (8) and also of a critical distance (38, 53). If a system can dynamically reach the minimum, it will fuse. Tangential friction gives the additional possibility of jumping from the entrance-channel l-wave potential energy curve to another at a lower l wave due to angular momentum transfer from orbital rotation to nuclear spin. Thus, even if the entrance-channel l-wave potential does not have a pocket, the system may still fuse if friction brings it down to a lower l-wave potential with a pocket. These simple models have been extended by Deubler & Dietrich (24), Birkelund et al (10), and Gross (42).

A general theoretical model of fusion and reseparation, which allows the two nuclei to deform, to form a neck, and amalgamate, was presented by Swiatecki (105). His approach explicitly considers the degrees of freedom that allow two nuclei to fuse. Thus such a model provides the most suitable framework not only for the description of fusion but of reseparation as well. Unfortunately, the illustration of this model goes beyond the scope and limits of this work.

Even if the two nuclei do not fuse, they do, of course, interact. During the interaction phase, a variety of degrees of freedom are called into play (33, 54, 66, 68, 82, 97, 98, 112). The coupling of the relative motion to the internal, or collective, nuclear degrees of freedom leads to energy dissipation, the most manifest property of these reactions. In a similar manner, angular momentum is transferred from orbital motion to nuclear spin. The exchange of matter between the two fragments may lead to an evolution along the mass asymmetry degree of freedom as well as to a redressing of a possible imbalance in the neutron-to-proton ratio of the two fragments in contact.

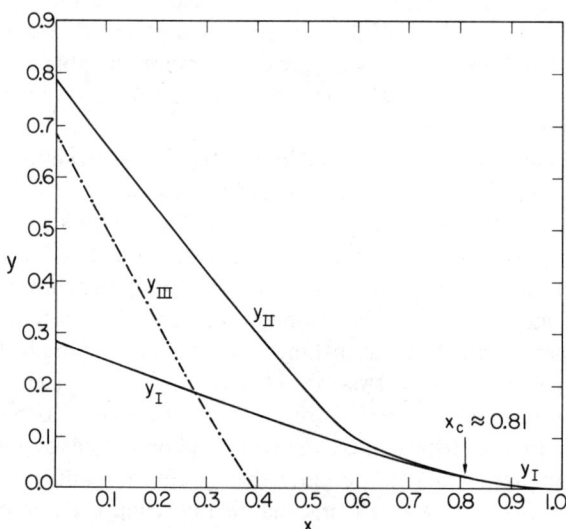

Figure 1 A classification of rotating systems according to the rotational parameter y and the fissility parameter x. With increasing angular momentum, rotating systems first deform into a flat shape (below Y_I), then into a triaxial shape (between Y_I and Y_{II}), and finally the fission barrier vanishes along curve Y_{II} (19).

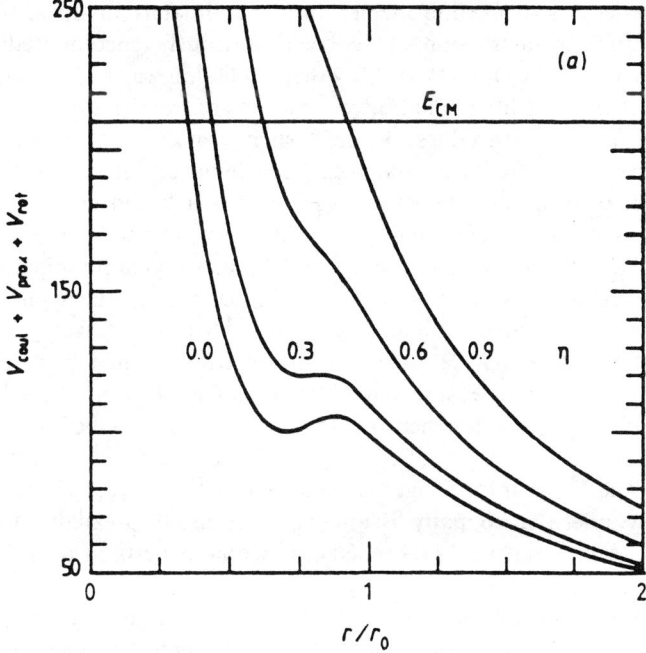

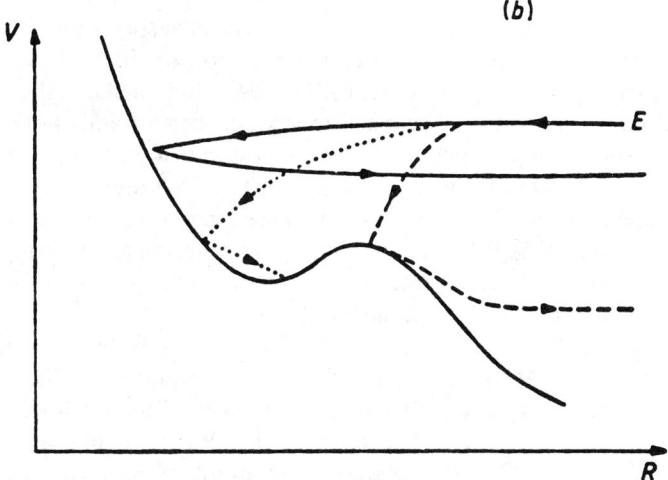

Figure 2 (a) Potential energies as a function of r/r_0 for various values of $\eta = l/l_{max}$. (b) Schematic showing the trajectories in the presence of weak (*full curve*), moderate (*dotted curve*), and strong (*dashed curve*) radial friction (68).

These degrees of freedom, which are so intimately involved with the nucleus-nucleus interaction, have been previously encountered in the fission process. The attempt to study them in fission was not too successful because of the inability of the experimenter to control the initial conditions of some relevant variables. Rather, such conditions are statistically determined at the fission saddle point, and thus very little variety can be observed. In deep inelastic processes, one has a broad choice of initial conditions for all the relevant degrees of freedom. The collision energy can be varied greatly and so can the entrance-channel angular momentum. The overall mass and charge of the system is limited only by the availability of stable or nearly stable isotopes (e.g. ^{238}U, ^{248}Cm or heavier). The mass asymmetry can be easily explored in its whole range from extreme asymmetry (p, α + nucleus) to symmetry (Ca + Ca, Nb + Nb, Pb + Pb, etc). The equilibration of the neutron-to-proton ratio can be explored by exploiting either the variety of isotopes available for a given atomic number Z (e.g. ^{40}Ca, ^{48}Ca) or the progressive increase of the neutron excess with Z. It is no wonder that so many fission experts found the availability of such reactions a godsend and proceeded to become experts in deep inelastic processes.

The experimental observations provided a very tantalizing picture of the evolution of such degrees of freedom. While some modes seemed to relax all the way to the equilibrium limit, others seemed hardly to evolve away from their initial conditions. So it was soon apparent that energy was dissipated with inelasticities covering the whole range from elastic energies to Coulomb energies (5, 43, 65, 115). Yet, at every inelasticity the dissipated energy was divided between the fragments proportionally to their mass as required by statistical equilibrium (31, 40, 45, 94, 106).[2] Similarly, the angular momentum was found to be partitioned between orbital motion and fragment spin from pure orbital motion all the way down to rigid rotation (29, 39, 76). Also, while the neutron-to-proton ratio was found to equilibrate rapidly (33, 35, 36, 47), the mass asymmetry mode seemed to develop extremely slowly (66). Whatever the degree of relaxation happened to be for a given mode, the mean value was observed to be accompanied by a corresponding development of the width.

This state of affairs led naturally to the treatment of the time evolution of the involved variables in terms of transport equations. The earliest suggestions along this line were made by Nörenberg (78–80), who described the energy dissipation in terms of a Fokker-Planck equation, and by Moretto & Sventek (70), who treated the evolution of the mass asymmetry degree of freedom by means of the Master equation. The two approaches

---

[2] Some very recent work (123) indicates that at small Q values the dissipated energy is distributed more equally than implied by statistical equilibrium.

are related to one another and are suitable for the description of the time evolution of distribution functions.

The Master equation for a single variable describes the time evolution of a population $f(x,t)$ as follows:

$$\dot{f}(x,t) = \int dx' [\Lambda(x,x')f(x',t) - \Lambda(x',x)f(x,t)], \qquad 1.$$

where $\dot{f}$ is the time derivative of f; $\Lambda(x,x')$ and $\Lambda(x',x)$ are the direct and inverse transition probabilities. The first term in the square bracket is called the gain term while the second is the loss term. The simplicity of this equation arises from the assumption that the transition probabilities are local in space and are time independent. Under these conditions the Λ's must obey the relations required by microscopic reversibility:

$$\Lambda(x,x') = \lambda(x,x')\rho_x \qquad 2.$$

$$\Lambda(x',x) = \lambda(x',x)\rho_{x'} \qquad 3.$$

$$\lambda(x,x') = \lambda(x',x), \qquad 4.$$

where ρ_x, $\rho_{x'}$ are the level densities at x, x'. This guarantees that the system will approach equilibrium. In fact, for $\dot{f} = 0$,

$$f(x', t = \infty)/f(x, t = \infty) = \rho_{x'}/\rho_x. \qquad 5.$$

The Fokker-Planck equation can be simply derived from the Master equation. If one sets $x' = x + h$ and expands all the quantities in powers of h about x, one obtains

$$\dot{f}(x,t) = -\frac{\partial}{\partial x}(\mu_1 f) + \frac{1}{2}\frac{\partial^2}{\partial x^2}(\mu_2 f) + \ldots, \qquad 6.$$

where μ_1, μ_2 are the first and second moments of the transition probabilities:

$$\mu_1 = \int h \Lambda(x,h)\, dh; \qquad \mu_2 = \int h^2 \Lambda(x,h)\, dh. \qquad 7.$$

The physics is of course contained in the Λ's or alternatively in μ_1, μ_2.

The justifications and the applications of these equations have been widely discussed (82). It may suffice here to recall the two most general approaches. The first, introduced by Nörenberg (79), calculates the transition probabilities in terms of shell-model matrix elements; the second, employed by Randrup (88) and Feldmeier (32), is based upon the one-body assumption following which the long nucleonic mean free path allows one to deal with the nucleons as independent particles. In this framework mass, charge, energy, and angular momentum are exchanged and/or dissipated through nucleon exchange between the interacting nuclei.

An even simpler approach to the description of heavy-ion reactions is the concept of conditional equilibrium (60). As discussed above, certain degrees of freedom equilibrate very fast while others are very slow. In this regime the former modes are essentially at equilibrium and smoothly follow the evolution of the slow modes. These fast modes can be treated within the framework of equilibrium statistical mechanics. The advantage of this approach is the minimal amount of information needed to predict the distribution of the variables under discussion. This approach has been applied with success to the problem of energy partition between fragments (71), to the distribution of charge at fixed mass asymmetry (35, 36), and to the distribution of angular momentum among the angular-momentum-bearing modes (67). The latter problem is discussed extensively in Section 3.

So far we have attempted to paint in broad strokes the picture of heavy-ion reactions. It may already be apparent that angular momentum plays a very pervasive role in this class of reactions, starting from the definition of the range in which these processes become dominant with respect to fusion on the one hand and direct reactions on the other, and ending with detailed effects on most of the variables that have been explored so far.

In what follows we consider the role of angular momentum in those modes whose excitation is associated with their ability to carry angular momentum, as well as in other modes, such as the mass asymmetry mode, that do not carry angular momentum but whose effective potential energy is directly affected by angular momentum through the associated rotational energy.

We begin by discussing the effect of angular momentum on angular distributions and we show how it is still possible to retain the concept of a deflection function in this class of reactions. We then proceed to discuss how the entrance-channel angular momentum is distributed or fractionated along the mass asymmetry degree of freedom. In this way the dependence of the angular distribution on mass asymmetry can be understood. The introduction of the rotational degrees of freedom associated with the dinuclear complex allows us to discuss their thermal or statistical excitation and to calculate the first and second moments of the angular momentum associated with the fragments.

The misalignment in fragment angular momentum that results from fluctuations is considered in its effects upon the angular distribution of the gamma rays, alpha particles, and fission fragments sequentially emitted by the primary deep inelastic fragments. The experimental evidence on the first and second moments of the fragment angular momentum is discussed in detail.

In Section 4, Time-Dependent Hartree-Fock, the excitation of surface modes, and transport models are discussed briefly. Finally the spin-spin correlation is discussed theoretically and the results of dynamical and equilibrium calculations are compared.

2. ROLE OF ENTRANCE-CHANNEL ANGULAR MOMENTUM IN EXIT-CHANNEL DISTRIBUTIONS

2.1 *Angular Distributions and the Classical Deflection Function*

In general, the angular momentum plays an essential role in determining the angular distributions of reaction products. In the case of complete decoupling between entrance and exit channel (e.g. compound nucleus decay), the magnitude of the orbital angular momentum carried by the emitted particle together with the total angular momentum determines, generally speaking, the sharpness of the angular distribution. For instance, in compound nucleus decay, evaporated neutrons and protons can carry little angular momentum and are nearly isotropic, while fission fragments carry the lion's share of angular momentum and are strongly forward peaked, like $1/\sin\theta$. In the case of elastic scattering, such a decoupling does not exist, and a unique relation can be established between the entrance-channel angular momentum l (or the impact parameter, b) and the exit-channel angle θ. This relation can be obtained directly from energy and angular momentum conservation. From energy conservation, we have

$$\frac{1}{2}m\dot{r}^2 = E - V(r) - \frac{l^2}{2mr^2}. \qquad 8.$$

From angular momentum conservation we have

$$l = mr^2\dot{\theta}. \qquad 9.$$

By eliminating the time, we obtain

$$d\theta = -\frac{l}{mr^2}\left\{\frac{2}{m}\left(E - V(r) - \frac{l^2}{2mr^2}\right)\right\}^{-1/2} dr. \qquad 10.$$

Setting $w = b/r$ and the impact parameter $b = l(2mE)^{-1/2}$, one obtains the deflection function Θ:

$$\Theta = \pi - 2\int_0^{b/r_{min}} \frac{dw}{\sqrt{1 - V(w)/E - w^2}}. \qquad 11.$$

For the Coulomb potential, for instance,

$$\Theta = 2\arctan\left(\frac{Z_1 Z_2 e^2}{2Eb}\right). \qquad 12.$$

The angular distribution is given by

$$\frac{d\sigma}{d\Omega} = \frac{b}{\sin\theta}\left|\frac{db}{d\Theta}\right|. \qquad 13.$$

The presence of energy dissipation and the ability of nuclei to pick up angular momentum in the form of spin can also be dealt with within the framework of classical mechanics. The most general formulation of the problem requires the definition of a set of dynamical variables $q, \dot{q}$, of the corresponding Lagrangian $L(q, \dot{q}) = T - V$, and of the Raleigh dissipation function $R = \frac{1}{2}\sum \gamma_{\mu\nu}\dot{q}_\mu\dot{q}_\nu$, where $\gamma_{\mu\nu}$ is the dissipation tensor. The equation of motion is then

$$\frac{d}{dt}\frac{\partial L}{\partial \dot{q}} - \frac{\partial L}{\partial q} + \frac{\partial R}{\partial \dot{q}} = 0. \qquad 14.$$

The simplest way to deal with such a problem is to consider the case of two rigid spheres interacting through the Coulomb field and the short-range nuclear force. The friction tensor must be defined empirically or determined from some microscopic model (78). Such a simple formulation of the problem allows one to define the maximum angular momentum for trapping with eventual compound nucleus formation, to determine the deflection function for the nontrapped orbits, and to establish the exit-channel fragment spin.

The transfer of angular momentum from the orbital motion to the fragments rotation can be divided into three regimes (54, 109). Initially the two nuclei "slide" on top of each other and they are brought to rest relative to one another by tangential friction. At this stage the nuclei roll and are slowed down by "rolling" friction. When the nuclei stop rolling, they stick, the orbital rotation and the nuclear rotation share the same angular velocity, and the dinuclear system rotates rigidly. It is interesting, but not too useful, to appreciate that in the absence of rolling friction one obtains the *rolling limit*; for spheres, this implies the following relation

$$l_{\text{rolling}} = \tfrac{5}{7}l_{\text{initial}} \qquad 15.$$

independent of mass asymmetry. The angular momentum of each nucleus is defined by

$$\frac{I_1}{R_1} = \frac{I_2}{R_2}, \qquad 16.$$

where R_i and I_i are the nuclear radius and spin, respectively. The rigid rotation limit, on the other hand, gives for the spin of one fragment:

$$I_{1,2} = \frac{\mathscr{I}_{1,2}}{\mathscr{I}_1 + \mathscr{I}_2 + \mu d^2} l_{\text{initial}}, \qquad 17.$$

where $\mathscr{I}_1$ and $\mathscr{I}_2$ are the moments of inertia, μ is the reduced mass, and d is the distance between centers of the fragments. For two touching spheres the spin of one fragment goes from $\frac{1}{2}l_{\text{initial}}$ to l_{initial} as one goes from a symmetric system ($A_1 = A_2$) to a progressively more asymmetric system. A completely fused system is, of course, made up of a single nucleus and its angular momentum must be equal to l_{initial}. Such a dependence on mass asymmetry can be readily written down for two touching spheres:

$$I_1 = \frac{\frac{2}{5}M_1 R_1^2}{\frac{2}{5}(M_1 R_1^2 + M_2 R_2^2) + \frac{M_1 M_2}{M_1 + M_2}(R_1 + R_2)^2} l_{\text{initial}}. \qquad 18.$$

This dependence of the transferred spin on mass asymmetry has been frequently used to verify rigid rotation (7, 39, 76, 99, 100).

The calculated deflection functions show two possible regimes: near-side scattering and far-side scattering or orbiting. It is possible to verify such predictions experimentally. A useful way to examine the experimental data is to plot the doubly differential cross section $d^2\sigma/dE\,d\theta$ as contour lines in the plane E, θ. When this is done (114), as shown in Figure 3, one notices that, rather than observing a unique relation between E and θ as predicted by the classical models, one obtains a broad spread of energies for a given angle and vice versa. This means that the deflection function as well as the energy loss are not well defined and that fluctuations, quantal or statistical in nature, dominate the picture. Still, at times it is possible to observe ridges that suggest that both regimes, near-side scattering and orbiting, can occur. For instance Figure 3a gives the distinct impression that the two ridges observed in the cross section do correspond to near-side scattering and far-side scattering for the high-energy and low-energy branch, respectively. Yet other reactions seem to indicate that near-side scattering alone is occurring. For this second class of reactions the angular distribution is side peaked at all energy losses (Figure 3b). These two classes of reactions are qualitatively distinguished by the deflection function, which dives to negative angles in the former case and remains confined to positive angles in the latter.

It has been shown (56, 66) that the product of the interaction time and the angular velocity, which determines the amount of forward swinging of the dinuclear system, is related to the ratio of the center-of-mass (c.m.) energy and the Coulomb barrier E/B. Typically, for $E/B > 1.5$, one observes forward peaking in the angular distribution associated with negative angle scattering and the corresponding two branches in the Wilczyński diagram. For $E/B < 1.5$ one observes side peaking in the angular distribution or near-side scattering and only one branch in the Wilczyński diagram.

The definitive way to establish the above picture is to measure the

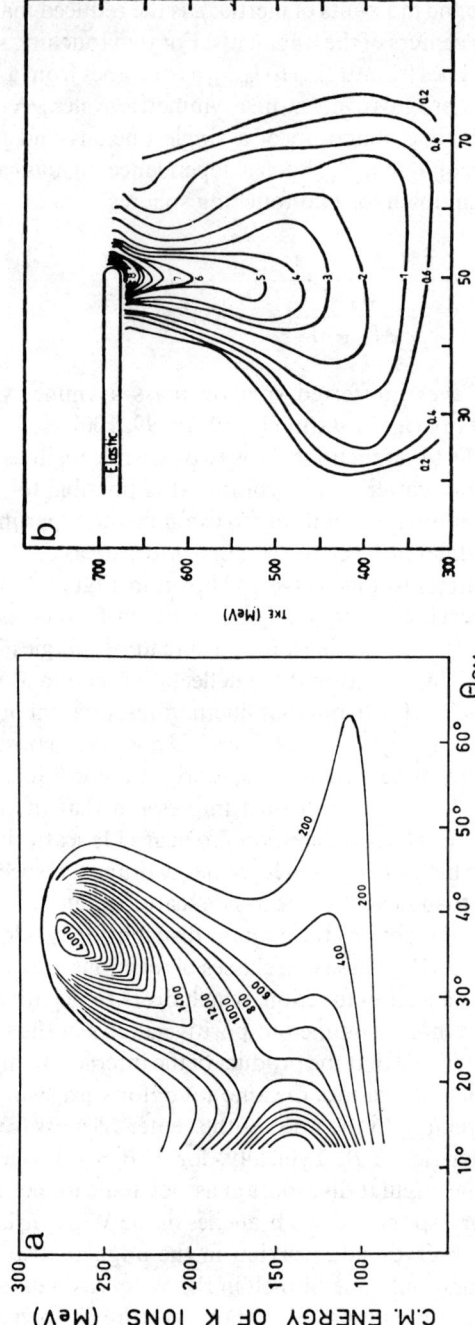

Figure 3 Contours of constant cross section for (*a*) potassium ions produced in the reaction 1130-MeV ^{136}Xe + ^{209}Bi (96), and (*b*) all products produced in the reaction 388-MeV ^{40}Ar + ^{232}Th (114).

polarization of the fragments' spins. The polarization is defined as

$$P = \left\langle \frac{m}{I} \right\rangle = \left\langle \sum P(m)m/I \right\rangle,\qquad 19.$$

where I is the spin, m is its Z projection, and $P(m)$ is the distribution of m values. If the spin transfer is induced by ordinary macroscopic friction, then P is positive for far-side scattering and negative for near-side scattering (28), as shown in Figure 4. The spin polarization can be determined by measuring the circular polarization of the emitted γ rays. Experiments performed so far are in general agreement with the above picture, provided one allows at times for some simultaneous near- and far-side scattering due to fluctuations (51, 107, 108). Some data obtained for the reaction ^{58}Ni+100-MeV ^{16}O are shown in Figure 5. For oxygen ejectiles one observes a large negative polarization in the quasi-elastic peak (near-side scattering) evolving into a large positive polarization in the deep inelastic peak, as expected. Similar results have been obtained with a different technique (46).

Two different techniques have been developed in order to obtain an empirical deflection function. Wolschin (116) observed that the deflection angle Θ is related to the rotation angle $\Delta\theta$ by

$$\Theta(l_i) = \pi - \theta_i - \theta_f - \Delta\theta,\qquad 20.$$

where θ_i and θ_f are the Coulomb angles in the entrance and exit channels and l_i is the entrance-channel angular momentum. The above equation can be written as

$$\Theta(l_i) = \Theta_C(l_i) - \Theta_N(l_i),\qquad 21.$$

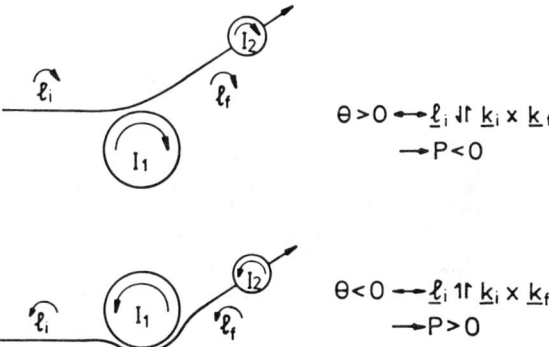

Figure 4 Relation between the spin polarization and the sign of the deflection angle in a frictional collision (28).

where Θ_C is the Rutherford deflection function and Θ_N is the nuclear part of the deflection function. The latter is parameterized as

$$\Theta_N(l_i) = \beta \theta_C^{gr} \frac{l_i}{l_{gr}} \left(\frac{\delta}{\beta}\right)^{l_i/l_{gr}}, \qquad 22.$$

where θ_C^{gr} is the grazing angle; l_{gr} is the grazing angular momentum, and β, δ are two free parameters. The quantity δ determines the deviation of the actual grazing angle from the Coulomb grazing angle, while β determines the depth of the deflection function. The two parameters δ and β are fitted to the experimental gross angular distribution through the relation

$$\frac{d\sigma}{d\theta} = \frac{2\pi}{K^2} \sum l \left|\frac{d\theta}{dl}\right|^{-1}. \qquad 23.$$

Examples of deflection functions extracted in this manner are shown in Figure 6.

Another approach is based on the assumption that the dependence between entrance-channel angular momentum and exit-channel kinetic energy is monotonic (96). Thus the reaction cross section can be divided into energy bins and, from the nearly triangular distribution of the cross section with l, one can assign an average angular momentum to each energy

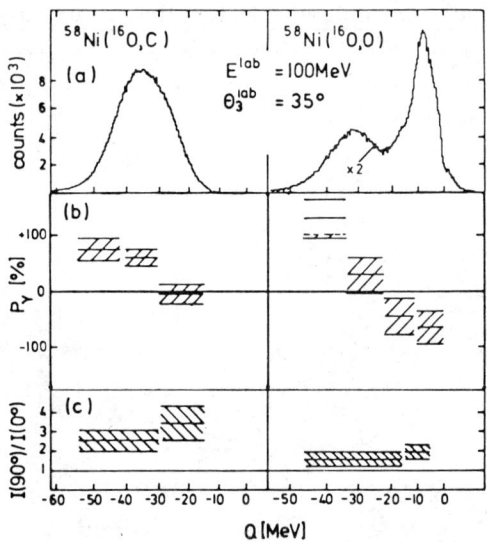

Figure 5 (a) Particle spectra in coincidence with γ rays, (b) circular polarization of the energy-integrated γ radiation, and (c) average in-plane/out-of-plane intensity ratio of stretched quadrupole transitions (51).

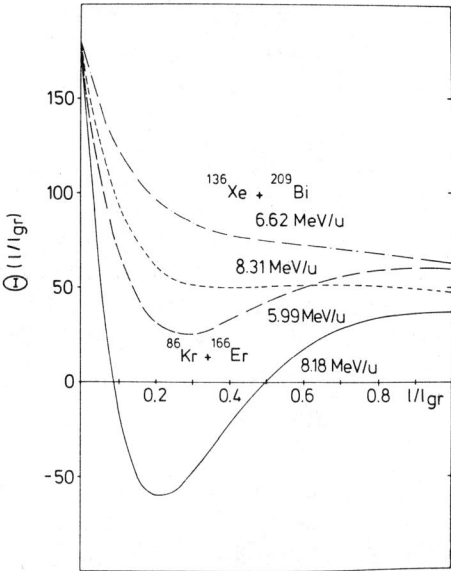

Figure 6 Mean deflection for Xe + Bi and Kr + Er at two different incident energies (116).

bin. At the same time, the angular distribution in each energy bin gives the most probable angle to be associated with that angular momentum. The combination of the two quantities defines an empirical deflection function, of which an example is given in Figure 7.

It is clear that both approaches suffer from several basic shortcomings.

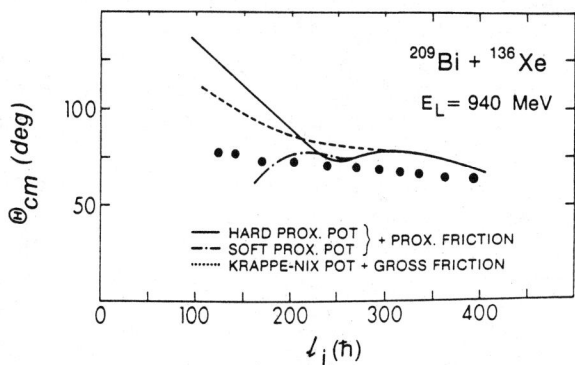

Figure 7 The experimentally deduced deflection function as a function of the angular momentum is compared to classical trajectory calculations, using potential and friction form factors as indicated. No deformations are taken into account (113).

Not only do they not allow for fluctuations, but they also assume that the deflection function is independent of the mass of the actual fragments. Furthermore, the second method leaves an ambiguity in the way one should set the energy windows. In order to be correct, one should be able to trace lines of constant entrance-channel angular momentum for the map of the cross section vs energy and atomic number. Unfortunately, that would be equivalent to knowing the answer at the outset, so different approaches have been suggested. On the one hand, straight cuts have been used (97, 111); on the other, lines parallel (66, 93) to the Coulomb-like line of the peaks of the deep inelastic bump. Neither are justified by a theoretical simulation (103). Therefore the conclusions obtained in such a way are to be considered only approximate.

2.2 Width of the Angular Distributions: Quantal or Statistical Fluctuations?

As discussed in the section above, the use of a classical Lagrangian leads to the definition of an energy distribution function $E = E(l)$ and of a deflection function $\Theta = \Theta(l)$ that relate the entrance-channel angular momentum l with the exit-channel energy E and angle θ. The two observables, E and θ, can be correlated, as can be seen in the so-called Wilczyński diagrams. Experimentally the cross section $\partial^2 \sigma / \partial \theta \, \partial E$ sometimes shows ridges that remind us of the theoretical predictions, especially for small energy losses and for masses near that of the entrance channel. However, even under the best conditions it is apparent that fluctuations substantially smear the angular distribution; under the worst, they wash out any indication of a deterministic trajectory. Typically the experimental angular distributions vary from side peaked to forward peaked, frequently in the same reaction, as the system moves toward ever more negative Q values. Such a remarkable evolution is associated with a progressive increase in the interaction time, on the one hand, and a corresponding increase in the width of the orbital angular momentum distribution on the other. The width of the distribution can arise from two very different contributions, as was pointed out by Strutinsky (102). We can expect a diffractive contribution arising from the width of the l packet associated with the process, and a "dynamical" contribution associated with the classical deflection function. The "diffractive" component can be estimated from the indetermination principle:

$$\sigma_{\theta,\text{diff}} = \frac{1}{2\sigma_l}. \qquad 24.$$

In other words, as one progressively narrows the l window, one observes a broadening of the angular range covered by the angular distribution. In particular, for a single l wave one expects to observe an angular distribution spread out over an angular range of 2π.

A purely classical width of dynamical origin is associated with the variation in scattering angle, which is associated with the variation in l:

$$\sigma_{\theta,\mathrm{dyn}} = \frac{d\Theta}{dl}\sigma_l. \qquad 25.$$

In this case an increase in the width of the l distribution produces a corresponding increase in the width of the angular distribution. The two widths can be combined in quadrature:

$$\sigma_\theta^2 = \frac{1}{4\sigma_l^2} + \left(\frac{d\Theta}{dl}\right)^2 \sigma_l^2. \qquad 26.$$

Which of the two terms dominates? It depends, of course, on the size of σ_l. A calculation (17) performed with a purely classical Fokker-Planck approach is unable to reproduce the width of the angular distribution at small energy losses, as shown in Figure 8a. The deficiency of this approach is related to the predicted smallness of σ_l at small energy losses. In this model the variance σ^2 is proportional to the temperature T, which in turn depends on the Q value. Consequently σ_l^2 tends to zero for Q tending to zero. However, the very smallness of σ_l^2 at small energy losses implies a large diffractive width. The inclusion of this quantal effect improves the theoretical picture quite dramatically (Figure 8b) and makes it almost indistinguishable from experiment.

Interestingly enough, this is not the end of the story. A different calculation (32) based upon a one-body diffusion model is quite realistic even at small Q values despite the absence of quantal fluctuations, as shown in Figure 9. The introduction of quantal fluctuations hardly changes the picture, since the angular distributions are not appreciably broadened. The reason for this apparently contradictory behavior lies in the large dynamical σ_l predicted by the one-body diffusion model even for small Q values. This large σ_l generates a large "dynamical" contribution to σ_θ. However, the same large σ_l implies a small "quantal" fluctuation. Therefore the introduction of quantal effects changes the picture almost imperceptibly.

For large energy losses, thermal fluctuations are almost certainly responsible for a great part of the effect. At the very least, they are simple to estimate. One of the possible fluctuations is associated with the angular momentum trade-off between orbital and intrinsic rotation (69). Given a total angular momentum I and a given orbital angular momentum l, the rotational energy for a dinucleus composed of two equal touching spheres is

$$E_R = \frac{l^2}{2\mathscr{I}^*} + \frac{I^2}{4\mathscr{I}} - \frac{Il}{2\mathscr{I}} \quad \text{where} \quad \frac{1}{\mathscr{I}^*} = \frac{1}{\mu d^2} + \frac{1}{2\mathscr{I}}, \qquad 27.$$

$\mathscr{I}$ being the moment of inertia of one of the two spheres, μ the reduced mass of the system, and d the distance between the centers of the two spheres. The

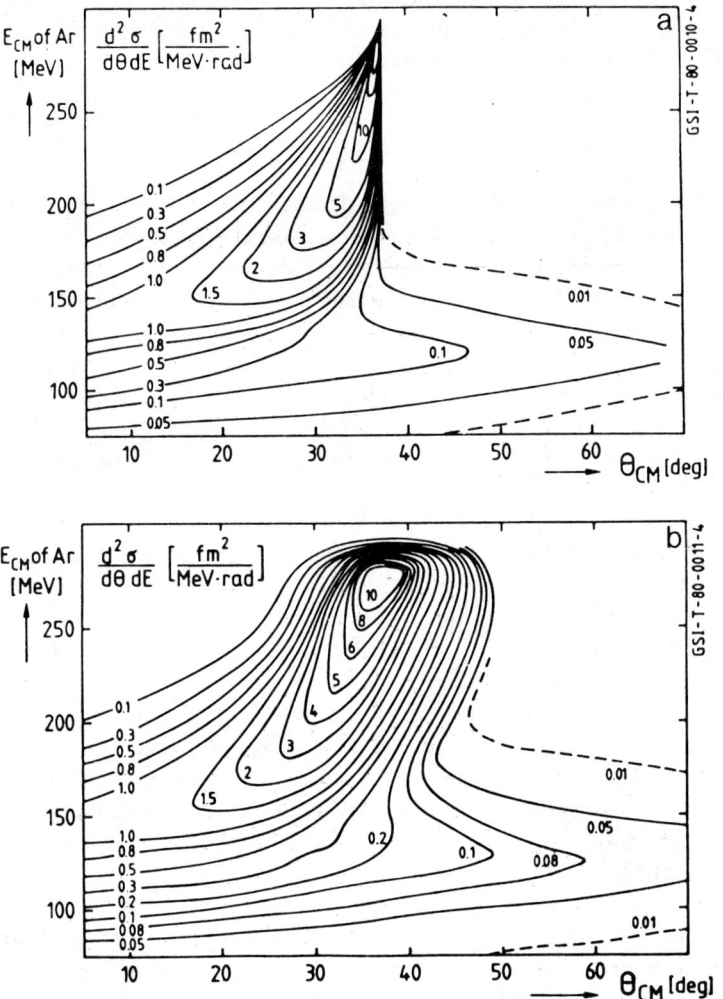

Figure 8 Wilczyński plot calculated (*a*) without and (*b*) with quantal fluctuations (17).

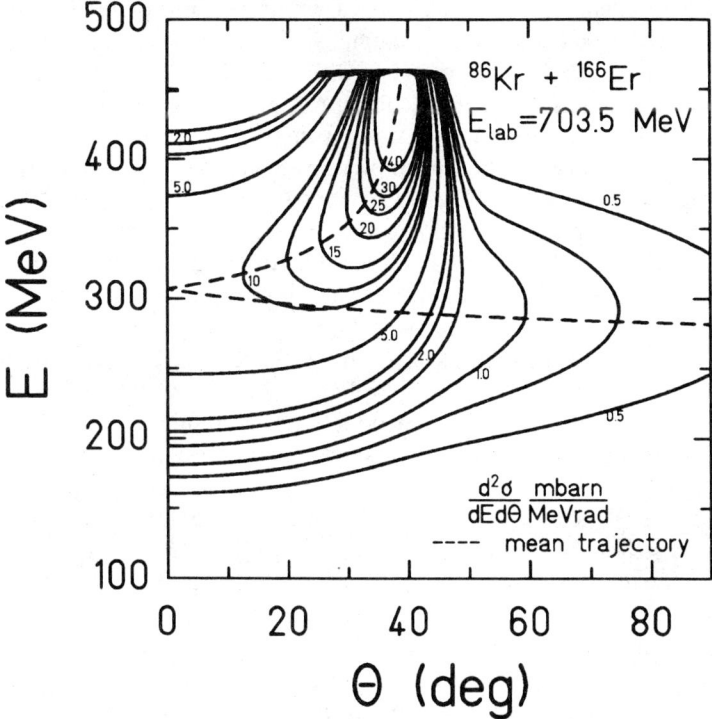

Figure 9 Wilczyński plot calculated from a one-body diffusion model without quantal fluctuations (32).

statistical distribution in l at fixed I at a given temperature T is

$$P(l)\,dl = (2\pi\mathscr{I}^*T)^{-1/2} \exp - \left[\frac{l^2}{2\mathscr{I}^*T} - \frac{Il}{2\mathscr{I}T} + I^2\frac{\mathscr{I}^*}{8\mathscr{I}^2T} \right]. \qquad 28.$$

This distribution at fixed I has the following variance

$$\sigma^2 = \frac{10}{7}\mathscr{I}T. \qquad 29.$$

Additional l fluctuations at fixed I can arise if the two fragments are allowed to deform (69).

2.3 Angular Distributions as a Function of Mass Asymmetry

One of the more complex and less understood problems in heavy-ion collisions is the distribution or fractionation of the entrance-channel angular momentum along the mass asymmetry coordinate. This distribution is also reflected in the angular distributions of the fragments as a function of mass asymmetry, although in a very complex way. Experimentally the angular distributions are seen to be either forward peaked (62) or side peaked (43). At times there is a smooth evolution from side peaking to forward peaking in the same reaction (5) as one moves away from the entrance-channel mass asymmetry or toward greater energy dissipation. An example of this evolution is given in Figure 10 for the reaction 620-MeV Kr + Au (64). The angular distributions evolve from side peaked near $Z = 36$ to forward peaked as one moves away, either way, from $Z = 36$.

A simple approach employing the Fokker-Planck equation nicely illustrates most of the important physics. For heavy systems and for asymmetries between the Businaro-Gallone mountains, the mass asymmetry potential is approximately parabolic for a broad range of angular momenta (66). Thus one can readily make use of the analytical solution of the Fokker-Planck equation in a parabolic potential (59) to calculate the charge distribution $\phi(Z, t)$.

Both the mass asymmetry potential energy and the diffusion rate depend upon the interpenetration of the fragments as well as their shapes. Also, calculations of the interaction time τ require knowledge of the dynamics. In the absence of detailed information concerning the time evolution of the system, we limit ourselves to an extremely simplistic approach, which nevertheless closely reflects the experimental data (59).

Let us first assume that the time-dependent curvature of the mass asymmetry potential can be replaced by a time-independent quantity that reflects the average shape of the system. We make this assumption because we know that the mass asymmetry potential for interpenetrating spheres can qualitatively explain many of the experimental features. The curvature is then easily obtained from a parabolic fit to the ridge line potential as calculated from the liquid drop model. The diffusion rate can be estimated in terms of the nucleonic fluxes.

The radial potential can be written as

$$V(D) = V_{\text{prox}} + \frac{Z(Z_T - Z)e^2}{d} + \frac{\hbar^2 l^2}{2\mathscr{I}(l)}, \qquad 30.$$

$\mathscr{I}(l)$ being the appropriate moment of inertia.

It is not very clear how much the fragments must interpenetrate before the above equation breaks down. This makes it difficult to formulate the dynamical problem. We just use the above potential to calculate the average force $F_R(l)$ at the interaction distance d_{int}. Knowing the reduced mass μ, the radial velocity v_R, and the radial force F_R for each l value at the interaction radius, one can introduce the following two ansatz for the interaction time τ and the average interpenetration x of the fragments:

$$\tau(l) = \frac{2\mu v_R}{F_R} = \frac{2[2\mu(E-B)]^{1/2}}{F_R}\left(1 - \frac{l^2}{l_{max}^2}\right)^{1/2}, \quad \bar{x}(l) = \frac{\alpha}{2}\frac{\mu v_R^2}{F_R}. \qquad 31.$$

In a more serious attempt to fit the experimental data, one could resort to a

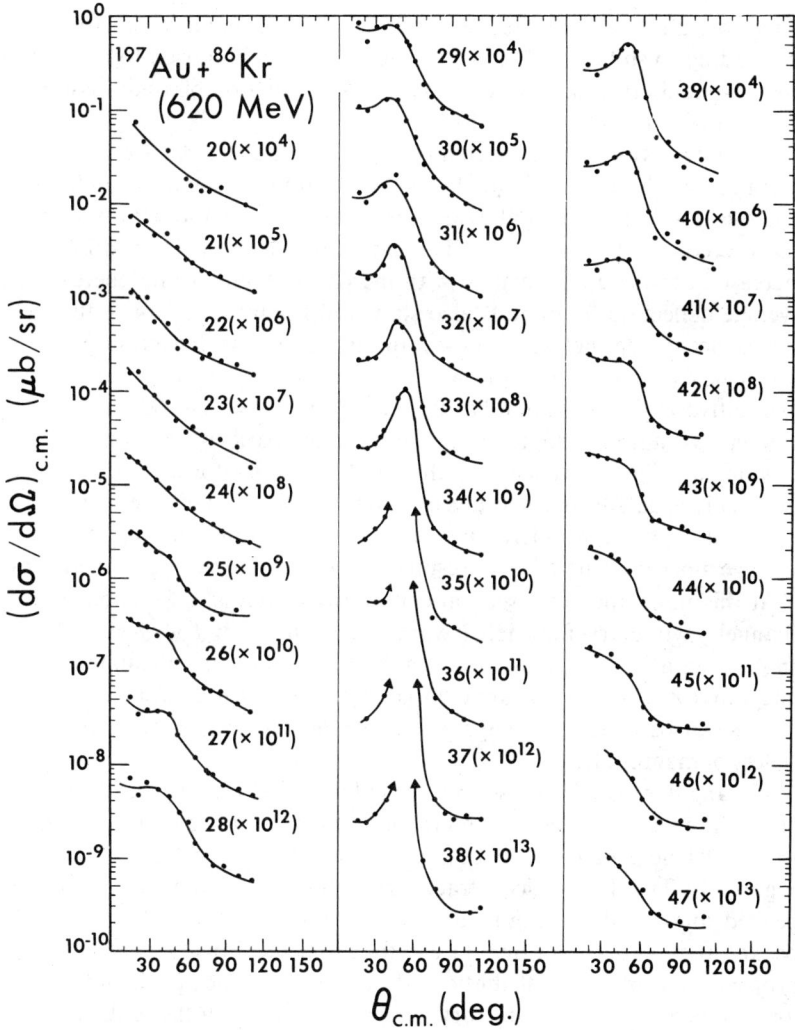

Figure 10 Center-of-mass angular distributions for products ($Z = 20$–47) from the reaction 620-MeV ^{86}Kr + ^{197}Au (64).

more detailed dynamical calculation. Obviously, it is a trivial matter to substitute the above ansatz with more reliable expressions. The diffusion along the asymmetry coordinate is then allowed to proceed with a form factor dependent upon $\bar{x}(l)$ for a time $\tau(l)$.

The tangential motion is treated assuming for the equation of motion the simple form

$$F_T = \mu\gamma(\omega_0 - \omega_{\text{rig}}), \qquad 32.$$

where ω_0 and ω_{rig} are the two limiting orbital angular velocities corresponding to sliding and sticking. The constant γ is chosen to reproduce approximately the mean kinetic energies as a function of angle, assuming that all of the radial energy is lost.

The interaction times calculated for the reaction Au + Kr at three energies are shown in Figure 11a as a function of angular momentum. There is good experimental evidence for the angular momentum dependence predicted by our ansatz. It is interesting to notice the rather mild increase in the average lifetime with increasing bombarding energy. The average deflection function is also shown in Figure 11a. Notice the well-pronounced deep inelastic rainbow (minimum in the deflection function, which yields a maximum in the angular distribution) moving from positive to negative angles as the bombarding energy increases. The 600-MeV curve predicts a rainbow angle of about 50°, in excellent agreement with experiment. The movement of the rainbow angle toward smaller and eventually negative angles results from the combination of three factors: (a) increasing lifetime, (b) increasing angular momentum, and (c) decreasing average moment of inertia as a result of the increasing average penetration.

At this point the cross section can be calculated as a function of exit-channel asymmetry for each l wave. Summing over l waves yields the angle-integrated charge distribution. In Figure 11b the calculated angle-integrated Z distributions are compared with experiment for the reaction Au + Kr at 620 MeV. The agreement is reasonable over more than two orders of magnitude.

The angular distributions can be calculated from the angular deflections of the fragments during the interaction and from their deflection in the Coulomb field. Angular distributions for the Kr + Au reaction are shown in Figure 11c. The theory nicely tracks the experiment in predicting forward-peaked angular distributions at small Z values that develop into side-peaked angular distributions close to the projectile. For Z values above the projectile, the angular distributions slowly lose their side peak and become forward peaked. The satisfactory agreement with both the Z distribution and the angular distribution shows that the calculated dependence of the interaction times and of the diffusion constant upon angular momentum

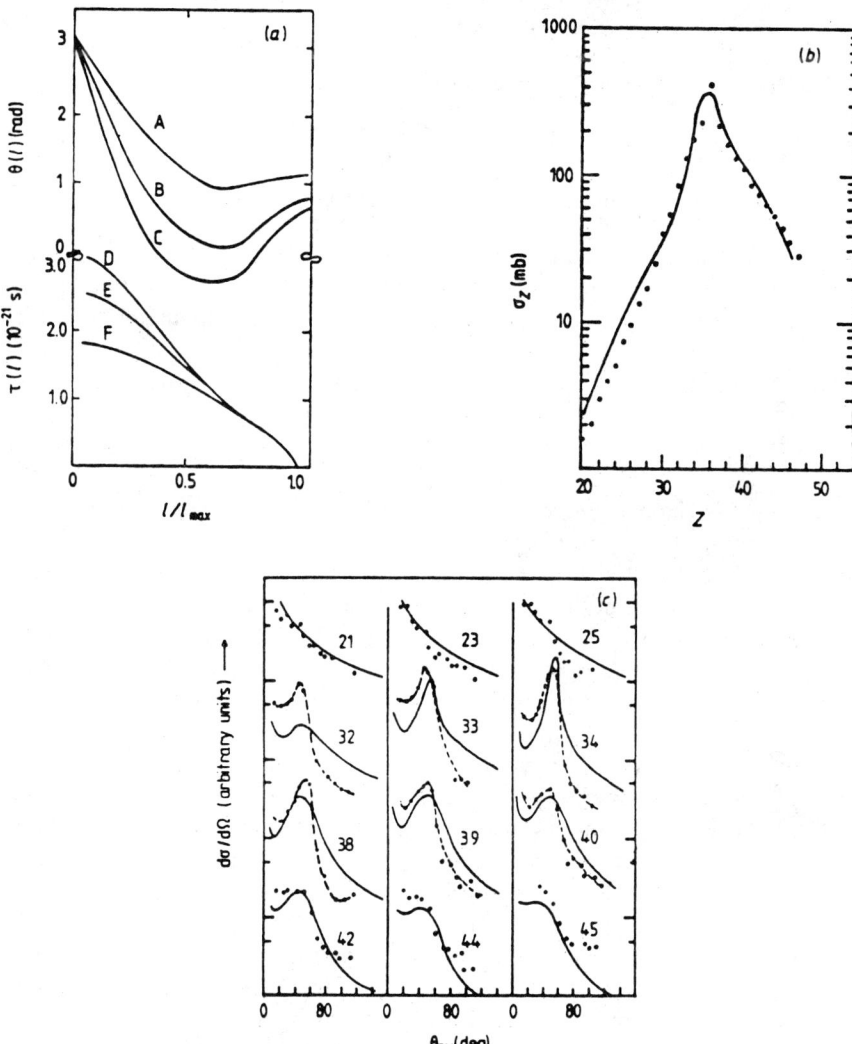

Figure 11 (a) Calculated deflection functions (A, 600 MeV; B, 800 MeV; C, 1000 MeV) and interaction times (D, 1000 MeV; E, 800 MeV; F, 600 MeV) for the reaction ^{86}Kr + ^{197}Au at three bombarding energies. For the 620-MeV ^{86}Kr + ^{197}Au reaction, (b) calculated (*full curves*) and experimental (*points*) angle-integrated charge distributions, and (c) c.m. angular distributions for selected atomic numbers. The broken curves are to guide the eye (59).

and radial velocity is reasonably good. Even better agreement should be expected with a more realistic treatment of the dynamics.

2.4 Rigid Rotation and Angular Momentum Fractionation Along the Mass Asymmetry Coordinate

The limit of rigid rotation can be tested by observing the dependence of either the spin of one fragment or the sum of the moduli of the spins of both fragments upon mass asymmetry. Historically this was the way in which rigid rotation was first demonstrated. Compound nucleus studies (25) have shown that, under optimal circumstances, the spin of a nucleus can be inferred from the number of γ rays emitted in any given reaction (γ-ray multiplicity). The reason for this lies in the fact that the majority of the angular momentum is removed by stretched E2 γ-ray transitions, so that each E2 transition accounts for two units of angular momentum. In general one obtains a relation of the form

$$I \approx 2(M_\gamma - \alpha), \qquad 33.$$

where I is the average angular momentum, M_γ is the measured average γ-ray multiplicity, and α is the number of "statistical" transitions that can be inferred from the γ-ray spectrum and that are weakly related to spin.

The measurement of the M_γ associated with heavy-ion reactions allows one to infer the *sum* of the moduli of the spins of both fragments. Early measurements of M_γ as a function of energy loss (Figure 12a) depicted an initial strong increase across the quasi-elastic region followed by saturation in the deep inelastic region (11, 83, 92). At large energy losses, Olmi et al (83) pointed out that the spin transfer exceeded the classical sticking limit (Figure 13), but was well described by a diffusion model that included statistical fluctuations of the dissipated angular momentum (119).

Measurement of M_γ as a function of the fragment mass or charge allows one to verify whether the limit of rigid rotation is in fact attained. Some measurements (39, 76) presented in Figure 14 show that, indeed, M_γ increases rapidly with increasing mass asymmetry, or decreasing atomic number of the detected fragment, thus verifying the rigid rotation limit. However, measurements on heavier systems (Figure 12b) show that M_γ is essentially constant with the atomic number of one of the fragments (4, 11, 18, 37, 83).

The lack of rise of the M_γ with increasing asymmetry may, at first, suggest that the rigid rotation limit is not established and that some intermediate relaxation stage is prevailing. This is not easy to accept in view of the extensive relaxation observed in the kinetic energy spectra away from the entrance-channel mass asymmetry. It is significant that the reactions exhibiting the rise of the M_γ typical of rigid rotation are associated with a narrow angular momentum window, while the other reactions are associated with a very broad angular momentum range. Therefore it is possible to explain the lack of rise of M_y and still retain the rigid rotation limit if, as is the case, the process is associated with a broad l window. In such cases a rather more stringent condition is required in order to obtain the rise in M_y signalling rigid rotation, namely that each l wave populates all mass asymmetries uniformly. This condition is not realized even in the equilibrium limit and it should not be expected to occur in the non-equilibrium regime that prevails for the mass asymmetry mode.

It is known from a variety of considerations that the interaction time (1, 2), τ, or the lifetime of the dinuclear system is a decreasing function of the entrance-channel angular momentum as shown, for instance, in Figure 11a. The diffusion along the mass asymmetry coordinate can be characterized in terms of a Gaussian whose centroid is drifting downhill on the potential energy surface and simultaneously spreading. If we consider the drift

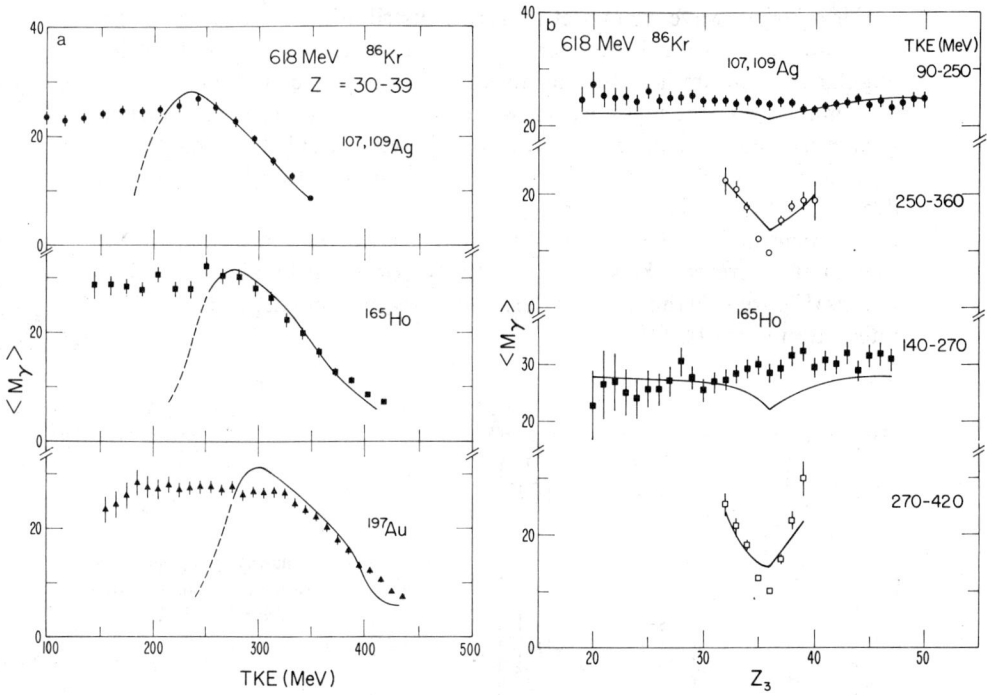

Figure 12 (a) M_y vs kinetic energy (TKE) for the reactions 618-MeV ^{86}Kr + ^{107,109}Ag, ^{165}Ho, ^{197}Au. (b) M_y vs atomic number for the reactions Kr + Ag and Kr + Ho. The full and open symbols are data for a TKE gate on the deep inelastic and quasi-elastic reactions, respectively. The full curves are diffusion model calculations (92).

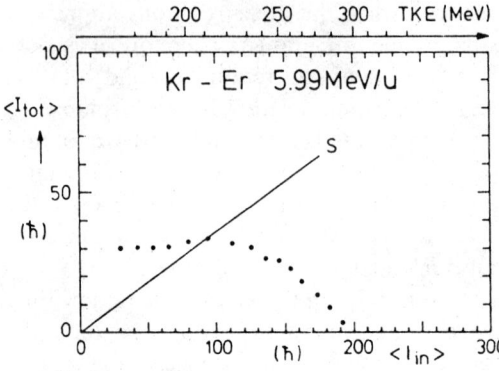

Figure 13 Internal spin of the fragments as a function of the entrance-channel angular momentum. The straight line indicates the sticking limit of two spheres for the Kr–Er exit channel (83).

negligible, as can be observed in several circumstances, the mass distribution for a given l wave can be described by a Gaussian with a variance

$$\sigma_A^2 = \sigma_A^2[\tau(l)]. \qquad 34.$$

For a large l wave we expect a narrow distribution while for a small l wave we expect a much broader distribution. Therefore a sample of the angular momentum near the entrance-channel asymmetry should show a predominance of the high l waves, while far from the entrance channel, at greater asymmetries, progressively lower l waves should be sampled.

As a consequence the mean total angular momentum is expected to decrease with increasing asymmetries. This decrease may well be sufficient to compensate for the rise in spin imposed by rigid rotation, as the asymmetry increases. This is quantitatively borne out by diffusion calculations (Figure 12b) that include the angular momentum dependence of the interaction time (92, 119).

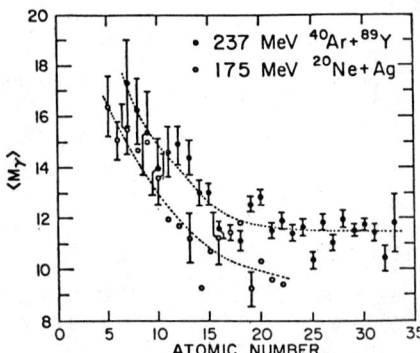

Figure 14 Gamma-ray multiplicity vs atomic number for the reactions 175-MeV ^{20}Ne+Ag (*open circles*) and 237-MeV ^{40}Ar+^{89}Yb (*filled circles*) (39, 76).

Angular momentum fractionation can be caused not only by the differential spreading of the various l waves, but also, in special cases, by the change in driving forces with increasing angular momentum. The total energy of the system, including the rotational energy as a function of mass asymmetry is given by

$$E(x, A, l) = E(x, A)_{LD} + \frac{l^2}{2\mathscr{I}(x)}. \qquad 35.$$

The driving force is

$$-\frac{dE}{dx} = -\frac{dE_{LD}}{dx} + \frac{l^2}{2\mathscr{I}(x)^2} d\frac{\mathscr{I}(x)}{dx}. \qquad 36.$$

If we define $x = (A_2 - A_1)(A_1 + A_2)^{-1}$ and notice that $\mathscr{I}(x)$ decreases monotonically in going from 0 to 1, we see that the centrifugal effect contributes a force that tends always to restore the system to symmetry.

The combination of the liquid drop and centrifugal terms is illustrated in Figure 15a, where the potential energy is given as a function of mass

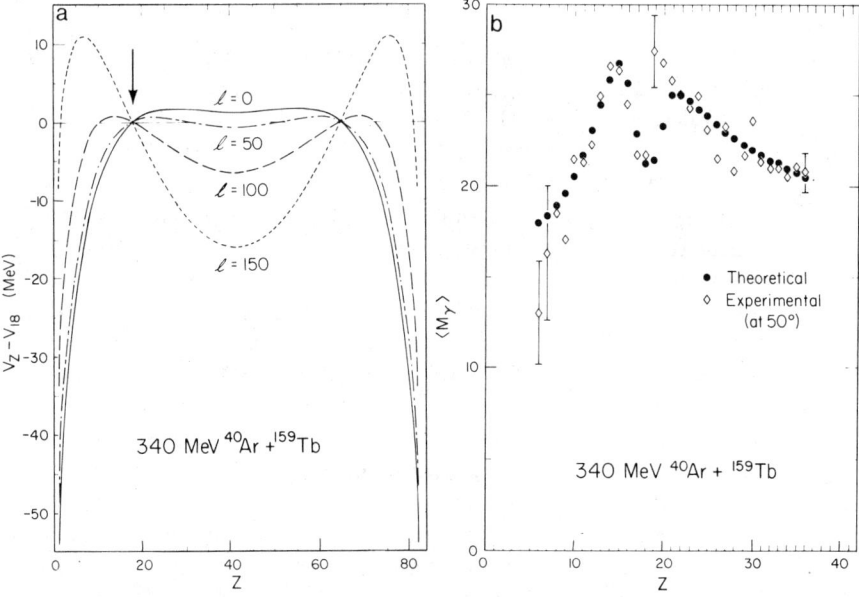

Figure 15 (a) Potential energy vs mass asymmetry (Z) for various l waves for the system 340-MeV ^{40}Ar + ^{159}Tb. (b) Gamma-ray multiplicity data (diamonds) as a function of mass asymmetry for the same system. The solid dots represent a diffusion model calculation using the potential energy surfaces in (a).

asymmetry for a number of l waves (61). The system chosen in the figure is Ar+Tb and all the curves are normalized at the entrance-channel asymmetry. It is evident that it would be quite difficult for the high l waves to populate the greater asymmetries to the left of the entrance-channel asymmetry, while it would become progressively easier for the lower l waves. The experiment (Figure 15b) illustrates this effectively. The gamma-ray multiplicity seems to rise going from right to left, as expected for rigid rotation (except for the sharp wedge at the entrance-channel asymmetry due to incomplete angular momentum relaxation), but actually it takes a steep plunge to the left of the entrance-channel mass asymmetry. This indicates that only low l waves are feeding that region. The same figure portrays a calculation (*solid dots*) that incorporates the physics discussed above and that fits the experimental data satisfactorily (61).

3. ANGULAR-MOMENTUM-BEARING MODES AND PARTITION OF ANGULAR MOMENTUM

3.1 *The Rotational Degrees of Freedom of the Dinuclear Complex*

Many of the degrees of freedom of the dinuclear system can carry angular momentum. If we simulate the dinuclear system with two equal touching spheres (67), these degrees of freedom can be easily identified (Figure 16). Let us fix a reference frame with the x axis coincident with the line of centers and the y and z axes perpendicular to it. The two "bending" modes correspond to a rotation of one fragment parallel to the y or z axis associated with an opposite rotation of the other fragment. The "twisting" modes correspond to a rotation of one fragment about the x axis associated with an opposite rotation of the other fragment. The two "wriggling" modes are rotations of both fragments parallel to the y or z axis compensated by a counter-rotation of the system as a whole about the same axis. Finally the "tilting" mode describes the inclination angle of the total angular momentum with respect to the x axis.

These modes may be excited through a variety of mechanisms. For instance, one of the two wriggling modes can be excited by a coupling with the relative motion mediated by tangential friction. Similarly multipole-multipole Coulomb and nuclear interactions, as well as particle exchange, can be responsible for this excitation. In general, since these collective modes are not exactly normal but are weakly coupled to the intrinsic modes, they can be "thermally" excited.

As an example, let us consider the relaxation of one wriggling mode that leads to the equilibration of the intrinsic rotation and the orbital rotation. If the total angular momentum is I and the fragment spin is s, the energy for an arbitrary partition between orbital and intrinsic angular momentum is

$$E(s) = \frac{(I-2s)^2}{2\mu r^2} + \frac{2s^2}{2\mathcal{I}} = \left(\frac{2}{\mu r^2} + \frac{1}{\mathcal{I}}\right)s^2 - \frac{2I}{\mu r^2}s + \frac{I^2}{2\mu r^2}. \qquad 37.$$

The first term is the orbital and the second the intrinsic rotational energy; $\mathcal{I}$ is the moment of inertia of one of the two equal spheres.

The partition function is given by

$$Z \propto \int \exp[-E(s)/T]\,ds = \sqrt{\frac{\pi \mu r^2 \mathcal{I} T}{2\mathcal{I} + \mu r^2}} \exp\left[-\frac{I^2}{2T(2\mathcal{I} + \mu r^2)}\right]. \qquad 38.$$

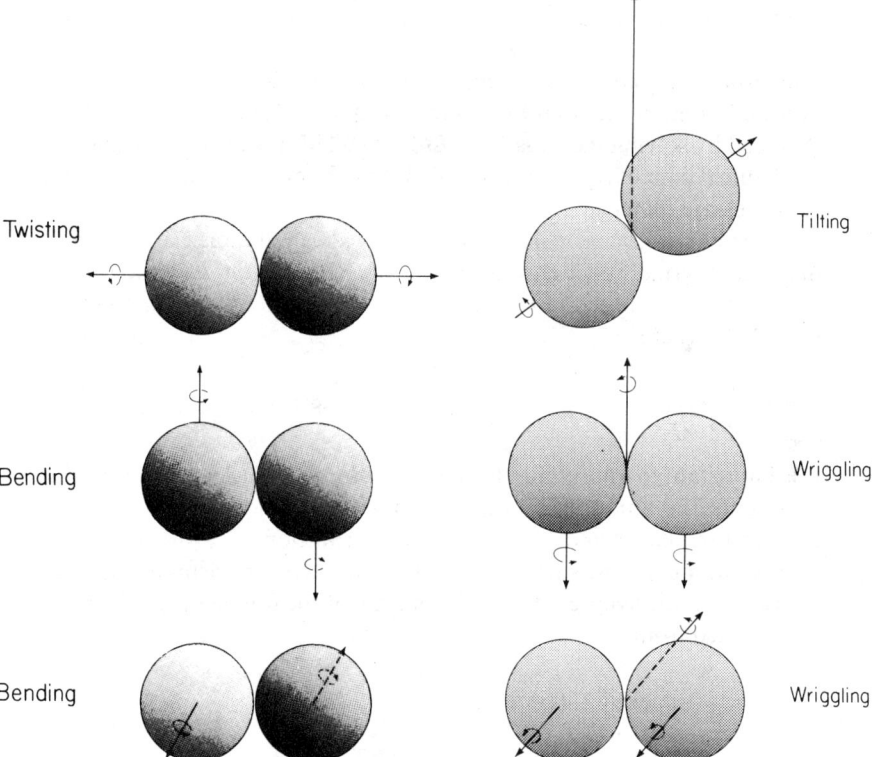

Figure 16 (*Left*) Schematic illustrating the twisting mode and the doubly degenerate bending modes for a two-equal-spheres model. In each case the spin vectors of the fragments (symbolized by arrows) are of equal length but point in opposite directions. (*Right*) Schematic illustrating the tilting mode and the doubly degenerate wriggling modes for the equal-spheres model. The short arrows represent the spin vectors of the fragments. The long arrows originating at the point of tangency of the two spheres represent the orbital angular momentum vectors (67).

The average spin for both fragments is given by

$$2\bar{s} = \frac{2\int s \exp[-E(s)/T]\,ds}{Z} = \frac{2\mathscr{I}}{\mu r^2 + 2\mathscr{I}} I = \frac{2}{7}I = 2I_R, \qquad 39.$$

where $I_R = I/7$ is the spin per fragment expected from rigid rotation. The second moment $\overline{s^2}$ is given by

$$4\overline{s^2} = \frac{2\mu r^2 \mathscr{I} T}{\mu r^2 + 2\mathscr{I}} + \frac{4I^2 \mathscr{I}^2}{(\mu r^2 + 2\mathscr{I})^2}. \qquad 40.$$

From this we obtain the standard deviation

$$4\sigma_s^2 = \frac{2\mathscr{I}\mu r^2 T}{\mu r^2 + 2\mathscr{I}} = \frac{10}{7}\mathscr{I}T. \qquad 41.$$

The result in Equation 39 is temperature independent, as one should have expected from the fact that Equation 37 is quadratic in s. It could in fact be obtained by solving the equation $dE/ds = 0$. This result corresponds to the mechanical limit of rigid rotation when the orbital and the intrinsic angular velocities are matched.

The result in Equation 41 could also have been obtained by recognizing that the thermal fluctuations about the average in Equation 39 are controlled by the second derivative of Equation 37 at the minimum, or $\sigma_s^2 = T/b$, where

$$b = \left(\frac{\partial^2 E}{\partial s^2}\right)_{\bar{s}}.$$

It is important to appreciate the meaning of Equation 41. The quantity $4\sigma^2$ represents the amount of angular momentum trade-off allowed by the temperature, between orbital and intrinsic rotation.

In some instances, such as in γ-ray multiplicity measurements, one is interested in the average sum of the moduli of the fragment spins. This can be obtained from

$$2|\bar{s}| = \int 2|s| \exp[-E(s)/T]\,ds/Z, \qquad 42.$$

which yields

$$2|\bar{s}| = 2\left(\left[\frac{\mu r^2 \mathscr{I} T}{\pi(\mu r^2 + 2\mathscr{I})}\right]^{1/2} \exp\left[\frac{-\mathscr{I}I^2}{\mu r^2 T(\mu r^2 + 2\mathscr{I})}\right]\right.$$
$$\left. + I\frac{\mathscr{I}}{\mu r^2 + 2\mathscr{I}} \operatorname{erf}\left\{I\left[\frac{\mathscr{I}}{\mu r^2 T(\mu r^2 + 2\mathscr{I})}\right]^{1/2}\right\}\right), \qquad 43.$$

or, in dimensionless form,

$$\frac{2|\bar{s}|}{\sqrt{\mathscr{I}^*T}} = 2\left[\frac{1}{\sqrt{\pi}}\exp(-x^2) + x\,\text{erf}(x)\right], \qquad 44.$$

where $x = I_R(\mathscr{I}^*T)^{-1/2}$ and $\mathscr{I}^* = \mu r^2 \mathscr{I}/(\mu r^2 + 2\mathscr{I})$. The above expression is plotted in Figure 17. In the limit of large I, one recovers Equation 39:

$$2|\bar{s}| = \frac{2\mathscr{I}I}{\mu r^2 + 2\mathscr{I}} = \frac{2}{7}I. \qquad 45.$$

For small I,

$$\frac{2|\bar{s}|}{\sqrt{\mathscr{I}^*T}} \approx \frac{2}{\sqrt{\pi}}(1+x^2) \qquad 46.$$

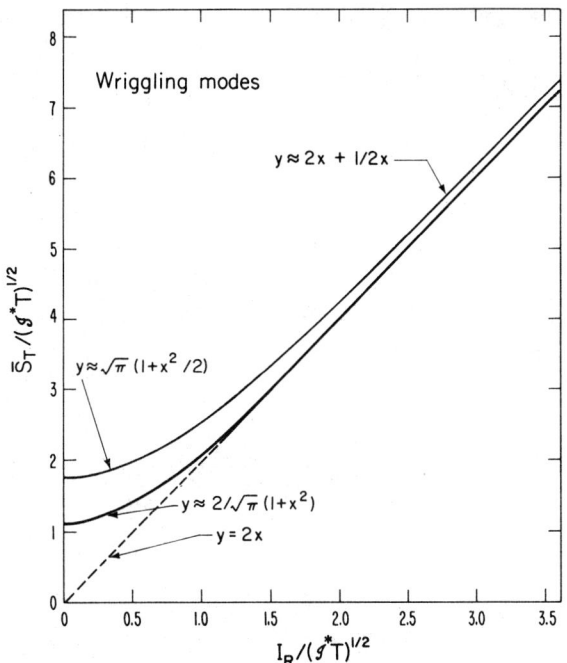

Figure 17 Total spin of the fragments arising from wriggling as a function of the spin arising from rigid rotation alone plotted in dimensionless form. The upper solid curve shows the result for both wriggling modes while the lower solid curve corresponds to the excitation of a single wriggling mode. The limiting behaviors for both small and large values of x are indicated in both cases (67).

to order x^2, so for $I = 0$ one obtains

$$2|\bar{s}| = 2\sqrt{\frac{\mathscr{I}T}{\pi}\left(\frac{\mu r^2}{\mu r^2 + 2\mathscr{I}}\right)^{1/2}} = 2\sqrt{\frac{5\mathscr{I}T}{7\pi}}. \qquad 47.$$

The second moment, still given by Equation 40, can be rewritten as $4\overline{s^2} = 2\mathscr{I}^*T + 4I_R^2$. Notice that the fragment angular momentum at zero total angular momentum arises from the excitation of a collective mode (wriggling) in which the two fragments spin in the same direction, while the system as a whole rotates in the opposite direction in order to maintain $I = 0$.

The overall statistical treatment of the angular-momentum-bearing modes allows us to describe the angular momentum of one of two fragments in terms of a tridimensional Gaussian distribution in the angular momentum components I_x, I_y, I_z:

$$P(\mathbf{I}) \propto \exp\left\{-\left[\frac{I_x^2}{2\sigma_x^2} + \frac{I_y^2}{2\sigma_y^2} + \frac{(I_z - \bar{I}_z)^2}{2\sigma_z^2}\right]\right\}, \qquad 48.$$

where $\bar{I}_z$ is the rigid rotation component:

$$\bar{I}_{z_i} = \frac{\mathscr{I}_i}{\mu d^2 + 2\mathscr{I}_i} I = \frac{1}{7}I \text{ for two equal touching spheres;} \qquad 49.$$

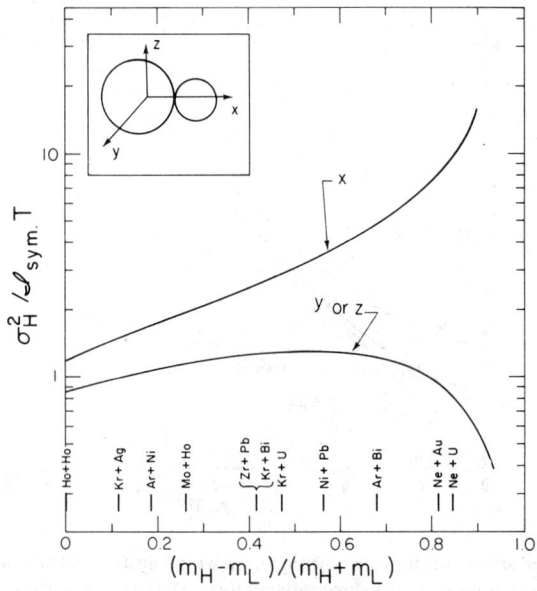

Figure 18 The variances of the normal modes of a dinuclear complex are shown as a function of mass asymmetry of the complex. The variances are shown in dimensionless units after division by $\mathscr{I}_{\text{sym}}T$, the moment of inertia of a mass symmetric spherical fragment times the temperature (73).

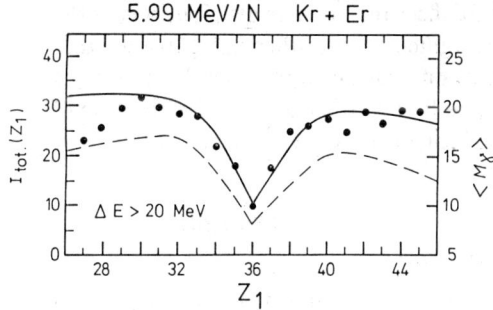

Figure 19 The total angular momentum $I_{tot}(Z_1)$ as a function of fragment atomic number and corresponding M_y data for ^{86}Kr (5.99 MeV per nucleon) + ^{166}Er. The solid curve includes statistical fluctuations and the dashed curve does not (119).

$$\sigma_x^2 = \sigma_{Tw}^2 + \sigma_{Ti}^2 = \frac{1}{2}\mathscr{I}T + \frac{7}{10}\mathscr{I}T = \frac{6}{5}\mathscr{I}T$$

$$\sigma_y^2 = \sigma_B^2 + \sigma_w^2 = \frac{1}{2}\mathscr{I}T + \frac{5}{14}\mathscr{I}T = \frac{6}{7}\mathscr{I}T$$

$$\sigma_z^2 = \sigma_B^2 + \sigma_w^2 = \frac{1}{2}\mathscr{I}T + \frac{5}{14}\mathscr{I}T = \frac{6}{7}\mathscr{I}T. \qquad 50.$$

In the case of an asymmetric system one obtains qualitatively similar results (95). The dependence of the three variances as a function of mass asymmetry is given in Figure 18. The presence of angular momentum fluctuations increases the average fragment spin over the value expected from simple transfer. An example of the role of fluctuations (119) is shown in Figure 19.

3.2 Angular Momentum Misalignment

The presence of fluctuating in-plane components in the fragment angular momentum leads to a misalignment of the fragment's spin, which fluctuates with a given amplitude about the normal to the reaction plane. The angular momentum alignment can be expressed quantitatively in terms of the polarization P and the alignment P_{zz}, given by

$$P = \left\langle \frac{m}{I} \right\rangle = \left\langle \sum_m P(m)m/I \right\rangle_I = \langle \cos\theta \rangle \qquad 51.$$

$$P_{zz} = \left\langle \frac{3}{I(2I-1)}\left[\sum_m P(m)m^2 - I(I+1)\right]\right\rangle_I \approx \frac{3}{2}\left\langle \frac{m^2}{I^2} \right\rangle - \frac{1}{2}, \qquad 52.$$

where $P(m)$ is the probability distribution of a given projection. In terms of the σ's we can obtain

$$P_{zz} \approx \frac{1}{1 + (\sigma_x^2 + \sigma_y^2 + \sigma_z^2)/\bar{I}_z^2}. \qquad 53.$$

Such misalignment is expected to affect the angular distributions of the particles and γ rays emitted by the fragments. It is rather straightforward to incorporate the angular momentum fluctuations into general expressions for the angular distributions. We deal first with the angular distribution of sequential α particles and fission fragments and later with the gamma-ray angular distributions.

3.3 Angular Distributions of Sequential Fission and of Sequential Light Particle Emission

We have shown elsewhere (58) that the angular distribution of fission fragments and of light particles emitted by a compound nucleus can be treated within a single framework. The direction of emission of a decay product (fission fragment, α particle, etc) is defined by the projection K of the fragment angular momentum on the disintegration axis. Simple statistical mechanical considerations show that the distribution in K values is Gaussian.

Specifically for any given K the particle width can be written as (60)

$$\Gamma_K^I \, dK = \Gamma^0 \exp\left[-\frac{\hbar^2 I^2}{2T}\left(\frac{1}{\mathscr{I}_\perp} - \frac{1}{\mathscr{I}_c}\right)\right] \exp\left(-\frac{K^2}{2K_0^2}\right)\frac{dK}{I} \qquad 54.$$

where Γ^0 is an angular-momentum-independent quantity; T is the temperature; $K_0^2 = \hbar^{-2}(1/\mathscr{I}_\parallel - 1/\mathscr{I}_\perp)^{-1} T$; $\mathscr{I}_\parallel, \mathscr{I}_\perp$ are the principal moments of inertia of the decaying system, with particle and residual nucleus just in contact, about an axis parallel and perpendicular to the disintegration axis, respectively; $\mathscr{I}_c$ is the moment of inertia of the compound nucleus.

Similarly, the neutron decay width, integrated over all the neutron emission directions, is

$$\Gamma_N \approx \Gamma_N^0 \exp\left[-\frac{\hbar^2 I^2}{2T}\left(\frac{1}{\mathscr{I}_N} - \frac{1}{\mathscr{I}_c}\right)\right]. \qquad 55.$$

In this expression $\mathscr{I}_N = \mathscr{I}_R + \mu R^2$, corresponding to $\mathscr{I}_\perp$ in Equation 54, is the sum of the moment of inertia of the residual nucleus after neutron decay and the orbital moment of inertia of the neutron at the surface of the nucleus.

Let us now express the particle decay width in terms of the emission angle α measured with respect to the angular momentum direction. Since $K = I \cos \alpha$ and $dK = I \, d(\cos \alpha) = I \, d\Omega$, we obtain

$$\Gamma^I(\alpha) \, d\Omega = \Gamma^0 \exp\left[-\frac{\hbar^2 I^2}{2T}\left(\frac{1}{\mathscr{I}_\perp} - \frac{1}{\mathscr{I}_c}\right)\right] \exp\left(-\frac{I^2 \cos^2 \alpha}{2K_0^2}\right) d\Omega. \qquad 56.$$

If the angular momentum has an arbitrary orientation with respect to our chosen frame of reference, defined by its components I_x, I_y, I_z, the angular distribution can easily be rewritten by noticing that

$$K = I\cos\alpha = \mathbf{I}\cdot\mathbf{n} = I_x \sin\theta\cos\phi + I_y\sin\theta\sin\phi + I_z\cos\theta, \qquad 57.$$

where $\mathbf{n}$ is a unit vector pointing along the direction of particle emission with polar angles θ, ϕ. If the orientation of the angular momentum is controlled by the distribution in Equation 48, we can integrate over the distribution of orientations and, if we drop angular-momentum-independent factors, we obtain (16, 63)

$$\Gamma^I(\theta,\phi)\,d\Omega \propto \exp\left[-\frac{\hbar^2 I^2}{2T}\left(\frac{1}{\mathscr{I}_\perp}-\frac{1}{\mathscr{I}_c}\right)\right]\frac{1}{S(\theta,\phi)}\exp\left[-\frac{I^2\cos^2\theta}{2S^2(\theta,\phi)}\right]d\Omega, \qquad 58.$$

where

$$S^2(\theta,\phi) = K_0^2 + (\sigma_x^2\cos^2\phi + \sigma_y^2\sin^2\phi)\sin^2\theta + \sigma_z^2\cos^2\theta. \qquad 59.$$

In Equation 58 we set $I_z = I$, in other words we averaged over the orientation but allowed the decay width to depend only upon the average angular momentum set equal to its z component. This expression should then be considered only as a high angular momentum limit ($\sigma/I \ll 1$).

The final angular distribution is obtained by integrating over the fragment angular momentum distribution, which is assumed to reflect the entrance-channel angular momentum distribution through the rigid rotation condition:

$$\frac{d\sigma}{d\Omega} \propto \int_{I_{\min}}^{I_{\max}} 2I\,dI\,\frac{\Gamma^I}{\Gamma_N}, \qquad 60.$$

where we have assumed $\Gamma_T \approx \Gamma_N$. More precisely:

$$W(\theta,\phi) \propto \int_{I_{\min}}^{I_{\max}} \frac{2I}{S}\exp\left[-I^2\left(\frac{\cos^2\theta}{2S^2}-\beta\right)\right]dI \qquad 61.$$

or

$$W(\theta,\phi) = \frac{1}{S}\left[\frac{I_{\min}^2}{A_{\min}}\exp(-A_{\min}) - \frac{I_{\max}^2}{A_{\max}}\exp(-A_{\max})\right]. \qquad 62.$$

If $I_{\min} = 0$, then

$$W(\theta,\phi) = \frac{1}{SA}[1-\exp(-A)], \qquad 63.$$

where

$$A = A_{\max} = I_{\max}^2\left(\frac{\cos^2\theta}{2S^2} - \beta\right); \quad A_{\min} = I_{\min}^2\left(\frac{\cos^2\theta}{2S^2} - \beta\right);$$

$$\beta = \frac{\hbar^2}{2T}\left(\frac{1}{\mathscr{I}_n} - \frac{1}{\mathscr{I}_\perp}\right). \qquad 64.$$

The quantity $\mathscr{I}_n$ is the moment of inertia of the nucleus after neutron emission, $\mathscr{I}_\perp$ is the perpendicular moment of inertia of the critical shape for the decay (e.g. saddle point). It is important to notice that the angular momentum dependence of the particle/neutron competition or fission/neutron competition is explicitly taken into account through β.

The final ingredient necessary for an explicit calculation of the angular distributions is the quantity K_0^2. This quantity can be expressed in terms of the principal moments of inertia of the critical configuration for the decay:

$$K_0^2 = \frac{1}{\hbar^2}\left(\frac{1}{\mathscr{I}_\parallel} - \frac{1}{\mathscr{I}_\perp}\right)^{-1} T = \mathscr{I}_{\text{eff}} T. \qquad 65.$$

For fission $\mathscr{I}_{\text{eff}}$ can be taken from liquid drop calculations (19). For light particle emission, the calculation of $\mathscr{I}_{\text{eff}}$ can be worked out trivially. Let m, M, A be the masses of the light, residual, and total nucleus. One obtains:

$$\mathscr{I}_\parallel = \frac{2}{5}MR^2 + \frac{2}{5}mr^2$$

$$\mathscr{I}_\perp = \frac{2}{5}MR^2 + \frac{mM}{A}(R+r)^2$$

$$\frac{\mathscr{I}_{\text{eff}}}{\mathscr{I}_{\text{sph}}} = \left(\frac{M}{A}\right)^{5/3}\left[1 + \frac{2}{5}\frac{A}{m}\left(\frac{R}{R+r}\right)^2\right], \qquad 66.$$

where r and R are the radii of the light particle and residual nucleus, respectively.

This result is adequate if $m \ll M$ and if the charge of the light particle is small. If the charge of the light particle is not negligible, one has to consider the shape polarization induced on the heavy fragment at the "ridge point," as discussed by Moretto (58). Since the shape polarization affects the asymptotic kinetic energy of the emitted particle as well, one can in principle utilize the particle kinetic energy spectra to verify that the shape of the system at the ridge point and its principal moments of inertia have been properly chosen. Again a more complete discussion on this point is available in Moretto's work (58).

Now we are in the position to calculate both in-plane and out-of-plane anisotropies (63). The in-plane anisotropy is given by

$$\frac{W(\phi = 90°)}{W(\phi = 0°)}\bigg|_{\theta=90°} = \left(\frac{K_0^2 + \sigma_x^2}{K_0^2 + \sigma_y^2}\right)^{1/2}. \qquad 67.$$

Since in most cases K_0^2 is fairly large, or at least comparable with σ_x^2 or σ_y^2, it is difficult to obtain a sizable in-plane anisotropy. Even by letting $\sigma_x = 0$ one needs $\sigma_y^2 = 3K_0^2$ just to obtain an anisotropy of 2! The out-of-plane anisotropy is somewhat more complicated. For a fixed angular momentum I one has:

$$\frac{W(\theta = 90°)}{W(\theta = 0°)}\bigg|_{\phi=0°} = \left(\frac{K_0^2 + \sigma_z^2}{K_0^2 + \sigma_x^2}\right)^{1/2} \exp\left[\frac{I^2}{2(K_0^2 + \sigma_z^2)}\right]. \qquad 68.$$

For the usual angular momentum distribution one obtains

$$\frac{W(\theta = 90°)}{W(\theta = 0°)}\bigg|_{\phi=0°} = \frac{1}{\beta}\left(\frac{K_0^2 + \sigma_z^2}{K_0^2 + \sigma_x^2}\right)^{1/2}\left[\beta - \frac{1}{2(K_0^2 + \sigma_z^2)}\right]$$

$$\times \frac{1 - \exp\beta I_{\max}^2}{1 - \exp I_{\max}^2\left[\beta - \frac{1}{2(K_0^2 + \sigma_z^2)}\right]}. \qquad 69.$$

At $\phi = 90°$ the anisotropy is obtained from the above equation by interchanging σ_x with σ_y.

3.4 Gamma-Ray Angular Distributions

Fragments with large amounts of angular momentum are expected to dispose of it mainly by stretched E2 decay (25). The relative amounts of dipole and quadrupole radiation depend mainly upon the ability of the nucleus to remain a good rotor over the whole angular momentum range. If the angular momentum of the fragment is aligned, the typical angular pattern of the quadrupole radiation should be observed. Any misalignment should decrease the sharpness of the angular distribution. If the distribution of the angular momentum components I_x, I_y, I_z is statistical, it is straightforward to derive an analytical expression for the angular distributions (63).

For a perfectly aligned system:

$$W(\alpha) = \frac{3}{4}(1 + \cos^2\alpha); \qquad W(\alpha) = \frac{5}{4}(1 - \cos^4\alpha). \qquad 70.$$
$$\text{for E1} \qquad\qquad\qquad \text{for E2}$$

If the angular momentum is not aligned with the z axis, one must express α in terms of θ and ϕ, which define the direction of the angular momentum vector. In particular,

$$\cos \alpha = \frac{\mathbf{I} \cdot \mathbf{n}}{I} = \frac{I_x \sin\theta \cos\phi + I_y \sin\theta \sin\phi + I_z \cos\theta}{(I_x^2 + I_y^2 + I_z^2)^{1/2}}. \qquad 71.$$

For any given I, the angular distribution is obtained by integration over the statistical distribution $P(\mathbf{I})$ of the angular momentum components:

$$W(\theta, \phi) = \int W(\alpha) P(\mathbf{I}) \, d\mathbf{I}. \qquad 72.$$

It is not possible to obtain an exact analytical expression for the general case. However, an expansion to order $\sigma_x^2/\bar{I}_z^2$, $\sigma_y^2/\bar{I}_z^2$, etc allows one to obtain expressions in closed form.

For the dipole decay,

$$W(\theta, \phi) = \frac{3}{4}(1 + \cos^2\theta)$$

$$+ \frac{3}{4}\left[(\sin^2\theta \cos^2\phi - \cos^2\theta)\frac{\sigma_x^2}{\bar{I}_z^2} + (\sin^2\theta \sin^2\phi - \cos^2\theta)\frac{\sigma_y^2}{\bar{I}_z^2}\right]$$

$$73.$$

Notice that there is no dependence upon σ_z^2. In the case in which $\sigma_x = \sigma_y = \sigma$, we obtain the simplified expression

$$W(\theta, \phi) = \frac{3}{4}(1 + \cos^2\theta) + \frac{3}{4}(\sin^2\theta - 2\cos^2\theta)\frac{\sigma^2}{\bar{I}_z^2}. \qquad 74.$$

A weak in-plane anisotropy is possible:

$$\left.\frac{W(\phi = 0°)}{W(\phi = 90°)}\right|_{\theta = 90°} = \frac{1 + \sigma_x^2/\bar{I}_z^2}{1 + \sigma_y^2/\bar{I}_z^2} \approx 1 + \frac{\sigma_x^2 - \sigma_y^2}{\bar{I}_z^2}. \qquad 75.$$

The out-of-plane anisotropy is

$$\frac{W(0°)}{W(90°)} = 2\frac{(1 - \sigma^2/\bar{I}_z^2)}{(1 + \sigma^2/\bar{I}_z^2)} \approx 2(1 - 2\sigma^2/\bar{I}_z^2). \qquad 76.$$

For the quadrupole decay, we have

$$W(\theta, \phi) = \frac{5}{4}(1 - \cos^4\theta) - \frac{5}{2}\bigg[(3\sin^2\theta \cos^2\theta \cos^2\phi - \cos^4\theta)\frac{\sigma_x^2}{\bar{I}_z^2}$$

$$+ (3\sin^2\theta \cos^2\theta \sin^2\phi - \cos^4\theta)\frac{\sigma_y^2}{\bar{I}_z^2}\bigg]. \qquad 77.$$

Again, no dependence upon σ_z^2 is predicted. If one assumes $\sigma_x = \sigma_y = \sigma$ as before, one obtains

$$W(\theta) = \frac{5}{4}(1-\cos^4\theta) - \frac{5}{2}(3\sin^2\theta\cos^2\theta - 2\cos^4\theta)\sigma^2/\bar{I}_z^2 \qquad 78.$$

$$\frac{W(0°)}{W(90°)} \approx 4\frac{\sigma^2}{\bar{I}_z^2}. \qquad 79.$$

For the in-plane anisotropy, we have

$$\left.\frac{W(\phi = 0°)}{W(\phi = 90°)}\right|_{\theta=90°} \approx 1 \qquad 80.$$

to order $\sigma^2/\bar{I}_z^2$. This can be easily understood. The rms misalignment is $\sim \sigma/I$; thus, at $\theta = 90°$

$$W(90°) = 1 - \cos^4\left(90° - \frac{\sigma}{I}\right) \approx 1 - \frac{\sigma^4}{I^4}. \qquad 81.$$

Thus, no second-order term exists. This result shows that it is very difficult to study anisotropies in the angular momentum misalignment by means of γ-ray angular distributions. The range of validity of the above expressions is rather limited because of the low-order expansion. In particular, the equations should not be trusted for $\sigma^2/\bar{I}_z^2 > 0.05$.

However, if we are willing to assume $\sigma_x^2 = \sigma_y^2 = \sigma_z^2 = \sigma$, then an exact result can be obtained. For the E1 distribution one obtains

$$W(\theta)_{E1} = \frac{3}{4}\{1 + \cos^2\theta + \beta^2[1 - D(\beta)](1 - 3\cos^2\theta)\}. \qquad 82.$$

For the E2 distribution one obtains

$$W(\theta)_{E2} = \frac{5}{4}\Big\{1 - \cos^4\theta - 2\beta^2\Big[3\sin^2\theta\cos^2\theta - 2\cos^4\theta$$

$$-\frac{3}{4}D(\beta)(\sin^2\theta - 4\cos^2\theta)\sin^2\theta\Big]$$

$$-3\beta^4(4\cos^4\theta + \frac{3}{2}\sin^4\theta - 12\sin^2\theta\cos^2\theta)[1 - D(\beta)]\Big\}. \qquad 83.$$

In these equations $\beta = \sigma/\bar{I}_z$ and $D(\beta) = \sqrt{2}\beta F(\sqrt{1/2}\beta)$ where

$$F(x) = e^{-x^2}\int_0^x e^{t^2}\,dt \qquad 84.$$

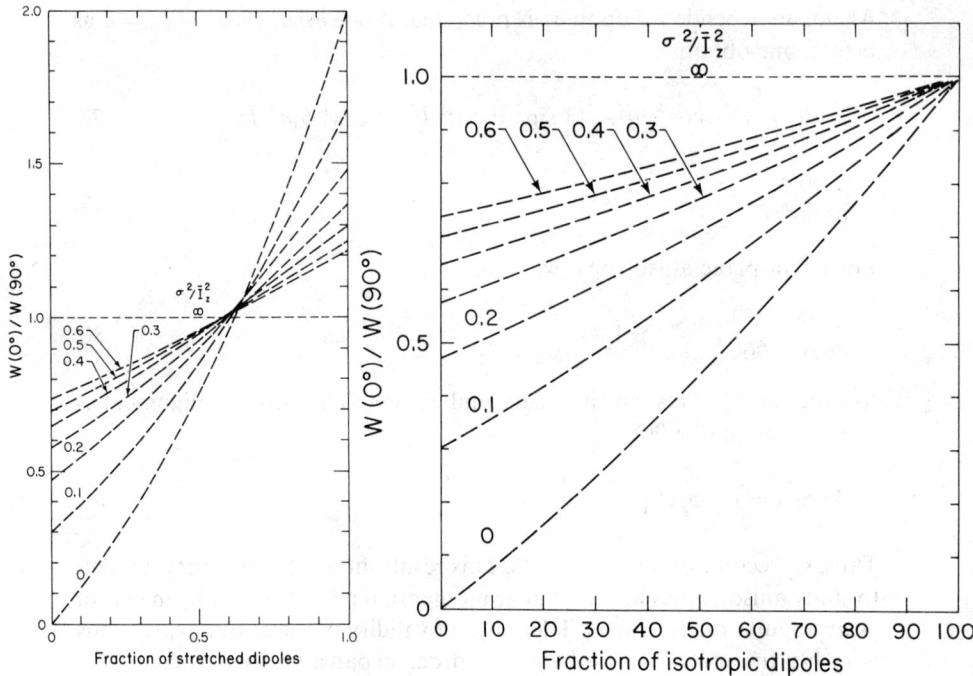

Figure 20 Calculated γ-ray anisotropies for mixtures of stretched E1 and E2 transitions as a function of the fraction of E1 radiation for various values of $\sigma^2/\bar{I}_z^2$ (60).

is the Dawson's integral. One can verify immediately that both expressions behave as expected in the limits of $\beta = 0$ and $\beta = \infty$. The anisotropy $W(0°)/W(90°)$ tends to 1 when β tends to infinity for both E1 and E2 transitions, while it tends to 0 for E2 and to 2 for E1 when $\beta = 0$.

These results are graphically summarized in Figure 20, where the anisotropy is plotted as a function of the fraction of E1 radiation for various values of $\sigma^2/\bar{I}_z^2$. The two extreme possibilities of stretched and nonstretched E1 decay are considered. If one has a fairly good experimental idea of the amount of E1 radiation to be expected from a given fragment and of its degree of stretching, the measurement of the antisotropy yields directly the value of $\sigma^2/\bar{I}_z^2$, which is of course the most direct information about the misalignment.

3.5 *Experimental Spin Alignment From Gamma-Ray Angular Distributions*

Discrete γ-rays in coincidence with deep inelastic fragments were observed to have large anisotropies, which indicates a high degree of spin alignment with the normal to the reaction plane (74, 85, 86, 110). Because the deep inelastic reaction strength is typically spread over a large number of final products, only low-lying γ-ray transitions have been observed at small Q values with modest statistics. These limitations and problems related to doppler broadening and side feeding have given impetus to continuum γ-ray studies.

Early attempts to observe continuum γ-ray anisotropies due to fragment spin alignment led to surprisingly small values of the alignment parameter P_{zz} (3, 9, 11, 22, 23, 51, 75, 76, 85, 86). The failure to observe substantial anisotropies had several causes: first, the presence of a sizable amount of "statistical" isotropic E1 transitions; second the possible presence of stretched M1 transitions under the E2 bump whose angular distribution is exactly out-of-phase with the stretched E2 transitions; third the integration over the deep inelastic part of the Q-value spectrum where the fluctuations are dominant and tend to decrease the alignment substantially. Once the above causes were understood, it was easy to choose suitable conditions under which strong anisotropies could be observed. In particular, it is important to maximize the number of stretched E2 transitions by selecting reaction products that are mostly good rotors, namely located in the region of the heavy rare-earth nuclei.

The continuum γ-ray anisotropy in heavy-ion reactions has been extensively studied (57, 120) for the reaction Ho + Ho at 8.5-MeV A^{-1} and extended (84) to Ho + Yb, Sm, Ag (always at the same energy). The Q-value spectrum was divided into a series of energy bins for which the γ-ray multiplicity, energy spectra, and anisotropy were measured.

The sum of the spins obtained from the γ-ray multiplicity as a function of Q value is shown in Figure 21. As in other reactions (11, 18, 83, 86, 92), an increase in energy loss leads to an initially rapid transfer of angular momentum to the fragments, followed by a relatively slow decrease as one moves toward the greatest inelasticities. These data (Figure 21), already corrected for the angular momentum removed by neutrons, show that the fragments can pick up as much as 35–40$\hbar$ of angular momentum each. Furthermore there is some evidence that, at least for large negative Q values, rigid rotation is obtained. The scaling of the Q value in terms of the common fragment temperature and of the spin in terms of the maximum spin expected for rigid rotation collapses all the data onto a single curve (Figure 22). This scaling suggests that rigid rotation is indeed reached.

The anisotropy of the γ rays (in the region of the γ-ray spectrum dominated by quadrupole radiation) as a function of Q value is shown in Figure 23. In all cases, but more visibly for Ho + Yb, the anisotropy rises initially with increasing energy dissipation to values as high as two, and then declines slowly with further energy dissipation.

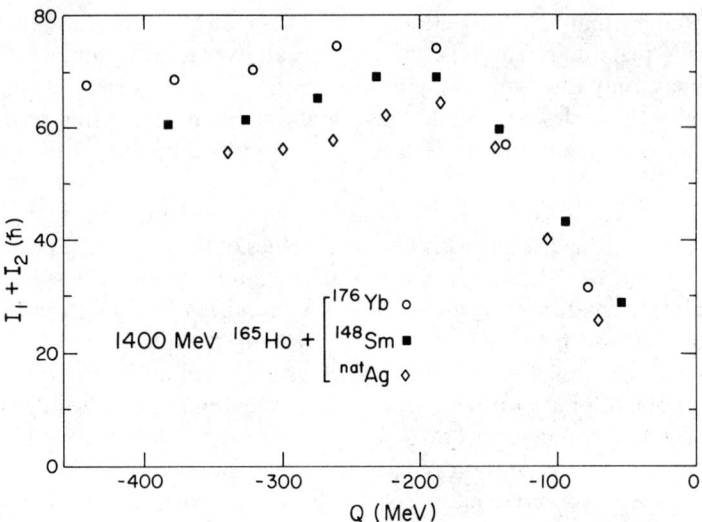

Figure 21 Sum of the spin magnitudes $(I_1 + I_2)$ as a function of Q value for three reaction systems (84).

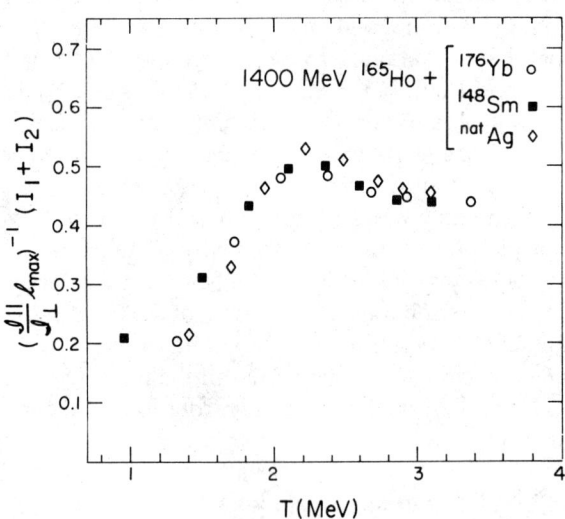

Figure 22 The sum of the spins in reduced units as a function of temperature. The angular momentum axis has been scaled according to the rigid rotation limit (84).

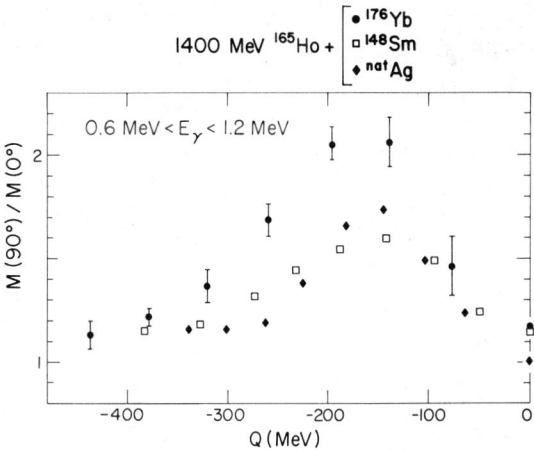

Figure 23 Gamma-ray anisotropy as a function of Q value, for heavy ions detected near the grazing angle. Error bars for the three systems are similar and are shown only for ^{165}Ho + ^{176}Yb (84).

Qualitatively, the rise and fall of the γ-ray anisotropy with increasing energy dissipation is easily understood if studied simultaneously with the spin transfer. For small energy dissipations there is a small amount of angular momentum transferred to the fragments, which in turn can be easily depolarized by in-plane components arising from detailed spectroscopic effects. As the energy dissipation increases, angular momentum is rapidly transferred to the fragments. This transferred angular momentum is aligned and is little perturbed by the in-plane thermally fluctuating components, which increase very slowly with excitation energy ($\sigma^2 \propto T \propto Q^{1/2}$). The resulting strong alignment is manifested in the substantial rise of the γ-ray anisotropy.

A further increase in the energy dissipation does not increase the transferred angular momentum but it increases the excitation energy and thus the thermal fluctuations of the in-plane components. As a consequence the total angular momentum becomes progressively more misaligned and the γ-ray anisotropy decreases.

Of course, there are additional sources of angular momentum misalignment, like particle evaporation from the primary fragments, but it appears that the main cause of angular momentum misalignment is the "thermal" excitation of the angular-momentum-bearing modes. This is shown in Figure 24, where all the sources of angular momentum misalignment are included with the exception of the thermal fluctuations. The calculated anisotropies clearly overestimate the corresponding experimental values.

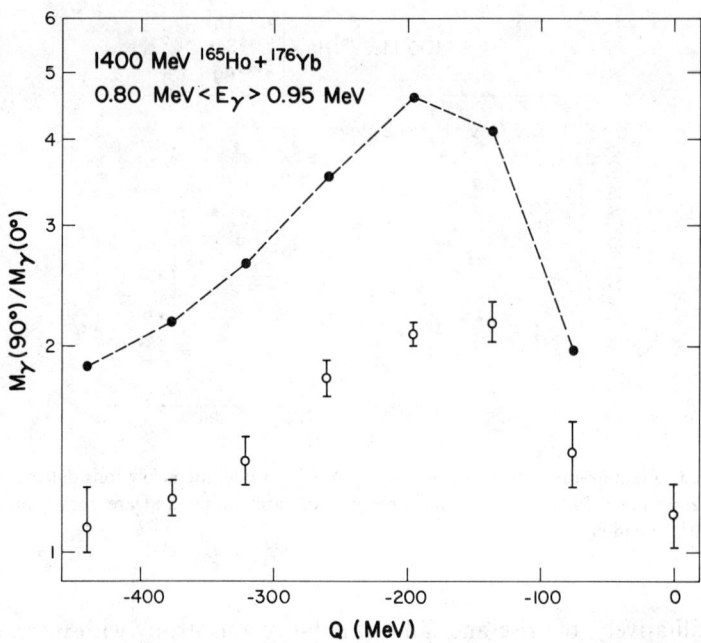

Figure 24 Comparison between the experimental anisotropies (*open symbols*) and a calculation (*solid symbols*) that does not include the effect of the thermal fluctuations (84).

On the other hand, the inclusion of thermal fluctuations provides us with a much improved picture, almost coincident with the experimental data, as seen in Figure 25. It should be pointed out that the calculation uses the experimental M_γ as an input for the fragment angular momentum and uses the theory only to calculate the σ's. In this way we can extend the use of the theory even in Q-value regions where the full equilibrium limit has not been attained, since it is well known that fluctuations tend to their equilibrium limit a great deal faster than the average values.

From the above analysis one can calculate the alignment for each individual fragment, although this decomposition is far less certain than the calculation of the anisotropy. In Figure 25 the alignment P_{zz} is shown for each of the two fragments. In general, alignments as great as 0.7 are observed, with the greatest alignment being associated with the heavier partner.

3.6 *Experimental Data From Sequential Decay*

3.6.1 SEQUENTIAL ALPHA EMISSION Inspection of Equations 65, 67, and 68 shows that if the σ's are comparable or smaller than K_0^2, it is difficult to differentiate between uncertainties in K_0^2 that are intrinsic to the sequential emission process and the angular momentum fluctuations arising from the deep inelastic process. In the case of alpha emission K_0^2 is indeed quite large in comparison with the σ's. As a consequence, the only quantity that one can hope to extract from angular distribution data is the spin of the emitting fragment (7, 50, 99, 100). In studies of the reaction Ar + Ni, Babinet et al (7) showed that the alpha particles from the Ni-like fragment could be isolated, that their angular distributions were peaked in-plane, and that the resulting spins agreed with the rigid rotation assumption (Figure 26). No account was taken of the angular momentum fluctuations. In a subsequent experiment using Kr + Ag, the out-of-plane angular distribution of the alpha particles was measured for a broad range of Ag-like fragments (99, 100). The angular distributions are shown in Figure 27a. It is apparent that the anisotropy of the alpha particles increases with increasing atomic number, which indicates an increase of the Ag-like fragment spin with increasing asymmetry. The fit to the angular distributions is shown in the same figure and the resulting spins are also shown in Figure 27c. The spin

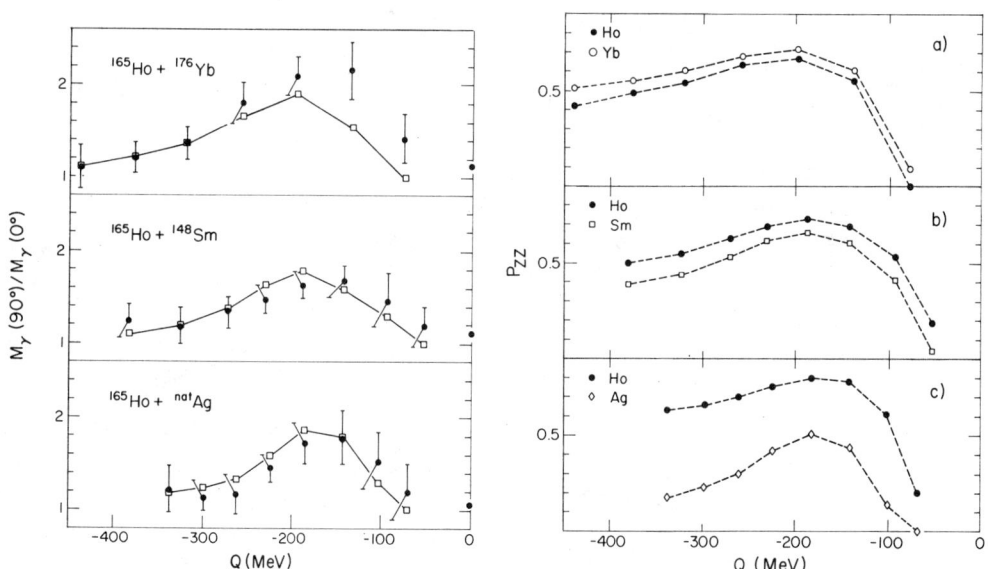

Figure 25 (*Left*) Comparison between experimental (*circles*) anisotropies of γ-rays ($E_\gamma = 0.80-0.95$ MeV) and a calculation based on the equilibrium statistical model (*squares*) as a function of Q value. Lines are drawn through the calculated points to guide the eye. (*Right*) Alignment parameter P_{zz} as a function of Q value, for each of the two deep inelastic fragments in the three reactions (84).

fluctuations were calculated on the basis of the thermal model described above. The measured spins are in excellent agreement with rigid rotation if one assumes that both fragments are prolate spheroids touching by their tips with a ratio of major to minor axes of $c/a = 2$. In this way it is also possible to reproduce the exit-channel kinetic energy distribution as shown in Figure 27b.

From the scission configuration we can reconstruct the sum of the two spins. This quantity can also be inferred from M_y, which was measured

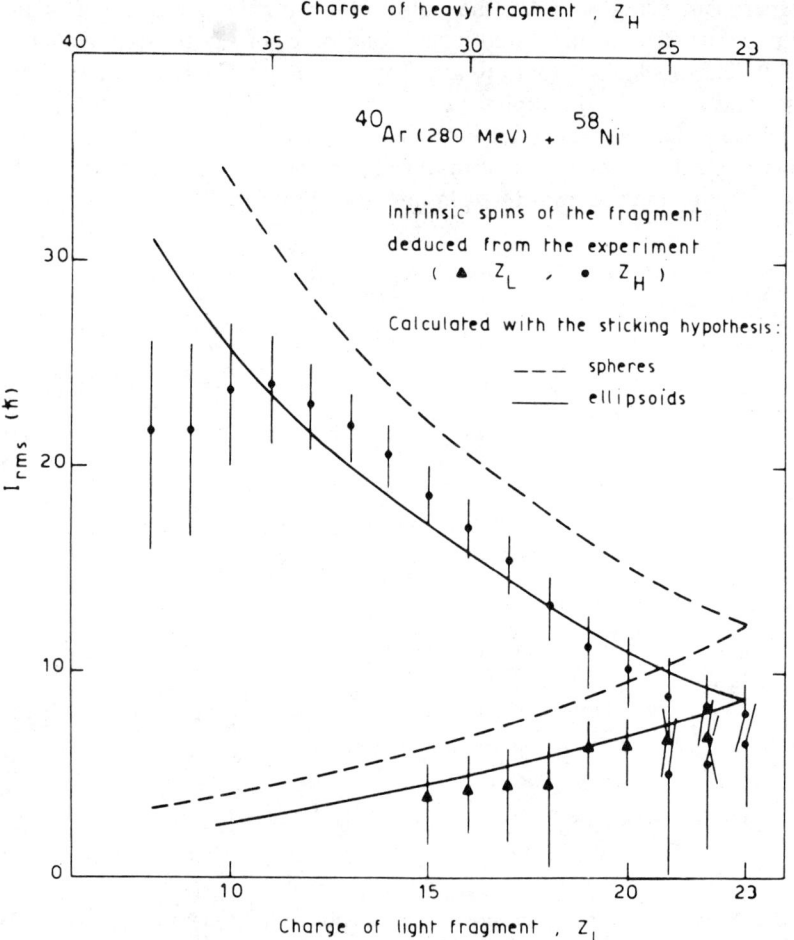

Figure 26 Experimental intrinsic spins of the individual fragments compared with the results of calculations for the sticking limit for rigid bodies (7).

simultaneously in the experiment. A comparison between the sum of the spins measured in both ways is shown in Figure 27c. The agreement implies that not only the trends but also the absolute values of the spins measured with either technique can be relied upon. Similar agreement has also been observed in studies of heavier systems (101).

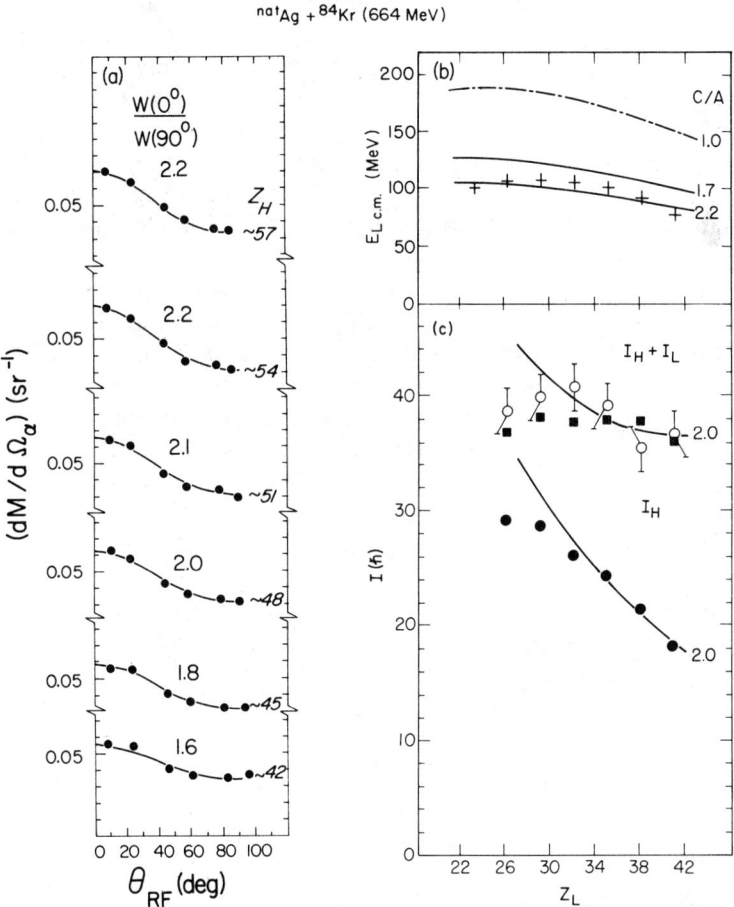

Figure 27 (a) Alpha-particle angular distributions (*points*) as a function of out-of-plane angle for several Z bins. Each bin is 3Z units wide and is labelled by the median Z value. These distributions are expressed in units of differential multiplicity and the solid curves are fits to the data. (b) Center-of-mass energies as a function of the charge of the light fragment. The curves are calculations for two equally deformed spheroids separated by 1 fm and are labelled by the ratio of axes. (c) The spin of the heavy-ion fragment extracted from the α-particle distributions (*solid circles*), the sum of the spins calculated from the α-particle data (*squares*), and M_γ data (*open circles*). The size of the solid symbols indicates the statistical error only (99).

3.6.2 SEQUENTIAL FISSION

In the case of fission, the quantity K_0^2 is comparable to or smaller than the three σ's so that sequential fission can provide information not only about the spin of one of the fragments, as in the case of the alpha emission, but about the degree of alignment as well (29, 121). The first experiment (29), performed for the reaction 610-MeV ^{86}Kr + Bi, showed that the sticking limit had been achieved. The out-of-plane anisotropy is quite strong as expected from the exponential in Equations 61 and 68. Many other systems (30, 44, 52, 72, 73, 87) have been studied, and alignments comparable to those extracted from the γ-ray work obtained (see Figure 28).

The angular distributions of γ rays and α particles are rather insensitive to differences in the in-plane projections of the random spin component. In contrast, the angular distributions of sequential fission fragments are more sensitive to such differences in that they can produce a substantial in-plane anisotropy. A complete measurement of the in-plane and out-of-plane distributions is necessary for this purpose. The anisotropies observed in-plane (29, 30, 52, 72, 73, 87) are much weaker than out-of-plane ones, as suggested by Equation 67. However, the measurement of in-plane anisotropies is quite worthwhile since it provides a sensitive test for the statistical model. In particular, as the mass asymmetry increases, the tilting mode becomes progressively softer, thus increasing σ_x^2 with respect to σ_y^2. The

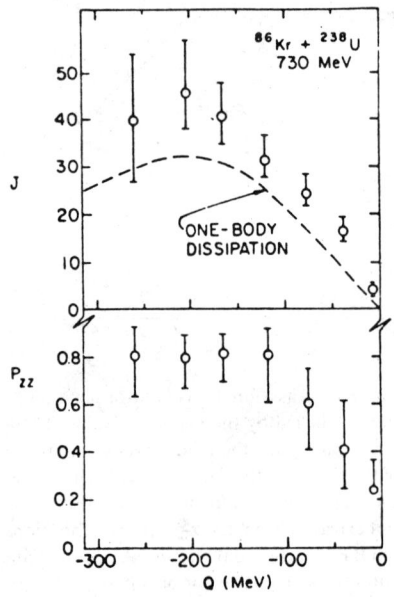

Figure 28 Dependence of the angular momentum J transferred to the heavy fragment and the alignment P_{zz} of the transferred angular momentum on Q value (87).

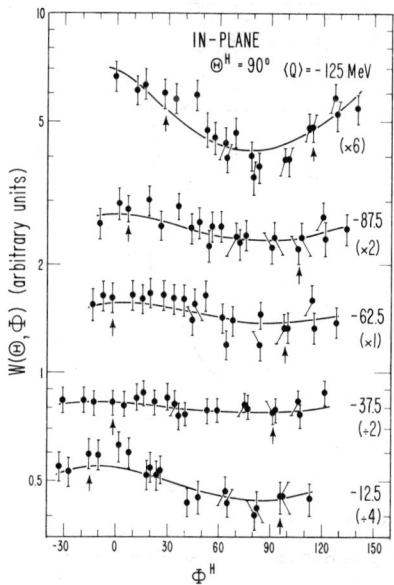

Figure 29 In-plane angular distributions of sequential fission fragments from the ^{20}Ne+^{238}U system for several different Q values (73).

dependence of the three variances with mass asymmetry is shown in Figure 18. The statistical model makes the following prediction for the in-plane angular distribution: namely, the anisotropy is expected to be small for symmetric exit channels and to become more pronounced for asymmetric ones. Some data are shown in Figure 29. For very negative Q values in the very asymmetric ^{20}Ne+^{238}U reaction, the prediction seems to hold; for more symmetric systems, smaller in-plane anisotropies have been observed (30, 52, 87, 122). More measurements are needed at intermediate mass asymmetries to determine whether or not the statistical model is successful even in this rather fine point.

4. MORE AMBITIOUS MODELS

4.1 Time-Dependent Hartree-Fock Model

The success of the self-consistent Hartree-Fock approach in describing low-lying nuclear stationary states prompted the attempt to adapt this kind of calculation to the time evolution of complex nuclear systems. The resulting theory is called time-dependent Hartree-Fock (20, 21, 77) and relies on the fact (or hope) that the complicated two-body reactions can be substituted by a time-dependent mean field on which the nucleons move as free particles. This approach, while it can be considered as fundamental as any in nuclear physics, unfortunately produces rather limited predictions. One

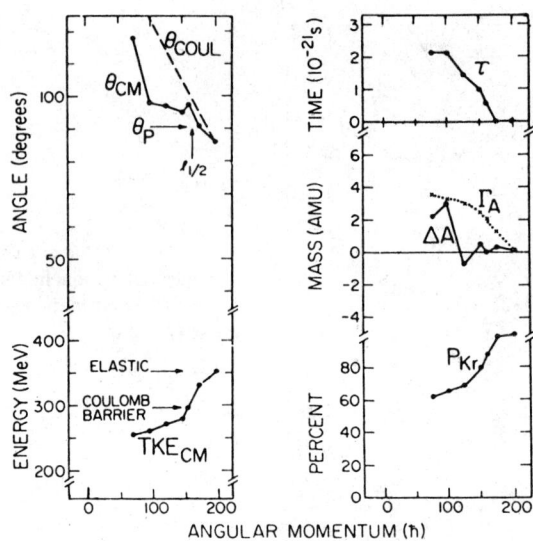

Figure 30 Quantitative summary of time-dependent Hartree-Fock results for the 494-MeV ^{84}Kr + ^{208}Pb reaction (20).

of its most serious limitations is its inability to produce dispersions in the relevant variables of a magnitude comparable to experiment, thus falling short of a qualitatively satisfying description. On the other hand, it appears that the mean energy dissipation and the mean angular momentum transfer are reproduced more or less adequately (Figure 30). The generalization of this model to include two-body interactions with the expected extensive thermalizations is not technically obvious. Still, the trained eye of the experimentalist can discern in the data the extended degree of thermalization that the present calculations have not yet been able to portray.

4.2 Coherent Surface Excitation Model

A program was undertaken by the Copenhagen group to verify the extent to which collective modes, excited through the mutual interaction of the two nuclei, could be responsible for energy and angular momentum transfer (13–15). In this approach, the surface modes are computed in the RPA approximation insofar as their strength distribution is concerned. These modes are further described in terms of equivalent damped oscillators. The results are mixed. Contrary to experiment, both fragments become loaded with approximately equal amount of energy: $Q_1/Q_2 \approx (A_2/A_1)^{1/3}$ and the overall energy dissipation is inadequate by a factor of ~ 2. The introduction of particle exchange together with the surface modes provides an adequate amount of energy dissipation and of angular momentum transfer as well. The latter result does not prove much. In a regime in which the radial energy is rapidly dissipated, the damping of the tangential energy goes hand in hand with angular momentum transfer. The merit of this model is that of stressing the role of collective modes and the important microscopic connection to standard nuclear theory.

4.3 Transport Models

We have already mentioned some phenomenological transport theories. The two theories that come close to satisfying most requirements (like the incorporation of a sufficient number of degrees of freedom, including that of relative motion, and like the microscopic qualification of the diffusion coefficients) are the one pioneered by Nörenberg (6, 79, 81) and Wolschin (55, 116–118), on the one hand, and those based upon one-body theory promoted by Randrup (88, 89, 91) and Feldmeier (32) on the other. Their success has also been mixed. Most problems regarding the mass drift and the partition of the dissipated energy have not reached a satisfactory solution as yet. Also, many of the successful predictions of the second moments and, at times, the first moments of dynamical variables have been preempted in their significance by the fact that the same quantities are predicted in the equilibrium limit. Thus no support can come to these theories from the reproduction of the equilibrium limit.

Insofar as the angular momentum partition is concerned, we have seen that the equilibrium statistical model does a good job in describing the experiment. However, some interesting theoretical results (90) show distinct differences between the equilibrium and the diffusion model in predicting the spin-spin correlation between the two fragments.

In order to attain statistical equilibrium of the spin distribution during a reaction, the appropriate relaxation times should be sufficiently short when compared with the reaction time. In the one-body picture, the primary excitations occur through the transfer of individual nucleons between the two reaction partners. A system of Fokker-Planck equations for the resulting evolution of the mean values and covariances of the spin distribution in the two nuclei can be derived.

The long-time limit solution to the equations is given by rigid rotation for the mean spin vectors, and by the statistical model for the variances and covariances. For a symmetric collision, one can obtain analytic expressions for the characteristic relaxation times as equilibrium is approached (26, 27). One should note that a different convention was chosen in this work. The line of centers of the dinuclear system and the orbital angular momentum were chosen to be the z and y axes, respectively. The relaxation times for the positive spin modes (in which the fragments are rotating in the same sense, wriggling and tilting) are denoted by t_{++} and t_{+z}, respectively. The relaxation times for the negative spin modes (in which the fragments are rotating in the opposite sense, bending and twisting) are denoted by t_{--}. Figure 31 shows the relaxation times calculated for the collision of 1400-MeV ^{165}Ho with ^{165}Ho. This theory predicts (see Figure 31) that the wriggling modes (t_{++}) are strongly excited and reach equilibrium in most cases, the negative modes (t_{--}) get partially excited, and the tilting mode (t_{+z}) receives only little excitation. These conclusions are also valid for results obtained by solving numerically the system of equations and performing the necessary transformations to the external coordinate system and to distributions not gated by impact parameters, but by total kinetic energy loss.

Figure 32 shows variances and covariances of the spins S^A and S^B of the projectile-like and target-like reaction products (for example $\sigma_{XZ}^{AB} \equiv \langle S_X^A S_Z^B \rangle - \langle S_X^A \rangle \langle S_Z^B \rangle$). The left-hand part of the figure shows the result obtained with the transfer-induced transport theory, and the right-hand part shows the predictions of the statistical model for the same collision. The principal

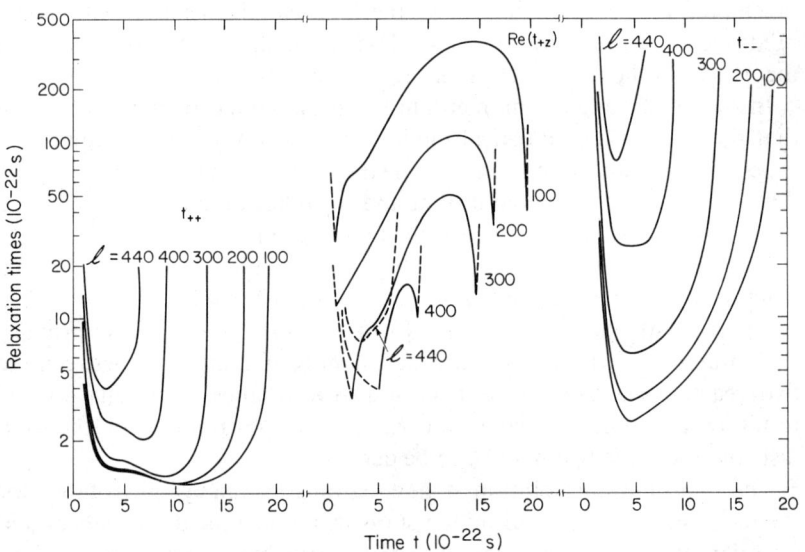

Figure 31 Calculated local relaxation times for the reaction 1400-MeV ^{165}Ho+^{165}Ho for various values of the total angular momentum l. The relaxation times for the two positive perpendicular modes (wriggling) are denoted t_{++}, while the one of the positive longitudinal mode (tilting) is denoted t_{+z}. The relaxation time for the three negative modes (bending and twisting) is denoted t_{--} (26).

variances σ_{ii}^{AA} within one nucleus all increase steadily with TKEL (except for a slight decrease of σ_{YY}^{AA} at the largest TKEL; this may be an artifact of the upper bound imposed on TKEL by the calculation). The covariance between the spins in the two nuclei along the normal direction, σ_{YY}^{AB}, is always positive because of the dominance of the positive wriggling mode for small TKEL, and because of the contributions from quite a wide range of l waves at large TKEL. The dependence of the relaxation times on l wave, as shown in Figure 31, is reflected in the dependence of the in-plane covariances σ_{XX}^{AB} and σ_{ZZ}^{AB} on TKEL. The larger in-plane component, σ_{XX}^{AB}, first increases to substantial positive values for small TKEL, as a result of the very short relaxation time for the wriggling mode. For large TKEL, σ_{XX}^{AB} decreases and finally becomes rather small as a result of the increasing excitation of the negative in-plane bending mode. For large l waves, the tilting relaxation time, t_{+z}, is smaller than the twisting relaxation time t_{--}, and the smaller in-plane component of the covariance, σ_{ZZ}^{AB}, attains small positive values for small TKEL. With decreasing l, the twisting relaxation time becomes smaller and the tilting time longer. Consequently, with

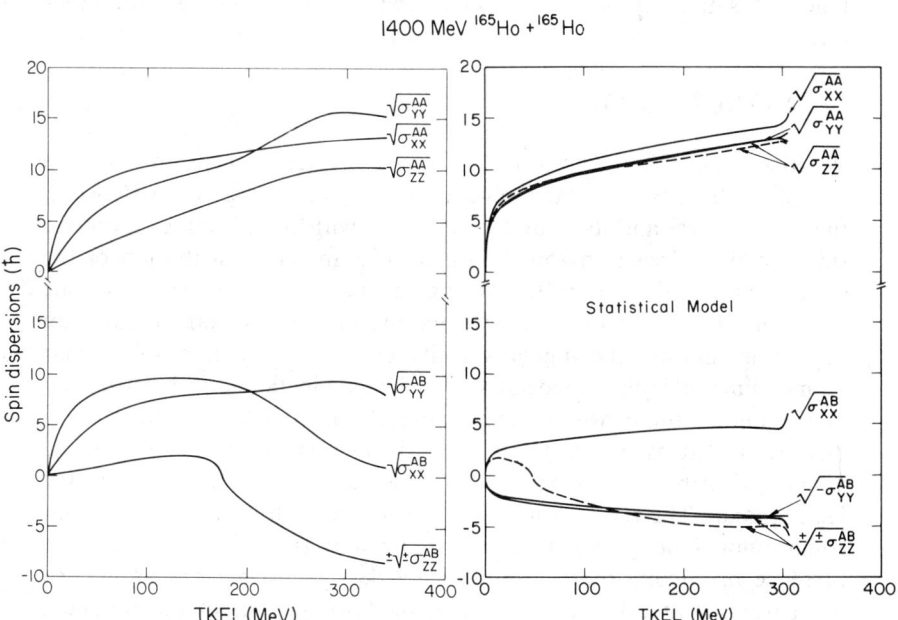

Figure 32 The spin dispersions along the principal directions as functions of the incurred energy loss TKEL, as calculated with the transfer-induced transport theory (*left*) and the statistical model (*right*) for the reaction 1400-MeV ^{165}Ho + ^{165}Ho by integrating over all l values (26).

increasing TKEL, σ_{ZZ}^{AB} changes sign and finally, for large TKEL, attains substantial negative values.

In contrast to this behavior, the dispersions calculated with the statistical model grow roughly as the fourth root of the total kinetic energy loss. For the normal variances and covariances, σ_{YY}^{AA} and σ_{YY}^{AB}, one should not attach so much significance to the difference between the two results, since part of the result in the dynamical case comes from integration over the impact parameter at fixed TKEL. A similar integration in the statistical model would diminish the difference between the two results. The in-plane variances and covariances, on the other hand, can be directly compared, and here the dynamical results display a much stronger anisotropy, and a quite different characteristic behavior as a function of TKEL, as compared to the statistical model result, which would be obtained in the dynamical calculations if the relaxation times were all very short.

Thus, the nucleon exchange transport model predicts strong correlations between the two fragment spins. The detailed structure of such spin-spin correlations can be probed in a fission-fission angular correlation experiment. Most notably, the existence of covariances between the two nuclear spins implies that the detection of fission from one nucleus breaks the reflection symmetry of the angular correlation of fission from the other nucleus.

5. CONCLUSION

We tried to show how important the role of angular momentum is in deep inelastic collisions. Angular momentum defines the range of existence of this very process and its boundaries as well, with complete fusion on one side and direct reactions on the other. The angular distribution of the ensuing products is controlled by angular momentum in more than one way. On the one hand the interaction time is angular momentum dependent, and so is the angular velocity. On the other, the orbital angular momentum that is dissipated and the dissipated kinetic energy both depend on the angular momentum. Furthermore, the different exit-channel asymmetries are fed by different l waves. All this contributes to making the angular distributions dependent on just about any variable in the Hamiltonian. The relaxation of angular momentum leads to the secular equilibrium of the system, which can rotate like a rigid body. However, the presence of angular-momentum-bearing degrees of freedom allows for fluctuations that can misalign the aligned component of the fragment angular momentum. The excitation of these modes can be studied through the angular distribution of particles and gamma rays emitted by the fragments.

The present understanding of the role of angular momentum has crystallized in a variety of theories ranging from conditional equilibrium statistical theories to assorted transport theories and including the time-dependent Hartree-Fock theory. At this stage it may be premature to say if any given theory satisfies all the details brought to light by experiment. It is clear that one ought to broaden the discussion to the overall field of heavy-ion collisions. Such an extension is beyond the scope of this publication. We ask forgiveness if our zeal in telling the whole story has been stronger at times than our commitment to a part of it, albeit an important one.

ACKNOWLEDGMENT

This work was supported by the Director, Office of Energy Research, Division of Nuclear Physics of the Office of High Energy and Nuclear Physics of the US Department of Energy under Contract DE-AC03-76SF00098.

Literature Cited

1. Agassi, D., Ko, C. M., Weidenmüller, H. A. 1978. *Phys. Rev. C* 18:223
2. Agassi, D., Ko, C. M., Weidenmüller, H. A. 1979. *Ann. Phys.* 117:407
3. Aguer, P., Schmitt, R. P., Wozniak, G. J., Habs, D., Diamond, R. M., et al. 1979. *Phys. Rev. Lett.* 43:1778
4. Aleonard, M. M., Wozniak, G. J., Glässel, P., Deleplanque, M. A., Diamond, R. M., et al. 1978. *Phys. Rev. Lett.* 40:622
5. Artukh, A. G., Gridnev, G. F., Mikheev, V. L., Volkov, V. V., Wilczyński, J. 1973. *Nucl. Phys. A* 215:91
6. Ayik, S., Wolschin, G., Nörenberg, W. 1978. *Z. Phys. A* 286:271
7. Babinet, R., Cauvin, B., Girard, J., Alexander, J. M., Chiang, T. H., et al. 1980. *Z. Phys. A* 295:153
8. Bass, R. 1974. *Nucl. Phys. A* 231:45
9. Berlanger, M., Deleplanque, M. A., Gerschel, C., Hanappe, F., Le Blanc, M., et al. 1976. *J. Phys. Lett.* 37:L323
10. Birkelund, J. R., Tubbs, L. E., Huizenga, J. R., De, J. N., Sperber, D. 1979. *Phys. Lett. C* 56:107
11. Bock, R., Fischer, B., Gobbi, A., Hildenbrand, K., Kohl, W., et al. 1977. *Nucleonika* 22:529
12. Bondorf, J. P., Sobel, M. I., Sperber, D. 1974. *Phys. Lett.* 15C:83
13. Broglia, R. A., Civitarese, O., Dasso, C. H., Winther, A. 1978. *Phys. Lett.* 73B:405
14. Broglia, R. A., Dasso, C. H., Pollarolo, G., Winther, A. 1978. *Phys. Rev. Lett.* 41:25
15. Broglia, R. A., Dasso, C. H., Winther, A. 1982. In *Proc. Int. Sch. Phys., "Enrico Fermi," Varenna, Italy, July 1979*, p. 327. Amsterdam: North-Holland
16. Broglia, R. A., Pollarolo, G., Dasso, C. H., Døssing, T. 1979. *Phys. Rev. Lett.* 43:1649
17. Cassing, W., Friedrich, H. 1981. *Z. Phys. A* 299:359
18. Christensen, P. R., Folkmann, F., Hansen, O., Nathan, O., Trautner, N., et al. 1980. *Nucl. Phys. A* 349:217
19. Cohen, S., Plasil, F., Swiatecki, W. J. 1974. *Ann. Phys.* 82:557
20. Davies, K. T. R., Maruhn-Rezwani, V., Koonin, S. E., Negele, J. W. 1978. *Phys. Rev. Lett.* 41:632
21. Davies, K. T. R., Koonin, S. E. 1981. *Phys. Rev. C* 23:2042
22. Dayras, R. A., Stokstad, R. G., Fulmer, C. B., Hensley, D. C., Halbert, M. L., et al. 1979. *Phys. Rev. Lett.* 42:697
23. Dayras, R. A., Stokstad, R. G., Hensley, D. C., Halbert, M. L., Sarantites, D. G., et al. 1980. *Phys. Rev. C* 22:1485
24. Deubler, H. H., Dietrich, K. 1975. *Phys. Lett.* 56B:241
25. Diamond, R. M., Stephens, F. S. 1980. *Ann. Rev. Nucl. Part. Sci.* 30:85
26. Døssing, T., Randrup, J. 1984. *Lawrence Berkeley Lab. Rep. No. LBL-16825.* Berkeley: LBL
27. Døssing, T., Randrup, J. 1984. *Lawr-*

ence *Berkeley Lab. Rep. No. LBL-16826.* Berkeley: LBL
28. Dünnweber, W. 1984. In *Proc. Int. Sch. Phys. "Enrico Fermi," Varenna, Italy, July 1982,* p. 389. Amsterdam: North-Holland
29. Dyer, P., Puigh, R. J., Vandenbosch, R., Thomas, T. D., Zisman, M. S. 1977. *Phys. Rev. Lett.* 39:392
30. Dyer, P., Puigh, R. J., Vandenbosch, R., Thomas, T. D., Zisman, M. S., Nunnelley, L. 1979. *Nucl. Phys. A* 322:205
31. Eyal, Y., Gavron, A., Tserruya, I., Fraenkel, Z., Eisen, Y., et al. 1978. *Phys. Rev. Lett.* 41:625
32. Feldmeier, H. 1984. See Ref. 28, p. 274
33. Galin, J. 1976. *J. Phys.* 37:C5–83
34. Galin, J., Guerreau, D., Lefort, M., Péter, J., Tarrago, X., Basile, R. 1970. *Nucl. Phys. A* 159:461
35. Gatty, B., Guerreau, D., Lefort, M., Pouthas, J., Tarrago, X., et al. 1975. *Z. Phys. A* 273:65
36. Gatty, B., Guerreau, D., Lefort, M., Tarrago, X., Galin, J., et al. 1975. *Nucl. Phys. A* 253:511
37. Gerschel, C., Deleplanque, M. A., Ishihara, M., Ngô, C., Perrin, N., et al. 1979. *Nucl. Phys. A* 317:473
38. Glas, D., Mosel, U. 1975. *Nucl. Phys. A* 237:439
39. Glässel, P., Simon, R. S., Diamond, R. M., Jared, R. C., Lee, I. Y., et al. 1977. *Phys. Rev. Lett.* 38:331
40. Gould, C. R., Bass, R., Czarnecki, J. v., Hartmann, V., Stelzer, K., et al. 1980. *Z. Phys. A* 294:323
41. Gross, D. H. E., Kalinowski, H. 1974. *Phys. Lett.* 48B:302
42. Gross, D. H. E. 1981. *Nukleonika* 26:985
43. Hanappe, F., Lefort, M., Ngô, C., Péter, J., Tamain, B. 1974. *Phys. Rev. Lett.* 32:738
44. Harrach, D. v., Glässel, P., Civelekoglu, Y., Männer, R., Specht, H. J. 1979. *Phys. Rev. Lett.* 42:1728
45. Hilscher, D., Birkelung, J. R., Hoover, A. D., Schröder, W. U., Wilcke, W. W., et al. 1979. *Phys. Rev. C* 20:576
46. Ishihara, M., Tanaka, K., Kammuri, T., Matsuoka, K., Sano, M. 1978. *Phys. Lett.* 73B:281
47. Jacmart, J. C., Colombani, P., Doubre, H., Frascaria, N., Poffé, N., et al. 1975. *Nucl. Phys. A* 242:175
48. Kaufmann, R., Wolfgang, R. 1961. *Phys. Rev.* 121:192
49. Kratz, J. V., Norris, A. E., Seaborg, G. T. 1974. *Phys. Rev. Lett.* 33:502
50. Kühn, W., Albrecht, R., Damjantschitsch, H., Ho, H., Ronningen, R. M., et al. 1980. *Z. Phys. A* 298:95
51. Lauterbach, C., Dünnweber, W., Graw, G., Hering, W., Puchta, H., Trautmann, W. 1978. *Phys. Rev. Lett.* 41:1774
52. Le Brun, C., Le Colley, J. F., Lefebvres, F., L'Haridon, M., Osmont, A., et al. 1982. *Phys. Rev. C* 25:3212
53. Lefort, M. 1975. *Lect. Notes Phys.* 33:275
54. Lefort, M., Ngô, C. 1979. *Riv. Nuovo Cimento* 2:1
55. Li, J. Q., Wolschin, G. 1983. *Phys. Rev. C* 27:590
56. Mathews, G. J., Wozniak, G. J., Schmitt, R. P., Moretto, L. G. 1977. *Z. Phys. A* 283:247
57. McDonald, R. J., Pacheco, A. J., Wozniak, G. J., Bolotin, H. H., Moretto, L. G., et al. 1982. *Nucl. Phys. A* 373:54
58. Moretto, L. G. 1975. *Nucl. Phys. A* 247:211
59. Moretto, L. G. 1978. *J. Phys. Soc. Jpn.* 44:Suppl. 7, p. 361
60. Moretto, L. G. 1981. *Lawrence Berkeley Lab. Rep. No. LBL-12596.* Berkeley: LBL
61. Moretto, L. G. 1982. In *Proc. Int. Conf. on Selected Aspects of Heavy-Ion Reactions, Saclay, France, LBL-14283.* Berkeley: LBL
62. Moretto, L. G., Babinet, R. P., Galin, J., Thompson, S. G. 1975. *Phys. Lett.* 58B:31
63. Moretto, L. G., Blau, S. K., Pacheco, A. J. 1981. *Nucl. Phys. A* 364:125
64. Moretto, L. G., Cauvin, B., Glässel, P., Jared, R., Russo, P., et al. 1976. *Phys. Rev. Lett.* 36:1069
65. Moretto, L. G., Heuneman, D., Jared, R. C., Gatti, R. C., Thompson, S. G. 1973. *Physics and Chemistry of Fission,* Vol. 2. Vienna: Int. Atomic Phys. Agency. 351 pp.
66. Moretto, L. G., Schmitt, R. P. 1976. *J. Phys.* 37:C5–109
67. Moretto, L. G., Schmitt, R. P. 1980. *Phys. Rev. C* 21:204
68. Moretto, L. G., Schmitt, R. P. 1981. *Rep. Prog. Phys.* 44:533
69. Moretto, L. G., Sobotka, L. G. 1981. *Z. Phys. A* 303:299
70. Moretto, L. G., Sventek, J. S. 1975. *Phys. Lett.* 58B:26
71. Morrissey, D. J., Moretto, L. G. 1981. *Phys. Rev. C* 23:1835
72. Morrissey, D. J., Wozniak, G. J., Sobotka, L. G., Pacheco, A. J., Hsu, C. C., et al. 1982. *Z. Phys. A* 305:131
73. Morrissey, D. J., Wozniak, G. J., Sobotka, L. G., Pacheco, A. J., McDonald, R. J., et al. 1982. *Nucl. Phys. A* 389:120

74. Mouchaty, G., Haenni, D. R., Nath, S., Garg, U., Schmitt, R. P. 1984. *Z. Phys. A* 316:285
75. Namoodiri, M. N., Natowitz, J. B., Kasiraj, P., Eggers, R., Adler, L., et al. 1979. *Phys. Rev. C* 20:982
76. Natowitz, J. B., Namboodiri, M. N., Kasiraj, P., Eggers, R., Adler, L., et al. 1978. *Phys. Rev. Lett.* 40:751
77. Negele, J. W. 1982. *Rev. Mod. Phys.* 54:913
78. Nörenberg, W. 1974. *Phys. Lett.* 53B:289
79. Nörenberg, W. 1975. *Z. Phys. A* 274:241
80. Nörenberg, W. 1976. *J. Phys.* 37:C5-141
81. Nörenberg, W. In *Proc. Predeal Int. Sch. Phys., Predeal, Romania, 1978. GSI Rep. 79-5*
82. Nörenberg, W., Weidnmüller, H. A. 1980. *Lecture Notes in Physics 51.* Berlin: Springer-Verlag
83. Olmi, A., Sann, H., Pelte, D., Eyal, Y., Gobbi, A., et al. 1978. *Phys. Rev. Lett.* 41:688
84. Pacheco, A. J., Wozniak, G. J., McDonald, R. J., Diamond, R. M., Hsu, C. C., et al. 1983. *Nucl. Phys. A* 397:313
85. Puchta, H., Dünnweber, W., Hering, W., Lauterbach, C., Trautmann, W. 1979. *Phys. Rev. Lett.* 43:623
86. Puigh, R. J., Doubre, H., Lazzarini, A., Seamster, A., Vandenbosch, R., et al. 1980. *Nucl. Phys. A* 336:279
87. Puigh, R. J., Dyer, P., Vandenbosch, R., Thomas, T. D., Nunnelley, L., Zisman, M. S. 1979. *Phys. Lett.* 86B:24
88. Randrup, J. 1978. *Nucl. Phys. A* 307:319
89. Randrup, J. 1979. *Nucl. Phys. A* 327:490
90. Randrup, J. 1981. *Phys. Lett.* 110B:25
91. Randrup, J. 1982. *Nucl. Phys. A* 383:468
92. Regimbart, R., Behkami, A. N., Wozniak, G. J., Schmitt, R. P., Sventek, J. S., Moretto, L. G. 1978. *Phys. Rev. Lett.* 41:1355
93. Rudolf, G., Gobbi, A., Stelzer, H., Lynen, U., Olmi, A., Sann, H., Stokstad, R. G., Pelte, D. 1979. *Nucl. Phys. A* 330:243
94. Schmitt, R. P., Bizard, G., Wozniak, G. J., Moretto, L. G. 1978. *Phys. Rev. Lett.* 41:1152
95. Schmitt, R. P., Pacheco, A. J. 1982. *Nucl. Phys. A* 379:313
96. Schröder, W. U., Birkelund, J. R., Huizenga, J. R., Wolf, K. L., Viola, V. E. Jr. 1978. *Phys. Lett. C* 45:301
97. Schröder, W. U., Huizenga, J. R. 1977. *Ann. Rev. Nucl. Sci.* 27:465
98. Scott, D. K. 1978. *Not. Fis. No. 5*, p. 1
99. Sobotka, L. G., Hsu, C. C., Wozniak, G. J., Morrissey, D. J., Moretto, L. G. 1981. *Nucl. Phys. A* 371:510
100. Sobotka, L. G., Hsu, C. C., Wozniak, G. J., Rattazzi, G. U., McDonald, R. J., et al. 1981. *Phys. Rev. Lett.* 46:887
101. Sobotka, L. G., McDonald, R. J., Wozniak, G. J., Morrissey, D. J., Pacheco, A. J., Moretto, L. G. 1983. *Phys. Rev. C* 28:219
102. Strutinsky, V. M. 1973. *Phys. Lett.* 44B:245
103. Sventek, J. S., Moretto, L. G. 1978. *Phys. Rev. Lett.* 40:697
104. Swiatecki, W. J. 1972. *J. Phys.* 33:C5-45
105. Swiatecki, W. J. 1982. *Nucl. Phys. A* 376:275
106. Tamain, B., Chechik, R., Fuchs, H., Hanappe, F., Morjean, M., et al. 1979. *Nucl. Phys A* 330:253
107. Trautmann, W., Dahme, W., Dünnweber, W., Hering, W., Lauterbach, C., et al. 1981. *Phys. Rev. Lett.* 46:1188
108. Trautmann, W., de Boer, J., Dünnweber, W., Graw, G., Kopp, R., et al. 1977. *Phys. Rev. Lett.* 39:1062
109. Tsang, C. F. 1974. *Phys. Scr.* 10A:90
110. Van Bibber, K., Ledoux, R., Steadman, S. G., Videbaek, F., Young, G., Flaum, C. 1977. *Phys. Rev. Lett.* 38:334
111. Vandenbosch, R., Webb, M. P., Thomas, T. D., Zisman, M. S. 1976. *Nucl. Phys. A* 269:210
112. Volkov, V. V. 1978. *Phys. Lett. C* 44:93
113. Wilcke, W. W., Birkelund, J. R., Hoover, A. D., Huizenga, J. R., Schröder, W. U., et al. 1980. *Phys. Rev. C* 22:128
114. Wilczyński, J. 1973. *Phys. Lett.* 47B:484
115. Wolf, K. L., Unik, J. P., Huizenga, J. R., Birkelund, J. R., Freiesleben, H., Viola, V. E. 1974. *Phys. Rev. Lett.* 33:1105
116. Wolschin, G. 1979. See Ref. 15, p. 508
117. Wolschin, G. 1979. *Phys. Lett.* 88B:35
118. Wolschin, G. 1979. *Nucl. Phys. A* 316:146
119. Wolschin, G., Nörenberg, W. 1978. *Phys. Rev.* 41:691
120. Wozniak, G. J., McDonald, R. J., Pacheco, A. J., Hsu, C. C., Morrissey, D. J., et al. 1980. *Phys. Rev. Lett.* 45:1081
121. Wozniak, G. J., Schmitt, R. P., Glässel, P., Jared, R. C., Bizard, G., Moretto, L. G. 1978. *Phys. Rev. Lett.* 40:1436
122. Zisman, M. S., Puigh, R. J., Vandenbosch, R., Thomas, T. D., Nunnelley, L. 1982. *Lawrence Berkeley Lab. Rep. No. LBL-13366*, p. 93. Berkeley: LBL
123. Vandenbosch, R., Lazzarini, A., Leach, D., Lock, D.-K., Ray, A., Seamster, A. 1984. *Phys. Rev. Lett.* 52:1964

Ann. Rev. Nucl. Part. Sci. 1984. 34: 247–84

SUPERCONDUCTING MAGNET TECHNOLOGY FOR ACCELERATORS[1]

R. Palmer

Brookhaven National Laboratory, Upton, New York 11973

A. V. Tollestrup

Fermi National Accelerator Laboratory, Batavia, Illinois 60510

CONTENTS

1. INTRODUCTION	248
2. MAGNET TECHNOLOGY	250
2.1 Coil Geometry	250
2.2 Iron Cross Section	253
2.3 The Cryostat	255
3. CONDUCTORS	256
3.1 The B-J-T Surfaces	257
3.2 Stability	259
3.3 Hysteresis and Eddy Currents	262
3.4 Temperature Effects	264
4. FIELD QUALITY CONSIDERATIONS	264
4.1 Field Analysis	264
4.2 Field Precision Requirements	266
4.3 Coil and Iron Cross-Section Design	266
4.4 The Magnet Ends	269
4.5 Superconducting Persistent Currents	271
4.6 Iron Effects	272
4.7 Correction Coils	273
5. FORCES AND COIL PACKAGING	274
5.1 Magnet Forces	275
5.2 Preload	276

[1] The US Government retains a nonexclusive, royalty-free license to publish or reproduce the published form of this contribution, or allow others to do so, for US Government purposes.

6. QUENCH: THEORY AND MAGNET PROTECTION 277
   6.1 Quench Theory 277
   6.2 Training 279
   6.3 Quench Heating 280
   6.4 Quench Protection 283

1. INTRODUCTION

A review article on superconducting magnets for accelerators should first answer the question, "Why superconductivity?" The answer revolves around two pivotal facts: (a) fields in the range of 2 to 10 T can be achieved; (b) the operating cost can be less than that with conventional magnets.

The relative importance of these two factors depends on the accelerator. In the case of upgrading an existing accelerator, the ability to obtain fields higher than those obtained with conventional magnets leads directly to an increase in machine energy for the given tunnel. In the case of a new facility, both factors must be balanced for the most economical machine. How to achieve this optimum is not yet clear.

In opting for superconducting magnets, one also encounters problems that were once foreign to accelerator technology. These can be lumped under these main headings of cryogenics, quench protection, and field quality.

At present, commercial refrigerators exist for large liquid He systems (~ 5000 liters hr^{-1}). At operating temperatures between 4 and 5 K, the ultimate Carnot efficiency is low ($\sim 1.4\%$) and the relative efficiency of a large plant ($\sim 10^5$ W at 4.2 K) may approach 30%. Thus, one needs 300–500 watts at room temperature for every watt of 4.2 K load. This places strict requirements on the cryostat design, forcing one to use intermediate heat shields, superinsulation, good insulating vacuum, and a minimum number of current leads between the cold conductor system and the outside world.

Other difficulties are imposed because of the cryogenic system. The continuity of the vacuum and cryogenic systems makes access to the beam difficult. Correction coils, pickup electrodes, etc all require careful attention to details since the system must have a low heat leak in spite of many vacuum feedthroughs. Failure of even minor components means warming up an entire string of magnets, which may take several days. The original installation relies on developing techniques to insure leak tightness of thousands of vacuum seals cooled to 4.2 K. This may sound difficult—it is! But techniques have been successfully developed to overcome these problems.

A second new regime of problems comes with "quench protection." A magnet quenches when some piece of the conductor is driven up in temperature until it loses its superconductivity; at that point, the I^2R losses from the current in the conductor continue to increase the temperature. If the cooling is sufficient to overcome the heating, the conductor will cool and return to its superconducting state; the magnet is said to be cryostable. However, it is hard to achieve this degree of cooling in an accelerator magnet where a very high current density is generally desired, although superconducting coils for bubble chambers are readily designed to be cryostable.

The temperature of the conductor may be driven up to the quench point by beam loss, frictional heat from the motion of the coil, or lack of sufficient refrigeration at coil support points. A successful design must protect every centimeter of conductor in a machine that may have several hundreds of kilometers of cable in the magnets. The problem is complicated by the inductance of the magnet, which makes it difficult to reduce the current to zero in a short time. Quench protection complicates the design, but solutions are now known.

Finally comes the question of field quality. An iron magnet has its field primarily controlled by the shape of the iron pole face. Laminations can be stamped to a precision and smoothness of better than 0.001 in. Classically, magnet design has centered on shaping the iron correctly.

However, a superconducting magnet relies on accurate placement of the conductor to achieve the desired field. Conductors must be held in place to an accuracy of about 0.001 in. in order to achieve accelerator quality fields ($\delta B/B \approx 10^{-4}$). This is difficult because of the large magnetic forces and differential thermal stresses. A successful magnet will have a uniform field region for between one third and two thirds of its available aperture. The space between the beam and the coil acts as a filter to remove the high harmonics introduced by individual wire strands and errors in their placement. Quality control of the components is required during the assembly, and a dedicated magnet measuring facility is the crucial final monitor of overall magnet quality. At the present level of technology, each magnet must be measured before installation.

Superconductivity was discovered by K. Onnes in 1911 and explained with a microscopic theory by Barden, Cooper, and Schrieffer in 1957. Type II superconductors using Nb_3Sn and NbTi have been developed since 1961. Thus, a long and difficult theoretical and experimental program led to the availability of conductors in the early 1970s that could be manufactured economically and could carry current densities high enough to be considered for use in accelerator magnets. Close collaboration between the conductor fabricators and the national laboratories resulted in the production of successful magnets.

The program described above culminated in the successful operation of the Tevatron early in 1983. The operational information to be obtained from this pioneering effort is crucial to the design of future accelerators. The basic magnet design used in the Tevatron has been reproduced at DESY, Serpukov, Saclay, KEK, LBL, and BNL. Many variations and improvements are being investigated, and one can expect future machines, HERA and UNK, to have magnets with better quality and higher fields (~ 5 T). Such magnets were demonstrated at BNL for the CBA.

Thus 75 years after the initial discovery of superconductivity, we can say that it has become a well-engineered technology with a wealth of well-understood solutions for the unique problems encountered. It is still difficult to predict beforehand if a particular variation of a magnet will "work." However, it can be stated that the sources of difficulties are well understood, and the necessary research and development program to produce a successful magnet is now well defined.

It is within this framework that we are approaching the greatest challenge—a superconducting collider in the 20 × 20 TeV range. It is here that the ultimate optimization between complexity, size, and cost will be made. We now turn to details of magnet technology.

2. MAGNET TECHNOLOGY

2.1 *Coil Geometry*

Magnets for accelerators require, in general, a vertical magnetic field in a long horizontal cylindrical volume. The coils of such magnets consist of long loops with conductor passing down one side of the magnet and returning up the other side. The magnet type can be described by the typical cross section of such a magnet; the ends we discuss later. We divide magnets into three types: low-field, ion-dominated magnets; high-field, rectangular magnets; and high-field magnets with circular cross sections.

For fields below 2 Tesla it is possible to design a magnet in which the field shape is dominated by the shape of magnetic poles above and below the field region. Such magnets are sometimes referred to as "superferric." Figure 1*a* shows (1) an H magnet in which the conductor is relatively far from the field region, and the field is truly dominated by the shape of the magnetic poles. In this type, variation in position of the conductor will have essentially no effect on the magnetic fields. The poles have to be specifically shaped to obtain a good field in the central region, and saturation effects in these shaped poles usually limit the field to well below 2 Tesla. Figure 1*b* shows (2) a superferric picture-frame magnet in which the conductor is in the form of current sheets to the left and to the right of the field region. No shaping of the poles is needed, and the entire space between the coils is, in principle, good field. The field is, however, to some extent dependent upon

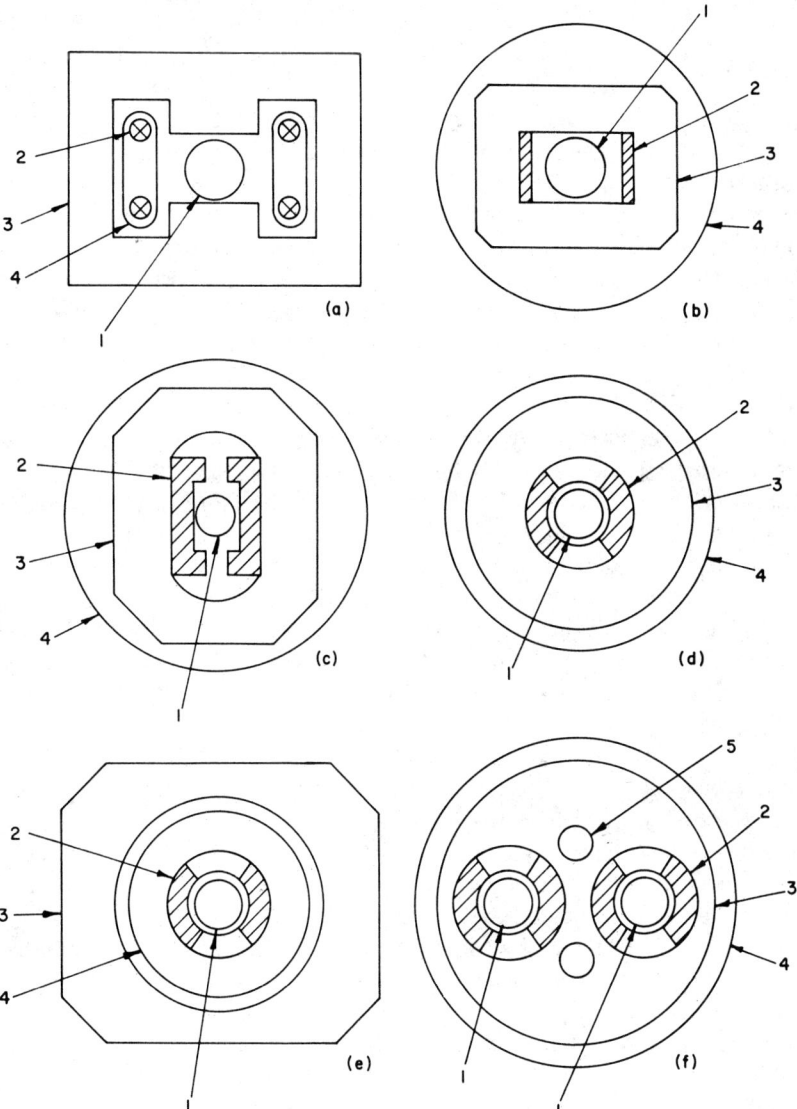

Figure 1 Magnet types: (*a*) warm iron superferric, (*b*) cold iron superferric, (*c*) high-field picture frame, (*d*) cold iron cos θ, (*e*) warm iron cos θ, and (*f*) 2-in-1 cos θ. 1. Beam pipe. 2. Conductor. 3. Iron yoke. 4. Cryostat. 5. Hole to correct saturation.

the conductor placement. The advantages are that less iron is required to return the field, and the design allows higher fields without large saturation effects. In neither of these cases is superconductor magnetization (persistent superconducting currents) a problem. The superconductor becomes uniformly "magnetized" in a vertical direction, and the resulting field lines are returned in the iron without affecting the central field.

If fields significantly above 2 Tesla are required, then the iron in the poles of a magnet saturates and that part of the field above 2 Tesla has to be generated by the conductors themselves. Consider again the picture-frame magnet shown in Figure 1b. As higher currents are passed in the two current sheets, the resulting field may be conceptualized as a uniform field from the iron plus a field generated by the coils alone. The field from the coils will not be uniform but will be higher at the left and right sides of the aperture and less in the center, which is further from the conductors. In order to correct this nonuniformity of field, extra conductors have to be placed above and below the aperture as shown (3) in Figure 1c. In this example, it is also seen that the aspect ratio of the coil has been changed to be higher relative to its width in order to reduce the aforementioned nonuniformity of field. It is also seen that the cross section of conductor is greatly increased. In this example, a field of 6 Tesla could easily be achieved, with ~ 4 Tesla of that field generated directly by the coils and ~ 2 Tesla contributed by the iron. The change of field shape as the field is increased, i.e. saturation effect, can be controlled by varying the relative current through the main coils and through the correction coils (sextupole correction coils) or self-correction can be achieved by changing the pole shape from flat to an arch, as shown in Figure 1c. This magnet, although conceptually based on the simple picture frame of Figure 1b, is now a hybrid between picture-frame designs and circular designs, which we discuss next.

Since in high-field magnets most of the field derives from the conductor placement, it is convenient to consider designs in which there is no iron present and all the field is so derived.

First we consider a current sheet of uniform radial thickness surrounding a cylindrical field volume. The current density in this current sheet is arranged to vary as the cosine of the azimuth with maximum currents at the left and right, and zero currents at the top and bottom. It can be easily seen that such a current sheet will generate a uniform vertical field in the field volume. In practice the varying current density is approximated by a finite number of conductors and wedges (see Figure 2a).

As an alternative, instead of varying current density in a current sheet of the uniform thickness as described above, one can keep current density constant and vary the radial thickness of the conductor. The resulting cross section, illustrated in Figure 2b, is the space between two intersecting

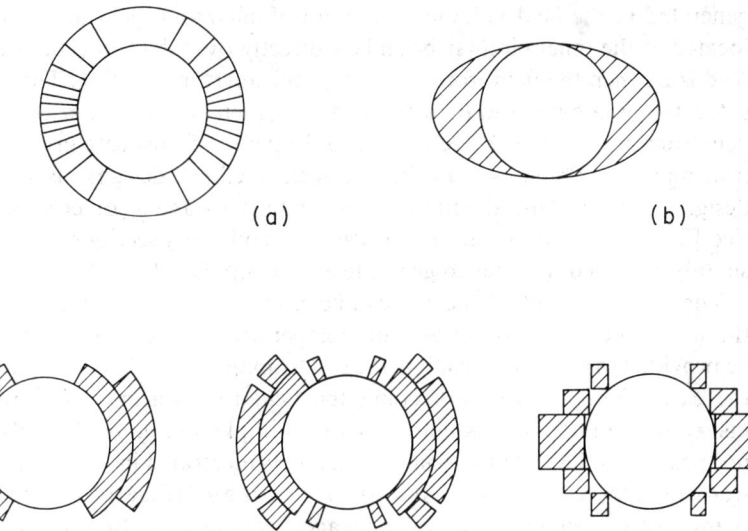

Figure 2 Approximations to cos θ coil: (*a*) pure cos θ, (*b*) intersecting ellipse, (*c*) two-layer approximation, (*d*) two-layer with wedges, (*e*) rectangular block.

ellipses. In practice this shape is approximated by one or more layers of conductor of fixed radial thickness, as indicated in Figure 2c. If three or more layers are used, a good approximation to the ideal field is generated. If only two layers are used, then the good field region is restricted to only about 50% of the diameter within the coils. The field can be improved by the addition of wedges without conductor placed in both the inner and outer layers (see Figure 2d). Since in the intersecting ellipse design the conductors are placed as close as possible to the field region, this design is both the most efficient in terms of conductor cross section for given fields, and also the one for which conductor placement is most critical. Magnetization effects at low field are also the greatest, and the design requires a conductor of a keystone cross-sectional shape; this may be considered a complication compared to rectangular shapes. There is, of course, an arrangement (Figure 2e) in which the intersecting ellipse cross section is approximated by rectangular conductors, and we see again the basic similarity to the picture-frame concept.

2.2 *Iron Cross Section*

In the cosine theta design, it can be shown that, if the coil is surrounded by a concentric tube of iron and if the iron is placed at sufficient distance to preclude saturation in that iron, then a strictly uniform field will still be generated in the field volume. The point of maximum field in the iron is located at the inner circular boundary directly over the coil center. If this field is B_i, then the field from the iron in the aperture is 0.5 B_i. Thus, if no saturation is allowed and B_i restricted to 2 T, then the maximum additional field from the iron is ~ 1 T (compare to the picture-frame case in which the iron may contribute ~ 2 T). In intersecting ellipse designs, as in cos θ designs, the iron is usually in the form of a pipe around a coil cross section (see Figures 1d and e). But in this case the coil cross sections have to be slightly modified in order to generate a pure dipole field.

The iron surrounding the coil can be either at the same temperature as the superconducting coil or at room temperature. If the latter, space must be provided to thermally insulate the coil from the iron. This can be done in the case of Figure 1a by keeping the field iron warm and providing insulation around the conductors. In the examples of Figures 1b and c, it is not possible to have warm iron. For magnets illustrated in Figure 2, either warm or cold iron is possible. If warm iron (4) is used (Figure 1e), insulation is introduced between the coil package and the iron. In this case, the contribution to the magnetic field from the iron, because of its greater distance from the field region, is relatively less and saturation effects may be avoided. If the iron is cold (5a,b), i.e. at the same temperature as the coils, then it may be brought in very close to the conductors (Figure 1d) with a resultant increase in field, but with the disadvantage of increased saturation effects that have to be corrected by either correction coils or the use of crenellation (2) or holes suitably placed in the iron. Such holes are illustrated in Figure 1f. A disadvantage of the use of cold iron is the large mass that must be cooled and the greater time and cooling capacity needed to perform this. Against this disadvantage must be weighed the higher fields achieved and the simplification resulting from the use of the iron both to retain the coils and to return the flux.

The radial thickness of iron required to return the flux can be approximated assuming a saturation field of 2 Tesla in the iron. This thickness rises linearly with both the field aperture of the magnet and the magnetic field, whereas the mass rises almost as the square of these parameters. This mass can be reduced in the special case of magnets built for proton-proton, colliding-beam machines. For such machines, two magnetic apertures are required: one with the field up and the other with the field down. If these two apertures are placed side by side (6a,b) in a common iron yoke, then field lines from one aperture are returned not through the midplane but across to the other magnetic aperture. The total weight of iron in this 2-in-1 design is approximately the same for the double magnet as would be required for each of the single magnets (see Figure 1f).

2.3 The Cryostat

In order to cool and maintain the superconductor at helium temperatures, thermal insulation must be provided between it and the surrounding warm material. If the iron is warm, this thermal insulation must be provided in the relatively small space between the superconductor and the iron. However, in these designs the surface area that must be insulated is relatively small. In the cold iron designs, space is not a problem but the surface to be insulated is larger. In both cases in order to maintain efficiently the superconductor at helium temperatures, it is desirable to introduce an intermediate shield, cooled either by liquid nitrogen or by the returning intermediate-temperature helium gas.

For a two-beam collider with coupled iron, only one cryostat is, of course, required for both beams and this reduces costs. Even if uncoupled magnets are used, they can share a common cryostat.

Cooling of the conductor itself is usually provided by placing the conductor inside a helium-filled cryostat with a flow of helium along the length of the magnet. In principle, and in simple test dewars, this helium is in the liquid phase at atmospheric pressure and heating causes the liquid to boil and introduce a gaseous phase in the upper parts of the vessel. The danger that the presence of this gaseous phase will provide inadequate cooling to the upper coils has led designers to prefer a single phase in the coil cryostat. This may be liquid helium at a somewhat elevated pressure or helium at a much higher pressure, above the critical point, in which case it is referred to as a supercritical helium. In either case, the helium is made to flow along the length of the magnet exchanging heat with the coil. The temperature of this helium will rise as a result of heat losses or heat generated in the coil until it is either replaced by new helium (Figure 3a) or passes through a heat exchanger (Figure 3c), exchanging heat with two-phase helium. One attractive solution (7) used in the Tevatron (Figure 3b) is to distribute such a heat exchanger along the outside of the magnet cryostat itself. Two-phase helium in an annular space in contact with the cryostat flows in the opposite direction and heat exchanges with the single-phase helium in contact with the coil. The temperature is held relatively constant, rising only as the pressure changes in the two-phase return manifold.

Supports must be provided to hold the coils and other members that are at helium temperature with respect to the room temperature structures. These supports must be rigid enough to maintain the required positional accuracy but should have a low heat leak. The problem is complicated as a little thought will show that rigidity and heat conduction tend to be proportional to each other for given materials.

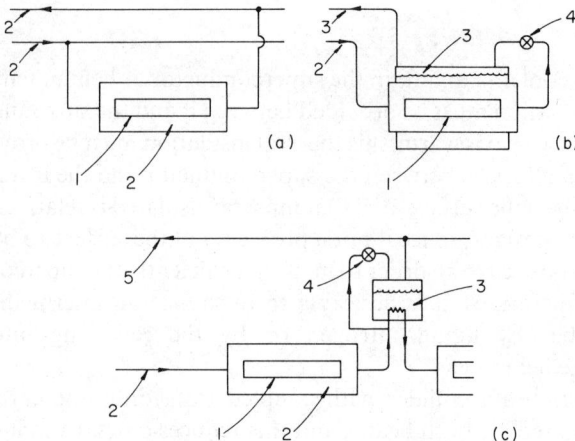

Figure 3 Cryogenic systems: (*a*) force flow, (*b*) two-phase jacket, (*c*) force flow with separate two-phase exchanger. 1. Magnet. 2. High-pressure single phase. 3. Two phase. 4. Joule-Thompson (JT) valve. 5. Low-pressure gas.

In the case of a warm iron, low-field design (Figure 1*a*), no great positional accuracy is required and little weight must be held, but little space is available to introduce long support rods. In a warm iron, high-field design the position tolerances are very tight, and again there is little space for long support rods. One solution to the tolerance problem used in the Tevatron (8) is to supply supports that can be adjusted from the outside. The required positioning of the coil package with respect to the iron is then achieved after the magnet is operating by observing the field distributions.

In the case of cold iron designs, the weight to be supported is far greater than in the warm iron designs. However, there is adequate space to use relatively long support rods and thereby keep the thermal losses low despite the larger cross sections required. Also the position accuracy required (at least in the case of the dipole bending magnets) is much lower in this cold iron case.

3. CONDUCTORS

At present the most widely used and commercially available superconductor is an alloy of niobium (50%) and titanium (50%), although Nb_3Sn was the first type II superconductor to be shown capable of carrying high current densities. Other alloys use components that are prohibitively expensive, or they have undesirable mechanical properties that make the conductor or magnet fabrication difficult or impossible. In what follows, we confine our attention to NbTi and Nb_3Sn.

3.1 The B-J-T Surfaces

A superconductor behavior is specified by the current density, J; the magnetic field, B; and the temperature, T. In particular there is a critical current density for any B and T, beyond which the superconductor becomes normal. Such a three-dimensional surface for Nb_3Sn is shown in Figure 4 and the projections are shown on the J, B plane in Figure 5a. Figure 5b compares NbTi and Nb_3Sn at 4.2 K (9).

The maximum field at which superconductivity is exhibited is called the critical field. This is about 11 T for NbTi and 20 T for Nb_3Sn at 4.2 K. The latter shows an advantage in reaching fields greater than 10 T or in operating at temperatures higher than 4.2 K. However, it is brittle, which leads to difficulties in magnet fabrication. NbTi is cheap, readily available, has good mechanical properties, and has now been extensively used for

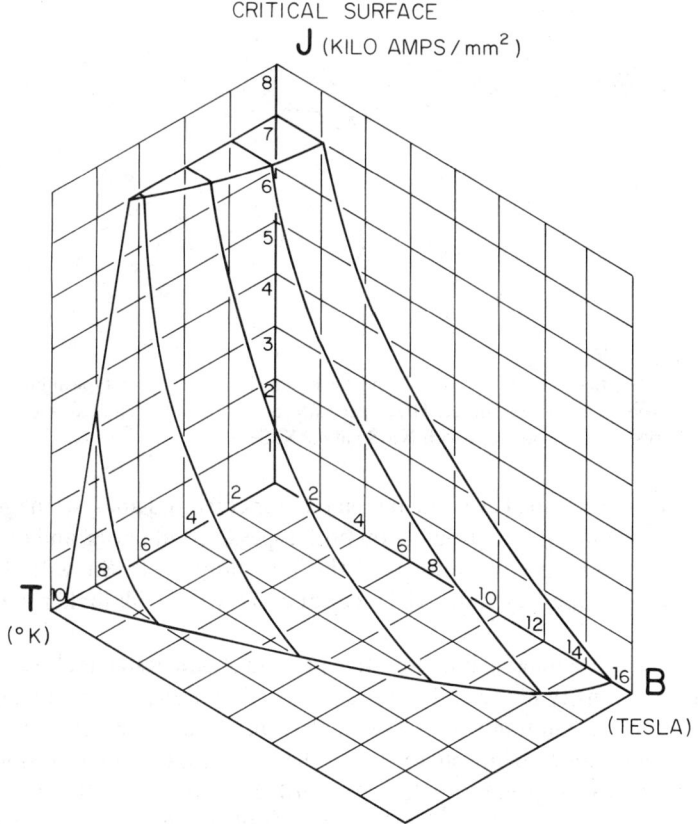

Figure 4   B-J-T surface for NbTi.

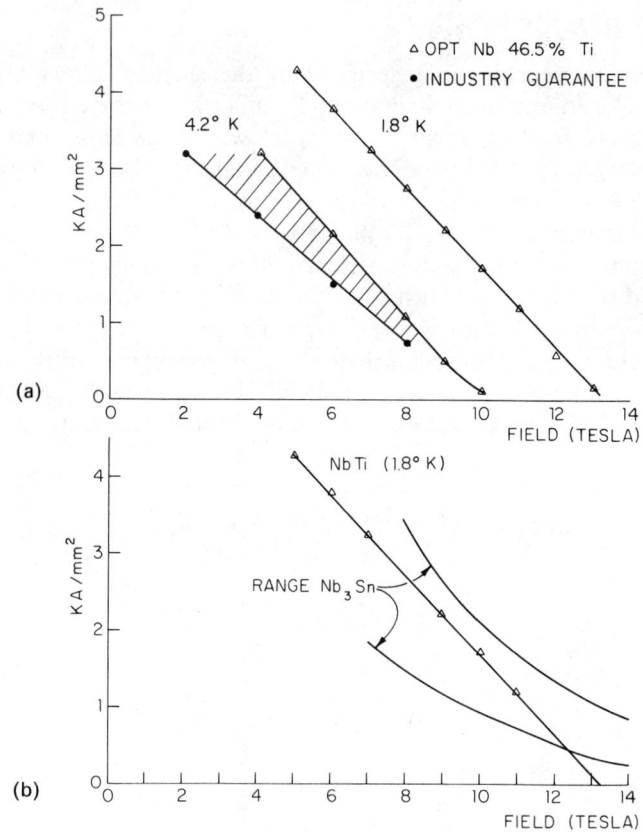

Figure 5 (*a*) B-J curves for NbTi at 4.2 and 1.8 K. The lower curve is a typical guaranteed minimum from an industrial source. The upper curve is what may be achieved optimally (9). (*b*) Comparison of NbTi at 1.8 K with Nb$_3$Sn at 4.2 K. See (9).

magnet construction (Tevatron, fusion reactor coils, and analysis magnets). Active development is in progress on both types of conductors, and one can expect, through careful control of the metallurgy, Nb$_3$Sn with better mechanical properties and NbTi conductors with still higher current density.

The design of a magnet is an iterative process. A coil design is chosen, and the current density in the coil required to reach a desired B_0 is calculated. The field in the winding will be higher than B_0 generally by 10–20%. One then checks the J, B graph to see if the operating point is inside the superconducting region by a safe margin (10–20%). If not, the coil geometry is changed, and the calculation repeated until a satisfactory solution is found.

3.2 Stability

We must now consider the details of conductor fabrication, because it is much more complicated than discussed above. There are two reasons for this: The first concerns whether or not a conductor is in a stable state at a given B and I, and the second concerns whether or not the conductor destroys itself because of I^2R heating.

A simplified picture of current flow in a conductor was proposed by Bean and is useful to help understand the effects we discuss. Figure 6a shows a very simple example of this model. A slab of type II superconductor is shown immersed in a magnet field $\mathbf{H} = \mathbf{e}_y H_0$. We assume the critical current is a constant, j_c, for a moment. As the field is increased, shielding currents start to flow in a manner that tries to keep the field in the conductor equal to zero. However, the current density cannot be greater than j_c, and so the field penetrates a depth Δ such that $H_0 = j_c \Delta$. As H_0 increases, the field penetrates further into the sheet until $\Delta = a$, the half width of the sheet. This is the maximum amount of shielding field the conductor can produce. After this point, the field increase inside follows that of the outside field. Note that

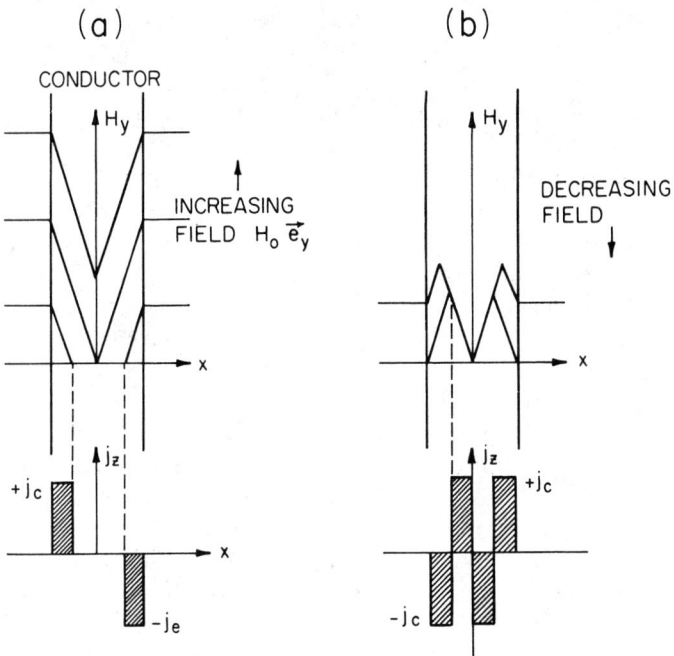

Figure 6 Shielding currents and field in a slab of superconductor exposed to (a) increasing and (b) decreasing fields according to the Bean model.

since flux is moving into a region there will be an Emf = $\partial \phi / \partial t$ that produces an E_z. This E_z combined with H_y gives $S_x = E_z H_y$, and energy flows into the conductor. A simple calculation shows that more energy flows into any region than is necessary to supply $W = \mu_0 B^2/2$, the field energy, and the difference appears as heat, i.e. the moving field boundary is doing work against the pinning forces.

When the field is decreased, the same action takes place as above, except now the currents that flow are $\mathbf{J}_z = -\mathbf{e}_z j_c$, and the resulting field and current patterns are shown in Figure 6b. Note that when the outside field reads zero, that fields and magnetizing current still exist in the slab. The conductor is behaving very much like a piece of magnetized iron.

If now we consider the slab to be an idealization of a small piece of a conductor in a magnet, we can complete the picture. The field is the coherent effect of all of the current elements in the magnet, and the resulting shielding currents must be added to the transport currents to obtain the total current density. In general the coherent field is much larger than the local self-field from current in the conductor, and for this reason we can consider j_c to be approximately constant over the conductor. We shall see that the shielding field (i.e. the ΔB across a filament) is of the order of 0.1 T, which is generally small compared to the field in the magnet.

We next describe the fabrication of commercial superconducting cable and then use the model we have just described to understand why it has been necessary to develop such a complicated technology.

NbTi rods about 3 mm in diameter and 65 cm long are first fabricated. These rods are inserted into specially purified copper tubes with a round hole in the center and a hexagonal shape outside. The copper-to-superconductor ratio is determined by the outside dimensions of the copper hexagon and may vary from 1:1 to 2:1. The composite rods are stacked inside a 10-in. diameter copper shell and in vacuum a nose cone and base plate are welded on. This structure, looking like an artillery shell, is then heated and extruded in an enormous press (10^6 psi). The extruded conductor of about 7-cm diameter is drawn through a succession of dies that reduce its diameter to final strand size, perhaps 1/2 to 1 mm. An electron micrograph of a wire used in the Tevatron is shown in Figure 7. Its end has been etched to remove the copper and reveal the filaments, which are 10 μm in diameter; the outer diameter (o.d.) is 0.6 mm. This incredible process produces over 100 km of wire with 2000 NbTi filaments encased in pure copper! Next, 20 to 30 of these strands are made into a cable, which can carry 5000 A at 5 T.

Nb_3Sn is made with the same extrusion techniques. However, the alloy itself is very brittle and cannot be extruded. Hence, Nb and Sn are coextruded in various configurations. After the cable is fabricated, it must be heated to ~700°C for a number of hours; this allows the Sn to migrate into the Nb and form the superconducting alloy. After this treatment the cable may not be bent to radii less than about 5 cm without loss of capacity. Sometimes, therefore, the reaction is performed after the coil is wound. This is a difficult fabrication technique as the dimensions of the coil change during the heat treatment. Subsequently, it must be handled very carefully during magnet assembly, and insulation has to be impregnated into the coil in its wound form.

The small filaments are necessary for several reasons. The first is called the adiabatic stability criterion and arises from the following reasoning. Flux penetrates a type II superconductor as we explained using Bean's simplified model. The current density at each point is given by the critical

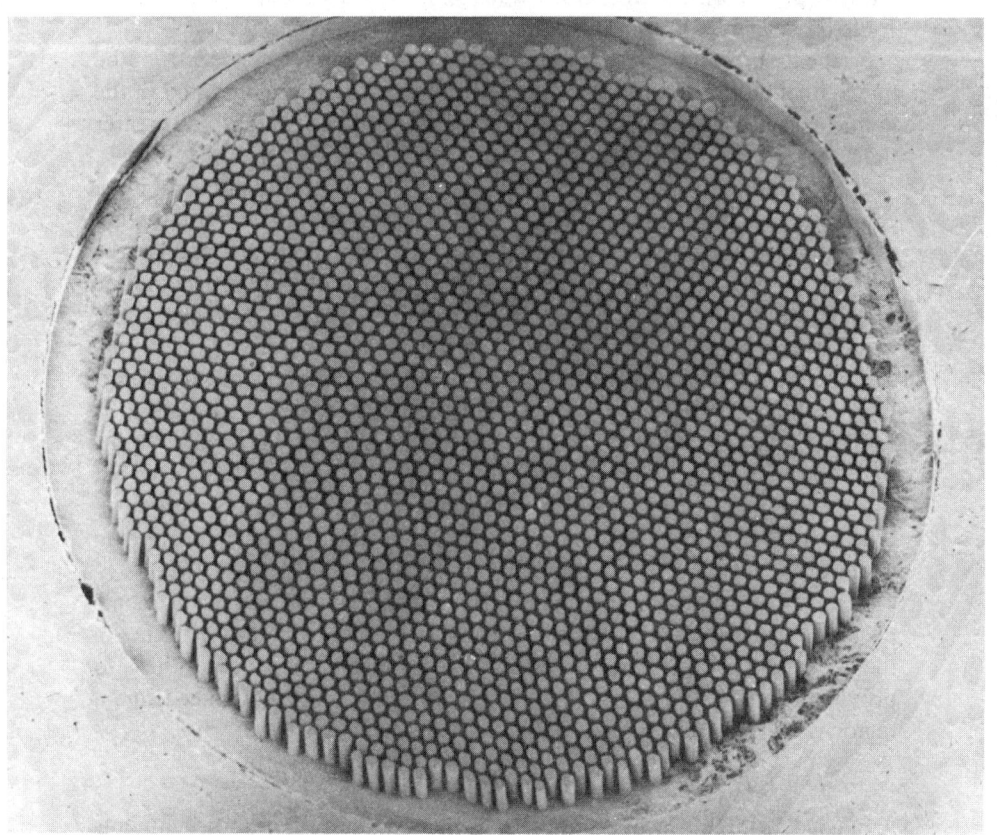

Figure 7 Photomicrograph of the end of a strand of a superconductor cable. The copper matrix has been etched back to expose the NbTi filaments, which are about 8 μm across.

current, $j_c(H)$, for the local field. If the temperature in the filament rises by δT_i for some reason, the value of the critical current decreases and the field in the filament must penetrate deeper. This change in the field causes a local deposition of heat in the filament that will remain in the NbTi because the diffusion times for heat are much longer than the characteristic times for changes of field. This heat raises the temperature of the conductor by δT_f and if $\delta T_f/\delta T_i > 1$, a runaway condition will exist. Since the magnetic energy stored in a filament is a function of its size, one can calculate (22) that for stability:

$$J_c^2 d^2 < -\frac{3}{\mu_0}(C_p \rho) J_c (\partial J_c/\partial T)^{-1},$$

where C_p is the heat capacity, ρ the density, and $\partial J_c/\partial T$ the change in critical current with temperature. $J_c(\partial J_c/\partial T)^{-1}$ is 5 K, and $C_p \rho$ is $\sim 10^{-3}$ Joules cm^{-3}. Substituting in the above equation gives $d < 30$ μm.

So the first reason for the copper matrix is for mechanical support of the fine filaments. It also aids in transporting away any heat generated in the filament, as its heat conductivity is almost three orders of magnitude greater than that of the superconductor.

The second reason the copper is necessary is to protect the magnet if the conductor should be driven normal. The resistivity of the superconducting filament once it goes normal is so high that it essentially transfers all of the current to the copper. The copper carries the current until either the He reduces the temperature to less than T_c or until the magnet is de-energized. If the cooling is greater than the $I^2 R$ heating, the conductor will recover its superconducting state, and the magnet is said to be cryogenically stable. Bubble chamber magnets are in this category because their stored energy is so large that it would be impractical to remove it in a short time. Accelerator magnets require such a high current density that space for copper and He in order to achieve cryogenic stability is not available. Hence, these magnets must have either active or passive quench protection systems.

In the Nb_3Sn conductor, copper is also introduced between the filaments to separate and support them, but in the course of the reaction with the tin the copper is converted into bronze, the resistance of which is too high to provide the needed stability. Therefore, a barrier layer is introduced, and a region of pure copper is maintained outside this barrier.

3.3 Hysteresis and Eddy Currents

A magnet with 100 turns of 23-strand cable has several million filaments carrying the magnet current. A filament in a uniform field has an induced dipole moment because of the shielding currents. These dipoles act coherently in much the same way as the atomic current in a piece of iron, i.e. there is a magnetization dipole moment per unit volume. This magnetization effect in a pulsed magnet causes the actual field to lag the field produced by the transport currents and in accelerator magnets causes dipole and sextupole terms to appear in the field expansion. The magnitude of these magnetized fields can be as large as 0.001 T, and they are important in determining the field shape. The effect is proportional to filament diameter.

These same currents as indicated earlier also result in an energy loss when a magnet is pulsed in a cyclical manner. Since the B-H curve is open, energy is deposited in the magnet each cycle and is proportional to the filament diameter for a fixed coil geometry and is also roughly proportional to B_{max}.

A second source of perturbing fields and ac losses is from the eddy currents in the cable structure. Consider the following simple picture. Flux threads a loop of resistance R and area A. If B rises linearly from zero to B_0 in time T_0 and back to zero in the same time, we can calculate the energy dissipated in heat:

$$\frac{W}{\text{cycle}} = \left(\frac{B_0}{T_0} A\right)^2 \frac{1}{R} 2T_0 = \frac{2B_0 \dot{B}_0 A^2}{R}.$$

In this equation, the term in parenthesis is the induced EMF, which generates a constant power for a time $2T_0$. The equation shows that the losses are proportional to B_0 and $\dot{B}_0$. We saw that the hysteresis losses were proportional to B_0 so we can write an expression the total loss per cycle of a linearly ramped magnet

$$\frac{W}{\text{cycle}} \approx \underbrace{K_1 B_0}_{\text{hysteresis}} + \underbrace{K_2 B_0 \dot{B}_0}_{\text{eddy current}}.$$

These losses may be separated by ramping the magnet to different B_0 with different $\dot{B}_0$. For the Tevatron (10) the hysteresis loss is about 200 J per cycle when the magnets are ramped to 4 T. (These magnets are 6 m long with a bore of 7.5 cm). The eddy current loss is negligible. However, some of the very early models that were constructed with a cable filled with solder had eddy current losses of >1 kJ per cycle.

There is a phenomenon closely related to eddy currents that can affect the performance of a magnet. We pose the question: How does one know that all of the strands in a cable are carrying the same current and, even more specifically, how does one know the filaments in a strand are equal participants? Current sharing is governed as follows. Consider two strands through a magnet. They are connected at each end and along their length by resistive contact between strands in the cable. During ramping the strands can be represented as two inductively coupled loops. To equalize the currents, it is important that there be no net flux from the main field that links the area between the two strands. If this is the case, the net inductance of the loop comprising the two strands will be very low, and the contact resistance, or resistance of the matrix, will allow equalization of any small current differences that may exist.

The strands are decoupled by either braiding them into a cable as was done for ISABELLE or spiraling them into a cable as in the Fermilab-Rutherford cables. In addition, when the cable is fabricated, each strand is given a twist about its own axis every inch or so to decouple the individual filaments. The cable is a much more complicated structure than casual examination discloses!

One may ask why not insulate the individual strands in the cable? This leads to very erratic magnet performance, which has been traced to unequal strand currents since it is difficult to decouple completely the interstrand circuits from the main field. On the other hand, good contact between strands leads to high eddy current losses and a magnet whose field shape is dependent on $\dot{B}$. The Fermilab-Rutherford cable has solved this problem by coating every other strand with CuO, which has enough resistance to control the losses without being so high that current sharing is upset.

3.4 Temperature Effects

It has been mentioned previously that Nb_3Sn has a higher critical field than NbTi. However, the critical field for NbTi can be increased by decreasing the temperature. Figure 5b compares the two conductors at different temperatures. The interesting prospect here is to investigate the behavior of a conductor below 2.17 K, the λ-point for HeII. Large accelerator magnets have been successfully operated at 1.8 K, with the expected increase in field due to the higher J_c. The superfluid properties of HeII show remarkable effects on the magnet stability, presumably because of the improved cooling from the superfluid. This approach is being studied extensively at LBL (11).

A question that remains to be answered is whether or not the decreased Carnot efficiency at the lower temperature will be a practical limitation. At present the overall efficiency is about 400 W/W at 4.2 K, i.e. large refrigeration plants have an overall efficiency of only about 15% of Carnot. A point that may favor Nb_3Sn is that even at present fields it can operate at higher temperature with resulting higher efficiency and lower operating costs for the refrigerator.

4. FIELD QUALITY CONSIDERATIONS

4.1 Field Analysis

In the central part of a long magnet the fields have no axial components. It is conventional to express the distribution of transverse magnetic fields on the horizontal midplane as a quadratic expansion of the distance from the magnet center. The fields at any point (r, θ) in the vertical direction (B_y) and horizontal direction (B_x) are then given by:

$$B_y = B_0 \left[1 + \sum_{n=1}^{\infty} \left(\frac{r}{r_0}\right)^n (b_n \cos n\theta - a_n \sin n\theta) \right]$$

$$B_x = B_0 a_0 + \sum_{n=1}^{\infty} \left(\frac{r}{r_0}\right)^n (a_n \cos n\theta + b_n \sin n\theta),$$

where r_0 is a reference radius usually taken at a given fraction (2/3) of the coil inner radius.

The radial and tangential components of these fields have a periodic variation in the azimuthal angle θ, where the number of poles $m = 2n+2$. They thus represent various multiple components of the field. The coefficient of these multipoles a_n and b_n are determined by the current density distribution. If the current density in a shell can be written as being proportional to $K_n^{(e)} \cos n\phi + K_n^{(o)} \sin n\phi$, then $a_n \approx K_{n+1}^{(o)}$ and $b_n \approx K_{n+1}^{(e)}$. Thus, the symmetry of the current distribution is reflected in the values of a_n, b_n. A coil that is symmetrical both up and down and right and left has only even b_n. It helps to visualize these fields if one notes the absolute value of field for a given multipole, $|B_n|$, is constant on any circle about the origin, and that this vector rotates $n+1$ times during a traversal of such a circle. Its magnitude is $\sim r^n$.

The use of such a multipole formulation has various advantages. Many specific behaviors of the accelerator are related to specific multipole components. The dipole (B_0), for instance, determines the equilibrium orbit of particles, and the quadrupole moment (b_1) affects the focusing of the beam. The sextupole provides momentum-dependent focusing and so on. A second advantage of this formulation is that it allows an easy estimation of the effect of different field errors as a function of the radial size of the beam. It is also convenient to design correction coils with explicit multipole contributions that can be used to correct unwanted components present in the main magnets.

The situation at the end of a magnet is more complicated; the conductors make the turn from passing up one side of the magnet to pass down the other side. The multipoles tend to vary rapidly in this region, and the magnetic fields contain axial field components. However, the effect on the beam is dependent primarily on the integral of the transverse fields and is very little affected by these axial fields or the details of variations of the transverse fields. The end contributions to these integrals can be obtained by calculating or measuring the multipole integrals for the entire magnet and subtracting the multipole contributions from the central part of the magnet multiplied by some suitably chosen effective magnet length.

4.2 Field Precision Requirements

The field requirements depend, of course, on the details of the machine in which they are to be used and on the beam size stored in that lattice. Field requirements thus cannot be specified without knowing these details. Nevertheless, it is convenient to give at least order-of-magnitude values that have been found to be required. If sufficient correction elements are included in the design, variations in the dipole field components may be allowed to vary about one part in 10^3. Quadrupole and sextupole components over a defined good field region should be within 2×10^{-4}. For higher multipoles, correction is clearly not possible and allowed errors should be less than 10^{-4}. In some cases, e.g. a stacking ring, the "good field region" may extend over as much as two thirds of the coil inner diameter. In other cases if injection, for instance, is at a high energy and the beam is already small, the required good field region may be one third, or even less, of the coil inner diameter.

When a magnet is energized some changes are observed in measured multipoles. In practice these changes are very small in the so-called unallowed multipoles, namely, all the multipoles except in the even b_n terms. Several effects can, however, generate changes in the even b_n terms. Magnetic forces cause small motion of the conductors and create changes in these multipoles. Changes in multipoles can be due to persistent currents and saturation. These effects are discussed below, but they are not a problem as long as they are the same for all magnets.

4.3 Coil and Iron Cross-Section Design

In the simple picture-frame magnet (Figure 1b) a uniform field is achieved from the intrinsic design. In all other cases the iron or coil shapes must be designed specifically to achieve the uniform field. If good field is only required over, say, one third of the aperture, then very simple approximations to the correct iron or coil shapes may be used (Figure 2c, for example). If good field is required over as much as two thirds of the coil inner diameter, then more complicated approximations (such as Figure 2d) must be used.

In order to understand the effects of errors of coil placement on the magnetic fields, we examine the multipoles that would be generated by random displacements of coil blocks in a particular case. The calculations were performed for a design of the type represented by Figure 2a but are applicable for any circular magnet. In each case the coil is divided into 24 (6 per quadrant) equal parts, and each part is allowed to have a random displacement in two directions of 10^{-3} times the coil inner radius.

Table 1 shows the resulting calculated variations of the various multi-

Table 1 Calculated and observed errors in CBA dipoles

	Variation ($\times 10^4$)			Corresponding displacement error (μm)
	Calculated	Observed	Ratio	
b_0		6.10		
b_2	2.1	1.76	0.84	55
b_4	1.0	0.73	0.73	48
b_6	0.5	0.24	0.48	32
b_1	3.8	0.84	0.22	15
b_3	1.3	0.24	0.18	12
b_5	0.7	0.35	0.5	33
a_2	2.1	0.45	0.21	14
a_4	1.0	0.19	0.19	13
a_6	0.5	0.10	0.20	13
a_1	3.8	2.44	0.64	42
a_3	1.3	0.73	0.56	37
a_5	0.7	0.37	0.53	35

poles together with the observed (12) variations in a set of 10 CBA magnets whose coil inner diameter was 6.6 cm (i.e. the random motions in this case were 66 μm). The table also shows the ratio of observed to calculated fluctuations of the multipole coefficients.

One can see that the ratios of observed to calculated errors are systematically higher for even b_n and odd a_n terms than they are for odd b_n and even a_n terms. These systematic effects indicate that block placement errors are correlated. It is instructive to study cases of particular coil errors that will generate particular lower multipoles. In Table 2 we give the

Table 2 Calculated effect of displacement errors

Type of error	Term	Effect from 1 mrad (66 μm) ($\times 10^{-4}$)			Error from Table 1	
		Inside layer	Outside layer	Total effect	Observed	rms μm
(a) Both poles to right	Δb_1	2.6	1.5	4.1	0.84	14
(b) Both midplanes up	Δa_1	2.2	1.7	4.0	2.4	40
(c) One midplane up, other down	Δa_2	0.8	1.1	1.8	0.45	16
(d) Coil nonuniformity	Δb_2	2.6	0.9	3.5	1.8	34
(e) Radius of all coils	Δb_0	6.8	2.1	8.9	6.1	45

magnitude of the induced lowest multipole corresponding to a particular set of 10^{-3} errors. We also give the magnitude of such errors that would produce the observed variations given in Table 1.

The particular cases given are:

(a) A 1-milliradian horizontal shift of both pole spacers. This compresses all coils on one side and expands those on the other. Such an error introduces a left-right asymmetry and generates only odd b_n terms. In the example such errors were not greater than 14 μm rms.

(b) A vertical displacement of the horizontal midplane. The coils in the upper half are uniformly smaller and those in the lower half uniformly larger. This gives a pure up-down asymmetry and induces only odd a_n terms. This is seen to be a more serious type of error; in the example it is of the order of 40 μm rms.

(c) An asymmetric vertical displacement of the horizontal midplane: higher on one side, lower on the other. Such an error would arise if there were left-right asymmetries in the coil curing fixtures. In the example this error is not greater than 16 μm rms.

(d) The conductors are assumed to have current density varying uniformly from the midplane to the pole in such a way as to produce a maximum azimuthal displacement of 10^{-3} radians half way up the coil (the displacement is always toward the midplane). Such an effect is produced by variations in the friction either during cure or assembly of a coil. It is a four-fold asymmetric effect and can only induce even b_n terms. In the example this is a relatively large effect: 34 μm rms. We must note, however, that variation in persistent currents or iron saturation could also cause fluctuation in this term.

(e) The coils are assumed to be located on a radius larger or smaller than designed, the error being one part in 10^3. The primary effect is to vary the central field (B_0). The observed variation in the example indicates radial errors somewhat less than 10^{-3}, corresponding to 45 μm, but it must be remembered that B_0 variation can also come from magnet length and other errors.

These observations suggest that the required field accuracy is not severely limited by the position of the poles or the radial dimensions, but is dominated by errors in coil size and conductor uniformity. Since such errors will scale down in a smaller magnet, we may hope that the multipole fluctuations arising from them will also scale, i.e. the even b_n's and odd a_n's will not greatly increase in a smaller magnet. The odd b_n's and even a_n's, which may already be dominated by pole position errors, would be expected to rise in inverse proportion to the coil size.

We now discuss the precautions taken in the construction of these magnets in order to achieve these high accuracies.

First, it is necessary to control the dimensions of all the components supporting the external dimensions of the coils. The iron laminations must be accurately punched, and the pole spacers must be accurate as must all the insulating materials on the outside and at the poles of the coils. All these dimensions can be held to approximately 25 μm. The conductor wire insulation and epoxy glass covering must also be controlled as well as possible. However, even when these dimensions are held to 5 μm per turn, the accumulated error in a coil stack of 60 turns is as large as 300 μm and clearly unacceptable. In order, therefore, to control the dimension of a final stack, the following procedure can be followed. The coils are first cured at a pressure such as to give dimensions slightly larger than those finally required. The coils are then measured at their final precompression and using this data the coils are recured to a final dimension chosen to produce coils of the correct size. A second measurement at final precompression is used to check that the size within 25 μm; a second recuring can be undertaken to correct any deviation. The coils are finally assembled between the yoke supports with teflon slip planes placed between the coils and their support and between the two coils to keep friction low.

A somewhat different philosophy was adopted for the Tevatron magnets. Great emphasis was placed upon constructing a factory that would produce a very uniform product, but absolute coil shape was given secondary importance. If one assumes for the moment that there are no random errors, then systematic errors can be corrected by means of small shims placed in the coil in such a way that they cancel the undesired moments. These shims are calculated on the basis of magnetic measurements made on the production stream flowing from the factory. This closes a feedback loop around the factory that can correct for any slowly changing systematic errors in the production stream. For this method to work, random errors must be constrained by good quality control, and the delay between production and measurement must be kept small. Nearly 800 dipole magnets and over 200 quadrupole magnets have been measured for the Tevatron. The characteristics of these magnets are in the literature (13a,b).

4.4 *The Magnet Ends*

If the magnet ends are designed with engineering considerations in mind and no special care is taken, then it is likely that significant multipoles will be contributed by these end regions. Such multipoles can be corrected in an integral sense by slight modification in the design of the central part of the magnet. Since the magnet is straight and since the beam bends, this correction is not simple. It may be desirable to design ends that contribute no unwanted multipoles. As an example, we consider a very simple cos θ

design, as illustrated by Figure 8a, which shows a z plot of the end of the conductors, i.e. what the conductors would look like if they were unwrapped from the beam pipe and laid out onto a flat surface. A section (yy) through the end shows that the conductors are now concentrated nearer to the poles, yielding relatively higher fields in those regions and relatively lower fields at the midplane. Such a field has a strong negative sextupole component. The addition of spacers within the end windings (Figure 8b) can correct these multipole components.

The accuracy required for locating the conductors in a magnet end are less stringent than those for placing them in the center of the magnet because the ends contribute less to the total magnetic field. There are usually no problems in such accuracy.

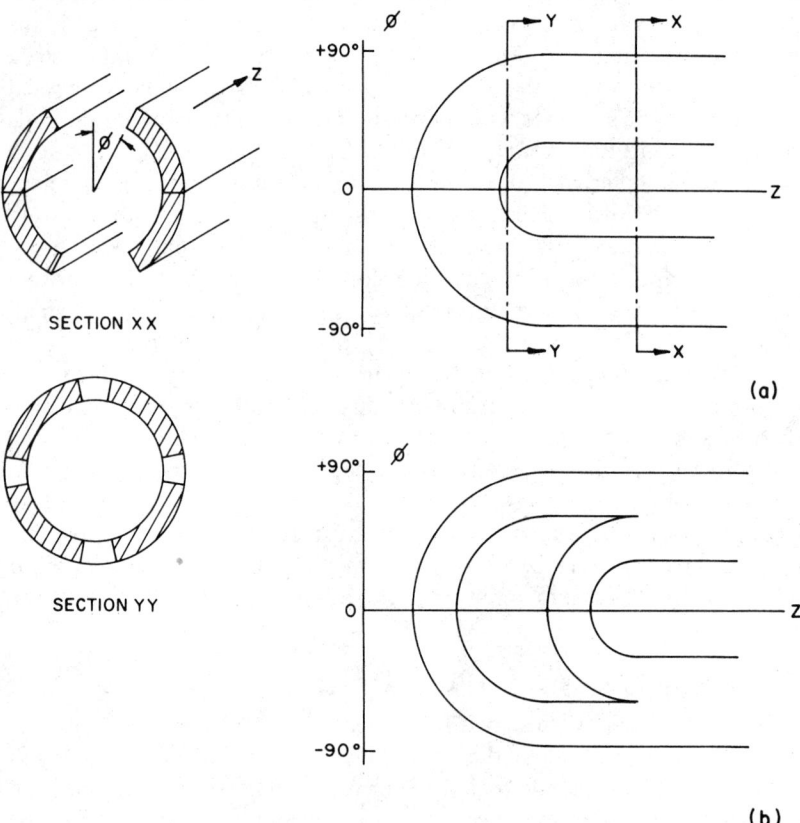

Figure 8 End sextupole effect. (a) $\theta - z$ plot of normal coil end; section xx shows central cross section, section yy shows end cross section with blocks moved toward poles creating sextupoles. (b) End with wedge to remove effect.

4.5 Superconducting Persistent Currents

Eddy currents in a superconductor have, of course, an infinite time constant and produce an antiferrimagnetic magnetization behavior. When the magnet is first excited these currents develop and rapidly increase until they reach the critical current within the superconductor. They are then damped by the onset of resistivity in the superconductor and become further damped as the field rises and the superconducting critical current falls. When the field in the magnet is reduced, eddy currents are generated in the opposite direction and again follow a bound set by the critical current behavior of the conductor.

As stated above, in a simple picture-frame magnet these magnetization effects do not produce changes in the fields in the good field region. However, in other geometry designs they do perturb the good field. Figure 9a shows the effects on the sextupole component (b_2) in a magnet in which the iron is relatively far from the coil so that there are negligible saturation effects. The lower line indicates the sextupole component observed while the current was rising and the upper line that observed while the current

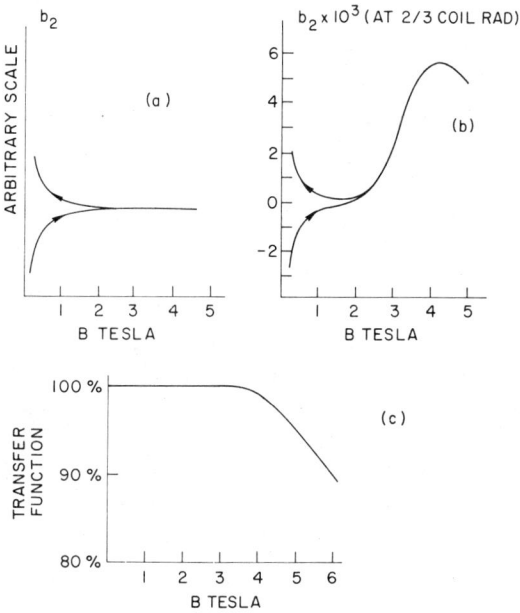

Figure 9 Field-dependent effects. (a) Sextupole in nonsaturating iron magnet showing persistent current effect, (b) as above with saturation effects also, (c) saturation effect on transfer function.

was falling. The difference between the two lines indicates a contribution from the magnetization. Such effects are corrected by trim coils and present no problems in principle. However, variations in the superconducting material in the cables can produce variations in these magnetization effects magnet-to-magnet, and correction of these effects is more difficult. Variations in niobium-titanium cable are probably of the order of 5% (although there is some evidence (14) that the true fluctuations are significantly less). Thus it is necessary to keep the magnetization contribution to less than 20 times the precision requirements specified above. The effect when expressed as a ratio to the central field is, of course, far worse as the central field reduces, since not only the denominator decreases but magnetization effects in the conductor increase roughly in inverse proportion to the field due to the increase of j_c. The effect is worse for a higher field magnet since more superconductor or better superconductor is employed. In practice these two effects result in a dependency on the ratio of injection to final field, and little dependence on the absolute fields. It is worse for smaller aperture magnets where the good field region is closer to the conductor. It can be reduced by reducing the sizes of individual filaments within the superconducting cable (see Section 3). The effect can be expressed:

$$\frac{\Delta B_1}{B_1} = \text{const} \, \frac{F_1}{F_2}\left(\frac{B_2}{B_1}\right)^2 \frac{\delta}{r} \approx \text{const} \left(\frac{B_2}{B_1}\right)^{1.5},$$

where F is the pinning force of the conductor, i.e. the product of the critical current times the local magnetic field; B_2 is the field of the magnet, δ the filament diameter, and r the coil inner radius. The subscripts refer to the injection (1) and final (2) fields. The constants depend on the geometry of the coil used.

4.6 Iron Effects

Below 2 Tesla variations in iron permeability (μ) and coercivity will dominate the field of a magnet. Iron can be obtained (9) with specified variations of $1/(\mu-1)$ of 0.2×10^{-4} rms. The effect of this on the central field is typically multiplied by a factor of five, which yields central field fluctuations of the order of 1×10^{-4}. The coercive force can be specified to within ± 0.2 Oersted rms and generates typically central field changes of ± 0.6 Gauss, thus producing field errors of only 1.5×10^{-4} at a typical injection field of 4000 Gauss.

At fields above 2 Tesla and if the iron yoke is relatively close to the coils such that the fields in the iron significantly exceed 2 Tesla, then saturation effects in the iron produce a reduction in the transfer function (the ratio of central field to coil current, see Figure 9c). The effect is due to the sharp reduction of μ at high fields, given approximately by

$$\frac{1}{\mu-1} = a + \frac{H}{M_s},$$

where a is a constant, H is the field in the iron, and M_s is the saturation induction. In a CBA magnet at 6 Tesla, the dependence of the central field (B_0) on the iron saturation induction (M_s) was

$$\frac{\Delta B_0}{B_0} \approx \frac{1}{4} \frac{\Delta M_s}{M_s}.$$

The observed rms variation in M_s was 0.13%, which produced central field variations of 3.3×10^{-4}.

Saturation also generally produces significant changes in the multipole content within a magnet. If, for instance, we look at the sextupole component b_2 for a magnet with the iron close to the coil (Figure 9b), and compare this with Figure 9a, then we observe changes in the sextupole moment above 2 Tesla fields that arise from these saturation effects. This high-field magnetization can be qualitatively understood as an initial rise in the sextupole moment due to saturation of the iron in the immediate vicinity of the pole, followed by an eventual fall in the sextupole component, which is caused by effects from the saturation of the iron in the magnetic return on either side of the coil package. A balancing of these two effects at the maximum field can be used to reduce the strength of correction coils required. Like the magnetization effects, these saturation effects are not intrinsically a problem provided they are the same for all magnets. The observed magnitude of fluctuations in these effects was 1.2×10^{-4} at 2/3 aperture.

Saturation effects can be reduced by shaping of the magnetic pole, by crenellation, or by the introduction of holes within the body of the yoke. An interesting example of the use of holes is provided in the design of 2-in-1 magnets. In these magnets saturation effects occur not only in the even harmonics of the field, but also in the odd harmonics, since left-right symmetry is not maintained. In particular, a quadrupole moment is induced that, while it can be corrected, requires additional coils. With the introduction of a hole placed between the two coils, as indicated in Figure 1f, all significant quadrupole saturation effects can be eliminated.

4.7 Correction Coils

In almost any accelerator, correction coils will be required, correction dipoles will be used to control the central orbit of the machine, correction quadrupoles will be employed to adjust the tune, and sextupoles will be used to control the chromaticity, that is, the variation of tune with momentum. If all magnets were identical, these correction coils could be the same at all points around the ring. However, in order to correct for errors both in surveying and in the magnet it is usually necessary to provide at least the dipole correctors with separate power supplies for each correction coil. A more limited number of power supplies are usually used for the quadrupole and sextupole correctors. The current in these correction coils must be adjusted as the main field in the machine is increased so as to correct the persistent current saturation, coil motion effects, and saturation effects. Since the persistent current and some magnetization effects at low field are dependent on the previous history of the magnetic fields, it is necessary to cycle the machine in a prescribed manner and to determine the correction currents required for this particular cycle. It should be also noted that the effects of persistent currents are strongly dependent on the temperature of the superconductor and thus good control of this temperature is required if more complicated corrections are to be avoided.

Ideally a correction coil to correct a defect in a particular magnet such as saturation or magnetization should be located at its source, i.e. it should be located within the magnet itself. In large-bore magnets, where there is no shortage of space between the coils and the beam tube, this can be done. Coils with different multipole moments can be wound in layers on the bore tube and such coils can be designed to show little training and to achieve nearly full, short-sample, current-carrying capacity.

An alternative to placing the trim coils within the magnets themselves is to wind special short trim magnets and place them relatively more sparsely around the ring. This while not theoretically so ideal is almost certainly a more economical course even though a slightly larger proportion of the machine azimuth must be devoted to this purpose. Such a solution is used in the Tevatron (16a,b).

5. FORCES AND COIL PACKAGING

A conductor in a field of 5 T and carrying 5000 A experiences a magnet force of 25 kg cm^{-1}. Such currents and fields are typical in the magnets we are discussing. Since a magnet may have 100 such conductors in its cross section, it is immediately evident that the total forces that must be contained by the support structures are enormous. Containment of these forces is complicated by the fact that the structure is at 4.2 K and must have a limited heat leak to the outside environment. In addition, the quality of the field is determined by the accuracy of the conductor placement and, consequently, the support must be very rigid so that the conductors do not move as the magnetic field changes. Finally, the large thermal contractions of the different components must track each other in order for the package to be mechanically stable when cold. To appreciate the magnitude of this latter problem, we remark that the coil package for the Tevatron magnets shrinks 0.75 in. axially when cold, and the diameter changes by 0.020 in.

5.1 Magnet Forces

Let us begin by considering the magnetic forces. The simplest picture has a field with a magnetic pressure of $B^2/2\mu_0$, which is equal to 100 kg cm^{-2} for 5 T. This force tries to blow the magnet apart both longitudinally and transversely. In the axial direction, the magnet is held together by the conductor itself. Transversely the forces must be absorbed in a cold rigid structure called a collar or yoke. If one examines $F = j \times B$ in cylindrical coordinates, one finds both an azimuthal and radial component for the force. The azimuthal force compresses the coil package, and Figure 10 shows these two components for the conductors in the two shells of the winding for the Tevatron magnets.

The collars must be strong enough to resist the sum of the radial forces without deflecting more than a few tenths of a millimeter. Small elastic deflections lead in first order to a sextupole moment that is a function of I^2 and may be compensated by dynamic correction coils (which are required for many reasons). For a pulsed machine, a more important consideration than deflection of the collars is their fracture by fatigue (17). This limits the maximum stress to less than about 30% of the yield point. Magnets with cold iron (CBA), of course, don't face this trouble as the support is massive and in storage rings the number of cycles during the life of the machine is not large.

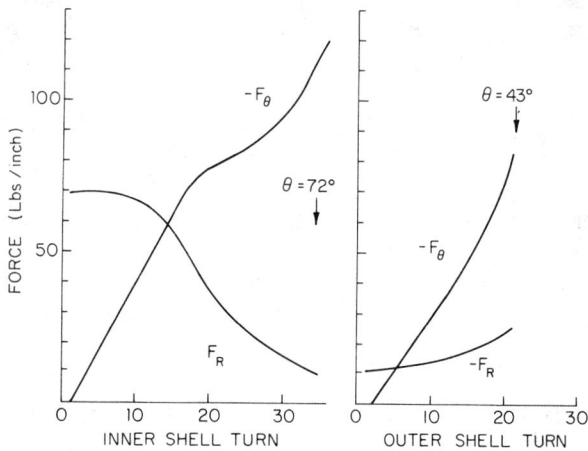

Figure 10 Force per linear inch of conductor in the Tevatron two-shell magnet at 4T.

We now consider azimuthal forces. Note that the largest forces occur in the turns nearest the poles decreasing to zero at the midplane. In a two-shell magnet, these forces are largest for the inner shell. Since there is no constraint on the winding in the azimuthal direction (i.e. winding can slip relative to the collar surface), the forces of the wires sum as one approaches the equator. The top half of the coil balances the bottom half. In the Tevatron, the azimuthal forces sum at the midplane to 250 kg cm^{-1} in a direction to compress the coil. If the elastic constants of the top and bottom half of the coil are the same, there will be no motion of the midpoint, but there is an elastic compression at $\sim \pm 30°$ toward the midplane and this induces a sextupole moment. If the elastic constants are not equal, the midplane can move, and quadrupole moments result.

5.2 Preload

We are now in a position to discuss a difficult problem in magnet construction. If the azimuthal forces of the last section were allowed to act on a coil supported only in the radial direction, the coil would compress itself as it was excited and the angles of the shell would change. If these change symmetrically, a sextupole moment is induced and, if asymmetrically, quadrupole terms appear as well. The field is enormously sensitive to these angles—they must be maintained to an accuracy of ~ 25 μrad for adequate field quality. The forces are so large that, with the elastic modulus available in the insulated coil packages ($E = 10^6$ psi), the compression of the coil would far exceed this limit. As a result, when the coil is constructed, it is preloaded in the azimuthal direction to the extent that the elastic forces are greater than the magnetic forces. This ensures that the boundaries of the coil package will stay in contact with the collars during excitation. The Tevatron coils (18) were assembled in a large press, and the CBA magnets were bolted together with a similar pressure. Elastic motion of the coil relative to its support can still take place but at a much reduced level. (It is similar to fixing two ends of a loaded beam, compared to fixing only one end.) Elastic motion can also be reduced by making the elastic modulus high. The group at LBL has had successes in achieving $E \approx 4 \times 10^6$ psi, which is perhaps four times larger than achieved in the Tevatron magnet; the CBA coils had 2×10^6 psi.

The problem of adequate preload is greatly complicated by the fact that when the temperature is reduced all the components become smaller. Stainless steel has a fractional change of 3×10^{-3}. The coil package decrease is much larger and also less reproducible or predictable because it is a composite of superconductor, copper, and insulation in a configuration that must be porous to allow He penetration. As a consequence, the preload established at room temperature is reduced as the temperature is lowered.

This effect must be carefully measured and compensated for at assembly time. If the contraction is α and the modulus is E, then the total pressure at room temperature is $p_0 = \alpha E + p_M$, where p_M is the magnetic pressure. Short circuits may develop if p_0 is so high that the insulation is crushed. Successful packaging must involve an elaborate development and test program where the designer is empirically feeling his way through a maze of unfriendly materials problems. Our discussion above has assumed the elastic modulus is linear, which it is not! As the operating fields increase, the preload must increase, and new higher strength coil packages must be developed.

6. QUENCH: THEORY AND MAGNET PROTECTION

A magnet is said to quench when some portion of the conductor goes normal for whatever reason, and the ensuing I^2R heat is greater than the cooling so that still more of the conductor is driven normal. Accelerator magnets require very high current densities to achieve fields of 4–5 T in the bore, and there is never enough cooling to return the conductor to its normal state. For instance, in the Tevatron magnets, which carry 4000 amps, if the cable goes normal, the I^2R heating is 16 W cm^{-1} of cable, whereas the cooling by the He is at most a few tenths of a watt. Thus, the area spreads. The velocity along the cable is 1–10 m s^{-1} but because of the insulation and He between turns the time to jump to adjacent turns is milliseconds. As discussed below, the current must be rapidly reduced to zero (~ 1 second) or the conductor will destroy itself at the point the quench started since that is the hottest spot in the magnet. We begin by describing various quench mechanisms.

6.1 Quench Theory

Figure 11 is an attempt to elucidate the energy diagram of an excited magnet. The curve is the enthalpy of 1 cm of 23-strand Tevatron cable as a function of temperature. This approximates a T^3 dependence. The normal operating point is at 4.8 K. If the field were zero, the temperature would have to be driven up to $T_c \approx 10.7$ K before the normal transition would occur. However, an accelerator magnet operates at nearly 90% of j_c at peak excitation. Since $j_c = j_0 (1 - T/T_c)$, it follows that $\delta T = +0.6$ K will cause a quench to start. This corresponds to a change of enthalpy per cm of cable of ~ 60 μJ!

In a cable, there is about 5–10% of open space which will be filled with liquid He. The helium is under some additional pressure to keep it from forming bubbles that would inhibit heat transfer from the conductor. The specific heat of liquid helium is ~ 5 J g^{-1} K^{-1}, and there is about 10^{-3} g of liquid per centimeter of cable. The effect of this added heat capacity is shown in the figure. For very short times (<100 s), the heat conduction at liquid He to metal interfaces has been observed to be very high, ~ 10 W cm^{-2} °C^{-1}. This decreases to ~ 0.2 W cm^{-2} K in the steady state. Since there is a strand surface area of about 5 cm^2 cm^{-1} in a 23-conductor cable, it is apparent that the helium is well coupled to the conductor. If the helium vaporizes, the latent heat is available, and this appears as a vertical jump in heat capacity. It is difficult to transfer this quantity of heat except over long periods of time, and it probably plays no role in stabilizing the conductor.

We now examine the heat sources. They are (a) eddy current heating, (b) magnetization losses, (c) heating from beam losses, (d) friction from wire

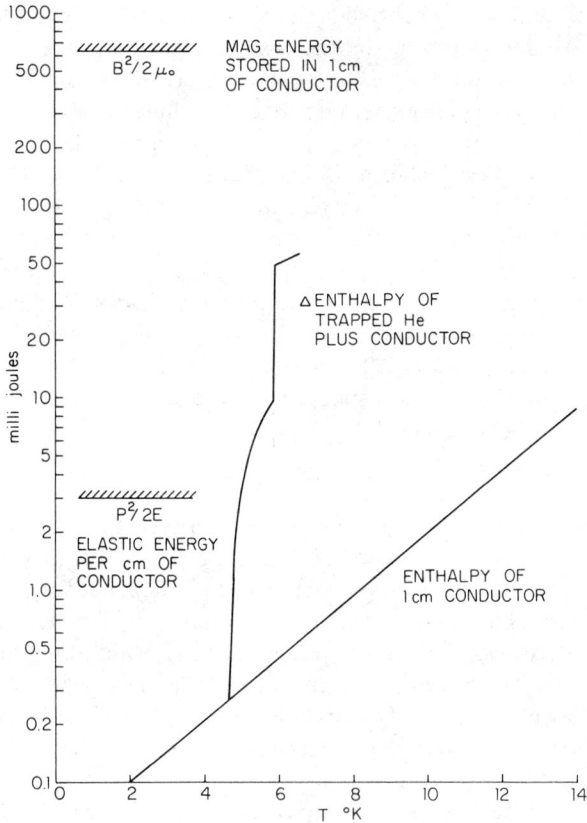

Figure 11 This diagram shows the various energies contained in a volume equal to that of 1 cm of cable in a Tevatron dipole at 4T. The operating point is at 4.6 K, and the short-sample limit would be reached at 5.6 K.

motion under $j \times B$ forces, and (e) microscopic fracture of the support matrix.

The first two correspond to part of the load on the central refrigerator and hence determine how rapidly the magnet can be cycled. Beam losses present a special problem. As we discuss in Section 6.3, there is always enough energy in the beam of the accelerator so that (if it is deposited in critical spots) it will raise the temperature of the conductor sufficiently to subsequently destroy itself by means of Joule heating before the current can be reduced to zero. As a consequence, protecting the magnet from this scenario involves early detection of beam instability and aborting the beam before it gets out of control and hits the magnets. This can be done in a single turn, which is a short time compared to how long it takes for any magnetic field to change enough to cause the beam to leave the aperture, even under fault conditions.

The frictional heating of the wire is controlled by having a very solidly constrained coil package. If the wire must move, then low friction "slip planes" can be intentionally inserted in the structure.

Microscopic cracking of the support matrix is controlled by having sufficient preload to keep the winding always under compression. A very important discovery has been that epoxy must never be allowed in direct contact with the superconductor. Apparently thermal stress and mechanical stresses always cause some fracture at the interface of these materials with a resultant quench of the conductor. The elastic energy density in the coil matrix, $P^2/2E$, is also shown in Figure 11. The strain energy per unit volume is many times the energy necessary to quench the conductor. Of course, only a very small fraction of this energy is available from microfailures in the matrix, and even that is isolated from the conductor by the Kapton wrapping, as indicated in Section 3.

6.2 Training

We come now to the subject of "training" (19). When a magnet is constructed and first cooled down, it has large elastic and thermal stress within the winding. As the magnet is excited, the $j \times B$ forces come into play. If motion occurs within the winding, heat is generated and the magnet may quench before the critical current density is reached in the conductor. The second time the excitation is applied, a somewhat higher current may be reached before quenching until finally the full design current is achieved. This process is referred to as "training." A well-constructed magnet may show very little of this phenomena, i.e. one or two quenches to full field, whereas a poorly constructed magnet may take hundreds of quenches. An extreme example of a magnet that had epoxy in direct contact with the superconductor is shown in Figure 12.

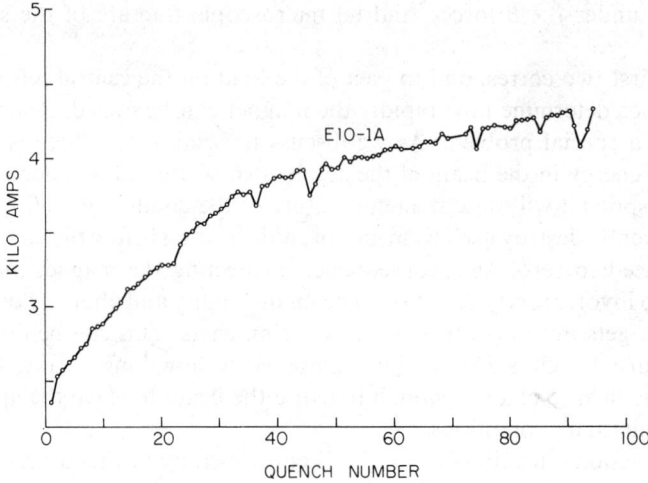

Figure 12 Example of a magnet with extremely bad training. A well-built magnet will show no changes after the first or second quench.

6.3 Quench Heating

When a quench starts for whatever reasons at a particular point in the coil, the wire develops a finite resistance and the temperature begins to rise. Since a certain percentage of the conductor cross section is usually made of high-purity copper, the resistance is initially low but rises steadily as the temperature increases. At the same time, the quench spreads from the point of origin with a certain velocity and new regions of conductor go normal and begin heating and developing greater resistivity. This rising resistivity coming from both the increased temperature and total length involved opposes the initial current. If no active quench protection system is used this resistance eventually stops the current and terminates further heating.

The temperature (T) reached at any particular region of the coil can be expressed as a function of the integral of the current density squared with respect to time $[f(t) = \int i^2 \, dt]$. The quantity $f(t)$ can be easily obtained experimentally providing the time when the quench occurred is known. A relationship between the temperature (T) and $f(T)$ is obtained by assuming that all the heat produced remains in the conductor, i.e. there is no cooling from the helium. This is a conservative assumption. We can then write

$$\frac{i^2 \rho(T)}{\alpha} \, dt = C_p(T) \, dT$$

$$f(T) = \int_0^t i^2 \, dt = \int_{T_i}^T \alpha \frac{C_p(T)}{\rho(T)} \, dT,$$

where $C_p(T)$ is the volumetric specific heat of the copper plus superconductor, $\rho(T)$ is the electrical resistivity of the copper, i is the current density in the conductor, and α is the fraction of the cross section that is copper. Figure 13a shows a plot of this relationship for a particular niobium titanium conductor (Tevatron/CBA type).

The dotted line in the figure shows an approximation to this relationship (20) assuming that the resistivity is linearly related with temperature, and the specific heat rises as the square root of temperature, i.e.

$$C_p = C_0(T/T_0)^{1/2}$$

$$\rho = \rho_0(T/T_0)$$

$$f(T) = f_0\alpha(T/T_0)^{1/2},$$

where $f_0 = 2C_0T_0/\rho_0$ and where C_0 and ρ_0 are the total mean volume specific heat and copper resistivity, each at room temperature T_0. Figures 13b and c show the observed and approximated dependence on temperature. This approximation, while being very crude, allows an analytic

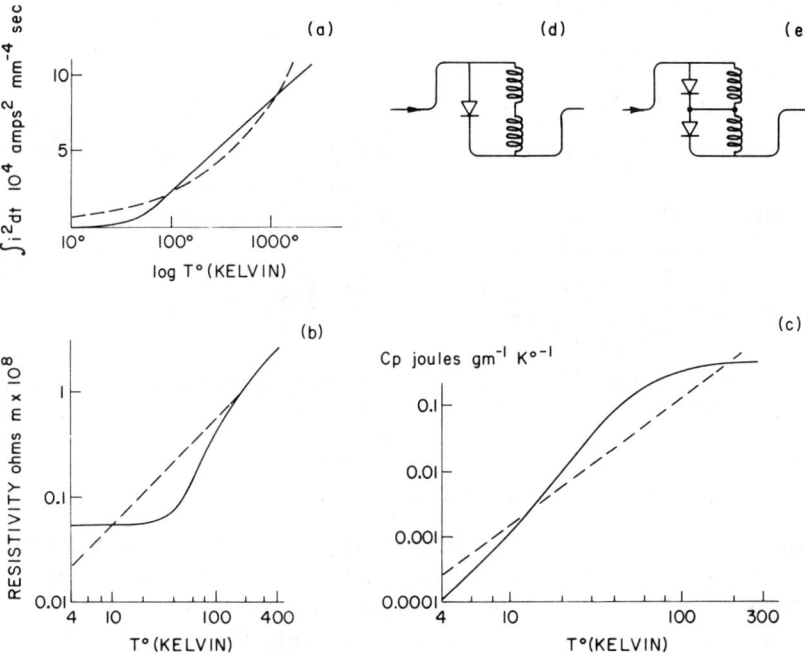

Figure 13 Quench protection. Graphs (a), (b), and (c) represent functions of temperature, with solid lines showing measured values and dashed lines showing simple approximations: (a) $\int i^2 \, dt$, (b) electrical resistivity of copper, (c) specific heat of copper. Diagrams (d) and (e) show single- and double-diode protection circuits.

solution to the rise in resistivity as a function of time and gives a qualitative feel for the effect.

The total resistance is obtained by integrating over elements of conductor that quenched at different times t' after the initial quench. These elements are at progressively larger distances from the point of quench origin in an axial and transverse direction:

$$R = \frac{4\rho_0 i^4 v_a v_l}{\alpha^3 A f_0^2 \Delta} \int_0^t (t-t')^2 t' \, dt'$$

$$= \frac{\rho_0 i^4 v_a v_l t^4}{3\alpha^3 A \Delta f_0^2}$$

where R is the total resistance of the copper and superconductor with total area A. In this approximate form the current i is assumed to be constant with time. The terms v_l and v_a are the quench propagation velocities in the longitudinal and azimuthal directions respectively; Δ is the conductor thickness in the azimuthal direction.

If the total stored energy is E, then the time (τ) before this energy has been dissipated by the resistance is given by

$$\tau = \left\{ \frac{15 E f_0^2 \alpha^3}{w \rho_0 i^6 v_l v_a} \right\}^{1/5}.$$

In reality, of course, the current would not be constant up to this time; nevertheless, the equation gives a good approximation to the time when the current has fallen to half its value, and it allows a calculation of the maximum temperature reached (T_{max}):

$$T_{max} = T_0 \left\{ \frac{15 E i^4}{w \rho_0 v_l v_a f_0^3 \alpha^2} \right\}^{2/5}.$$

Ignoring cooling effects, the velocity of quench propagation is given by (23)

$$v_{l,a} = \frac{i}{C_p} \left\{ \frac{\rho k_{l,a} \beta}{\alpha (i_q - i)} \right\}^{1/2},$$

where i_q is the quench current density and β is the dependence of quench current with temperature (di_q/dT). For the longitudinal case, k_l is the thermal conductivity of the wire along its length, and in this case we note that ρk_l is, from the Vedaman-Franz-Lorentz law, approximately a constant, $\rho k_l \approx K$. The longitudinal quench velocity is thus independent of the resistivity of the wire. In the transverse direction, however, k_a is the mean thermal conductivity in the azimuthal direction and is dominated by the insulating layers between turns. For this case the velocity increases as the square root of the electrical resistance of the wire.

Experimental measurements have been made on the damage thresholds both for the electrical insulation and for the superconducting properties of the wire. No electrical degradation was observed even with values of the $\int i^2 \, dt$ as high as 1.2×10^5 A^2 mm^{-4} s, whereas slight reductions of the superconducting properties of niobium-titanium cable were observed at approximately 1.0×10^5 A^2 mm^{-4} s. Both values correspond to extremely high temperatures (above 1000 K). In an experiment on an actual magnet connected in series to other magnets with no protection provided, no damage was observed with an $\int i^2 \, dt = 1.1 \times 10^5$ A^2 mm^{-4} s, but substantial damage was done at a value of 1.4×10^5 A^2 mm^{-4} s. Again we note that these numbers correspond to extremely high temperatures, and we must remember that the calculations assume negligible cooling either by conduction or by gas convection during the quench process. In the case of the actual magnet test, it is clear that some cooling must be present.

6.4 Quench Protection

The simplest form of protection is to provide no protection at all and have the resistance in the magnets absorb all available energy. But if no protection of any kind is provided, a single magnet will be required to absorb the energy of an entire string of magnets assuming that these are connected in series. Something has to be done.

The easiest protection scheme is to place a high-current, solid-state diode across each magnet with a direction such as to allow the main current to bypass the magnet in the forward direction through the diode (see Figure 13d). If a typical solid-state diode is operated at nearly 4 K, then a relatively large forward voltage (of the order of 4 volts) is required before conduction begins. Providing the excitation rate does not require a voltage across the magnet greater than about 4 volts, the diode will not conduct unless a quench occurs. When the magnet quenches, however, a large voltage is induced, the diode becomes conducting, and the main current can pass through the diode instead of through that magnet. As a result, the magnet is only required to absorb its own quench energy and not that of other magnets in the string. The requirements for this were described above but may in some cases result in very high temperatures being reached. These are somewhat reduced if the upper and lower coil halves each have a separate diode (as was done in CBA, Figure 13e). The simplicity and reliability of this protection scheme are its greatest attractions.

An alternative protection scheme used in the Tevatron (21) involves sensing the presence of a quench electronically and then putting a pulse of current into an electrical heater, which more generally distributes the quench in the magnet or magnets of a string. When this is done, the maximum temperature in the magnets can be kept relatively low, considerably lower than in the diode scheme. The disadvantage of such an active system is that a failure in the system could cause the loss of magnet, and great care has to be exercised in the detailed design. It is not clear whether or not such an active system can be extrapolated to a system as extended as the SSC. However, quench protection will be a dominant consideration for large accelerators using high-field (>4 T) magnets.

Literature Cited

1. Boom, R. W., McIntyre, P., Mills, F. E. 1983. *Workshop on Technical Discussions on the SSC Magnets*, Argonne, informal report
2. Wilson, R. R. 1982. *Proc. DPF Summer Study on Elementary Particle Physics and Future Facilities*, Snowmass, pp. 330–34. Batavia, Ill: Fermilab
3. Allinger, J. A., Danby, G. T., Jackson, J. W., Prodell, A. 1977. *High Precision Superconducting Magnets, IEEE Trans. Nucl. Sci.* NS-24:1299
4. Lundy, R. A. 1981. *IEEE Trans. Magn.* MAG-17(1):709–12
5a. Palmer, R. B., Cottingham, J. G., Fernow, R. C., Goodzeit, C. L., Kelly, E. R., et al. 1981. *BNL Internal Rep.*, TLM No. 23. Upton, NY: Brookhaven Natl. Lab. (BNL)
5b. Palmer, R. B., Baggett, N. V., Dahl, P. F. 1983. *IEEE Trans. Magn.* MAG-19(3) Part I:189–94
6a. Palmer, R. B. 1982. *BNL (CBA), Tech. Note 427*. Upton, NY: BNL
6b. CBA Newsletter No. 2. 1982. *2-in-1 Magnet*, p. 27. Upton, NY: BNL
7. Biallas, G., Finks, J. E., Strauss, B. P., Kuchnir, M., Hanson, W. B., et al. 1979. *IEEE Trans. Magn.* MAG-15(1):131–33
8. Cooper, W. E., Fisk, H. E., Gross, D. A., Lundy, R. A., Schmidt, E. E., et al. 1983. *IEEE Trans. Magn.* MAG-19(3)
9. Larbalestier, D. C. 1983. Proc. Part. Accel. Conf., Santa Fe. *IEEE Trans. Nucl. Sci.* In press
10. Tollestrup, A. V. 1979. *IEEE Trans. Magn.* MAG-15(1):647–53
11. Althaus, R., Caspi, S., Gilbert, W. S., Hassenzahl, W., Mauser, R., et al. 1981. *IEEE Trans. Nucl. Sci.* NS-28(3):3280–82
12. Kahn, S. A., Engelmann, R., Fernow, R. C., Greene, A. F., Herrera, J. C. 1983. See Ref. 9
13a. Hanft, R., Brown, B. C., Cooper, W. E., Gross, D. A., Michelotti, L., et al. 1983. See Ref. 9
13b. Schmidt, E. E., Brown, B. C., Cooper, W. E., Fisk, H. E., Gross, D. A., et al. 1983. See Ref. 9
14. Blesser, E. J., Thompson, P. A., Wanderer, P., Willen, E. 1983. *Accel. Phys. Issues for a Superconducting Supercollider Workshop Ann Arbor, Mich.*, UMHC84-1
15. Tannenbaum, M. J., Ghosh, A. K., Robins, K. E., Sampson, W. B. 1938. See Ref. 9
16a. Ciazynski, D., Mantsch, P. 1981. *IEEE Trans. Nucl. Sci.* NS-28(3):3280–82
16b. Ciazynski, D., Mantsch, P. 1981. *IEEE Trans. Magn.* MAG-17(1):165–67
17. Tollestrup, A. V., Peters, R. E., Koepke, K., Flora, R. H. 1977. *IEEE Trans. Nucl. Sci.* NS-124(3):1331–33
18. Koepke, K., Kalbfleisch, G., Hanson, W., Tollestrup, A. V., O'Meara, J., et al. 1979. *IEEE Trans. Magn.* MAG-15(1):658–61
19. Tollestrup, A. V. 1981. *IEEE Trans. Magn.* MAG-17(1):863–72
20. Wilson, M. N. 1983. *Superconducting Magnets*. London: Clarendon. (Note that although this reference uses this approximation, the following treatment is not equivalent, but follows that by Alan Stevens, private communication.)
21. Koepke, K., Rode, C., Tool, G., Flora, R., Jostlein, H., et al. 1981. *IEEE Trans. Magn.* MAG-17(1):713–15
22. Brechna, H. 1973. *Superconducting Magnet Systems*. New York: Springer-Verlag. (This is an excellent overall reference.)
23. Tollestrup, A. V. 1982. *Am. Inst. Phys. Conf. Proc.* 87:699–804. (This is a general reference. See p. 764 for the equation for quench wave velocity.)

HIGH-RESOLUTION ELECTRONIC PARTICLE DETECTORS

G. Charpak and F. Sauli

CERN, 1211 Geneva 23, Switzerland

CONTENTS

1.	INTRODUCTION	285
2.	THE PHYSICAL MESSAGE	286
3.	GASEOUS DETECTORS	290
	3.1 Introduction	290
	3.2 Drift and Diffusion of Charges in Gases	291
	3.3 The Multiwire Proportional Chamber	296
	3.4 Drift Chambers	310
	3.5 Parallel-Plate and Multistep Chambers	321
	3.6 Condensed Noble-Gas Ionization Detectors	325
4.	SOLID-STATE DETECTORS	328
	4.1 Introduction	328
	4.2 The Depletion Region in Rectifying Structures	329
	4.3 Position-Sensitive Detectors	334
	4.4 Microstrip Detectors	334
	4.5 Charge-Coupled Devices	338
	4.6 Detection by Charge Storage in Shallow Levels in Semiconductors	343
	4.7 The Silicon Drift Chamber	345
5.	CONCLUSION	346

1. INTRODUCTION

High-energy physics currently presents a heavy demand on high-accuracy particle detectors, in the μm or tens of μm range. As is almost always the case in particle-detector development, this demand from physics stimulates the conception of new ideas or, more often, the perfection of well-known techniques whose potential properties for the highest possible accuracy were not fully exploited. Very illustrative is the spectacular rejuvenation of the bubble chamber, which has stopped developing in the direction of mammoth dimensions to become a live target with an accuracy of 5 μm, thanks to the use of holographic methods for the retrieval of the bubble position. A similar development is observed for streamer chambers, where the combination of high pressure and refined read-out methods has been used to improve greatly the localization accuracies of the device. In the two cases the interest in these developments has been fostered by the need to obtain clear information on the interaction vertices where short-lived particles are produced and decay over path lengths of μm or tens of μm. While these detectors—which provide an almost complete picture of the most complex vertices—have gained a new popularity with the research on charmed particles, a considerable effort has been undertaken in the development of electronic detectors capable of withstanding high fluxes to bring them to the same level of accuracy. The study of short-lived particles is not the only goal in these searches. The improvement of accuracy in the measurement of momenta of charged particles in detectors being used around high-energy colliders would result in a serious reduction of size in some expensive systems. The most important developments—those permitting a realistic projection in the near future for the design of experiments—are based on solid-state or gaseous electronics and are the subject of this review.

2. THE PHYSICAL MESSAGE

Charged particles, traversing a gaseous or condensed medium, release energy in various ways. Of all interactions, however, only the electromagnetic one provides a physical message that can be exploited for detection in thin layers of matter, since it is several orders of magnitude more probable than nuclear or weak collisions. The interaction between the electromagnetic fields of the particle and of the medium can result, for energy transfers exceeding several eV, in the excitation or ionization of the molecules; the ionization message is the one exploited in most position-sensitive detectors.

The probability of a given collision is a fast decreasing function of the energy transfer involved. As an example, Figure 1 gives the computed number of ionizing collisions in argon for fast singly charged particles, as a function of the energy of the ejected electron (1). Typically, 80% of all ionization encounters lead to the production of an electron with energy below the average that is needed for further ionizations (around 25 eV in argon). Despite their low probability, however, large-energy transfer encounters whose outcome is an energetic electron capable of ionizing the medium (a delta electron) dominate the statistics of the energy-loss processes. Indeed in 1 cm of argon at normal temperature and pressure (NTP) there is a 5% probability of emission of a 2-keV electron, which doubles the energy loss (2 keV/cm on an average) and, with its 150 μm of range, offsets the ionization trail.

The total number of ion pairs released in the medium can be evaluated by dividing the fractional average energy loss by the average energy per ion pair, w. This phenomenological quantity depends on the nature and energy of the ionizing radiation, and equals roughly twice the lowest ionization potential of the medium. Values of w measured under different conditions and in various materials are given, for example, by Franzen & Cochran (2) and Christophorou (3).

The localization accuracy in detectors is often dominated by the physical extension of the ionization trail in the medium due to the nonresolved, heavily ionizing, slow electrons ejected in the primary encounter at energies in the keV region. Precise range-energy curves can be computed; however, because of the very large scattering angles suffered by low-energy electrons in their collisions with molecules, the integrated range does not represent well the physical extension of the released ionization. It is customary to define a practical range as the extrapolation to zero intensity in an absorption curve; the ratio between integral path length and practical range is energy-dependent, and is about a factor of 2 around 10 keV. Figure

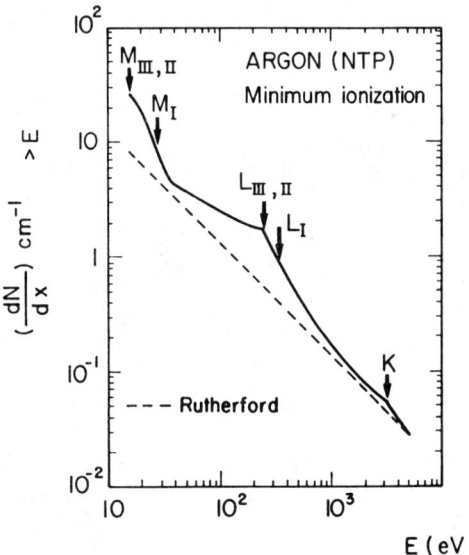

Figure 1 Collisions per unit length, in argon at NTP, with an energy transfer $\geq E$ as a function of the energy (1).

2 (4) shows values of practical range measured in solids between 0.1 and 10 keV and given in g/cm²; the range in cm can be computed by dividing these values by the corresponding density. Although data are scarce for gases in this energy domain, it is assumed here that the average values from Figure 2 can be used.

Analytical and statistical methods have recently been developed to allow a detailed description of the energy-loss process of fast charged particles (1, 5, 6). Although accurately computing the number and distribution of the ionization clusters, the quoted works unfortunately do not include an estimate of the physical track width. Figure 3 (6) shows the remarkable agreement between computed and experimental values of differential energy loss for fast particles in 1 cm of argon at NTP. Additional complications arise in the calculation of energy loss for thin layers of condensed materials because of the delta electrons escaping from the layer. Still an excellent agreement can be obtained with experimental measurements, as shown in Figure 4 (7).

Soft x-ray generators or sources are often used to check efficiency and localization accuracies of detectors, since they are easily collimated. The interaction of protons with matter consists in a single localized encounter

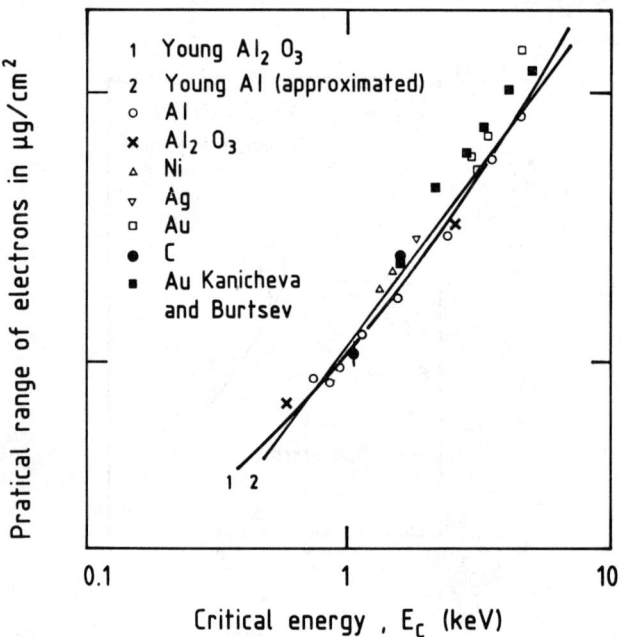

Figure 2 Practical range-energy relation for slow electrons (4).

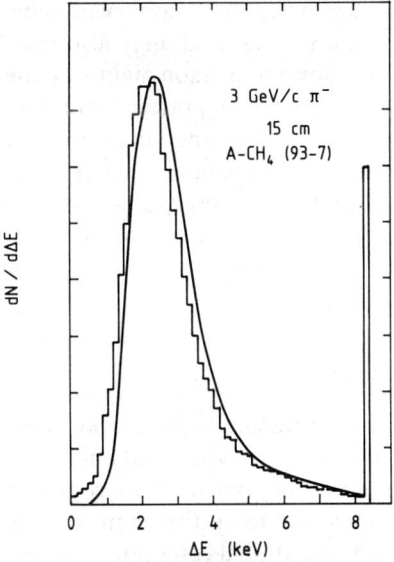

Figure 3 Differential energy loss of minimum ionizing particles in argon at NTP compared with a model calculation (6).

generally involving large transfers. In the energy domain we are concerned with, a few keV to a few tens of keV, the main absorption process is due to photo-ionization with the emission of an electron, accompanied either by a lower-energy photon or by an Auger electron. The range in the medium of the emitted electron or electrons is one of the factors determining the

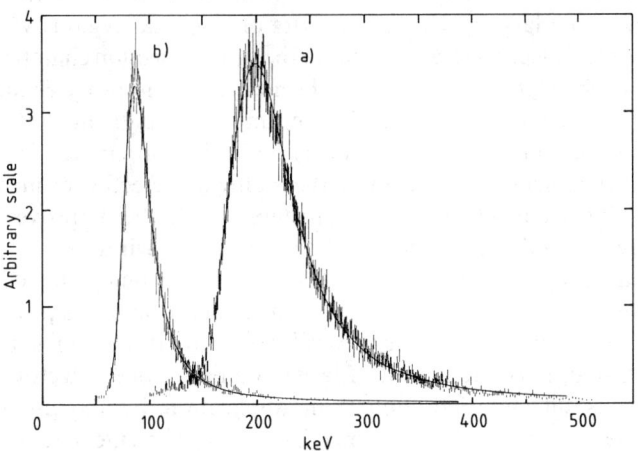

Figure 4 Experimental and computed differential energy loss in silicon (7) for 0.736 (*a*) and 115 GeV/*c* (*b*) protons.

achievable localization accuracy of the detector. Reconversion—within the active volume—of a fluorescence photon, if not resolved, may also spoil localization. For example, a 5.9-keV x-ray converts in argon mainly on the K shell (3.2 keV); the 2.7-keV photoelectron has a practical range of 30 $\mu g/cm^2$ or 200 μm at NTP. With 88% probability, another electron of energy around 3 keV (250 μm range) is ejected; in the remaining cases, a 3.2-keV fluorescence photon is produced with a mean absorption length of 40 mm. This illustrates the physical limits in the attainable localization accuracy for soft x-rays using gaseous detectors.

3. GASEOUS DETECTORS

3.1 *Introduction*

Position-sensitive gaseous detectors based on the multiwire proportional chamber (MWPC) are widely used in particle physics. Submillimeter localization accuracies in the detection of ionizing radiation are routinely achieved and there have been many attempts to further improve the accuracy by using suitable geometries and operational conditions. Among the resolution-limiting factors, the major ones are the width of the physical message due to emission of long-range delta electrons in the gas and to the diffusion of charges before collection. In drift chambers localization accuracies around 20 μm rms have been achieved, about the same as intrinsically allowed by the center-of-gravity avalanche localization method.

For energetic charged particles, not suffering too much from Coulomb scattering, multiple sampling can be realized to improve localization accuracy, using large stacks of detectors or thick gas regions with many independent collecting electrodes (as in the time projection chamber). In the absence of correlations, one expects the accuracy to improve as the square root of the number of samples. At the limit, each individual electron or cluster in the ionization trail could be located, as attempted in the time expansion chamber; replacement of the multiwire detection element with a parallel-plate chamber, single or multistep, may further improve the issue by removing the dispersive effects of discrete anode wires.

Both the physical ionization width and the diffusion can be reduced by increasing the gas pressure. However, while operation of gas proportional chambers at pressures above several hundred atmospheres has been demonstrated, implementation of large, multielectrode detectors at high pressures is difficult and limited in application by the use of thick containment vessels. This has encouraged experimenters to turn their attention to liquid or solid ionization devices; however, since electron transport is observed over distances of interest only in ultrapure condensed noble gases, the necessity of using cryogenic containment and purification, for the time being, seriously limits the usefulness of such an approach. Recently in the field of calorimetry research, several room-temperature liquids capable of reasonable electron transport were found and may change the issue in the near future.

3.2 Drift and Diffusion of Charges in Gases

Charges released in a gas by the ionizing encounters diffuse uniformly in the absence of external fields, having multiple elastic collisions with the molecules until they eventually get neutralized by mutual collision, capture, or collision at the walls. However, in the presence of an applied electric field, a global motion in the direction of the field superimposes on the random thermal motion; the average displacement of the swarm per unit time is called drift velocity.

For positive ions, and up to high values of field (several kV/cm · atm) the drift velocity increases linearly with the field; the ratio $\mu = w/E$ is called mobility. The average energy of ions remains in this region equal to its thermal value kT. Detailed theories of ion motion and experimental values of mobility can be found, for example, in McDaniel & Mason (8).

The much lighter electrons, on the other hand, can easily be accelerated between collisions with the molecules by the external field, reaching average energies far exceeding the thermal one even at moderate fields. The behavior of drift velocity and diffusion depends very strongly on the gas nature, through the detailed structure of the elastic and inelastic electron cross section of molecules. Because of the very different values and energy dependence of the cross section in different gases, addition of even very small quantities of one species to another may substantially modify the drift properties; often, in mixtures of noble gases with organic quenchers (which increase the maximum gain attainable by proportional amplification) a maximum or saturated value of drift velocity is reached at fields around 1 kV/cm · atm, typically of several cm/μs.

The general requirements for a gas mixture optimal for obtaining high resolutions depend on the detector design. Very high drift velocity is desired whenever collection time plays a role, while a slow gas is preferred if a measurement of drift time is performed to provide space coordinates, in order to ease the electronics requirements. In this case, it is also desirable to use a gas with velocity saturation at high fields, as this generally reduces the dependence of drift velocity on variations of field, temperature, pressure, and gas composition. For example, Figure 5 shows the field dependence of drift velocity and of its variance with temperature, for an argon-methane 90-10 mixture at NTP (9). Electron diffusion, setting an ultimate limit to the obtainable accuracy, has to be kept small, thus favoring the choice of "cool" gas mixtures (where the energy of electrons remains close to thermal at high fields); since diffusion is inversely proportional or the square root of the gas density, increasing the gas pressure improves accuracy. Needless to say, one has to preserve the other requirements of the detectors (for example, proportional or semiproportional avalanche multiplication, small electron attachment, and so on). Extensive compilations of cross sections and drift velocities in pure gases and in some mixtures can be found in the literature (3, 10–14). Theories and experimental methods are reviewed in the quoted references and elsewhere (15–17).

Recently, and specifically in the development of gas detectors, electron transport theories have been reviewed by several authors and applied to the gas mixtures commonly used in MWPCs and drift chambers (18–21). The agreement between theory and experiment is remarkable, and the quoted calculation methods can be used to tailor a gas mixture to specific needs or to compute its drift properties, e.g. in the presence of magnetic fields.

In the case of a spherically symmetric, Gaussian diffusion (this assumption is not verified for most gases at large fields, see below), the dispersion

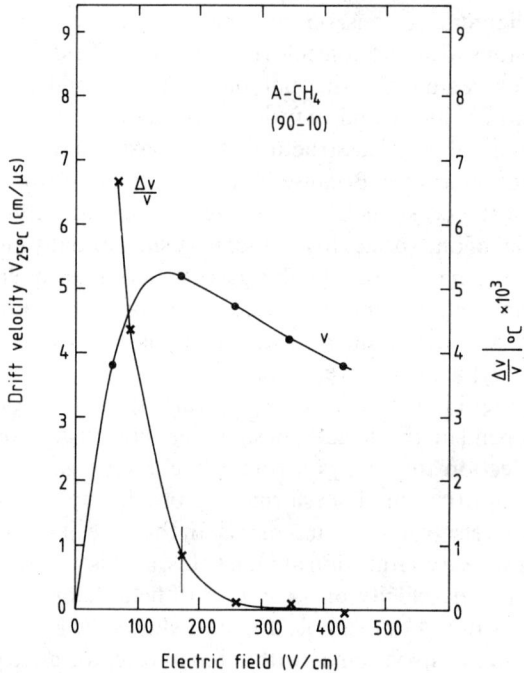

Figure 5 Electron drift velocity and its temperature dependence, as a function of electric field, in argon–methane (90-10) at NTP (9).

with time of an originally point-like charge distribution is described by a field-dependent diffusion coefficient D such that

$$\sigma_x = \sqrt{2Dt} \qquad\qquad 1.$$

represents the rms of the swarm along any direction x after a time t. It is customary to define a quantity named characteristic energy ε_K as the ratio between diffusion coefficient and mobility; one can write then

$$\sigma_x = (2\varepsilon_K x/eE)^{1/2} = (2\varepsilon_K x/EP^{-1})^{1/2} P^{-1/2}, \qquad\qquad 2.$$

where the last expression shows explicitly the pressure dependence of space diffusion at given values of the reduced field EP^{-1}.

For an ideal gas, where the energy distribution of electrons remains thermal at any value of the electric field, the characteristic energy equals its classic value kT and the space diffusion is given by

$$\sigma_x = (2kTx/eE)^{1/2}, \qquad\qquad 3.$$

a quantity often called the thermal limit.

In "hot" gases, where their average energy is greatly increased at high fields, electrons can have diffusions that are orders of magnitude larger than the thermal limit [see Figure 6, computed for 1 cm of drift in gases at NTP (18)]. In carbon dioxide, on the other hand, electrons remain thermal up to rather high fields; this is partly true also for mixtures containing a large fraction of CO_2. It has been found recently that dimethylether $(CH3)_2O$ presents thermal diffusion properties at even higher fields, see Figures 7a and 7b (22); at the same time, this gas has a low drift velocity, which makes it interesting in some applications.

As mentioned, diffusion is not always symmetric, especially at high fields, but tends to be smaller in the direction of drift. One has therefore to consider a transverse diffusion coefficient D_T (as given by the previous considerations), and a longitudinal one D_L. Although a detailed theory has been developed on the subject (23) the above-mentioned calculation methods have so far not been extended to include the effect, and provide only the transverse values. Longitudinal diffusions, on the other hand, have been measured in several gas mixtures commonly used in proportional chambers (24–26); Figure 8 is a compilation from the quoted sources, while Figure 9 shows the comparison of computed transverse and measured longitudinal diffusions for a particular gas mixture (24).

Other effects may appear to partly invalidate the previous considerations. As an example, Figure 10 (27) gives the dependence on electric field of the longitudinal diffusion measured for a 16-mm drift length in pure CO_2 at various pressures in an experimental high-accuracy drift chamber. At each pressure, there is an optimum field value for minimum diffusion.

Notice, however, that the results shown in the figure violate the usual E/P scaling law, probably because of residual attachment of electrons to CO_2 molecules, increasing with pressure.

Any search for high-resolution detectors has to take into account the dispersive role of diffusion; the quoted values, however, provide the expected localization accuracy for single electrons. Very often the physical message consists instead of many electrons, localized or scattered along an ionization trail. Calculation of the resulting resolution depends then on the way the detector handles the message; for the ideal case of a localized

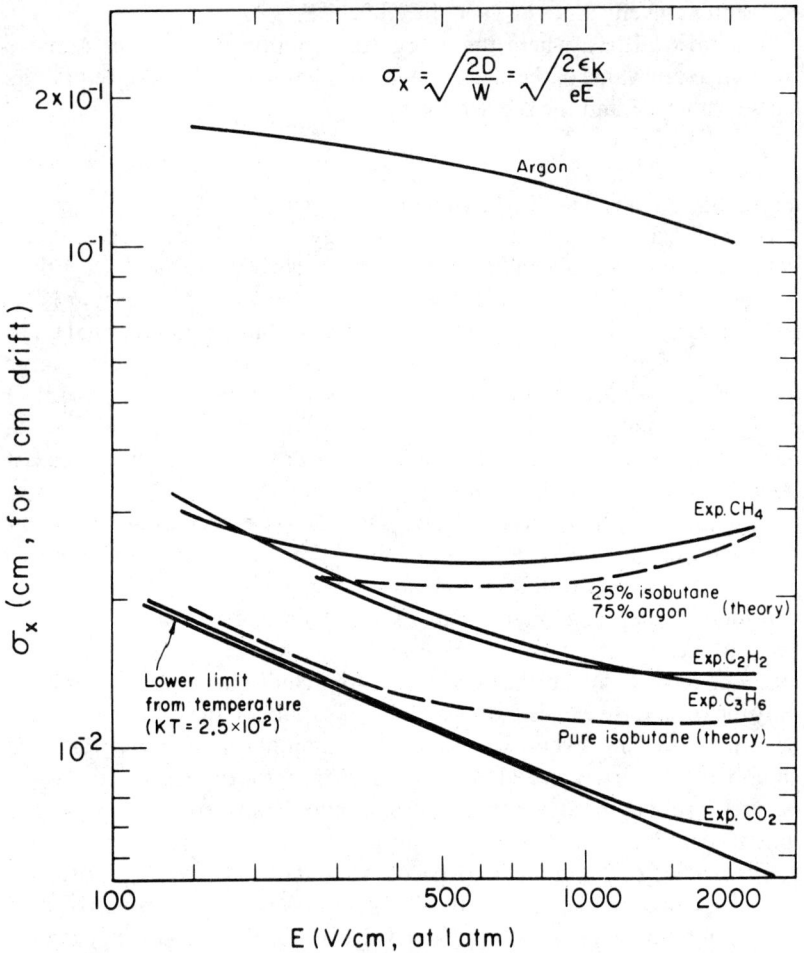

Figure 6 Single-electron diffusion rms for 1 cm drift, as a function of field for various gases at NTP (18).

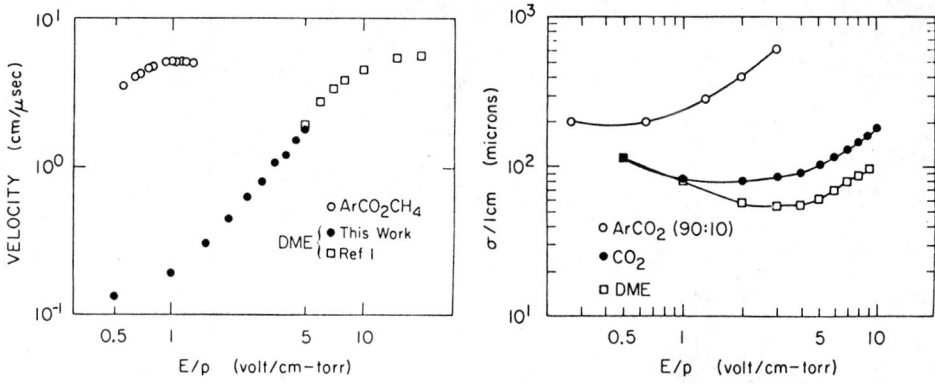

Figure 7 Drift velocity and diffusion, for 1 cm drift, in dimethylether (DME) (22).

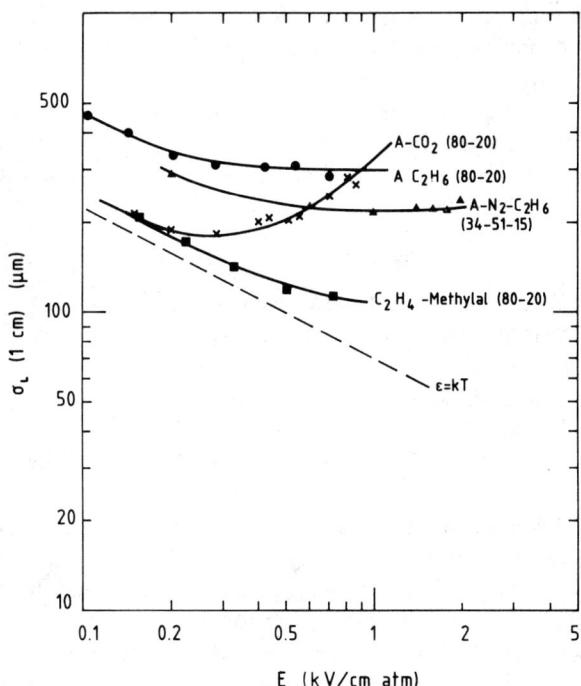

Figure 8 Experimental measurements of single-electron longitudinal diffusion for 1 cm drift, in various gases at NTP (24–26).

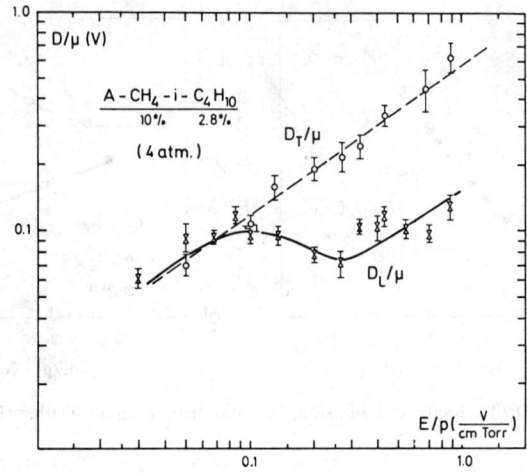

Figure 9 Comparison between transverse and longitudinal diffusion coefficients in a mixture of argon, methane, and isobutane (24).

cluster consisting of N electrons (a soft x-ray conversion in high-pressure gases), and if one can somehow determine the center of gravity of the detected swarm, the expected accuracy will be $\sigma_x(N)^{-1/2}$.

For more complex topologies of energy loss and electric fields, as is the case for drift chambers in the detection of charged tracks, the limiting resolution can either be computed using Monte Carlo simulations, or be directly measured (see the next sections).

3.3 The Multiwire Proportional Chamber

The MWPC was conceived fifteen years ago as a fast, high-resolution detector of charged particles and soft x-rays [Charpak et al (28)]. Since then, an impressive amount of work has been done to improve the general operational characteristics of the device, as documented in many review articles (9, 29–34).

In its basic design, the MWPC consists of a set of parallel, evenly spaced, thin anode wires between two equally spaced cathode planes. Typical values for the anode wire spacing range between 1 and 5 mm, the anode to cathode distance (the gap) being 5 to 10 mm. The operation gets increasingly difficult at smaller wire spacing, a fact that has generally prevented taking this obvious direction for obtaining higher resolutions. The chamber is filled with a gas mixture suitable for charge collection and proportional multiplication, in a range of pressures between a few Torr and tens of atm, depending on experimental needs.

Applying a potential difference between anode and cathodes, field lines

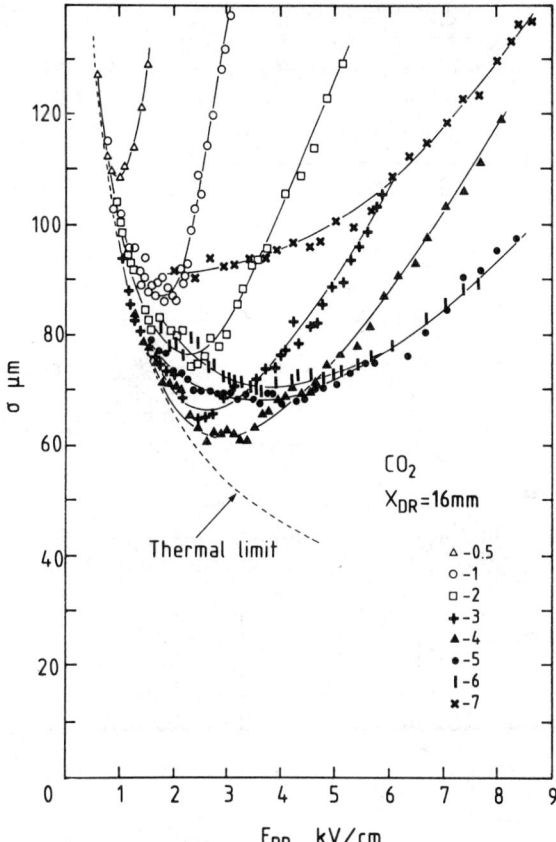

Figure 10 Experimental single-electron diffusion width (rms for 1 cm drift), in carbon dioxide at various pressures in atmospheres (27).

and equipotentials develop as shown in Figure 11. In most of the gas volume, the electric field is such as to simply drift charges: electrons to the anodes, and positive ions to the cathodes. In the immediate surroundings of the thin anode wires, the sharp $(1/r)$ increase of the field strength imparts to electrons enough energy to allow inelastic collisions with the gas molecules, causing, as a consequence, both excitations (often followed by photon emission) and ionizations. The creation of electron-ion pairs in the collisions is at the origin of avalanche multiplication, and is exploited as a signal amplification mechanism in gas counters; excitations and subsequent photon emissions participate in the avalanche spread processes, and can be detected directly by optical means.

The typical avalanche growth around a thin wire is depicted in Figure 12.

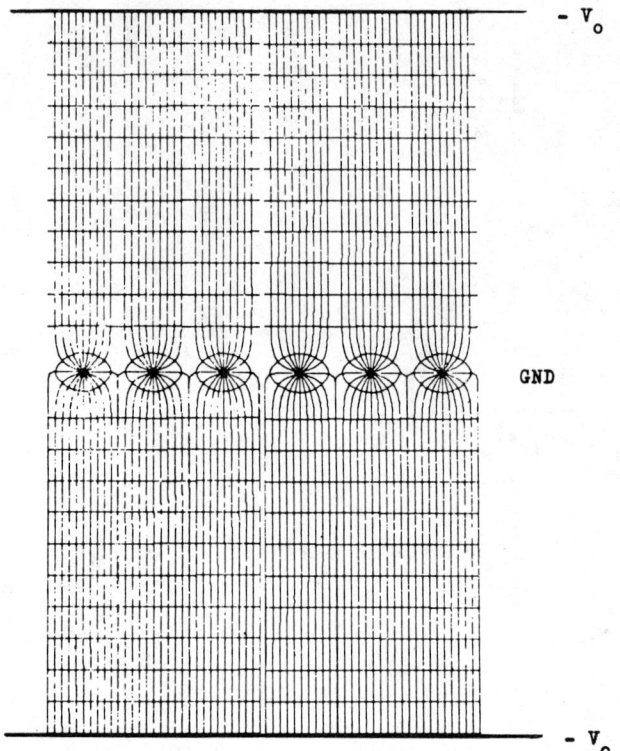

Figure 11 Electric field and equipotentials in the multiwire proportional chamber (29).

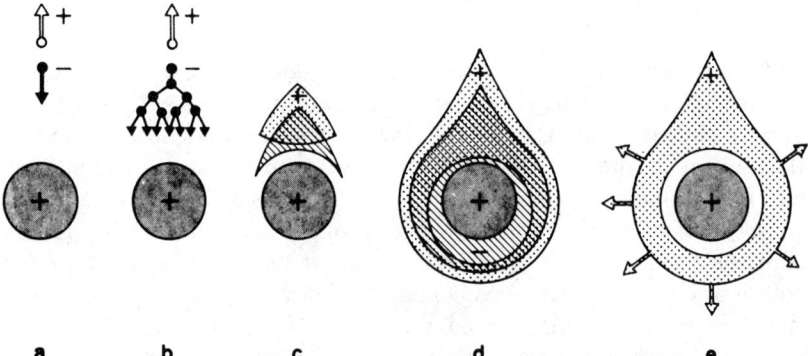

Figure 12 Schematic picture of the time development of the avalanche around the anode wire of a proportional counter (29).

Approaching the anode wire, electrons multiply in cascade, leaving behind an exponentially increasing number of positive ions. Both mechanical and photon-mediated diffusion contribute to the spread of the growing avalanche, which may completely surround the anode; the whole process, which begins a few wire radii from the anode, is over after a fraction of a nanosecond, leaving the cloud of positive ions receding from the anode at decreasing speed as a consequence of the decrease in field strength. This motion is responsible for the largest fraction of charge signals detected on the anode induced by the avalanche and on all surrounding electrodes; a measurement of the charge profile induced on the cathodes, suitably segmented, allows bi-dimensional localization of the ionizing event (35) to be achieved. This is the center-of-gravity method, which allows attainment of the highest localization accuracies in MWPCs (see Figure 13); if Y_i is the measured charge on the strip of central coordinate y_i, the average position of the avalanche along the direction y is estimated as

$$\bar{y} = \sum Y_i y_i / \sum Y_i. \qquad 4.$$

An example of induced charge profile, recorded in a MWPC with a 5-mm gap and 5-mm wide cathode strips, is shown in Figure 14. It has a FWHM of about 10 mm, twice the anode to cathode distance.

The computed center of gravity of induced pulses has a continuous dependence on the position of the original ionization only in one direction, that of the anode wires. For the perpendicular coordinate the dependence reflects the quantizing effect of the anode wires; should the avalanche symmetrically surround the anodes, the localization accuracy in this

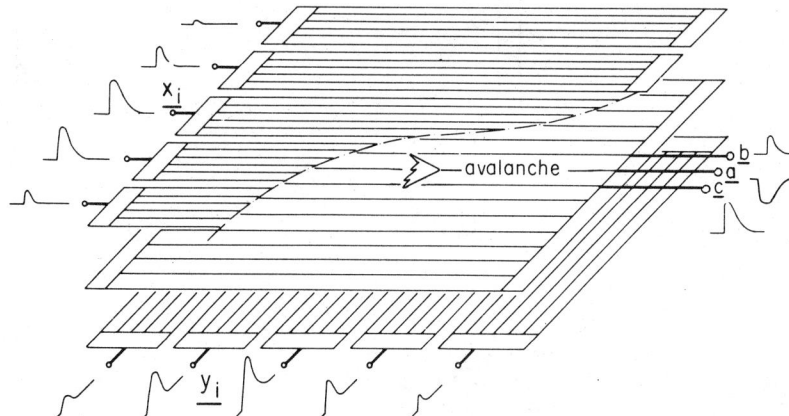

Figure 13 Principle of the center-of-gravity localization method by read-out of the induced charges on cathode planes (43).

direction would be limited to the wire spacing. As we see below, this is not necessarily the case and indeed, with a proper choice of operating conditions and reconstruction algorithms, almost equal localization accuracies can be obtained in both directions, at least in soft x-ray detection.

A precise knowledge of the induced charge distribution is necessary to obtain the highest resolutions. Indeed, while the distribution for a localized avalanche is intrinsically symmetric (and therefore its weighted average is the best estimate of the real position), for practical reasons the charge is integrated over cathode strips or pads with finite width, thus introducing a nonlinear correlation between the real position and the computed center of gravity. The distribution of induced charge on the cathode planes has been computed by many authors with various assumptions; the results of some model calculations are compared with experimental points in Figure 15 (36). Model D (37), overestimating the width, consists in a simple solid-angle dependence of charges induced by a point charge at an anode position, neglecting all other electrodes. Curve C (38) assumes a continuous anode plane at a very small distance from the avalanche; curve B (39) considers the presence of both cathode planes but neglects the adjacent

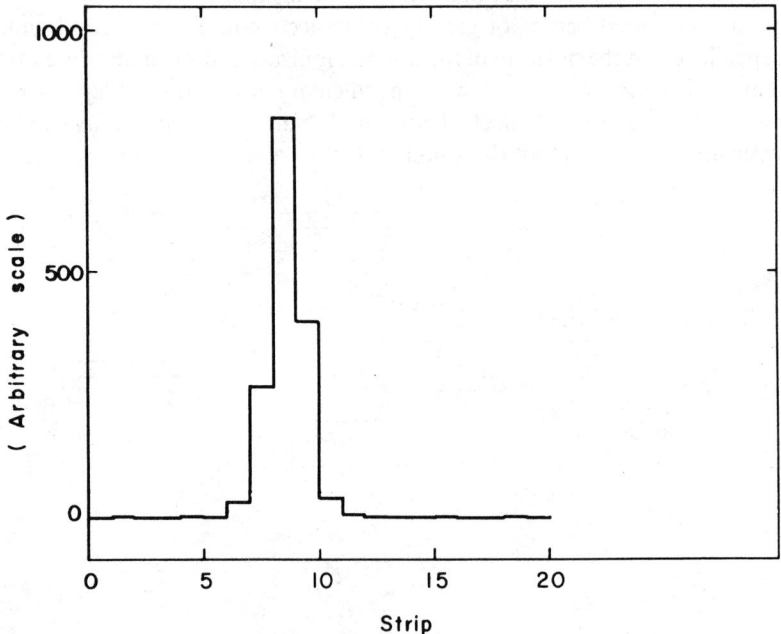

Figure 14 Typical charge profile induced on cathode strips by an avalanche; strip width and MWPC gap are 5 mm (43).

anodes. Curve A is the result of a very detailed and rigorous electrostatic calculation (40). While this obviously provides the best match with the experiment, the simplified expressions of the other models are often used for practical reasons. The effect of finite sampling width has been analyzed by many authors (36–41); in particular, Gatti et al (40) estimated theoretically both the integral and differential dispersions due to finite sampling size, to mutual capacitance between strips and from strips to ground, and to amplifier noise. Figure 16 shows one of the results, the figure of merit of avalanche localization, as a function of the ratio a/D ($2a$ being the cathode strip width and D the anode to cathode distance or gap). The result is given for three positions of the avalanche, in front of the center of a strip ($\lambda = 0$), between two strips ($\lambda = a/D$), or in an intermediate position. The result depends on the anode wire spacings ($b = s/D$ in the figure), but obviously

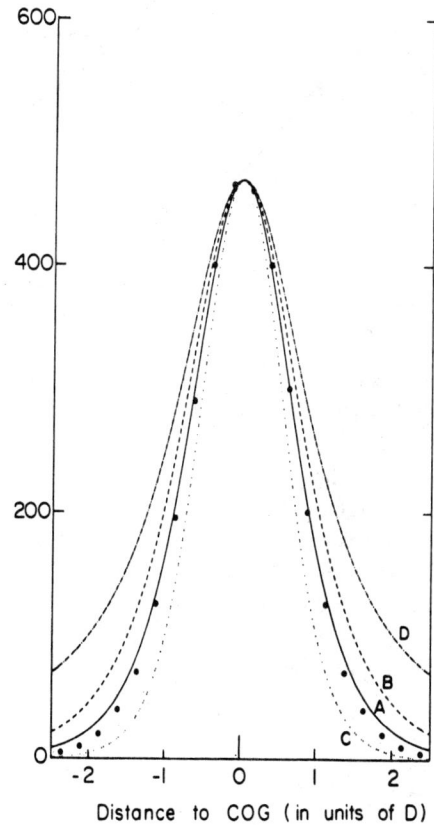

Figure 15 Experimental (*full points*) and computed cathode charge profile, according to various models (36).

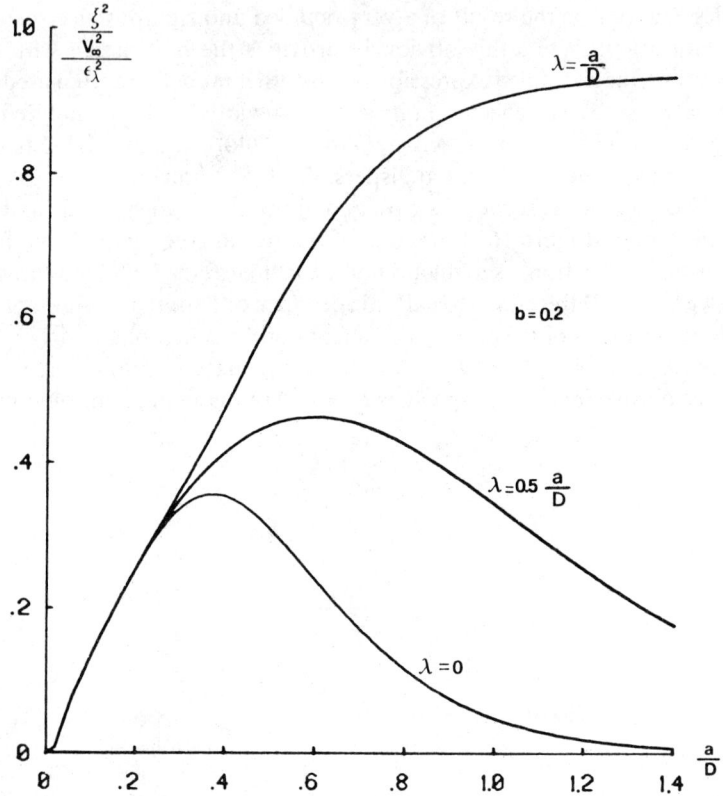

Figure 16 Figure of merit for avalanche localization, as a function of the strip width/gap ratio, for different avalanche relative positions (40).

there is an optimum in the region around $a/D = 0.5$, which is for the cathode strip width equal to the MWPC gap, as is already well known experimentally (42). This is a natural choice for a high-accuracy detector; a smaller strip width increases the number of samples, but this advantage is spoiled by the decrease of detected charge on each strip, with the resulting increase of both cross-talk and noise effects.

The simplest way to reduce the dispersive effects of noise and finite sampling size is the subtraction of a bias level to individual strip amplitudes before averaging (43):

$$\bar{y} = \sum (Y_i - B) y_i / \sum (Y_i - B). \qquad 5.$$

A good choice of the constant B appears to be a fraction of the total amplitude for each event:

$$B = b \sum Y_i. \qquad 6.$$

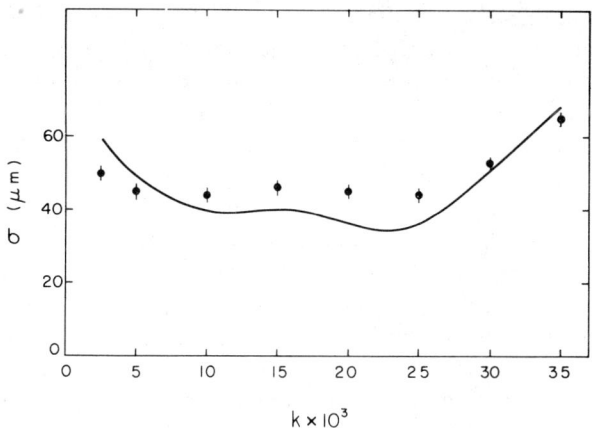

Figure 17 Influence of the choice of the bias constant (Equation 6) on localization accuracy (43).

Figure 17 (43) shows the dependence on b of the localization accuracy in the direction of the anode wires, measured for a minimum ionizing particle beam perpendicular to the MWPC plane. The choice of the value of b is not very critical, and produces consistently around 40 μm rms for the localization accuracy. Figure 18 (36) shows the systematic nonlinearity error in units of the half gap D, as a function of avalanche position, for several choices of the bias constant b [computed with the model of (40)].

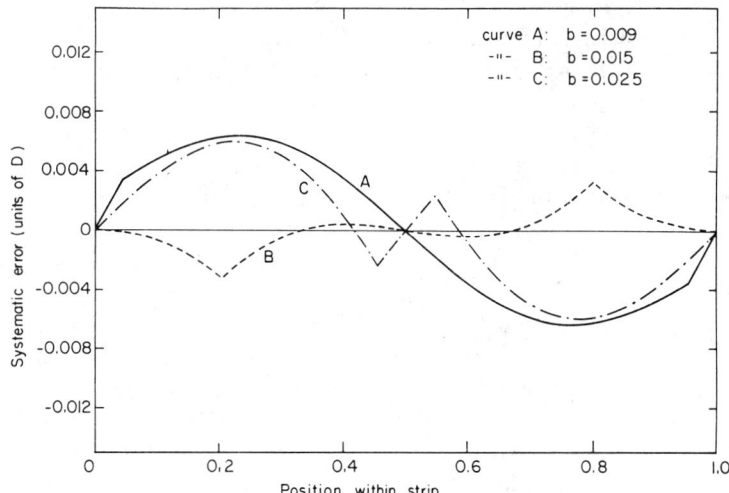

Figure 18 Systematic nonlinearity error as a function of avalanche position, for several choices of the bias constant (36).

The error appears as a periodic modulation, with a wavelength corresponding to the cathode-strip width. For the particular case of large cathode strips or pads, such as used in the time projection chambers, simple algorithms have been developed to restore the linearity of the response in reconstruction using a functional weighted average of the detected charges.

Mutual capacitance between strips and electronics cross-talk, producing a reinjection of charge in adjacent channels, may cause considerable distortions if not taken into account, as shown for example by Piuz et al (36). As noted by these authors, systematic effects of this kind may be masked by the experimental procedure that often involves the use of three identical, aligned chambers in a perpendicular beam.

Gain differences between channels, mostly due to the electronic amplifiers, also cause local nonlinearities. From Equation 4, and for a typical induced charge distribution such as that shown in Figure 14, one can estimate that a 10% gain difference in one strip results in a position error corresponding to 2% of the strip width (40 μm in the example). Rather straightforward calibration procedures can be used to determine the relative gains of all channels, applying then the corresponding scale correction at the analysis level. A convenient method of calibration consists in uniformly irradiating the detector with a monochromatic soft x-ray source (such as ^{55}Fe producing a 5.9-keV line), and recording the complete charge profile in each event. For each channel, a plot is then constructed containing the recorded charge only for the events in which the computed center of gravity lies within the size of the corresponding strip: this guarantees a good enough energy resolution to allow the determination of the average relative gain with an accuracy of a few percent.

Localization accuracies of 35 μm rms have been reported in the detection of soft x-rays (44, 45). Figure 19 (45) shows the reconstructed center-of-gravity distributions measured for three positions of a collimated 1.5-keV x-ray beam, 200 μm apart; the localization accuracy (in the direction of the anode wires) is 35 μm rms. The MWPC used for the measurement had an 8-mm gap, 3-mm cathode strips, and was operated in a xenon-isobutane mixture in order to reduce the photoelectron range in the gas (see Section 2.1). Figure 20, from the same reference, shows the linearity of the correlation between real and computed positions of the source.

Positioning accuracy is limited in the detection of charged particles by the physical extension and by the non-uniform density of the ionized trail, both due to emission of delta electrons. The second effect is largely dominant for tracks nonperpendicular to the MWPC plane. This is illustrated in Figure 21 (43). The localization accuracy for fast, charged particles is given for four incidence angles ($\theta = 0°$ meaning a beam perpendicular to the MWPC), and is shown in each plot for several

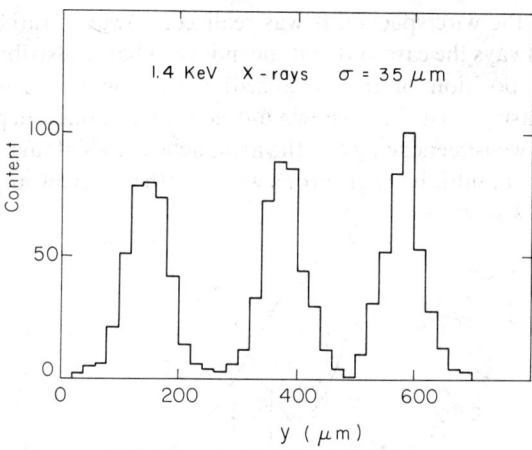

Figure 19 Position resolution, measured with the center-of-gravity method, in the direction parallel to the anode wires for three positions, 200 μm apart, of a collimated source (45).

intervals of actual energy loss. Resolution tends to deteriorate for large energy losses at each angle, and moreover the average accuracy goes from 50 μm rms at $\theta = 0°$ to 300 μm at 30°. For large losses, the resolution distribution has considerable non-Gaussian tails; this is very apparent when the experimental conditions allow for very high intrinsic resolutions, as shown for example in Figure 22 (46).

Should the avalanche process symmetrically surround the anode wire, the localization accuracy of a MWPC would be necessarily limited in one

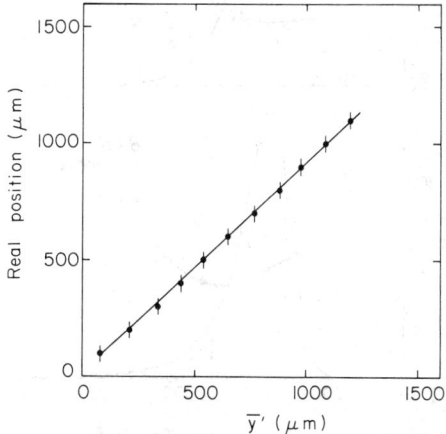

Figure 20 Linearity of the correlation between the computed center of gravity and the real position, in the direction of the anode wires (45).

direction by the wire spacing. It was realized, however, rather early that such is not always the case and that the induced charge distribution reflects the original position of the (localized) ionization between wires (47). Detailed measurements have shown indeed that at moderate proportional gains and in well-quenched gases the avalanche grows along the field lines of approach of multiplying electrons with a rather narrow angular spread

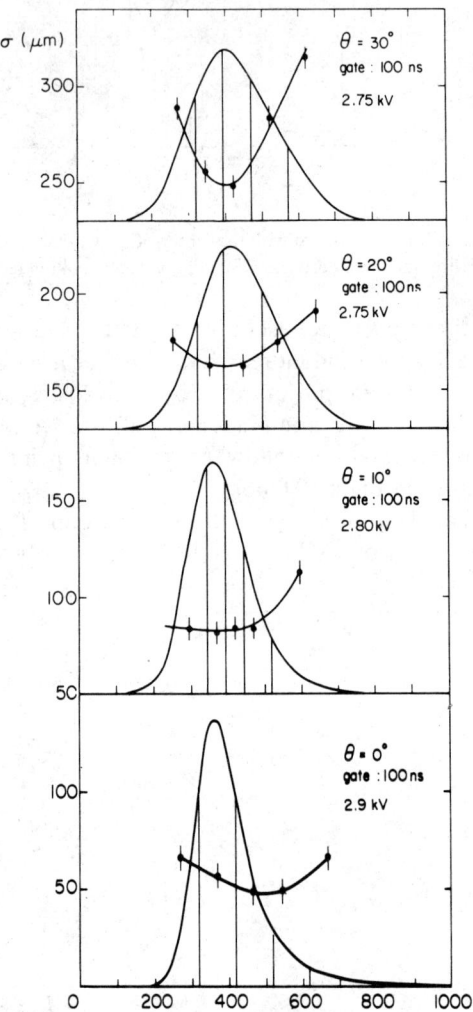

Figure 21 Experimental accuracy in the detection of minimum ionizing particles at different angles of incidence. The accuracy is computed for different slices in the energy loss as shown (43).

(44, 45, 48–52). This can be exploited for obtaining localization accuracies far better than the wire spacing.

The angular spread of avalanches around the anode has been measured both directly in special counters and indirectly by observation of the induced charge ratios on cathodes. The spread depends strongly on the counter gain and, to a smaller extent, on the gas mixture and anode diameter. Figure 23 (49) shows the measured spread, FWHM, as a function of avalanche size in argon-methane mixtures and for several anode wire diameters. Figure 24 instead gives the measured correlation between real position and center of gravity, computed from charges induced on cathode strips parallel to the anode wire, for a collimated 5.9-keV x-ray source (45). A 2-mm displacement of the source is reflected by a change of about 150 μm in the center of gravity. Despite this scale compression, using a plot such as the one in the figure one can deconvolute the measurements to achieve along the direction perpendicular to the anode wires almost the same accuracy as for the crossed coordinate. This is apparent in Figure 25, a bi-dimensional plot obtained by irradiating a MWPC with soft x-rays through a mask with letters cut out (45); the real size of the image is 4 × 2 mm. A single anode wire was hit, located horizontally in the center of the figure, and the appropriate transformation was applied to restore the vertical scale.

The sensitivity of the induced charge distributions to the position and angular spread of the avalanche has been computed in detail (53); Figure 26

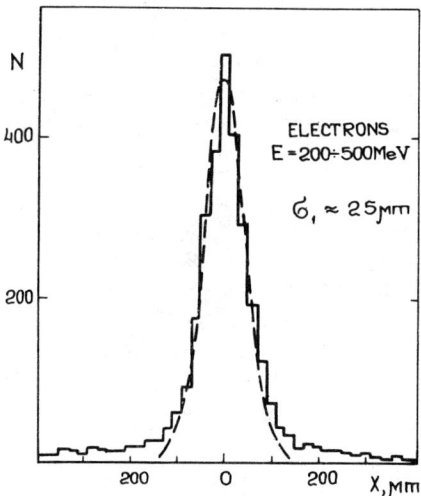

Figure 22 Localization accuracy distribution for an electron beam perpendicular to the chamber, showing the extended tails at large deviations (46).

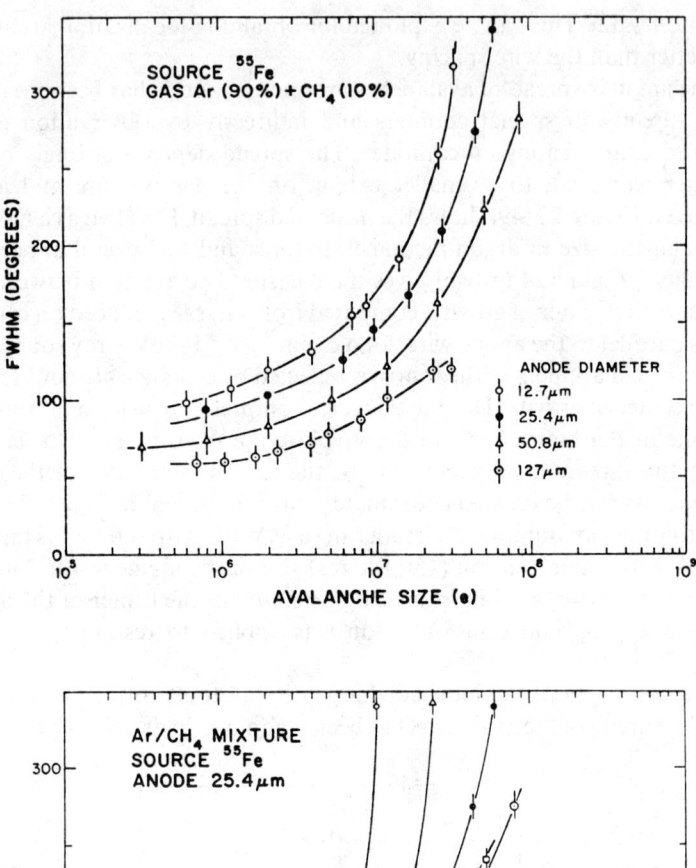
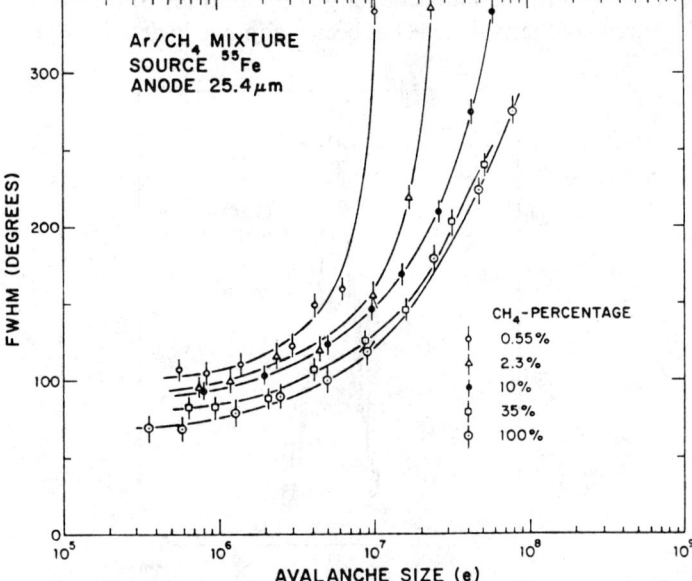

Figure 23 Avalanche width measured in argon–methane mixtures as a function of gain and for various wire diameters (48).

HIGH-RESOLUTION DETECTORS 309

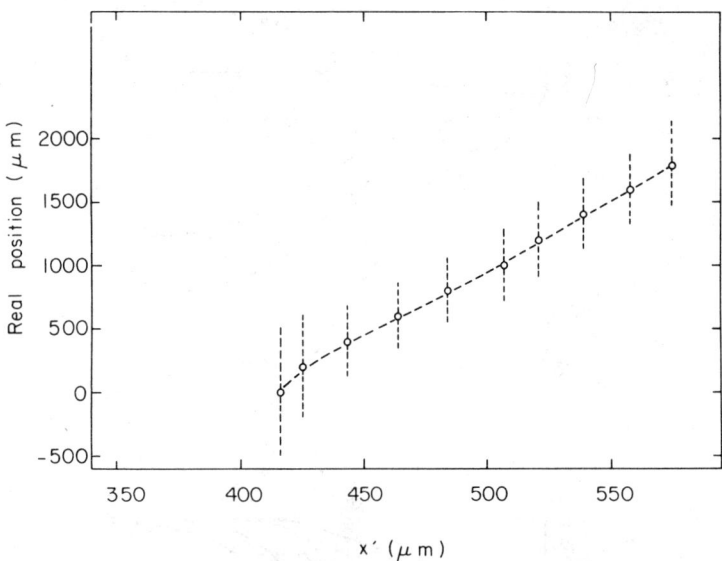

Figure 24 Real and measured coordinates in the direction perpendicular to the anode wires (47).

is an example of time dependence of the charge induced on two adjacent cathode strips, one of which is facing the anode, for several average angular locations of the avalanche (90° meaning an average avalanche growth perpendicular to the anode plane). The apparent displacement of the center of gravity is clearly time dependent, as experimental results obtained with

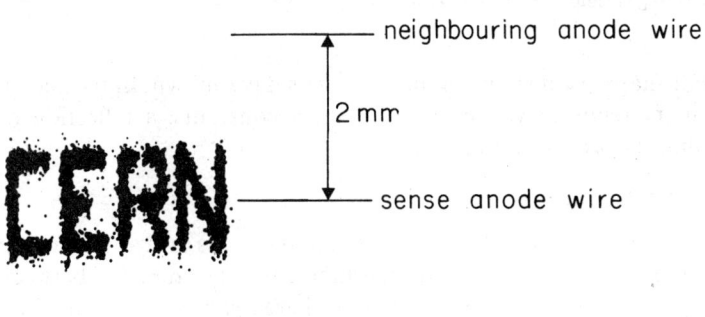

Figure 25 Bi-dimensional image obtained with a high-accuracy MWPC with cathode readout using a collimator with letters cut out; the size of the mask is 4×2 mm^2 (45).

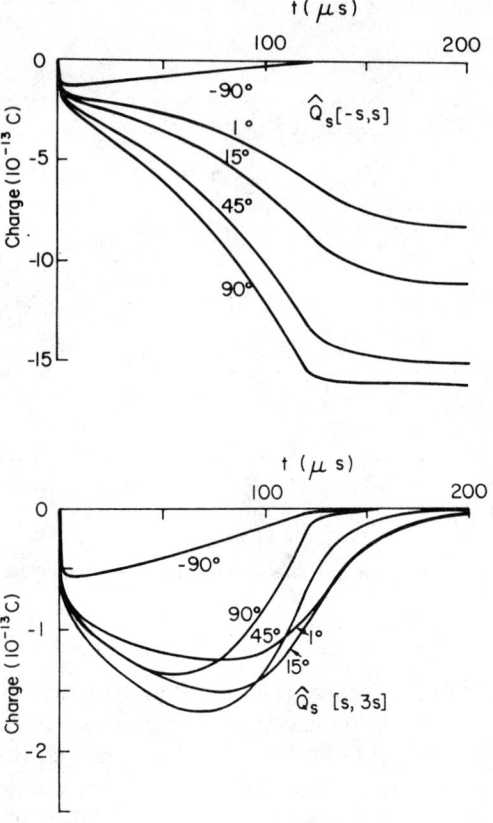

Figure 26 Computed time dependence of the charge induced on adjacent cathode strips for several average avalanche positions (53).

different shaping constants in the amplifiers have shown. In particular, the asymmetry tends to vanish for very short constants, a reflection of the screening effect of the anode wire.

3.4 Drift Chambers

As indicated in Section 2.1, electrons released in a gas drift and diffuse under the action of an electric field. A measurement of the time lag between the ionizing encounter and detection of a charge signal at the anodes in a suitable structure provides the average distance, along the field lines, of the ionization, as pointed out already in the early works on the MWPC (35). Of course, this can be done only if the interaction time is known. The first detectors specifically designed to exploit the drift process for localization were operated by Saudinos et al (54, 55) and Walenta et al (56); with various improvements, the drift chamber is still a very popular detector in high-energy physics because of the rather good position accuracy that can be obtained over large surfaces at moderate cost. Large-volume imaging detectors, such as the time projection chamber (57–59), have been developed by fully exploiting the drift properties in gases. For a review of existing drift chamber systems see the references quoted in the previous section; we are concerned here only with the high-resolution detectors.

The localization accuracy that can be achieved in a drift is limited mainly by the following effects:

1. the physical extension and statistical spread of the original ionization mostly due to energetic delta electron emission;
2. the value and stability of the electron drift velocity as well as the detailed knowledge of the drifting field, both in direction and strength; and
3. the dispersion due to diffusion, which depends on the gas nature and pressure and on the field strength, and increases with the drifting distance.

A structure optimized to achieve high resolutions in detecting charged tracks is shown in Figure 27 (60). The closely spaced cathode wires receive from an external network a uniformly decreasing potential, from ground (for the wire facing the anode) to a high negative value in front of the field wires; the electric field is then uniform across most of the cell, with the exception of the region around the anode. This design, together with the choice of a gas mixture exhibiting velocity saturation, allows one to obtain a very stable operation (61, 62). Figure 28 (63) shows an example of positioning accuracy obtained with a set of small drift chambers detecting a beam of minimum ionizing particles perpendicular to the wire planes, as a function of drift distance. Various contributions to the dispersion are also indicated, and the decrease in accuracy, due to diffusion, is apparent; in the argon-isobutane mixture used for this measurement, the single electron diffusion is about 200 μm for 1 cm of drift (see Figures 6 and 8).

One way to reduce the dispersion caused by diffusion and electronics sampling errors is to use a gas in which electrons remain thermal and with small drift velocity up to very high electric fields. Villa (22), operating with dimethylether at atmospheric pressure, measured a localization accuracy of 16 μm rms for minimum ionizing tracks drifting over about 1 cm (see also Figure 7).

The most obvious way to improve localization accuracy is to increase the gas pressure (see Equation 2). This both reduces diffusion and increases the primary ionization yield. Figure 29 shows measurements in a drift chamber at increasing pressures of a propane-ethylene gas mixture (64); Figure 30 is

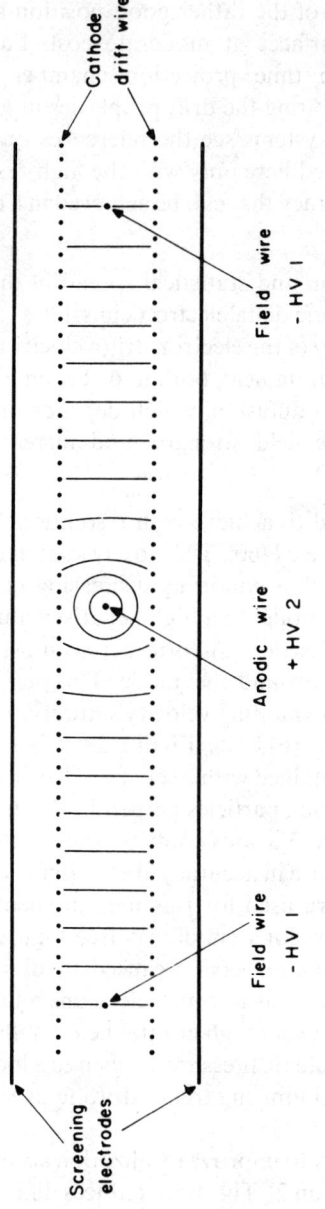

Figure 27 Principle of construction of the high-precision drift chamber (60).

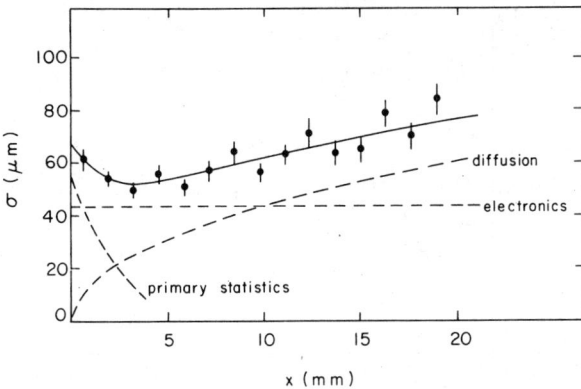

Figure 28 Experimental localization accuracy measured with a drift chamber as a function of the distance of tracks to the anode wire (63).

the result obtained in xenon at 20 atm (65). Figure 31, from the same work, shows the actual accuracy distribution plot for a 20-mm drift distance; the standard deviation of the distribution is 16 μm, but one should notice the non-Gaussian tails due to long-range delta electrons. They contain a non-negligible fraction of the events—about 6%. This information does not appear in plots such as the ones in Figures 28 and 29, and should be taken into proper account when designing high-resolution detectors; redundancy in the number of recorded points for each track allows one to disregard the measurements exceeding a given confidence level in the fit. For this

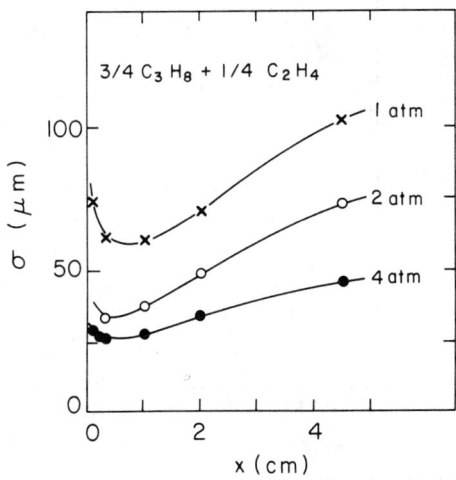

Figure 29 Influence of gas pressure on localization accuracy of a drift chamber (64).

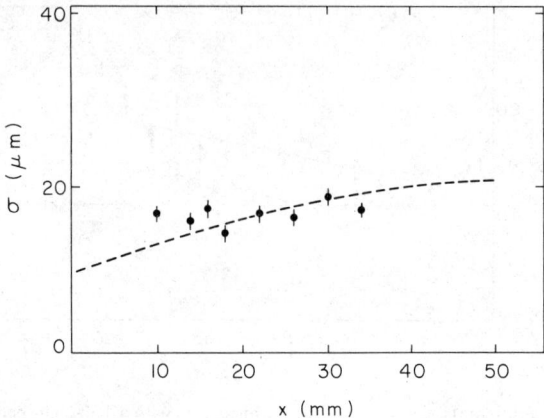

Figure 30 Localization accuracy measured with a scintillating drift chamber operated in xenon at 20 atm (65).

measurement, the authors used the so-called scintillating drift chamber, a device in which charges are detected by a photomultiplier through the light emitted by secondary gas scintillation in the high field around the anodes (66, 67). The intrinsically faster response of a photomultiplier, as compared to electronic amplifiers, and the absence of space-charge distortion (light

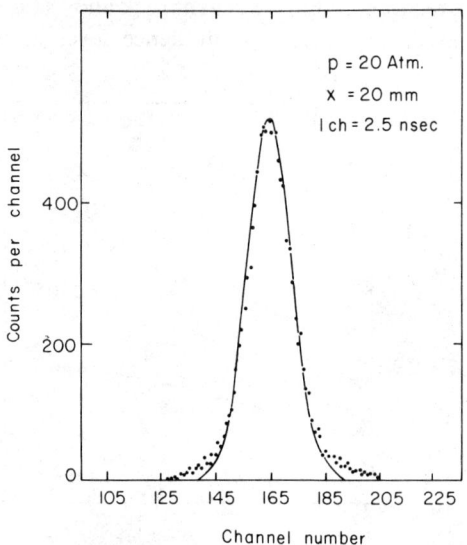

Figure 31 Accuracy distribution in a high-pressure drift chamber, for 20 mm of drift; the tails due to delta electrons are apparent (65).

emission can be obtained without large charge multiplication) encourage this approach for high-resolution, high-rate detectors.

Various ways have been proposed for coupling scintillating gas detectors to photomultipliers using scintillating rods, wavelength shifters, or optical filters (68), but no operational system based on this approach has emerged so far. Operation of drift chambers at ultra-high pressures (up to 200 atm) has been investigated (69); recent results (27) suggest, however, that the $1/P$ decrease in diffusion may not continue in some gases at high pressures, probably because of enhanced electron attachment and detachment (see also Figure 10).

In the devices described so far, tracks were mostly perpendicular to the drift direction, a particularly favorable case. This is not so in the large-volume drift detectors such as the time projection chamber (57–59), the jet chamber (24), and the imaging chamber (70). In these devices ionized trails form within the gas volume, as a result of external or internal interactions, and a suitable electric field drifts the tracks to a single plane of detection, generally a modified MWPC having field wires interleaved with anode wires. Track segments are continuously sampled and recorded electronically, using systems of analog memories [charge-coupled devices (CCDs)] or fast flash encoders. The time information is then used to compute the coordinate in the drift direction. In the time projection chamber the coordinates in the direction of the anode wires are determined by recording the induced charges on rows of cathode pads and computing the center of gravity by one of the algorithms discussed in the previous section. In other imaging devices this coordinate is obtained either by current division on the anodes or by assembling different modules at a small stereo angle; although less accurate than the center of gravity, these methods allow a more flexible construction geometry. Most of the detectors described are mounted inside a large magnet to allow momentum analysis.

Va'vra (71, 72) recently analyzed both experimentally and theoretically, in great detail, the physical and geometrical factors limiting the attainable accuracy in a drift chamber, in view of building a high-resolution vertex detector. His microjet chamber prototype design is shown in Figure 32; it is essentially a miniature drift chamber with thin anode wires (7.8 μm) operated at high pressures, with tracks developing parallel to the wire plane. Figure 33, generated by a Monte Carlo simulation program, illustrates the different sources of dispersion at detection. Because of the electric field structure, segments of track at equal distance from the anodes reach the wires with a characteristic (and unavoidable) U-shaped distribution. In the absence of clustering, i.e. local increases of ionization due to delta electrons, the distortion does not necessarily introduce a localization error, since the average time of recorded charge corresponds to the average track distance. A large cluster, on the other hand, may largely offset the time measurement because of the local increase in detected signal. Diffusion helps to reduce the cluster effects by smearing the charge, but of course represents another source of dispersion. The author extensively describes the best strategy for determining the timing of track segments, reaching the conclusion that first electron timing (obtained using large chamber gains and low discrimination thresholds) is the simplest and most accurate method. Figure 34 shows a comparison between computed and measured accuracies, as a function of distance of drift; the result was obtained by operating the microjet prototype in argon-methane 90-10 at 6.1 atm.

To overcome the intrinsic limitation in resolution due to the described clustering effect, various chamber geometries have been proposed, either focusing or otherwise limiting in extension the accepted segments of ionized tracks (27, 73); Figure 35 shows the design of the precision drift imager (27) operated at high pressures and in which only an 0.8-mm long segment of track contributes to the detected signal to each wire. Although of course the same performance could be obtained with a conventional drift chamber having small wire spacing, operation of such a detector at high pressures would require very thin wire diameters with the associated safety and reliability problems. Figure 36 shows the accuracy measured with the detector in a high-energy hadron beam, using as gas filling a carbon dioxide-isobutane mixture at 4 atm; Figure 37 shows instead the measured accuracy for a fixed drift distance, as a function of pressure, in pure CO_2. As already mentioned, the resolution does not improve indefinitely with increasing pressures, as would be expected from Equation 2, but an optimum is reached around 4–5 atm. The decrease in accuracy above this value could either be due to charge asymmetries produced by clustering (the probability of obtaining energetic electrons in a given length increases with gas density) or, as suggested by the authors, to residual electronegative

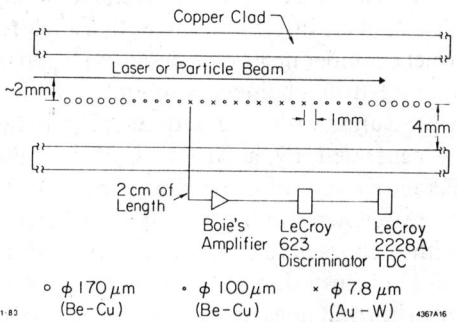

Figure 32 Principle of construction of the microjet chamber (71).

HIGH-RESOLUTION DETECTORS 317

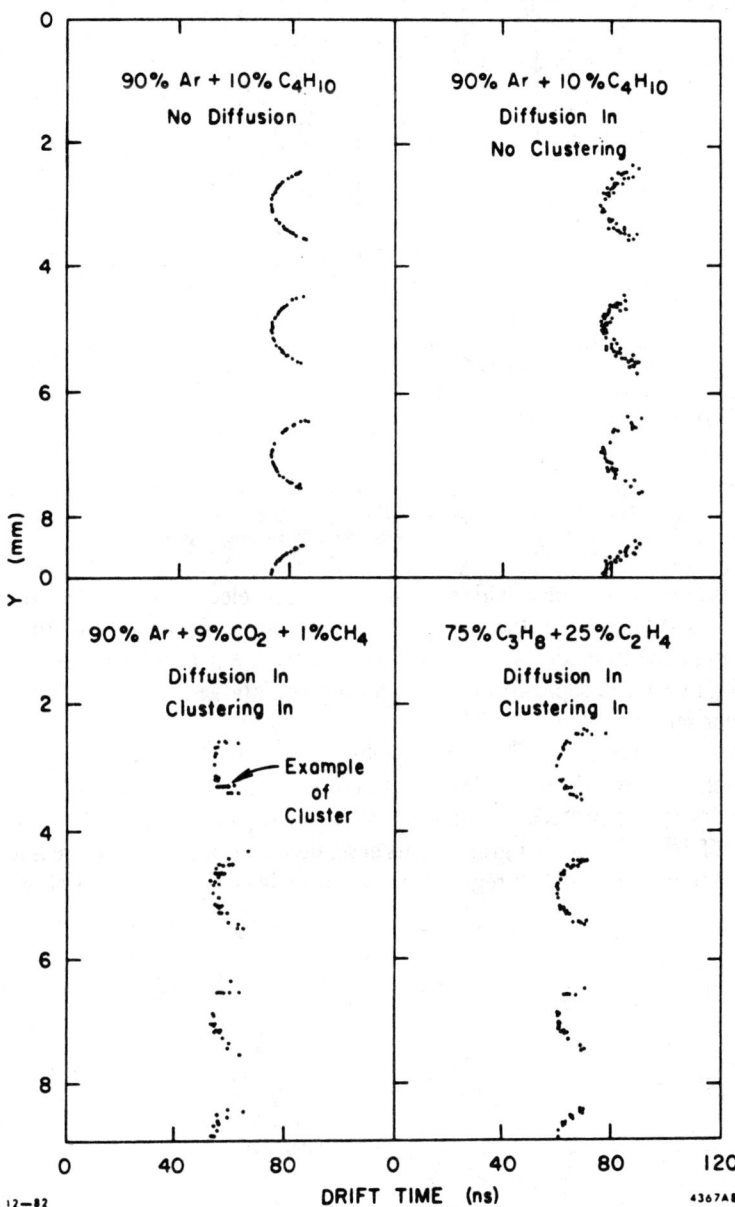

Figure 33 Computed time-shape of detected electrons in the microjet chamber, under various assumptions (72).

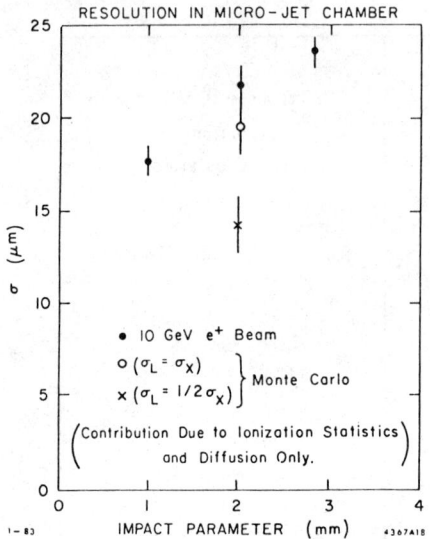

Figure 34 Computed and measured resolution in the microjet chamber (72).

pollutants in the gas, which attach and lose electrons, thus producing delayed afterpulses at the wires. The low diffusion and narrow sampling in the described detectors results also in a very good two-track resolution; 100% two-track separation has been obtained above 700 μm of distance, a rather remarkable result (73).

A different approach for obtaining high resolutions has been proposed by Walenta with the time expansion chamber, see Figure 38 (74). As in a standard time projection chamber, the drift region is separated from the amplifying section by a grid. In this case, however, the gas mixture and the field strength in the drift region are chosen such as to obtain a very low drift

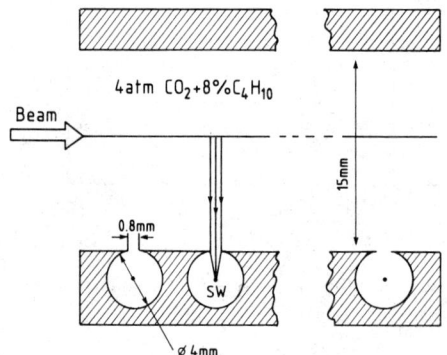

Figure 35 Principle of construction of the high-accuracy drift imager (27).

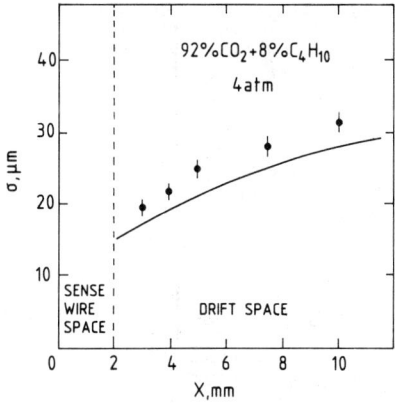

Figure 36 Localization accuracy measured with the drift imager operated at high pressure (27).

velocity (typically 0.5 cm/μs), while in the amplification region a normal fast collection is restored. A relatively coarse time measurement is then sufficient to obtain good spatial resolutions. Using a recording electronics system of fast flash ADCs, with 10-ns binning (corresponding in the drift region to a sampling length of 50 μm) and computing the center of gravity of the recorded time information one can reach resolutions that are mostly diffusion-limited. Figure 39 shows the accuracy measured with a time expansion chamber, operated in a mixture of CO_2 and isobutane in the ratio 80–20 and detecting tracks parallel to the wire plane (75).

The chamber geometry and operating conditions can be optimized in order to attempt a measurement of drift-time for each individual electron or localized cluster in the track segments; such a piece of information contains the full history of the ionizing event and, in principle, allows one to extract the best estimate of its coordinates. With a typical 20–30-ns duration for

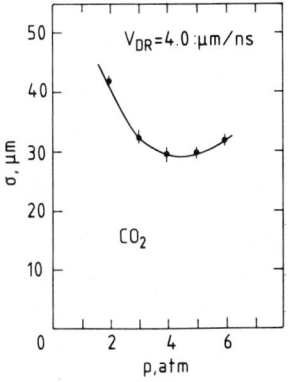

Figure 37 Best localization accuracy measured in the drift imager chamber as a function of gas pressure (27).

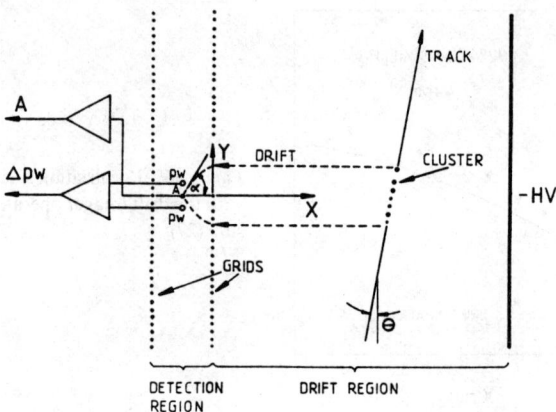

Figure 38 Principle of construction of the time expansion chamber (74).

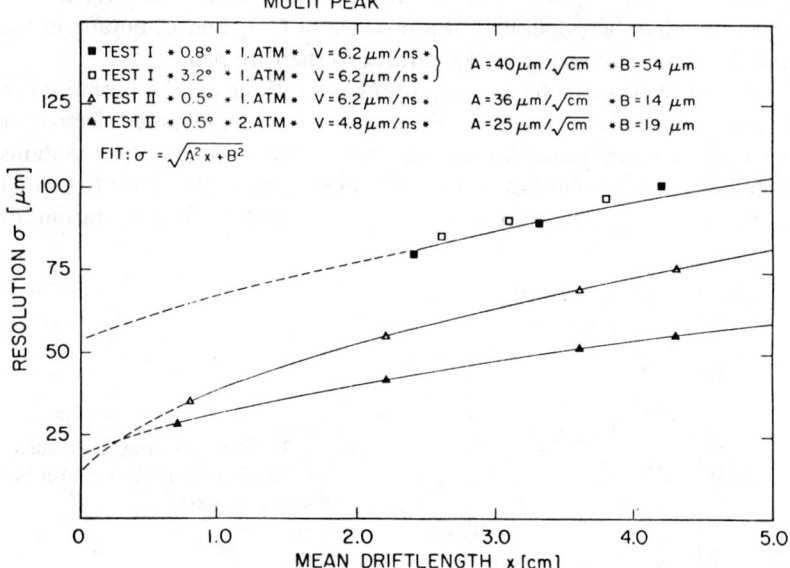

Figure 39 Localization accuracy measured with the time expansion chamber, for tracks parallel to the wire plane (75).

each detected pulse, obtainable using fast, low-impedance amplifiers, clusters with an average distance of 200–300 μm can be resolved (34). Obviously, in order for the clusters to maintain individuality the electron diffusion has to be kept to a value well below the average cluster separation; whence the need for using a gas mixture having low diffusion and low drift velocity.

For inclined tracks a good localization along the drift direction is obviously not sufficient to locate the cluster, because of the difference in the drift lengths depending on the transverse position. To obviate this problem, two field wires on each side of the anode are connected to a difference amplifier, providing additional information on the direction of approach of each cluster (see the discussion in the previous section on the avalanche localization).

3.5 Parallel-Plate and Multistep Chambers

The finite distance between anode wires has a distorting effect on localization along the coordinate perpendicular to the wire, because of the different length of field lines. Although the center of gravity of the induced charge reflects, to some extent, the original position of ionization, this is only true at moderate proportional gains (see Section 3.2), which seems a major limitation to the possibility of locating individual clusters of one or a few electrons, as attempted in the time expansion chamber. Except for tracks parallel to the wire planes, the different trajectory and therefore drift-time for clusters at equal distance from a wire, if not corrected for, is obviously a source of large errors. In the presence of external strong magnetic fields, the deflection of the electron swarm introduces also large dispersions.

A device that, in principle, allows these problems to be avoided is the parallel-plate avalanche chamber, as it has a uniform electric field in the multiplication region. Generally realized with thin meshes or metal foils, parallel-plate chambers can be built using printed-circuit boards, or otherwise patterned electrodes, for localization. Proportionality and imaging capability can be obtained by separating with a mesh the drift from the multiplication regions, much in the same way as in normal time projection chambers.

In the absence of other effects (space charge, gap distortions, etc.) the overall charge gain in a gap of thickness d is given by [see, for example, Raether (76)]

$$Q = Q_0 \exp(\alpha d), \qquad 7.$$

where α is the (constant) first Townsend coefficient pertinent to the gas and field conditions. Quite differently from the MWPC case, the fast electron signal represents a considerable fraction of the total induced charge, and is given by

$$Q_e = Q_0 \exp(\alpha d)/\alpha d \qquad 8.$$

or about 10% of the total signal at gains around 10^5 (as compared with 1% in MWPCs). Moreover, since the ions move in the uniform field with constant velocity, a simple differentiation at the amplifiers' input removes their contribution leaving a very fast detected signal. This makes the parallel-plate chamber a rather promising device to obtain good time and multitrack resolutions in time projection chambers (77–79).

Although widely used in nuclear physics as a detector of heavily ionizing projectiles (80–85), the parallel-plate chamber has not been much considered so far in high-energy physics because of the difficulty of obtaining large and uniform gains over extended surfaces. The discovery, however, that two or more parallel-plate amplifying elements can be combined in the same structure can modify the picture [Charpak & Sauli (86)].

Figure 40 shows the simplest two-step chamber design. Charges produced by ionizing encounters in the upper region drift into a region of very high field, the preamplification region, separated from the previous one by a thin mesh. Here, avalanche multiplication occurs; a fraction of the electrons generated in the avalanche escape into the following, lower field transfer region and continue toward the second element of amplification, another parallel-plate chamber in the drawing. With typical amplification factors in both multiplying gaps of 2×10^3, and 20% transfer efficiency,

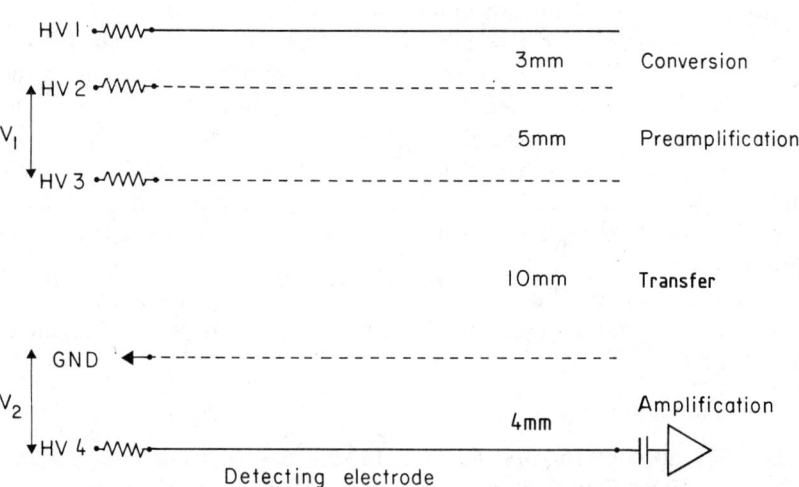

Figure 40   Principle of construction of the multistep avalanche chamber (86).

overall stable gains around 10^6 are attained, quite sufficient for single-electron detection. Using a gated two-step device (87), the authors detected minimum ionizing tracks perpendicular to the electrodes with a localization accuracy of about 150 μm and a two-track separation of about 1.5 mm (Figure 41). The single-electron imaging capability of multistep devices has also been demonstrated (88–90) in Cherenkov ring imaging, a technique devised for particle identification through the detection and localization of photons radiated by fast charged particles in suitable media.

It appears that the presence of a transfer low-field region, originally thought to help in suppressing photon and ion feedback between the two amplifying elements, may not be necessary to obtain large gains, at least in well-quenched gases. Peisert (77) demonstrated that a double-step structure, where the multiplying gaps follow each other, has a stable gain at least 20 times that of each gap individually, for reasons that are not very clear. The detector structure is shown in Figure 42; a region of drift is followed by two multiplying gaps, 4 and 1 mm thick, realized using crossed wire meshes or thick parallel wires at a small pitch; setting the operational conditions such as to have a multiplication factor of at least 20 or 30 in the last gap, a very fast electron signal can be detected. The last anode, realized with a printed-circuit board, contains the read-out structure, which is made of an alternance of strips, 500 μm apart, and rows of 1 mm^2 pads in a geometry rather similar to that of the time projection chamber except for the size of the elements. Figure 43, from the same reference, shows that the electron avalanche measured on the strips of the last electrode has a FWHM of

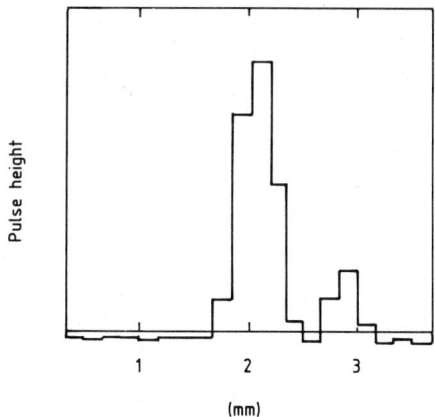

Figure 41 Avalanche profile measured on the last electrode of the multistep chamber, for a two-track event; the bin width is 500 μm (87).

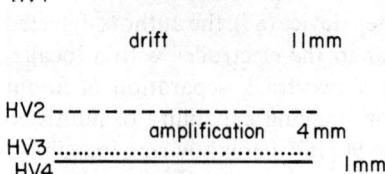

Figure 42 A two-step avalanche chamber as end-cap detector of a time projection chamber (77).

1 mm, just the right value to obtain a good localization by center-of-gravity measurement on the strips sharing the charge; using a low-impedance amplifier, with fast time constant, only the electron signal with 8-ns full width is detected (see Figure 44), with the consequent very good intrinsic two-track resolution in the drift direction (typical drift times are around 20 ns/mm).

For the reasons indicated, we expect the parallel-plate structure, simple or multistep, to take an important part in the future detector development because of its intrinsic superior characteristics and simplicity of construction.

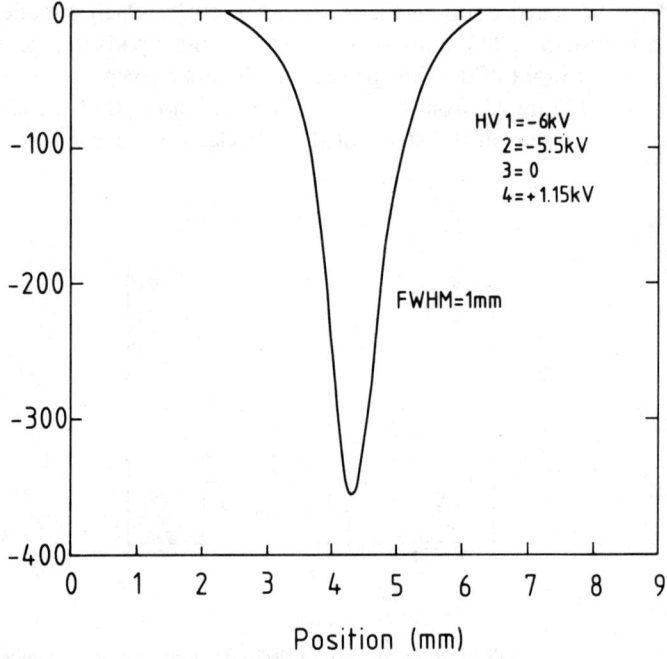

Figure 43 Avalanche profile measured in the two-step chamber shown in Figure 42 (77).

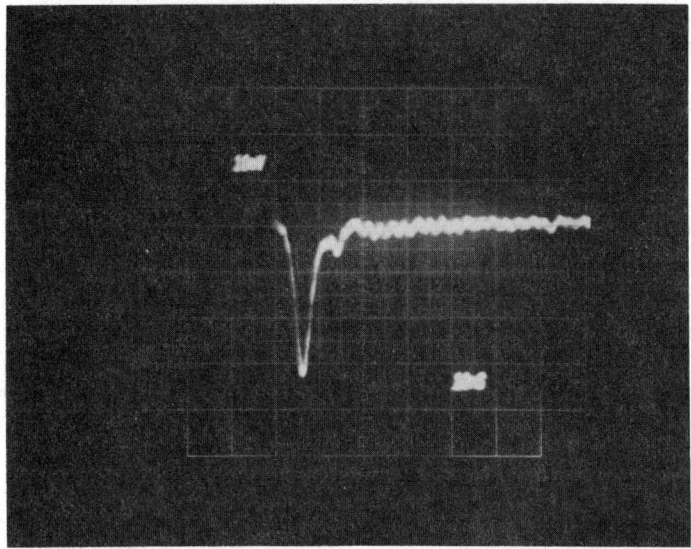

Figure 44 Typical signal detected with the two-step chamber using a fast current amplifier: only the fast electron component is seen (77).

3.6 Condensed Noble-Gas Ionization Detectors

Both diffusion and physical track width are reduced by increasing the density of the medium, a fact naturally leading to the search for suitable liquids or solids as detectors. Most liquids have, however, very large attachment coefficients, leading to a quick loss of drifting electrons, the exception being liquified noble gases with exceedingly low impurity levels.

More than a decade ago Derenzo et al (91) demonstrated electron drift and detection in a multistrip ionization chamber filled with liquid argon. Attempts to obtain proportional amplification around thin wires in the same conditions were, however, only partly successful, preventing at the time the use of the device as a detector of minimum ionizing particles. With the present availability of inexpensive charge-sensitive amplifiers with a typical noise of 0.2 fC rms (92–94) there is no real need for amplification within the detector, as minimum ionizing particles lose about 1.5 fC/mm.

A liquid-argon microstrip ionization chamber with resolutions better than 10 μm has been operated by Deithers et al (95). It consists of a 2-mm wide gap with a flat cathode and an anode made with vacuum-deposited metal strips, 10-μm wide at 20-μm intervals. The active area of the detector is about 3×0.4 mm^2, each individual strip being read out with a low-noise charge amplifier connected to analog-to-digital converters.

Because of diffusion and of the large gap, several strips (8 to 10) collect charge in each event, and the tracks' coordinates is computed by the usual center of gravity method. A stack of several aligned identical detectors allow the measurement of the localization accuracy by a straight-line fitting algorithm; Figure 45 shows the residuals' distribution, having a width of 8.5 μm rms.

In the described detector electrons are collected over a few millimeters and the requirements on gas purity are not very stringent (0.3 ppm residual oxygen is quoted by the authors). It was proposed some time ago (96) to use liquid argon in a detector with the time projection chamber geometry, having several tens of centimeters drift length; in this case the required impurities should be reduced to the ppb level.

A 50-liter liquid-argon time projection chamber has been operated successfully (97–99); see Figure 46. A large cryogenic vessel contains the actual detector, consisting of a drift volume of variable thickness; a grid separates the volume from the segmented anode electrode. After extensive purification, an attentuation length for electrons of about 30 cm at a field of 1 kV/cm has been measured, quite appropriate for track detection. The pulse profile within one unit of drift time and for a number of adjacent strips is shown in Figure 47, for two cosmic-ray events. Because of the modest sampling frequency of the waveform digitizers employed (1 MHz, cor-

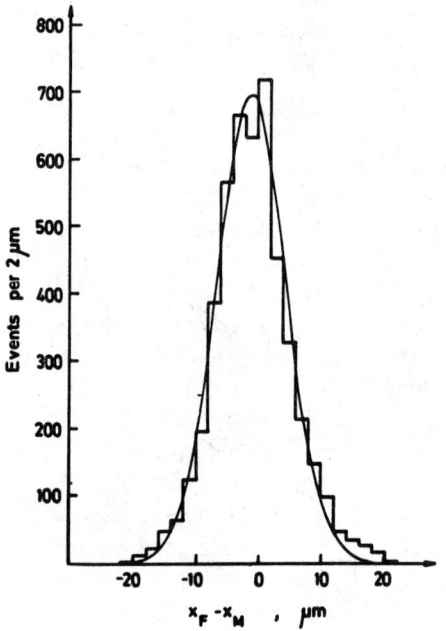

Figure 45 Localization accuracy measured with a stack of microstrip ionization chambers operated in liquid argon (95).

HIGH-RESOLUTION DETECTORS 327

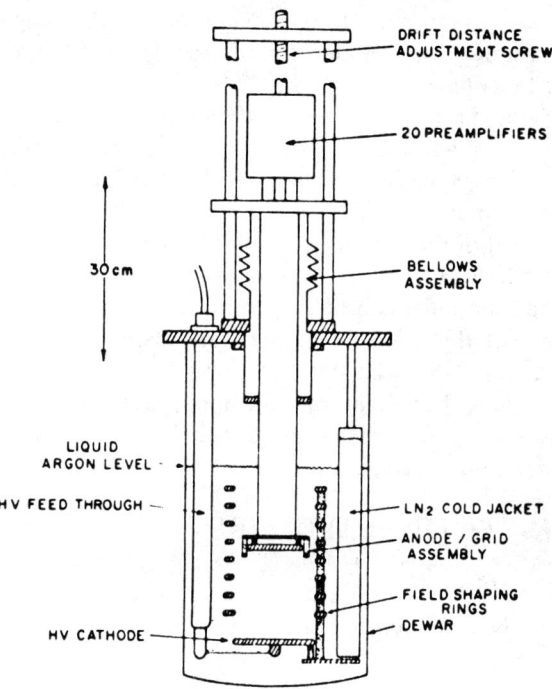

Figure 46 Schematics of the liquid-argon time projection chamber assembly (97).

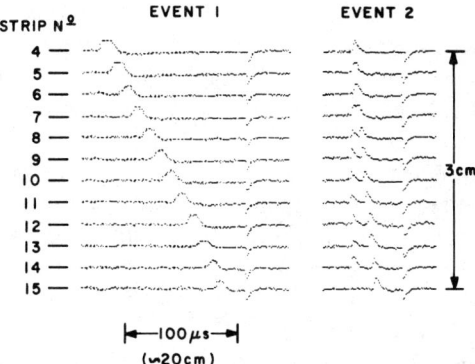

Figure 47 Pulse-height profile as a function of drift time, recorded in the liquid-argon time projection chamber, for a single- and double-track event (99).

responding to 2-mm samplings), the data are far from reaching the ultimate localization accuracy that should be permitted by the reduced diffusion in the condensed medium.

Various methods of obtaining bi-dimensional localization on the detector plane have been investigated by the authors, as mosaics of pads and interleaved strips. Because of the small detected charge, however, some of these methods may turn out to be impractical. Further progress is also certainly required in the amplifiers' design to increase the signal-to-noise ratio and the bandwidth.

Even a small amplification factor would be rewarding in this kind of game. Because of the difficulty of obtaining gains in the liquid phase, Dolgoshein et al (100) successfully attempted to drift tracks, produced within the condensed medium, into an upper layer of gas where a conventional MWPC would work. This very challenging approach does not seem to have been pursued further.

4. SOLID-STATE DETECTORS

4.1 Introduction

While most of the basic phenomena underlying the gaseous detectors are often well understood by the users, this is not the case with solid-state detectors. The reason is not connected with the complexity of the phenomena in semiconductors but with the fact that most experimental groups in particle physics have had to build their own gaseous detectors and thus became familiar with the underlying physics. Solid-state detectors require high-technology devices built by specialists and appear as black boxes with unchangeable characteristics. The situation is changing since solid-state devices so clearly present perfect features for the solution of pressing problems that some teams of high-energy physicists have started to play with these devices and there is a strong pressure in some laboratories to acquire expensive facilities of the type used by the manufacturers of integrated circuits. An extensive introduction to this subject can be found in recent reviews (101–104).

The solid-state detectors rely on technologies developed for the semiconductor industry. While silicon and germanium present favorable features for particle detection, only silicon allows operation at normal temperature. This is important for devices aiming at the localization of charged particles where the material of the cryogenic tanks would be unacceptable in many cases.

Silicon has four valence electrons. In the crystalline structure each atom is equidistant from the other atoms. Each valence electron is coupled to an electron of a neighboring atom in a covalent bond. In a pure crystal at absolute zero all the electrons are bound and cannot conduct electricity. The valence band is filled. It is separated from the conduction band by an energy gap of 1.1 eV for silicon and 0.7 eV for germanium.

At higher temperatures the thermal energy can break valence bonds, thus allowing electrons to reach the conduction band and become current carriers (thermal generation). The vacancy left behind behaves like a current carrier of opposite sign. An electron from a neighboring atom can fill the hole by breaking its own valence bond and jump over to the first atom. It appears as if the hole has moved. At room temperature the thermal energy is sufficient for a large number of free electrons, of the order of 10^{11} cm^{-3}, to exist in a perfectly pure crystal ("intrinsic semiconductor").

In most applications a controlled amount of impurities is added to the pure crystals ("extrinsic semiconductors"). The impurities are of two types: an element that has three valence electrons (p-type) and one that has five (n-type). The impurity atoms replace semiconductor atoms in the lattice. With a p-type impurity one of the four covalent bonds will not be formed. This makes it easy for an electron in a bond in the neighboring semiconductor atoms to fill this vacant bond, leaving a hole behind it. Common p-type impurities are boron, gallium, and indium. With an n-type impurity the fifth electron is in excess and can easily be removed, becoming free. Typical impurities are antimony, phosphorus, and arsenic.

In intrinsic semiconductors the number of holes equals the number of free electrons (with no applied voltage). In extrinsic semiconductors, at room temperature, the number of current carriers is usually much greater than the number that would be present in the undoped material, but the product of the concentration of electrons n and holes p is a constant $n \cdot p = N_c N_v \exp(-E_g/kT)$, where E_g is the band gap, k is Boltzmann's constant, N_c and N_v are the numbers of allowed energy levels in the conduction band and the valence band, respectively; typically, at 300 K, $N_c = 2.8 \times 10^{19}$ and $N_v = 1.04 \times 10^{19}$, for silicon.

The most important material for charged-particle detection is silicon doped with an n-type impurity. Most of the current is carried by free electrons in the conduction band. They are called the majority carriers. With p-type impurities the holes are majority carriers. Table 1 shows some intrinsic properties of silicon and germanium, at room temperature. It illustrates the advantages of silicon for operation at room temperature.

4.2 The Depletion Region in Rectifying Structures

Particle detection exploits the properties of rectifying structures made of the junction of materials of different types, for instance an n-type and a p-type, which we take as examples.

The charges in an unbiased p–n junction are shown schematically in

Table 1 Some intrinsic properties of silicon and germanium at room temperature

Property	Ge	Si	Unit
Electron mobility	4×10^3	1.4×10^3	$cm^2/V \cdot s$
Hole mobility	1.9×10^3	500	$cm^2/V \cdot s$
Intrinsic resistivity	65	2×10^5	$\Omega \cdot cm$
Energy gap	0.7	1.1	eV

Figure 48. The silicon atoms are not shown. The holes and electrons are free to move. There will be a current generated by the diffusion of holes from the p region to the n region and vice-versa. The electrons and holes recombine. The current stops because each of the regions was initially neutral and the departure of the majority carrier leaves a charge, produced by the fixed ions

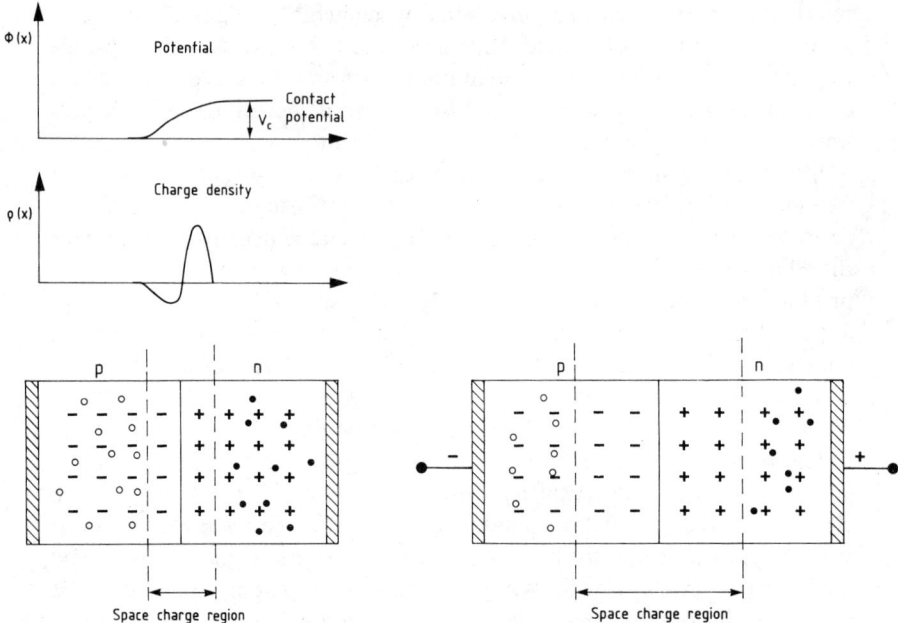

Figure 48 Schematic view of a p-n junction; − and + represent the fixed ionized acceptor and donor atoms. ○ and ● represent the mobile charge carrying holes and electrons. The substrate Si or Ge atoms are not represented. (*a*) No applied external potential. The holes and electrons recombine in the space-charge or depletion region; the extension of this region is prevented by the potential produced by the fixed ions. (*b*) External potential in the reverse direction. More atoms are "uncovered" by the field. The depletion region is extended to a thickness depending on the external field strength.

in the lattice. The p region becomes negative and the n region positive. This build-up of space charge prevents further diffusion. It is only near the junction that the current carriers have disappeared. This region is not electrically neutral; it is called the "space-charge region" or the "depletion region."

If a voltage source is applied across the junction in such a way as to pull away the majority carriers from the junction, more fixed donor and acceptor ions are "uncovered," more space charge is built up, and the potential barrier between the two regions is increased; the only current flowing in such a structure is due to minority carriers and is called the saturation current of the rectifying structure.

When ionizing radiations liberate charges in this depletion region the strong internal electric field drifts them away and produces a detectable current. This current has to be larger than the noise that is due to the fluctuations of the saturation current. This is where silicon presents its main advantage over germanium, at room temperature, because of its high resistivity and its larger energy gap.

The art of construction of the detecting silicon diodes for high-accuracy detectors in high-energy physics consists in having a depletion layer of sufficient thickness for an efficient detection of minimum ionizing particles and a minimum thickness of the dead region necessary to produce the rectifying structure and carry the currents. The materials of the two types can have very different impurity concentrations leading to different resistivities ρ_p and ρ_n; if $\rho_p \ll \rho_n$ the active region extends primarily into the n-type side of the device and is referred to as a p^+n structure. In a p^+n device the charge mobility is essentially the mobility of the electrons, which can be an advantageous feature for the speed of the device. A depleted region can also be obtained by the junction of an n-type crystal with a metal, such as a thin layer of gold (Schottky diode). Silicon has the advantage of growing a natural oxide that serves as an insulator, as a passivation layer, and also as a diffusion mask. The various types of detecting diodes are described in (101, 105). Figure 49a shows a practical structure with the various manufacturing steps only to give an idea of the complexity of the construction.

The width X_D of the depletion layer depends on the doping density and the applied bias voltage V_B:

$$X_D = [2\varepsilon\mu_e\rho(V_0 + V_B)]^{1/2}. \qquad 9.$$

V_0 is the voltage created by the space charge in the junction, ε is the permittivity ($\varepsilon = 1.054 \times 10^{-12}$ F/cm), μ_e is the electron mobility.

The resistivity of n-type silicon can be 20 k$\Omega \cdot$ cm and a depletion layer of 1 mm thickness can be obtained with $V_B = 300$ V. A depletion layer X_D has

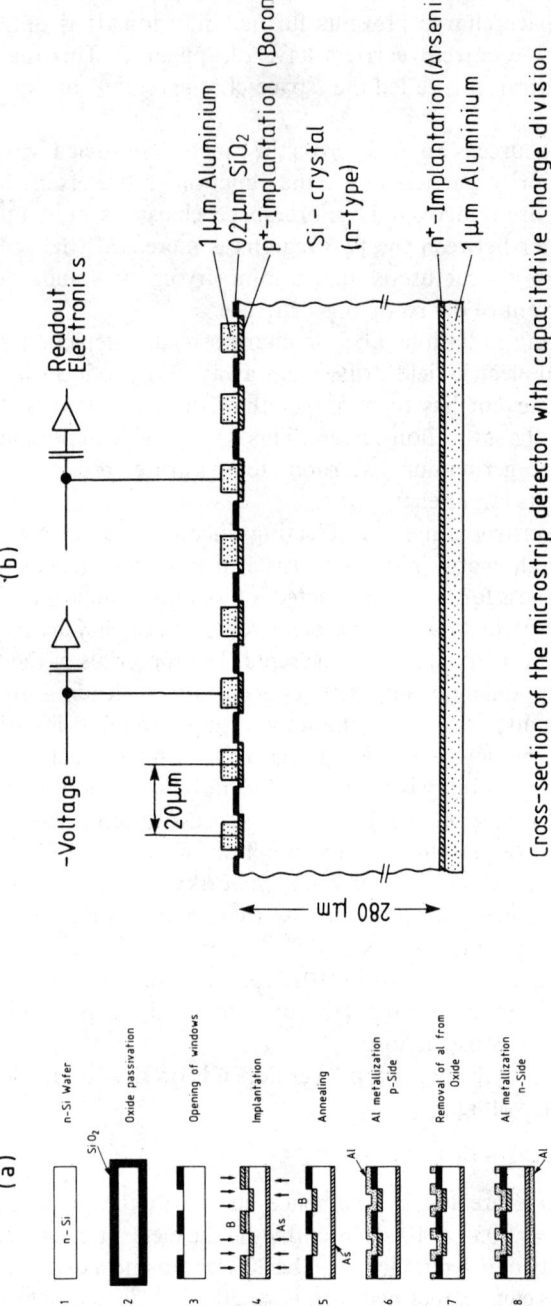

Figure 49 Successive steps of the manufacturing process of passivated ion-implanted silicon detectors. (*a*) The manufacturing steps. (*b*) Details of the final product (103).

a capacitance, per unit area, of $C_D = \varepsilon/X_D$. Above a certain voltage the capacitance becomes constant and the complete thickness of silicon is depleted, the electrical field extending from the rectifying contact to the ohmic rear contact. This is the situation reached in most detectors, which then behave like a solid ionization chamber.

The typical detecting thicknesses vary between 20 μm in charge-coupled devices (CCDs) and a few hundred μm in silicon detectors. With liberated charges of 80 electron–hole pairs per μm and, as we discuss below, with structures allowing for an energy sharing between several electrodes, the signals to detect can vary between a few hundred electron charges to a thousand times more. The various sources of noise, intrinsic to the solid-state detector, vary from a few tens of electron charges to a few thousands in silicon detectors and can relatively easily be made negligible compared to the signals. We refer the reader to the articles discussing the various sources of noise (101–105) and the matching of the external read-out electronics. The main source of noise is due to the statistical fluctuations in the number of carriers, leading to changes in conductivity, and it is proportional to the current flow:

$$\langle i^2 \rangle = 2qI\Delta f, \qquad \qquad 10.$$

where I is the dark current, q is the charge of the carrier, and Δf is the bandwidth. Optimizing the electronics as discussed by Radeka (in 101), and making the capacitance C_d of the detector equal to the input capacitance C_a of the amplifier, leads to an equivalent noise charge ENC,

$$\text{ENC} \simeq 8kTC_d/f_T\lambda, \qquad \qquad 11.$$

where f_T is the frequency for unity gain of the amplifier and λ is the time constant of the filter. For typical values of $f_T = 1$ GHz and $\lambda = 100$ ns, ENC $\simeq 3 \times 10^2$ [$C_d^{1/2}$, C_d being in pF] rms electron charges. For $C_d = 10^{-2}$ pF, which is more than the capacitance of the detecting elements of a CCD, ENC is only $30e$; for $C_d = 1$ pF, ENC is equal to $300e$; and even for parasitic capacitances of 100 pF, ENC $\simeq 1000e$. With present amplifiers, where the input noise is of the order of $1000e$, the contribution of the solid-state detector is not the dominant factor. And it is easy, with the various structures, to obtain charge signals from minimum ionizing particles larger than the noise. In many applications, however, where high accuracy is desirable, the thickness is also limited by the errors introduced by multiple scattering. This contribution is very dependent upon the geometry of the experiment. With a radiation length of 9 cm, the most promising silicon detectors, with a thickness of about 300 μm, do not contribute much more than gaseous detectors if they are operated in vacuum.

The finite thickness leads also to another source of error. While the initial column of charges produced by the ionizing particle is estimated to have an extension of less than 1 μm, it broadens by diffusion when the charges are drifting to the electrodes. For silicon and a drift length of 1 mm, the diffusion leads to a spread of about 30 μm (FWHM).

4.3 Position-Sensitive Detectors

There are many concepts for position sensing with silicon detectors, of which we mention five:

1. Charge transfer from the whole detector with current division to localize the initial charge deposit. The energy loss of minimum ionizing particles in thin layers of silicon, say 300 μm, liberates about 25,000 charges. For a surface of a few cm^2, with a capacitance around 100 pF and a high-resistivity electrode, the total noise is equivalent to 50 keV. This does not allow accuracies in the micrometer range to be reached over a length of the order of cm, and the method is used only for particles of a few MeV stopping in the detector where 1% accuracy is possible.
2. Arrays of narrow strips, read out individually or using interpolation methods to reduce the number of strips.
3. Charge-coupled devices where a mosaic of small read-out elements is used to localize the ionizing event.
4. Charge storage in P–I–N diodes.
5. Silicon drift chambers.

We describe these different approaches, which are, at present, at quite different levels of development.

4.4 Microstrip Detectors

The breakthrough for the high accuracy was the demonstration in 1983 by Hyams et al (105) that resolutions of 5 μm rms can be achieved with strips 20-μm wide, read out every 60 μm, while a resolution of 8 μm was achieved with a read-out of 120 μm, exploiting capacitive charge division between adjacent strips. A large-scale experiment has confirmed the operational character of this device (105–108). Figure 49b shows the detailed structure of the detector. The sensitive area of the counter is a rectangle of 24 mm × 36 mm with 1200 strips of 12 μm × 36 mm and 20-μm pitch.

A relativistic particle of charge one traversing the detector produces ~25,000 electron-hole pairs, which are collected within <10 ns at the electrodes. The intermediate strips are kept at the potential of the read-out strip. The impedance between read-out strips is much greater than the input impedance of the electronics in order to avoid cross-talk. The interstrip capacitance is greater than the strip to ground capacitance. The charge collected at intermediate strips can be divided among the neighboring read-out strips. The position of the impinging particles is found by computing the center of gravity of the collected charges.

Figure 50 shows the strip layout of the detector. The results obtained with such a detector are illustrated by Figure 51, which shows that a resolution of 4.5 μm can be achieved. A two-particle separation of $\sim 120\ \mu$m has been obtained. Detectors of this type have now been employed for several years in an experiment and constitute one of the best candidates for detectors in this range of accuracy. It should be mentioned that the width of the charge distribution arriving at the strip is approximately 6 μm and that one would expect the charges to be collected by no more than 2 strips. However 1/3 of all clusters originating from a single particle involve at least 3 strips. This effect, spoiling the particle separation, is so far unexplained. The counting rate achieved was 10^6 per second per strip, limited only by the electronics.

Belau et al (106) have studied the behavior of such a detector as a function of applied voltage and magnetic field. Their results are summarized in Figure 51. A simple model permits the prediction of the observed spatial resolution. A magnetic field perpendicular to the direction of the drifting charges and parallel to the read-out strip increases the average number of the collecting strips, as expected from the effect of the Lorentz force. A field of 1.68 T shifts the measured coordinate by an amount of the order of 10 μm and increases the width of the collected charge from 5 to about 12 μm. The accuracy of 5 μm is not considered the lowest limit by Belau et al. Fluctuations in the center of gravity of the deposited charge, of the order of only 1 μm, are expected from delta electrons. These authors calculate that with a read-out pitch of 20 μm an accuracy of 2.8 μm could be reached, with the diode arrangement on the strip side. This has to be compared with the observed resolution of 4.5 μm, with the read-out of 60 μm, while their model

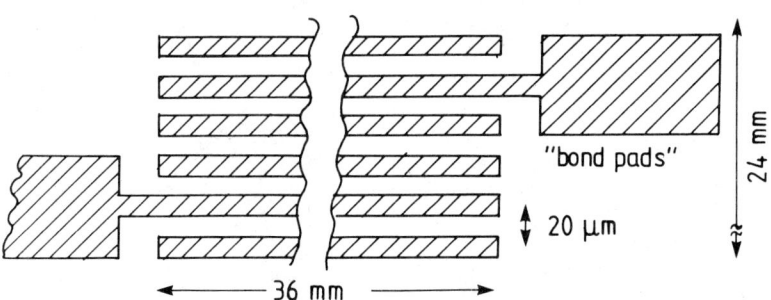

Strip pattern of the microstrip detector

Figure 50 Strip pattern of a microstrip detector (105).

gave 3.6 μm. With a read-out pitch of 20 μm and a diode arrangement opposite to the strips, their model gives an accuracy of 0.8 μm, i.e. the limit of accuracy where the delta electrons enter into the game.

The understanding of the microstrip silicon detectors and the experience acquired in their operation give support to the estimate that they are at present good candidates for accurate measurements of relativistic particles in the range of 5 μm and probably down to 2 μm.

It should be stressed that the possibility of individual read-out of strips spaced by 20 μm, as demonstrated by Belau et al (106), which leads to the same accuracy as that reached so far by capacitive read-out, is an attractive

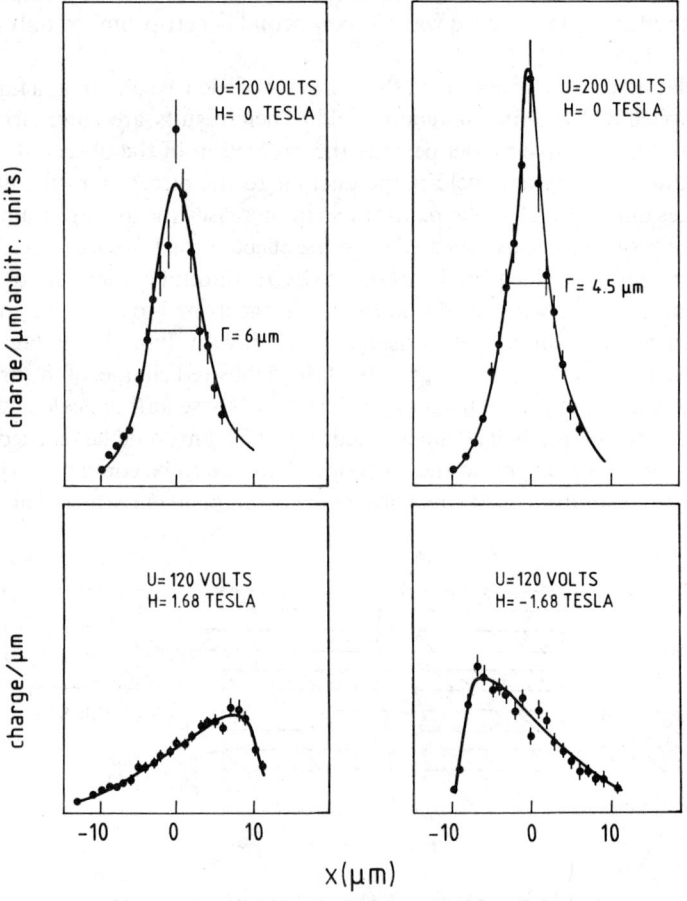

Figure 51 Position resolution of a microstrip detector, as a function of voltage and magnetic field: U = applied voltage, H = magnetic field. The magnetic field is parallel to the strip (106).

solution for the future. The shaped pulse width can be of only 50 ns at the basis, against 800 ns with current division, thus permitting higher rates. This fast response is of value for applications where the microstrip chambers are used for the triggering on preselected multiplicities. This is permitted by the resolution, close to 20 keV (FWHM), for the energy loss of minimum ionizing particles (109). The two particles' resolution can be 40 μm, which is also a very attractive feature for high-multiplicity reactions.

Large-scale integration of the associated electronics on the same chips as the microstrips would make a considerable difference in the prospects for a generalization in the use of this technique. The present ratio of the area of the electrons to the area of the detector is 300, while with a suitable integration it could possibly reach one. Such a development could permit the construction of the moderate surfaces needed for the vertex detectors close to the interaction regions at storage rings. Figure 52 shows a detector planned for LEP, with only a partial integration of the electronics (108).

The commercial cost of a microstrip detector is only $1000 for an area of about 5 cm². The main cost is in the associated electronics and large-scale

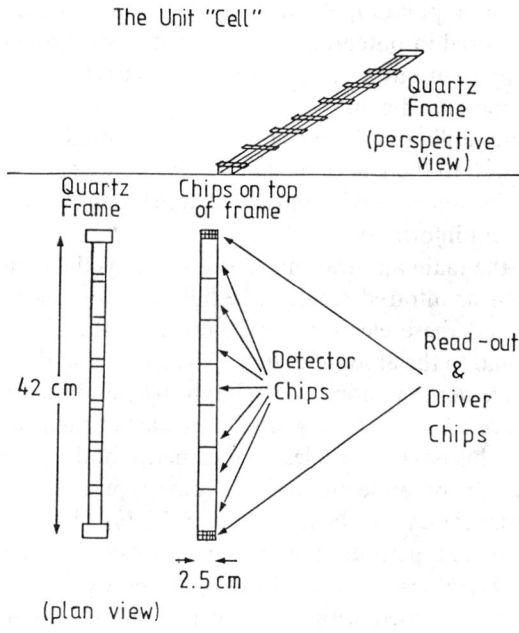

Figure 52 Projected large-surface microstrip detector. A combination of the unit "cell" made of microstrip detectors should permit the construction of high-accuracy vertex detectors for colliding beams. Part of the electronics is integrated on the detecting chips (108).

integration of the whole device could make it very attractive. Present technology could permit the construction of chips of 7×7 cm^2.

The radiation damage could be a worry for some applications. It is very dependent upon the nature of the radiation involved. Sizeable changes take place at silicon surface barriers irradiated on the front surface by fast electrons at above 10^{14} cm^{-2}, 10^{13} cm^{-2} for protons, 10^{11} cm^{-2} for α particles, about 3×10^8 cm^{-2} for fission fragments and x-ray doses of about 10^4 Gy (103).

Passivated, ion-implanted, silicon junction detectors were exposed to doses of high-energy protons at CERN up to fluences of 8.3×10^{13} particles cm^{-2}, and a considerable increase in leakage current was observed. Nevertheless full charge collection can be maintained by linear increase of the electric field with the fluence (110). Also, for extreme cases, different methods of manufacture or choice of the silicon resistivity offer possibilities to decrease the sensitivity to radiation damage.

4.5 Charge-Coupled Devices

The CCD is, like the silicon diode, a concept of semiconductor electronics. However, it does not rely mainly on the modulation of electrical currents but on the manipulation of information. This information is in the form of electrons stored in potential wells, located in very small volumes of a few μm^2 of apparent surface, constituting a matrix of tens of thousands of elements per silicon chip.

Charge-coupling is the collective transfer of all the charge stored in a potential well to the adjacent similar element, by external manipulation of voltages. The quantity of the stored charge is conserved in this transfer and can represent information.

One of the main applications of this device is the imaging of light in the visible or near infrared region. The light liberates electrons in the silicon elements and these electrons are trapped in the wells, the charge being proportional to the amount of light. The transfer of the charges to external read-out elements is done very simply by manipulating collectively the potentials of a few electrodes, which are part of each storage element. The interest of this device for television cameras, and its extension to particle detection, can be understood from some typical characteristics: (a) The quantum efficiency can be as high as 80%. (b) The low capacity for each storage element permits a very low intrinsic noise of the transferable charges. Overall rms noise of 10 electrons can be achieved. Excellent pictures are obtained with CCDs where charges as small as about 100 electrons each are transferred through a centimeter of silicon.

This is not the only application of CCDs. They can be used as analogue memory elements. The charges are then introduced through diodes and gates and they then operate as discrete delay lines, with the timing

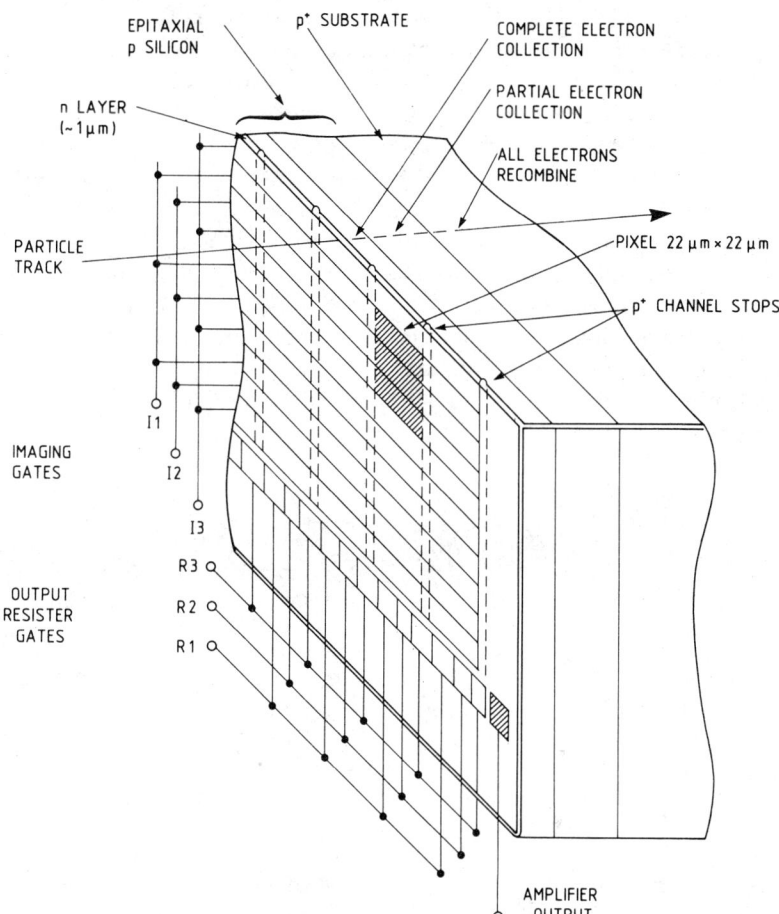

Figure 53 A CCD structure. A corner of the CCD enlarged to show details of the pixel (storage element) structure (111).

properties controlled by the clocking potential applied to the transferring electrodes.

Before discussing the use of CCDs for particle detection (111, 112) a summary description of their physical structure may be appropriate. This device illustrates beautifully the huge potential for high-accuracy particle detection that exists in solid-state electronics.

Figure 53 shows the structure of one type of CCD used by Damerell et al.[1] The CCD is fabricated on an epitaxial p-type layer (dopant concentration 5×10^{14} cm^{-3}) of thickness 25 μm. This layer is grown on a p$^+$

---
[1] Charge Coupled Device P 8600, English Electric Valve Co. Chelmsford, England.

substrate (dopant 5×10^{18} cm^{-3}), which is inactive from the point of view of particle detection. The surface of the epitaxial silicon is converted, by ion implantation, to an n-type, with dopant concentration of 10^{16} cm^{-3} to a depth of approximately 1 μm. Above a thin oxide layer are deposited transparent electrodes, called the imaging gates, which are insulated from one another and form the substrate. Holding the I_2 gates high (~ 10 V) and the I_1 and I_3 low (0 V) creates a matrix of potential wells (pixels) with minima near the upper interface (buried channel), defined in the x direction by the I_2 gates and in the y direction by narrow p$^+$ implants. The CCD has 576×385 such pixels in a sensitive area of 12.7 mm × 8.5 mm. With an interval between pixel centers of 22 μm in the x and y directions, the CCD is essentially a very precise rectangular matrix of potential wells that act to trap electrons. Figure 54 shows a typical distribution of potentials in a CCD

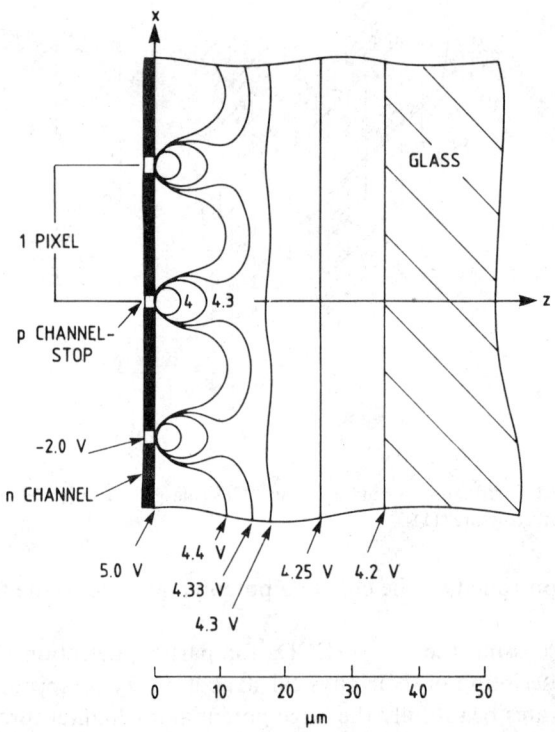

Figure 54 Map of the electrostatic potentials in the CCD substrate when all gate voltages are held to zero (109). The electrons generated in the depletion region diffuse to the n channel. Their drift is limited in the x direction by the p$^+$ stops. In the y direction the spread among several channels improves the accuracy of the centroid-finding methods. The application of a pulse of a few volts to the gates I_2 freezes the charges inside the pixels.

slightly different from that represented in Figure 53 and the first investigated by Damerell et al (111, 112).

At room temperature the potential wells rapidly fill with electrons, owing to thermal generation. This is reduced to a negligible level at moderately low temperatures, which require the CCD to be kept in a cryostat in thermal contact with a bath of liquid nitrogen. In the applications to optical imaging the read-out of the stored image is obtained by shifting the stored electrons in the y direction, from I_2 to I_3, I_3 to I_1, I_1 to I_2 (for a three-phase CCD) [Figure 55, from (113)]. At this step the charges have moved by one pixel in the x direction. Those charges that were in the bottom row are shifted to the R register, which provides 385 R triplets to read out the first I triplet, by moving the charges, on the same principle, in the orthogonal direction, sideways. The charge in each pixel is sensed at the output by an analog circuit. Once the first row has been read out, there follows another I triplet and 385 R triplets as the second row is read. This system has been invented to read out an optical image where most of the pixels have a stored bucket and for speeds of about 25 images per second.

When the particle ionizes the depletion region, the initial column of liberated electrons, of about 1 μm thickness, starts diffusing toward the

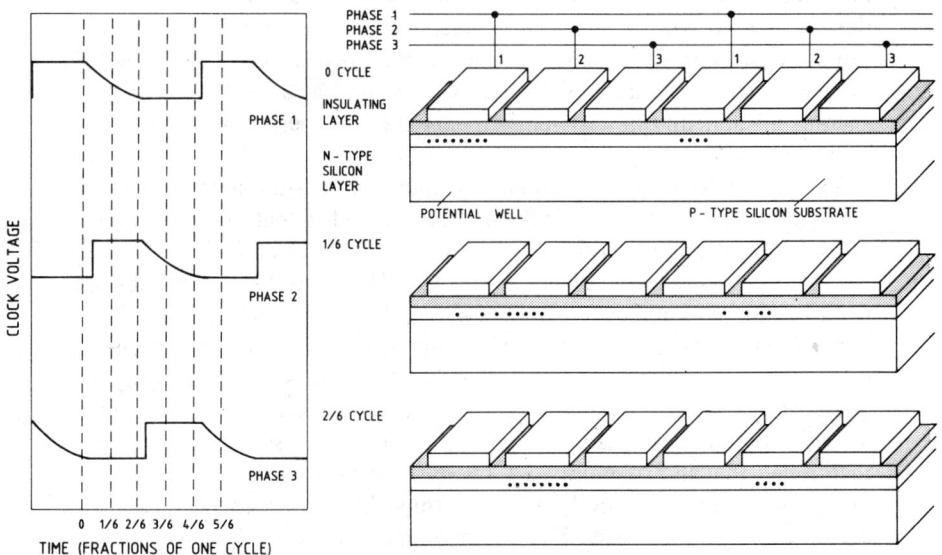

Figure 55 The read-out of a three-phase CCD. A full cycle of the clock voltages represented at the left shifts the electrons by one pixel (3 strips). The electrons in the end row are moved to the R registers (Figure 53), where the electron buckets are moved sideways, on the same principle, and their contents are measured, one after the other (113).

region of minimum potential, which is part of the depletion region closest to the I gates. It is collected in a time of the order of 30 ns, during which it diffuses transversely by about 10 μm. The diffusion is stopped in the x direction by the p$^+$ stops.

In the y direction the electrons would continue to diffuse laterally if all the I gates were at the same potential. The initial idea of the first investigators of the CCD was to apply a pulsed voltage on every third I gate, say I_2, 100 ns after having obtained a signal from auxiliary counters detecting an interesting event. This would have frozen the electrons in a matrix of potential wells which could then only be read out by the conventional potential manipulation of the I gates. Because of the diffusion the electrons would be distributed among several elements of this matrix in the y direction, thus permitting an accurate measurement of the charge centroid. In order to avoid beam tracks being registered during the read-out procedure, a fast kicker magnet switching off the beam was foreseen.

The experience gained during the study of this method has led to another approach. The I_2 gates are polarized as soon as the accelerator is active and the potential matrix is then permanent. The events are clocked out continuously, at a rate of 3 MHz, to the R registers. For the most remote events this leads to a clearing time of a few hundred μs during which accidental beam tracks traverse the detectors. However, by having a second device back to back with the first one, a precise geometrical relation is imposed on every track. The great number of pixels, close to 0.25×10^6, compensates for the lack of time resolution, and at a beam rate of 10^5/s there is no difficulty in tagging the beam tracks superimposed on a good event.

The sensitivity of the system is very high. A minimum ionizing particle generates 1300 electrons while the noise is only a few tens of electrons. The pulse-height distribution from x-rays of 5.9 keV shows a 12% (FWHM) resolution (Figure 56a). Figure 56b shows the pulse-height distribution from minimum ionizing particles for which an efficiency of 98 ± 2% is reached with a spatial accuracy of 4.3 μm and 6.1 μm in two orthogonal directions and a two-track resolution of 40 μm in space (with only 2% overlap). The authors claim that a spatial precision better than 2 μm can be achieved by fabricating the CCDs on high-resistivity silicon. It should be mentioned that in astronomy, CCDs have permitted the tracking of star images with an accuracy of 0.2 μm, thus illustrating the considerable usefulness of this device. The main advantage of CCDs over silicon microstrip detectors is their better two-track resolution and the two-dimensional read-out from every plane. They are much more limited in counting rate.

This work illustrates beautifully the fact that semiconductor technology

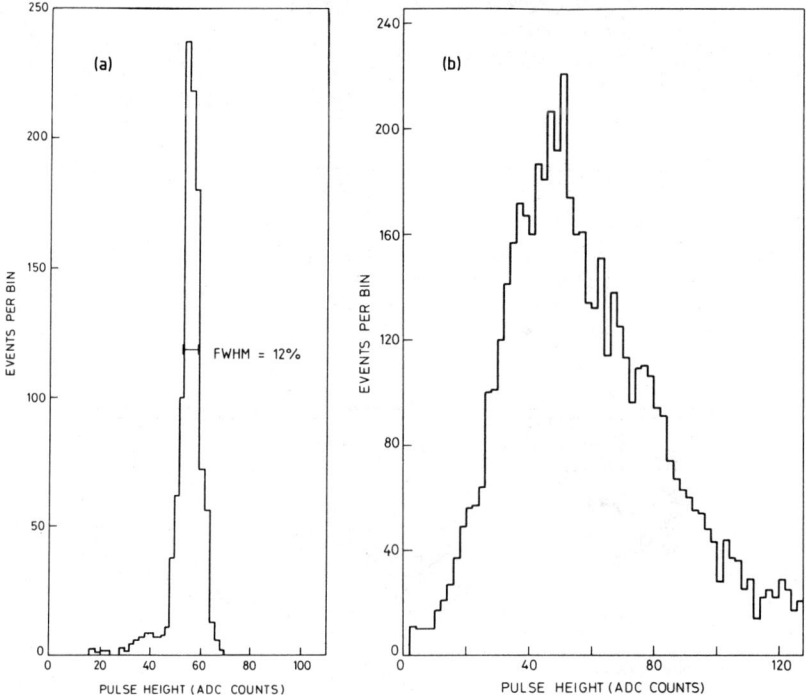

Figure 56 Energy resolution of a CCD. (*a*) Pulses from 5.9-keV x rays. (*b*) Energy loss of a minimum ionizing particle (111).

has the potential properties for high-accuracy, high-resolution detection of charged particles. However, because of the considerable investment in money required for the development of this technology, physicists have so far been forced to use commercially available chips, even if they do not perfectly fit their needs. The time is ripe for the elaboration of specific systems.

4.6 *Detection by Charge Storage in Shallow Levels in Semiconductors*

In 1978 it was observed in so-called P–I–N diodes, which in our notation is a pnn^+ structure, that charges liberated by ionizing particles can be stored and subsequently released by the application of an electric field (Figure 57) [Shepard, in (101)]. This was interpreted as being due to the capture of the carriers by the ionized impurities. The storage charge could be released by applying an electric field of a few $V/\mu m$. Figure 58 illustrates the process. The *I* region is silicon-doped with an n-type impurity. The trapped charges

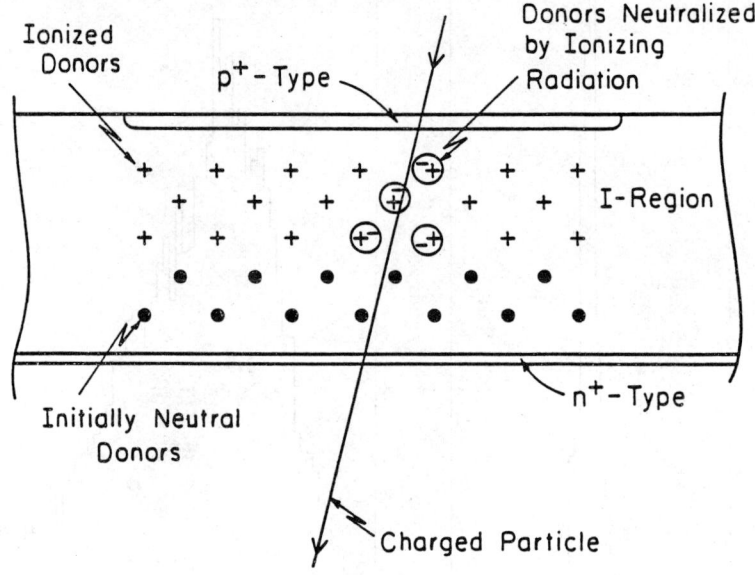

Figure 57 Schematic view of P–I–N structure suitable for cryogenic charge storage produced by ionizing radiation. [From Shepard, in (101) page 73.]

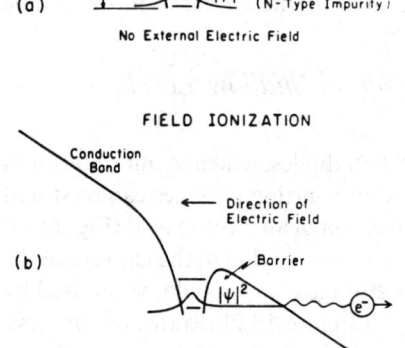

Figure 58 Schematic view of the potential energy levels and a ground-state wave function near a donor impurity site in (*a*) the absence of an external electric field and (*b*) the presence of an electric field of sufficient strength to produce field ionization by quantum mechanical tunnelling. [From Shepard, in (101) page 73.]

are tunnelled out from their bound levels by the applied electrical field (Figure 58b). It was established that at 4.2 K the trapping lifetime is $\gg 10^5$ s.

Under the influence of an electric field at 2 V/μm the charge is extracted in times of about 10 ps from phosphorus impurities in silicon. The tests were performed in commercial diodes with a thickness of 50 μm for the I region doped with phosphorus at a concentration of $\sim 10^{13}/\text{cm}^3$. The authors estimate that the capture occurs within a characteristic distance of the primary ionizing particle of only 1 μm, which sets the ultimate accuracy limit.

The preliminary test permitted successfully trapping and releasing of ionization electrons, the best results being obtained with a diode of 5×10^{-4} cm^{-2}. The authors extrapolate these observations into the conception of structures capable of exploiting the intrinsic properties into the few μm range of accuracies. The read-out relies on a strip structure, of 10 μm width on the n$^+$ side, in a diode, of a total width 0.5 mm. Each strip surface is only 10 times smaller than the tested commercial diode, and the calculations show that the accuracy should be about 3 μm in one dimension, with a perfect separation between signals from minimum ionizing particles and from the noise.

4.7 The Silicon Drift Chamber

The silicon drift chamber, introduced by Gatti & Rehak (114), relies on the idea that a flat pnp structure can be depleted from the side over a long distance (Figure 59a).

Until complete depletion is reached, the undepleted region around the median plane is a conductor at the potential of the side electrode. Thus the whole volume of the wafer is depleted at the voltage that would be required

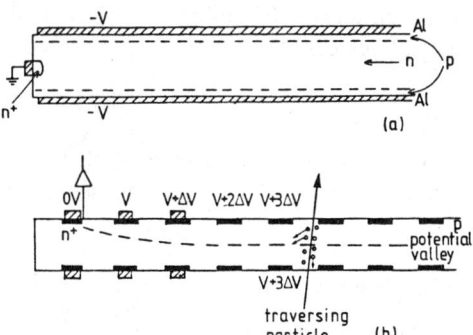

Figure 59 The silicon drift chamber (114). (a) Principle. The whole volume is depleted except the median plane, which acts as a conductor at the potential of the side electrode. (b) Practical scheme with a sidewise superimposed drift field.

to deplete a wafer of half this thickness constructed as a conventional planar diode.

The positive charges left after free electrons are removed by depletion give rise to a potential minimum in the median plane. Figure 59b shows how one may construct a practical drift chamber by introducing strip electrodes to give a transverse field. Electrons liberated by a charged track will drop to the potential minimum and then drift to the positive electrode; collection times in the range 100 ns to 10 μs can be envisaged. Preliminary tests of this device have been reported; it can aim at the same accuracy as the microstrip detectors, with much simpler electronics and mechanics.

5. CONCLUSION

We have reviewed the efforts being undertaken to improve the accuracy recently achieved in tracking high-energy particles. The great flexibility and relative ease of construction of gaseous detectors has encouraged a sizeable effort in this field. It seems, however, that in practical cases it will be hard to achieve accuracies better than 30–40 μm. This is compensated partly by the fact that multisampling along a track is easy.

Solid-state detectors offer considerable potentialities and two satisfactory structures emerge: (a) the multistrip detectors, which have demonstrated a one-dimensional accuracy of 5 μm (rms) with 120 μm multitrack resolution; and (b) the CCDs, which offer an accuracy of 5 μm, with a multiparticle resolution of 40 μm. The first device permits high counting rates and has been tested up to 10^6 counts/s on a single strip. The second has considerable limitations connected with its read-out system but has been used successfully up to about 10^5 counts/s · cm². These two devices can still be improved. The other approaches that we mentioned illustrate the considerable potentialities of solid-state electronics and may well produce new practical devices.

ACKNOWLEDGMENT

The authors are grateful to Drs. C. Damerell, B. D. Hyams, and P. G. Rancoita, for illuminating comments on the last section of this article.

Literature Cited

1. Lapique, F., Piuz, F. 1980. *Nucl. Instrum. Methods* 175:247
2. Franzen, W., Cochran, L. W. 1971. In *Nuclear Instruments and Their Use*, ed. A. H. Snell, p. 1. New York: Wiley
3. Christophorou, L. G. 1971. *Atomic and Molecular Radiation Physics*. New York: Wiley
4. Kanter, H. 1961. *Phys. Rev.* 121:461
5. Talman, R. 1979. *Nucl. Instrum. Methods* 159:198
6. Allison, W. W. M., Cobb, J. H. 1980. *Ann. Rev. Nucl. Part. Sci.* 30:253
7. Hancock, S., James, F., Movchet, J., Rancoita, P. G., Van Rossum, L. 1983. *Phys. Rev. A* 28:615

8. McDaniel, E. W., Mason, E. A. 1973. *The Mobility and Diffusion of Ions in Gases*. New York: Wiley
9. Sauli, F. 1983. In *Techniques and Concepts of High Energy Physics*, ed. T. Ferbel, Vol. 11, p. 301. New York: Plenum
10. Brown, S. C. 1959. *Basic Data of Plasma Physics*. New York: Wiley
11. Brown, S. C. 1966. *Basic Data of Plasma Physics*. Cambridge, Mass: MIT Press
12. Varney, R. N., Fisher, L. H. 1968. In *Methods of Experimental Physics*, ed. L. Marton. New York: Academic
13. Huxley, L. G., Crompton, R. W. 1974. *The Diffusion and Drift of Electronics in Gases*. New York: Wiley
14. Fehlmann, J., Viertel, G. 1983. *Compilation of Data for Drift Chamber Operation*. Zurich: ETH
15. Loeb, L. B. 1961. *Basic Processes of Gaseous Electronics*. Berkeley: Univ. Calif. Press
16. McDaniel, E. W. 1964. *Collision Phenomena in Ionized Gases*. New York: Wiley
17. Massey, H. S. W., Burhop, E. H. S. 1969. *Electronic and Ionic Impact Phenomena*. Oxford: The Univ. Press
18. Palladino, V., Sadoulet, B. 1975. *Nucl. Instrum. Methods* 128:323
19. Schultz, G., Gresser, J. 1978. *Nucl. Instrum. Methods* 151:413
20. Mathieson, E., el Hakeem, N. 1979. *Nucl. Instrum. Methods* 158:489
21. Ramanantsizehena, P. R. 1979. Thesis and Rep. *No. CRN-HE 79-13*. Strasbourg: Univ. Louis-Pasteur
22. Villa, F. 1983. *Nucl. Instrum. Methods* 217:273
23. Parker, J. H., Lowke, J. J. 1969. *Phys. Rev.* 181:302
24. Drumm, H., Ganz, B., Heintze, J., Heinzelmann, G., Heuer, R. D., et al. 1980. *Nucl. Instrum. Methods* 176:333
25. Piuz, F. 1982. *Nucl. Instrum. Methods* 205:425
26. Farilla, A., Sauli, F., Ropelewski, L. 1983. Search for a low drift velocity, low diffusion gas. Unpublished
27. Bobkov, S., Cherniatin, V., Dolgoshein, B., Evgrafov, G., Kalinovsky, A., et al. 1983. Preprint CERN-EP/83-81. Submitted to *Nucl. Instrum. Methods*
28. Charpak, G., Bouclier, R., Bressani, T., Favier, J., Zupančič, Č. 1968. *Nucl. Instrum. Methods* 62:235
29. Charpak, G. 1970. *Ann. Rev. Nucl. Sci.* 20:195
30. Rice-Evans, P. 1974. *Spark, Streamer, Proportional, and Drift Chambers*. London: Richelieu
31. Sauli, F. 1977. *Rep. CERN 77-09*. Geneva: CERN
32. Charpak, G., Sauli, F. 1979. *Nucl. Instrum. Methods* 162:405
33. Fabjan, C. W., Fischer, H. G. 1980. *Rep. Prog. Phys.* 43:1003
34. Walenta, A. H. 1983. *Nucl. Instrum. Methods* 217:65
35. Charpak, G., Rahm, D., Steiner, H. 1970. *Nucl. Instrum. Methods* 80:13
36. Piuz, F., Roosen, R., Timmermans, J. 1982. *Nucl. Instrum. Methods* 196:451
37. Fischer, H. G., Pech, J. 1972. *Nucl. Instrum. Methods* 100:515
38. Lee, D. E., Sobottka, S. E., Thiessen, H. A. 1972. *Nucl. Instrum. Methods* 104:179
39. Endo, I., Kawamoto, Y., Mizuno, Y., Ohsugi, T., Taniguchi, T., Takeshita, T. 1981. *Nucl. Instrum. Methods* 188:51
40. Gatti, E., Longoni, A., Okuno, H., Semenza, P. 1979. *Nucl. Instrum. Methods* 163:83
41. Fancher, D. L., Schaffer, A. C. 1979. *IEEE Trans. Nucl. Sci.* NS-26:150
42. Breskin, A., Charpak, G., Demierre, C., Majewski, S., Policarpo, A., et al. 1977. *Nucl. Instrum. Methods* 143:29
43. Charpak, G., Melchart, G., Petersen, G., Sauli, F. 1979. *Nucl. Instrum. Methods* 167:455
44. Charpak, G., Petersen, G., Policarpo, A., Sauli, F. 1978. *IEEE Trans. Nucl. Sci.* NS-25:122
45. Charpak, G., Petersen, G., Policarpo, A., Sauli, F. 1978. *Nucl. Instrum. Methods* 148:471
46. Bondar, A. E., Onuchin, A. P., Penin, V. S., Telnov, V. I. 1983. *Nucl. Instrum. Methods* 207:379
47. Charpak, G., Sauli, F. 1973. *Nucl. Instrum. Methods* 113:381
48. Fisher, J., Okuno, H., Walenta, A. H. 1978. *Nucl. Instrum. Methods* 151:451
49. Okuno, K., Fisher, J., Radeka, V., Walenta, A. H. 1979. *IEEE Trans. Nucl. Sci.* NS-26:160
50. Harris, T. J., Mathieson, E. 1978. *Nucl. Instrum. Methods* 154:183
51. Mathieson, E., Harris, T. J. 1979. *Nucl. Instrum. Methods* 159:483
52. Kochelev, N. I., Telnov, V. I. 1978. *Nucl. Instrum. Methods* 154:407
53. Erskine, G. A. 1982. *Nucl. Instrum. Methods* 198:325
54. Saudinos, J. 1970. *Proc. Topical Seminar on Interactions of Elementary Particles on Nuclei*, p. 313. Trieste: INFN
55. Saudinos, J., Duchazeaubeneix, J. C., Laspalles, C., Chaminade, R. 1973. *Nucl. Instrum. Methods* 111:77
56. Walenta, A. H., Heintze, J., Schürlein, B. 1971. *Nucl. Instrum. Methods* 92:373

57. Allison, W. W. M., Brooks, C. B., Bunch, J. N., Cobb, J. H., Lloyd, J. L., Pleming, R. W. 1974. *Nucl. Instrum. Methods* 119:499
58. Clark, A. R., et al. 1976. Proposal for a PEP facility based on the Time Projection Chamber, PEP-4. Stanford, Calif: SLAC
59. Nygren, D., Marx, J. 1978. *Phys. Today* 31, No. 10
60. Charpak, G., Sauli, F., Duinker, W. 1973. *Nucl. Instrum. Methods* 108:413
61. Breskin, A., Charpak, G., Gabioud, B., Sauli, F., Trautner, N., et al. 1974. *Nucl. Instrum. Methods* 119:9
62. Breskin, A., Charpak, G., Sauli, F., Atkinson, M., Schultz, G. 1975. *Nucl. Instrum. Methods* 124:189
63. Filatova, N., Nigmanov, T., Pugachevich, V., Riabtsov, V., Shafranov, M., et al. 1977. *Nucl. Instrum. Methods* 143:17
64. Farr, W., Heintze, J., Hellenbrand, K. H., Walenta, A. H. 1978. *Nucl. Instrum. Methods* 156:283
65. Baskakov, V. Y., Chernjatin, V. K., Dolgoshein, B. A., Lebedenko, V. N., Romanjuk, A. S., et al. 1979. *Nucl. Instrum. Methods* 158:129
66. Charpak, G., Majewski, S., Sauli, F. 1975. *Nucl. Instrum. Methods* 126:381
67. Mathis, K. D., Simon, M., Henkel, M. 1983. *Siegen Univ. Preprint SI-83-19*
68. Anderson, D., Charpak, G. 1982. *Nucl. Instrum. Methods* 201:527
69. Chernyatin, V. K., Dolgoshein, B. A., Golutvin, I. A., Kaftanov, V. S., Kalinovskii, A. N., et al. 1979. *Proc. Second ICFA Workshop, Les Diablerets, 1979*, p. 320. Geneva: CERN
70. Barranco Luque, M., Calvetti, M., Dumps, L., Girard, C., Hoffmann, H., et al. 1980. *Nucl. Instrum. Methods* 176:175
71. Va'vra, J. 1983. *Nucl. Instrum. Methods* 217:322
72. Va'vra, J. 1983. *Stanford preprint SLAC-PUB-3131* (Presented at the 2nd Pisa Meeting on Advanced Detectors, Castiglione, 1983). Stanford, Calif: SLAC
73. Bettoni, D., Dolgoshein, B., Evans, M., Fabjan, C. W., Hoffman, H., et al. 1983. In *Proc. Int. Conf. on High Energy Physics, Brighton, 1983*, p. 424
74. Walenta, A. H. 1979. *IEEE Trans. Nucl. Sci. NS-26*:73
75. Vertex chamber group (Aachen-Siegen-Zurich), LEP experiment L3, 1983 (private communication from G. Viertel)
76. Raether, H. 1964. *Electron Avalanches and Breakdown in Gases*. London: Butterworth
77. Peisert, A. 1983. *Nucl. Instrum. Methods* 217:229
78. Hilke, H. 1983. *Nucl. Instrum. Methods* 217:189
79. Peisert, A., Charpak, G., Sauli, F., Viezzoli, G. 1984. *IEEE Trans. Nucl. Sci. NS-31*:125
80. Hempel, G., Hopkins, F., Schatz, G. 1975. *Nucl. Instrum. Methods* 131:445
81. Jared, R. C., Glaessel, P., Hunter, J. B., Moretto, L. G. 1978. *Nucl. Instrum. Methods* 150:597
82. Harrach, D. V., Specht, H. J. 1979. *Nucl. Instrum. Methods* 164:477
83. Breskin, A., Zwang, N. 1977. *Nucl. Instrum. Methods* 146:461
84. Van der Plicht, J. 1980. *Nucl. Instrum. Methods* 171:43
85. Fabris, D., Fortuna, G., Gramegna, F., Prete, G., Viesti, G. 1983. *Nucl. Instrum. Methods* 216:167
86. Charpak, G., Sauli, F. 1978. *Phys. Lett.* 78B:523
87. Breskin, A., Charpak, G., Majewski, S., Melchart, G., Peisert, A., et al. 1979. *Nucl. Instrum. Methods* 161:79
88. Bouclier, R., Charpak, G., Cattai, A., Million, G., Peisert, A., et al. 1983. *Nucl. Instrum. Methods* 205:403
89. Adams, M., et al. 1983. *Nucl. Instrum. Methods* 217:237
90. Charpak, G., Sauli, F. 1983. *Preprint CERN-EP/83-128* (Presented at 2nd Pisa Meet. on Adv. Detectors, Castiglione, 1983). Geneva: CERN
91. Derenzo, S. E., Kirschbaum, A. R., Eberhard, P. H., Ross, R. R., Solmitz, F. T. 1974. *Nucl. Instrum. Methods* 122:319
92. Willis, W. J., Radeka, V. 1974. *Nucl. Instrum. Methods* 120:221
93. Karlovac, N., Mayhugh, T. L. 1977. *IEEE Trans. Nucl. Sci. NS-24*:327
94. Gatti, E., Hrisoho, A., Manfredi, P. F. 1983. *IEEE Trans. Nucl. Sci. NS-30*:319
95. Deithers, K., Donat, A., Lanius, K., Leiste, R., Roeser, U., et al. 1981. *Nucl. Instrum. Methods* 180:145
96. Rubbia, C. 1977. *Preprint CERN-EP/77-8*. Geneva: CERN
97. Chen, H. H., Doe, P. J. 1981. *IEEE Trans. Nucl. Sci. NS-28*:454
98. Doe, P. J., Mahler, H. J., Chen, H. H. 1982. *Nucl. Instrum. Methods* 199:639
99. Mahler, H. J., Doe, P. J., Chen, H. H. 1983. *IEEE Trans. Nucl. Sci. NS-30*:86
100. Dolgoshein, B. A., Kruglov, A. A., Lebedenko, V. N., Miroshnichenko, V. P., Rodionov, B. U. 1973. *Sov. J. Part. Nucl.* 4:70
101. Ferbel, T., ed. 1982. *Proc. Fermilab*

Workshop on Silicon Detectors for High-Energy Physics, Batavia, 1981. Batavia: FNAL
102. Stefanini, A. 1983. *Miniaturization of High-Energy Physics Detectors*, Ettore Majorana International Science Series. New York: Plenum
103. Rancoita, P. G., Seidman, A. 1982. *Riv. Nuovo Cimento* 5: No. 7
104. Heijne, H. M. 1983. *Rep. CERN 83-06.* Geneva: CERN
105. Hyams, B., Kötz, U., Belau, E., Klanner, R., Lutz, G., et al. 1983. *Nucl. Instrum. Methods* 205:99
106. Belau, E., Klanner, R., Lutz, G., Neugebauer, E., Seebrunner, H. J., et al. 1983. *Nucl. Instrum. Methods* 214:253
107. Hyams, B., Kötz, U. 1983. *Nucl. Instrum. Methods* 205:9
108. Hyams, B., private communication
109. Rancoita, P. G. 1984. *J. Phys. G* 10: No. 3, 299
110. Borgeaud, P., McEwen, J. G., Rancoita, P. G., Seidman, A. 1983. *Nucl. Instrum. Methods* 211:363
111. Damerell, C. J. S., Farley, F. J. M., Gillman, A. R., Wickens, F. J. 1981. *Nucl. Instrum. Methods* 185:33
112. Bailey, R., Damerell, C. J. S., English, R. L., Gillman, A. R., Lintern, A. L., et al. 1982. *Nucl. Instrum. Methods* 213:201
113. Amelio, G. F. 1974. *Sci. Am.* 230:22
114. Gatti, E., Rehak, P. 1983. *Brookhaven preprint BNL 33523* (Presented at 2nd Pisa Meeting on Advanced Detectors, Castiglione, 1983). Brookhaven, NY: BNL

HYPERON BETA DECAYS

Jean-Marc Gaillard and Gilles Sauvage

Laboratoire de l'Accélérateur Linéaire, 91405 Orsay, France

CONTENTS

1. INTRODUCTION 351
2. PROPERTIES OF THE BARYON MATRIX ELEMENT 359
   2.1 The Baryon Matrix Element in the Cabibbo Theory 359
   2.2 Observable Effects of the V−A Structure 364
   2.3 q^2 Dependences of the Form Factors 366
3. EXPERIMENTAL DATA ON HYPERON DECAYS 367
   3.1 Low-Energy Experiments 367
   3.2 Hyperon Beam Experiments 372
4. RADIATIVE CORRECTIONS 382
5. FITS TO THE CABIBBO MODEL 385
   5.1 Fits without Radiative Corrections 386
   5.2 Fits with Radiative Corrections 388
   5.3 Discussion and Comparison with Other Results 388
6. SU(3)-BREAKING EFFECTS 390
7. GENERALIZATION OF THE CABIBBO MODEL 392
8. $\Omega^- \to \Xi^0 e \nu$ DECAY 395
9. CONCLUSIONS 395
   APPENDIX 397

1. INTRODUCTION

The theory of weak interactions was born 50 years ago with the neutrino hypothesis of Pauli (1) and Fermi's theory of beta decay (2). Fermi's description of the first known example of baryon semileptonic decay, $n \to p e^- \bar{\nu}_e$, was analogous to that of the emission of a photon in radiative decay. He replaced the electromagnetic potential A_μ by a four-vector constructed from the electron and the neutrino fields. Fermi wrote the beta-decay interaction as the product of two currents:

$$H_{int} \sim (\bar{u}_p \gamma_\mu u_n)(\bar{u}_e \gamma_\mu u_\nu),$$

assuming that the weak interaction was a *local* interaction, i.e. the four fermions interacted at the same space-time point.

Twenty-five years later the discovery of parity nonconservation (3) led to the V–A theory of Feynman & Gell-Mann (4) in which the weak transitions arise from the self-coupling of a single charged current J_μ^W:

$$H_w = (G/\sqrt{2})J_\mu^W J_\mu^{W+} + \text{h.c.} \qquad 1.$$

The current J_μ^W is the sum of the current operators for the leptons J_μ^ℓ and for the strongly interacting particles J_μ^h. Each current is a linear combination of vector and axial terms. The weak interaction is a very short-range force, so that the currents were still taken as interacting only at the same space-time point.

The explicit V–A form of the lepton current in terms of the lepton fields is

$$J_\mu^\ell = \sum [\bar{\ell}\gamma_\mu(1+\gamma_5)v_\ell], \qquad 2.$$

where the sum extends over the three types of leptons, $\ell = e, \mu, \tau$, which have been observed and their associated neutrinos. The third lepton τ was only discovered later; at the time of the Feynman & Gell-Mann formulation of the V–A theory, only the electron and the muon had been observed. The assumption that each type of lepton contributed to J_μ^ℓ with the same coefficient was called the e–μ *universality* hypothesis. The matrix element of J_μ^ℓ between physical lepton and neutrino states is

$$\ell_\mu = \bar{u}_\ell(p_\ell)\gamma_\mu(1+\gamma_5)u_{v_\ell}(p_v), \qquad 3.$$

where the $u(p)$ are Dirac spinors.

All the observed properties of purely *leptonic transitions* corresponding to a charged weak current (e.g. $\mu^- \to e^- \bar{v}_e v_\mu$) are well represented by the matrix element

$$M = (G/\sqrt{2})\ell_\mu \ell^\mu, \qquad 4.$$

where G is the Fermi weak coupling constant, and the factorization of the two lepton terms is a consequence of the assumed locality of the weak interaction. For leptonic transitions, the corrections to the first-order V–A weak amplitude are small: electromagnetic or higher order in the weak interaction.

The measurement of the muon decay rate, taking into account radiative corrections, gives for the universal weak coupling constant the value (5)

$$G = G_\mu = (1.16632 \pm 0.00004) \times 10^{-11} \text{ MeV}^{-2}. \qquad 5.$$

The other two classes of charged-current weak transitions are (a) *the semileptonic interactions* in which hadrons take part together with leptons;

beta radioactivity and hyperon semileptonic decays $\Lambda \to pe\nu$, etc, belong to this class; and (b) *the nonleptonic interactions* in which only hadrons take part; kaon decays into mesons, e.g. $K \to \pi\pi$, and hyperon decays into mesons and baryons, e.g. $\Lambda \to N\pi$.

The systematic study of the baryon semileptonic decays

$$B_1 \to B_2 \ell \bar{\nu}$$

provides rather complete information about the properties of the hadronic weak current J_μ^h. The semileptonic current-current interaction

$$H_{SL} = (G/\sqrt{2})J_\mu^h J_\ell^\mu + \text{h.c.} \qquad 6.$$

gives for those decays a matrix element of the form

$$M = (G/\sqrt{2})\langle B_2|J_\mu^h|B_1\rangle \ell^\mu. \qquad 7.$$

The structure of the baryonic part of the matrix element is complicated because hadrons interact strongly. For purely leptonic transitions the intrinsic V–A structure of the weak current is almost unaltered by the small electromagnetic corrections. In contrast, the large perturbations, due to strong interaction effects, drastically modify the matrix element of the weak hadronic current, which is nevertheless expected to remain a combination of V and A terms.

The approach taken to specify the expected properties of the weak hadronic current has been based upon the assumed flavor SU(3) symmetry of the strong interactions. The baryons are characterized by their masses and their quantum numbers: baryonic number B, spin J, parity P, charge Q, strangeness S, isospin I and its third component I_3. With the exception of the Ω^- hyperon ($S = -3$ and $J^P = 3/2^+$), whose semileptonic decay is discussed separately, all the hyperons (Λ, Σ, Ξ) and the nucleons belong to the lightest $J^P = 1/2^+$ SU(3) octet of baryons. Figure 1 is the representation of that baryon octet in the (I_3, S) plane.

Strangeness S is a conserved quantum number in strong and electromagnetic interactions, whereas it is not necessarily conserved in weak transitions. All observed semileptonic decays may be classified into two groups according to the quantum numbers carried by the hadronic current:

$\Delta S = 0$ transitions $\Delta Q = \Delta I_3 = \pm 1$

$n \to pe\nu_e$ $\pi^\pm \to \pi^0 e\nu_e$

$\Sigma^\mp \to \Lambda e\nu_e$ $\pi^\pm \to \ell\nu_\ell$

$\Sigma^- \to \Sigma^0 e\nu_e$

$|\Delta S| = 1$ transitions	$\Delta Q = \Delta S = \pm 1, \Delta I_3 = \pm 1/2$
$\Lambda \to p \ell \nu_\ell$	$K^\pm \to \pi^0 \ell \nu_\ell$
$\Sigma^- \to n \ell \nu_\ell$	$K^\pm \to \ell \nu_\ell$
$\Xi^- \to (\Lambda, \Sigma^0) \ell \nu_\ell$	$\bar{K}^0 (K^0) \to \pi^{+(-)} \ell \nu_\ell$
$\Xi^0 \to \Sigma^+ \ell \nu_\ell$	$K^\pm \to \pi^+ \pi^- \ell \nu_\ell$

where $\ell =$ e or μ, and to avoid complications in the notation we have omitted the distinction between leptons and antileptons. Transitions with $|\Delta S| > 1$ or with $\Delta S = -\Delta Q$ have not been observed.

The charged weak hadronic current J_μ^h is therefore decomposed into $\Delta S = 0$ and $|\Delta S| = 1$ components, which are both sums of vector and axial vector terms:

$$J_\mu^h(\Delta S = 0) = V_\mu^+ + A_\mu^+,$$
$$J_\mu^h(\Delta S = 1) = v_\mu^+ + a_\mu^+,$$

8.

and the quantum numbers of the currents must satisfy the selection rules given above for the two classes of decays.

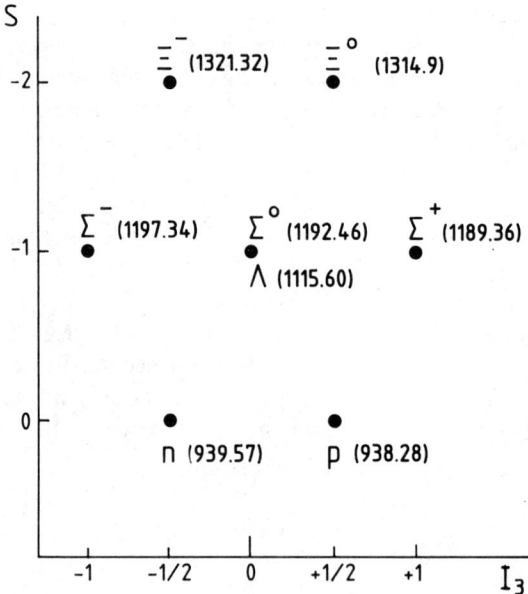

Figure 1 Representation of the lightest $J^P = 1/2^+$ octet of baryons in the (I_3, S) plane. The masses of the particles are given in MeV.

It is convenient for further discussion of the properties of the currents to adopt a formulation in terms of quarks, the assumed constituents of hadrons. The baryons, bound states of three quarks, correspond to the product of three triplet SU(3) representations:

$$3 \otimes 3 \otimes 3 = 10 \oplus 8 \oplus 8 \oplus 1.$$

The lightest eight $J^P = 1/2^+$ baryons constitute one of the octets, whereas the Ω^- hyperon was predicted before its discovery as the $S = -3$ member of the $J^P = 3/2^+$ decuplet. Similarly, the eight $J^P = 0^-$ mesons (π, K, η^0), quark–antiquark bound states, are members of the octet obtained from the product

$$\bar{3} \otimes 3 = 8 \oplus 1.$$

The quantum number assignments of the three types (flavors) of quarks u, d, s—building blocks of the $J^P = 1/2^+$ baryons and $J^P = 0^-$ mesons—are given in Table 1.

The quantum numbers of each of the three quarks satisfy the relation

$$Q = I_3 + (S+B)/2. \qquad 9.$$

These quantum numbers are additive; therefore the relation is also satisfied by the hadronic states. Examples of the quark composition of the hadrons are p = uud, n = udd, $\Sigma^- $ = sdd. The pieces of the hadronic current contributing to the semileptonic decays n → pe$^-\nu$ and $\Sigma^- \to $ ne$^-\nu$ correspond to the quark weak transitions d → u and s → u, respectively. With the corresponding transitions between antiquarks $\bar{d} \to \bar{u}$ and $\bar{s} \to \bar{u}$, these are the only charged $|\Delta Q| = 1$, quark weak transitions, as can easily be verified from Table 1.

The isospin and strangeness quantum numbers carried by the $\Delta S = 0$ and $|\Delta S| = 1$ hadronic weak charged currents can be identified with those of charged mesons belonging to an octet of SU(3):

$$\begin{array}{ll} d \to u & \pi^+ = \bar{d}u \\ s \to u & K^+ = \bar{s}u \\ \bar{d} \to \bar{u} & \pi^- = d\bar{u} \\ \bar{s} \to \bar{u} & K^- = s\bar{u} \end{array}$$

Table 1 Quark quantum numbers

	B	Q	I	I_3	S
u	1/3	2/3	1/2	+1/2	0
d	1/3	−1/3	1/2	−1/2	0
s	1/3	−1/3	0	0	−1

These properties of the quantum numbers of the current led to the first assumption of the Cabibbo theory (6): the hadronic weak currents belong to a single self-conjugate representation of SU(3). More precisely, the components of the current are the charged members of an SU(3) octet.

Using the Wigner–Eckart theorem, all the matrix elements of an octet operator between octet states are linear combinations of two reduced matrix elements for that operator. The Cabibbo assumption therefore implies definite relations between the baryon matrix elements for the various transitions.

A more detailed discussion of the structure, conservation laws, and other properties of the hadronic current is given in Section 2. Here we just note that flavor SU(3) is only an approximate symmetry for the strong interactions. Since SU(3) is broken by only about 15% in the baryon masses, one may reasonably hope that the corrections to baryon semileptonic decays may not be too large. Several attempts at estimating the SU(3)-breaking effects are discussed in Section 6.

The formulation in terms of quark currents leads naturally to the classification given above for the observed transitions. In addition, it specifies the more restrictive $|\Delta I|$ properties of the current, which are $|\Delta I| = 1$ for $\Delta S = 0$ transitions and $|\Delta I| = 1/2$ for $\Delta S = 1$ transitions. Both the baryon and the meson decays have been listed since it is clear, for instance from the quark formulation of the current, that the Cabibbo assumptions must apply to all of them. Within the scope of the present article we limit the discussion of meson decays to a few crucial pieces of information and to comparisons. Detailed discussions of meson decays may be found in several review articles (7).

Early measurements of the hyperon semileptonic decay rates and of the ratio between the $K^+ \to \mu^+ \nu$ and $\pi^+ \to \mu^+ \nu$ decay rates gave much lower values than were expected from the V–A theory of Feynman & Gell-Mann. In addition, the measured value of the neutron beta-decay coupling constant was about 3% lower than G_μ. Cabibbo made the assumption— also proposed by Gell-Mann & Lévy (8)—that the hadronic weak current had a "unit length"

$$J_\mu^h = \cos\theta_C J_\mu^h(\Delta S = 0) + \sin\theta_C J_\mu^h(|\Delta S| = 1)$$

and was coupled to the leptonic current with the universal coupling strength $G = G_\mu$. The data suggested a value of $\theta_C \simeq 0.25$. With that last hypothesis the quark form of the hadronic current, very similar to the leptonic current of Equation 3, is

$$\bar{u}\gamma_\mu(1+\gamma_5)d', \qquad\qquad 10.$$

with

$$d' = d \cos \theta_C + s \sin \theta_C.$$

The Cabibbo theory has been very successful at describing the properties of the semileptonic decays.

The theoretical developments and the experimental discoveries of the last 15 years have led to the SU(3) × U(1) gauge theory of Glashow, Weinberg & Salam (9) as the Standard Model of electroweak interactions. The model provides a unification of weak and electromagnetic interactions and thus gives definite predictions for the masses of the bosons $W^\pm$ and Z^0 assumed to mediate the charged and neutral weak currents. The SU(2) × U(1) model has recently received striking confirmation with the discovery of the $W^\pm$ and Z^0 particles at the CERN $p\bar{p}$ Collider. The values of the masses are in excellent agreement with the theoretical predictions. The model incorporates the V–A theory of the charged weak current, as well as the neutral weak current discovered in 1973.

How do the new theoretical developments affect the postulates of the Cabibbo theory? In the SU(2) × U(1) model, the charged weak interaction corresponds to transitions between *left-handed* doublets of leptons and quarks; it is therefore a V–A theory. It is normal to associate quarks and leptons in the doublets:

$$\begin{pmatrix} \nu_e \\ e \end{pmatrix}_L \begin{pmatrix} \nu_\mu \\ \mu \end{pmatrix}_L \begin{pmatrix} \nu_\tau \\ \tau \end{pmatrix}_L \quad \begin{matrix} \text{Charge} \\ 0 \\ -1 \end{matrix}$$

$$\begin{pmatrix} u \\ d \end{pmatrix}_L \begin{pmatrix} c \\ s \end{pmatrix}_L \begin{pmatrix} t \\ b \end{pmatrix}_L \quad \begin{matrix} 2/3 \\ -1/3 \end{matrix}$$

These include both the old leptons and quarks and the recently discovered states with new quantum numbers: charm c, beauty b, lepton τ. States corresponding to the top quark t have not yet been observed. The elementary weak transition corresponding to the decay n → peν is shown in Figure 2.

Nonlocal effects due to the W boson propagator are negligible for processes in which the square of the four-momentum transfer q^2 is small compared to $m_W^2 \cong (80 \text{ GeV}/c^2)^2$. For hyperon decays the maximum value of q^2 is $(m_{\Sigma^-} - m_n)^2 = (0.23 \text{ GeV}/c^2)^2$, so the assumption of locality is well justified. The postulates of the Cabibbo theory based upon the SU(3) symmetry properties of the strong interaction perturbations are also not affected by the new developments of the weak interaction theory.

Gauge invariance requires the coupling of the W boson with any quark

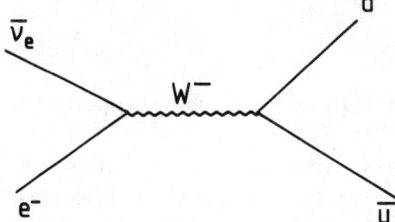

Figure 2 Elementary weak transition corresponding to the decay n → peν.

or lepton doublet to have the same strength. For the leptons this justifies the e–μ universality assumption and its natural extension to the τ lepton. For quarks the situation is more complex than is reflected in the quark doublets introduced above, since, for example, the $|\Delta S| = 1$ semileptonic decays correspond to transitions between members of different quark doublets (s → u). It is therefore assumed that the states with definite weak coupling are different from the quark states with definite masses. The assumption of universality for the weak coupling strength implies that the two sets of states are connected by a unitary transformation. Defining the quark doublets with definite weak couplings as

$$\begin{pmatrix} u \\ d' \end{pmatrix}_L, \begin{pmatrix} c \\ s' \end{pmatrix}_L, \begin{pmatrix} t \\ b' \end{pmatrix}_L, \qquad 11.$$

the universality assumption implies

$$\begin{pmatrix} d' \\ s' \\ b' \end{pmatrix} = U \begin{pmatrix} d \\ s \\ b \end{pmatrix}, \qquad 12.$$

where U is a 3×3 unitary matrix.

In this context the weak hadronic current corresponding to the doublet (u, d') takes the form

$$\bar{u}\gamma_\mu(1+\gamma_5)d', \qquad 13.$$

where $d' = \alpha d + \beta s + \gamma b$ and $|\alpha|^2 + |\beta|^2 + |\gamma|^2 = 1$. The success of the Cabibbo theory, which assumed (Equation 10) that $\alpha^2 + \beta^2 = 1$, implies that the b-quark mixing parameter γ is small. The matrix U was first proposed by Kobayashi & Maskawa (10). It is a function of three real numbers, the weak mixing angles between the quarks, and a CP-violating phase, which provides a possible source for the CP violation observed in K^0 decays.

Any fundamental understanding of the mixing between the three quark families (Equation 12) requires an accurate determination of the mixing parameters in the matrix U. This re-emphasizes the importance of measurements of semileptonic hadron decays in general, and hyperon decays in particular.

This article presents a detailed comparison of the experimental information on baryon (hyperon and neutron) semileptonic decays with the Cabibbo model. Section 2 deals with the properties of the baryon matrix element and of the observable quantities within the framework of the Cabibbo theory. Section 3 presents the experimental data. In Section 4 we briefly discuss the effects due to radiative corrections. The confrontation between the experimental data and the Cabibbo model, including a discussion of the SU(3)-breaking effects, is set out in Sections 5 and 6. Other information on charged-current weak mixing angles is discussed in Section 7. The status of the Ω^- semileptonic decay is briefly reviewed in Section 8 and Section 9 summarizes our conclusions.

2. PROPERTIES OF THE BARYON MATRIX ELEMENT

2.1 *The Baryon Matrix Element in the Cabibbo Theory*

The most general form of the baryon matrix element for the decay $B_1 \to B_2 \ell \nu$ is

$$\langle B_2 | J_\mu^h | B_1 \rangle = C \bar{u}_{B_2}(p_2) [f_1(q^2)\gamma_\mu + i[f_2(q^2)/M_1]\sigma_{\mu\nu}q^\nu + [f_3(q^2)/M_1]q_\mu$$
$$+ \{g_1(q^2)\gamma_\mu + i[g_2(q^2)/M_1]\sigma_{\mu\nu}q^\nu + [g_3(q^2)/M_1]q_\mu\}\gamma_5] u_{B_1}(p_1), \quad 14.$$

where $u_{B_1}(p_1)$, p_1, $M_1[\bar{u}_{B_2}(p_2), p_2, M_2]$ are the Dirac spinor, the four-momentum, and the mass of the initial [final] baryon, and $q = p_2 - p_1$.

It can be seen that, a priori, each semileptonic decay mode depends on six independent functions of q^2, called form factors: $f_1(q^2)$, $f_2(q^2)$, $f_3(q^2)$ for the vector part of the weak interaction, and $g_1(q^2)$, $g_2(q^2)$, $g_3(q^2)$ for the axial part. The factor C has been extracted from the matrix element as a prelude to the introduction of the Cabibbo angle already mentioned in Section 1. Time reversal invariance requires that the form factors be real.

The Cabibbo assumptions reduce the number of independent form factors: namely, all the form factors will be a function of only three independent parameters not predicted by the theory.

We now review the different assumptions of the Cabibbo theory and discuss their consequences.

Assumption 1: *The components of the weak hadronic current belong to a single self-conjugate representation of SU(3).* The components of the current are the charged members of an SU(3) octet, and the isospin and strangeness quantum numbers carried by the weak hadronic current correspond to those of the charged mesons of the $J^P = 0^-$ octet:

$$\Delta S = 0 \quad \begin{cases} \Delta Q = \Delta I_3 = +1 & |\Delta I| = 1 \quad \pi^+ \\ \Delta Q = \Delta I_3 = -1 & |\Delta I| = 1 \quad \pi^- \end{cases}$$

$$\Delta S = +1 \qquad \Delta S = \Delta Q = +1 \qquad |\Delta I| = 1/2 \quad K^+$$

$$\Delta S = -1 \qquad \Delta S = \Delta Q = -1 \qquad |\Delta I| = 1/2 \quad K^-$$

The self-conjugation implies that the currents with $\Delta Q = 1$ and $\Delta Q = -1$ belong to the same octet.

The $J^P = 1/2^+$ baryon states are identified in terms of SU(3) octet components (11). The matrix element of an operator—itself belonging to an octet—between two octet states is a linear combination of two reduced matrix elements, on account of the symmetric and antisymmetric octets that appear in the direct product of two octets in SU(3) ($8 \otimes 8 = 1 \oplus 8a \oplus 8b \oplus 10 \oplus \overline{10} \oplus 27$).

More precisely, any of the six form factors $f_m(q^2)$, $g_m(q^2)$ previously introduced is given by

$$\begin{cases} f_m(q^2) = aF_m(q^2) + bD_m(q^2), \\ g_m(q^2) = aF_{m+3}(q^2) + bD_{m+3}(q^2), \quad (m = 1, 3), \end{cases} \qquad 15.$$

where $F_i(q^2)$ and $D_i(q^2)$ with $i = 1, 6$ are different functions of q^2 for each of the six form factors. The constants a and b in Equations 15 are generalized Clebsch–Gordan coefficients whose values are given in Table 2 for the various transitions.

Table 2 Clebsch–Gordon coefficients in hyperon decays

Transition	a	b
$n \to p$	1	1
$\Sigma^\pm \to \Lambda$	0	$\sqrt{2/3}$
$\Sigma^- \to \Sigma^0$	$\sqrt{2}$	0
$\Xi^- \to \Xi^0$	-1	1
$\Lambda \to p$	$-\sqrt{3/2}$	$-\sqrt{3/2}$
$\Sigma^- \to n$	-1	1
$\Xi^- \to \Lambda$	$\sqrt{3/2}$	$-\sqrt{3/2}$
$\Xi^- \to \Sigma^0$	$1/\sqrt{2}$	$1/\sqrt{2}$
$\Xi^0 \to \Sigma^+$	1	1

The first assumption of the Cabibbo theory, by making each form factor a linear combination of two independent functions of q^2, has reduced the number of independent functions to 12. Equations 15 rely on exact SU(3) symmetry, and later we discuss the consequences of the SU(3) breaking of the strong interactions. As the mass splitting for hyperons is small ($\sim 15\%$) compared to mesons, one may expect the predictions of exact SU(3) to hold better in the case of hyperons.

Assumption 2: *Universality*. The leptonic current is coupled to a single hadronic current of unit length:

$$J_\mu^h = \cos\theta_C J_\mu^h(\Delta S = 0) + \sin\theta_C J_\mu^h(|\Delta S| = 1)$$

with the universal coupling $G = G_\mu$. The constant C of Equation 14 is equal to $\cos\theta_C$ for $\Delta S = 0$ transitions and to $\sin\theta_C$ for $|\Delta S| = 1$ transitions.

Assumption 3: *Generalized conserved vector current (CVC) hypothesis*. The CVC hypothesis (8, 12) states that the vector part $J_\mu^\pm$ of the $\Delta S = 0$ weak current is a conserved current like the electromagnetic current. Furthermore, CVC implies that $J_\mu^\pm$ form an isotriplet with the isovector part of the electromagnetic current. Generalized CVC assumes that the vector part of the $\Delta S = 0$ and $|\Delta S| = 1$ weak currents and the electromagnetic current are members of the same SU(3) octet. The electromagnetic (e.m.) form factors and the weak vector form factors are therefore directly related. We observe that, historically, the CVC hypothesis was an important step toward the unification of the electromagnetic and weak forces.

The proton and the neutron e.m. form factors are given by the relations:

$$f_m^p(q^2) = F_m(q^2) + \tfrac{1}{3}D_m(q^2); \qquad f_m^n(q^2) = -\tfrac{2}{3}D_m(q^2).$$

Therefore the six functions $F_m(q^2)$ and $D_m(q^2)$ for the vector form factors in Equations 15 are completely determined by the equations:

$$\begin{aligned} F_m(q^2) &= f_m^p(q^2) + \tfrac{1}{2}f_m^n(q^2), \\ D_m(q^2) &= -\tfrac{3}{2}f_m^n(q^2). \end{aligned} \qquad 16.$$

At $q^2 = 0$, the e.m. form factor $f_1(0)$ is equal to the electric charge of the baryon, therefore $F_1(0) = 1$ and $D_1(0) = 0$.

The e.m. form factor $f_2(q^2)$ is related to the anomalous magnetic moment, μ. Defining $f_2'(q^2) = [(M_1 + M_2)/M_1]f_2(q^2)$, we have $f_2'^{p,n}(0) = \mu_{p,n}$, which gives $F_2'(0) = \mu_p + \tfrac{1}{2}\mu_n$ and $D_2'(0) = -\tfrac{3}{2}\mu_n$. The weak f_2 form factor is often called the weak magnetism term.

Finally the conservation of the electromagnetic current requires that

$f_3^{p,n}(q^2)$ be equal to zero, implying $F_3(q^2) = D_3(q^2) = 0$. Therefore the form factor $f_3(q^2)$ vanishes for all the semileptonic decays.

We have shown how all the vector form factors for baryon decays are determined by the generalized CVC hypothesis and are related, at $q^2 = 0$, to the electric charges and the anomalous magnetic moments of the nucleons. In the limit of exact SU(3) symmetry, the q^2 dependences of the weak vector form factors are also given by the q^2 dependences of the electromagnetic form factors of the nucleons. In Section 2.3 we discuss the specific forms adopted for the q^2 dependences.

Assumption 4: *Absence of second-class currents*. There remain, at this stage, the three axial-vector form factors, g_1, g_2 and g_3, each one being a linear combination of two unknown functions for the various decays. To further reduce the number of form factors, a supplementary hypothesis is needed.

Let us consider the properties of the combined operation of charge conjugation C and rotation by 180° about the second axis I_2 of the isospin space, which is called G parity. When applied to the neutron beta-decay matrix, it interchanges the neutron and the proton twice, and gives back the initial matrix up to possible changes of sign. Under G parity the weak vector (f_1) and axial-vector (g_1) currents transform as

$$V_\mu \to V_\mu,$$
$$A_\mu \to -A_\mu.$$

17.

Currents that transform as Equation 17 are called currents of the *first class*; those that transform with an opposite sign are called *second class*. The terms f_1, f_2, g_1, and g_3 correspond to currents of the first class, whereas the f_3 and g_2, terms are second class. Since the strong interactions are invariant under C and transformations in isospin space, the above properties of the currents are unaltered by the effects due to strong interactions.

We note that at the quark level (see Equation 10) there are only currents of the first class for the d → u transition, occurring in neutron beta decay, and that the term f_3 has already been eliminated by the generalized CVC hypothesis.

The same reasoning applies to the (Ξ^-, Ξ^0) isospin doublet for the decay $\Xi^- \to \Xi^0 \ell \bar{\nu}$. The absence of currents of the second class gives:

$$g_2^{n,p} = F_5(q^2) + D_5(q^2) = 0; \qquad g_2^{\Xi^-,\Xi^0} = D_5(q^2) - F_5(q^2) = 0,$$

and so

$$F_5(q^2) = D_5(q^2) = 0.$$

Therefore all the pseudotensor form factors g_2 vanish in all decays up to symmetry-breaking effects.

In addition, in the expression of the differential decay rate, all the terms involving the form factors f_3 and g_3 are multiplied by a factor $(m_\ell/M_1)^2$. As accurate experiments have been performed only on the electronic hyperon decays, $g_3(q^2)$ can be safely ignored. Finally, the matrix element M for $B_1 \to B_2 \ell \nu$ (Equation 14) is reduced to

$$M = (G/\sqrt{2})C\bar{u}_{B_2}(p_2)\{\gamma_\mu f_1(q^2) + i[f_2(q^2)/M_1]\sigma_{\mu\nu}q^\nu + \gamma_\mu\gamma_5 g_1(q^2)\}u_{B_1}(p_1) \times \bar{u}_\ell(p_\ell)\gamma^\mu(1+\gamma_5)u_\nu(p_\nu), \qquad 18.$$

where

$$C = \begin{pmatrix}\cos\theta_C \\ \sin\theta_C\end{pmatrix} \quad \text{for} \quad |\Delta S| = \begin{pmatrix}0 \\ 1\end{pmatrix}$$

and where $f_1(q^2)$, $f_2(q^2)$ are predicted and $g_1(q^2)$ is a linear combination of two unknown functions $F(q^2) = F_4(q^2)$ and $D(q^2) = D_4(q^2)$. The values at $q^2 = 0$ of the form factors for all the semileptonic hyperon decays are given in Table 3, where $f_2'(0) = [(M_1 + M_2)/M_1]f_2(0)$, and F, D stand for $F(0)$, $D(0)$.

All the decays are therefore described by three parameters, the Cabibbo angle θ_C and two coupling constants F and D. We take here the sign convention that $g_1/f_1 = F+D$ is positive for the neutron decay, which fixes all other signs. This sign convention is used by Bender et al (13), whereas Linke (14) and the Particle Data Group (5) have taken the opposite convention. The determination of the Cabibbo angle requires the measure-

Table 3 Parameters of the baryon weak matrix

Transition	C	$f_1(0)$	$f_2'(0)$[a]	$g_1(0)$
$n \to p e \nu_e$	$\cos\theta_C$	1	$\mu_p - \mu_n$	$F+D$
$\Sigma^\pm \to \Lambda e \nu_e$	$\cos\theta_C$	0	$-\sqrt{3/2}\mu_n$	$\sqrt{2/3}D$
$\Sigma^- \to \Sigma^0 e \nu_e$	$\cos\theta_C$	$\sqrt{2}$	$\sqrt{2}[\mu_p + (\mu_n/2)]$	$\sqrt{2}F$
$\Lambda \to p \ell \nu$	$\sin\theta_C$	$-\sqrt{3/2}$	$-\sqrt{3/2}\mu_p$	$-\sqrt{3/2}(F+D/3)$
$\Sigma^- \to n \ell \nu_\ell$	$\sin\theta_C$	-1	$-(\mu_p + 2\mu_n)$	$-(F-D)$
$\Xi^- \to \Lambda \ell \nu_\ell$	$\sin\theta_C$	$\sqrt{3/2}$	$\sqrt{3/2}(\mu_p + \mu_n)$	$\sqrt{3/2}(F-D/3)$
$\Xi^- \to \Sigma^0 \ell \nu_\ell$	$\sin\theta_C$	$1/\sqrt{2}$	$1/\sqrt{2}(\mu_p - \mu_n)$	$(F+D)/\sqrt{2}$
$\Xi^0 \to \Sigma^+ \ell \nu_\ell$	$\sin\theta_C$	1	$\mu_p - \mu_n$	$F+D$
$\Xi^- \to \Xi^0 \ell \nu_\ell$	$\cos\theta_C$	1	$\mu_p + 2\mu_n$	$F-D$

[a] The values of the anomalous magnetic moments are: $\mu_p = 1.793$, $\mu_n = -1.913$.

ment of decay rates, whereas F and D can be determined also by studying the distributions of various kinematical variables, as discussed in the next section and in the Appendix.

The results given in Table 3 are valid at $q^2 = 0$ and neglect the SU(3) symmetry-breaking effects. The q^2 dependences of the vector form factors $f_1(q^2)$, $f_2(q^2)$, and of the axial vector form factor $g_1(q^2)$ are discussed in Section 2.3. As the q^2 values involved in the hyperon decays are relatively small, it is expected that the q^2 dependence will only introduce small modifications of the predictions summarized in Table 3. The breaking of SU(3) symmetry by strong interactions may introduce large modifications of the results of Table 1, and possible estimates of these effects are discussed in Section 6.

2.2 Observable Effects of the V–A Structure

The leptonic decay rate of a baryon B_1 is given by

$$\Gamma = \frac{1}{(2\pi)^5} \frac{1}{2E_1} \int \frac{dp_2}{2E_2} \frac{dp_\ell}{2E_\ell} \frac{dp_\nu}{2E_\nu} \delta^4(p_1 - p_2 - p_\ell - p_\nu) \frac{1}{2} \sum_{\text{spins}} |M|^2. \qquad 19.$$

The numerical integration of Γ is given for the various decays in the article of Bender et al (13). Contributions to Γ of terms other than f_1^2 and g_1^2 are of the order of 1%, and to a very good approximation

$$\Gamma = G^2 \begin{bmatrix} \cos^2 \theta_C \\ \sin^2 \theta_C \end{bmatrix} \Delta m^5 / 60\pi^3 (f_1^2 + 3g_1^2)(1 - 3\delta), \qquad 20.$$

where $\Delta m = M_1 - M_2$ and $\delta = (M_1 - M_2)/(M_1 + M_2)$.

There are several distributions and correlations that can be used to determine the decay form factors, such as the differential decay rate $\partial^2 \Gamma / \partial E_\ell \partial E_2$ and the angular correlation between the lepton and the antineutrino. Other distributions that make use of the polarization information for the initial and/or final baryons are especially relevant for the determination of the relative sign of the form factors f_1 and g_1. The complete expressions are obtained by inserting in the matrix element the suitable spin projection operators, which are $\frac{1}{2}[1 + \gamma_5(S_{B_1}\gamma)]$ and $\frac{1}{2}[1 + \gamma_5(S_{B_2}\gamma)]$ for the initial and final baryons, respectively.

As discussed in Section 2.1, the basic simplicity of the V, A structure of the interaction is somewhat obscured by finite momentum transfer effects for the physical decays. However, in the limit of zero momentum transfer (i.e. equal baryon masses and zero lepton mass), the effects of the V, A structure in the decay distributions can be represented in a very concrete way. Consider the decay of a completely polarized baryon $B_1 \to B_2 \ell \bar{\nu}$. The vector (Fermi) and axial-vector (Gamow–Teller) transitions correspond to

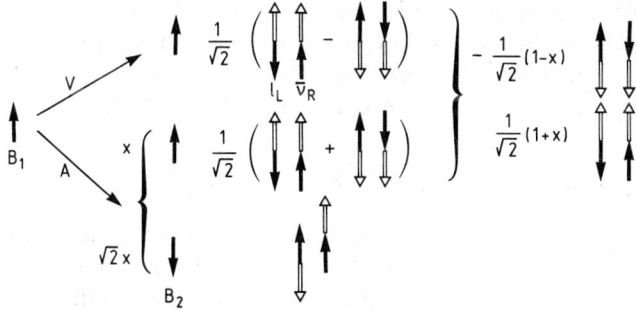

Figure 3 The process $B_1 \to B_2 + \ell + \bar{v}$: solid arrows indicate the spin orientation; open arrows indicate the momentum direction; $x = g_1/f_1$ is the axial-vector to vector form-factor ratio.

the $J = 0$ singlet and $J = 1$ triplet spinor states, respectively, for both $\bar{B}_2 B_1$ and $\ell \bar{v}$ systems. The helicities of the lepton (ℓ_L) and of the antineutrino ($\bar{v}_R$) are fixed and the decay proceeds as indicated in Figure 3, where $x = g_1/f_1$.[1]

It is a simple matter of counting states to determine several observable quantities from Figure 3. The total decay rate is proportional to $(1+3x^2)$. Counting the lepton–antineutrino states with parallel and antiparallel momentum directions gives the correlation $\alpha_{\ell \bar{v}}$ between the lepton and antineutrino directions:

$$\alpha_{\ell \bar{v}} = [\tfrac{1}{2}(1-x)^2 + \tfrac{1}{2}(1+x)^2 - 2x^2]/(1+3x^2) = (1-x^2)/(1+3x^2).$$

Both the decay rate and $\alpha_{\ell \bar{v}}$ are independent of the polarization.

The correlation α_ℓ between the lepton direction and the polarization of B_1 is given by

$$\alpha_e = [\tfrac{1}{2}(1+x)^2 - \tfrac{1}{2}(1-x)^2 - 2x^2]/(1+3x^2) = [-2x(x-1)]/(1+3x^2).$$

Similarly, for the antineutrino one gets

$$\alpha_v = [\tfrac{1}{2}(1+x)^2 + 2x^2 - \tfrac{1}{2}(1-x)^2]/(1+3x^2) = [2x(x+1)]/(1+3x^2).$$

We emphasize again that these formulae are only valid for zero momentum transfer. They are given to indicate how, without complete and complex calculations, one can get a concrete idea of the way in which the observable effects arise. At finite q^2 the simple formulae given in this section have to be completed with recoil corrections and the contributions of other form factors. In addition, the q^2 dependences of the form factors have to be introduced.

In the Appendix we give a detailed discussion of the distributions

---

[1] Note that, with the convention we have used, (V–A) corresponds to $x = +1$.

accessible to experiments. Since the complete expressions are in all cases very lengthy, we instead tend to give convenient but accurate approximate formulae. The complete expressions that should be used to analyze the experiments can be found in several articles, for example Bender et al (13) and Linke (14).

2.3 q^2 Dependences of the Form Factors

The q^2 dependences of the form factors, which are determined by the dynamics of the strong interactions, can only be obtained from the phenomenology of these interactions. A possible method, based on dispersion relations, is to include only the effects of the nearest poles, assuming that they dominate the behavior of the form factors at low q^2. However, it seems more reasonable to use directly the experimental information obtained from electroproduction and neutrino experiments.

Dipole formulae have been shown to give good fits to the electroproduction and the neutrino experiments involving $\Delta S = 0$ transitions. From electroproduction one gets

$$G_V^E(q^2) = G_V^E(0)[1/(1+q^2/m_V^2)^2],$$

with

$$m_V = (0.84 \pm 0.04) \text{ GeV}/c^2,$$

and CVC predicts the same q^2 dependence for the vector form factor in the $\Delta S = 0$ decays (15).

For the axial form factor dependence, the neutrino experiments (16) give

$$G_A(q^2) = G_A(0)[1/(1+q^2/m_A^2)^2],$$

with

$$m_A = (1.08 \pm 0.08) \text{ GeV}/c^2.$$

For the baryon semileptonic decays, extrapolations of these dependences in the negative q^2 region are used, and since the q^2 values involved are always small we retain only the first term of the expansion. Thus

$$f_1(q^2) = f_1(0)(1+2q^2/m_V^2),$$
$$g_1(q^2) = g_1(0)(1+2q^2/m_A^2).$$

For the $\Delta S = 0$ transitions, the values to be used for m_V and m_A are the ones given above. For the $\Delta S = 1$ transitions, it seems reasonable to adopt a renormalization procedure that accounts for the SU(3) mass splitting effects. For the $\Delta S = 0$ vector transition, the nearest pole is the ρ meson ($m_\rho = 0.77 \text{ GeV}/c^2$), whereas for $\Delta S = 1$ transitions the nearest pole is the

K* ($m_{K^*} = 0.89$ GeV/c^2). Thus for $\Delta S = 1$ transitions the following values are used:

$$m'_V = [0.84 \times (0.89/0.77)] \text{ GeV}/c^2 = 0.98 \text{ GeV}/c^2,$$

[handwritten correction: 0.97]

and

$$m'_A = [1.08 \times (0.89/0.77)] \text{ GeV}/c^2 = 1.25 \text{ GeV}/c^2.$$

The sensitivity of g_1/f_1 to the exact choice of vector and axial-vector masses in the q^2 expansions given above is small. For example, in the case $\Sigma^- \to ne\nu$, which corresponds to the largest value of q^2_{max}, varying m'_V or m'_A by ± 0.15 GeV/c^2 causes a relative change $\Delta(g_1/f_1)/(g_1/f_1) = \pm 2\%$. However, ignoring the q^2 dependence altogether would cause a shift of 17% in g_1/f_1. The q^2 dependence of f_2 can be safely neglected.

3. EXPERIMENTAL DATA ON HYPERON DECAYS

The experiments may be classified in two groups according to the experimental techniques used for producing the hyperons.

For all the early experiments, and for a recent one with polarized Σ^-, the hyperons were produced by secondary pion and kaon beams. Typical production reactions were $\pi^+ + N \to \Lambda + K^+$, $K^- + p \to \Sigma^\pm + \pi^\mp$, $K^- + p \to \Xi^- + K^+, \ldots$. In all those experiments the hyperon momenta were low ($\lesssim 1$ GeV/c). The statistics were limited by the (available or tolerable) beam intensity, the production rate, and the acceptance of the apparatus. On the other hand, in several cases the choice of the beam energy and the production reaction could be such as to obtain a large polarization of the hyperon—a very valuable tool for the form-factor analysis in the decays, as we saw in the previous section.

With the advent of high-energy hyperon beams directly produced by the interactions of primary proton beams, a new round of experiments was initiated. In the most recent ones, performed in hyperon beams of 100–200-GeV/c momenta, fluxes of several thousand hyperons per machine cycle have been used and high-statistics information on hyperon semileptonic decays obtained. In the most complete experiment using a hyperon beam at the CERN Super Proton Synchrotron (SPS), five different semileptonic decays have been analyzed in a single experiment.

3.1 Low-Energy Experiments

The information collected in the early experiments is summarized in Table 4, taken from the 1974 edition of the Particle Data Group (PDG) (17). These set the stage for later hyperon beam experiments. The average values of the branching ratio and of the form-factor ratio g_1/f_1 (f_1/g_1 for $\Sigma^\pm \to \Lambda e\nu$, expected to be a pure axial-vector transition according to CVC) are given for each decay mode.

The experimental findings in the early 1960s, that the decay rates were a lot smaller than the naive V–A predictions, played a decisive role in the formulation of the Cabibbo assumptions introduced in Section 2.1. The table shows the importance of the early work on hyperon semileptonic decays. The experimental results already showed an impressive general agreement with the basic hypotheses of the Cabibbo theory, although there were inconsistencies between the assumptions made in the analysis of separate experiments. A fit to all the experimental data available at the end of 1970 was made by Ebenhöh et al (18). The value used at that time for the form factor ratio in neutron beta decay was $g_1/f_1 = 1.231 \pm 0.010$. The values of the parameters of the Cabibbo theory obtained from the fit were

$$\theta_C = 0.239 \pm 0.005,$$
$$F = 0.451 \pm 0.019, \qquad\qquad 21.$$
$$D = 0.777 \pm 0.021.$$

We now discuss in detail a few important aspects of these low-energy data.

3.1.1 $\Lambda \to pe\nu$ DECAYS This is the decay that was studied the most extensively during the pre-hyperon-beam era. Several experiments have measured the $\Lambda \to pe\nu$ branching ratio and the electron–antineutrino correlation, obtaining values of $|g_1/f_1|$. In other experiments with polarized Λ, the angular distributions of the decay particles relative to the Λ polarization provided additional information, such as the magnitude and the sign of g_1/f_1 (see Section A2.2 in the Appendix). The consistency among the various determinations of g_1/f_1 is a means of testing theoretical

Table 4 Hyperon semileptonic data from low-energy experiments[a]

Transition	Branching ratio $\times 10^3$	g_1/f_1 or $\|g_1/f_1\|$[b]
$\Lambda \to pe\nu$	0.813 ± 0.029	$+0.66 \pm 0.05$
$\Lambda \to p\mu\nu$	0.16 ± 0.04	
$\Sigma^- \to ne\nu$	1.082 ± 0.038	0.21 ± 0.12
$\Sigma^- \to n\mu\nu$	0.45 ± 0.04	
$\Sigma^- \to \Lambda e^- \nu$	0.060 ± 0.006	-0.45 ± 0.20[c]
$\Sigma^+ \to \Lambda e^+ \nu$	0.020 ± 0.005	
$\Xi^- \to \Lambda(\Sigma^0)e\nu$	0.68 ± 0.22	

[a] Average values taken from Particle Data Group 1974 (17) except for $|g_1/f_1|$ in $\Sigma^- \to ne\nu$ (see Section 3.1.3).
[b] The sign of g_1/f_1, when measured, is shown by $+/-$ in front of the value.
[c] Value of f_1/g_1 ($f_1 = 0$ expected) for the combined $\Sigma^\pm \to \Lambda e\nu$ data sample.

assumptions, such as, for example, the absence of second-class current g_2 terms.

The reaction $\pi^{\pm} + N \to \Lambda + K$ just below ΣK threshold ($p_\pi \sim 1$ GeV/c) was used as a source of highly polarized Λ in several experiments. We discuss the results of three experiments (19–21), each with several hundred beta decays and Λ polarizations $\gtrsim 70\%$.

The results are summarized in Table 5. The values of $|g_1/f_1|$ were obtained from the electron–antineutrino correlation and/or from the proton recoil spectrum, taking $g_2 = 0$ and for f_2 the value expected from CVC. As an illustration of the sensitivity of the α_e, α_v, and α_p asymmetry parameters, their variations as a function of g_1/f_1 are shown in Figure 4. Using the value $g_1/f_1 = 0.73$ obtained from a Cabibbo fit to the data for all the decays (see Section 5), we find the expected values for α_e, α_p, and α_v, which are also listed in Table 5. The agreement with the experimental results is acceptable. The table shows the wealth of information that has been furnished by the large polarization of the Λ's and the measurements of the asymmetry parameters. Similar measurements in Λ hyperon beams, which have smaller polarization but higher statistics, are extremely desirable.

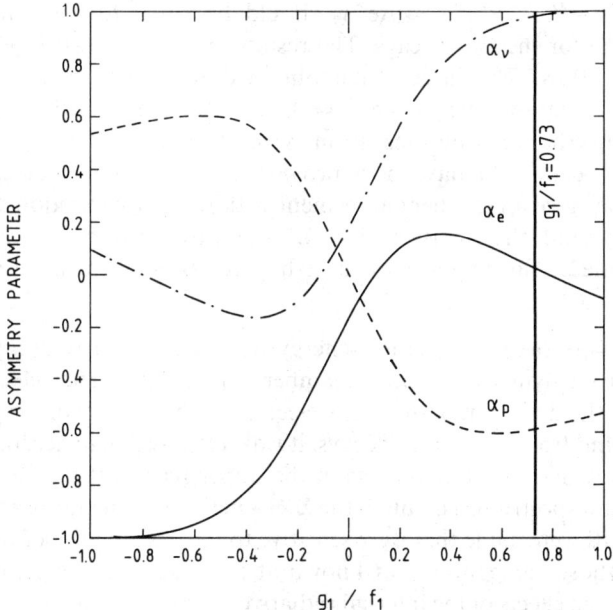

Figure 4 Variation of the electron, neutrino, and proton asymmetry parameters as a function of g_1/f_1 for the $\Lambda \to pev$ decay mode.

Table 5 Polarized Λ beta-decay experiments

	Λ Polarization	$|g_1/f_1|$	α_e	α_p	α_ν
Althoff et al (19)	0.75±0.03	0.64±0.06	0.15±0.09	−0.51±0.09	0.89±0.08
Lindquist et al (20)	0.78±0.04	$0.60^{+0.13}_{-0.10}$	0.05±0.12	−0.47±0.12	0.72±0.12
Burnett et al (21)	0.70±0.02	0.52±0.10		−0.56±0.15	0.75±0.15
Cabibbo fit		0.73	0.02	−0.59	0.98

3.1.2 $\Sigma^\pm \to \Lambda e\nu$ DECAYS An analysis of the combined $\Sigma^\pm \to \Lambda e^\pm \nu$ data samples obtained in three different bubble chamber experiments has been presented by the groups involved (22). The combined sample, which represents the totality of the low-energy data, consists of 163 Σ^- and 23 Σ^+ decays. According to the CVC hypothesis, $\Sigma^\pm \to \Lambda e\nu$ is a pure axial transition ($f_1 = 0$), except for the small contribution of the weak magnetism term f_2. In addition, in the absence of second-class currents, the ratio $R = \Gamma(\Sigma^+ \to \Lambda e^+\nu)/\Gamma(\Sigma^- \to \Lambda e^-\bar{\nu})$ should be equal to 0.61, the phase-space ratio for the two decays. The results are $f_1/g_1 = -0.45 \pm 0.20$ and $R = 0.67 \pm 0.18$. The large value obtained for f_1/g_1, more than two standard deviations away from $f_1 = 0$, probably indicates the difficulty of combining the distributions from several low-statistics experiments. Hyperon beam data have now provided a large sample of $\Sigma^- \to \Lambda e\nu$ decays that give an excellent agreement with $f_1 = 0$ (see Section 3.2.4). On the other hand, there exist no new Σ^+ data, owing to the strong limitations on the Σ^+ flux in positive charge-hyperon beams, which are explained in Section 3.2.1.

3.1.3 $\Sigma^- \to n e\nu$ DECAYS The low-energy data on $\Sigma^- \to n e\nu$ decays comes mainly from hydrogen bubble chamber experiments. The selection was based on the detection of the decay electron, and, because of the potential background from $\Sigma^- \to n\pi^-$ decays, it was not possible to accept $\Sigma \to n e\nu$ events in which the electron was in the upper part of the center-of-mass momentum spectrum. To obtain the $\Sigma^- \to n e\nu$ branching ratio, acceptance corrections were made that were sensitive to the exact choice of weak form factors whose values were not known at the time of the experiments. In addition, the effects of the internal radiative corrections on the acceptances were not taken into account. For those reasons the average of the $\Sigma^- \to n e\nu$ branching ratio given in Table 4 is very difficult to use. A rough evaluation indicates that the value would have to be shifted downward by about 4%.

In Table 4 we have given an average value for $|g_1/f_1|$ obtained by measuring the electron–antineutrino angular correlation $\alpha_{e\nu}$ with the small fraction of the events (5% of the $\Sigma^- \to ne\nu$ decays) where the neutron scattered in the bubble chamber giving a proton recoil. There were also three experiments that attempted to measure the sign of g_1/f_1 and that gave conflicting results. The 1974 PDG listing (17), referring to the $\Sigma^- \to ne\nu$ decay, stated that "since the sign of g_1/f_1 is undetermined the average of the six values is meaningless." Ten years later the sign of g_1/f_1 in $\Sigma^- \to ne\nu$ is still an open question. Four low-energy experiments (23–26) have now measured the electron asymmetry parameter α_e in the decay of polarized Σ^-. We discuss their results below. In addition, from a study of the electron spectrum in the SPS hyperon beam experiment (see Section 3.2.4) it was concluded that the negative sign of g_1/f_1, which is expected within the Cabibbo framework ($g_1/f_1 = F-D$), was favored by at least 2.6 standard deviations.

Table 6 summarizes the information about the four low-energy experiments that measured the asymmetry parameter α_e of the electron in the decay of polarized Σ^-. For the K^- experiments, with beam momenta around 400 MeV/c, the Σ^- polarization (50–60%) was calculated by using partial-wave analysis results. For the π^- experiment, the Σ^- polarization was measured to be (32 ± 11)% in a separate experiment with the same beam momentum of 1.13 GeV/c. Although for the first two K^- experiments a few $\Sigma^- \to n\mu\nu$ were also analyzed, we have chosen to give only the values of the electron asymmetry, since for muons the background was large. On the other hand, for both of those experiments the background in the $\Sigma^- \to ne\nu$ data sample was estimated to be negligible. Ellis et al (25) have given a background estimate of (6 ± 12)%, whereas in the experiment of Keller et al (26) the 193-event sample included a 36% calculated background contribution (29% K_{e_3} and 7% $\Sigma^- \to n\pi^-$). In the latter case α_e was corrected for

Table 6 Experiments with polarized Σ^-

	Beam	Detector	Events	α_e	α_-
Gershwin et al (23)	K^-	HBC	53	-0.26 ± 0.37	-0.071 ± 0.012
Bogert et al (24)	K^-	HBC	44	$+0.23 \pm 0.44$	-0.067 ± 0.011
Ellis et al (25)	π^-	counters + chambers	43	$+0.39^{+1.9}_{-0.53}$	$+0.27^{+0.88}_{-0.21}$
Keller et al (26)	K^-	counters + chambers	193	$+0.35 \pm 0.29$	-0.073 ± 0.055

the background effect, the asymmetry before correction being $+0.35\pm0.25$. The values of the $\Sigma^- \to n\pi^-$ asymmetry parameter α_- have been measured in the same experiment or in the same bubble chamber exposure. They should be considered as checks of possible instrumental asymmetries for hadrons in the case of the two counter experiments (25, 26).

The variation of α_e as a function of g_1/f_1 in the $\Sigma^- \to ne\nu$ case is very similar to that shown in Figure 4 for $\Lambda^- \to pe\nu$, and negative values of g_1/f_1 around -0.3, which are required by any standard Cabibbo fit, correspond to α_e large and negative. To be more specific, the Cabibbo fit of Section 5, which does not make use of the $\Sigma^- \to ne\nu$ α_e information, gives $g_1/f_1 = -0.28$, which corresponds to $\alpha_e = -0.54$. The set of experimental results given in Table 6 is not in agreement with that value of α_e and favors a positive sign for g_1/f_1. The problem of the sign of g_1/f_1 has now become the keystone of the Cabibbo theory. It appears to require a decisive experimental clarification, which we hope will come from the new experiment under way in the charged hyperon beam at Fermilab.

3.1.4 $\Xi^- \to \Lambda(\Sigma^0)e\nu$ DECAYS In a fine counter experiment (27) performed at CERN in a K^- beam of 1.65 GeV/c, 23 Ξ^- leptonic decay candidates were observed. The background was estimated to be 6.3 ± 0.7 events. In that experiment the decay modes $\Xi^- \to \Lambda e\nu$ and $\Xi^- \to \Sigma^0 e\nu$ could not be separated. The result based on 17 signal events was $(0.68\pm0.22) \times 10^{-3}$ for the sum of the branching ratios of the two decay modes.

3.1.5 MUONIC DECAYS The $\Sigma^- \to n\mu\nu$ branching ratio was measured in several bubble chamber experiments. To identify the muon and to avoid background contamination, only a small fraction of the μ^- spectrum was detected $(20 \lesssim p_\mu \lesssim 80 \text{ MeV}/c)$. Therefore the calculated rate depended upon the values assumed for the form factors. The average value of the branching ratio, for $g_1/f_1 \sim -0.3$, was $(0.45\pm0.04) \times 10^{-3}$, in agreement with the expectation from μ-e universality that $\Gamma(\Sigma^- \to n\mu\nu)/\Gamma(\Sigma^- \to ne\nu) = 0.45$. The average of the $\Lambda \to p\mu\nu$ branching ratio was $(0.16\pm0.04) \times 10^{-3}$.

3.2 Hyperon Beam Experiments

The construction of hyperon beams, first at the Brookhaven AGS and the CERN PS, and more recently at the CERN SPS and at Fermilab, has allowed high-statistics experiments to be performed, with 10^3 to 10^4 events for most decay channels, and up to 10^5 events in the case of the $\Lambda \to pe\nu$ decay channel.

We first give the main features of the charged and neutral hyperon beams realized so far. Then the results obtained by the different experiments are presented and discussed. A more complete description of hyperon beams can be found in the review article of Lach & Pondrom (28).

3.2.1 HYPERON BEAMS Hyperon beams are secondary beams of particles, produced by striking a target with the primary proton beam of an accelerator. These beams contain a useful flux of hyperons together with a background of the more copiously produced particles: mainly π^- for negative beams, π^+ and protons for positive beams, n and γ for neutral beams. The main design problem is the short decay length of the hyperons. For example, at 100 GeV/c the decay lengths (in meters) are, respectively, 7.1, 2.0, 3.7, 6.6, 3.7, and 1.5 for $\Lambda, \Sigma^+, \Sigma^-, \Xi^0, \Xi^-$, and Ω^-. This means that one has to build beams that are as short as possible with minimal dead spaces, and with maximal value of the magnetic field for the magnet components of the secondary beam, and that no particle separation is possible.

There is an advantage in going to higher energies: the decay length grows linearly with the energy whereas the length of shielding necessary to absorb the hadronic cascades produced by dumping the primary proton beam grows only logarithmically. Consequently the length of the hyperon beam channels will not scale with the hyperon energy. Since the fractional yield of hyperons at the target depends on x_F (Feynman x) rather than on the energy, the hyperon flux available at the exit of the channel is enhanced in a spectacular way. For example, between the first-generation beams at the Brookhaven AGS and at the CERN PS (15–20 GeV) and the second-generation beams at the CERN SPS and at Fermilab (100–200 GeV), the fluxes of hyperons have been increased by about two orders of magnitude, and fluxes of 5×10^3 Σ^-, 5×10^2 Ξ^-, or 5×10^4 Λ per pulse have been achieved.

Another interesting feature is the existence of the polarization of the produced hyperons. In the case of Λ production, the polarization grows linearly with p_T, being zero at $p_T = 0$ and 25% at $p_T = 2$ GeV/c (28). This polarization enables the observation, in leptonic decays, of asymmetries that are very sensitive to the sign of the g_1/f_1 ratio. However, up to now, no published data have explicitly used this polarization. Two high-statistics $\Lambda \to pe\nu$ experiments using polarized Λ beams are being analyzed. The Fermilab charged-hyperon beam measured also the polarization of the Σ^- at 250 GeV/c, and a high-statistics $\Sigma^- \to ne\nu$ experiment is taking data to settle the question of the sign of g_1/f_1 in this particular decay. Hyperons in the CERN SPS beam are not polarized; however, in all decays involving a Λ in the final state the asymmetries of the decay products of the Λ are also very sensitive to the sign of the g_1/f_1 ratio. The Λ's produced by the Ξ^- decay are longitudinally polarized, and therefore in the $\Lambda \to pe\nu$ channel there are asymmetries that permit a determination of the sign of g_1/f_1. Table 7 summarizes the decay channels studied and the number of events observed for each hyperon beam.

The Ω^- semileptonic decay has recently been observed for the first time at the CERN SPS. On the other hand, as the Σ^+ flux is about 100 times smaller than the Σ^- flux, owing to the shorter life time and the large proton component in the positive beam, no Σ^+ leptonic decay experiment has so far been performed with a hyperon beam.

3.2.2 EXPERIMENTS WITH CHARGED HYPERON BEAMS AT THE BROOKHAVEN AGS AND CERN PS

The first such experiment, done by Tanenbaum et al (29), studied the leptonic decays of the Σ^- and Ξ^-. From a sample of 3507 reconstructed $\Sigma^- \to ne\nu$ and 55 $\Sigma^- \to \Lambda e\nu$ they determined the axial-to-vector form-factor ratio g_1/f_1 for both channels. The experimental set-up of Tanenbaum et al (29) is shown in Figure 5. Beam particles of mass less than 1 GeV/c^2 are vetoed by a threshold Cherenkov counter, which is part of the beam channel. The direction and momentum of the outgoing hyperons are measured by a set of high-pressure, high-resolution, magnetostrictive spark chambers. A spectrometer equipped with magnetostrictive chambers located after a 2.9-m decay region measures the direction and momentum of the charged particles (π^-, e^-, p). A second spectrometer is used to improve the proton momentum measurement. The electron signature is provided by a gas Cherenkov counter. The direction of the neutron coming from $\Sigma^- \to n\pi^-$ or $\Sigma^- \to ne\nu$ decay is determined by measuring the neutron shower coordinates in a neutron counter and the Σ^- decay point. For the $\Sigma^- \to ne\nu$, the measured quantities—the Σ^- and e^- momenta and the neutron direction—yield a zero constraint fit with two solutions for the neutron energy. The major background for the leptonic triggers was due to

Table 7 Summary of decay channels

AGS charged beam	first experiment: $\Sigma^- \to ne\nu$ 3507 ev; $\Sigma^- \to \Lambda e\nu$ 55 ev; $\Xi^- \to \Lambda e\nu$ 11 ev	
	second experiment: $\Sigma^- \to \Lambda e\nu$ 119 ev; $\Xi^- \to \Lambda e\nu$ 14 ev	
AGS neutral beam	$\Lambda \to pe\nu$	10,039 ev
		(100,000 events under analysis)
PS charged beam	$\Sigma^- \to ne\nu$	519 ev
SPS charged beam	$\Sigma^- \to \Lambda e\nu$	1649 ev
	$\Sigma^- \to ne\nu$	4456 ev
	$\Lambda \to pe\nu$	7111 ev
	$\Xi^- \to (\Lambda, \Sigma^0 e\nu)$	3011 ev
	$\Xi^- \to \Lambda e\nu$	2608 ev
	$\Xi^- \to \Sigma^0 e\nu$	154 ev
	$\Omega^- \to \Xi^0 e\nu$	branching ratio measured
Fermilab neutral beam	$\Lambda \to pe\nu$	($\sim 55{,}000$ events under analysis)
Fermilab charged beam	$\Sigma^- \to ne\nu$	data taking (10^5 events foreseen)

accidentals between a background muon entering the Cherenkov counter and a normal $\Sigma^- \to n\pi^-$ or $\Xi^- \to \Lambda\pi^-$ decay. The extensive use of the Cherenkov information is not sufficient to eliminate all the hadronic background, and a cut on the Σ^- mass, interpreting the event as a $\Sigma^- \to n\pi^-$ decay, is done to obtain the final leptonic sample of 3507 events.

The axial-to-vector form-factor ratio has been determined by comparing experimental and Monte Carlo (MC) neutron energy spectra in the Σ^- rest frame. To take into account the double solution arising from the 0C fit, both solutions are plotted in the energy spectrum. A scatter plot of the lower solution versus the higher solution has also been made in order to obtain maximum information in a bias-free manner, and this scatter plot has been compared with the equivalent MC scatter plot. The data have been taken at two values of the spectrometer field. The result found is $|g_1/f_1| = 0.435 \pm 0.035$ neglecting the q^2 dependence of the form factors and the radiative corrections.

The analysis of the $\Sigma^- \to \Lambda e \nu$ is done in a similar way by requiring a $\Lambda \to p\pi^-$ instead of a neutron. The $\Xi^- \to \Lambda\pi^-$ background is eliminated by a

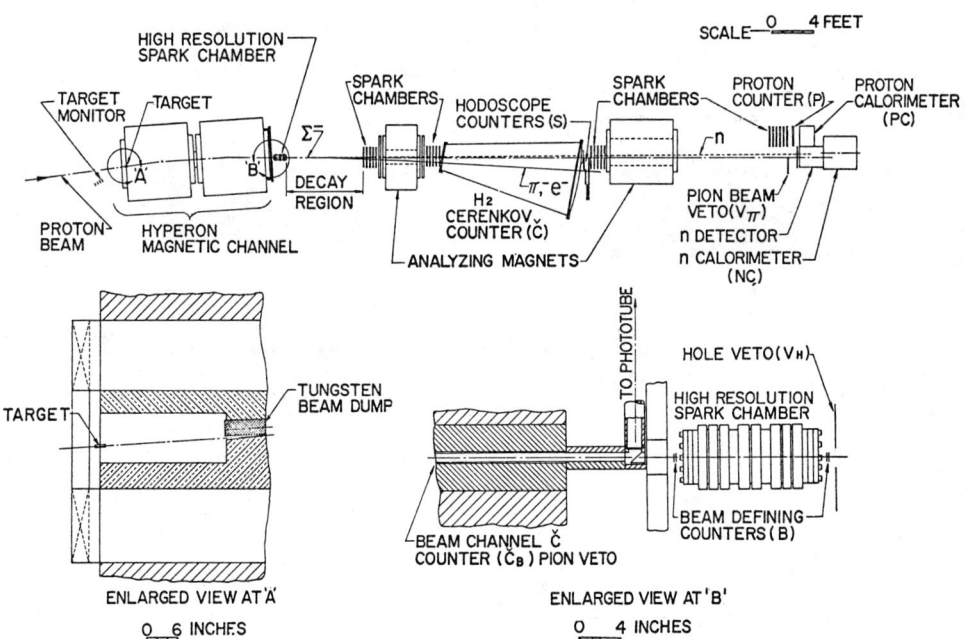

Figure 5 Apparatus of Tanenbaum et al (29) for detecting Σ^- and Ξ^- semileptonic decays.

kinematic cut, which leaves a final sample of 55 events. The value of g_1/f_1 is extracted from the Λ kinetic energy in the Σ^- rest frame. The angular distributions of the proton depend on the Λ polarization and allow a determination of the sign of f_1/g_1. As f_1 is predicted to be zero for this channel, one has the ratio f_1/g_1 instead of the usual g_1/f_1. The result is $f_1/g_1 = 0.17 \pm 0.35$, assuming that f_2 has the CVC value. A sample of 11 $\Xi^- \to \Lambda e \nu$ events with a residual contamination of three or four $\Xi^- \to \Lambda \pi^-$ events has also been obtained.

A second experiment, using the same beam and a similar set-up, has been performed by Thompson et al (30) and has studied the $\Sigma^- \to \Lambda e \nu$ and the $\Xi^- \to \Lambda e \nu$ channels. The two branching ratios have been determined as well as the ratio f_1/g_1 for the $\Sigma^- \to \Lambda e \nu$ channel. The electron signature was also given by a gas Cherenkov counter. The Cherenkov information was not sufficient to eliminate the $\Xi^- \to \Lambda \pi^-$ background, and a cut on the $\Lambda \pi^-$ mass has been done (interpreting the electron candidate as a pion). The $\Sigma^- \to \Lambda e \nu$ and $\Xi^- \to \Lambda e \nu$ channels are separated according to the value of the $(\Lambda e \nu)$ mass.

There are 119 $\Sigma^- \to \Lambda e \nu$ events with a background of 5 ± 1.5 events and 14 $\Xi^- \to \Lambda e \nu$ events with a background of 3.4 ± 1.7 events. The corresponding values of the branching ratios are

$$B(\Xi^- \to \Lambda e \nu / \Xi^- \to \Lambda \pi^-) = (0.30 \pm 0.13) \times 10^{-3},$$

and

$$B(\Sigma^- \to \Lambda e \nu / \Sigma^- \to n \pi^-) = (0.63 \pm 0.11) \times 10^{-4}.$$

A search for $\Xi^- \to \Sigma^0 e \nu$ and $\Xi^- \to \Lambda \mu \nu$ has given the following upper limits:

$$B(\Xi^- \to \Sigma^0 e \nu / \Xi^- \to \Lambda \pi^-) < 0.14 \times 10^{-3} \quad (90\% \text{ C.L.}),$$

$$B(\Xi^- \to \Lambda \mu \nu / \Xi^- \to \Lambda \pi^-) < 2.3 \times 10^{-3} \quad (90\% \text{ C.L.}).$$

The f_1/g_1 and f_2/g_1 ratios have been determined using the electron and Λ kinetic energy distributions in the Σ^- rest frame and the angular distribution of the proton in the Λ rest frame. The values are

$f_1/g_1 = +0.32 \pm 0.30$,

$f_2/g_1 = +0.6 \pm 3.6$,

and

$\text{Im}(f_1/g_1^*) = -0.2 \pm 0.7$,

consistent with time reversal invariance, which predicts f_1/g_1 to be real. For both experiments the sign of g_1 in the $\Sigma^- \to \Lambda e \nu$ decay quoted here has been reversed to be consistent with the sign convention defined in Section 2.

Decamp et al (31) studied the $\Sigma^- \to ne\nu$ decay in the charged PS hyperon beam, and collected a sample of 519 events. The Σ^- were selected by a DISC counter. The direction and momenta of the charged decay particles were measured in two streamer chambers placed upstream of and inside the spectrometer magnet. A gas Cherenkov counter located in front of the spectrometer magnet identified the electron. A neutron counter composed of iron-plate optical spark chambers was placed downstream. As in the AGS experiment, the main background was caused by accidental coincidence between a $\Sigma^- \to n\pi^-$ and a particle triggering the Cherenkov counter. This background was reduced to an acceptable level by making tight cuts on the Cherenkov information and a cut on the Σ^- mass. The $|g_1/f_1|$ ratio was extracted from the neutron kinetic energy distributions where both solutions were used. A procedure was developed to recover 25% of the events where the discriminant that gives rise to the two solutions is negative owing to measurement errors. The value found, $|g_1/f_1|$ = $0.17^{+0.07}_{-0.09}$, is more than three standard deviations away from the result of Tanenbaum et al (29).

For all these initial experiments with charged-hyperon beams the analysis was done neglecting the q^2 dependence of the form factors and radiative corrections. In the case of the $\Sigma^- \to ne\nu$ decay, which corresponds to the largest value of q^2_{max}, neglecting the q^2 dependence introduces a large shift of the value of g_1/f_1, as seen in Section 2.4. This shift depends upon the exact acceptance of each experiment, and it is therefore difficult to combine data where this correction has not already been done at the analysis level.

3.2.3 EXPERIMENTS WITH NEUTRAL BEAMS   At the AGS Wise et al (32) have collected about 10^5 events in the channel $\Lambda \to pe\nu$. The present analysis, based on a sample of 10,000 events, provides measurements of the branching ratio and the g_1/f_1 and f_2/f_1 ratios. The experimental apparatus is similar to that of Figure 5 except that it uses a neutral beam. The momentum spectrum of the beam is determined by using the hadronic $\Lambda \to p\pi^-$ decay. The spectrometer is composed of two analyzing magnets to measure the proton and electron momenta with the same accuracy. An excellent electron-pion separation is achieved by using two gas Cherenkov counters for the electron identification and two different ones for the pion identification. The remaining contamination of $\Lambda \to p\pi^-$ after applying cuts mainly based on the Cherenkov information is $2.5 \pm 0.5\%$. There is a $0.7 \pm 0.3\%$ K_{e3} contamination. Other sources of background such as $K^0 \to 3\pi$, $K^0 \to \pi\mu\nu$, or $\Lambda \to p\pi$, followed by $\pi \to e\nu$ decay in flight, etc, contribute to less than 1% of the signal. The final Λ beta-decay sample, after correcting for these effects contains 10,039 events.

The $|g_1/f_1|$ ratio has been determined from the electron-neutrino angular correlation, which is independent of the initial Λ polarization. There are two solutions for the magnitude of the Λ momentum coming from the 0C fit $\Lambda \to pev$. For the analysis the value closer to the mean beam momentum has been used for the data events and for the Monte Carlo simulation. The results are

$|g_1(0)/f_1(0)| = 0.734 \pm 0.031$

$\Gamma(\Lambda \to pev)/\Gamma(\text{all}) = (0.843 \pm (0.017)\, 10^{-3} \times [1+(0.80-|g_1/f_1|) \times 0.057]$.

The q^2 dependences have been taken into account by using the parametrization given in Section 2.3 but with different values of the masses: $m'_V = 0.84$ GeV/c^2 and $m'_A = 0.89$ GeV/c^2. If one uses instead the values of m'_A and m'_V given in Section 2.3, the value of $|g_1/f_1|$ is increased by about 0.03. The same correction applies to all the measurements of g_1/f_1 in that experiment.

The acceptance of the apparatus for the Λ beta-decay mode depends weakly on the value of $|g_1/f_1|$ as shown in the above formula. For the branching-ratio determination, the change in the shape of the density distribution over the Dalitz plot, which is due to radiative correction, has been taken into account for the acceptance calculation. The quoted errors include the statistical errors and the systematic uncertainties due to the background subtraction, the Cherenkov inefficiencies, and the Λ^0 momentum spectrum determination.

A more complete analysis of the same sample has recently been presented (33). By studying the Dalitz plot distribution, the sign of g_1/f_1 and the value of the weak magnetism form factor are obtained:

$g_1(0)/f_1(0) = +0.715 \pm 0.026 \quad \text{and} \quad f_2/f_1 = 1.1 \pm 0.2.$

The value of $g_1(0)/f_1(0)$ has diminished by less than one standard deviation and the value of f_2 is in agreement with the CVC prediction $f_2/f_1 = 0.97$. The value of $|g_1/f_1|$ was obtained assuming $g_2 = 0$ and f_2 equal to the value predicted by CVC. The dependence of $|g_1/f_1|$ on the value of g_2 is:

$|g_1/f_1| = 0.715 + 0.25\, g_2/f_1$

The analysis of the full sample, including the study of the proton and electron asymmetries with respect to Λ polarization, has not yet been completed. It requires a very careful study of the various systematic effects.

Another experiment (34) has been performed at Fermilab with a polarized Λ hyperon beam. A sample of about 55,000 Λ beta decays has been collected and the analysis is in progress.

3.2.4 EXPERIMENT IN THE CERN SPS CHARGED-HYPERON BEAM The charged-hyperon SPS beam has been used by Bourquin et al (35) to collect data on the Σ^-, Ξ^-, and Λ semileptonic decays in a single experiment, in order to minimize possible biases when comparing different decay channels. The numbers of events in the different channels are large:

$\Sigma^- \to \Lambda e \nu$ 1649 events

$\Sigma^- \to n e \nu$ 4456 events

$\Lambda \to p e \nu$ 7111 events

$\Xi^- \to \Lambda e \nu$ 2608 events

$\Xi^- \to \Sigma^0 e \nu$ 154 events

The $\Lambda \to p e \nu$ sample is obtained from the decay of $\Xi^- \to \Lambda \pi^-$ followed by a Λ beta decay.

The experimental apparatus is shown in Figure 6. The Σ^- and Ξ^- are identified in parallel by a DISC counter placed at the exit of the magnetic channel, which selects 100-GeV/c negative particles. A set of multiwire proportional chambers (MWPCs) located upstream and downstream of the DISC counter enables the direction of the beam particles and their momentum to be measured. After a 10-m decay region a spectrometer, comprising an analyzing magnet and two sets of four drift chambers, is used to measure the tracks of the charged particles upstream and downstream of the magnet. Several electron detectors are used: a lead-glass array, two transition radiation detectors, and a gas Cherenkov counter. To detect low-energy electrons a lead/scintillator shower counter is placed on one side of the lead-glass array. Only the lead-glass array and the shower counters are used in the trigger. For the $\Sigma^- \to n e \nu$ data a neutron counter, made of iron plates and proportional chambers, is located in front of the lead-glass array. Small-angle γ rays are detected in the lead-glass array and large-angle γ rays in the γ-ray hodoscope placed at the end of the decay region.

As the number of leptonic events is very large, a careful study of the Monte Carlo simulation of the experiment has been made. The hadronic decays $\Xi^- \to \Lambda \pi^-$, $\Sigma^- \to n \pi^-$, and $\Sigma^+ \to p \pi^0$ have been used to check the resolutions, the detection efficiencies, and the absolute momentum calibration.

The data reduction requires the presence in the event of a $\Lambda \to p \pi^-$ for the $\Xi^- \to \Lambda(\Sigma^0) e \nu$ and $\Sigma^- \to \Lambda e \nu$ decay modes, or of a missing Λ, $[\Xi^- \to \Lambda \pi^-]$, for the $\Lambda \to p e \nu$ mode. The electron selection is done by using the lead-glass and the transition radiation detector information. At that stage of the analysis there remain two main sources of background: accidental coincidences between a beam electron and a hadronic Ξ^- decay, and the Σ^- and Ξ^- interactions. To eliminate both sources of background, a set of kinematic cuts tailored to each decay has been applied. The remaining $\Xi^- \to \Lambda \pi^-$ contamination is estimated to be negligible for the

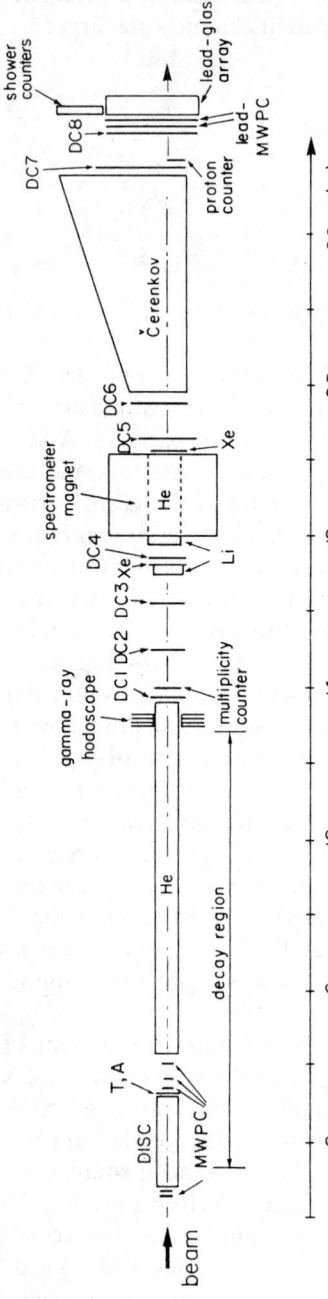

Figure 6 Plan view of the apparatus of the CERN SPS hyperon experiment. He: helium bags; Li: lithium radiators; Xe: xenon proportional chambers; DC: drift chambers.

$\Sigma^- \to \Lambda e \nu$ channel, and 7.5% and 12% for the $\Lambda \to p e \nu$ and the $\Xi^- \to \Lambda e \nu$ channels, respectively. The results on g_1/f_1 and the branching ratio are obtained either by adding to the MC events the right amount of hadronic events, or by eliminating these events with a cut on the Λ mass. The two methods give consistent results.

For the $\Xi^- \to \Sigma^0 e \nu$ channel, the presence of a γ ray detected in the lead-glass array is required. The low statistics and the background due to spurious γ rays in the γ chamber has allowed only the branching ratio to be determined.

The $\Sigma^- \to n e \nu$ decay is less constrained than the other decays; therefore the kinematic cuts used to eliminate the background due to Σ^- interactions and beam electrons have to be different. The remaining $\Sigma^- \to n\pi^-$ background is $(5\pm1)\%$ for the branching ratio sample. The contaminations from the other leptonic decays of Ξ^- and Σ^- have been calculated and they amount to $(6.6\pm2.2)\%$ in total. These backgrounds are smaller for the form factor sample. In the MC simulation of the $\Sigma^- \to n e \nu$ the corresponding amount of each type of background has been included. As in the other $\Sigma^- \to n e \nu$ experiments, there are two possible solutions for the neutron momentum:

$$p_{1,2} = p_0 \pm \sqrt{D}.$$

For about 30% of the events the discriminant D is negative owing to measurement errors. The form-factor study is done either with the events having $D > 0$, using p_1 and p_2 as independent entries, or with all the events using the "pseudo" momentum p_0.

To extract the g_1/f_1 ratio from the data, the following distributions have been studied: baryon kinetic energy, lepton kinetic energy, Dalitz plot, and electron–neutrino angular correlation. When a Λ is involved in the decay, the asymmetries of the final baryon have also been studied. The q^2 dependences of the form factors, as defined in Section 2, have been applied. The radiative corrections are important only for the branching ratio determination. At this stage, only the change in the experimental acceptances due to radiative corrections is taken into account. The radiative corrections (see Section 4) will be applied for the Cabibbo fit in Section 5. The results are summarized in Table 8.

In the case of the $\Sigma^- \to n e^- \nu$ decay the sign of the form-factor ratio is only given by the electron energy spectrum. The sign given here, in agreement with the Cabibbo theory, is favored by at least 2.6 standard deviations. A careful study of the systematic errors has been made, and the quoted errors in Table 8 take into account both the statistical and systematic errors.

The values of g_1/f_1 have been obtained assuming $g_2 = 0$ and $f_2 =$

Table 8 Results of the SPS experiment on hyperon semileptonic decays

Decay	Branching ratio	g_1/f_1 ($g_2 = 0$)	$\Delta g_1/\Delta g_2$	$\Delta(g_1/f_1)$ ($g_2 \neq 0$)[b]
$\Sigma^- \to \Lambda e \nu$	$(0.561 \pm 0.031) \times 10^{-4}$	$+0.03 \pm 0.08$[a]		
$\Sigma^- \to n e \nu$	$(0.96 \pm 0.05) \times 10^{-3}$	-0.34 ± 0.05	0.24	-0.013
$\Xi^- \to \Lambda e \nu$	$(5.64 \pm 0.31) \times 10^{-4}$	$+0.25 \pm 0.05$	0.15	$+0.007$
$\Xi^- \to \Sigma^0 e \nu$	$(0.87 \pm 0.17) \times 10^{-4}$			
$\Lambda \to p e \nu$	$(8.57 \pm 0.36) \times 10^{-4}$	$+0.70 \pm 0.03$	0.20	$+0.021$

[a] f_1/g_1 is quoted in this case.
[b] The estimates of g_2 are from (43).

f_2 (CVC). The measured dependences $\Delta g_1/\Delta g_2$ of g_1 on g_2 are given in Table 8. The weak magnetism form factors, and the quantity Im (f_1/g_1^*) have also been determined and are listed in Table 9.

4. RADIATIVE CORRECTIONS

Experiments on several hyperon semileptonic decays have now reached a level of precision where the radiative corrections are expected to be significant compared with the experimental errors. It is therefore important to take into account these corrections when comparing the experimental information with the predictions of the Cabibbo model. Radiative corrections to muon decay and to neutron beta decay have been discussed extensively in the past because of their relevance to the problem of universality of weak interactions. A theoretical difficulty in the old V–A theory was that the corrections were logarithmically divergent for the neutron decay, whereas the corresponding quantities for muon decay were finite and well defined. In practice, the divergence problem was avoided by introducing an arbitrary cut-off Λ into the calculation. Clearly this situation was very unsatisfactory from a theoretical point of view, in particular because of the shadow cast on the universality concept by the uncertainties associated

Table 9 Weak magnetism form factors and Im $(f_1 g_1^*)$

	f_2/f_1 (meas)	f_2/f_1 (CVC)	Im $(f_1 g_1^*)$
$\Sigma^- \to \Lambda e \nu$[a]	1.82 ± 1.82	1.21	-0.10 ± 0.14
$\Sigma^- \to n e \nu$	-1.0 ± 0.34	-1.14	
$\Xi^- \to \Lambda e \nu$	-0.24 ± 0.25	-0.07	0.009 ± 0.076
$\Lambda \to p e \nu$	1.32 ± 0.81	0.97	0.098 ± 0.148

[a] In the case of $\Sigma^- \to \Lambda e \nu$ the quoted numbers are f_2 and Im (f_1/g_1^*).

with the arbitrariness of the cut-off Λ. Theoretical progress culminated with the demonstration by Sirlin (36) that within the framework of the standard $SU(2) \times U(1)$ gauge model of electroweak interactions the ultraviolet divergences cancel to order α relative to the basic weak process. The radiative corrections are finite for the semileptonic baryon decays, without the need for an arbitrary cut-off, as well as for the muon decay. The new calculation introduces additional terms for the muon decay radiative corrections. However, for the purpose of discussing universality and for the comparison with Cabibbo theory, the exact definition of the weak coupling constant is not important provided that any renormalization factor is applied in an identical manner to every process. Therefore, since for convenience the "old" V–A definition of the universal weak coupling constant G_μ is kept,

$$\tau_\mu^{-1} = (G_\mu^2 m_\mu^5/192\pi^3)[(1-(8m_e^2/m_\mu^2)]\{1+(\alpha/2\pi)[(25/4)-\pi^2]\},$$

the additional radiative corrections to muon decay are subtracted for all the other processes.

In the case of baryon semileptonic decays, it is convenient to identify three separate contributions to the total radiative correction (δ_a, δ_b, and δ_c), so that the square of the matrix element can be written as

$$M'^2 = M^2(1+\delta_a)(1+\delta_b)(1+\delta_c),$$

where M is the matrix element defined in Section 1.

In this formulation, δ_a is a "classical" Coulomb term present in neutral baryon decays and describing the charge attraction between the electron and the final-state charged baryon. It is given by the expression $\delta_a = \pi\alpha/\beta$, where β is the velocity of the electron.

The term δ_b contains all the contributions that depend explicitly on the electron energy, with the exception of the Coulomb correction δ_a. It does not depend on the details of the weak decay or of the strong interactions. It is given (36) by the expression $\delta_b = (\alpha/2\pi)g(E, E_M)$, where E is the electron energy and E_M is the maximum value of E. This correction modifies both the shape of the electron spectrum and the decay rate. For example, averaging over an electron spectrum of the form $dN \simeq \beta E^2(E_M-E)^2 dE$, a good approximation in the case of neutron beta decay, we obtain $\tilde{g}(E_M) = 3 \ln (m_{ch}/2E_M) + 81/10 - 4\pi^2/3$, where m_{ch} is the mass of the charged baryon in the decay.

The term δ_c is electron energy independent, and contains all the leftover corrections. It is often called the model-dependent term since, in contrast with δ_b, it does depend on the detailed anatomy of the weak interaction and it also includes small model-dependent effects induced by the strong interactions; δ_c is the term that was logarithmically divergent in the old

V–A derivation. In the framework of the standard SU(2) × U(1) model of electroweak interactions, it is given by the expression

$$\delta_c = \alpha/2\pi[3 \ln (m_Z/m_p) + 6\bar{Q} \ln (m_Z/M) + 2C + A_S], \qquad 22.$$

where $m_Z \cong 90$ GeV/c^2 is the mass of the neutral weak intermediate boson Z^0, m_p is the proton mass, and $M \cong 1$ GeV/c^2. Here $\bar{Q}$ is the mean charge of the quarks, $\bar{Q} = (2/3 - 1/3)/2 = 1/6$. The term $2C$ is a nonasymptotic piece contributed by the weak axial current. Estimates indicate that $|2C| \simeq 1$, a value about 20 times smaller than the contribution of the sum of the two leading logarithmic terms in δ_c. The contribution A_S induced by the strong interactions is estimated to be about 50 times smaller than the leading contribution.

The effects of the model-independent corrections δ_a and δ_b on the decay rates are obtained by taking an average over the electron spectrum for each decay. The results are shown in Column 1 of Table 10. For the model-dependent term δ_c taking $M = 1$ GeV/c^2 and neglecting the C and A_S contributions in Equation 22, one obtains $\delta_c = 2.1\%$. Column 2 of Table 10 shows the total radiative correction obtained by applying a further correction of 2.1% to the results given in Column 1 for each of the decays.

As indicated at the beginning of this section the uncertainties in the calculation of the radiative corrections constitute limitations on probing the hypothesis of the universality of weak interactions, which relates the "bare" coupling constants of the various decays. The dominant contributions to the uncertainties within the framework of recent calculations come from the model-dependent term δ_c. The contributions of the terms that have been neglected, C and A_s, may amount to $\delta(\delta_c) \approx 0.15\%$. The uncertainty corresponding to the choice of the value of M is negligible because of the logarithmic dependence.

An independent calculation by Toth is reported by the SPS hyperon group (35d), with the results listed in column 3 of Table 10. The work of Sirlin (36) has been extended by considering the corrections as a function of both the electron and the final-state baryon energies and by refining the

Table 10 Radiative correction to the semileptonic decay rates (in %)

Decay	1	2	3
$\Lambda \to pe\nu$	+1.4	+3.50	+3.9
$\Sigma^- \to ne\nu$	−0.81	+1.29	+1.1
$\Sigma^- \to \Lambda e\nu$	−0.23	+1.87	+1.6
$\Xi^- \to \Lambda e\nu$	−0.50	+1.60	+1.3
$n \to pe\nu$	+4.9	+7.10	+7.3

calculations of δ_c. In the latter case Sirlin's approach, starting with free quarks, has been adopted at large values of k^2 (where k is the photon energy). However, for $k^2 < 10$ (GeV/c)^2 the quarks have been treated collectively by using the baryon weak form factors g_1, f_1, and f_2 to parametrize the matrix element. The differences between the results derived from Sirlin's expressions and those obtained by Toth are between 0.2% and 0.4% and may be considered as yet another way of gauging the uncertainties in the calculation of the radiative corrections. For all the decays, the corrections to the values of the form factor ratio g_1/f_1 deduced from the decay distributions are negligible.

5. FITS TO THE CABIBBO MODEL

Within the framework of the Cabibbo model the analysis of the experimental results is made in terms of three parameters θ_C, F, and D. The relations between the form factors at $q^2 = 0$, $f_1(0)$, $g_1(0)$, and $f_2(0)$ have been given in Table 3.

The purpose of a Cabibbo fit is to determine the values of θ_C, F, and D corresponding to the best agreement between the experimental results and the theoretical expectations for all the observed decays. Conclusions about the validity of the model are inferred from the quality of the agreement, e.g. value of the χ^2 for a least-square fit.

In previous tests of the Cabibbo model it has been necessary to combine data from many different experiments. These experiments had been analyzed under a variety of assumptions that affect the values obtained for the weak form factors and for the leptonic decay branching ratios. For example, the dependences of the form factors on q^2 has often been neglected. With the advent of the high-statistics experiments on hyperon semileptonic decays, the complete consistency of the assumptions used in the experimental analyses has become crucial for the significance of the fit procedure. For these reasons we use as hyperon input data for the fit only the results of the SPS experiment (Section 3.2.4) in which five hyperon semileptonic decays have been measured in a consistent way. In that high-statistics experiment, the q^2 dependences of the form factors (following the recipe given in Section 2.3) and the effects of radiative corrections have been included in the experimental analysis. It is possible to perform a worthwhile Cabibbo analysis using hyperon data alone; however, most of the fits will include results on neutron beta decay, which provides powerful additional constraints. The current status of neutron decay has recently been reviewed (35d). The absolute value of g_1/f_1 obtained from the neutron lifetime measurements taking into account the radiative corrections is

$$|g_1/f_1|_{\tau_n} = 1.239 \pm 0.009.$$

The decay correlation measurements yield an independent value of g_1/f_1:

$(g_1/f_1)_{\text{corr.}} = 1.258 \pm 0.009$.

These two values of g_1/f_1 differ by 1.5 standard deviations, which is not particularly significant by itself. But, as we see below, the quality of the fits does change significantly, depending on which of the two values for $(g_1/f_1)_{n \to p}$ is used.

We have calculated all the phase-space coefficients from the formulae of Bender et al (13) using the most recent values of the particle masses (5) and the universal weak coupling derived from the muon lifetime,

$G = (1.16632 \pm 0.00004) \times 10^{-11} \text{ MeV}^{-2}$.

For the particle lifetimes we have used (5):

$\tau_{\Sigma^-} = (1.482 \pm 0.011) \times 10^{-10}$ s,

$\tau_\Lambda = (2.632 \pm 0.020) \times 10^{-10}$ s,

$\tau_{\Xi^-} = (1.641 \pm 0.016) \times 10^{-10}$ s.

5.1 Fits without Radiative Corrections

We first consider fits in which radiative corrections to the hyperon semileptonic branching ratios have been omitted. Radiative corrections have been applied to the neutron lifetime, however, since they are important in relation to the small error on this quantity.

Three fits have been made. The first uses the SPS data alone, which are given in Table 8; the second includes the neutron lifetime; and the third fit includes both the neutron lifetime and neutron decay correlation data. The results are summarized in the first three columns of Table 11, where the residuals defined as $R = (x_{\text{fit}} - x_{\text{meas}})/\delta x_{\text{meas}}$ are given for each measured quantity used in the corresponding fit. The input data values are also given in the table.

The fit made with the SPS data alone (Column 1) is good with a χ^2 of 3.9 for 5 degrees of freedom (DOF). The value obtained for the Cabibbo angle corresponds to $\sin \theta_C = 0.2365 \pm 0.0038$. Including the neutron lifetime in the fit increases the χ^2 significantly, but the fit is still good with $\chi^2 = 6.9$ for 6 DOF. The partial χ^2 values are quite uniformly distributed with no single measurement contributing more than 2 to the total. The values of F, D, and $\sin \theta_C$ obtained from the two fits are similar; however, introducing the neutron lifetime constraint reduces the error on $\sin \theta_C$ by 20% and the errors on F and D by 30%. Column 3 of the table shows the result of the fit made using both neutron lifetime and decay correlation information. The fit is poorer, but still acceptable, with $\chi^2 = 10.6$ for 7 DOF.

Table 11 Fits to the Cabibbo model[a]

		Input values	Additional features	1	2	3	4	5	6
			n lifetime		*	*		*	*
			$(g_1/f_1)_{n\to p e\nu}$			*			*
			Rad. corr.				*	*	*
g_1/f_1	$\Lambda \to pe\nu$	$+0.70 \pm 0.03$	Residual	$+0.1$	$+1.0$	$+1.3$	-0.1	$+1.0$	$+1.2$
	$\Xi^- \to \Lambda e\nu$	$+0.25 \pm 0.05$		-0.7	-0.5	-0.4	-0.8	-0.5	-0.4
	$\Sigma^- \to n e\nu$	-0.34 ± 0.05		$+1.3$	$+1.2$	$+1.2$	$+1.3$	$+1.2$	$+1.2$
	$n \to pe\nu$	$+1.258 \pm 0.009$		—	—	-1.4	—	—	-1.5
Br. ratio ($\times 10^3$)	$\Sigma^- \to \Lambda e\nu$	0.0561 ± 0.0031	Residuals	-0.1	$+1.0$	$+1.3$	-0.0	$+1.3$	$+1.6$
	$\Lambda \to pe\nu$	0.857 ± 0.036		$+0.5$	$+0.9$	$+1.2$	$+0.7$	$+1.2$	$+1.5$
	$\Sigma^- \to n e\nu$	0.96 ± 0.05		$+0.4$	-0.1	-0.1	$+0.3$	-0.3	-0.3
	$\Xi^- \to \Sigma e\nu$	0.087 ± 0.017		-0.2	-0.0	$+0.0$	-0.3	-0.1	-0.0
	$\Xi^- \to \Lambda e\nu$	0.564 ± 0.031		-1.1	-1.4	-1.3	-1.3	-1.5	-1.5
	$n \to pe\nu$	1.239 ± 0.009[b]		—	-0.6	$+0.7$	—	-0.7	$+0.7$
Results of the fits	F			0.458	0.478	0.483	0.453	0.477	0.483
				± 0.016	± 0.011	± 0.011	± 0.016	± 0.012	± 0.011
	D			0.734	0.756	0.762	0.729	0.756	0.762
				± 0.017	± 0.011	± 0.011	± 0.017	± 0.011	± 0.011
	$\sin \theta_C$			0.237	0.233	0.233	0.235	0.231	0.230
				± 0.004	± 0.003	± 0.003	± 0.004	± 0.003	± 0.003
	χ^2			3.9	6.9	10.6	4.5	8.8	13.0
	DOF			5	6	7	5	6	7

[a] The basic fit, shown in Column 1, uses the SPS data alone, without radiative corrections. The additional features included in each of the other fits are indicated by *. The χ^2 for the fit is given by $\chi^2 = \Sigma R_i^2$, where R_i are the residuals of the measured quantities, defined as $R_i = [x_i(\text{fit}) - x_i(\text{meas})]/\delta x_i(\text{meas})$.
[b] In the case of the neutron decay the value of $|g_1/f_1|$ extracted from the lifetime is given.

5.2 Fits with Radiative Corrections

In addition, we now apply to the hyperon semileptonic branching ratios the radiative corrections calculated by Toth for the SPS experiment (see Table 10) and we repeat the three fits of the previous section. The results are shown in Columns 4, 5, and 6 of Table 11, respectively, and may be compared with the fits in Columns 1, 2, and 3, which were made with radiative corrections to the neutron lifetime alone.

The fit to the SPS data alone is almost unchanged when the radiative corrections are applied. When the neutron lifetime constraint is added, the χ^2/DOF is increased to 8.8/6 which is still quite acceptable. However, the fit including both the neutron lifetime and the decay correlation data gives a χ^2/DOF of 13.0/7, with radiative corrections. We attribute this rather poor fit to the inconsistency between the two types of neutron data, and include only the neutron lifetime constraint in most of the fits discussed in the following section. The results of the fits are insensitive to the set of radiative corrections (Toth or Sirlin) used.

5.3 Discussion and Comparison with Other Results

The values of θ_C, F and D given by the last two fits in the previous section are very similar. The results of the fit (called FIT 5 below) with the neutron lifetime measurements are $F = 0.477 \pm 0.012$, $D = 0.756 \pm 0.011$, and $\sin \theta_C = 0.231 \pm 0.003$.

The extent to which the various decays constrain the fit is graphically illustrated in Figure 7, which shows the bands of allowed F and D values deduced both from the Dalitz plot distributions and the branching ratio measurements. Apart from the n → peν correlation measurement (not used in this particular fit) and the neutron lifetime, all the data on this plot are from the SPS experiment. It can be seen that the bands, whose widths represent the experimental errors, share a common region of overlap as predicted by the standard Cabibbo model.

The good fit obtained is in sharp contrast with the results of several recent Cabibbo analyses (37–39), which use data obtained by averaging results from several hyperon decay experiments. This disagreement highlights the dangers of combining results from separate experiments analyzed under different assumptions (see Section 2.3).

For the fit we have taken the sign of g_1/f_1 in $\Sigma^- \to $ neν decay to be negative, in agreement with the result of the SPS experiment. It is obvious from Figure 7 that this sign is constrained to be negative within the framework of a Cabibbo fit to the hyperon semileptonic decay data. However, as discussed in Section 3.1.3, several low-energy experiments attempting to measure the electron asymmetry α_e in the decay of polarized

Σ^- have observed indications that g_1/f_1 is positive. If this result were confirmed, it would cause serious problems for the Cabibbo model.

The $\Lambda \to pe\nu$ high-statistics experiment (32, 33) at the AGS has been analyzed using assumptions consistent with those of the SPS experiment analysis, except for a difference in the q^2 dependence parametrization. They give (33) $g_1/f_1 = 0.715 \pm 0.026$, a value that is increased by 0.03 if the q^2 parametrization of Section 2.3 is used. This result is in excellent agreement with the FIT 5 result $g_1/f_1 = 0.73$. The measured value of the branching ratio is only 6% lower than the FIT 5 value, which, however, amounts to three times the quoted error.

Finally, the values of the asymmetry parameters α_e, α_p, and β_ν in the beta decay of polarized Λ, measured in three low-energy experiments, are in fair agreement with the FIT 5 results (see Table 5).

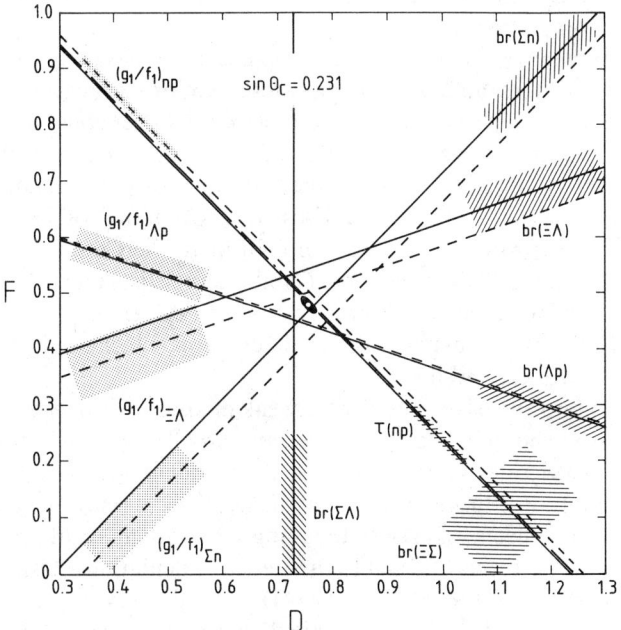

Figure 7 A plot of allowed values of F and D showing the constraints on the Cabibbo fit imposed by the various measurements. The solid lines were obtained from the branching ratio measurements assuming $\sin \theta_C = 0.231$, and the dashed lines correspond to the direct (g_1/f_1) determinations. [The $(\Xi\Sigma)$ branching ratio constraint is indicated by a dashed-dotted line to distinguish it from the nearby neutron lifetime line.] The shaded bands indicate the experimental errors. With the exception of the results on neutron decay, all the data on the plot come from the SPS experiment. The black ellipse shows the region within one standard deviation of the F and D values given by the fit of Column 5 in Table 11.

6. SU(3)-BREAKING EFFECTS

The results of the good fit presented in the previous section have been obtained within the framework of the Cabibbo model, which assumes SU(3) invariance of the strong interactions. This is only an approximate symmetry, and, for instance, it is broken by about 15% in the baryon masses. Effects of SU(3) breaking are also expected to modify the baryon matrix element. Ademollo & Gatto (40) showed that for the vector part (f_1 and f_2) the first-order symmetry-breaking effects vanish. An equivalent theorem does not exist for the axial-vector part of the hadronic weak current. Therefore the effects due to the breaking of SU(3) are expected to be more important in the axial-vector terms than in the vector terms. As a consequence, the predictions on g_1 may be modified, and finite g_2 terms may appear. However, as the theoretical calculations of such effects are very much model dependent, we first investigate them from a purely phenomenological point of view.

As the symmetry-breaking effects are expected to affect mainly the axial-vector form factors, an empirical way to search for such effects may be to perform a fit in which the g_1 form factors are no longer constrained by the Cabibbo hypotheses. The parameters of the fit are then the axial-vector form factors g_1 for each decay and the Cabibbo angle θ_C. The fit gives the values of g_1/f_1, $\sin\theta_C$, and χ^2/DOF listed in Column 2 of Table 12. For comparison, the results of FIT 5, the final fit to the Cabibbo model, are given in Column 1. The value of the Cabibbo angle is almost the same but the error has increased by a factor of 1.6. The values of g_1/f_1 differ from those of FIT 5 by at most one standard deviation, which is why the simple Cabibbo fit works so well.

The Cabibbo angle is essentially the weak mixing angle between the s and d quarks and should therefore be the same in all semileptonic processes. The K_{e_3} decay rates can provide an independent measurement of the Cabibbo angle. For these decays, only the weak vector current contributes. Therefore the effects of SU(3) breaking are rather small and appear to be within the reach of reliable theoretical calculations. Using the K_{e_3} experimental data, Shrock & Wang (41) derived a value of the Cabibbo angle $\sin\theta_C = 0.219 \pm 0.003$. A detailed and careful new analysis of the effects of SU(3) breaking in K_{e_3} decays by Leutwyler & Roos (42) gives $\sin\theta_C = 0.220 \pm 0.002$. Although corrections have been applied to obtain this value of $\sin\theta_C$, it is not clear at this stage whether the difference between this and the hyperon value, $\sin\theta_C = 0.231 \pm 0.003$, comes from SU(3)-breaking effects in the meson or in the baryon decays. Using the meson value $\sin\theta_C = 0.220 \pm 0.002$ as an additional constraint, the fit to the hyperon data gives the results listed in Column 3 of Table 12. The fit is good,

Table 12 Values of g_1/f_1 in various fits

	FIT 5	Fit with g_1 free[a]	Fit with g_1 free[a] and sin θ_C constrained[b]	Fit with SU(3) breaking[c]
$\Sigma^- \to \Lambda e\nu$[d]	0.62	0.595	0.594	0.62
$\Lambda \to pe\nu$	0.73	0.70	0.73	0.80
$\Xi^- \to \Sigma^0 e\nu$	1.233	1.26	1.32	1.36
$\Xi^- \to \Lambda e\nu$	0.22	0.28	0.31	0.24
$\Sigma^- \to n e\nu$	-0.28	-0.31	-0.35	-0.33
$n \to pe\nu$	1.233	—	—	1.227
sin θ_C	0.231 ± 0.003	0.231 ± 0.005	0.221 ± 0.002	0.229 ± 0.003
F	0.477 ± 0.012	—	—	0.466 ± 0.010
D	0.756 ± 0.011	—	—	0.761 ± 0.010
χ^2/DOF	8.8/6	1.0/2	4.2/3	21.8/6

[a] For the $\Sigma^- \to \Lambda e\nu$ and $\Xi^- \to \Sigma^0 e\nu$ channels the inputs for the fits are only the branching ratio values, and the variation of g_1 between Columns 2 and 3 simply compensates the change due to the Cabibbo angle.
[b] The input value of sin $\theta_C = 0.220 \pm 0.002$ is taken from K_{e3} data (42).
[c] All the SU(3)-breaking effects as computed by Donoghue & Holstein (43) have been included. The experimental values of g_1/f_1 are modified owing to the nonzero g_2 terms as indicated in Table 8.
[d] In the case of $\Sigma^- \to \Lambda e\nu$ the quoted numbers are the values of g_1.

with a χ^2/DOF of 4.2/3. For the $|\Delta S| = 1$ decays, the general increase of the fitted values of $|g_1/f_1|$ with respect to FIT 5 is a consequence of the 5% decrease of sin θ_C. The average relative increase $\Delta(g_1/f_1)/(g_1/f_1)$ is about 15%. If one relies upon the corrected value of sin θ_C for K_{e3} decays, these changes may be attributed to SU(3) breaking.

Finally, as an example of a theoretical attempt to calculate the effects of SU(3) breaking in hyperon semileptonic decays, we take the results of Donoghue & Holstein (43). These authors have used the MIT bag model to calculate the g_2 terms and the modifications to f_2, f_1, and g_1 induced by SU(3) breaking. For all the $|\Delta S| = 1$ decays the expected magnitude of g_2/g_1 is similar,[2] about $+0.20$, and renormalization factors of 0.97 and 1.08 are obtained for f_1 and g_1, respectively. For $\Delta S = 0$ decays, g_2/g_1 is much smaller and there are no renormalization effects.

In order to perform a fit that includes all these calculated effects, the experimental values of g_1/f_1 for $g_2 = 0$ given in Column 2 of Table 8 have been modified as shown in Column 4 of the same table by using the dependence on g_2 measured in the experiment. The results of the fit are shown in Column 4 of Table 12. The χ^2 of 21.8 for 6 DOF is very poor,

[2] Note that the definition of g_2 used in the present review differs from that of Donoghue & Holstein (43) by a factor $M_1/(M_1 + M_2)$.

showing a strong disagreement between the data and this particular model of SU(3) breaking. We note, however, that the disagreement is entirely due to the renormalization factors for f_1 and g_1. In fact a fit with the calculated g_2 terms alone shows a slight improvement of the χ^2 from 8.8 to 8.1 with respect to FIT 5. This variation of χ^2 is too small to be considered significant, indicating that the data are not yet sufficiently precise to establish the presence of g_2 terms.

In conclusion, although the good fit obtained to the simple Cabibbo model does not call for sizeable SU(3)-breaking effects, there is certainly room for such effects as shown by the fit using the K_{e_3} value of the Cabibbo angle, which is about 5% lower than that obtained from the hyperon data alone. However, the theoretical ground rules for hunting the SU(3)-breaking game seem far from being established.

7. GENERALIZATION OF THE CABIBBO MODEL

Initially the Cabibbo model was dealing with the three quarks, u, d, and s, and the weak hadronic charged current was written:

$$j_\mu^+ = \bar{u}\gamma_\mu(1+\gamma_5)d' \qquad \text{(see Equation 10)}$$

with

$$d' = d\cos\theta + s\sin\theta.$$

The generalization of Equation 10 to the case of the six quarks u, d, c, s, t, b has been described in the Introduction, where the connection between the quarks with definite mass and the quarks with definite weak couplings (d', s', b') have been given through a unitary transformation:

$$\begin{pmatrix} d' \\ s' \\ b' \end{pmatrix} = U \begin{pmatrix} d \\ s \\ b \end{pmatrix}$$

The matrix U, first proposed by Kobayashi & Maskawa (10), may be written as[3]

$$U = \begin{pmatrix} c_1 & s_1 c_3 & s_1 s_3 \\ -s_1 c_2 & c_1 c_2 c_3 + s_2 s_3 e^{i\delta} & c_1 c_2 s_3 - s_2 c_3 e^{i\delta} \\ -s_1 s_2 & c_1 s_2 c_3 - c_2 s_3 e^{i\delta} & c_1 s_2 s_3 + c_2 c_3 e^{i\delta} \end{pmatrix},$$

where there are three angles $\theta_1, \theta_2, \theta_3$ ($c_i = \cos\theta_i$, $s_i = \sin\theta_i$) to describe the mixing between the quarks and a CP-violating phase δ. The quark state d'

---

[3] Watch for the different sign conventions. The form given here is that used by the Particle Data Group (5).

in Equation 10 is now given by

$$d' = c_1 d + s_1 c_3 s + s_1 s_3 b. \qquad 23.$$

The universality of weak interactions is still preserved, but now $|u_{11}|^2 + |u_{12}|^2 = c_1^2 + s_1^2 c_3^2$ is smaller than 1. In other words, the constant C of the baryon matrix element (Equation 14) is equal to

c_1 for $\Delta S = 0$ transitions, instead of $\cos \theta_C$,

$s_1 c_3$ for $|\Delta S| = 1$ transitions, instead of $\sin \theta_C$.

The success of the simple Cabibbo fit implies a small value of the mixing angle θ_3.

In principle the value of s_3 may be determined by performing a four-parameter fit $(F, D, \theta_1, \theta_3)$ to the SPS data and the neutron lifetime. However, in such a fit, although the product $s_1 c_3$ is constrained to remain close to the value 0.23 found for $\sin \theta_C$, c_3 can become large, and s_1 small. The corresponding minute changes in c_1 are compensated by very small changes in the quantity $(F+D)$. Therefore, to determine a limit on θ_3, nuclear beta-decay data are used to impose a further constraint on θ_1 in the four-parameter fit. The additional constraint is $c_1 = G_V (1 + \delta_c)^{-1/2}/G$, where G is the weak coupling constant obtained from the muon lifetime, G_V is the beta-decay constant determined from the ft values of $0^+ \to 0^+$ superallowed Fermi transitions, and δ_c is the model-dependent radiative correction to the ratio of the squares of the β and μ decay coupling constants discussed in Section 4.

In order to extract G_V and G from the experimental measurements, it is necessary to apply radiative corrections of the type δ_b. These may be estimated reliably, especially in the case of muon decay. However, a further correction must be applied to the nuclear ft values due to electromagnetic breaking of the isospin analogue symmetry. This nucleus-dependent radiative correction introduces a significant theoretical uncertainty in the value obtained for G_V.

A recent estimate (44) of the superallowed ft value, extrapolated to zero nuclear charge, is $(\text{ft})_{z=0} = (3084.4 \pm 1.9)$ s. The result is obtained from a fit to measurements made on eight different nuclei, with $\chi^2/\text{DOF} = 3.3$; thus the error should be scaled up, giving

$$(\text{ft})_{z=0} = (3084.4 \pm 3.5) \text{ s},$$

which yields

$$G_V = (1.14730 \pm 0.00064) \times 10^{-11} \text{ MeV}^{-2}.$$

Using the value for G given in Section 5 and Sirlin's value of 2.10% for δ_c, we obtain $c_1 = 0.9735 \pm 0.0005$, including only the experimental error.

Uncertainties in the nuclear radiative corrections are at least partly taken into account in the scaling factor we have applied to the error on $(ft)_{z=0}$. However, the model-dependent correction δ_c is not included in this procedure and it is difficult to estimate the uncertainty on this quantity, which is computed starting with free quarks. We have chosen to assign an error of $\pm 0.2\%$ to $(1+\delta_c)$, which leads to an additional uncertainty on c_1 of 0.0010. Adding this error linearly to the experimental error, the result $c_1 = 0.9735 \pm 0.0015$ is obtained. We emphasize that the estimate of this error is somewhat arbitrary, since it includes an attempt to quantify the inadequacy of the understanding of the theoretical corrections.

Performing the four-parameter fit using the SPS results, the neutron lifetime, and the constraint $c_1 = 0.9735 \pm 0.0015$, gives the results

$F = 0.477 \pm 0.011,$

$D = 0.755 \pm 0.011,$

$s_1 c_3 = 0.231 \pm 0.003,$

$s_1 = 0.228 \pm 0.006,$

$c_3 = 1.012 \pm 0.032.$

with $\chi^2/\text{DOF} = 8.7/6$. This is in good agreement with the corresponding three-parameter fit of Column 5 in Table 3. Using the one-standard-deviation lower limit on c_3 of 0.980, we can set an upper bound on the modulus of s_3: $|s_3| < 0.20$.

It should be noted that the error on c_3 is dominated by the theoretical error on the value of c_1 obtained from nuclear beta-decay data rather than the error on $s_1 c_3$ from hyperon decay measurements. Another limitation of the significance of the upper bound obtained for θ_3 comes from possible SU(3)-breaking effects, as discussed in the preceding section. One may assume, for example, that the Cabibbo angle $\sin \theta_C$, or more precisely the product $s_1 c_3$, is given by the semileptonic meson decays, and that the difference with the value obtained in the semileptonic baryon decays is due to SU(3)-breaking effects not taken into account in the normal Cabibbo fit. Then the two constraints $c_1 = 0.9735 \pm 0.0015$ and $s_1 c_3 = 0.220 \pm 0.002$ give $c_3 = 0.962 \pm 0.029$. The upper bound on s_3 is now only $|s_3| < 0.36$.

In conclusion, to measure the value of s_3 by this method will require more accurate hyperon decay measurements as well as a better theoretical understanding of the nuclear radiative corrections and of the SU(3)-breaking effects.

Additional information on s_3 can be obtained from beauty decay measurements. The standard model prediction for the beauty hadron decay rate is (45)

$$1/\tau_B \cong 1.08 \times 10^{14}(m_b/5 \text{ GeV})^5 \times (2.75|U_{cb}|^2 + 7.69|U_{ub}|^2) \text{ s}^{-1}, \qquad 24.$$

where $U_{cb} = (c_1c_2s_3 - s_2c_3e^{i\delta})$ and $U_{ub} = s_1s_3$ are the terms of the matrix U that are relevant for the decay, and m_b is the mass of the beauty quark.

From a study of the decay lepton spectra at the Cornell storage ring CESR, the CLEO and CUSB Collaborations (46) have obtained

$$|U_{ub}|/|U_{cb}| < 0.2 \qquad 25.$$

at the 90% confidence level, by making plausible assumptions on the masses of the decay products. Recent measurements of the beauty hadron lifetime by the MAC and Mark II Collaborations (47, 48) at the storage ring PEP give $\tau_B = (18 \pm 6 \pm 4) \times 10^{-13}$ s and $\tau_B = (12^{+4.5}_{-3.6} \pm 3) \times 10^{-13}$ s, respectively.

From these values for the lifetime, using Equations 24 and 25 and assuming $m_b = 5$ GeV, one gets $|U_{ub}| < 0.01$, which, with $s_1 = 0.23$, gives the stringent limit $|s_3| < 0.05$. It is clearly important that this limit be confirmed with a better control of the assumptions made and more accurate measurements of the beauty hadron lifetime.

8. $\Omega^- \to \Xi^0 e\nu$ DECAY

In a recent experiment (49) at the CERN SPS charged-hyperon beam, a sample of 16,000 Ω^- decays was collected at an Ω^- momentum of 131 GeV/c. In an extensive study of the various decay modes, the $\Omega^- \to \Xi^0 e\nu$ branching ratio was measured:

$$R = \Gamma(\Omega^- \to \Xi^0 e\nu)/\Gamma(\Omega^- \to \text{all}) = (0.56 \pm 0.28) \times 10^{-2}.$$

The Ω^- semileptonic decay rate is expected to be roughly one order of magnitude larger than the other $|\Delta S| = 1$ hyperon semileptonic decay rates because of the larger phase space available. But, since the transition occurs between a $J = 3/2^+$ decuplet state (Ω^-) and an octet state (Ξ^0), there is no Cabibbo prediction here. However, the matrix element for $\Omega^- \to \Xi^0 e\nu$ is related by SU(3) to that for $\nu + N \to \ell + \Delta$. With the measured Ω^- lifetime, $\tau_{\Omega^-} = (0.823 \pm 0.013) \times 10^{-10}$ s (49), a theoretical calculation (50) using the ν-induced Δ production information gave $R \cong 1 \times 10^{-2}$.

9. CONCLUSIONS

The SU(2) × U(1) gauge theory of Glashow, Weinberg, and Salam is now considered to be the Standard Model of electroweak interactions. In particular, the Kobayashi-Maskawa six-quark version of the theory has gained wide acceptance since it incorporates the charm and beauty quarks and the τ lepton, in addition to the lighter quarks and leptons. In this extension of the Cabibbo model the fermion eigenstates of the weak gauge group are related to the mass eigenstates by three mixing angles of fundamental significance. In its original form the Cabibbo model only accommodated the three lightest quarks and had a single mixing angle θ_C. Nevertheless it meets with remarkable success in fitting the most recent and complete data on hyperon semileptonic decays, showing that the effects of both SU(3) breaking and the mixing of higher-generation quarks are small.

The advent of a new generation of hyperon decay experiments using high-energy hyperon beams has provided high-statistics data samples for almost every type of semileptonic decay. This new situation requires a greater consistency between the experimental analyses, especially in the ways of handling effects such as q^2 dependences and radiative corrections, both of which had often been neglected in the analysis of earlier experiments.

Using recent experimental results on hyperon semileptonic decays, combined with neutron lifetime measurements to achieve a better sensitivity, an excellent fit to the Cabibbo model is obtained. The fit gives the parameter values

$F = 0.477 \pm 0.012,$

$D = 0.756 \pm 0.011$

$\sin \theta_C = 0.231 \pm 0.003.$

However, the controversial question of the sign of the form-factor ratio g_1/f_1 in $\Sigma^- \to ne\nu$ still needs definitive experimental clarification. Effects due to the breaking of SU(3) have yet to be uncovered in the decays.

With the additional constraint of the value of G_V, the beta-decay coupling measured in $0^+ \to 0^+$ superallowed transitions, a limit on the weak mixing angle of the beauty quark in the SU(2) × U(1) model is obtained:

$|\sin \theta_3| < 0.20.$

It is an open challenge to reduce both the theoretical and experimental uncertainties sufficiently so that the beauty quark mixing angle can be determined. This is perhaps the benchmark by which future hyperon experiments should be measured.

ACKNOWLEDGMENTS

We would like to thank John Ellis and Stuart Tovey for their critical reading of the manuscript.

Appendix

The matrix element M for the decays

$$B_1 \to B_2 + \ell + \bar{\nu}_\ell,$$

where B_1 and B_2 are baryons and where ℓ and ν denote the lepton and the neutrino, respectively, is given by the product of the matrix elements of the baryonic weak current and of the lepton current:

$$M = (G/\sqrt{2}) \langle B_2 | J_\mu^h | B_1 \rangle \bar{u}_\ell(p_\ell) \gamma^\mu (1+\gamma_5) u_\nu(p_\nu),$$

where G is the universal weak coupling constant [$G = (1.16632 \pm 0.00004) \times 10^{-11}$ MeV^{-2}], and p_ℓ and p_ν are the lepton and the antineutrino four-momenta.

The hadronic weak current is expressed in terms of form factors that take into account the effects of the strong interactions. There are different conventions for the normalization of some form factors, which alter only the expected numerical values but not the theoretical description. Here we follow the convention adopted by Bender et al (13) and by Linke (14). The baryon term is written as:

$$\langle B_2 | J_\mu^h | B_1 \rangle = C\bar{u}_{B_2}(p_2)(f_1(q^2)\gamma_\mu + i[f_2(q^2)/M_1]\sigma_{\mu\nu}q^\nu + [f_3(q^2)/M_1]q_\mu$$
$$+ \{g_1(q^2)\gamma_\mu + i[g_2(q^2)/M_1]\sigma_{\mu\nu}q^\nu + [g_3(q^2)/M_1]q_\mu\}\gamma_5)u_{B_1}(p_1),$$

where

$$C = \begin{pmatrix} \cos\theta_C \\ \sin\theta_C \end{pmatrix} \quad \text{for} \quad \Delta S = \begin{pmatrix} 0 \\ 1 \end{pmatrix} \text{ transitions,}$$

θ_C being the Cabibbo angle p_1 and p_2 the B_1 and B_2 four-momenta, $q = p_1 - p_2$, and M_1 and M_2 the masses of the initial and final baryon. Within the framework of the Cabibbo model (see Section 2) $f_3 = g_2 = 0$. The g_3 term, which is multiplied by a factor $(m_\ell/M_1)^2$, is neglected.

A.1 DECAY RATE, DALITZ PLOT, AND ELECTRON-NEUTRINO CORRELATION

A1.1 *Decay Rate*

The leptonic decay rate of a baryon B_1 is given by

$$\Gamma = \frac{1}{(2\pi)^5} \frac{1}{2E_1} \int \frac{dp_2}{2E_2} \frac{dp_\ell}{2E_\ell} \frac{dp_\nu}{2E_\nu} \delta^4(p_1 - p_2 - p_\ell - p_\nu) \frac{1}{2} \sum_{\text{spins}} |M|^2. \qquad \text{A.1}$$

The numerical integration of Γ is given for the various decays in the article of Bender et al (13). Contributions to Γ of terms other than g_1^2 and f_1^2 are of the order of 1%, and to a very good approximation

$$\Gamma = G^2 \left(\frac{\cos^2 \theta_C}{\sin^2 \theta_C}\right) \frac{\Delta m^5}{60\pi^3} (f_1^2 + 3g_1^2)(1 - 3\delta), \qquad \text{A.2}$$

where $\Delta m = M_1 - M_2$ and $\delta = (M_1 - M_2)/(M_1 + M_2)$.

A1.2 Dalitz Plot

The differential decay rate $\delta^2\Gamma/\delta E_\ell \delta E_2$ is obtained by integrating Equation A.1 over p_ν only, and is given in the article of Bender et al (13). This expression has been used for the Dalitz plot fits to extract g_1/f_1 (or f_1/g_1 for $\Sigma^- \to \Lambda e\nu$ since CVC predicts $f_1 = 0$) and to study the effects of the other form factors. However, the expression is lengthy and we do not reproduce it here. The two projections of the Dalitz plot give the lepton energy (E_ℓ) spectrum and the baryon kinetic energy ($T_2 = E_2 - M_2$) spectrum. To illustrate the information that can be obtained from these spectra, we give an approximate expression for the differential decay rate, keeping only the main form factors f_1, g_1, and f_2, and terms up to first order in δ:

$$\delta^2\Gamma/\delta E_\ell \delta T_2 \simeq (1+x^2)(1-L^2) - B(1-x^2) - 4\delta(1+y)xLB, \qquad \text{A.3}$$

where

$B = 1 - T_2/T_2^{\max}$

$L = (E_\ell - E_\nu)/(E_\ell - E_\nu)_{\max}$,

$x = g_1/f_1$ and $y = f_2/f_1$.

To obtain the baryon spectrum, Equation A.3 is integrated over L, which varies between two limits that have opposite signs and the same absolute value. The last term of Equation A.3, which is odd in L, does not contribute, and thus the baryon spectrum depends only upon $x^2 = |g_1/f_1|^2$. In the integral over B, the last term contributes and thus the lepton spectrum depends upon both x^2 and x allowing, in principle, a determination of the sign of g_1/f_1. However, the dependence is weak: for example, in the $\Sigma^- \to n e\nu$ decay, if $|g_1/f_1|$ has the value expected from a Cabibbo fit, the effect of reversing the sign of this term is to cause a shift of about 4 MeV (2% of $E_{\ell\,\max}$) in the central region of the spectrum.

A1.3 Lepton–Neutrino Correlation

The distribution of the angle between the lepton and the antineutrino in the c.m.s. is

$$W(\cos\theta_{\ell\bar\nu}) = \tfrac{1}{2}(1 + \alpha_{\ell\bar\nu} \cos\theta_{\ell\bar\nu}).$$

The coefficient $\alpha_{\ell\bar\nu}$ depends upon the form factors and is computed using the general matrix element. To again illustrate the sensitivity of this distribution to the form factors, we give an expression for $\alpha_{\ell\bar\nu}$ obtained using the same approximations as in the previous section (51):

$$\alpha_{\ell\bar{\nu}} = (f_1^2 - g_1^2)/(f_1^2 + 3g_1^2) - 2\delta.$$

Thus the lepton–neutrino correlation determines only the absolute value of g_1/f_1.

A.2 ASYMMETRIES

The V–A structure of the interaction manifests itself also in the angular distributions of the decay products. These distributions are constructed either with respect to the spin of the incoming baryon when it is polarized ($\Lambda \to pe\nu$ case) or with respect to the spin of the outgoing baryon, since the Λ, which is present in all the other hyperon decays studied (except $\Sigma^- \to ne\nu$), analyzes its own polarization via the decay $\Lambda \to p\pi^-$. The complete formulae are given in the article of Linke (14). We again restrict ourselves to defining the distributions that can be studied and giving simple approximate expressions for the expected asymmetries.

A2.1 Unpolarized Initial Baryon

The polarization of the decay baryon (Λ in our case) is

$$\mathbf{P}_{B_2} = A\hat{\boldsymbol{\alpha}} + B\hat{\boldsymbol{\beta}} + N\hat{\mathbf{n}},$$

where $\hat{\boldsymbol{\alpha}}, \hat{\boldsymbol{\beta}}, \hat{\mathbf{n}}$ form an orthonormal basis,

$$\hat{\boldsymbol{\alpha}} = [1/\sqrt{2(1+\cos\theta)}] (\hat{\mathbf{p}}_\ell + \hat{\mathbf{p}}_\nu),$$

$$\hat{\boldsymbol{\beta}} = [1/\sqrt{2(1-\cos\theta)}] (\hat{\mathbf{p}}_\ell - \hat{\mathbf{p}}_\nu),$$

$$\hat{\mathbf{n}} = (1/|\sin\theta|)\hat{\mathbf{p}}_\ell \times \hat{\mathbf{p}}_\nu;$$

$\hat{\mathbf{p}}_\ell$ and $\hat{\mathbf{p}}_\nu$ are unit vectors in the decay baryon c.m.s., θ is the angle between $\hat{\mathbf{p}}_\ell$ and $\hat{\mathbf{p}}_\nu$, and A, B, N are complicated functions of the form factors.

This polarization is reflected in the projected distributions of the proton from the $\Lambda \to p\pi^-$ decay, which are given by

$$W[\cos(\hat{\mathbf{p}}_p, \hat{\boldsymbol{\alpha}})] = \tfrac{1}{2}(1 + \alpha_\Lambda A \hat{\mathbf{p}}_p \hat{\boldsymbol{\alpha}})$$

and similar formulae for the other projections, with the Λ asymmetry parameter $\alpha_\Lambda = 0.642 \pm 0.014$ (5).

For zero momentum transfer, the asymmetry parameters A, B, and N take the simple forms (14, 52):

$$A = 8/3 \ \mathrm{Re}(f_1 g_1^*)/(f_1^2 + 3g_1^2),$$

$$B = 8/3 \ |g_1|^2/(f_1^2 + 3g_1^2),$$

$$N = 1/2\pi \ \mathrm{Im}(f_1 g_1^*)/(f_1^2 + 3g_1^2),$$

which show that the sign of f_1/g_1 can be obtained from the first distribution. In addition, as the form factors should be real by T invariance, one expects $N = 0$.

A2.2 Polarized Initial Baryon

For polarized hyperons the parity-violating correlation between the directions of the decay particles and the hyperon polarization $\mathbf{P}_{B_1}$ is described by the distribution

$$I(\hat{q}) = 1/2(1 + \alpha_i \mathbf{P}_{B_1} \cdot \hat{q}),$$

where $\hat{q}$ and α_i are the unit momentum and the asymmetry parameter for the relevant decay product.

To first order in δ and retaining only the f_1, g_1, and f_2 terms, the asymmetry parameters are (53)

$$\alpha_\ell = \{2(f_1 g_1 - g_1^2) - 1/3\delta[2(f_1 + g_1)^2 + 2f_2'(f_1 + g_1)]\}/(f_1^2 + 3g_1^2)$$

$$\alpha_\nu = \{2(f_1 g_1 + g_1^2) + 1/3\delta[2(f_1 - g_1)^2 + 2f_2'(f_1 - g_1)]\}/(f_1^2 + 3g_1^2)$$

$$\alpha_{B_2} = \{-5/2 f_1 g_1 + 5/6\delta[2(f_1 + f_2')g_1]\}/(f_1^2 + 3g_1^2),$$

where

$$f_2' = (M_1 + M_2) f_2 / M_1.$$

Another type of analysis can also be performed using an orthonormal system $(\hat{\boldsymbol{\alpha}}, \hat{\boldsymbol{\beta}}, \hat{\mathbf{n}})$ as in the unpolarized case (see Section A2.1). The unit vectors $\hat{\mathbf{p}}_\ell$, $\hat{\mathbf{p}}_\nu$, and $\mathbf{p}_{B_2}$ are defined in the parent baryon c.m.s. for the present case, and the distributions are

$$W[\cos(\mathbf{P}_{B_1}, \hat{\boldsymbol{\alpha}})] = 1/2(1 + A\mathbf{P}_{B_1}\hat{\boldsymbol{\alpha}}),$$

$$W[\cos(\mathbf{P}_{B_1}, \hat{\boldsymbol{\beta}})] = 1/2(1 + B\mathbf{P}_{B_1}\hat{\boldsymbol{\beta}}),$$

$$W[\cos(\mathbf{P}_{B_1}, \hat{\mathbf{n}})] = 1/2(1 + N\mathbf{P}_{B_1}\hat{\mathbf{n}}).$$

For zero momentum transfer the asymmetry parameters are (14)

$$A = 8/3 \, \text{Re}(f_1 g_1^*)/(f_1^2 + 3g_1^2),$$

$$B = -8/3 \, |g_1|^2/(f_1^2 + 3g_1^2),$$

$$N = 1/2\pi \, \text{Im}(f_1 g_1^*)/(f_1^2 + 3g_1^2).$$

For the polarized baryon case there are several distributions that are sensitive to the sign of g_1/f_1.

Literature Cited

1. Pauli, W. 1930. Address to group on radioactivity (Tübingen, December 4, 1930). *Rappts septième conseil phys. Solvay, Bruxelles, 1933.* Paris: Gauthier-Villars, 1934
2. Fermi, E. 1934. *Nuovo Cimento* 11:1; *Z. Phys.* 88:161–77
3. Lee, T. D., Yang, C. N. 1956. *Phys. Rev.* 104:254–58; Ambler, E., Hayward, R. W., Hoppes, D. D., Hudson, R. P., Wu, C. S. 1957. *Phys. Rev.* 105:1413–15; Garwin, R. L., Lederman, L. M., Weinrich, M. 1957. *Phys. Rev.* 105:1415–17; Friedman, J. I., Telegdi, V. L. 1957. *Phys. Rev.* 105:1681–82
4. Feynman, R. P., Gell-Mann, M. 1958. *Phys. Rev.* 109:193–98
5. Particle Data Group. 1982. Review of Particle Properties. *Phys. Lett.* 111B
6. Cabibbo, N. 1963. *Phys. Rev. Lett.* 10:531–33
7. Chounet, L. M., Gaillard, M. K. 1970. *Phys. Lett.* 32B:505–9; Chounet, L. M., et al. 1972. *Phys. Rep.* 4C:199–323; Pondrom, L. G. 1976. Weak decay processes, *Proc. Particle and Fields 1976*, BNL Rep. 50598, pp. C1–C24; See also Ref. 5, pp. 73–76
8. Gell-Mann, M., Lévy, M. 1960. *Nuovo Cimento* 16:705–26
9. Glashow, S. 1961. *Nucl. Phys.* 22:579–88; Weinberg, S. 1967. *Phys. Rev. Lett.* 19:1264–66; Salam, A. 1968. *Proc. 8th Nobel Symp., Stockholm, 1968*, pp. 367–77, ed. N. Svartholm. Stockholm: Almquist & Wiksells
10. Kobayashi, M., Maskawa, T. 1973. *Prog. Theor. Phys.* 49:652–57
11. Gell-Mann, M. 1961. *Phys. Rev.* 125:1067–84; Gell-Mann, M. 1961. *The Eightfold Way*, CALTECH Rep. CTSL-20. Pasadena: Calif. Inst. Technol.
12. Gershtein, S. S., Zel'dovich, J. B. 1956. *Sov. Phys. JETP* 2:576–78
13. Bender, I., Linke, V., Rothe, H. J. 1968. *Z. Phys.* 212:190–212
14. Linke, V. 1969. *Nucl. Phys. B* 12:669–93
15. Bartoli, R., et al. 1972. *Rivista del Nuovo Cimento* 2:241–302
16. Cnops, A. M., et al. 1978. *Proc. Top. Conf. on Neutrino Physics, Oxford, 1978*, p. 62. Didcot: Rutherford Publ. No. RL-78-081
17. Particle Data Group. 1974. Review of Particle Properties. *Phys. Lett.* 50B:1–198
18. Ebenhöh, H., et al. 1971. *Z. Phys.* 241:473–79
19. Althoff, K. W., et al. 1973. *Phys. Lett.* 43B:237–39
20. Lindquist, J., et al. 1977. *Phys. Rev. D* 16:2104–14
21. Burnett, T. H., et al. 1976. *Nuovo Cimento* 34A:14–20
22. Franzini, P., et al. 1972. *Phys. Rev. D* 6:2417–23
23. Gershwin, L. K., et al. 1968. *Phys. Rev. Lett.* 20:1270–72; Gershwin, L. K. 1969. *UCRL Rep.* 19246 (thesis)
24. Bogert, D., et al. 1970. *Phys. Rev. D* 2:6–29
25. Ellis, R. J., et al. 1972. *Nucl. Phys. B* 29:77–105
26. Keller, P., et al. 1982. *Phys. Rev. Lett.* 48:971–74
27. Duclos, J., et al. 1971. *Nucl. Phys. B* 32:493–512
28. Lach, J., Pondrom, L. G. 1979. *Ann. Rev. Nucl. Part. Sci.* 29:203–42
29. Tanenbaum, W., et al. 1975. *Phys. Rev. D* 12:1871–83
30. Thompson, J. A., et al. 1980. *Phys. Rev. D* 21:25–44
31. Décamp, D., et al. 1977. *Phys. Lett.* 66B:295–99
32. Wise, J., et al. 1980. *Phys. Lett.* 91B:165–68 and 98B:123–26
33. Jensen, D., et al. 1983. *Proc. Int. Conf. High-Energy Physics, Brighton, 1983*, p. 255. Didcot: Rutherford Appleton Lab.
34. Dworkin, J. S. 1983. *A High-Statistics Study of Lambda Beta-Decay* (thesis). Univ. Michigan, Ann Arbor
35. Bourquin, M., et al. *Measurement of Hyperon Semi-Leptonic Decays at the CERN SPS.*
35a. I "The $\Sigma \to \Lambda e\nu$ Decay Mode. 1982. *Z. Phys. C* 12:307–21
35b. II "The $\Lambda \to pe\nu$, $\Xi^- \to \Lambda e\nu$ and $\Xi^- \to \Sigma^0 e\nu$ Decay Modes. 1983. *Z. Phys. C* 21:1–15
35c. III "The $\Sigma^- \to ne\nu$ Decay Mode. 1983. *Z. Phys. C* 21:17–26
35d. IV "Tests of the Cabibbo Model. 1983. *Z. Phys. C* 21:27–36
36. Sirlin, A. 1978. *Rev. Mod. Phys.* 50:573–605 and references therein; 1982. *Nucl. Phys. B* 196:83–92; 1974. *Nucl. Phys. B* 71:29–51 and references therein
37. Donoghue, J. F., Holstein, B. R. 1982. *Phys. Rev. D* 25:2015–18
38. Paschos, E. A., Turke, U. 1982. *Phys. Lett.* 116B:360–64
39. Garcia, A., Kielanowski, P. 1982. *Phys. Rev. D* 26:1090–1102
40. Ademollo, M., Gatto, R. 1964. *Phys. Rev. Lett.* 13:264–66
41. Shrock, R. E., Wang, L.-L. 1978. *Phys. Rev. Lett.* 41:1692–95
42. Leutwyler, H., Roos, M. 1984. Determination of the Elements V_{us} and V_{ud} of the

Kobayashi–Maskawa Matrix, preprint TH.3830–CERN. Submitted to *Z. Phys. C*
43. Donoghue, J. F., Holstein, B. R. 1982. *Phys. Rev. D* 25:206–12
44. Wilkinson, D. H. 1978. *Symmetries and Nuclei, Nuclear Physics with Heavy Ions and Mesons* Vol. II, ed. R. Balian, M. Rho, G. Ripka, pp. 877–1017. Amsterdam: North-Holland
45. Gaillard, M. K., Maiani, L. 1979. *Proc. 1979 Cargese Summer Inst. on Quarks and Leptons*, ed. M. Lévy, et al., pp. 433–514. New York: Plenum
46. Gittelman, B., CLEO Coll. 1982. *Proc. 21st Int. Conf. on High-Energy Physics, Paris. J. Phys.* 43, Suppl. 12, *C* 3:110–13; Franzini, P., CUSB Coll. 1982. *Proc. 21st Int. Conf. on High-Energy Physics, Paris. J. Phys.* 43, Suppl. 12, *C* 3:114–16; See also Stone, S. 1983. *Proc. Int. Symp. on Lepton and Photon Interactions at High Energies, Cornell*, ed. D. G. Cassel, D. L. Kreinick, pp. 203–43. Cornell: Newman Lab. Nucl. Studies
47. Fernandez, E., et al. 1983. *Phys. Rev. Lett.* 51:1022–25
48. Lockyer, N. S., et al. 1983. *Phys. Rev. Lett.* 51:1316–19
49. Bourquin, M., et al. 1984. *Nucl. Phys. B* 241:1–47
50. Finjord, J., Gaillard, M. K. 1980. *Phys. Rev. D* 22:778–86
51. Frampton, P. H., Tung, W. K. 1971. *Phys. Rev. D* 3:1114–21
52. Alles, W. 1962. *Nuovo Cimento* 26:1429–33
53. Garcia, A. 1971. *Phys. Rev. D* 3:2638–48

RECENT PROGRESS IN UNDERSTANDING TRINUCLEON PROPERTIES[1]

J. L. Friar and B. F. Gibson

Theoretical Division, Los Alamos National Laboratory, Los Alamos, New Mexico 87545

G. L. Payne

Department of Physics and Astronomy, University of Iowa, Iowa City, Iowa 52242

CONTENTS

1. INTRODUCTION 404
2. THREE-NUCLEON FORCES 405
   2.1 *Introduction and Definitions* 405
   2.2 *Types of Three-Body Forces* 406
   2.3 *Evidence for Three-Nucleon Forces* 409
   2.4 *The Two-π Three-Nucleon Force* 413
3. MESON-EXCHANGE CURRENTS 416
   3.1 *Introduction* 416
   3.2 *Experimental Evidence* 420
4. EFFECTS OF THE TENSOR FORCE 424
   4.1 *Nucleon-Nucleon Force Properties* 424
   4.2 *Bound States* 424
   4.3 *Electromagnetic Reactions* 428
   4.4 *Elastic and Inelastic Scattering* 429
5. SUMMARY 430

---

[1] The US Government has the right to retain a nonexclusive, irrevocable royalty-free license in and to any copyright covering this paper.

1. INTRODUCTION

The traditional approach of nuclear physics describes nuclei by means of a model in which nonrelativistic nucleons interact via two-body or pair-wise forces; subnuclear degrees of freedom are ignored. Although deviations from this approach have been sporadically pursued for the last fifty years, the simplifications inherent in the traditional model are enormous and have resulted in a semiphenomenological description of nuclear physics that has enjoyed considerable success. There is little theoretical justification for these assumptions other than a rough consistency between theoretical predictions and experimental data. The inability to calculate observables accurately for a given two-body force model throughout the periodic table has been a serious impediment to the development of nuclear physics.

For these reasons much of the insight we have gathered about details of the nuclear mechanism has come from the few-nucleon systems. Unfortunately, the common two-nucleon observables must be used as input to fix parameters in the semiphenomenological forces, which are needed because of our inability to calculate nuclear properties beginning with an underlying field theory. Little is known with certainty beyond the long-range pion-exchange parts of the force. Therefore, consistency of description is the primary measure of success in the two-body problem. The three- and four-nucleon systems offer nontrivial tests of the traditional model, however. There is no guarantee that a force model constructed to reproduce the deuteron binding energy and nucleon-nucleon phase shifts will reproduce the trinucleon observables.

The seminal work of Faddeev (1) on techniques for implementing the boundary conditions for the scattering of three nucleons also provided the original impetus for accurate calculations of the ^{3}He and ^{3}H ground states. These positive parity, spin-$\frac{1}{2}$ and isospin-$\frac{1}{2}$ states were originally investigated by means of the Rayleigh-Ritz variational principle; the strong short-range repulsion of the two-nucleon force and the strong noncentral, primarily tensor, force make such calculations extremely difficult, although progress is still being made (2, 3). Computational sophistication has dramatically improved in the past decade, to the point where we can now challenge experimentally the theoretical predictions of the traditional model.

In this brief review we concentrate entirely on the three- and four-nucleon systems, with emphasis on three-nucleon calculations using realistic forces. We focus our discussion on advances in three topics that have preoccupied researchers recently: (*a*) meson-exchange currents and the trinucleon charge and magnetic densities and form factors (4, 5); (*b*) three-nucleon forces and their effect on observables (6–10); (*c*) the effect of the nucleon-nucleon tensor force on observables. Clearly, a comprehensive study of any of these topics is beyond our resources. Fortunately, several related themes recur and unite these disparate topics, and we concentrate on these. References (4–10) are recent reviews.

2. THREE-NUCLEON FORCES

2.1 *Introduction and Definitions*

The study of three-body forces and their effect on observables is not new in nuclear and other branches of physics. These forces depend in an irreducible way on the simultaneous coordinates, spins, momenta, and internal quantum numbers of three particles rather than the usual two. The word "irreducible" is an essential one; the most challenging technical difficulty in any calculation of such forces is distinguishing two successive two-body interactions between three particles and a "true" three-body force. Clearly, $V = V_{12} + V_{13}$ is a potential that depends on the coordinates of particles 1, 2, and 3; as written it is merely the sum of pair-wise interactions between particles 1 and 2, and between 1 and 3, and is not a three-body force. The problem is complicated further by the fact that few three-body or many-body forces are fundamental. Most are artifacts of the theorist's imagination and, in particular, his chosen method for performing a calculation. The difficulty lies in frozen degrees of freedom, the simplification introduced into a calculation in order to make it tractable. Internal degrees of freedom, when frozen, always lead to many-body forces, but seldom do when fully incorporated into a wave function. Thus our operating definition of three-nucleon forces will be this: *Forces that depend in an irreducible way on the simultaneous coordinates of three nucleons when only nucleon degrees of freedom are taken into account.* Meson and subnucleon degrees of freedom are generally regarded as frozen. It is clear that one theorist's three-body force could be another's two-body force imbedded in a more complicated model space.

These concepts are best illustrated by example. The archetype of three-body forces is the polarization force caused by the presence of two systems distorting a third. Typical is the Axilrod-Teller (11) three-atom force caused by the simultaneous distortions of the electron clouds of three atoms when the electron degrees of freedom are frozen. Such forces usually depend on the relative angular orientation of the three systems as well as their spatial separations. This peculiarity of three-body forces is perhaps their most notable feature.

The best-known and oldest example of three-body forces is the classical electromagnetic force of Primakoff & Holstein (12). It arises from the electromagnetic interactions in the nonrelativistic Schrödinger equation or classical Hamiltonian: $(\mathbf{p}-e\mathbf{A}/c)^2/2m$, where c is the speed of light and m is the particle mass. The diamagnetic $\mathbf{A}^2/2m$ term leads to the emission of two photons from one particle, which are subsequently absorbed by the other two through the $\mathbf{p}\cdot\mathbf{A}/m$ convection current interaction. This classical and nonrelativistic quantum mechanical argument leads to a force that is a second-order relativistic correction $(1/c^4)$ and consequently very weak; it is also momentum dependent and the form depends on the gauge used for the virtual photons. Alternatively, in a relativistic formalism, part of the sequential interaction of electromagnetic two-body forces from two different particles on the third can be viewed as creation and annihilation of particle-antiparticle "pairs"; this process leads directly to the diamagnetic term written above. Thus, in a Dirac formalism the Primakoff-Holstein force is merely the sum of two-body interactions, while in nonrelativistic formalisms it is a "real" three-body force. To the best of our knowledge this force has never been seen experimentally.

The pion-exchange three-body forces commonly considered in nuclear physics have the schematic form V_π^2/Mc^2, where M is the nucleon mass and V_π is the two-body, static, one-pion-exchange potential (OPEP). This has the form of a relativistic correction. On the average, a nucleus is weakly bound and should be nonrelativistic, since the binding energy is not more than 1% of the rest mass of the nucleons. Concomitantly, a typical nuclear momentum is 100–200 MeV c^{-1}, and $(v/c)^2$ is typically a few percent. We therefore expect that the size of any three-body force contribution to the energy should be on the order of 1% of the total potential energy (roughly 50 MeV), or 0.5 MeV.

We summarize this section by giving a litmus test for three-body forces: freezing out degrees of freedom leads to three-body forces, while including such degrees of freedom in one's wave function generally does not. Three-body forces are usually dependent on the relative angular orientation of the three particles. Three-body forces in nuclei generally are of order $(1/c^2)$, and therefore a small part of the total potential energy in a weakly bound system like a nucleus.

2.2 Types of Three-Body Forces

Over the years many different three-body forces have been developed from disparate models and with varying degrees of physically constrained input. Several of these are illustrated in Figure 1. We have concentrated on the pionic processes, because the strong short-range repulsion of the two-nucleon force produces "holes" in the nuclear wave function that suppress the contribution of other operators with short range. This dominating effect is discussed further in the subsequent section on meson-exchange currents.

Most attention has been directed at the two-pion-exchange, three-nucleon (2π-3N) force. This longest-range force is depicted in Figure 1a, which shows a π^+ being emitted by the left-most nucleon (changing from a proton to a neutron), the subsequent rescattering and charge exchange of the pion on the middle nucleon, and ultimately the absorption of the π^0 on the remaining nucleon. Clearly many other processes involving charged pions are possible in different orders; the 2π-3N force has a complex isospin structure. The key element is the shaded ellipse, which represents the π-N scattering amplitude *off-shell*. This amplitude involves processes such as that in Figure 1b, which must be removed since it represents the iteration of the usual two-body OPEP. Although the most general calculational approach, discussed below, does not separate the amplitude into distinct contributions, they can be succinctly categorized as in Figures 1c–f.

The process shown in Figure 1c is the genesis of the august Fujita-Miyazawa force (13); the strong N-Δ transition involving p-wave pions scattering from the central nucleon was the original motivation for this force. It has been the most commonly used three-nucleon force, and

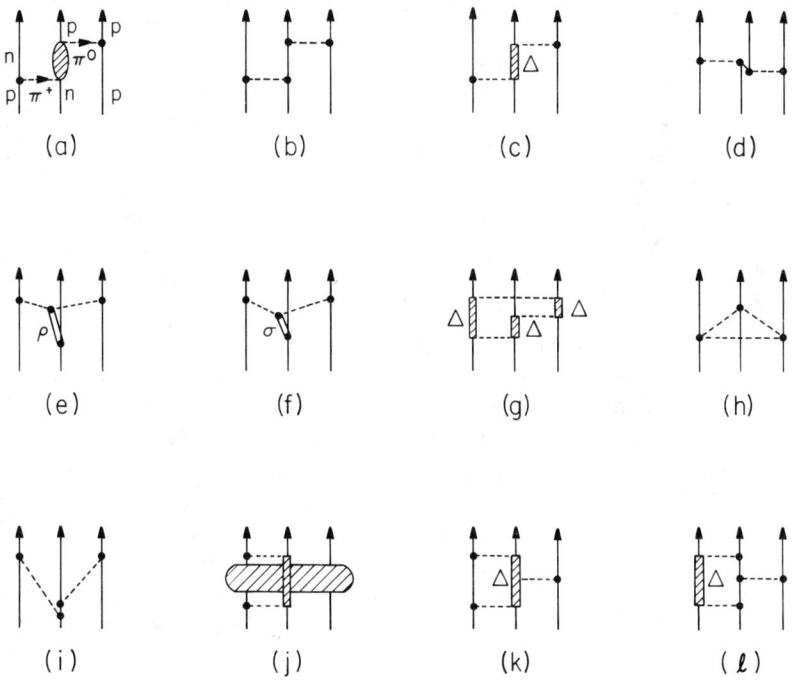

Figure 1 Various physical processes contributing to three-nucleon forces. Solid, dashed, shaded, and double lines depict nucleons, pions, isobars (nucleon excited states), and heavy mesons, respectively. See text.

conceptually it arises from the polarization of the nucleon. An s-wave pion can also contribute through the "pair" diagram in Figure 1d, and the σ term in Figure 1f, while p-wave pions are generated by the σ term and the ρ-meson diagram in Figure 1e. As pointed out originally by Brown et al (14, 15) and implemented by the Tucson-Melbourne group (16–18), it is extremely important in constructing the potentials to respect two principles: 1. the on-shell π-N amplitude is well known experimentally and is strongly constrained by (approximate) chiral symmetry through PCAC and current algebra; and 2. the pions are necessarily off-shell and the off-shell amplitude has important chiral constraints. Using arbitrary models that involve Δ's, ρ's, and σ's is inadequate, as are ad hoc phenomenological prescriptions. Much of the older work did not adhere to these principles. The primary problem is that the pseudoscalar (PS) model of π-N coupling produces unphysically large s-wave π-N scattering amplitudes, due largely to "pair" terms such as shown in Figure 1d. These are compensated by the σ-meson pole terms (Figure 1f), which exactly cancel the dominant part of the pair term. Alternatively one can use pseudovector (PV) π-N coupling, which has much smaller pair terms. However one proceeds, the s-wave parts of the 2π-3N potential (19–21) are not easy to calculate accurately.

Another approach to the p-wave 2π-3N force is the isobar constituent model (22, 23). This method does not freeze out the isobars; rather, it includes them as explicit components of the wave function. The process shown in Figure 1c can be viewed as a pion flipping a nucleon's "isobaric spin," creating a Δ, which then flips back to a nucleon. This approach has the advantage of more easily incorporating such effects as the isobar-nucleon interaction and multiple isobar formation. Its disadvantage is the additional complexity required to keep track of the new wave function components. The only extensive study of the three-nucleon force using this method is the calculational *tour de force* by Hajduk & Sauer (22, 23).

The 3π-3N forces have suffered from neglect. Two processes that contribute are shown in Figures 1g and 1h. The nuclear analogue (24) of the atomic Axilrod-Teller force is represented by the 3-Δ graph in Figure 1g, while the s-wave π-N contribution from pair terms (25) is shown in Figure 1h. In view of our warning about pair terms, calculation of the latter process must incorporate the appropriate constraints. These processes have a range corresponding to two pion masses, which can be seen by vertically snipping the pion lines; one always cuts two, unlike the graph in Figure 1a. For this reason they may be less important in the triton. More quantitative work needs to be performed on these forces.

Recently, more attention has been focused on short-range three-nucleon forces mediated by two mesons: $\pi\rho$-3N and $\rho\rho$-3N forces. The Tucson-Melbourne group (26) has extended its program of studying chiral constraints on pion-nucleon scattering to incorporate these processes. Graphically one replaces one or both pions in Figure 1a by a ρ meson. A similar step in this direction has been taken by Martzolff et al (27). The isobar constituent calculations by Hajduk & Sauer allow excitation and de-excitation of the Δ by ρ mesons as well as pions; they find this to be a non-negligible correction to the dominant pion-exchange potential.

Finally, several other three-nucleon forces have been developed. Time ordering the exchange of two pions also generates a three-body force residue (28), when the iteration of the static two-body OPEP is subtracted out. These forces, one of which is depicted in Figure 1i, have not been investigated quantitatively in the three-nucleon system. "Dispersion" in the isobar constituent process (29) displayed in Figure 1j subsumes Figures 1k and 1l. This mechanism is discussed further in Section 2.4.

In summary, the most thoroughly investigated and possibly most important three-nucleon force is the longest-range 2π-3N force, an important component of which involves virtual isobar excitation. Chiral symmetry constraints *must* be incorporated in any quantitatively reliable calculation of this force. Recent improvements have incorporated $\pi\rho$-3N and $\rho\rho$-3N components. Other forces have been largely neglected.

2.3 Evidence for Three-Nucleon Forces

Most of the evidence for three-nucleon forces comes from the three-nucleon systems, and all of this evidence is circumstantial, but nontrivial. Quantitatively accurate calculations of binding energies for a variety of realistic nonrelativistic force models have been performed recently using the Faddeev method, all of which fall short of the necessary binding and produce too large a charge radius (30–32). In order to produce a tractable calculational framework this approach decomposes the potential as an infinite sum of nonlocal potentials, each effective in a specific two-body partial wave or channel, and solves the Faddeev equations exactly for a fixed (finite) number of partial waves. It is exactly equivalent to solving the Schrödinger equation for that set of potentials. Calculations typically involve either the [$^1S_0$, $^3S_1$-$^3D_1$] partial waves (5 channels), or all partial waves with total $J \leq 2$ (18 channels). The angular momentum barrier and the relatively weak binding in the trinucleon suppress the contributions of the higher partial waves. Typically, the 5-channel and 18-channel binding energies differ by 200 keV or less, with a negligible contribution from the odd-parity forces (32). Binding energies, E_B, for the Reid Soft Core (33) (RSC), Paris (34), Super Soft Core C (35) (SSCC), and Argonne V_{14} (36) (AV14) potential models and the experimental datum are 7.2, 7.4, 7.5, 7.6, and 8.5 MeV, respectively. The corresponding rms charge radii for ^{3}He are 2.04, 2.04, 2.03, 2.02, and 1.88(2) fm, respectively, and for ^{3}H they are 1.83,

1.83, 1.83, 1.82, 1.67(3) fm, respectively. The binding energies were calculated in the 18-channel approximation and verified as upper bounds using the Rayleigh-Ritz technique. The radii include the nucleons' finite size. The Faddeev approach to calculating wave functions has proven superior to the use of analytic trial functions with the variational technique, and to other methods. If the tails of the wave functions dominate the calculation of the rms radius, $\langle r^2 \rangle^{1/2}$, one expects to find $\langle r^2 \rangle^{1/2} \sim E_B^{-1/2}$, as in the deuteron, and underbinding will produce the excessively large radii seen above.

Several obvious flaws exist in these calculations: 1. we assumed a nonrelativistic model; 2. we neglected three-nucleon forces; 3. we assumed that our "realistic" force models are, indeed, realistic. The most difficult flaw to assess is the third since only the outer part of the potential has significant theoretical underpinnings. The best approach, though hardly optimal, is to calculate with as many different potential models as possible and to correlate the results. We are helped by the fact that the relative diffuseness of the triton causes the long-range parts of the potential to dominate. Relativistic corrections occur as additions to the kinetic energy (attractive), as additions to the two-body potential energy (partly repulsive), and as the three-body potentials that were discussed above. Although few calculations of the first two types of corrections have been made, most yield a small net attraction (37). It would seem appropriate to pursue this problem in more detail.

Given that three-body forces may account for part (or most) of the binding defect, are there other problems in the trinucleon that might be corrected by a three-body force? A recurring problem for ^{3}He has been the charge form factor (F_{ch}), or Fourier transform of the charge density [$\rho_{ch}(r)$], in the region of the secondary maximum. Because the charge density is a real scalar quantity, the form factor can exhibit diffraction structure, becoming alternately positive and negative. The magnitude of the latter will then exhibit cusp-like structures at diffraction minima, where it vanishes. Such behavior is seen in Figure 2, where both experimental data and an impulse approximation calculation (23) based on the Paris potential are plotted as a function of momentum transfer, q. Diffraction minima for the latter are apparent at $q^2 \approx 13.6$ fm^{-2} and $q^2 \approx 60$ fm^{-2}. The behavior at large momentum transfers is clearly outside the province of nonrelativistic calculations, but the behavior for $q^2 < 25$ fm^{-2} may not be.

Alternatively we can transform both data and theoretical calculations to configuration space in the form of $\rho_{ch}(r)$, and simultaneously remove the effect of the nucleon form factors. The latter arise from the intrinsic charge distributions of the proton and neutron; before an external electric field can probe the nuclear wave function, it must first "grab" the nucleons, which are the carriers of charge. Unfortunately, this procedure involves considerable theoretical extrapolation and has uncertain reliability, since large momentum transfers contribute non-negligibly for small r. Nevertheless, the region of moderate momentum transfer ($q^2 \leq 50$ fm^{-2}) tends to dominate. The result of this process, popularized by Sick (38), is shown in Figure 3.

The "experimental" charge density has a large hole not exhibited by the theoretical calculations (39) for the RSC potential, with and without a Coulomb interaction between the protons. This hole is essentially the same problem we saw in the previous figure: the experimental strength in the secondary (negative) maximum of the form factor is larger than theoretical predictions. The connection is easily seen from the Fourier transform relationship

$$\rho_{ch}(0) = \frac{1}{2\pi^2} \int_0^\infty F_{ch}(q^2) q^2 \, dq,$$

where large negative contributions to F_{ch} contribute to the hole in ρ_{ch}. The value of $\rho_{ch}(0)$ has a simple physical interpretation in impulse approxi-

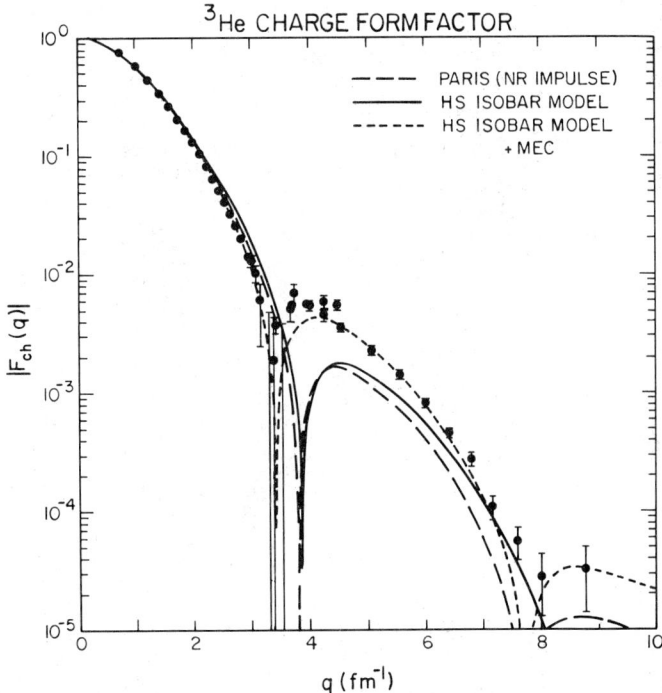

Figure 2 Magnitude of ^{3}He charge form factor versus momentum transfer. Three calculations (23) are shown together with data.

mation. The coordinate r in $\rho_{ch}(r)$ is the distance from the trinucleon center of mass to one of the protons. Thus, setting r to zero corresponds to a collinear configuration of nucleons. Theoretical calculations therefore produce too large a value of this part of the wave function. It should be noted that the strong short-range repulsion when two nucleons overlap does not play an important role, and that less than 1% of the charge of ^{3}He will fill the hole. The effect is thus much smaller than it appears to be; only small Fourier components are involved.

The angular dependence of three-body forces might accommodate all of the evidence presented above (40). It is entirely possible for the three-body force to be repulsive for collinear configurations and suppress those wavefunction components that contribute to $\rho_{ch}(0)$, while at the same time being attractive for isosceles configurations, which dominate the binding energy. Most 2π-3N forces have this property.

Another area of concern is the doublet neutron-deuteron scattering length: a_2. The n-d system has both doublet and quartet spin configurations. The latter scattering length has a large value (6.4 fm) determined

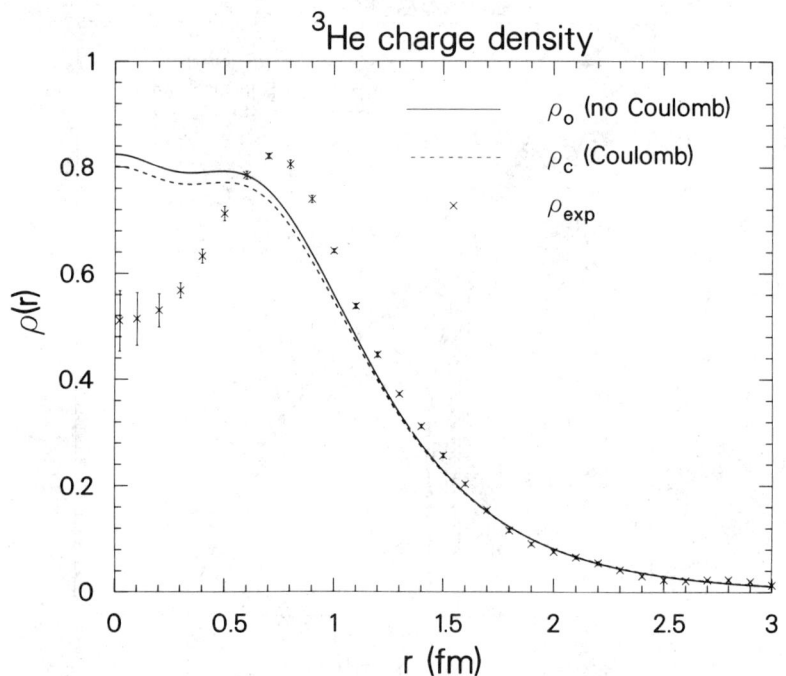

Figure 3 ^{3}He charge density for point protons. Two representative calculations (39) using the RSC potential model and the quasi-experimental data are shown.

primarily by the deuteron binding energy. Predictions for all realistic force models are commensurate. The doublet calculations on the other hand tend to lie along a narrow band, called the Phillips line (41), when plotted versus the triton binding energy. Recent calculations using a variety of force models, both realistic and otherwise (31, 42, 43), produce a curve that passes through the experimental datum. Curing the underbinding of the ground state will likely produce the correct value of a_2.

The remaining reaction with possible indications of significant three-nucleon force contributions is $n+d \rightarrow (n+n)+p$, which is used to determine the n-n scattering length in two ways: 1. an energetic proton is knocked out [ko], leaving the neutrons nearly at rest with small relative energy; 2. an energetic neutron is picked up [pu] (with small relative energy), leaving the proton nearly at rest. Analysis of this process with two-body force models produces two values for a_{nn}: $-20.7(20)$ fm [ko] and $-16.7(5)$ fm [pu]. The process $\pi^- + d \rightarrow \gamma + n + n$ also allows determination of a_{nn}: $-18.6(5)$ fm. Slaus et al (44) argue that the three-nucleon force should reduce the magnitude of the [ko] value of a_{nn} while increasing the magnitude of the [pu] value. This tantalizing result is based on an approximate scattering calculation. Better calculations are now necessary.

We summarize this section by noting that three-nucleon forces could reduce the trinucleon binding energy discrepancy, reduce $\rho_{ch}(0)$, lower the charge radii, reduce a_2, and bring the various "experimental" values of a_{nn} closer together. This evidence for a significant three-nucleon force effect is largely circumstantial.

2.4 The Two-π Three-Nucleon Force

We mentioned earlier that the 2π-3N force was believed by many to be the dominant three-body force in the trinucleon because of its long range; it is certainly the best studied. Much of the older work did not respect chiral constraints and is obsolete. We now discuss three recent approaches. Because constraints given by simple spin and isospin arguments determine the form of the result, disagreements by various groups largely concern the numerical coefficients in various terms. Critical discussions of the values of these parameters exist in (18, 45).

The Tucson-Melbourne group utilized current algebra and PCAC constraints without a detailed model expressed in terms of diagrams such as those of Figure 1. This approach has the advantage of not assuming a detailed model and making maximal use of general principles and data. The Brazilian group (Saõ Paulo–Recife) used (45) a phenomenological Lagrangian incorporating these constraints and also reproducing the π-N data; it is an approach familiar to nuclear physicists and is conceptually simpler. An excellent description of the physics input in both these approaches is given by McKellar & Glöckle (9). An additional, purely phenomenological 2π-3N force has been developed by the Urbana group (2, 3).

Calculations of the complete 2π-3N force effect in the trinucleon have largely depended on perturbation theory (46–48) or the variational method (49). The Monte Carlo variational calculations using the Urbana force (2, 3) found a large (~ 1 MeV) increase in binding, while the perturbation theory calculations for the Tucson-Melbourne potential found that projected 5-channel Faddeev basis functions produce large and nearly complete cancellation between the s- and p-wave components of the force (each roughly 1 MeV in size). Recently, a first-order perturbation theory Monte Carlo calculation (50) using complete 5-channel Faddeev wave functions has qualitatively confirmed both results. The Tucson-Melbourne force produced relatively small additional binding: -0.41 MeV for the RSC case, -0.14 MeV for the AV14 case, and -0.01 MeV for the SSCC case. The corresponding cases for the Brazilian force are: -1.10 MeV (RSC), -1.01 MeV (AV14), and -0.92 MeV (SSCC), with virtually identical results for the Urbana force.

The trinucleon wave function contains a maximum of 10 wave-function components determined by permutation symmetry and the requirements of the Pauli principle: 3 S-state components, 4 P-state components, and 3 D-state components. The results for the Tucson-Melbourne force calculated with a RSC wave function can be further broken down by turning off selected parts of the wave function: -0.59, -0.04, and -0.41 MeV, respectively assuming S-state only (S), S+D-state only (S+D), and the complete wave function (S+P+D). Somewhat surprising is the large effect of the tiny P-states ($\sim 0.1\%$), which are nearly as important as the larger wave-function components. Not surprising is the large D-state contribution. This calculation confirms (22, 23, 46, 47) that quantitatively accurate calculations are impossible unless small components of the wave function are included. Moreover, the results are sensitive to details of the three-nucleon force and to the short-range behavior mediated by the pion-nucleon form factor (discussed below).

The final model we examine is the isobar constituent calculation of Hajduk & Sauer (22, 23). This impressive calculation, which does not include the processes illustrated in Figures 1d–f, concentrates instead on the isobar processes and involves 33 Faddeev channels—18 with nucleons only and 15 with a single isobar each. It is undoubtedly the most ambitious three-nucleon calculation to date. Excitation and de-excitation of isobar components of the triton are accomplished by π and ρ exchange. Because the isobars generate a component of the two-body force, the purely nucleonic part of that force must be modified to accommodate the isobars.

This introduces a conceptually complicated "dispersion" effect. Hajduk and Sauer find a small net enhancement of binding (-0.3 MeV) from the cancellation of a repulsive two-body dispersive effect ($\Delta E \approx 0.6$ MeV) and an attractive three-body isobar force ($\Delta E \approx -0.9$ MeV) of the Fujita-Miyazawa type. The 2π-3N force by itself contributes -1.4 MeV. The effect on $\rho_{ch}(0)$ or the charge form factor of ^{3}He in the region of the secondary maximum is rather small, as indicated in Figure 2. The effect is larger in the region of higher q^2, and is larger in ^{3}He than in ^{3}H. The dispersive effect on the three-body force is also large, reducing the result by a factor of two.

The primary conclusion of this work is that the dispersive effect is extremely important and not included in other approaches. In order to understand this phenomenon one needs to perform a schematic perturbation theory calculation (29) in the trinucleon with a single isobar. For simplicity we imagine that an isobar, once created, propagates with a mass equivalent to that of the nucleon (a 25% error) and that it interacts with the nucleons via a potential equivalent to the nucleon-nucleon one. Hajduk and Sauer ignored the Δ-N interaction because of the additional computational complexity; isobar-hole calculations (53, 54) of π-nucleus scattering find wide variations of the strength of the Δ-N potential ranging from half to the full N-N value. Defining Δ to be the Δ-N mass difference and O_{jk} to be the two-body transition operator that causes an isobar to be created on nucleon line k by absorption of a pion emitted from line j, we obtain the energy shift in second-order perturbation theory:

$$\langle 0|\Sigma O_{kj}(E-\Delta-H)^{-1}O_{ki}|0\rangle,$$

where $H|0\rangle = E|0\rangle$ and H is both the NNN Hamiltonian and the ΔNN Hamiltonian (less the mass shift, Δ).

The terms with $i=j$ correspond to the aforementioned two-body dispersive terms. If one expands the energy denominator about the static value (Δ), the leading-order correction involves only commutators of H with O_{ij}. The kinetic energy part of H generates two-body terms, which must be eliminated in order to maintain the same two-body potential energy. The remaining potential terms generate three-body forces of the type shown in Figures 1k and 1l. If the potential were assumed to commute with O, the entire two-body dispersive effect would vanish in this approximation. Our analysis is analogous (55) to removing the energy dependence in the OPEP problem. Our conclusions are threefold. The two-body dispersive effect can be reinterpreted as a three-body force effect. The size of the effect will depend on details of the Δ-N interaction. The effect should be smallest when the N-N and Δ-N interactions are identical. More work is needed on this problem.

In summary, a recent calculation has found a substantial contribution to trinucleon binding from the Brazilian and Urbana 2π-3N forces, and substantially less from the Tucson-Melbourne force, with small wavefunction components producing substantial contributions. The impressive isobar constituent calculation of Hajduk and Sauer leads to a small 2π-3N effect. Most calculations find a small effect on $\rho_{ch}(0)$.

3. MESON-EXCHANGE CURRENTS

3.1 *Introduction*

In classical and quantum mechanics the motion of any charged particle creates a convection current: $e\mathbf{v}/c$, where e is the charge and $\mathbf{v}$ is the velocity of the particle. The total current density $\mathbf{J}(\mathbf{x})$ and charge density $\rho_{ch}(\mathbf{x})$ for a system with Hamiltonian H must satisfy the current continuity constraint: $\mathbf{\nabla} \cdot \mathbf{J}(\mathbf{x}) = -i[H, \rho_{ch}(\mathbf{x})]/c$, which states that no net charge can be created or destroyed. Separating H into kinetic (T) and potential (V) energies, the convection current is conserved using T alone. Particles with spin also possess a spin-magnetization current whose divergence vanishes. Together these currents for the individual nucleons comprise the nonrelativistic impulse approximation: $\mathbf{J}_0(\mathbf{x})$. Because the nuclear potentials contain isospin (charge) degrees of freedom and momentum dependence, current continuity requires additional currents, $\mathbf{J}_{ex}$, called variously interaction, potential-dependent, two-body, or meson-exchange currents. They can be interpreted as currents arising from the motion of virtual charged mesons in the nucleus (55).

Previously we categorized contributions to the Hamiltonian as either nonrelativistic or relativistic corrections, on the assumption that the *average* nuclear velocity is small compared to c. We categorize currents in the same way. The nonrelativistic current, $\mathbf{J}_0$, is of order $(1/c)$, and corrections are of order $(1/c^3)$ (i.e. relative $1/c^2$ corrections) and higher; corrections to the nonrelativistic charge operator (order $1/c^0$) are of order $(1/c^2)$ and higher. The rules of scale we developed previously can be applied here (37), as well: relativistic corrections should be of the order of a few percent. Because a nucleus is weakly bound, the potential and kinetic energies are roughly equal in magnitude and opposite in sign. Consequently, we expect that the potential-dependent currents, $\mathbf{J}_{ex}$, are comparable to the impulse current, $\mathbf{J}_0$.

Isovector transitions, which involve the flow of charged mesons from one nucleon to another, are expected to generate large exchange currents. Unfortunately, in magnetic transitions of this type the isovector nucleon magnetic moment, $\mu_v^0 = 4.7$, makes $\mathbf{J}_0$ "abnormally" large. Our weak-binding argument that $\mathbf{J}_{ex}$ is roughly 50–100% of $\mathbf{J}_0$ gets scaled down by μ_v^0

to 10–20%, which is observed. Electric transitions are primarily sensitive to $\rho_{ch}(\mathbf{x})$; they are relatively insensitive to exchange currents.

Given our desire to calculate these currents, how can we avoid the time-worn problem of being able to calculate only in perturbation theory, which does not converge for the strong interactions? Fortunately, nature comes to our rescue: the potential between two nucleons is very repulsive when they are close together and the corresponding part of the wave function is very small. Consequently, when evaluating matrix elements of currents, short-range operators are suppressed. This is indicated in Figure 4, which depicts the isoscalar trinucleon two-body correlation function C_s for two potential models, with the ranges (Compton wavelengths) of π, 2π, and ρ exchange indicated. The experimental and theoretical evidence suggests that OPE operators dominate, with a range determined by the pion mass, m_π.

The traditional approach to calculating exchange currents is similar to that of the three-nucleon forces: one assumes a model for photopion production from nucleons (56) that enforces the known constraints and

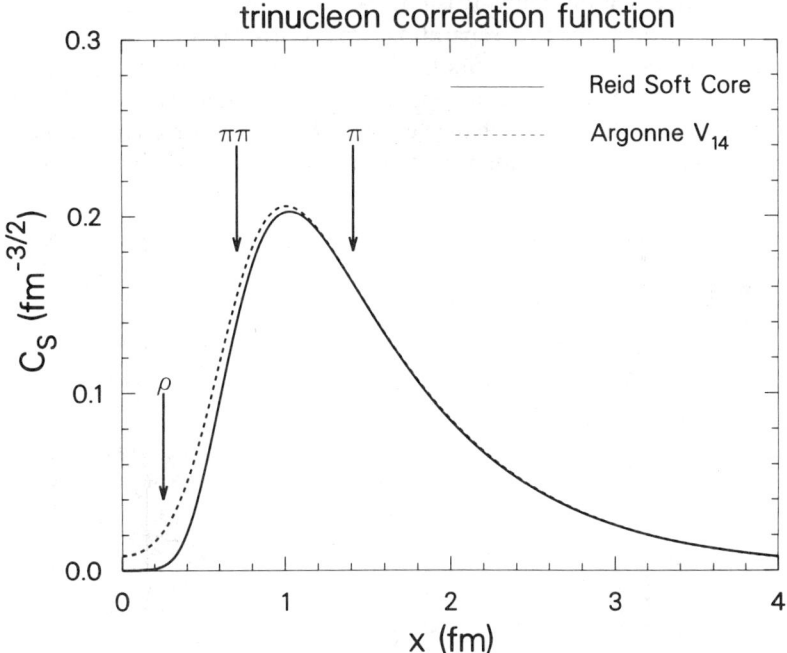

Figure 4 Isoscalar trinucleon correlation function versus two-nucleon separation. Representative calculations for two potential models and the ranges of π, 2π, and ρ exchange are shown.

adequately reproduces experimental data, and then freezes the meson degrees of freedom in the nucleus (57). This leads to currents of the types shown in Figure 5, induced by the electromagnetic interaction and produced by virtual pions being exchanged between nucleons. The dominant process is illustrated by the "seagull" diagram in Figure 5a, which incorporates the correct threshold (s-wave) pion photoproduction for both PS and PV pion-nucleon couplings. In the former it is the nucleon-antinucleon pair process depicted in Figure 5b that forms the larger dot in Figure 5a. In the latter scheme the seagull is a fundamental "gauge" term. This equality applies only to the static, leading-order current. Higher-order (in $1/c^2$) corrections to **J** and ρ_{ch} in the two models are physically different. Only the PV-Born terms reproduce threshold pion photo- and electro-production (56). The true exchange process, whose semiclassical origin is the overlapping pion clouds of the two nucleons, is illustrated in Figure 5c. Isobar photoproduction of p-wave pions, which are subsequently absorbed, leads to the current illustrated in Figure 5d. The processes shown in Figures 5b and c are essentially model independent in the static approximation, while that in Figure 5d is not (4).

The technical problem of separating two successive pion exchanges from the true three-body force recurs in Figures 5e and f in a different guise: separating successive external electromagnetic interactions and pion exchanges from the true-exchange current. Folklore correctly identifies a meson exchange before or after the external interaction as being part of the

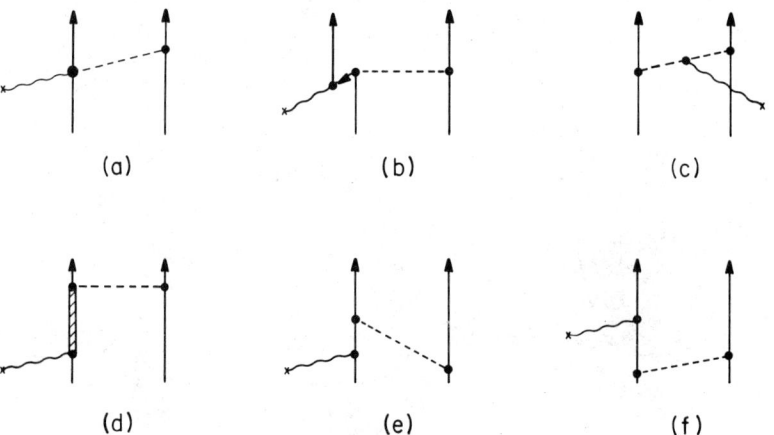

Figure 5 Various physical processes leading to meson-exchange currents. Solid, dashed, shaded, and wiggly lines depict nucleons, virtual pions, isobars, and the external electromagnetic interaction, while dots represent form factors.

wave function, not the current operator, but does not resolve the "recoil" graph, Figure 5e.

A rather complex analysis (57) leads to the following conclusions: Figure 5f defines the OPEP to be used in calculating wave functions. The static part (bulk) of the recoil graph disappears if this OPEP is chosen to be independent of energy (Hermitian). The nonstatic residue is of relativistic order ($1/c^2$). The appropriate OPEP must contain relativistic corrections for consistency and is nonlocal. The recoil current residue and the relativistically corrected OPEP are only uniquely defined up to a common unitary transformation. The common "realistic" potentials do not correspond to any of these OPEPs, and a consistent calculation using the former is not possible.

Most practitioners (58–65) use the Chemtob-Rho (66) approach to calculating transition operators, but this approach is neither unique nor does it possess any particularly desirable properties. It can be shown that the aforementioned (two-body) unitary transformation, when applied to the static OPEP, generates a 2π-3N potential identical in form to those parts of the Tucson-Melbourne force not determined by virtual isobars. A consistent treatment of relativistic corrections thus requires simultaneous analysis of the nonstatic OPEP, the charge or current operator, and the three-nucleon force. This is a manifestation of warnings issued in a slightly different context (67, 68).

Another source of uncertainty in the π-N coupling is the size of the form factor $F_{\pi NN}(q^2)$ indicated by the small dots in Figure 5. Although this is a short-range effect, it is not negligible. Theoretical and experimental efforts to determine this size have been inconclusive (69); most of our experimental information comes from the Goldberger-Trieman (GT) discrepancy, which is assumed to be proportional to $F_{\pi NN}(0)-1$, and experimentally is -0.06. A monopole form factor, $(\Lambda^2 - m_\pi^2)/(\Lambda^2 - q^2)$, then gives $\Lambda \approx 4m_\pi$. A thorough discussion of this problem by Coon & Scadron (69) suggests that only half of the GT discrepancy arises from the form factor, a result consistent with many theoretical efforts to understand $F_{\pi NN}$. That assumption leads to $\Lambda = 5.8\ m_\pi$. There is little to recommend the commonly used value $\Lambda = 8.5\ m_\pi$.

Although shorter-range currents (70) should be suppressed by the mechanism we discussed above, their effects are not negligible. Typical contributions are generated by replacing the pions in Figure 5 by ρ mesons, for example. A distinct type is the isoscalar ($\pi\rho\gamma$) current or the isovector ($\pi\omega\gamma$) current calculated by replacing one of the pion legs in Figure 5c by a ρ or ω meson. A very readable discussion of these processes is given by Riska (5).

The interaction of an electromagnetic field with a hadron is mediated by another type of form factor, indicated by the dots in Figure 5 attached to the wavy lines. In general they are all different. Simple Born-term models are current conserving only if these various form factors are taken to be identical. The problem lies in additional physical processes not included in these models. Magnetic transitions can be shown to depend only on current components that are solenoidal. There is therefore no compelling reason to equate these form factors. The preferred PV coupling leads naturally to the use of the nucleon axial-vector form factor, G_A, in the seagull current rather than the more commonly used charge form factor, G_E.

We summarize this section by noting that current conservation implies exchange currents between nucleons, mediated by charged mesons. In isovector magnetic transitions these currents should be large. The strong short-range repulsion between nucleons dictates the dominance of long-range currents in general and OPE currents in particular.

3.2 Experimental Evidence

Much of the interest in exchange currents has centered on magnetic moments. The trinucleon magnetic moments can be written as isoscalar $[\mu_s = \mu(^3\text{He}) + \mu(^3\text{H})]$ and isovector $[\mu_v = \mu(^3\text{He}) - \mu(^3\text{H})]$ combinations:

$$\mu_s = \mu_s^0 - 2P(D)(\mu_s^0 - \tfrac{1}{2}) + \Delta\mu_s = 0.85131, \qquad 1.$$

$$\mu_v = -\mu_v^0[1 - \tfrac{4}{3}P(S') - \tfrac{2}{3}P(D)] + \Delta\mu_0 + \Delta\mu_v = -5.10641, \qquad 2.$$

where $\Delta\mu_0$ is the small isovector orbital magnetic moment; $\Delta\mu_s$ and $\Delta\mu_v$ are the contributions from relativity and meson currents; terms proportional to the percentage of D state, $P(D)$, and mixed symmetry S state, $P(S')$, originate in impulse approximation. The numerical values are experimental ones. The expression for μ_s is similar to that of the deuteron and is determined by the isoscalar nucleon magnetic moment, $\mu_s^0 = 0.88$. Using typical values of $P(S') \approx 1.5\%$ and $P(D) \approx 9.0\%$, we find that $\Delta\mu_s$ is 0.05, while $\Delta\mu_v/\mu_v^0$ is -16.5%. A very small value of $P(D) = 3.9\%$ in Equation 1 would make the impulse approximation value equal to the experimental value. The large isovector correction is typical of pion-exchange currents. The consensus of contributions (65, 71–75) from the "pair" or seagull-exchange current is approximately -0.6 or $-0.14 \mu_v^0$. The true-exchange process is opposite in sign to the seagull and isobar contributions, and tends to cancel the latter. Several researchers (65, 71–75) find a small residue from the true-exchange and isobar contributions. The short-range contributions are roughly 10% of those of pion range. The overall conclusion, however, is clear. The massive failure of the impulse approximation to account for the isovector magnetic moment is alleviated by the large seagull-exchange current of pion range. Other contributions are smaller, and tend to cancel.

Related in impulse approximation to the isovector magnetic moment is the β-decay amplitude of ^{3}H, which proceeds through a mixture of Fermi and Gamow-Teller processes, the smaller Fermi terms being accurately calculated assuming a conserved vector current. The Gamow-Teller operator has a spin-flip, isospin-flip character and its matrix element is given by (4)

$$\frac{|M_A|}{\sqrt{3}} = 1 - \tfrac{4}{3}P(S') - \tfrac{2}{3}P(D) + \Delta M_A = 1 - 0.042(8),$$

where the numerical value is experimental (76). The exchange-current corrections, ΔM_A, are expected to be of relativistic order (a few percent); using our previous values of $P(S')$ and $P(D)$, we estimate $\Delta M_A \approx 0.04(2)$. A variety of small and largely cancelling one-pion-exchange, two-pion-exchange, isobar-mediated, and other mechanisms (73, 76–80) are consistent with this estimate in several calculations, although the individual calculations of each are not. The isoscalar trinucleon and deuteron magnetic moments and the β-decay rate of ^{3}H are both exceptionally difficult problems, with many comparable mechanisms making small contributions.

The charge and magnetic trinucleon form factors have received much attention lately (81–84). These measurements together with older data lead to: $\langle r^2 \rangle_{\mathrm{mag}}^{1/2} = 1.84(3)$ fm for ^{3}He; $\langle r^2 \rangle_{\mathrm{mag}}^{1/2} = 1.72(6)$ fm for ^{3}H; $\langle r^2 \rangle_{\mathrm{ch}}^{1/2} = 1.88(2)$ fm for ^{3}He; $\langle \dot{r}^2 \rangle_{\mathrm{ch}}^{1/2} = 1.67(3)$ fm for ^{3}H. The experimental values include the intrinsic sizes of the neutron and proton. Extensive data exist only for the ^{3}He charge and magnetic form factors. Recent measurements (81) of the latter have shown the existence of a diffraction minimum. Although its existence was predicted using the impulse approximation, the actual minimum occurs at a much higher value of q, as shown in Figure 6. The Hajduk-Sauer isobar model (73) is qualitatively similar to other calculations, and the explicit pair and exchange currents and implicit isobar currents of that model account for the differences seen in the solid and long-dashed curves. The pair current in the former was multiplied by the nucleon isovector charge form factor G_E^v, while the short-dashed curve used F_1^v, a component of G_E^v; the difference is a relativistic correction. The large cancellation between the pionic seagull current and the impulse approximation creates this sensitivity near the minimum. We advocate the use of G_A, rather than G_E, although the experimental uncertainties in the former would also lead to large uncertainties. Nevertheless, the dominance of the seagull graph, the large secondary role of the isobar process near the minimum, and the relatively small contribution of short-range processes again confirms the importance of pion degrees of freedom in the trinucleon. The ^{3}H magnetic form factor is similar to that of ^{3}He, but no data exist for $q^2 > 8$ fm^{-2}. Planned experiments will remedy this situation.

We saw in Figure 2 that the charge form factor of ^{3}He was deficient in the region of the secondary maximum, and that the three-body forces of the Hajduk-Sauer isobar model did not help appreciably. Adding part of the pion-exchange charge operator (short-dashed curve) with PS coupling according to the Chemtob-Rho prescription (66) produces agreement between experiment and theory. Using PV coupling reduces the effect, although the role of uncalculated (nonlocal) terms has not yet been assessed. Also, different prescriptions lead to different results when used with common, "realistic" potentials. Thus, this apparent success cannot be taken too seriously. The triton charge form factor shows only a limited pion-exchange influence. This is an artifact of having nearly identical

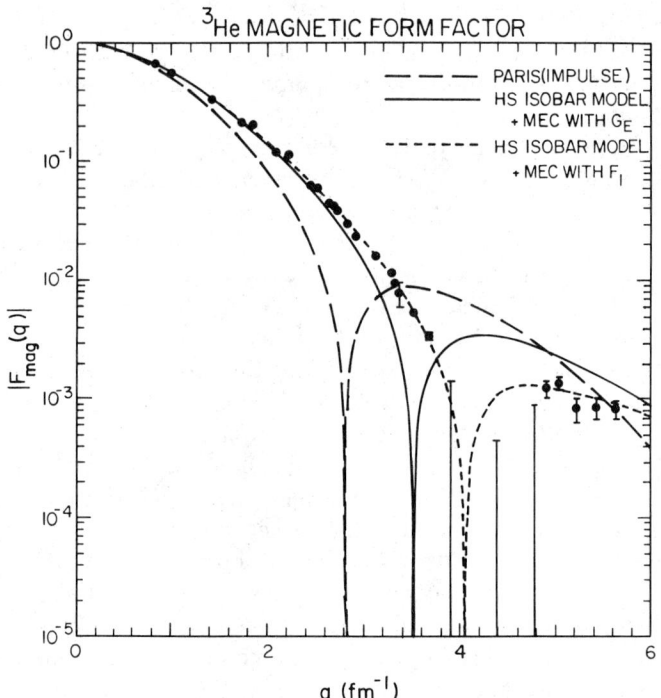

Figure 6 Magnitude of ^{3}He magnetic form factor versus momentum transfer. Three calculations (73) are shown together with data. See text.

isovector and isoscalar pion-exchange contributions, which interfere constructively in the ^{3}He case and destructively in the ^{3}H case (64).

The n-d system at thermal energies can radiatively decay to the triton ground state: $n+d \to {}^3H+\gamma$. Both magnetic dipole and electric quadrupole processes compete, the former dominating and the latter occurring only for the quartet initial-spin configuration. The doublet transition in impulse approximation can be considered an analogue of the magnetic moment, given by Equations 1 and 2, if we replace percentages of various wave-function components by overlaps of the ground and scattering state. Because the scattering and ground-state wave functions are orthogonal, one drops the leading μ_s^0 and μ_v^0 terms. Consequently, in impulse approximation the transition proceeds via small components of the wave function and by meson-exchange currents; this leads to a large suppression relative to the n-p radiative capture process. Our analysis of the magnetic moments indicates that the impulse and meson-exchange current contributions should be comparable. The most recent measurement (85) of the transition rate for thermal neutrons combined with the previous measurement (86) by a different technique gives 0.515(9) mb, 600 times smaller than the corresponding n-p capture rate.

Several earlier theoretical calculations (87, 88) that were consistent with this rate have been superseded by the extensive calculations of Torre & Goulard (89). Their calculations for the RSC, SSCC, and Malfliet-Tjon (90) potentials were shown to scale with the trinucleon binding energy and were extrapolated to the physical value of the latter quantity. They find that in impulse approximation the quartet contribution to the cross section is 20% larger than that of the doublet. Exchange currents lower the former by 20% while raising the latter by a factor of 5. The net result is that the impulse approximation cross section of 0.2 mb is raised by exchange currents to 0.6 mb. Uncertainties of approximately 20% in their calculation make this result consistent with experiment. The pion-range currents dominate those of shorter range, with the seagull, isobar, and true-exchange currents contributing roughly in the ratio $3:2:-1$.

In summary, the magnetic properties of the trinucleon ground state receive large contributions from the pion-exchange seagull process. The isovector magnetic moment requires (roughly) a 16% exchange current contribution, most of it originating from the seagull graph. The exchange currents shift outward the diffraction minimum in the ^{3}He magnetic form factor, and this is observed. The thermal n-d radiative capture rate is strongly influenced by the pion-exchange currents, which raise the cross section by a factor of three. Agreement with experiment is adequate. Exchange contributions to the ^{3}H β-decay rate and to the charge form factors are of relativistic order and are more poorly defined.

4. EFFECTS OF THE TENSOR FORCE

4.1 Nucleon-Nucleon Force Properties

Physical effects stemming from the tensor nature of the nuclear force have been the object of experimental search since the discovery that the deuteron had a quadrupole moment. Measurement of the D-state probability of the deuteron (P_D), the triton, etc was attempted until Amado (91) and Friar (92) pointed out that P_D was not an observable. The emphasis then switched to investigation of the asymptotic normalization constants (C_S and C_D) and their ratio η. Still, the value of P_D remains a convenient model characterization, as can be seen by examining Table 1, where binding energy, P_D, and C_D/C_S are tabulated for three "realistic" potential models: the RSC (33), and AV14 (36), and the SSCC (35). The D-state probability varies much more than either asymptotic property—binding energy or asymptotic D/S ratio η. We also include properties of two separable potential models IJ4 and IJ7 (93), which are used to illustrate the point that simple models yield results for asymptotic properties similar to those of "realistic" local potentials.

4.2 Bound States

Although there are several situations in which the tensor nature of the nuclear force is apparent, extracting a definitive measure of its size is difficult. Consider the bound-state properties of the trinucleon system. (The effect of a tensor force in ^{4}He has been investigated only in the case of the binding energy via perturbation theory.) Even though the rms radius does vary as a function of P_D, the correlation is with the ^{3}H binding energy: the tensor force has a longer range than a central force and is consequently less effective in binding the more compact triton than the deuteron. Therefore, increasing P_D while holding the deuteron binding fixed decreases the triton binding and this produces a concomitant expansion of the radius. Because the alpha particle is even more compact than the triton, the tensor force is even less effective in binding ^{4}He than ^{3}H; consequently, except for overall

Table 1 Properties of selected local and separable NN potentials

Model	Binding energy, B_2 (MeV)	$\eta = C_D/C_S$	P_D (%)
RSC	2.2246	0.0262	6.45
AV14	2.225	0.0266	6.08
SSCC	2.224	0.0255	5.45
IJ7	2.226	0.0254	7.0
IJ4	2.226	0.0289	4.0

scale, details of the tensor force are less apparent in the bound-state properties of ^{4}He. The $A = 3$ theoretical charge densities do show some difference between the s-wave [Malfliet-Tjon (90)] model and the tensor-force RSC model. However, much of the structure that appears in the point-charge isoscalar and isovector densities is washed out after folding in the finite size of the nucleons. Still, the RSC model predicts a small hole in ^{3}H, unlike ^{3}He—the result of the D-state component of the wave function (39). To test our modeling of the trinucleon system, it is absolutely essential to obtain adequate ^{3}H charge form factor data.

The trinucleon asymptotic normalization constants provide direct evidence for the existence of tensor forces: $C_D(^3\text{H}) \neq 0$. Physically, C_S and C_D echo the internal dynamics of the wave function through the overall normalization. [In fact, the deuteron S-state asymptotic normalization is determined uniquely by its binding energy and the triplet effective range, r_t (94): $C_S(^2\text{H}) = (1 - \sqrt{MB_2 r_t})^{-1/2}$.] This has even led to the suggestion that the ^{3}H and ^{3}He asymptotic normalization constants be accorded the same status as other trinucleon properties such as the binding energy and charge radius (95). However, because the experimental determination of these parameters is incomplete, it has not yet been possible to use them to discriminate among candidate nucleon-nucleon interaction models. No data exist for ^{4}He.

Asymptotic normalization constants are defined such that their value is unity when the effective interaction in the asymptotic channel is of zero range (96):

$$\lim_{y_1 \to \infty} \Psi_{^3\text{H}}(\mathbf{y}_1, \mathbf{x}_1) \to C_S N_{ZR} \frac{\exp(-\beta y_1)}{y_1} \{[Y_0(\hat{y}_1) $$

$$\times \chi^{1/2}(1)]^{[1/2]} \times \Phi^{[1]}(\mathbf{x}_1)\}^{[1/2]} \frac{\eta'}{\sqrt{2}}$$

$$+ C_D N_{ZR} \frac{\exp(-\beta y_1)}{y_1} \left(1 + \frac{3}{\beta y_1} + \frac{3}{\beta^2 y_1^2}\right)$$

$$\times \{[Y_2(\hat{y}_1) \times \chi^{1/2}(1)]^{[3/2]} \times \Phi^{[1]}(\mathbf{x}_1)\}^{[1/2]} \frac{\eta'}{\sqrt{2}},$$

where $\beta^2 = 4M[B_3 - B_2]/3$, $N_{ZR} = \sqrt{2\beta}$ is the zero-range normalization, $Y_l(\hat{y})$ is a spherical harmonic (m suppressed due to coupling), $\chi^{1/2}$ is a spin-$\frac{1}{2}$ function, $\Phi^{[1]}$ is the deuteron wave function, and η' is the isospin-$\frac{1}{2}$ function for three particles in which particles 2 and 3 (the deuteron) are coupled to isospin 0. For ^{3}He (96), one should replace $\exp(-\beta y_1)$ by the Whittaker function $W_{-\kappa,1/2}(2\beta y_1)$, replace $\exp(-\beta y_1)[1 + 3/(\beta y_1) + 3/(\beta y_1)^2]$ by

$W_{-\kappa,5/2}(2\beta y_1)$, and replace N_{ZR} by $N_W = N_{ZR}[\frac{1}{2}\Gamma(3+\kappa)\Gamma(2+\kappa)/{_3}F_2(\kappa, 2, 1+\kappa; 3+\kappa, 2+\kappa; 1)]^{1/2}$, where $\kappa = 2M\alpha/(3\beta)$. With these definitions, one recovers the ^{3}H constants (C_S and C_D) from the ^{3}He constants (C_S^C and C_D^C) in the limit that $\kappa \to 0$ (i.e. in the limit that the fine structure constant $\alpha \to 0$):

$$W_{-\kappa,1/2}(2\beta y_1) \xrightarrow[y_1 \to \infty]{} \frac{\exp(-\beta y_1)}{(2\beta y_1)^\kappa} \xrightarrow[\kappa \to 0]{} \exp(-\beta y_1) \quad \text{and} \quad N_W \xrightarrow[\kappa \to 0]{} N_{ZR}.$$

In Table 2 we list selected model values of the asymptotic normalization constants. Compare the MT I-III model C_S (96, 97) with that of the RSC3 model (96, 98), a three-channel calculation in which the tensor force does not act on the spectator nucleon (y_1 coordinate) so that $C_D \equiv 0$. Approximately 8% of the 11% difference is accounted for by the reduction of the S-state probability to 92%; the remainder is due to the difference in binding energy between the two models. Solving the full 5-channel problem for the RSC potential (RSC5) yields little change in C_S compared to the RSC3 value (31, 96, 99). The fact that $C_D \ll 1$ simply reflects the relative size of the S-state and D-state nucleon-deuteron wave functions; recall that C_D is defined relative to the S-state zero-range function. The IJ model parameters (100) result from the separable potentials in Table 1 for P_D values of 4% and 7%. They demonstrate that separable potential models do yield parameters that are indeed comparable to those of realistic potential models. The differences between the RSC5 numbers and the IJ7 numbers are consistent with the binding energy and P_D variations. [We return to the IJ model results below in our discussion of the ^{3}He(e, e'p)d reaction.] The distorted-wave parameter D_2 is approximately related to the C_D/C_S ratio: $D_2 \approx -C_D/(\beta^2 C_S)$ and $D_2^C \approx -C_D^C/(\beta^2 C_S^C) f(\kappa)$, where $f(\kappa) \xrightarrow[\kappa \to 0]{} 1$. ($D_2$ is defined more precisely below in Equation 4.) Because C_D has not been determined experimentally, it is D_2 that directly reflects the tensor nature of the nuclear force in the ^{3}H $\to$ n+d (or ^{3}He $\to$ p+d) vertex. Comparing

Table 2 Binding energy and asymptotic constants for the MT I-III, RSC3, RSC5, and separable-potential models

Model	Ref.	Binding energy, B_3 (MeV)	$C_S(C_S^C)$	$C_D(C_D^C)$	$D_2(D_2^C)$ (fm^2)
^{3}H/MT I-III	96, 97	8.54	1.96	—	—
^{3}H/RSC3	91	6.38	1.74	—	—
^{3}H/RSC5	91, 96, 99	7.02	1.76	0.066	−0.243
^{3}He/RSC5	96	6.39	1.77	0.061	−0.240
^{3}H/IJ-7	100	7.64	1.79	0.069	−0.220
^{3}H/IJ-4	100	8.58	1.84	0.092	−0.233

RSC5 model values for ^{3}H and ^{3}He shows that Coulomb effects should not lead to significant differences in any experimental determination of C_S and C_S^C or D_2 and D_2^C.

Experimental values of the asymptotic normalization constants of ^{3}H and ^{3}He have been obtained by several different means. [For an excellent review of various definitions and relationships between asymptotic normalization constants and vertex constants, see Locher & Mizutani (101).] Methods include (a) forward dispersion relation analysis (FDR and FDRC) (102, 103), (b) partial-wave dispersion relation analysis (PWDR) (104), and (c) fits to tensor analyzing powers for the $(\vec{d}, ^3\text{H})$ and $(\vec{d}, ^3\text{He})$ reactions (105–107). In Table 3 we list what are considered to be the most reliable values (96). There is no clear consensus for C_S; neutron experiments are intrinsically difficult. The value of C_S^C is much better determined; it clearly indicates that the MT I-III model (90) with no tensor force has an unacceptable C_S value. The values for the RSC3 and RSC5 models are consistent with the PWDR result. In contrast, the one-boson-exchange model (OBE) yields (108) a much smaller value of 1.61, consistent with the FDR result. Clearly, more experimental work is needed on D_2 and D_2^C before strong conclusions can be drawn. However, separable potential model studies of D_2 indicate that P_D in the range of 5–8% is acceptable, whereas such studies of the analogous D_0 parameter suggest limits on P_D lying between 4 and 6% (108–110).

In summary, the trinucleon binding energies, form factor data (including rms radii), and asymptotic normalization parameters all yield evidence supporting the existence of the tensor nature of the nuclear force. However, experimental data available now appear to be consistent with any reasonable model description of the nucleon-nucleon data that yields a value of P_D in the range of 4–7%.

Table 3 Experimental values of asymptotic constants

Parameter	Ref.	Value	Method
$C_S^2(^3\text{H} \rightarrow \text{nd})$	102	2.6 ± 0.3	FDR
$C_S^2(^3\text{H} \rightarrow \text{nd})$	104	3.3 ± 0.1	PWDR [B_3 and a_2 fixed at experimental values]
$(C_S^C)^2(^3\text{He} \rightarrow \text{pd})$	103	3.3 ± 0.4 3.19 ± 0.24	FDRC [corrected for Coulomb zero-range function not normalized to unity (96)]
$D_2(^3\text{H} \rightarrow \text{nd})$	105	-0.279 ± 0.012 fm^2	$(\vec{d}, ^3\text{H})$
$D_2^C(^3\text{He} \rightarrow \text{pd})$	105	-0.339 fm^2	$(\vec{d}, ^3\text{He})$
	106	-0.37 fm^2	$(\vec{d}, ^3\text{He})$
	107	-0.22 fm^2	$(\vec{d}, ^3\text{He})$

4.3 Electromagnetic Reactions

The asymptotic normalization constants can also be extracted from coincidence experiments like ^{3}He(e, e'p)d (100). This structural information is contained in the momentum distribution amplitudes $f_0(q)$ and $f_2(q)$, which are the $l = 0$ and 2 components of the overlap of the ground-state wave function with the wave function for a nucleon of relative momentum **q** moving freely relative to a deuteron. The general form of the coincidence cross section is given by

$$\frac{d\sigma}{d\Omega_e dE_e d\Omega_p} = \frac{d\sigma}{d\Omega_{ep}}(\text{k.f.})\frac{3}{2}\{[f_0(q)]^2 + 2[f_2(q)]^2\},$$

where $d\sigma/d\Omega_{ep}$ is the "half-off-shell" ep cross section and (k.f.) represents a kinematic factor. The integral d^3q over the quantity in brackets{...} is usually called the spectroscopic factor. The asymptotic normalization constants are given by

$$C_l = i^l \left\{ 2\pi i \beta^{1/2} \lim_{q \to i\beta} (q - i\beta) f_l(q) \right\}.$$ 3.

The measured distorted-wave parameters are obtained from the $q = 0$ values of the $f_l(q)$; in particular, one has

$$D_2 = \lim_{q \to 0} \left[\frac{-f_2(q)}{q^2 f_0(q)} \right],$$ 4.

which is only approximately related to the pole value $\tilde{D}_2 \equiv -C_D/(\beta^2 C_S)$. Unfortunately, the available data (111) are consistent with any model value of P_D between 4 and 7% (100). In this range, D_2 decreases linearly with P_D, and the approximation of D_2 by $\tilde{D}_2$ is not unreasonable. Thus, the data require a tensor force but are not sufficiently sensitive to distinguish among candidate nuclear force models.

At one point it appeared that the E2 multipole component of the ^{3}He(γ, p)d cross section might be primarily the result of the tensor force. Early experiments (112) indicated an E2 cross section as large as 10% of the total in the region of the peak ($E_\gamma \approx 12-15$ MeV). In contrast, a separable-potential model calculation (113) using only s-wave potentials found the E2 cross section to be only about 2%. However, a later study (114) firmly established that the E2 strength below $E_\gamma \approx 25$ MeV is less than 2%. Recent angular distribution data (115) from the ^{2}H(p, γ)^{3}He capture reaction for incident proton energies from 6.5 → 16 MeV have been combined with data from earlier experiments (114, 116, 117) to investigate the size of tensor force effects. An optical model analysis employing spectroscopic factors from separable-potential Faddeev calculations indicates that P_D in the range of 4–7% can explain the magnitude and systematics of the data (100). A word of caution is in order: when examining angular distributions to extract small cross-sectional components (such as the higher multipole coefficients), one should include the plane-wave Born result for all multipoles not otherwise contained in the analysis (118); otherwise, extraneous values of the small parameters will likely be generated by the partial wave truncation.

We summarize by recalling that electromagnetic interactions primarily explore the one-body density properties of the nucleus. The (e, e′) coincidence measurements can be directly related to the asymptotic normalization constants. Multipole data from the photodisintegration reaction can be compared with sophisticated model calculations. Within the accuracy of the available data, any reasonable nucleon-nucleon force model with P_D in the 4–7% range appears acceptable.

4.4 Elastic and Inelastic Scattering

Below 50 MeV, polarization effects in N-d scattering are an order of magnitude larger than polarization effects in N-N scattering. Polarization effects are also significant in the $A = 4$ scattering problem. However, because one expects vector quantities to be most sensitive to p-wave forces and tensor quantities to tensor forces, and because one must utilize the $(\vec{d}d)$ system to investigate tensor quantities, we concentrate on the $A = 3$ system. The polarization data prior to 1980 were admirably reviewed by Grüebler (119) at the Eugene Few-Body Conference. In addition, Kloet (120) summarized the cross section and polarization situation at the Santa Fe Polarization Conference. Since then, more precise nd and pd data have appeared, such as the differential and total cross-section measurements from Karlsruhe (121) and the ETHZ/SIN (122) facilities. Very low-energy measurements (below threshold for breakup of the deuteron) have also been made (123). The question remains whether quantitative constraints on the tensor nature of the nucleon-nucleon force result.

Theoretical calculations for elastic scattering include complex, separable-potential models (124) and local, realistic RSC and SSCC potential models (125). The advantage of the former lies in the flexibility that permits testing the sensitivity of the calculation to details of the N-N interaction model; the parameters of the local potentials are fixed and their results either fit the data or do not. The calculations of Stolk & Tjon (125) lead one to the following conclusions: (a) the details of the s-wave forces are the most important aspect for most observables, although comparison with experimental data may indicate that P_D for the RSC model is too large; and

(b) several observables are as sensitive to d-wave forces as to the p-wave or tensor forces.

The more extensive calculations of Doleschall have been used to analyze a variety of data (126). The largest difference in the tensor analyzing power calculations arises in comparing results for a two-term $^3S_1$-$^3D_1$ potential model with those for a four-term model; the ε_1 mixing parameter is not realistic for the two-term potential, although both models have the same 4% deuteron D state. It is perhaps unfortunate that all these models fit the tensor analyzing powers as well as they do. While one must have a tensor ($^3S_1$-$^3D_1$) force to obtain a nonzero result, any reasonable model with 4% $< P_D <$ 7% appears to provide an acceptable qualitative fit. If so, very precise data will be required before one obtains a significant constraint upon the nonasymptotic nature of the tensor force.

Before closing this section, we note that it has been suggested that one can look at the Nd → Nnp breakup reaction for evidence of tensor force effects (127). Using Doleschall's code, it was found that a significant filling in of a minimum occurred in the triple differential cross section for a two-term 7% P_D model compared to a two-term 4% P_D model. An earlier perturbative treatment of the Faddeev equations found significant differences among several available potential models for various physical observables in the breakup reaction (128). The sensitivity of the breakup reaction to off-shell effects apparently should be larger than that of the elastic scattering process. Hence, the extraction of quantitative constraints on the tensor nature of the nuclear force would be less reliable. In addition, pion-exchange three-body forces are primarily tensor in character and would add further to the uncertainty in extracting definitive information from the breakup reaction, as well as from the elastic scattering process.

In summary, the nonspin observables are most sensitive to the details of the s-wave component of the nuclear force. Moreover, several spin observables are as sensitive to the p-wave and d-wave components of the nuclear force as they are to the tensor components. However, a tensor force is required by the data. Yet, any reasonable nucleon-nucleon force model with P_D in the range of 4–7% appears to provide an adequate representation of the available data.

5. SUMMARY

Strong circumstantial evidence exists for three-nucleon forces in the trinucleon. The two-pion-exchange part of this force is the best studied and is attractive. Technical problems with the theory are still an impediment.

Excellent experimental evidence exists for meson-exchange currents in the trinucleon. The strong short-range repulsion in the nucleon-nucleon force makes the long-range currents dominate in general and the one-pion-exchange currents in particular.

Unambiguous experimental evidence exists for the tensor force in the trinucleon systems, but much of it is insensitive to details of this force.

ACKNOWLEDGMENTS

The authors acknowledge the invaluable assistance of D. R. Lehman, D. M. Skopik, P. U. Sauer, and R. B. Wiringa. The work of GLP was supported in part by the US Department of Energy; that of JLF and BFG was performed under the auspices of the US Department of Energy.

Literature Cited

1. Faddeev, L. D. 1961. *Sov. Phys. JETP* 12:1014
2. Carlson, J., Pandaharipande, V. R., Wiringa, R. B. 1983. *Nucl. Phys. A* 401:59
3. Wiringa, R. B. 1983. *Nucl. Phys. A* 401:86
4. Ivanov, E. A., Truhlik, E. 1981. *Sov. J. Part. Nucl.* 12:198
5. Riska, D. O. 1983. *Univ. Helsinki Rep. Ser. in Phys. HU-P-224*
6. Friar, J. L., Gibson, B. F., Payne, G. L. 1983. *Comments Nucl. Part. Phys.* 11:51
7. McKellar, B. H. J., Rajaraman, R. 1979. In *Mesons in Nuclei*, ed. M. Rho, D. Wilkinson, p. 358. Amsterdam: North Holland
8. Friar, J. L. 1983. *Proc. Workshop on the Interaction Between Medium Energy Nucleons in Nuclei*, p. 378. New York: Am. Inst. Phys.
9. McKellar, B. H. J., Glöckle, W. 1984. *Proc. 10th Int. Conf. on Few Body Problems in Physics* (Karlsruhe, 1983). *Nucl. Phys. A* 416:435c
10. Glöckle, W., Sauer, P. U. 1984. *Europhys. News* 15-2:5
11. Axilrod, B. M., Teller, E. 1943. *J. Chem. Phys.* 11:299
12. Primakoff, H., Holstein, T. 1939. *Phys. Rev.* 55:1218
13. Fujita, J.-I., Miyazawa, H. 1957. *Prog. Theor. Phys.* 17:360
14. Brown, G. E., Green, A. M., Gerace, W. J. 1968. *Nucl. Phys. A* 115:435
15. Brown, G. E., Green, A. M. 1969. *Nucl. Phys. A* 137:1
16. Coon, S. A., Scadron, M. D., Barrett, B. R. 1975. *Nucl. Phys. A* 242:467
17. Coon, S. A., Scadron, M. D., McNamee, P. C., Barrett, B. R., Blatt, D. W. E., McKellar, B. H. J. 1979. *Nucl. Phys. A* 317:242
18. Coon, S. A., Glöckle, W. 1981. *Phys. Rev. C* 23:1790
19. Loiseau, B. A., Nogami, Y., Ross, C. K. 1971. *Nucl. Phys. A* 165:601
20. Yang, S.-N. 1974. *Phys. Rev. C* 10:2067
21. Ueda, T., Sawada, T., Takagi, S. 1977. *Nucl. Phys. A* 285:429
22. Hajduk, C., Sauer, P. U. 1979. *Nucl. Phys. A* 322:329
23. Hajduk, C., Sauer, P. U., Strueve, W. 1983. *Nucl. Phys. A* 405:581
24. Fujita, J.-I., Kawai, M., Tanifuji, M. 1962. *Nucl. Phys.* 29:252
25. Drell, S. D., Hwang, K. 1953. *Phys. Rev.* 91:1527
26. Ellis, R. G., Coon, S. A., McKellar, B. H. J. 1984. *Nucl. Phys. A.* In press
27. Martzolff, M., Loiseau, B., Grangé, P. 1980. *Phys. Lett.* 92B:46
28. Hasagawa, K. 1963. *Prog. Theor. Phys.* 30:827
29. Hajduk, C., Sauer, P. U., Yang, S.-N. 1983. *Nucl. Phys. A* 405:605
30. Hajduk, C., Sauer, P. U. 1981. *Nucl. Phys. A* 369:321
31. Benayoun, J. J., Gignoux, C., Chauvin, J. 1981. *Phys. Rev. C* 23:1854
32. Chen, C. R., Payne, G. L., Friar, J. L., Gibson, B. F. 1984. *Phys. Rev. C.* In press
33. Reid, R. V. Jr. 1968. *Ann. Phys. (NY)* 50:411
34. LaCombe, M., Loiseau, B., Richard, J. M., Vinh Mau, R., Côté, J., et al. 1980. *Phys. Rev. C* 21:861
35. de Tourreil, R., Rouben, B., Sprung, D. W. L. 1973. *Nucl. Phys. A* 201:193
36. Wiringa, R. B., Smith, R. A., Ainsworth, T. A. 1984. *Phys. Rev. C* 29:1207
37. Friar, J. L. 1981. *Nucl. Phys. A* 353:233c

38. Sick, I. 1978. *Lecture Notes in Physics* 87:236. Berlin: Springer
39. Friar, J. L., Gibson, B. F., Tomusiak, E. L., Payne, G. L. 1981. *Phys. Rev. C* 24:665
40. Fabre de la Ripelle, M. 1979. *C. R. Acad. Sci. (Paris)* 288:325
41. Phillips, A. C. 1977. *Rep. Prog. Phys.* 40:905
42. Torre, J., Benayoun, J. J., Chauvin, J. 1981. *Z. Phys. A* 300:319
43. Gibson, B. F., Payne, G. L., Friar, J. L. 1984. *Phys. Rev. C.* In press
44. Slaus, I., Akaishi, Y., Tanaka, H. 1982. *Phys. Rev. Lett.* 48:993
45. Coelho, H. T., Das, T. K., Robilotta, M. R. 1983. *Phys. Rev. C* 28:1812
46. Bömelburg, A. 1983. *Phys. Rev. C* 28:403
47. Bömelburg, A., Glöckle, W. 1983. *Phys. Rev. C* 28:2149
48. Muslim, Kim, Y. E., Ueda, T. 1983. *Nucl. Phys. A* 393:399
49. Sato, M., Akaishi, Y., Tanaka, H. 1981. *Prog. Theor. Phys.* 66:930
50. Wiringa, R. B., Friar, J. L., Gibson, B. F., Payne, G. L., Chen, C. R. 1984. *Phys. Lett.* In press
51. Glöckle, W. 1982. *Nucl. Phys. A* 381:343
52. Das, T. K., Coelho, H. T., Fabre de la Ripelle, M. 1982. *Phys. Rev. C* 26:2288
53. Horikawa, Y., Thies, M., Lenz, F. 1980. *Nucl. Phys. A* 345:386
54. Freedman, R. A., Miller, G. A., Henley, E. M. 1981. *Phys. Lett.* 103B:397
55. Villars, F. 1947. *Helv. Phys. Acta* 20:476
56. Friar, J. L., Gibson, B. F. 1977. *Phys. Rev. C* 15:1779
57. Friar, J. L. 1977. *Ann. Phys. (NY)* 104:380
58. Haftel, M., Kloet, W. M. 1977. *Phys. Rev. C* 15:404
59. Kloet, W. M., Tjon, J. A. 1976. *Phys. Lett.* 61B:356
60. Hadjimichael, E. 1978. *Nucl. Phys. A* 294:513
61. Katayama, T., Akaishi, Y., Tanaka, H. 1980. *Prog. Theor. Phys.* 63:2127
62. Hajduk, C., Sauer, P. U., Arenhövel, H., Drechsel, D., Giannini, M. M. 1981. *Nucl. Phys. A* 352:413
63. Hadjimichael, E., Bornais, R., Goulard, B. 1982. *Phys. Rev. C* 26:294
64. Dreschsel, D. 1983. *Nuovo Cimento* 76A:388
65. Hadjimichael, E., Goulard, B., Bornais, R. 1983. *Phys. Rev. C* 27:831
66. Chemtob, M., Rho, M. 1971. *Nucl. Phys. A* 163:1
67. Saenz, A. W., Zachary, W. W. 1975. *Phys. Lett.* 58B:13
68. Brayshaw, D. D. 1976. *Phys. Rev. C* 13:1024
69. Coon, S. A., Scadron, M. D. 1981. *Phys. Rev. C* 23:1150
70. Sato, T., Hyuga, H., Ohtsubo, H. 1980. *Prog. Theor. Phys.* 63:516
71. Barroso, A., Hadjimichael, E. 1975. *Nucl. Phys. A* 238:422
72. Riska, D. O. 1980. *Nucl. Phys. A* 350:227
73. Strueve, W., Hajduk, C., Sauer, P. U. 1983. *Nucl. Phys. A* 405:620
74. Maize, M. A., Kim, Y. E. 1983. *Nucl. Phys. A* 407:507
75. Maize, M. A., Kim, Y. E. 1984. *Nucl. Phys. A* 420:365
76. Bargholtz, C. 1982. *Phys. Lett.* 112B:193
77. Jaus, W. 1976. *Nucl. Phys. A* 271:495
78. Jaus, W. 1976. *Helv. Phys. Acta* 49:475
79. Ciechanowicz, S., Truhlik, E. 1984. *Nucl. Phys. A* 414:508
80. Towner, I. S., Khanna, F. C. 1983. *Nucl. Phys. A* 399:334
81. Cavedon, J. M., Frois, B., Goutte, D., Huet, M., Leconte, P., et al. 1982. *Phys. Rev. Lett.* 49:986
82. Dunn, P. C., Kowalski, S. B., Rad, F. N., Sargent, C. P., Turchinetz, W. E., et al. 1983. *Phys. Rev. C* 27:71
83. Retzlaff, G. A., Skopik, D. M. 1984. *Phys. Rev. C* 29:1194
84. Beck, D. H., Kowalski, S., Schulz, M., Turchinetz, W. E., Lightbody, J. W. Jr., et al. 1984. *Phys. Rev. C.* In press
85. Jurney, E. T., Bendt, P. J., Browne, J. C. 1982. *Phys. Rev. C* 25:2810
86. Merritt, J. S., Taylor, J. G. V., Boyd, A. W. 1968. *Nucl. Sci. Eng.* 34:195
87. Phillips, A. C. 1972. *Nucl. Phys. A* 184:337
88. Hadjimichael, E. 1973. *Phys. Rev. Lett.* 31:183
89. Torre, J., Goulard, B. 1983. *Phys. Rev. C* 28:529
90. Malfliet, R. A., Tjon, J. A. 1969. *Nucl. Phys. A* 127:161
91. Amado, R. D. 1979. *Phys. Rev. C* 19:1473; 1981. *Comments Nucl. Part. Phys.* 10:131
92. Friar, J. L. 1979. *Phys. Rev. C* 20:325
93. Ioannides, A. A., Johnson, R. C. 1978. *Phys. Rev. C* 17:1331
94. Schiff, L. I. 1955. *Quantum Mechanics*, p. 312. New York: McGraw-Hill. 2nd ed.
95. Kim, Y. E., Tubis, A. 1974. *Ann. Rev. Nucl. Sci.* 24:69
96. Friar, J. L., Gibson, B. F., Lehman, D. R., Payne, G. L. 1982. *Phys. Rev. C* 25:1616
97. Borbely, I., König, V., Grüebler, W., Jenny, B., Schmelzbach, P. A. 1981. *Nucl. Phys. A* 351:107

98. Sasakawa, T., Sawada, T., Kim, Y. E. 1980. *Phys. Rev. Lett.* 45:1386
99. Kim, Y. E., Muslim. 1979. *Phys. Rev. Lett.* 42:1328
100. Gibson, B. F., Lehman, D. R. 1984. *Phys. Rev. C* 29:1017
101. Locher, M. P., Mizutani, T. 1978. *Phys. Rep.* 46:43
102. Bornand, M., Plattner, G. R., Voillier, R. D., Alder, K. 1978. *Nucl. Phys. A* 294:492
103. Plattner, G. R., Bornand, M., Voillier, R. D. 1977. *Phys. Rev. Lett.* 39:127
104. Giraud, B., Fuda, M. G. 1979. *Phys. Rev. C* 19:583
105. Knutson, L. D., Colby, P. C., Hichwa, B. P. 1981. *Phys. Rev. C* 24:411; Sen, S., Knutson, L. D. 1982. *Phys. Rev. C* 26:257
106. Brandon, M. E., Haberli, W. 1977. *Nucl. Phys. A* 287:213
107. Roman, S., Basak, A. K., England, J. B. A., Nelson, J. M., Sanderson, N. E., et al. 1977. *Nucl. Phys. A* 289:269
108. Harper, E. P., Lehman, D. R., Prats, F. 1980. *Phys. Rev. Lett.* 44:237
109. Ioannides, A. A., Nagarangan, M. A., Shyam, R. 1981. *Nucl. Phys. A* 363:150
110. Borbély, I., Doleschall, P. 1982. *Phys. Lett.* 113B:443
111. Jans, E., Barreau, P., Bernheim, M., Finn, J. M., Morgenstern, J., et al. 1982. *Phys. Rev. Lett.* 49:974
112. Skopik, D. M., Weller, H. R., Roberson, N. R., Wender, S. A. 1979. *Phys. Rev. C* 19:601
113. Barbour, I. M., Hendry, J. E. 1972. *Phys. Lett.* 38B:151
114. Skopik, D. M., Asai, J., Beck, D. H., Dielschneider, T. P., Pywell, R. E., Retzlaff, G. A. 1983. *Phys. Rev. C* 28:52
115. King, S. E., Roberson, N. R., Weller, H. R., Tilley, D. R. 1983. *Phys. Rev. Lett.* 51:877
116. Belt, B. D., Bingham, C. R., Halbert, M. L., van der Woude, A. 1970. *Phys. Rev. Lett.* 24:1120
117. Matthews, J. L., Kruse, T., Williams, M. E., Owens, R. O., Savin, W. 1974. *Nucl. Phys. A* 223:221
118. Gibson, B. F., O'Connell, J. S. 1970. *Phys. Lett.* 32B:331
119. Grüebler, W. 1981. *Proc. 9th Int. Conf. on the Few-Body Problem,* p. 31c. Amsterdam: North-Holland; 1981. *Nucl. Phys. A* 353:31c
120. Kloet, W. M. 1981. *Proc. 5th Int. Symp. on Polarization Phenomena in Nuclear Physics,* p. 1132. New York: Am. Inst. Phys.
121. Schwarz, P., Klages, H. O., Doll, P., Haesner, B., Wilczynski, J., et al. 1983. *Nucl. Phys. A* 398:1
122. Grüebler, W., König, V., Schmelzbach, P. A., Sperisen, F., Jenny, B., et al. 1983. *Nucl. Phys. A* 398:445
123. Weber, J. 1981. *Helv. Phys. Acta* 54:547; Huttel, E., Arnold, W., Berg, H., Krause, H. H., Ulbricht, J., Clausnitzer, G. 1983. *Nucl. Phys. A* 406:435
124. Doleschall, P. 1973. *Nucl. Phys. A* 201:264; 1974. *A* 220:491; Bruinsma, J., van Wageningen, R. 1977. *Nucl. Phys. A* 282:1
125. Stolk, C., Tjon, J. A. 1978. *Nucl. Phys. A* 295:384; 1976. *Phys. Rev. Lett.* 35:985; Benayoun, J. J., Chauvin, J., Gignoux, C., Laverne, A. 1976. *Phys. Rev. Lett.* 36:1438
126. Doleschall, P., Grüebler, W., König, V., Schmelzbach, P. A., Sperisen, F., Jenny, B. 1982. *Nucl. Phys. A* 380:72
127. Svenne, J. P., Birchall, J., McKee, J. S. C. 1982. *Phys. Lett.* 119B:269
128. Stolk, C., Tjon, J. A. 1977. *Nucl. Phys. A* 319:1; 1979. *Phys. Rev. Lett.* 39:395

NUCLEAR REACTION TECHNIQUES IN MATERIALS ANALYSIS

G. Amsel

Groupe de Physique des Solides de l'Ecole Normale Superieure, Tour 23, 2, place Jussieu, 75005 Paris, France

W. A. Lanford

Joseph Henry Physics Building, Department of Physics, State University of New York at Albany, Albany, New York 12222 USA

CONTENTS

INTRODUCTION	436
PRINCIPLES FROM NUCLEAR PHYSICS	440
Q Values and Selectivity	440
Cross Sections and Resonances	442
Optimizing Selectivity	444
MEASUREMENT METHODS	444
Overall Near-Surface Contents	444
Nonresonant Depth Profiling	447
Resonant Depth Profiling	448
Recoil Depth Profiling	452
Channeling Measurements	453
TECHNICAL ASPECTS	453
APPLICATIONS	455
Nuclear Analysis of Hydrogen Isotopes	455
Storage of Ultracold Neutrons in Material Bottles	456
Hydrogen on Surfaces	457
Hydrogen in Metals	457
Reaction Between Water and Glass	457
SUMMARY	458

INTRODUCTION

Since the earliest experiments in low-energy nuclear physics, reviewed by Livingston & Bethe (1), nuclear reactions induced by charged particles have been studied by bombarding thin targets of well-known composition with various ions of well-defined energy. From the beginning, target purity was a crucial experimental problem: "Many elements were found to give alpha particles (from p,α reactions), but as was suggested at that time, they were later identified as due to impurities of boron" (Ref. 1, p. 308). Thus the observation of known groups of reaction products was used to infer the presence of elements or isotopes in a target of *a priori* unknown or even unexpected trace contamination.

Nuclear reaction microanalysis (NRA) is just the reversal of the aims of low-energy nuclear physics: the corpus of data accumulated in the frame of the latter is put to advantage for analytical purposes. In nuclear physics known targets are bombarded to study unknown reactions; in NRA unknown targets are bombarded and known reactions observed. In the latest, most sophisticated analytical experiments, the samples are sometimes produced *in situ*, sometimes in ultra-high-vacuum chambers, often with the ability to control rigorously temperature, vacuum, etc, in order to carry out kinetic studies.

While in low-energy nuclear physics a single target is bombarded for long times, in NRA many targets are bombarded in turn routinely for times as short as possible. For NRA the accelerator, detectors, spectrum recording, and interpretation must be reliable, simple, and fast. Hence, NRA could become a standard analytical tool only when these conditions were met to a reasonable extent. Such a situation was reached after 1960. Robust 2–3-MV Van de Graaff accelerators became common enough to be accessible for non-nuclear work. Fast, low-noise, and stable electronics as well as multichannel analyzers became commercially available. Above all, the new semiconductor detectors brought a revolution in high-resolution, charged-particle detection techniques because of their low cost, reliable operation, small size, large accessible solid angle, and nondispersive, linear, high-resolution energy response at high counting rates.

The same evolution took place for the three other MeV ion-beam analytical techniques that are complementary to NRA: Rutherford backscattering (RBS), proton-induced x-ray emission (PIXE), and the more recent method of elastic recoil detection (ERD). While NRA and to some extent RBS and ERD arose from low-energy nuclear physics, PIXE's mother is atomic physics, its father the Si(Li) solid-state x-ray detector. These four techniques are now grouped under the generic term "ion-beam analysis," to which an international conference has been devoted every second year since 1975. Recently, thermal-neutron-induced nuclear reactions have also been used for near-surface analysis and should be included in NRA, although experimental conditions are radically different since high-flux reactors are required as a neutron source.

In fact, each time a body of data is accumulated on some process induced by a beam of ions, neutrons, electrons, photons, etc, the corresponding techniques and data may be used for analysis. Varying the energy range of both incident and detected particles (from eV to MeV energies) leads to an alphabet soup of analytical procedures: SIMS, ESCA, EXAFS, etc. We do not discuss such techniques here.

In the case of NRA, the lower limit of the beam energy range is set by Coulomb barrier penetrabilities and is about 0.3 MeV for protons bombarding light nuclei like ^{19}F or ^{23}Na. The upper limit is reached when the reaction products are too numerous for efficient analysis. Most experiments are carried out below 2 MeV for proton and deuteron beams bombarding nuclei in the ^{2}H to ^{58}Ni range. Higher energies, up to 5 MeV, may be useful for ^{3}He- and ^{4}He-induced reactions, while 7 MeV are required for ^{6}Li-induced reactions and for the particular case of ^{1}H depth profiling with ^{15}N ions. Heavier projectiles and higher energies are seldom used in NRA because small machines are preferred over large tandems for routine work. Detected particles are usually p, d, t, ^{3}He, α, and γ rays. Neutrons are seldom observed because their detection does not benefit from the simplicity of semiconductor detectors. NRA is thus a specific tool for determining the depth profile of light nuclei in the near-surface regions of solids, to depths set by the slowing-down process of the beam particles and by reaction cross section variation with energy.

NRA is called a "prompt" method because the reaction products are directly detected, in contrast to activation analysis techniques where the beam-induced radioactivity is monitored. We do not discuss the latter in this review.

The relationship between NRA and low-energy nuclear physics (and to some extent atomic, solid-state, and semiconductor physics) has far reaching consequences on several levels (Figure 1).

1. By its very nature NRA is multidisciplinary. Many physicists engaged in NRA are converts who started in nuclear physics (as did the present authors) or in semiconductor detector research. They can do a good job only if they become acquainted with the field to which they apply NRA. Those who stem from the field of application in which they use NRA should acquire enough knowledge of nuclear and atomic physics to be able to have critical insight into their experimental procedures and results.

2. Many of the pioneer groups in NRA developed in nuclear laboratories with all the advantages of this environment and all the conflicts that are

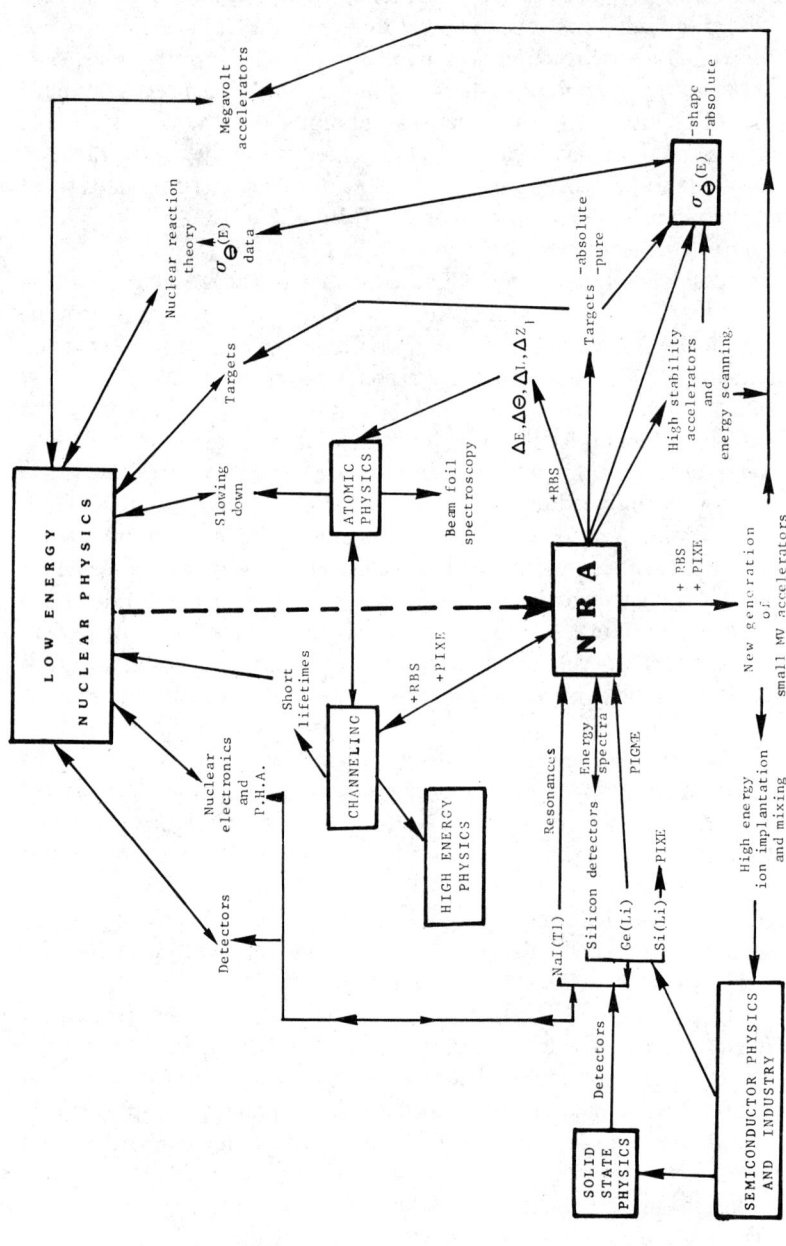

Figure 1 Schematic relationships between low-energy nuclear physics and the development of NRA methodology.

unavoidable between "parents" and "child," from accelerator time problems to budgeting of "non-nuclear" activities. Since the late 1970s the situation has changed rapidly. The amount of nuclear physics work done with small accelerators has decreased drastically and small machines are steadily spreading in the non-nuclear environment in both universities and industrual research laboratories. The latter spread emphasizes that NRA and more generally ion-beam analysis has proven its usefulness in many areas. Most of the nuclear authorities recognize the advantages of maintaining (at relatively low cost) ion-beam analysis activities in a nuclear laboratory but some do not. We hope that this paper will contribute to further understanding between the nuclear physics and the growing ion-beam analysis communities.

3. The relationship between nuclear physics and NRA is indeed mutually beneficial. NRA contributes to its parent discipline in a variety of ways. For example, very thin, well-known, pure and absolutely calibrated, natural or isotopically enriched targets are routinely produced for NRA and supplied to nuclear physicists. Consider, for example, the experimental and theoretical study of charged-particle slowing-down processes: energy-straggling ΔE, angular multiple scattering $\Delta \theta$, lateral spread ΔL, and charge state fluctuations ΔZ have all been extensively developed for the interpretation of analysis measurements in general and for NRA results in the particular case of small energy loss, i.e. very thin targets. This leads to marked progress in the evaluation of nuclear physics data especially for studies related to narrow resonances and for designing and interpreting beam foil spectroscopy experiments. Last but not least, the need in NRA for high-precision differential cross-section curves $\sigma_\theta(E)$, both in shape and absolute values, produced the most precise set of cross-section curve data ever recorded and led to the discovery of numerous very narrow resonances, as illustrated below. These results were reached through a significant improvement of the stability of MV accelerators. Thus NRA further increased the corpus of data of low-energy nuclear physics, which was the very starting point of its development.

4. The spread of small MV accelerators outside the nuclear community gave rise to a new, most dynamic, accelerator building industry. A new generation of small accelerators was thus created, in particular 1-MV to 2.5-MV tandems. The latter recently found new uses in high-energy ion implantation and mixing studies that connect this outcome of RBS and NRA back to the semiconductor industry and solid-state physics, which triggered their birth through the invention of the semiconductor particle detectors.

In conclusion NRA is an offspring of low-energy nuclear physics and it still contributes to it, especially in the field of high-precision measurements.

NRA, being a high-energy technique, is sensitive only to the nuclei present in the sample. Hence, it yields information on the elemental or isotopic distribution within the target and no direct information on chemical bonding. While this is a limitation, it also leads to one of the most important of its advantages: it is easily made quantitative by comparison with universal high-precision reference standard samples. Most of the low-energy methods are sensitive to poorly understood atomic, chemical, or solid-state phenomena making quantitative analysis difficult if not impossible. Moreover, NRA is highly sensitive, it presents in many cases good depth resolution in the first few micrometers of the sample, and it is nondestructive. It is an excellent tool for tracing stable isotopes. The applications of NRA proceed from these basic properties. They pertain to a large variety of fields including solid-state physics, semiconductor technology, thin-film deposition and growth mechanisms, ion implantation, laser annealing, solid-state electrochemistry, corrosion science, catalysis, crystallography, metallurgy, geology, biology and archeology.

In what follows we present the basic principles and methods of NRA and illustrate its use with analytical examples. Some specific applications are stressed.

Limitation of length prevents us from presenting in detail the literature on this subject. Our aim is to be didactic rather than exhaustive. The reader may find in (2–13) some basic, well-documented books, proceedings, and papers on NRA and more generally on ion-beam analysis. The birth and spread of ion-beam analysis techniques have been studied in detail recently with scientometric methods (14).

PRINCIPLES FROM NUCLEAR PHYSICS

Q Values and Selectivity

Let us consider, for example, the nuclear analysis of the oxygen isotopes ^{16}O and ^{18}O (natural abundances 99.758 and 0.204%) often used in isotopic tracing experiments with ^{18}O-enriched compounds (15). For ^{18}O determinations the best reaction is ^{18}O(p,α)^{15}N with a Q value of $+3.970$ MeV; because the first level of ^{15}N is at 5.28 MeV, a single α group is emitted for beam energies E below the threshold at 1.3 MeV. The deuteron-induced stripping reaction ^{16}O+d leads to proton groups p_i from ^{16}O(d,p_i)^{17}O, which correspond to the ground state of ^{17}O ($Q_0 = 1.919$ MeV) and to its first excited state at 0.871 MeV ($Q_1 = Q_0 - 0.871 = 1.048$ MeV). The second excited state of ^{17}O at 3.058 MeV can be reached only for $E > 1.2$ MeV. ^{16}O is usually measured with the ^{16}O(d,p_1)^{17}O* reaction, the cross section of which is the higher (15).

The ^{16}O(p,α)^{13}N reaction has a strongly negative Q value, $Q = -5.208$

MeV, and MeV protons induce no reaction on ^{16}O: protons are blind to ^{16}O, while they see ^{18}O very well. On the other hand, ^{18}O may be measured with deuteron beams either with the ^{18}O(d,p)^{19}O reaction or with ^{18}O(d,α)^{16}N(15); ^{18}O(p,α)^{15}N is, however, usually preferred for several reasons. (a) ^{18}O+d reactions produce complex spectra; (b) their cross sections are low; (c) while to use two projectiles (i.e. two runs) for measuring ^{18}O and ^{16}O is a complication, the total selectivity of proton beams for ^{18}O is a basic advantage—no overlap from ^{16}O peaks can occur even at the highest ^{16}O/^{18}O ratios.

Such radical selectivity properties are typical of NRA and are in contrast to mass selection (RBS, mass spectroscopy) or isotopic shift (infrared absorption) measurement techniques, where overlap from strong neighboring peaks may completely mask weak signals from nuclei at low relative concentrations. Moreover, for mass spectroscopy and in particular secondary ion mass spectroscopy (SIMS) used for near-surface analysis, neither of the mass-16 or mass-18 peaks can be identified unambiguously because polyatomic ions may severely blur the spectra. Finally, the only proton-induced α groups with energies above that corresponding to ^{18}O are those from ^{15}N, ^{19}F, ^{11}B, and ^{7}Li. All of them except ^{11}B have quite different energies from that of ^{18}O. Thus the only nucleus that may interfere with the proton-beam measurement of ^{18}O is ^{11}B.

A similar selectivity holds for the medium- and high-Z nuclei in the target. The reaction cross sections are negligible at MeV energies because of the Coulomb barrier, except for the (p,γ) type reactions on medium-Z nuclei considered below.

We may conclude this section by noting that, while NRA cannot determine in a single measurement all the nuclei in the near-surface region of a target, it may do so very efficiently and unambiguously for various light nuclei using well-chosen nuclear reactions. The selectivity between different nuclei, including isotopes, is excellent and often total. Table 1 lists some reactions ordered according to decreasing associated Q values; it gives a preliminary idea of total selectivity relationships between light nuclei, as discussed in (16).

For many (p,α) reactions the alpha particles have energies too low for efficient detection. The γ rays associated with the α's are then detected, the reaction being noted (p,αγ). This is the case for the important reactions ^{19}F+p, which yield in practice ^{19}F(p,α)^{16}O and ^{19}F(p,αγ)^{16}O, and ^{15}N+p leading to ^{15}N(p,α)^{12}C and ^{15}N(p,αγ)^{12}C ($E\alpha = 4.43$ MeV). Inelastic scattering reactions of the (p,p'γ) or (α,α'γ) type can also be detected only through the associated γ rays as well as the capture reactions of the (p,γ) type. For all these reactions full selectivity cannot be achieved because of the very nature of the γ-ray detection process. Using Ge(Li) high-resolution

Table 1 Q values for nuclear reactions induced by protons and deuterons on some light isotopes (from 16)

Isotope	Q (MeV)	Isotope	Q (MeV)	Isotope	Q (MeV)
		(p,α) reactions:			
^{7}Li	17.347	^{6}Le	4.02	^{9}Be	2.125
^{11}B	8.582	^{18}O	3.970	^{3}P	1.917
^{19}F	8.119	^{37}Cl	3.030	^{27}Al	1.594
^{15}N	4.964	^{23}Na	2.379	^{17}O	1.197
				^{10}B	1.147
		(d,α) reactions:			
^{3}He	18.352				
^{10}B	17.819	^{11}B	8.022	^{32}S	4.890
^{6}Li	22.36	^{15}N	7.683	^{18}O	4.237
^{7}Li	14.163	^{9}Be	7.152	^{30}Si	3.121
^{14}N$_{(\alpha_0)}$	13.579	^{25}Mg	7.047	^{16}O	3.116
^{19}F	10.038	^{23}Na	6.909	^{26}Mg	2.909
^{17}O	9.812	^{27}Al	6.701	^{24}Mg	1.964
^{14}N$_{(\alpha_1)}$	9.146	^{29}Si	6.012	^{28}Si	1.421
^{31}P	8.170	^{13}C	5.167	^{12}C	<0
		(d,p) reactions:			
^{3}He	18.352				
^{10}B	9.237	^{17}O	5.842	^{26}Mg	4.212
^{25}Mg	8.873	^{27}Al	5.499	^{12}C	2.719
^{14}N$_{(p_0)}$	8.615	^{24}Mg	5.106	^{16}O	1.919
^{29}Si	8.390	^{6}Li	5.027	^{18}O	1.731
^{32}S	6.418	^{23}Na	4.734	^{14}N$_{(p_5)}$	1.305
^{28}Si	6.253	^{9}Be	4.585	^{11}B	1.138
^{13}C	5.947	^{19}F	4.379	^{15}N	0.267
^{31}P	5.712	^{30}Si	4.367	^{7}Li	<0

γ-ray detectors may improve selectivity at the expensive of detection efficiency as compared to large NaI(Tl) detectors. However, the resonant character of many of these reactions may improve selectivity, as seen below.

Cross Sections and Resonances

To each particle group or γ ray associated with the various levels of the residual nucleus corresponds a different differential cross section $\sigma_\theta(E)$ called an excitation function. A typical $\sigma_\theta(E)$ is shown in Figure 2 for the ^{18}O(p,α)^{15}N reaction recorded at the laboratory angle $\theta = 150°$. We observe a slowly varying component with a practically exponential rise (15) and a strong $\Gamma = 2.1$ keV wide resonance at 629 keV (17), shown in great detail. This is a typical example of $\sigma_\theta(E)$ shape measurement with much higher precision than is usual in low-energy nuclear physics. The peak-to-valley ratio changes rapidly with θ.

Other high-precision and detailed data of this type arising from NRA may be found in (2, 4). Let us quote some recent typical results. $^{19}F(p,\alpha)^{16}O$ presents some medium wide resonances while numerous narrow ($\Gamma = 1$ to 4 keV) resonances and one very narrow resonance occur in $^{19}F(p,\alpha\gamma)^{16}O$ (18). The very narrow resonance is at 1088 keV and is $\Gamma = 150 \pm 50$ eV wide; its former reported value was 700 ± 300 eV (19). The very narrow resonance in $^{15}N(p,\alpha\gamma)^{12}C$ at 429 keV was shown to be $\Gamma = 120 \pm 30$ eV wide (20); its former reported value was $\Gamma = 900$ eV (19). This resonance, which plays a basic role in hydrogen depth profiling as illustrated below, was also shown to follow a Breit-Wigner law over four decades (21). A $\Gamma = 50$ eV wide resonance was measured in both the $^{18}O(p,\alpha)^{15}N$ and $^{18}O(p,\alpha)^{19}F$ reactions at 1167 keV (22) corresponding to a state in ^{19}F with isotopic spin $T = 3/2$ and spin $7/2^-$. As a final example the cross-section curve for the $^3He + D$ reaction was remeasured with absolute precision when needed for ^3He and D depth profiling experiments (23).

Let us notice here that while elastic Rutherford scattering processes pertain to RBS techniques, elastic nuclear scattering processes belong to the

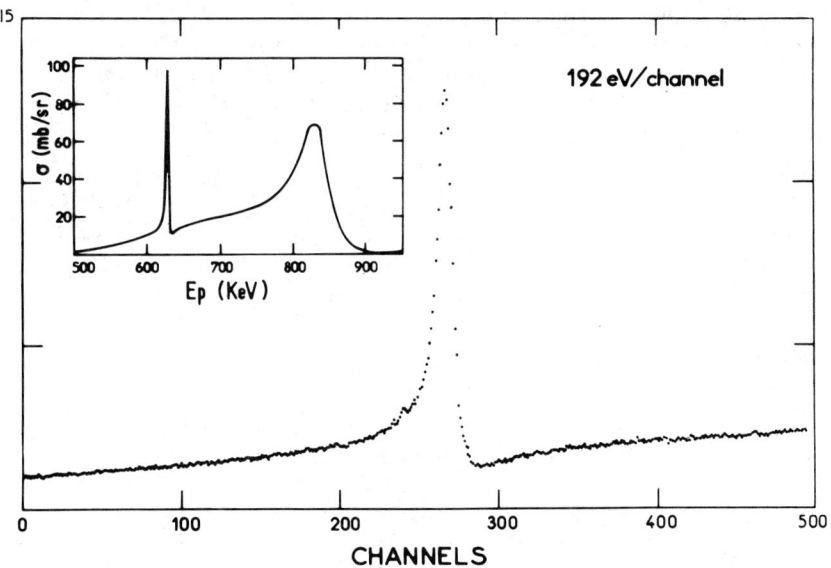

Figure 2 Master excitation curve for the 629-eV, $\Gamma = 2.1$ keV wide resonance at $\theta = 150°$ and $^{18}O(p,\alpha)^{15}N$ recorded with a 50 Å thick, 75% ^{18}O-enriched Ta_2O_5 target; $\Omega = 0.08$ sr, 20 μC per channel. Beam energy spread is 200 eV FWHM. The insert shows the position of the resonance in a broad cross-section curve (from 17).

field of NRA: $\sigma_\theta(E)$ may exhibit large positive and negative interferences depending critically on θ. Some very strong resonances of this type are widely used in RBS (the acronym should then be reworded as resonance back scattering), in particular $^{16}O(\alpha,\alpha)^{16}O$ at 3.05 MeV, $\Gamma = 30$ keV (10, p. 436; 11, pp. 303–12) and $^{12}C(\alpha,\alpha)^{12}C$ at 4.26 MeV, $\Gamma = 35$ keV (24).

In conclusion, the mass of cross-section data needed in NRA is large and increases steadily. The corresponding compilation of the numerical data creates unsolved problems (how does one transmit the detailed master excitation curve in Figure 2?). This subject was discussed recently in (3). In practice the physicists who need a particular $\sigma_\theta(E)$ curve must very often measure it again; this emphasizes the urgent need for an easily accessible and reliable $\sigma_\theta(E)$ data base.

Optimizing Selectivity

It may happen in unfavorable cases that two particle groups corresponding to two nuclei tend to overlap, and no better nuclear reaction is available. Proton and α peaks may be unscrambled by suitably choosing the thickness of the depletion zone in the surface barrier detector, and by using well-chosen absorbers (16). The different rates of variation of the energies of various particle groups with E and θ, as calculated from the kinematics of the reactions, may also be put to beneficial use. Finally, advantage may be taken of the usually very different shapes of the $\sigma_\theta(E)$ curves corresponding to the two reactions for optimizing the counting ratios of nearby peaks through a skillful choice of E and θ. Note that no such possibilities exist in RBS, where all the cross sections vary in the same proportion with E and θ, except for the nuclear scattering processes of the type quoted above. Similarly the selectivity for (p,$\alpha\gamma$), (p,pγ) and (p,γ) reactions may be enhanced by operating near well-chosen resonances in $\sigma_\theta(E)$.

Hence, it appears that the ultimate selectivity of NRA in difficult cases depends on the knowledge of all the $\sigma_\theta(E)$ curves involved and on the skill of the physicist who carries out the experiment. Figure 3 shows a typical optimized spectrum for the unfavorable case of trying to determine traces of oxygen in the presence of large amounts of nitrogen in a thin film (25). The paper from which this figure was taken (25) discusses selectivity optimization and illustrates well the above considerations.

MEASUREMENT METHODS

Overall Near-Surface Contents

Let us choose the bombarding energy in the vicinity of an energy E_0 such that for a given nucleus $\sigma_\theta(E)$ varies slowly with E between $E_0 - \Delta E$ and E_0, for instance around 750 keV for ^{18}O in Figure 2. If, in a thin film of thickness

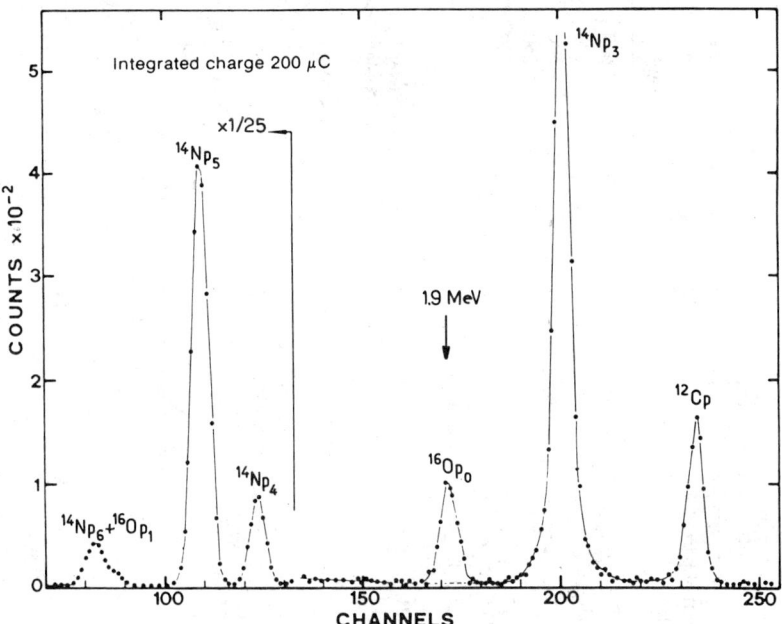

Figure 3 Proton spectrum from the reactions $^{16}O(d,p)^{17}O$ and $^{14}N(d,p)^{15}N$ recorded at $E = 0.61$ MeV for an oxygen-contaminated, 1700 Å thick Si_3N_4 film on a silicon backing containing 10^{18} ^{14}N per cm^2 and 2×10^{16} ^{16}O per cm^2; $\Omega = 0.052$ sr, $\theta = 150°$; 13 μm thick Mylar absorber; 10 keV per channel. The ^{16}O peak P_0 is well resolved although the $^{16}O/^{14}N$ ratio was only 2% (from 25).

X, the energy lost by the beam is below ΔE, the area of the corresponding peak in the spectrum is proportional to the total number per cm^2 of these nuclei in the film, irrespective of their concentration profile over X. Typical values of X are some thousands of Å. Absolute values of nuclei per cm^2 are obtained by comparison to universal thin-film reference targets calibrated once and for all. The construction of such standards for each light isotope is a central technical problem in NRA. The state of the art was reviewed in detail recently (26). Absolute precisions around 3% are common while relative precisions from sample to sample are limited only by statistics.

Nuclei that may be determined in this way are D, ^{3}He, ^{6}Li, ^{7}Li, ^{10}B, ^{11}B, ^{9}Be, ^{12}C, ^{13}C, ^{14}N, ^{15}N, ^{16}O, ^{17}O, ^{18}O, ^{19}F (and some heavier nuclei) using reactions induced by p, d, and ^{3}He at energies below 2 MeV. Table 2 gives an idea of the counting rates in typical conditions. Thus a limiting sensitivity for ^{18}O is of the order of 10^{13} atoms per cm^2 (15 counts per min.), i.e. 10^{-2} monolayer or 3×10^{-12} g on a 1 mm diameter beam spot.

This ability of NRA to determine unambiguously in a single, non-

Table 2 Typical experimental conditions for nuclear microanalysis of absolute amounts of nuclei per cm^2

Nucleus	Reaction	Bombarding energy (keV)	Energy of emitted particles (MeV)	Thickness of Mylar absorber (μm)	Counting rate[a]
^{16}O	^{16}O(d,p$_1$)^{17}O*[d]	830	1.52	19	1,600
^{18}O	^{18}O(p,α)^{15}N	730	3.38	12	4,600
^{14}N	^{14}N(d,α_1)^{12}C*[d]	1300	6.76	19	500
^{19}F	^{19}F(p,α)^{16}O	1260	6.93	26	250
	^{19}F(p,α)^{16}O	1340[c]	6.97	31	1,400
	^{19}F(p,$\alpha\gamma$)^{16}O	870[c]	$\left.\begin{array}{c}7.12\\6.92\\6.13\end{array}\right\}\gamma$	—	~14,000[b]
^{12}C	^{12}C(d,p)^{13}C	1000	3.01	19	~20,000
^{2}H	D(d,p)T	550	2.45	6	~2,000
	D(^{3}He,α)^{1}H / D(^{3}He,p)^{4}He	700	$\begin{cases}\alpha: 1.98\\ p: 13\end{cases}$	5	~20,000
^{3}He	^{3}He(d,α)^{1}H / ^{3}He(d,p)^{4}He	700	$\begin{cases}\alpha: 2.29\\ p: 13.46\end{cases}$	5	~20,000
^{6}Li	^{6}Li(d,α)^{4}He	1560	9.25	26	900
^{7}Li	^{7}Li(p,α)^{4}He	1000	7.84	19	400

[a] For a film containing 10^{16} atoms per cm^2, a 1-μA beam, per min. The surface barrier detector solid angle is 0.12 steradian, at a 150° detection angle. The counting rates may be multiplied by 3, using three identical detectors, when necessary.
[b] 3 in. × 3 in. NaI(Tl) scintillation detector at 7 cm at θ = 90°.
[c] For films of equivalent thickness <10 keV.
[d] Reaction leading to the first excited state (^{17}O* or ^{12}C*) of the residual nucleus.

destructive measurement the total absolute amount of light nuclei near the surface of solids is unparalleled by any other analytical technique and has many useful applications.

Nonresonant Depth Profiling

Let us consider the same experiment as before, for a medium thick target. Figure 4 shows the trajectory of a beam particle hitting a nucleus at M and the trajectory of the emitted particle, assumed to be charged, toward the detector. Because of the energies lost by the particles over y and l, we obtain $E(x) = E_0 - kx$, where k depends on E and θ and on the relevant stopping powers. Usually k is positive, i.e. the contribution of deep-lying nuclei broadens the peak toward lower energies in a way very similar to what happens in RBS depth profiling. The peak shape reflects the depth distribution $C(x)$ with a depth resolution depending on (a) the stopping powers and the detector energy resolution; (b) the energy-straggling processes arising during particle slowing down; and (c) the tilting of the target. In the so-called glancing geometry either y or l or both may be considerably increased for given x, thus improving depth resolution. The ultimate resolution in glancing geometry is limited by energy-straggling and angular multiple scattering processes.

Figure 5 shows a typical depth profile measured in this way in an ^{18}O-tracing experiment in which the growth mechanism of Al_2O_3 films during anodic oxidation was studied (22). Typical depth resolutions near the surface range from 500 to 2000 Å in perpendicular geometry.

It should be emphasized that as long as $\sigma_\theta(E)$ does not vary greatly along y, the area of the whole peak remains proportional to the total amount of

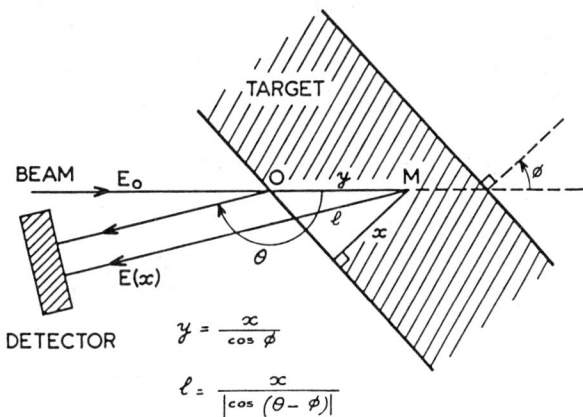

Figure 4 Geometry of the particle trajectories in the target (from 16).

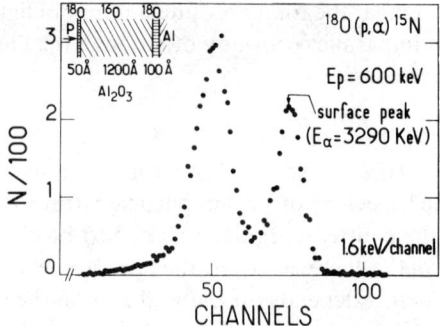

Figure 5 Spectrum of α particles from $^{18}O(p,\alpha)^{15}N$ recorded at $E_0 = 0.6$ MeV at $\theta = 165°$ and the target perpendicular to the beam ($\phi = 0$). The ^{18}O regions are enriched to 80% ^{18}O (from 22).

nuclei in the surface region, i.e. to the integral of $C(x)$. For thick films the spectra should be corrected for the variation of $\sigma_\theta(E)$ along y.

Note that for very light nuclei k may be negative: the tail of the spectra extends then toward higher energies, as a result of inverted kinematics at backward angles. Examples are the $D(d,p)T$, $D(^3He,\alpha)^1H$, and $D(^3He,p)^4He$ reactions. Such behaviors cannot be encountered in RBS.

Resonant Depth Profiling

The depth resolutions in nonresonant depth profiling are not very high. Moreover, in many cases there is no region where $\sigma_\theta(E)$ is large enough and flat enough to allow one to use this method. Finally, spectrum analysis yields no depth profile when the observed particles are γ rays.

High-resolution depth profiling may be achieved by recording the excitation curve for an unknown target in the vicinity of a narrow resonance of the relevant nuclear reaction. The first use of this technique dates back to 1962 (27), as applied to ^{18}O and ^{27}Al. It is described in detail in (17) from a theoretical point of view, particularly for the role of energy-straggling processes and for data interpretation. High resolution and technical aspects were discussed in detail recently (28). The technique has been intensively used during the recent years for hydrogen depth profiling as described in (29) and we use this example for illustrating its principle.

The reaction used is $^{15}N + {}^1H \rightarrow {}^{12}C + {}^4He + 4.43$-MeV γ ray. We mentioned above that the width of the corresponding isolated resonance in $^{15}N(p,\alpha\gamma)^{12}C$ at 429 keV is $\Gamma = 120 \pm 30$ eV. This reaction may be used for isotopic tracing and depth profiling of ^{15}N (20). However, it may also be used in its reversed version as $^1H(^{15}N,\alpha\gamma)^{12}C$ for depth profiling hydrogen, the target being bombarded with a ^{15}N beam. In this case the resonance occurs at 6.385 MeV and the width is $\Gamma = 1.8$ keV.

Figure 6 illustrates the principle of the technique: if the sample is bombarded at the resonance energy E_R, the γ-ray yield is proportional to the hydrogen on the surface of the sample. If the nominal (i.e. average) ^{15}N beam energy E_0 is raised, the yield from the surface tends to vanish as the resonance energy E_R is reached only in the vicinity of a depth x_0, where the average energy lost by the beam is just $E_0 - E_R$. Now the γ-ray yield is proportional to the hydrogen concentration at this depth. Thus the higher is the energy E_0 above E_R, the deeper the resonance is "pushed" into the target: the excitation curve $N(E_0)$ is hence an image of the profile $C(x)$.

This procedure is illustrated in Figure 7, which shows the hydrogen profile in a silicon wafer ion implanted with 10^{16} H per cm^2 at 40 keV. The curve in this figure represents both the raw data (γ-ray yield vs beam energy) and the final result [$C(x)$, the hydrogen concentration vs depth]. In fact, the conversion of raw data to a profile is here trivial because straggling effects may be neglected with respect to the experimental energy resolution. This is due to a most favorable ratio between stopping power and straggling for 7-MeV ^{15}N, dE/dx being near its maximum. The depth scale is given by $x = (E_0 - E_R)/(dE/dx)$ and the concentration scale is calculated from the resonance cross section and width, dE/dx, and the overall detection efficiency if they are known. If not, a single calibration is enough for all the experiments.

A number of features of this profiling method are illustrated in Figure 7: the narrow but intense peak of surface hydrogen (contamination) demonstrates the good near-surface depth resolution; the background (mainly cosmic rays) corresponds to 100 ppm atomic hydrogen concen-

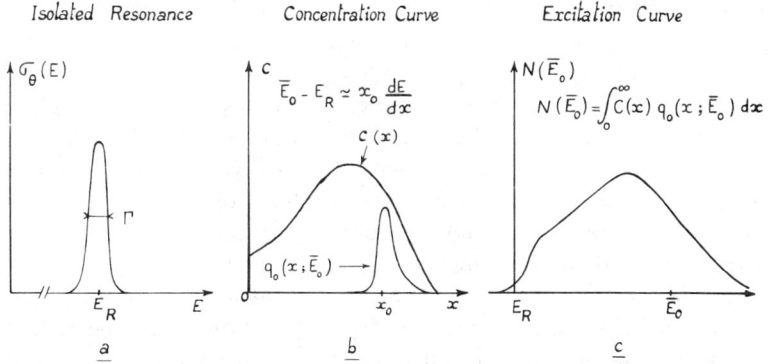

Figure 6 Principle of depth profiling with the resonance technique (from 16).

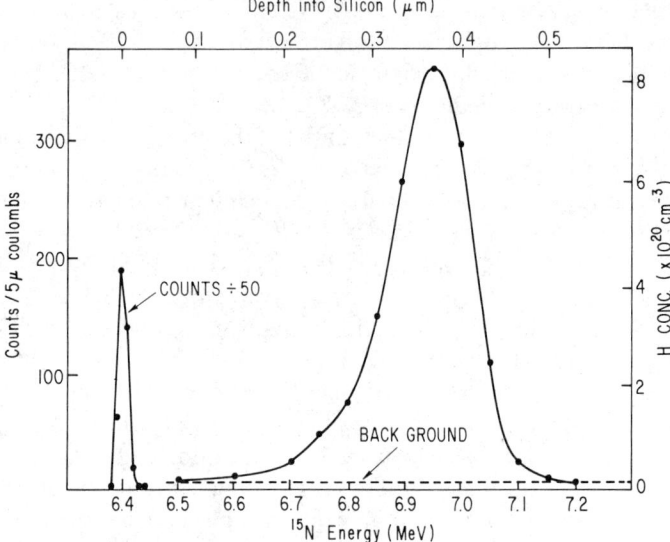

Figure 7 The hydrogen concentration vs depth of a silicon wafer ion implanted with 10^{16} H per cm^2 at 40 keV measured with the ^{15}N(p,α)^{12}C resonance.

tration (without sophisticated optimization); the method is quantitative. Such profiles are precise enough to measure not only mean projected range and rms straggle width of the implantation profile but also skewness. One of the first uses of nuclear reaction analysis of hydrogen was to test and refine theories of ion implantation.

In the case of light projectiles, mainly protons, straggling effects dominate the interpretation of the excitation curves. Because of the beam energy spread, resonance width, and straggling, the depths x at which the particles of a beam of average energy E_0 (the accelerator energy setting) may react are spread around the mean depth $x_0 = (E_0 - E_R)/(dE/dx)$ according to $q_0(x; E_0)$. The latter distribution, which is actually the depth resolution function, broadens with x_0 (i.e. $E_0 - E_R$) roughly as $x_0^{1/2}$. Thus depth resolution deteriorates with depth. The yield curve $N(E_0) = \int C(x) q_0(x; E_0) \, dx$ is hence an integral transform of $C(x)$. The stochastic theory of slowing down must be used for calculating $q_0(x; E_0)$ and interpreting the data accurately, as developed in detail in (17).

This phenomenon does not, however, prevent us from depth profiling light and medium-Z nuclei with high-depth resolution. In fact the better the instrumental resolutions the stronger is the influence of straggling. Resonance widths are often between 50 and 150 eV for proton energies from 300 to 1200 keV, while the corresponding beam energy spreads may be comparable (28). This leads to near-surface depth resolutions in the tens of Å range. The latter may be improved further by using grazing incidence geometry. If such high resolutions are not needed or if wider (>1 keV) resonances are used (such as in Figure 2), the deterioration of the resolution with depth is not as fast with respect to its near-surface value.

Lack of space does not allow us to illustrate in detail this most powerful technique. This method is even more important because SIMS techniques are especially unreliable in the first tens of Å of a sample. Among nuclei that may be depth profiled in this way, mainly with (p,γ) or $(p,\alpha\gamma)$ resonances, let us cite ^{13}C, ^{15}N, ^{18}O, ^{19}F, ^{20}Ne, ^{23}Na, ^{27}Al, ^{30}Si, ^{52}Cr, and ^{58}Ni.

In Figure 8 we show one example of high-resolution excitation curves, recorded using the $\Gamma = 50$ eV wide resonance at 1167 keV in ^{18}O$(p,\gamma)^{19}$F. The beam energy spread in this experiment was 170 eV FWHM; the targets, which consisted of uniform Ta$_2$O$_5$ layers enriched to 90% ^{18}O and were of increasing thicknesses, were covered by a hydrocarbon contamination layer of probable composition CH$_4$ and thickness estimated to be 33 Å. The stochastic theory of energy loss described in (17) was used to fit the data. Care was taken to include in the calculations the additional apparent beam energy spread resulting from the thermal-vibration-induced Doppler effect (17, 28), which amounts here to 140 eV FWHM.

Two striking features of the results are noteworthy: the sharp rise (over <200 eV, 10–90% of the plateau) followed by the strong overshoot called

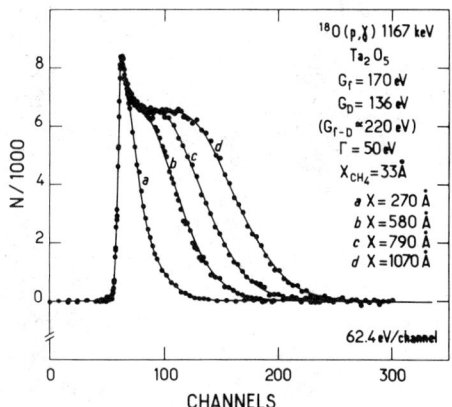

Figure 8 Excitation curves for 90% ^{18}O-enriched uniform Ta$_2$O$_5$ layers of increasing thicknesses recorded near the 1167 keV resonance in ^{18}O$(p,\gamma)^{19}$F; 100 μC per channel; 3 in. × 3 in. NaI(Tl) γ-ray detector near $\theta = 0°$ at 1 cm. Background is substracted. Beam energy spread is 170 eV, Doppler broadening is 140 eV FWHM. The solid lines are best fits using stochastic energy loss theory. Note the strong, well-fitted Lewis effect overshoot. Equivalent depth scale: 20 Å per channel (from 17).

the Lewis effect. The latter is observed only for very high-resolution excitation curves. It is of purely statistical origin. During their slowing down in the target, the energy of the protons is lost in discrete steps of random magnitude occurring at random along their trajectory. For $E = E_R$ all the protons may resonate; for $E > E_R$ they may "jump over" the resonance by losing in a single encounter an energy much larger than Γ. This induces the overshoot (17, 22, 28). The second feature is the increasingly slow decay of the curves corresponding to the straggling induced in the targets of increasing thicknesses. This illustrates the depth resolution that may be expected at various depths.

It should be emphasized that the resonance width $\Gamma = 50\,\text{eV}$ was actually extracted from these data (22). This illustrates the dual aspects of these techniques; for unknown targets they lead to NRA depth profiling; for known targets they represent sophisticated excitation curve fitting procedures that allow one to extract precise, high-resolution $\sigma_\theta(E)$ shapes from the data. This duality was considered in detail in (28).

Recoil Depth Profiling

We mention the relatively recent ERD, which is not an NRA method, only because it competes with NRA for very light ions, in particular for hydrogen isotopes. These experiments are similar to RBS except that the recoiling target atoms are detected in forward directions instead of the backscattered particles. The target must therefore be tilted strongly toward the beam ($\phi \approx 75°$ in Figure 4) while the particles are detected at angles $\theta \approx 30°$.

High-energy heavy ions, such as 30-MeV ^{35}Cl, are needed for detecting isotopes from ^{6}Li to ^{16}O, as illustrated in Figure 9. Sophisticated detection techniques may be used for unambiguously recording concentration profiles of several light elements even when, as a result of scatterings at different depths, the atoms of different mass reach the detector with the same energy. While extremely efficient and practical, as demonstrated by the Montreal group (31), high-energy ERD requires large machines, has a good isotope selectivity for very light nuclei only, and is not background free. The grazing geometry is demanding for target smoothness. Nevertheless high-energy ERD is likely to become more widely utilized and to compete in some cases with NRA for nuclei below oxygen.

More recently, several groups demonstrated that 2.5-MeV ^{4}He beams may be used for ERD analysis of ^{1}H (32) and D (33). These techniques, which require only a small machine, are extremely promising especially for D profiling, for which the usual D(^{3}He,α)^{1}H reaction is neither very sensitive nor very precise for depth resolution. ^{15}N-induced NRA is still more sensitive to ^{1}H traces, because of the lower background and presents much higher depth resolutions. However, the extremely high efficiency of ERD allows one to carry out measurements with very low beam fluences

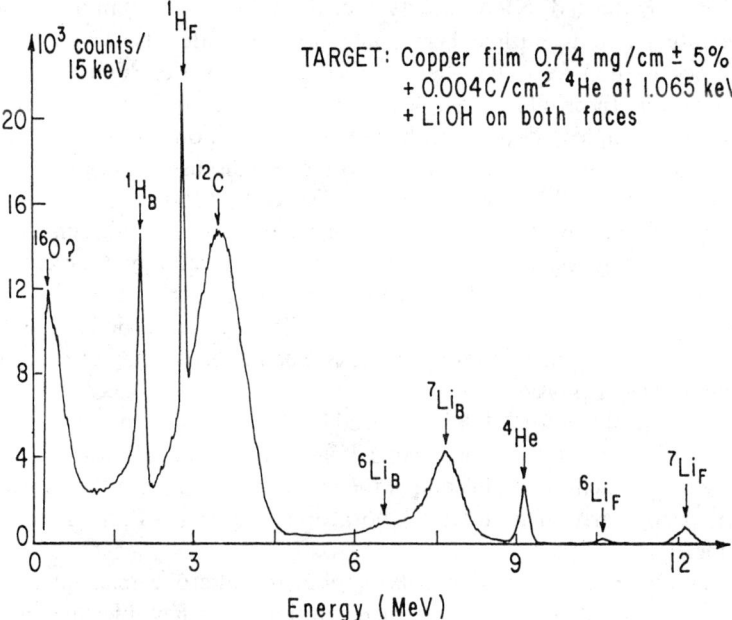

Figure 9 The ERD spectrum recorded for 30-MeV Cl beam bombarding a thin copper film coated on each face with LiOH and ion-implanted He at 1 keV. Note peaks for both ^{6}Li and ^{7}Li on both the front (F) and back (B) as well as peaks resulting from H, ^{4}He, and C (from 30).

and in very short times. Both techniques will probably go on being used for ^{1}H profiling, each one retaining its specific domain of application.

Channeling Measurements

Because of space limitations, we do not review the wealth of techniques and results arising from NRA combined with channeling processes in single-crystal samples. NRA with channeling was used for fundamental studies on channeling processes, for crystallographic site determinations of impurities in metals and semiconductors, for damage and amorphization studies in ion implanataion, for studying solid solutions and metastable phases, in surface physics, etc. One of the main specific interests of NRA is its ability to be coupled with channeling for site-locating light nuclei. A deep insight in this wide field is given by Feldman et al (34).

TECHNICAL ASPECTS

The experimental techniques are basically those of low-energy nuclear physics and are not detailed here. Reference 16 presents, in addition to the general principles of NRA, a description of the detection techniques used and of a dedicated NRA facility based on a 2-MV Van de Graaff accelerator. Let us outline here some nontrivial details. The types of machines needed for NRA were described above. For high-resolution resonance depth profiling, energy stability and spread must be optimized, using, for example, a double feedback system based on slit and capacitive pick-off plate control as shown in (28), which also discusses energy spread in Tandem machines. Stabilities of 30 eV and spreads of 100 eV, both FWHM, are routinely achieved for 1-MeV protons. An automatic, high-precision, energy scanning system based on electrostatic deflection may be coupled readily to any slit-controlled accelerator (35), and radically simplifies detailed measurements of excitation curves. These are recorded like simple spectra, as illustrated by Figures 2 and 8. Such a device may also be very useful in nuclear physics.

A major problem for NRA is the need for a heavy antineutron shielding when working with deuteron beams. This is unfortunately not available around many facilities and is one of the main factors limiting the spread of NRA, particularly in industrial laboratories where RBS and PIXE are routine.

A major application of NRA is its coupling to a microbeam facility. Light nuclei may be mapped with lateral resolutions of a few micrometers, as discussed at length at the conference devoted mainly to this subject (13).

The targets must be protected (a) against damage and excessive heating by the beam, perhaps a delicate problem in some special cases (hydrogen, sodium, organic targets); (b) against electrical breakdown if they are of insulating material; and (c) against carbon contamination. This may be improved by careful liquid nitrogen trapping. Detection, electronics, and data acquisition are standard. The elastically backscattered beam particles are stopped in well-chosen Mylar absorbers, as shown in Table 2.

As for the scattering chambers, the only difference between NRA and nuclear physics is that a large throughput of samples must be foreseen. Automation based on a stepping motor is widespread as for RBS and PIXE. In fact, these three ion-beam analysis techniques, NRA, RBS, and PIXE, are very often used in conjunction in the same chamber, sometimes on the same samples simultaneously. If channeling processes are studied, goniometric chambers are required. Because of the ease with which high-energy beams may be handled, a large variety of NRA measurements may be carried out "on line." Low- and high-temperature processes may be studied in situ as well as phenomena taking place under moderate gas pressures (oxidation processes, plasma-sample interaction, etc), using specially designed, differentially pumped chambers. The direct coupling to accelerators of ultra high-vacuum chambers for in situ surface studies, in parallel with classical instruments like LEED and Auger spectroscopy, is now routine (36).

APPLICATIONS

The applications of NRA stem from its basic and unique properties, which we may summarize as follows. NRA is nondestructive, sensitive, fast, insensitive to solid-state structure effects except for single-crystal structure (which allows channeling experiments to be carried out), may be used with finely focused microbeams, and, because of easy beam handling, allows one to perform in situ experiments at low and high temperatures, low vapor pressures, or in ultra high vacuum or on large objects. Moreover, NRA is specifically sensitive to light nuclei, which it can determine with high selectivity, precision, and in absolute amounts near the surface of solids. It may thus serve as a basic calibration tool for other analytical techniques; it may yield depth profiles of light nuclei; and it is fully isotope selective and is ideal for isotopic tracing experiments.

It is clear, however, that NRA gives neither chemical nor microscopic information on the samples. NRA is thus often used in conjunction with techniques such as SIMS, Auger, and ESCA. Moreover, the uniformity of the samples over the beam impact area must be checked; more and more often NRA studies are coupled to detailed electron microscope or microprobe observations so as to take into account the possible heterogeneous nature of the samples under study.

The main domains of application of NRA were enumerated at the end of the introduction section. In what follows we give, for illustrative purposes, some examples pertaining to the nuclear analysis of hydrogen isotopes, the latest and one of the most successful chapters of NRA.

Nuclear Analysis of Hydrogen Isotopes

The presence of hydrogen can have dramatic effects on the physical, chemical, and electrical properties of many materials. Hydrogen embrittlement of steels is one widely known example but there are also many other examples in both high-technology materials (e.g. thin-film superconductors, semiconductors, and insulators) and more traditional materials (e.g. metals, minerals, and glasses). H and D also play a major role in plasma-wall interactions in fusion reactors. Because hydrogen is invisible in most commonly available analytic methods, many of these hydrogen-related problems have been incomplete studies and are poorly understood. As a result of this widely recognized need for better techniques for hydrogen analysis, a number of nuclear reaction methods have been developed and are now regularly used in many laboratories around the world.

The two most widely used reactions for hydrogen studies (in addition to ERD) are $^1H(^{15}H,\alpha\gamma)^{12}C$ discussed in detail above and the $D+^3He$ reaction reported in Table 2. For overall amount measurements and channeling studies, D(^{3}He,p)^{4}He is mainly used; D(^{3}He,α)H is generally used for depth profiling. While deuterium-based experiments are carried out for isotopic tracing purposes, deuterium is often used instead of hydrogen for taking advantage of the large cross sections and ease of use of the D + ^{3}He reaction, which requires only 700-keV ^{3}He ions and measurements at a single energy.

Storage of Ultracold Neutrons in Material Bottles

Neutrons are generally thought of as very penetrating particles. However, as pointed out by Fermi over 40 years ago, at sufficiently low energy they should undergo total reflection from many solid surfaces and, hence, it should be possible to contain neutrons in material bottles. Such bottled neutrons would provide unique opportunities for a number of fundamental studies such as a more sensitive search for the electric dipole of the neutron (a test of time reversal invariance) and tests of quantum mechanics of heavy particles in previously unavailable energy regimes (10^{-8} eV). At these extreme energies, the neutron's wavelength becomes long enough (1000 Å) that tests of elementary predictions of quantum mechanics can be made in macroscopic systems, e.g. resonant barrier penetration of heavy particles through macroscopic barriers.

It was not realized until the early 1970s that reasonable fluxes of ultracold neutrons were actually available from high-flux reactors. The first experiments attempted by the various groups were typically transmission and containment measurements. All these experimenters were successful in containing neutrons in bottles but all also observed much larger loss rates than theoretically predicted. Losses are expected during a reflection because of finite probabilities that a neutron will undergo nuclear capture or be thermalized, but experimental losses were always observed to be $\sim 10^{-3}$ per reflection, which was two orders of magnitude greater than expected.

There were a number of suggestions for this large and persistent discrepancy, varying from questions about the validity of quantum mechanics at these very low energies to suspicions about "dirt" in the experiments. Because the presence of large amounts of hydrogen on the surfaces of the ultracold neutron bottles could explain this anamolous loss, a series of measurements were made of hydrogen on ultracold neutron bottle materials (37). These measurements showed that even technically "clean" materials typically have of order 10^{16} H per cm^2 within the first five nanometers of their surface. This result surprised the neutron community, which had expected at most an order of magnitude less, and was sufficient to explain essentially all of the discrepancy between predicted and measured ultracold neutron loss rates (37).

Hydrogen on Surfaces

The above example is one in which nuclear reaction analysis for hydrogen on surfaces is being applied to solve a problem in another area of physics. However, studies of hydrogen on surfaces is of fundamental importance in surface science and in related practical areas such as catalysis. There are a number of groups working in this area, and, while these studies are still at an early stage, it is clear that hydrogen plays a special role because of two unique features. First, hydrogen is normally by far the most mobile species in a solid or on a surface. Second, hydrogen is a universal contaminant in vacuum systems and, apparently, in most materials.

When a surface is made, for example by fracture in a vacuum system, bonds are broken and the resultant dangling bonds at the surface are chemically very reactive. Because of its special properties, in most situations hydrogen is the first element to reach these dangling bonds. This results in a monolayer of bonded hydrogen on the surface. In the words of L. Feldman, "hydrogen acts as a surface terminator" (38).

Hydrogen in Metals

Because even very low concentrations of hydrogen can embrittle most metals, there has been intense interest in and research on the role of hydrogen in metals for many years. Hydrogen can enter metals from the gas phase, by electrolysis, or as a result of aqueous corrosion. Its effects are both difficult to predict and potentially catastrophic. Because of the presence of hydrogen (and hydrogen embrittlement) in essentially all types of present or proposed energy schemes (from conventional fossil fuel, to geothermal, to various fission or fusion reactors), hydrogen embrittlement is of ever increasing technological concern.

While some of the first applications of nuclear reaction analysis for hydrogen were in hydrogen-metal systems, there was surprisingly little progress in this area for some years. However, in the past two or three years techniques have been developed and systematically applied that have dramatically increased our knowledge of hydrogen in solids (39). For example, by measuring the amount of hydrogen retained in a sample vs the temperature, it has been possible to determine the binding enthalpies of hydrogen to various defects. Further, by using channeling measurements (at various temperatures) to determine the lattice location of this trapped hydrogen, it has also been possible to unambiguously determine the nature of the binding defect (39).

Reaction Between Water and Glass

Reactions with water or aqueous solutions are among the most widespread and costly degrading processes for most materials. These reactions are important not only in everyday conventional materials but also in various high-technology applications, such as long-term durability of thin-film electronic and optical devices or the long-term consolidation of radioactive reactor wastes. Protection against unwanted reactions begins with an understanding of the reaction mechanisms. For water-based reactions, nuclear reaction measurements of elemental profiles of hydrogen, oxygen, and other elements provide unique, valuable data for establishing these reaction mechanisms. We describe the study of the reaction between water and glass as an illustration of the power of these techniques.

Glass is one of the most durable materials widely used by man. However, all glasses do undergo a slow reaction with water, either as a liquid or an atmospheric vapor, resulting in the modification (hydration) of the glass surface. While this reaction is undesirable from a durability of point of view, measurement of the extent of the surface modification is used by archaeologists to date artifacts made of obsidian (40) (a natural volcanic glass) and man-made glass (41) objects.

The mechanisms responsible for this reaction have been of interest for over one hundred years and, until recently, there was still considerable controversy over even the basic transport processes of the water penetrating the hydrated glass. While there were many variations, the fundamental question was whether the water (hydrogen) was carried in molecular or ionic form.

The ionic model made a number of specific predictions. The essential feature of this model is that the very mobile alkali ions present in all common glasses undergo an ionic exchange reaction with water on the surface of the glass:

$$Na^+(glass) + 2H_2O \rightarrow H_3O^+(glass) + NaOH,$$

and that the resulting flux of hydronium ions (H_3O^+) diffuses into the glass, replacing Na^+ ions diffusing out.

The most crucial test of this ionic exchange model is to see if the sodium is depleted in the surface regions of high hydrogen content. ^{23}Na profiles can be measured by nuclear reaction analysis and results of such a measurement are shown in Figure 10 (42). One sees from these profiles that the H and Na profiles are clearly complementary, which confirms the ionic exchange model.

SUMMARY

NRA was first noted indirectly in the earliest period of low-energy nuclear physics but it has expanded rapidly since the 1960s thanks to the spread of MV accelerators and the invention of semiconductor detectors. It interacts upstream with nuclear and atomic physics and downstream with the large number of fields where the determination of light nuclei in the near-surface

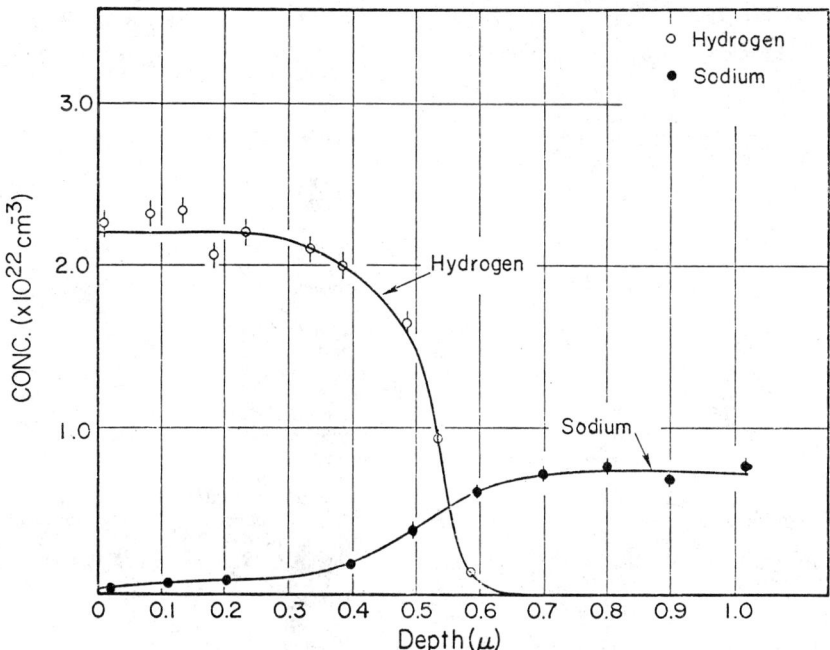

Figure 10 Hydrogen and sodium profiles of a sample of soda-lime glass exposed to water at 90°C. The Na profile was measured by the ^{23}Na(p,γ) resonant nuclear reaction method (from 42).

region of solids is of prime importance. Applications range from solid-state physics and electrochemistry, semiconductor technology, metallurgy, materials science, and surface science to biology and archeology. The nuclear processes involved in NRA are simple and the MeV particle slowing-down processes that intervene in data reduction are now reasonably well understood. Therefore NRA yields reliable, unambiguous, accurate, and quantitative analytical results. Not only is NRA in many cases the only available analytical tool, in particular when coupled with channeling studies, but it may also be used as a calibration and cross check technique for other analytical methods that, while less quantitative, have other advantages. The use of ion-beam analysis will continue to spread steadily both in academic and industrial research, where MV accelerators dedicated to ion-beam analysis are increasingly common.

ACKNOWLEDGMENTS

This work was partially supported by the French Centre National de la Recherche Scientifique under the program RCP No. 157 and by the US Army Research Office.

Literature Cited

1. Livingston, M. S., Bethe, H. A. 1937. *Rev. Mod. Phys.* 9:245–390
2. Jarjis, R. A. *Handbook of Nuclear Cross-Section Data for Surface Analysis.* New York: Academic. In press
3. Bird, J. R. 1980. *Nucl. Instrum. Methods* 168:85–91
4. Mayer, J. W., Rimini, E. 1977. *Ion Beam Handbook for Material Analysis.* New York: Academic. 488 pp.
5. Ziegler, J. F. 1975. *New Uses of Ion Accelerators.* New York: Plenum. 482 pp.
6. Thomas, J. P., Cachard, A. 1978. *Material Characterization Using Ion Beams.* London: Plenum. 517 pp.
7. Deconninck, G. 1978. *Introduction to Radioanalytical Physics (RBS, NRA, PIXE).* Amsterdam: Elsevier. 242 pp.
8. Chu, W. K., Mayer, J. W., Nicolet, M. A. 1978. *Backscattering Spectrometry.* New York: Academic. 384 pp.
9. Bird, J. R., Duerden, P., Wilson, D. J. 1983. *Nucl. Sci. Appl. B* 1:357–526
10. Mayer, J. W., Ziegler, J. F. 1973. *Ion Beam Surface Layer Analysis, Proc. IBA Conf., Yorktown Heights. Thin Solid Films* 19:1–463
11. Meyer, O., Linker, G., Kappler, F. 1975. *Ion Beam Surface Layer Analysis, Proc. IBA Conf., Karlsruhe,* Vols. 1, 2. New York: Plenum. 985 pp.
12. *Proceedings of IBA Conferences in Nucl. Instrum. Methods* 1978. Washington, Vol. 149; 1980. Aarhus, Vol. 168; 1981. Sydney, Vol. 191; 1983. Phoenix, Vol. 218
13. Demortier, G. 1982. *Proc. Int. Conf. in Namur, Nucl. Instrum. Methods* 197:1–258
14. Bujdoso, E., Lyon, W. S., Noszlopi, I. 1982. *J. Radioanal. Chem.* 74:197–238
15. Amsel, G., Samuel, D. 1967. *Anal. Chem.* 39:1689–98
16. Amsel, G., Nadai, J. P., d'Artemare, E., David, D., Girard, E., Moulin, J. 1971. *Nucl. Instrum. Methods* 92:481–98
17. Maurel, B., Amsel, G., Nadai, J. P. 1982. *Nucl. Instrum. Methods* 197:1–13 (in Ref. 13)
18. Dieumegard, D., Maurel, B., Amsel, G. 1980. *Nucl. Instrum. Methods* 168:93–103
19. Ajzenberg-Selove, F. 1972 and 1977. *Nucl. Phys. A* 190:1–196; *A* 281:1–148
20. Maurel, B., Amsel, G. 1983. *Nucl. Instrum. Methods* 218:159–64
21. Damjantschitsch, H., Weiser, W., Hensser, G., Kalbitzer, S., Mannsperger, H. 1983. *Nucl. Instrum. Methods* 218:129–40
22. Maurel, B. 1980. Thesis. Univ. Paris VII, October
23. Möller, W., Besenbacher, F. 1980. *Nucl. Instrum. Methods* 168:111–14
24. Östling, M., Petersson, C. S. 1983. *Nucl. Instrum. Methods* 218:439–44
25. Berti, M., Drigo, A. V. 1982. *Nucl. Instrum. Methods* 201:473–79
26. Amsel, G., Davies, J. A. 1983. *Nucl. Instrum. Methods* 218:177–82
27. Amsel, G., Samuel, D. 1962. *J. Phys. Chem. Solids* 23:1707–18
28. Amsel, G., Maurel, B. 1983. *Nucl. Instrum. Methods* 218:183–96
29. Lanford, W. A., Trautvetter, H. P., Ziegler, J., Keller, J. 1976. *Appl. Phys. Lett.* 28:566–68
30. Terreault, B., Martel, J. G., St.-Jacques, R. G., L'Ecuyer, J. 1977. *J. Vac. Sci. Technol.* 14:492–500
31. Moreau, C., Knystautas, E. J., Timsit, R. S., Groleau, R. 1985. *Nucl. Instrum. Methods* 218:111–15
32. Doyle, B. L., Peercy, P. S. 1979. *Appl. Phys. Lett.* 34:811–13
33. Scherzer, B. M. U., Langley, R. A., Möller, W., Roth, J., Schulz, R. 1982. *Nucl. Instrum. Methods* 134:497–500
34. Feldman, L. C., Mayer, J. W., Picraux, S. T. 1982. *Materials Analysis by Ion Channeling.* New York: Academic. 300 pp.
35. Amsel, G., d'Artemare, E., Girard, E. 1983. *Nucl. Instrum. Methods* 205:5–26
36. Narusawa, T., Kinoshita, K., Gibson, W. M. 1980. *Surf. Sci.* 1:673–781
37. Lanford, W. A., Golub, R. 1977. *Phys. Rev. Lett.* 39:1509–72
38. Feldman, L. 1981. Presented at 5th Int. Conf. on Ion Beam Analysis, Sydney, Australia, February 16–20, 1981
39. Möller, W. 1983. *Nucl. Instrum. Methods* 209/210:773–90
40. Taylor, R. E., ed. 1976. *Advances in Obsidian Glass Studies.* Pack Ridge, NJ: Noyes Press. 360 pp.
41. Lanford, W. A. 1977. *Science* 196:975–78
42. Lanford, W. A., Davis, K., LaMarche, P., Laursen, T., Groleau, R., Doremus, R. 1979. *J. Non-Crystalline Solid* 33:249–66

MAGNETIC MONOPOLES[1]

John Preskill[2]

California Institute of Technology, Pasadena, California 91125

CONTENTS

1. INTRODUCTION	462
2. THE DIRAC MONOPOLE	466
2.1 Monopoles and Charge Quantization	466
2.2 Generalizations of the Quantization Condition	468
3. MONOPOLES AND UNIFICATION	471
3.1 Unification, Charge Quantization, and Monopoles	471
3.2 Monopoles as Solitons	472
3.3 The Monopole Solution	474
4. MONOPOLES AND TOPOLOGY	477
4.1 Monopoles without Strings	477
4.2 Topological Classification of Monopoles	478
4.3 Magnetic Charge of a Topological Soliton	480
4.4 The Kaluza-Klein Monopole	484
4.5 Monopoles and Global Gauge Transformations	485
5. EXAMPLES	487
5.1 A Symmetry-Breaking Hierarchy	487
5.2 A Z_2 Monopole	491
5.3 The $SU(5)$ and $SO(10)$ Models	493
5.4 Monopoles and Strings	496
6. DYONS	498
6.1 Semiclassical Quantization	498
6.2 The Anomalous Dyon Charge	501
6.3 Composite Dyons	502
6.4 Dyons in Quantum Chromodynamics	504
7. MONOPOLES AND FERMIONS	506
7.1 Fractional Fermion Number on Monopoles	506
7.2 Monopole-Fermion Scattering	508
8. MONOPOLES IN COSMOLOGY AND ASTROPHYSICS	513
8.1 Monopoles in the Very Early Universe	513
8.2 Astrophysical Constraints on the Monopole Flux	517
9. DETECTION OF MONOPOLES	522
9.1 Induction Detectors	522
9.2 Ionization Detectors	524
9.3 Catalysis Detectors	527

[1] Work supported in part by the US Department of Energy under contract DEAC-03-81-ER40050.

[2] Alfred P. Sloan Fellow.

1. INTRODUCTION

How is it possible to justify a lengthy review of the physics of the magnetic monopole when nobody has ever seen one? In spite of the unfortunate lack of favorable experimental evidence, there are sound theoretical reasons for believing that the magnetic monopole must exist. The case for its existence is surely as strong as the case for any other undiscovered particle. Moreover, as of this writing (early 1984), it is not certain that nobody has ever seen one. What seems certain is that nobody has ever seen two.

The idea that magnetic monopoles, stable particles carrying magnetic charges, ought to exist has proved to be remarkably durable. A persuasive argument was first put forward by Dirac in 1931 (1). He noted that, if monopoles exist, then electric charge must be quantized; that is, all electric charges must be integer multiples of a fundamental unit. Electric charge quantization is actually observed in Nature, and no other explanation for this deep phenomenon was known.

Many years later, another very good argument emerged. Polyakov (2) and 't Hooft (3) discovered that the existence of monopoles follows from quite general ideas about the unification of the fundamental interactions. A deeply held belief of many particle theorists is that the observed strong and electroweak gauge interactions, which have three apparently independent gauge coupling constants, actually become unified at extremely short distances into a single gauge interaction with just one gauge coupling constant (4, 5). Polyakov and 't Hooft showed that any such "grand unified" theory of particle physics necessarily contains magnetic monopoles. The implications of this discovery are rich and surprising and are still being explored.

While Dirac had demonstrated the consistency of magnetic monopoles with quantum electrodynamics, 't Hooft and Polyakov demonstrated the necessity of monopoles in grand unified gauge theories. Furthermore, the properties of the monopole are calculable, unambiguous predictions in a given unified model.

All grand unified theories possess a large group of exact gauge symmetries that mix the strong and electroweak interactions, but these symmetries become spontaneously broken at an exceedingly short distance scale M_X^{-1} (or, equivalently, an exceedingly large mass scale M_X). The properties of the magnetic monopole, such as its size and mass, are determined by the distance scale of the spontaneous symmetry breakdown (the "unification scale"). The prediction that magnetic monopoles must exist does not depend on the *mechanism* of the symmetry breakdown; for example, it does not matter whether the Goldstone bosons associated with the symmetry breakdown are elementary or composite. Nor does it matter whether gravitation becomes unified with the other particle interactions at the unification scale.

The magnetic charge g of the monopole is typically the "Dirac charge" $g_D = 1/2e$. (Magnetic charge will be defined so that the total magnetic flux emanating from a charge g is $4\pi g$. Electric charge is defined so that the electric flux emanating from a charge e is e.) This magnetic charge is distributed over a core with a radius of order M_X^{-1}, the unification distance scale, and the mass of the monopole is comparable to the magnetostatic potential energy of the core.

The unification mass scale M_X varies from one grand unified model to another. But M_X can be calculated if we make a very strong assumption—the "desert hypothesis"—that is, if we assume that no unexpected new interactions or particles appear between present-day energies (of order 100 GeV) and the unification scale M_X. [This assumption is also the basis of the highly successful calculation (6) of the electroweak mixing angle $\sin^2 \theta_w$.] From the desert hypothesis follows the prediction $M_X \approx 10^{14}$ GeV (6); the properties of the monopole may then be summarized by

Charge: $\quad g = g_D = 1/2e$,

Core size: $\quad R \approx M_X^{-1} \approx 10^{-28}$ cm, $\qquad\qquad$ 1.

Mass: $\quad m \approx (4\pi/e^2)M_X \approx 10^{16}$ GeV.

Here $e^2/4\pi$ is the running coupling constant renormalized at the mass scale M_X, making it somewhat larger than $\alpha \approx 1/137$.

Of course, the desert hypothesis could easily be wrong, even if the general idea of grand unification is correct. So the size and mass of the monopole could be much different from the estimates in Equation 1. It is nonetheless interesting to note that one can reasonably expect the monopole to be an *extremely* heavy stable elementary particle; 10^{16} GeV $\approx 10^{-8}$ g $\approx 10^{6}$ J is comparable to the mass of a bacterium, or the kinetic energy of a charging rhinoceros. It is hardly surprising that magnetic monopoles have not been produced by existing particle accelerators.

We also see from Equation 1 that the size R of the core of the monopole is expected to be larger than its Compton wavelength by a factor of order $4\pi/e^2$. In this sense, the monopole is a nearly classical object; quantum mechanics plays an insignificant role in determining the structure of its core, if e^2 is small. In fact, magnetic monopoles appear in spontaneously broken unified gauge theories even in the classical limit, as stable time-independent solutions to the classical field equations.

The stability of the classical monopole solution is ensured by a topological principle to be explained in detail below. Loosely speaking, the monopole is a "defect" in the scalar field that acts as an order parameter for the spontaneous breakdown of the grand unified gauge symmetry. Trapped inside its core is a region in which the scalar field respects symmetries different from those respected by the vacuum state. This scalar field configuration is energetically unfavorable, so the core cannot expand. But the magnetostatic energy of the core prevents it from shrinking. So the core is stable.

While most of the mass of the monopole is concentrated in its tiny core of radius M_X^{-1}, the monopole has interesting structure on many different size scales (Figure 1). At distances less than $M_Z^{-1} \approx 10^{-16}$ cm from the center of the monopole, virtual W and Z bosons have important effects on its interactions with other particles. The monopole is also a hadron; it has a color magnetic field that extends out to distances of order 10^{-13} cm, and then becomes screened by nonperturbative strong-interaction effects. And, because of its large magnetic charge, the monopole is strongly coupled to a surrounding cloud of virtual electron-positron pairs, which extends out to distances of order $m_e^{-1} \approx 10^{-11}$ cm. In a grand unified theory in which new physics appears at energies below the unification scale M_X (so that the desert hypothesis does not apply), the structure of the monopole might be even more complicated.

The existence of magnetic monopoles is a very general consequence of the unification of the fundamental interactions. But it is one thing to say that monopoles must exist, and quite another to say that we have a reasonable chance of observing one. If monopoles are as heavy as we expect, there is no hope of producing monopoles in any foreseen accelerator. Our best hope is to observe a monopole in cosmic rays. But since no process occurring in the present universe is sufficiently energetic to produce monopoles, any

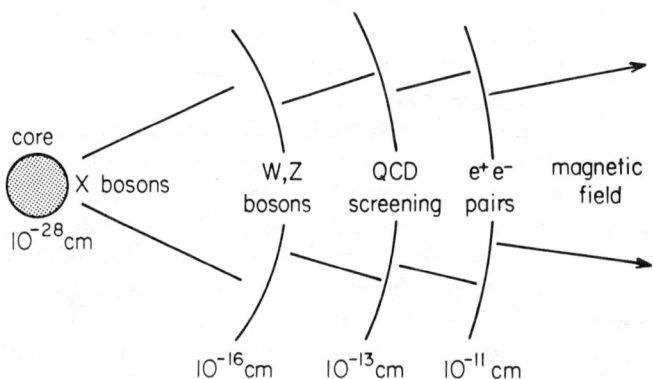

Figure 1 Structure of a grand unified monopole.

monopoles around today must have been produced in the very early universe, when higher energies were available. Thus, the abundance of magnetic monopoles is a cosmological issue (7–9).

In fact, estimates based on the standard cosmological scenario indicate that the monopole abundance should exceed by many orders of magnitude the current observational limits. Thus, our failure to observe a monopole is itself a significant piece of information, casting doubt on either the standard view of the evolution of the universe, or on cherished beliefs about particle physics at extremely short distances. This dilemma has led to revolutionary new developments in theoretical cosmology (10–12).

Significant as it may be not to see a monopole, it would be even more interesting to see one. But astrophysical arguments indicate that the flux of monopoles in cosmic rays is probably quite small (13, 14). Furthermore, if monopoles are very heavy, those bombarding the earth are likely to be moving relatively slowly, with velocities of order 10^{-3} c. Detection of these slow, rare monopoles is a challenging problem for experimenters.

If a magnetic monopole is ever discovered, it will be a momentous occasion, with many fascinating implications. For one thing, that there are any monopoles at all would be evidence that the universe was once extremely hot. And severe constraints would be placed on our attempts at cosmological model building, for the observed monopole abundance would have to be explained by any realistic cosmological scenario.

Detection of a monopole would also confirm a very fundamental prediction of grand unification. The mass of the monopole, if it could be measured, would reveal the basic symmetry-breaking scale at which electrodynamics becomes truly united with the other particle interactions. More could be learned about very short-distance physics by studying the interactions of monopoles with fermions. Remarkably, a charged fermion (e.g. a quark or lepton) incident on a monopole at low energy can penetrate to the core of the monopole, and probe its structure (15, 16). Thus monopoles could provide us with a unique window on new physics at incredibly short distances.

But even if nobody ever sees a magnetic monopole, there is surely much to be gained by studying the theory of monopoles. Already, marvelous insights into gauge theory and quantum field theory have been derived from this study. There is little reason to doubt that further surprising discoveries await the dedicated student of the magnetic monopole.

The main purpose of this article is to present the basic results of monopole theory. For the most part, the presentation is intended to be accessible to a reader with a minimal background in theoretical particle physics. In Section 2, the connection between magnetic monopoles and the quantization of electric charge is explained, and in Section 3 the classical monopole solution of 't Hooft and Polyakov is introduced. The theory of magnetic monopoles carrying nonabelian magnetic charge is developed in Section 4, and the general connection between the topology of a classical monopole solution and its magnetic charge is established there. Various examples illustrating and elucidating the formalism of Section 4 are discussed in Section 5. Section 6 is concerned with the properties of dyons, which carry both magnetic and electric charge. Aspects of the interactions of fermions and monopoles are considered in Section 7. In Section 8, the cosmological production of monopoles and astrophysical bounds on the monopole abundance are described. Some remarks about the detection of monopoles are contained in Section 9.

The reader who finds gaps in the present treatment may wish to consult some of the other excellent reviews of these topics. For a general review of grand unified theories see (17, 18). For more about some of the topics in Section 2–4, see (19–21); for Section 6, see (21); for Section 8, see (22–26); and for Section 9, see (27, 28).

2. THE DIRAC MONOPOLE

2.1 *Monopoles and Charge Quantization*

Measured electric charges are always found to be integer multiples of the electron charge. This quantization of electric charge is a deep property of Nature crying out for an explanation. More than fifty years ago, Dirac (1) discovered that the existence of magnetic monopoles could "explain" electric charge quantization.

Dirac envisaged a magnetic monopole as a semi-infinitely long, infinitesimally thin solenoid (Figure 2). The end of such a solenoid looks like a

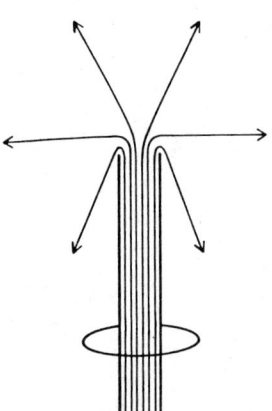

Figure 2   The end of a semi-infinite solenoid.

magnetic charge, but it makes sense to identify this object as a magnetic monopole only if no conceivable experiment can detect the infinitesimally thin solenoid.

We can imagine trying to detect the solenoid by doing an electron interference experiment (29); such an experiment gives a null result only if the phase picked up by the electron wave function, when the electron is transported along a closed path enclosing the solenoid, is trivial. Suppose a point monopole with magnetic charge g sits at the origin, so that the magnetic field is

$$\mathbf{B} = g \frac{\hat{r}}{r^2}, \qquad 2.$$

and that the solenoid lies on the negative z-axis. Then, in spherical coordinates, and in an appropriate gauge, the only nonvanishing component of the vector potential is

$$A_\phi = g(1 - \cos \vartheta), \qquad 3.$$

where A_ϕ is defined by $\mathbf{A} \cdot d\mathbf{r} \equiv A_\phi \, d\phi$. The electron interference experiment fails to detect the solenoid if

$$\exp[-ie\oint \mathbf{A} \cdot d\mathbf{r}] = \exp[-i4\pi eg] = 1, \qquad 4.$$

where $-e$ is the electron charge. Hence, we require the magnetic charge g to satisfy Dirac's quantization condition (1)

$$eg = \frac{n}{2}. \qquad 5.$$

The minimum allowed magnetic charge $g_D = 1/2e$ is called the Dirac magnetic charge.

Dirac's reasoning shows that it is consistent in quantum mechanics to describe a magnetic monopole with the vector potential Equation 3, even though it has a "string" singularity for $\vartheta = -\pi$. The string is undetectable. In fact, we formulate in Section 4.1 a different mathematical description of the monopole, in which the string is avoided altogether.

The quantization condition Equation 5 requires all magnetic charges to be integer multiples of the Dirac charge $g_D = 1/2e$. We can also turn this argument around, as follows: Suppose there exists a magnetic monopole with magnetic charge g_D. Then it is consistent for a particle with electric charge Qe (and vanishing magnetic charge) to exist only if $\exp[i4\pi Qeg_D] = 1$, or

$$Q = (1/2eg_D)n = n, \qquad 6.$$

where n is an integer. Therefore, the existence of a magnetic monopole implies quantization of electric charge.

2.2 Generalizations of the Quantization Condition

To derive the Dirac quantization condition (Equation 5), we used the electron charge $-e$. But we believe that quarks exist, and the electric charge of a down quark, for example, is $-e/3$. Will not the same argument as before, applied to a down quark instead of an electron, lead to the conclusion that the minimal allowed magnetic charge is $3g_D$ instead of g_D?

No, not if quarks are confined (30). For if quarks are permanently confined in hadrons, it makes sense to speak of performing a quark interference experiment only over distances less than 10^{-13} cm, the size of a hadron. It is true that, when the down quark is transported around Dirac's string, its wave function acquires the nontrivial phase

$$\exp[-i(e/3)\oint \mathbf{A}_{em} \cdot d\mathbf{r}] = \exp(-i2\pi/3) \neq 1 \qquad 7.$$

due to the coupling of the down quark to the electromagnetic vector potential, if the monopole carries the Dirac magnetic charge g_D. But we must recall that the down quark carries another degree of freedom, color. The string is not detectable if the monopole also has a *color-magnetic field*, such that the phase acquired by the down quark wave function due to the color vector potential compensates for the phase due to the electromagnetic vector potential, or

$$\exp[ie_c \oint \mathbf{A}_{color} \cdot d\mathbf{r}] = \exp(i2\pi/3), \qquad 8.$$

where e_c is the color gauge coupling.

The correct conclusion, then, if quarks are confined, is not that the minimal magnetic charge is g_D, but rather that the monopole carrying magnetic charge g_D must also carry a color-magnetic charge. The color-magnetic field of the monopole becomes screened by nonperturbative strong-interaction effects at distances greater than 10^{-13} cm (21, 31). We also conclude that there cannot exist both *isolated* fractional electric charges and monopoles with the Dirac magnetic charge, unless there is some other (as yet unknown) long-range field that couples to both the monopoles and the fractional electric charges (32).

To state the Dirac quantization condition in its most general form, we note that the vector potential of a magnetic monopole carrying more than one type of magnetic charge can in general be written (33)

$$\sum_a e^a T^a A^a_\phi = \tfrac{1}{2}M(1-\cos\vartheta), \qquad 9.$$

where M is a constant matrix. The sum over a runs over all the generators of the gauge group, and the gauge couplings e^a have been absorbed into M. By an argument similar to that invoked above (see also Section 4.2), we can derive the generalized Dirac quantization condition

$$\exp(i2\pi M) = 1. \qquad 10.$$

That is, M must have integer eigenvalues.

For example, in the SU(5) grand unified model, the electric charge generator may be written as a 5×5 matrix

$$Q_{\rm em} = {\rm diag}(\tfrac{1}{3}, \tfrac{1}{3}, \tfrac{1}{3}, 0, -1), \qquad 11.$$

where the diag$(-,-,-,-,-)$ notation denotes a diagonal matrix with the indicated eigenvalues. The eigenvalues of $Q_{\rm em}$ are the electric charges, in units of e, of the elements of the 5 representation of SU(5)—antidown quarks, in three colors, the neutrino and the electron. The color SU(3) generators are traceless 3×3 matrices acting on the quarks only; one of these is

$$Q_{\rm color} = {\rm diag}(-\tfrac{1}{3}, -\tfrac{1}{3}, \tfrac{2}{3}, 0, 0). \qquad 12.$$

A matrix M that satisfies Equation 10 is

$$M = Q_{\rm em} + Q_{\rm color} = {\rm diag}(0, 0, 1, 0, -1), \qquad 13.$$

and the magnetic charge of the monopole described by Equation 13 is the coefficient of $eQ_{\rm em}$ in $\tfrac{1}{2}M$, or $1/2e = g_{\rm D}$, the Dirac charge.

In the SU(5) model, a restatement of the criteria in Equations 10 and 13 for the existence of a magnetic monopole with the Dirac charge is

$$\exp[i2\pi Q_{\rm em}] = {\rm diag}[\exp(i2\pi/3), \exp(i2\pi/3), \exp(i2\pi/3), 1, 1] \equiv Z \qquad 14.$$

where Z is a nontrivial element of Z_3, the center of color SU(3). Equation 14 is just a fancy way of saying that objects that carry trivial color SU(3) triality have integer electric charge (in units of e), even though objects with nontrivial triality have fractional charge. That the U(1) group generated by $Q_{\rm em}$ contains the center of color SU(3) also has a topological significance, which is elucidated in Sections 4 and 5.

Another interesting generalization of Equation 5 applies to *dyons*, objects that carry both electric and magnetic charge. Consider the two dyons with electric and magnetic charges $(Q_1 e, M_1 g_{\rm D})$ and $(Q_2 e, M_2 g_{\rm D})$. Each dyon is unable to detect the string of the other if and only if (34)

$$Q_1 M_2 - Q_2 M_1 = n, \qquad 15.$$

where n is an integer. The minus sign in Equation 15 arises because transporting the first dyon counterclockwise around the string of the second is equivalent to transporting the second dyon clockwise around the string of the first.

The condition represented by Equation 15 requires all magnetic charges to be integer multiples of g_D, if there exists a particle with $Q = 1$ and $M = 0$. But magnetically charged objects are allowed to carry anomalous electric charges. Equation 15 is satisfied if Q and M for all dyons are related by

$$Q = n - \frac{\vartheta}{2\pi} M, \qquad 16.$$

where n is an integer, and ϑ is an arbitrary parameter defined modulo 2π (see Figure 3). For a dyon carrying more than one type of magnetic charge, there is a distinct ϑ for each type.

The significance of the parameter ϑ is discussed further in Sections 6 and 7. Here we merely note that the dyon charge spectrum (Equation 16) violates CP unless ϑ is 0 or π, because Q is CP odd, and M is CP even.

So far, we have taken the magnetic monopole to be pointlike; the magnetic field (Equation 2) is singular at the origin. But it is obvious that our derivation of the quantization condition will also apply to a nonsingular field that approaches the form of Equation 2 at large distances. Such a nonsingular monopole is constructed in Section 3.

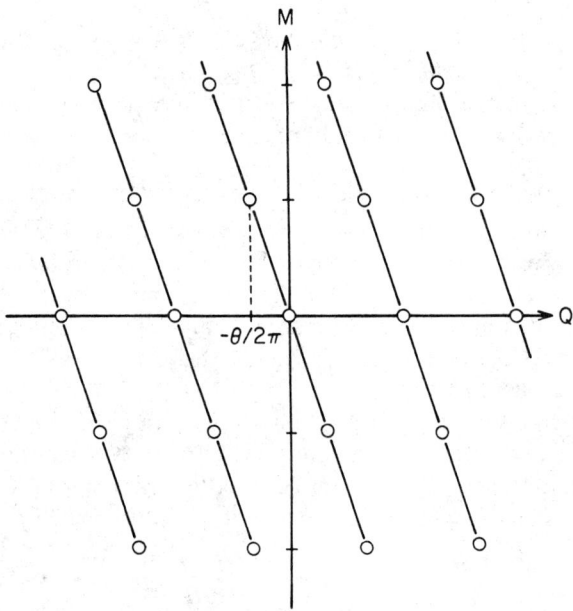

Figure 3 Electric charges (Q) and magnetic charges (M) allowed by the Dirac quantization condition.

3. MONOPOLES AND UNIFICATION

3.1 *Unification, Charge Quantization, and Monopoles*

Dirac showed that quantum mechanics does not preclude the existence of magnetic monopoles. Moreover, the existence of monopoles implies quantization of electric charge, a phenomenon observed in Nature. The monopole thus seems to be such an appealing theoretical construct that, to quote Dirac, "one would be surprised if Nature had made no use of it" (1).

Nowadays, we have another way of understanding why electric charge is quantized. Charge is quantized if the electromagnetic $U(1)_{em}$ gauge group is compact. But $U(1)_{em}$ is automatically compact in a *unified* gauge theory in which $U(1)_{em}$ is embedded in a nonabelian semisimple group. [Note that the standard Weinberg-Salam-Glashow (35) model is not "unified" according to this criterion.]

In other words, in a unified gauge theory, the electric charge operator obeys nontrivial commutation relations with other operators in the theory. Just as the angular momentum algebra requires the eigenvalues of J_z to be integer multiples of $\frac{1}{2}\hbar$, the commutation relations satisfied by the electric charge operator require its eigenvalues to be integer multiples of a fundamental unit. This conclusion holds even if the symmetries generated by the charges that fail to commute with electric charge are spontaneously broken.

These two apparently independent explanations of charge quantization are not really independent at all. Dirac found the existence of monopoles to imply charge quantization, but the converse, in a sense, is also true. Any unified gauge theory in which $U(1)_{em}$ is embedded in a spontaneously broken semisimple gauge group, and electric charge is thus automatically quantized, necessarily contains magnetic monopoles. The discovery of this remarkable result, by 't Hooft (3) and Polyakov (2), ushered in the modern era of monopole theory.

In contrast to Dirac's demonstration of the *consistency* of magnetic monopoles with quantum electrodynamics, t' Hooft and Polyakov demonstrated the *necessity* of monopoles in unified gauge theories. Furthermore, the properties of the monopole are calculable in a given unified model. In particular, its mass can be related to the masses of certain heavy vector bosons, while in Dirac's formulation of electrodynamics, the monopole mass must be regarded as an arbitrary free parameter.

There has been much speculation in recent years about "grand unified" models of elementary particle interactions, in which the standard low-energy gauge group $SU(3)_{color} \times [SU(2) \times U(1)]_{electroweak}$ is embedded in a simple gauge group that is spontaneously broken at a large mass scale. The simplest model of this type is the $SU(5)$ model (4). But the prediction that magnetic monopoles must exist applies to any grand unified model, and also to the even more ambitious models purporting to unify gravitation with the other particle interactions.

3.2 Monopoles as Solitons

In this section we show how magnetic monopoles arise in unified gauge theories as solutions to the classical field equations. A semiclassical expansion about the classical monopole solution can be carried out to arbitrary order in $\hbar$, but for now we confine our attention to the classical approximation. Some properties of the semiclassical expansion in higher order are discussed in Section 6.

Here we consider the simplest unified gauge theory containing a monopole solution (36). The generalization to more complicated models is described in Sections 4 and 5.

The model has the gauge group SU(2) and a Higgs field Φ in the triplet representation of the group; its Lagrangian is

$$\mathscr{L} = -\tfrac{1}{4}F^a_{\mu\nu}F^{\mu\nu a} + \tfrac{1}{2}D_\mu\Phi^a D^\mu\Phi^a - U(\Phi), \qquad 17.$$

where

$$U(\Phi) = \tfrac{1}{8}\lambda(\Phi^a\Phi^a - v^2)^2, \qquad 18.$$

$$D_\mu\Phi^a = \partial_\mu\Phi^a - e\varepsilon^{abc}W^b_\mu\Phi^c, \qquad 19.$$

$$F^a_{\mu\nu} = \partial_\mu W^a_\nu - \partial_\nu W^a_\mu - e\varepsilon^{abc}W^b_\mu W^c_\nu, \qquad 20.$$

and $a = 1, 2, 3$. The energy density can be written as

$$\mathscr{E} = \tfrac{1}{2}[E^a_i E^a_i + B^a_i B^a_i + D_i\Phi^a D_i\Phi^a] + U(\Phi), \qquad 21.$$

where

$$E^a_i = F^a_{0i}, \qquad B^a_i = \tfrac{1}{2}\varepsilon_{ijk}F^a_{jk}. \qquad 22.$$

Since $\mathscr{E} \geq 0$, the classical "vacuum" of this theory is a field configuration such that $\mathscr{E} = 0$. In the "unitary" gauge, the scalar field Φ may be written

$$\Phi = (0, 0, v+\varphi), \qquad 23.$$

and the vacuum configuration is

$$\varphi = 0, \qquad W^a_\mu = 0. \qquad 24.$$

To determine the perturbative spectrum in this gauge, we substitute Equation 23 into the Lagrangian. Since

$$\tfrac{1}{2}D_\mu\Phi^a D^\mu\Phi^a = \tfrac{1}{2}(\partial^\mu\varphi)^2 + \tfrac{1}{2}e^2(v^2 + \ldots)[(W^1_\mu)^2 + (W^2_\mu)^2], \qquad 25.$$

and

$$U(\Phi) = \tfrac{1}{2}\lambda v^2 \varphi^2 + \ldots, \qquad 26.$$

we find that the theory has undergone the Higgs mechanism; there is a massless "photon" W_μ^3 that couples to the unbroken $U(1)_{em}$ current, as well as charged vector bosons $W_\mu^\pm$ with mass

$$M_W = ev \qquad 27.$$

and a neutral scalar with mass

$$M_H = \sqrt{\lambda}\, v. \qquad 28.$$

To investigate the spectrum of this theory beyond perturbation theory, let us determine whether there is a stable time-independent solution to the classical field equations other than the vacuum solution. Equivalently, we seek a field configuration at a fixed time that is a local minimum of the energy functional $\int d^3 r\, \mathscr{E}$. Such a "soliton" configuration behaves like a particle in the classical theory, and can be expected to survive in the spectrum of the quantum theory.

Our search for a nontrivial local minimum of the energy functional will surely succeed if there are field configurations that cannot be continuously deformed to the vacuum configuration while the total energy remains finite. For if we start with such a configuration and deform it until a local minimum is obtained, the final configuration is guaranteed to be different from the vacuum solution.

Furthermore, it is easy to demonstrate the existence of such a configuration. For a field configuration of finite energy, the scalar field Φ is required to approach a minimum of the potential $U(\Phi)$ at large distances, but Φ is free to select different minima of U in different spatial directions. The asymptotic behavior of Φ defines a mapping $\Phi^a(\hat{r}) = \lim_{\alpha \to \infty} \Phi^a(\alpha \mathbf{r})$ such that

$$\Phi^a(\hat{r})\Phi^a(\hat{r}) = v^2; \qquad 29.$$

that is, a mapping from the sphere at spatial infinity to the sphere of minima of $U(\Phi)$.

Consider a "hedgehog" configuration such that the mapping $\Phi^a(\hat{r})$ is the identity mapping

$$\Phi^a(\hat{r}) = v\hat{r}^a. \qquad 30.$$

It is evident that there is no way of continuously deforming the mapping of Equation 30 to the trivial mapping $\Phi^a(\hat{r}) = $ constant, while preserving the finite-energy condition, Equation 29. The number of times the mapping

$\Phi^a(\hat{r})$ "wraps around" the manifold of minima of $U(\Phi)$ is an integer. Since an integer cannot change continuously, this "winding number" is preserved by continuous deformations; it is said to be a "topological invariant." But the hedgehog configuration has winding number 1, and the vacuum configuration has winding number 0. Therefore, the vacuum configuration cannot be obtained by any continuous deformation of the hedgehog configuration that is consistent with Equation 29.

It only remains to verify that there really is a hedgehog configuration that asymptotically approaches Equation 30 and has finite energy. The contribution $\frac{1}{2}\int d^3x (D_i\Phi^a)^2$ to the energy is finite only if $D_i\Phi^a$ approaches zero at large r sufficiently rapidly. We therefore require

$$D_i\Phi^a \sim 0 \qquad 31.$$

for large r, or

$$W_i^a \sim \frac{\varepsilon_{iak}\hat{r}_k}{er}, \quad \text{and} \quad B_i^a \sim \frac{\hat{r}_i\hat{r}^a}{er^2}. \qquad 32.$$

The long-range gauge field (Equation 32) is a $U(1)_{em}$ gauge field that carries magnetic charge $g = 1/e$ [where $U(1)_{em}$ is the subgroup of $SU(2)$ left unbroken by the scalar field, Equation 30]. The charge $1/e$ is really the Dirac magnetic charge in this model, since it is possible to introduce matter fields in the doublet representation of $SU(2)$ that carry electric charge $e/2$.

We thus conclude that there must be a stable, finite-energy, time-independent solution to the classical equations of motion such that the asymptotic scalar field configuration $\Phi^a(\hat{r})$ has winding number 1. Finiteness of the energy requires the long-range gauge field of this soliton to be the field of a Dirac magnetic monopole.

In general, we may consider field configurations such that the winding number is an arbitrary integer. Since time evolution is continuous, and the winding number is discrete, it must be a constant of the motion in the classical field theory. We have seen that this "topological conservation law" is equivalent to conservation of magnetic charge. The conservation law survives in quantum theory because the probability of quantum mechanical tunneling between configurations with different winding numbers vanishes in the infinite volume limit.

The above discussion of the topological charge and its connection with magnetic charge is reformulated in much more general language in Section 4.

3.3 *The Monopole Solution*

We have demonstrated the existence of a time-independent monopole solution to the classical field equations. Let us now consider how the solution can be explicitly constructed.

The task of finding an explicit monopole solution is greatly simplified if we make the plausible assumption that the solution is spherically symmetric. In a gauge theory, it is not sensible to demand more than spherical symmetry up to a gauge transformation; we say that the scalar field configuration $\Phi^a(\mathbf{r})$, for example, is spherically symmetric if the effect of a spatial rotation of $\Phi^a(\mathbf{r})$ can be compensated by a gauge transformation. The asymptotic behavior of Φ^a given by Equation 30 and of W_i^a given by Equation 32 is invariant under a simultaneous rotation and global SU(2) gauge transformation. Let us assume that this invariance, and also invariance under the "parity" transformation

$$\mathbf{r} \to -\mathbf{r}, \qquad \Phi^a \to -\Phi^a, \qquad W_i^a \leftrightarrow -W_i^a, \qquad\qquad 33.$$

hold for all $\mathbf{r}$. We thus obtain the ansatz (2, 3)

$$\Phi^a(\mathbf{r}) = v\hat{r}_a H(M_W r)$$

$$W_i^a(\mathbf{r}) = \frac{\varepsilon_{iak}\hat{r}_k}{er}[1 - K(M_W r)]. \qquad\qquad 34.$$

Finite-energy solutions will obey the boundary conditions

$$\begin{aligned} H = 0, \quad K = 1 \quad & (r = 0); \\ H = 1, \quad K = 0 \quad & (r = \infty). \end{aligned} \qquad\qquad 35.$$

H and K satisfying the classical field equations can now be obtained by numerical methods (20, 37). [In fact, an analytical solution is possible in the limit $\lambda = 0$ (38, 39).] Here we merely note a few general features of the solution.

The gauge field W_i^a rapidly approaches its asymptotic value outside a core with radius of order R_c; the heavy gauge fields are excited only inside the core. The size R_c is chosen to minimize the sum of the energy stored in the magnetic field outside the core and the energy due to the scalar field gradient inside the core. In order of magnitude these are

$$\begin{aligned} E_{\text{mag}} &\sim 4\pi g^2 R_c^{-1} \sim (4\pi/e^2)R_c^{-1}, \\ E_{\text{core}} &\sim 4\pi R_c v^2 \sim (4\pi/e^2)M_W^2 R_c, \end{aligned} \qquad\qquad 36.$$

so the core size is determined to be

$$R_c \sim M_W^{-1}. \qquad\qquad 37.$$

The energy of the solution, the monopole mass m in the classical approximation, does not depend sensitively on the scalar self-coupling λ;

one finds

$$m = \frac{4\pi}{e^2} M_W f(\lambda/e^2), \qquad 38.$$

where f is a monotonically increasing function such that (37)

$$f(0) = 1$$
$$f(\infty) = 1.787. \qquad 39.$$

The mass m becomes independent of λ for large λ because the scalar field approaches its asymptotic form outside an inner core with radius $R_H \sim M_H^{-1}$, and the scalar field energy stored in the inner core is of order

$$E_{\text{scalar}} \sim 4\pi\lambda v^4 R_H^3 \sim m(M_W/M_H), \qquad 40.$$

which becomes negligible for large λ.

Comparing Equations 37 and 38, we see that the size of the monopole core is larger by the factor $\alpha^{-1} = (4\pi/e^2)$ than the monopole Compton wavelength. As a result, the quantum corrections to the structure of the monopole are under control, if α is small. Even though the coupling $g = 1/e$ is large, the effects of virtual monopole pairs are small, because the monopole is a complicated coherent excitation that cannot be easily produced as a quantum fluctuation. (See Section 6.)

This situation should be contrasted with the quantum mechanics of a point monopole. Virtual monopole pairs have a drastic effect on the structure of the point monopole, for which g is a genuine strong coupling. In fact, the vacuum-polarization cloud of a point monopole must extend out to distances of order $(\alpha m)^{-1}$, because the magnetic self-energy of a monopole of that size is of order m. Thus, both the nonsingular monopole and the point monopole have a complicated structure in a region with radius of order $(\alpha m)^{-1}$. But for the nonsingular monopole, we have an explicit classical description of this structure, and quantum corrections are small and calculable if α is small. The point monopole, on the other hand, is a genuine strong-coupling problem. We cannot calculate anything.

We have shown how the magnetic monopole arises as a solution to the classical field equations in a simple SU(2) gauge theory. The discussion is generalized in Section 4 and various more complicated examples are cited in Section 5.

It turns out (40) that in many, but not all (41), more complicated examples it is possible to construct a monopole solution that satisfies a suitable generalization of the spherically symmetric ansatz, Equation 34. But nothing further is said here about the construction of explicit solutions.

4. MONOPOLES AND TOPOLOGY

4.1 *Monopoles without Strings*

The Dirac string is a considerable embarrassment in monopole theory. It is disconcerting to find that the vector potential that describes a Dirac monopole has a string singularity along which the magnetic field is formally infinite, even though we can argue that the string is undetectable. One is therefore encouraged to discover that it is possible to eliminate the string (42).

Let us consider the vector potential on a sphere centered at the monopole. (The monopole may be either pointlike or nonsingular; in the latter case we choose the radius of the sphere to be much larger than the core radius.) The trick by which we avoid the string is to divide the sphere into upper and lower hemispheres, and define a vector potential on each. For example, we may choose the nonvanishing component of **A** on each hemisphere to be

$$A_\varphi^U = g(1-\cos\vartheta), \quad \text{upper}\left(0 \le \vartheta \le \frac{\pi}{2}\right),$$

$$A_\varphi^L = -g(1+\cos\vartheta), \quad \text{lower}\left(\frac{\pi}{2} \le \vartheta \le \pi\right), \qquad 41.$$

where A_φ is defined by $\mathbf{A}\cdot d\mathbf{r} = A_\varphi d\varphi$. Both A^U and A^L are nonsingular on their respective hemispheres, and both have the curl

$$\mathbf{B} = g\frac{\hat{r}}{r^2}. \qquad 42.$$

On the region where the hemispheres intersect, the equator ($\vartheta = \pi/2$), we must require that A^U and A^L describe the same physics; therefore, they differ by a gauge transformation. And, indeed

$$A_\varphi^U\left(\vartheta = \frac{\pi}{2}\right) - A_\varphi^L\left(\vartheta = \frac{\pi}{2}\right) = 2g = \frac{1}{ie}(\partial_\varphi \Omega)\Omega^{-1}, \qquad 43.$$

where

$$\Omega(\varphi) = \exp[i2eg\varphi]. \qquad 44.$$

If Ω is not single-valued, then the change in the phase of the wave function of an electron, as the electron is transported around the equator, is ill defined. So we must demand

$$eg = \frac{n}{2}, \qquad 45.$$

where n is an integer. We have thus found an alternative derivation of the Dirac quantization condition, in which the string singularity makes no appearance.

It is easy to see that this quantization condition applies to any vector potential on the sphere, not just one with the special form of Equation 41. In general, if the nonsingular vector potentials A^U and A^L defined on the upper and lower hemispheres differ by a gauge transformation $\Omega(\varphi)$ at the equator, then we may interpret $\Omega(\varphi)$ as an object that detects the total magnetic flux Φ through the sphere. If $\Omega(\varphi = 0) = 1$, then $\Omega(\varphi = 2\pi)$ satisfies

$$\Omega(\varphi = 2\pi) = \exp[ie\oint d\mathbf{x} \cdot (\mathbf{A}^U - \mathbf{A}^L)] = \exp[ie(\Phi^U + \Phi^L)]$$
$$= \exp[ie(4\pi g)], \qquad 46.$$

where the line integral is taken along the equator, and g is the magnetic charge enclosed by the sphere. Single-valuedness of $\Omega(\varphi)$ again implies Equation 45.

The integer n, the magnetic charge of the monopole in Dirac units, is a winding number; it is the number of times $\Omega(\varphi)$ covers the $U(1)_{em}$ gauge group as φ varies from 0 to 2π. So we have discovered a topological basis for the Dirac quantization condition. Magnetic charge is quantized because the winding number must be an integer.

If we now allow the radius r of the sphere to vary, $\Omega(\varphi, r)$ and the winding number n are continuous functions of r as long as A^U and A^L are nonsingular. Since n is required to be an integer, it must be a constant, independent of r. If n is nonzero, we are forced to conclude that the magnetic charge g is contained in an arbitrarily small sphere; the monopole is a point singularity.

It is possible to avoid the singularity only if Ω is allowed to wander through a larger gauge group containing $U(1)_{em}$. This is precisely the option exercised by the nonsingular monopole described in Section 3, which has nonabelian gauge fields excited in its core.

4.2 Topological Classification of Monopoles

It is easy to generalize the above discussion to apply to magnetic monopoles with nonabelian, long-range gauge fields, and thus obtain a topological definition of magnetic charge appropriate for the nonabelian case (21, 43).

Let us consider gauge fields, defined on a sphere, in the Lie algebra of an arbitrary Lie group H. As before, we describe the gauge field configuration by specifying nonsingular gauge potentials A^U and A^L on the upper and lower hemispheres, and a single-valued gauge transformation $\Omega(\varphi)$, which relates A^U and A^L on the equator. The gauge transformation $\Omega(\varphi)$ is a "loop" in the gauge group H, a mapping from the circle into H. We define the magnetic charge enclosed by the sphere to be the winding number of $\Omega(\varphi)$. This is the natural nonabelian generalization of the abelian magnetic charge.

For example, suppose that the gauge group is $H = SO(3)$. It is well known that $SO(3)$ is topologically equivalent to a three-dimensional sphere with antipodal points identified. Therefore, there are closed paths in $SO(3)$, those beginning at one point of the three-sphere and ending at an antipodal point, which cannot be continuously deformed to a point. Such a path is said to have winding number 1. But a path that begins and ends at the same point of the three-sphere can be continuously deformed to a point; it has winding number 0. Thus, the winding number of a loop in $SO(3)$ can have only two possible values, 0 and 1, and the magnetic charge in an $SO(3)$ gauge theory can have only the values 0 and 1. In particular, a magnetic monopole is indistinguishable from an antimonopole.

In general, the closed paths in a Lie group H beginning and ending at the identity element of H fall into topological equivalence classes, called "homotopy" classes (44). Two paths are in the same class if they can be continuously deformed into one another. The classes are endowed with a natural group structure, since the composition of two paths may be defined to be a path that traces the two paths in succession. This group is called $\pi_1(H)$, the "first homotopy group" of H. According to the above remarks, $\pi_1[SO(3)]$ is Z_2, the additive group of the integers defined modulo 2.

The example $H = SO(3)$ exhibits all the essential features of the general case. Every Lie group H has a covering group $\bar{H}$, which is simply connected; that is, such that $\pi_1(\bar{H})$ is trivial. For $H = SO(3)$, the covering group is $\bar{H} = SU(2)$, which is topologically equivalent to the three-sphere itself, without antipodal points identified. The Lie group H is always isomorphic to the quotient group $\bar{H}/K$, where K is a subgroup of the center of $\bar{H}$. The center is a discrete subgroup of $\bar{H}$ that commutes with all elements of $\bar{H}$. For $\bar{H} = SU(2)$, the center is Z_2, consisting of the elements 1 and -1, and $SO(3)$ is isomorphic to $SU(2)/Z_2$. In general, we may think of the group H as the group $\bar{H}$, but with elements differing by multiplication by an element of K identified as the same element.

All paths in H that begin and end at the identity element of H correspond to paths in $\bar{H}$ that begin at the identity and end at an element of K. And the topological class of a path in H can be labeled by the end point of the corresponding path in $\bar{H}$, just as the class of a path in $SO(3)$ is determined by whether it ends at its starting point on the three-sphere or at the antipodal point. So we finally have

$$\pi_1(H) = \pi_1(\bar{H}/K) = K. \qquad 47.$$

For $H = U(1)$, which is covered by $\bar{H} = \mathbb{R}$, the additive group of the real numbers, K is Z, the integers. For $H = SO(3)$, K is Z_2, and for any simple Lie group H, K is Z_N, for some integer N.

Our topological definition of the nonabelian magnetic charge is sensible. As long as the gauge fields are nonsingular and Ω is an element of H, the winding number must be a constant, independent of the radius of the sphere. So the magnetic charge is not carried by the long-range field of a monopole; it resides on a point singularity (Dirac monopole) or a core in which gauge fields other than the H gauge fields are excited (nonsingular monopole). And this magnetic charge is obviously conserved. It is a discrete quantity. But time evolution is continuous, and a discrete quantity can be continuous only by being constant.

While other gauge-invariant definitions of magnetic charge are possible (33), only the topological definition, which requires a magnetic monopole to have a point singularity or a core, can guarantee the stability of a monopole. If we assign "magnetic charge" to an H gauge field configuration that is nonsingular everywhere in space, nothing can prevent this "magnetic charge" from propagating to spatial infinity as nonabelian radiation (21, 45).

So far we have only considered magnetic monopole configurations in classical gauge field theory. Eventually, we must worry about quantum mechanical effects on the magnetic field. There is really something to worry about, because we believe that nonabelian gauge field theories are confining and have no massless excitations. Therefore the magnetic field cannot survive at arbitrarily large distances, it must be screened at distances larger than the confinement distance scale (21, 31). The mechanism of magnetic screening is briefly discussed in Section 6.4.

Fortunately, since our definition of magnetic charge is topological, it can be applied in the quantum theory, and conservation of magnetic charge is still guaranteed. The gluon fluctuations about the classical long-range magnetic field, which cause the magnetic screening, cannot change the winding number of the classical field.

4.3 *Magnetic Charge of a Topological Soliton*

The object of this section is to generalize the discussion of topological solitons in Section 3 to an arbitrary gauge group, and to demonstrate the general connection between the topology of a soliton and its magnetic charge.

We consider an arbitrary gauge field theory, with gauge group G, which undergoes spontaneous symmetry breakdown to a subgroup H. Acting as an order parameter for this symmetry-breaking pattern is a multiplet of scalar fields Φ, transforming as some (in general reducible) representation of

G. The classical potential $U(\Phi)$ has many degenerate minima, and we identify one arbitrarily chosen minimum as Φ_0. H is the "stability group" of Φ_0, the subgroup of G that leaves Φ_0 invariant.

We find it convenient in this section to assume that G is simply connected, $\pi_1(G) = 0$. This assumption entails no loss of generality, because we may always consider G to be the covering group of a specified Lie group.

We wish to construct finite energy solutions to the classical field equations of this gauge field theory. Therefore, we restrict our attention to field configurations such that Φ approaches a minimum of $U(\Phi)$ at spatial infinity. Barring "accidental" degeneracy, degenerate minima of U not required by G symmetry, the manifold of minima of U is equivalent to the coset space G/H,

$$G/H = \{\Phi : \Phi = \Omega\Phi_0, \Omega \in G\}. \qquad 48.$$

Associated with each finite-energy field configuration is a mapping from the two-dimensional sphere S^2 at spatial infinity into the vacuum manifold G/H. As noted in Section 3.2, a field configuration is a topological soliton if this mapping cannot be continuously deformed to the trivial constant mapping that takes all points on S^2 to Φ_0.

By multiplying by an appropriate constant element of G, we may turn any mapping from S^2 into G/H into a mapping that takes an arbitrarily chosen reference point, the north pole, say, to Φ_0. [This procedure suffers from an ambiguity if H is not connected (19). A consequence of this ambiguity is explained in Section 5.4.] Mappings from S^2 into G/H that take the north pole to Φ_0 fall into topological equivalence classes, homotopy classes, such that mappings in the same class can be continuously deformed into one another. These classes are endowed with a natural group structure, since there is a natural way of composing two mappings that both take the north pole to Φ_0 (see Figure 4). This group is $\pi_2(G/H)$, the "second homotopy group" of G/H (44).

The group $\pi_2(G/H)$ is discrete; its elements are the possible "topological charges" of finite-energy field configurations. Since time evolution is continuous, the discrete topological charge must be a constant of the motion. The classical field theory has a "topological conservation law."

We found that the topological charge of the soliton constructed in Section 3 could be identified with its magnetic charge. We can now show that this identification applies in general (19).

Mappings from the sphere S^2 into the coset space G/H are not very easy to visualize. Fortunately, we can, by a trick, reduce the topological classification of these mappings to the topological classification of closed paths in H. That is, we can reduce the calculation of $\pi_2(G/H)$ to the calculation of $\pi_1(H)$, and we already know how to calculate $\pi_1(H)$.

The trick is to cut the sphere into two hemispheres, along the equator. Each point (ϑ, φ) on the sphere is mapped to some $\Phi(\vartheta, \varphi) \in G/H$. On the upper and lower hemispheres we can find smooth gauge transformations Ω_U and Ω_L that take Φ to Φ_0:

$$\Omega_U(\vartheta, \varphi)\Phi(\vartheta, \varphi) = \Phi_0, \quad \text{upper}\left(0 \leq \vartheta \leq \frac{\pi}{2}\right),$$

$$\Omega_L(\vartheta, \varphi)\Phi(\vartheta, \varphi) = \Phi_0, \quad \text{lower}\left(\frac{\pi}{2} \leq \vartheta \leq \pi\right). \qquad 49.$$

On the region where the hemispheres intersect, the equator ($\vartheta = \pi/2$), the gauge transformation $\Omega_U \Omega_L^{-1}$ is defined. It leaves Φ_0 invariant, and is therefore an element of H. So

$$\Omega_U\left(\vartheta = \frac{\pi}{2}, \varphi\right)\Omega_L^{-1}\left(\vartheta = \frac{\pi}{2}, \varphi\right) \equiv \Omega(\varphi) \in H \qquad 50.$$

defines a closed path in H. We have thus found a natural way of associating with each mapping from S^2 into G/H a closed path in H.

This association actually defines a group homomorphism from $\pi_2(G/H)$ to $\pi_1(H)$. If we choose the arbitrary reference point that is mapped to Φ_0 to be a point on the equator, instead of the north pole, then it is obvious that the composition of mappings from S^2 into G/H corresponds to the composition of loops in H, and the group structure is preserved.

The kernel of this homomorphism is trivial because, if $\Omega(\varphi)$ has winding number zero, then it can be continuously deformed to the trivial loop $\Omega(\varphi) = 1$. Therefore, there is a smooth gauge transformation in G, defined on the whole sphere, which takes $\Phi(\vartheta, \varphi)$ to Φ_0. Furthermore, it is known that $\pi_2(G) = 0$ for any compact Lie group G (46). Thus, this gauge transformation can be continuously deformed to a trivial gauge transformation, and the mapping $\Phi(\vartheta, \varphi)$ can be continuously deformed to Φ_0. Therefore, the

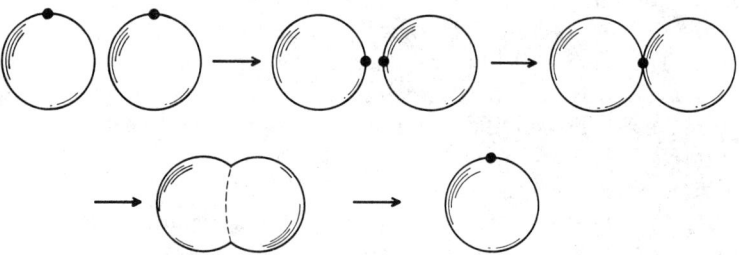

Figure 4 Composition of two mappings from S^2 into G/H that take the north pole *(black dot)* to Φ_0.

homomorphism takes only the identity element of $\pi_2(G/H)$ to the identity element of $\pi_1(H)$.

Moreover, if G is simply connected, we can show that this homomorphism is onto; every element of $\pi_1(H)$ is the image of some element of $\pi_2(G/H)$. Given any loop $\Omega(\varphi)$ in H we can find smooth gauge transformations Ω_U and Ω_L in G, defined on the upper and lower hemispheres, such that

$$\Omega_U\left(\vartheta = \frac{\pi}{2}, \varphi\right) = \Omega(\varphi),$$

$$\Omega_L\left(\vartheta = \frac{\pi}{2}, \varphi\right) = 1, \qquad 51.$$

because we may choose $\Omega_U(\vartheta, \varphi)$ to be the continuous deformation of the loop $\Omega_U(\vartheta = \pi/2, \varphi)$ to the point $\Omega_U(\vartheta = 0)$, which is guaranteed to exist if G is simply connected. Then

$$\Phi(\vartheta, \varphi) = \begin{cases} \Omega_U(\vartheta, \varphi)\Phi_0, & 0 \le \vartheta \le \frac{\pi}{2}, \\ \Phi_0 & \frac{\pi}{2} \le \vartheta \le \pi, \end{cases} \qquad 52.$$

is a smooth mapping from S^2 into G/H associated with the loop $\Omega(\varphi)$.

We have now established the group isomorphism

$$\pi_2(G/H) = \pi_1(H), \qquad 53.$$

which holds if G is simply connected. [It is easy to see, by slightly modifying the above argument, that the general result is $\pi_2(G/H) = \pi_1(H)/\pi_1(G)$.] As promised, we have found that the topological classification of mappings from S^2 into G/H is equivalent to the topological classification of loops in H.

Since we have already seen that the elements of the group $\pi_1(H)$ specify the possible magnetic charges of a configuration with long-range H gauge fields, we suspect, in view of Equation 53, that the topological charge of a finite-energy field configuration coincides with its magnetic charge. To verify this conjecture, we must consider the long-range gauge field of the soliton.

As we saw in Section 3, a finite-energy field configuration must obey

$$D_i\Phi = (\partial_i - ieA_i^a T^a)\Phi = 0 \qquad 54.$$

on the sphere at spatial infinity, where the T^as are the generators of G in the representation according to which Φ transforms. In the gauge constructed above, in which $\Phi = \Phi_0$ is a constant on the sphere, the only gauge fields that can be excited at large distances are the H gauge fields, those associated with the generators of G that annihilate Φ_0. The gauge transformation (Equation 49) is nonsingular except on the equator, so the gauge fields A^U and A^L defined on each hemisphere are nonsingular in this gauge. But on the equator they differ by the gauge transformation $\Omega(\varphi)$. The winding number of $\Omega(\varphi)$, which we have now seen is the topological charge of the soliton, is also the magnetic charge defined in Section 4.2. So topological charge equals magnetic charge.

We have now verified the claim in Section 3.1, that any unified gauge theory in which $U(1)_{em}$ is embedded in a spontaneously broken semisimple gauge group necessarily contains magnetic monopoles as topological solitons. There are topologically stable finite-energy solutions to the classical field equations associated with each element of $\pi_2(G/H)$. We have now learned that $\pi_2(G/H) = \pi_1(H)$, if G is chosen to be simply connected, and that $\pi_1(H)$ contains the integers if H has a $U(1)_{em}$ factor. Finally, we have found that the integer labeling an element of $\pi_2(G/H)$ is precisely the magnetic charge in Dirac units. [The Dirac magnetic charge is that corresponding to the minimal $U(1)_{em}$ charge occurring in a representation of the simply connected gauge group G.]

We discuss further applications for our topological formalism when we analyze the examples of Section 5.

4.4 The Kaluza-Klein Monopole

Kaluza-Klein theories (47), which unify Einstein's theory of gravitation with other gauge interactions, also contain topological solitons that can be identified as magnetic monopoles. The connection between the topological charges and magnetic charges of these objects are described here. This connection is closely analogous to, but not exactly the same as, that discussed in Section 4.3.

The basic hypothesis of Kaluza-Klein theory is that space-time is not really 4-dimensional, but $(4+n)$-dimensional. The $(4+n)$-dimensional space-time is endowed with a metric satisfying a $(4+n)$-dimensional generalization of Einstein's equations, but n dimensions have become spontaneously compactified to a manifold N with radii of order the Planck length. At low energies, far below the Planck mass, the effects of the microscopic compact dimensions cannot be perceived directly, but a remnant of the underlying $(4+n)$-dimensional theory may survive. The metric on the compact n-dimensional manifold N is typically invariant under some group H of isometries, and the massless fields of the theory include, in addition to the four-dimensional metric, spin-one gauge fields associated with H. These gauge fields are components of the $(4+n)$-dimensional metric that have managed to avoid acquiring large masses upon compactification. Low-energy physics is described by an effective four-dimensional field theory that is an H-gauge theory coupled to gravity.

The classical vacuum solution of the Kaluza-Klein theory is assumed to be $M^4 \times N$, the direct product of four-dimensional Minkowski space and the compact manifold N. Any classical field configuration that approaches the vacuum solution at spatial infinity thus defines an N "bundle" (44) over the sphere at spatial infinity S^2. This bundle has the *local* structure of a direct product $S^2 \times N$; that is, the manifold N sits on top of every point of S^2. But it need not be a direct product *globally*. If the N bundle over S^2 cannot be continuously deformed to the global direct product $S^2 \times N$, then the field configuration cannot be continuously deformed to the vacuum solution; it is a topological soliton.

To perform the topological classification of N bundles over S^2, we cut the sphere S^2 into two hemispheres, along the equator. The N bundles over the two hemispheres D^U and D^L are then easily deformed to direct product bundles, $D^U \times N$ and $D^L \times N$, by performing coordinate transformations on each hemisphere. Along the equator, these two coordinate transformations must differ by a transformation that leaves the geometry of the manifold N invariant; that is, an isometry of N. Thus we can associate with every N bundle over S^2 a loop in the isometry group H. The N bundle over S^2 is topologically nontrivial if and only if the loop in H has a nontrivial winding number. We conclude, as before, that all topological solitons have H-magnetic charges.

The Kaluza-Klein monopole solution has been explicitly constructed in the simplest Kaluza-Klein theory, the five-dimensional theory in which N is a circle and H is U(1) (48, 49). It has some interesting properties. In particular, the four-dimensional constant time slices of both the monopole and antimonopole solution have handles; therefore, a monopole-antimonopole pair has a different toplogy from the vacuum, and cannot annihilate classically.

Generally, one expects a Kaluza-Klein monopole to have a mass m of order $(1/e)M_{\text{Planck}}$. In the five-dimensional theory it has been found that

$$m = \tfrac{1}{4}\alpha^{-1/2} M_{\text{Planck}} \sim 5 \times 10^{19} \text{ GeV}. \qquad 55.$$

As we see in Section 8.2, this is an interesting mass from an astrophysical viewpoint.

4.5 *Monopoles and Global Gauge Transformations*

It was recently discovered (50, 51) that a global gauge transformation cannot be defined in the vicinity of a magnetic monopole with a nonabelian long-range magnetic field, unless the gauge transformation acts trivially on the long-range field. Implications of this result are considered in Section 6. Here we sketch the proof, which is simple and involves topological concepts that we have already encountered.

A classical field f on a sphere surrounding a magnetic monopole is defined by specifying smooth functions f_U and f_L on the upper and lower hemispheres, and the gauge transformation Ω, which relates f_U and f_L on the equator, where the hemispheres intersect:

$$f_U\left(\vartheta = \frac{\pi}{2}, \varphi\right) = \Omega(\varphi) f_L\left(\vartheta = \frac{\pi}{2}, \varphi\right), \qquad 56.$$

$\Omega(\varphi)$ is a loop in the gauge group H with a nontrivial winding number; the winding number is the magnetic charge of the monopole.

A local gauge transformation of f on the sphere consists of gauge transformations Ω_U and Ω_L on the two hemispheres that preserve the relations (Equation 56):

$$f_U(\vartheta, \varphi) \to \Omega_U(\vartheta, \varphi) f_U(\vartheta, \varphi), \qquad \text{upper}\left(0 \leq \vartheta \leq \frac{\pi}{2}\right),$$

$$f_L(\vartheta, \varphi) \to \Omega_L(\vartheta, \varphi) f_L(\vartheta, \varphi), \qquad \text{lower}\left(\frac{\pi}{2} \leq \vartheta \leq \pi\right),$$

$$\Omega_U\left(\vartheta = \frac{\pi}{2}, \varphi\right) = \Omega(\varphi) \Omega_L\left(\vartheta = \frac{\pi}{2}, \varphi\right) \Omega^{-1}(\varphi). \qquad 57.$$

To define an infinitesimal global gauge transformation on the sphere surrounding the monopole, we must specify a set of generators $\{T^a\}$ of the gauge group H at each point of the sphere. If the transformation is globally defined, the commutation relations satisfied by the generators must be independent of the position on the sphere, but we still have the freedom to perform a local redefinition of the generators of the form

$$T^a(\vartheta, \varphi) = \Sigma(\vartheta, \varphi) T^a \Sigma^{-1}(\vartheta, \varphi), \qquad 58.$$

where $\Sigma \in H$. The redefinition of the generators determined by Σ is called an inner automorphism of the Lie algebra **H** of the group H. The group aut **H** of inner automorphisms (which is the connected component of the group Aut **H** of all automorphisms preserving the Lie algebra **H**) is evidently isomorphic to H/K where K is the center of H, since the elements of K, and only the elements of K, define trivial automorphisms (52).

A global gauge transformation of f on the sphere must be compatible with Equation 56. Therefore, the generators have the form

$$T_U^a(\vartheta, \varphi) = \Sigma_U(\vartheta, \varphi) T^a \Sigma_U^{-1}(\vartheta, \varphi), \qquad \text{upper}\left(0 \leq \vartheta \leq \frac{\pi}{2}\right),$$

$$T_L^a(\vartheta, \varphi) = \Sigma_L(\vartheta, \varphi) T^a \Sigma_L^{-1}(\vartheta, \varphi), \qquad \text{lower}\left(\frac{\pi}{2} \leq \vartheta \leq \pi\right), \qquad 59.$$

where

$$T_U^a\left(\vartheta = \frac{\pi}{2}, \varphi\right) = \Omega(\varphi) T_L^a\left(\vartheta = \frac{\pi}{2}, \varphi\right) \Omega^{-1}(\varphi),$$

or

$$T^a = \Sigma_U^{-1}\left(\frac{\pi}{2}, \varphi\right) \Omega(\varphi) \Sigma_L\left(\frac{\pi}{2}, \varphi\right) T^a \Sigma_L^{-1}\left(\frac{\pi}{2}, \varphi\right) \Omega^{-1}(\varphi) \Sigma_U\left(\frac{\pi}{2}, \varphi\right). \qquad 60.$$

We see that $\Sigma_U^{-1}(\pi/2, \varphi)\Omega(\varphi)\Sigma_L(\pi/2, \varphi)$ defines a trivial automorphism, and is therefore an element of the center of H. If H is semisimple, its center is discrete, and we have

$$\Omega(\varphi) = \Sigma_U\left(\vartheta = \frac{\pi}{2}, \varphi\right) \Omega_0 \Sigma_L^{-1}\left(\vartheta = \frac{\pi}{2}, \varphi\right) \qquad 61.$$

where Ω_0 is a constant element of the center. If we now allow ϑ, the argument of $\Sigma_U(\Sigma_L)$ to vary smoothly from $\vartheta = \pi/2$ to $\vartheta = 0 (\vartheta = \pi)$ in Equation 61, we find that the loop $\Omega(\varphi)$ in H can be continuously deformed to a point, and therefore has winding number zero.

We are forced to conclude, if H is semisimple, that a global H transformation can be performed on a sphere only if the sphere encloses no H magnetic charge. In the vicinity of a nonabelian monopole, a global nonabelian gauge transformation cannot be implemented!

There is obviously no topological obstacle to defining globally the generators that commute with Ω. So global gauge transformations of the U(1) magnetic monopole can be performed. In general, we can define any global gauge transformation that acts trivially on the long-range gauge field, and hence leaves Ω intact.

5. EXAMPLES

5.1 *A Symmetry-Breaking Hierarchy*

In order to illustrate the topological principles developed in Section 4, we consider various examples of model gauge theories containing magnetic monopoles. In all these examples, it is possible (40) to construct explicit monopole solutions by using suitable generalizations of the spherically symmetric ansatz of Section 3.3. But here we note only the general properties of the monopoles, and do not exhibit explicit solutions.

Our first example illustrates the importance in monopole theory of the global structure of the unbroken gauge group. Consider a model with gauge group $G = SU(3)$ and a scalar field Φ transforming as the adjoint (octet) representation of G: Φ can be written as a hermitian traceless 3×3 matrix, which, under a gauge transformation $\Omega(x)$, transforms according to

$$\Phi(x) \to \Omega(x)\Phi(x)\Omega^{-1}(x). \qquad 62.$$

Suppose that Φ acquires the expectation value

$$\langle \Phi \rangle = \Phi_0 = (v)\,\text{diag}(\tfrac{1}{2},\tfrac{1}{2},-1),$$

where v is the mass scale of the symmetry breakdown, and the diag$(\tfrac{1}{2},\tfrac{1}{2},-1)$ notation denotes a diagonal matrix with the indicated eigenvalues.

The unbroken subgroup H of G, the stability group of Φ_0, is locally isomorphic to SU(2) × U(1). "Locally isomorphic" means that H has the same Lie algebra of infinitesimal generators as SU(2) × U(1). The generators of H are the SU(3) generators that commute with Φ_0. These are the SU(2) generators that mix the two degenerate eigenstates of Φ_0, and also the U(1) generator

$$Q = \text{diag}(\tfrac{1}{2},\tfrac{1}{2},-1), \qquad 63.$$

which is proportional to Φ_0, and obviously commutes with it. [The eigenvalues of Q are the U(1) electric charges of the members of the SU(3) triplet, in units of e.]

To perform the topological classification of monopole solutions in this model, we need to determine $\pi_2(G/H) = \pi_1(H)$. So it is not sufficient to know that H has the local structure of the direct product SU(2) × U(1); we must know its global structure. For this purpose, we check to see whether the U(1) subgroup of G generated by Q has any elements in common with the unbroken SU(2) subgroup, other than the identity. And, indeed

$$\exp(i2\pi Q) = \text{diag}(-1,-1,1) \qquad 64.$$

is the nontrivial element of the center Z_2 of SU(2). We conclude that

$$H = [\text{SU}(2) \times \text{U}(1)]/Z_2, \qquad 65.$$

where "=" denotes a global isomorphism; there are two elements of SU(2) × U(1) corresponding to each element of H.

The topologically nontrivial loops in H consist of loops winding around the U(1) subgroup of H, and also of loops traveling through the U(1) subgroup from the identity to the element in Equation 64, and returning to the identity through the SU(2) subgroup of H. If we had failed to recognize that H is not globally the direct product SU(2) × U(1), we would have missed the latter set of nontrivial loops, and thus missed half of the monopole solutions in this model.

The monopole with minimal U(1) magnetic charge defines a loop that winds only half-way around U(1); it necessarily also has a Z_2 nonabelian magnetic charge. We anticipated the existence of this solution in our discussion of the generalized Dirac quantization condition in Section 2.2. Equation 64 implies that objects with trivial SU(2) "duality" have integer U(1) charge, although objects with nontrivial duality can have half-integer charge. According to the discussion of Section 2.2, a monopole can exist with the Dirac U(1) magnetic charge $g_D = 1/2e$, provided that it also carries an SU(2) magnetic charge. Alternatively, the charge carried by this monopole can be regarded, in an appropriate gauge, as the U(1)' charge generated by

$$Q' = T^3 + Q = \mathrm{diag}(1, 0, -1), \tag{66}$$

where

$$T^3 = \mathrm{diag}(\tfrac{1}{2}, -\tfrac{1}{2}, 0), \tag{67}$$

is an SU(2) generator. The monopole with minimal U(1) magnetic charge defines a closed loop in U(1)'.

In realistic unified gauge theories, spontaneous symmetry breakdown typically occurs at two or more mass scales differing by many orders of magnitude. To illustrate the effect of such a symmetry-breaking hierarchy on magnetic monopoles, let us imagine that the $G = $ SU(3) gauge symmetry of our model breaks down in two stages, first to $H_1 = $ [SU(2) × U(1)]/Z_2 at mass scale v_1, then to $H_2 = $ U(1) at mass scale $v_2 \ll v_1$,

$$G = \mathrm{SU}(3) \xrightarrow{v_1} H_1 = [\mathrm{SU}(2) \times \mathrm{U}(1)]/Z_2 \xrightarrow{v_2} H_2 = \mathrm{U}(1). \tag{68}$$

The effect of the second stage of symmetry breakdown on the monopoles generated by the first stage depends on which U(1) subgroup of H_1 remains unbroken at the second stage (53).

First, suppose that H_2 is the U(1) subgroup generated by

$$Q_2 = Q' = \mathrm{diag}(1, 0, -1). \tag{69}$$

Since this is the same charge as that carried by the monopole associated with the $G \to H_1$ breakdown at mass scale v_1, the breakdown at the much lower mass scale v_2 has no significant effect on the monopole.

But if H_2 is the U(1) subgroup generated by

$$Q_2 = Q = \mathrm{diag}(\tfrac{1}{2}, \tfrac{1}{2}, -1), \tag{70}$$

the monopole is significantly affected, for the only monopole solutions now have twice the U(1) magnetic charge allowed by the $G \to H_1$ breakdown.

What would happen to the minimal G/H_1 monopole if we varied the parameters of the model so as smoothly to turn on the second symmetry-breaking scale v_2? This question is not entirely academic, because the H_1 symmetry is expected to be restored at sufficiently high temperature,

$T \gg v_2$. As the temperature is lowered, a phase transition occurs at $T \sim v_2$ in which H_1 becomes spontaneously broken. We might be interested in what happens to the minimal G/H_1 monopoles during this phase transition, especially since a phase transition like this one may have occurred in the very early universe.

A reasonable guess is that pairs of minimal G/H_1 monopoles or monopole-antimonopole pairs become connected by magnetic flux tubes, and form composite objects with either twice the minimal U(1) magnetic charge or zero magnetic charge. To verify this guess, we need a mathematical criterion to determine when such flux tubes occur.

A magnetic flux tube in three spatial dimensions, a static solution to the field equation with finite energy per unit length, may be regarded as a topological soliton in two spatial dimensions with finite energy. The topological classification of these solitons is very similar to the classification of monopoles in Section 4.3; so similar that we need only sketch the analysis.

In a gauge theory with gauge group G and unbroken group H, the finite-energy two-dimensional field configurations define mappings from the circle S^1 at (two-dimensional) spatial infinity into the vacuum manifold G/H, and are classified by the first homotopy group $\pi_1(G/H)$. To facilitate the calculation of $\pi_1(G/H)$, we cut the circle open at $\varphi = 0$, and find a gauge transformation $\Omega(\varphi) \in G$ that rotates the order parameter $\Phi(\varphi)$ to the standard values Φ_0 for all φ. The discontinuity of this gauge transformation at $\varphi = 0$ is an element of H,

$$\Omega(\varphi = 0)\Omega^{-1}(\varphi = 2\pi) = \Omega_0 \in H. \qquad 71.$$

We thus obtain a group homomorphism from $\pi_1(G/H)$ into a group called $\pi_0(H)$. The elements of $\pi_0(H)$ are equivalence classes of elements of H, defined such that $\Omega_1, \Omega_2 \in H$ are in the same class if there is a continuous path in H from Ω_1 to Ω_2. Group multiplication in H defines the group structure in $\pi_0(H)$.

It is easy to see that the homomorphism from $\pi_1(G/H)$ into $\pi_0(H)$ has a trivial kernel, if G is simply connected, and is onto, if G is connected. So we have the isomorphism

$$\pi_1(G/H) = \pi_0(H), \qquad 72.$$

which holds if G is connected and simply connected.

To apply this result to the symmetry-breaking pattern (Equation 68), with the H_2 generator Q_2 given by Equation 70, we note that the U(1) factor of H_1 is not affected by the second stage of symmetry breakdown, so that the flux tubes are classified by

$$\pi_1[\text{SU}(2)/Z_2] = \pi_0(Z_2) = Z_2. \qquad 73.$$

As we expected, there are Z_2 flux tubes, to which the nonabelian SU(2) magnetic flux becomes confined, generated by the spontaneous breakdown of the SU(2) gauge symmetry. The thickness and energy per unit length of the flux tubes are determined by the lower symmetry-breaking scale v_2; the thickness is of order $(ev_2)^{-1}$, and the energy per unit length is of order v_2^2 (54).

The flux tubes link each G/H_1 monopole with minimal H_1 magnetic charge to either another monopole or an antimonopole, since the monopole and antimonopole carry the same Z_2 charge. The bound pairs of monopoles have the minimal H_2 magnetic charge allowed by the Dirac quantization condition.

Finally, suppose that the unbroken U(1) group H_2 is generated by

$$Q_2 = T_3 = \mathrm{diag}(\tfrac{1}{2}, -\tfrac{1}{2}, 0). \qquad 74.$$

In this case H_2 is contained in $\mathrm{SU}(2) \subset H_1$ and the symmetry breakdown $H_2 \to H_1$ can be represented by

$$\begin{array}{ccc} H_1 = \mathrm{SU}(2) & \times & \mathrm{U}(1) \\ \downarrow & & \downarrow \\ H_2 = \mathrm{U}(1) & & 1 \end{array} \qquad 75.$$

The flux tubes associated with the breakdown of H_1 are classified by

$$\pi_1[\mathrm{U}(1)] = Z. \qquad 76.$$

These are Z flux tubes to which the U(1) magnetic flux becomes confined, and therefore no heavy monopoles with mass of order v_1/e can survive when v_2 turns on; all heavy monopoles become bound to antimonopoles by the flux tubes. Since $\pi_2(G/H_2) = Z$, there must still be stable, but light (mass of order v_2/e), monopoles associated with the symmetry breakdown $H_1 \to H_2$.

We see that magnetic monopoles generated at a large symmetry-breaking mass scale may be affected by a small symmetry-breaking mass scale in various ways. The monopoles may survive intact, may become bound by flux tubes into monopole-antimonopole pairs, or may become bound into both monopole-antimonopole pairs and clusters of n monopoles. And, of course, new monopoles might also be generated at the smaller mass scale.

5.2 *A Z_2 Monopole*

We encountered above a monopole carrying both a U(1) magnetic charge and a nonabelian magnetic charge. Of course, it is also possible for a monopole to carry only a nonabelian charge.

For example, consider a model with gauge group $G = \mathrm{SU}(3)$ and a scalar field Φ transforming as the symmetric tensor representation of G. Φ can be written as a symmetric 3 × 3 matrix, which, under a gauge transformation $\Omega(x)$, transforms according to

$$\Phi(x) \to \Omega(x)\Phi(x)\Omega^T(x). \qquad 77.$$

If Φ acquires the expectation value

$$\langle\Phi\rangle = \Phi_0 = v\mathbb{1}, \qquad 78.$$

then G is spontaneously broken to $H = SO(3)$. The monopoles of this model are classified by

$$\pi_2(G/H) = \pi_1[SO(3)] = Z_2. \qquad 79.$$

They are Z_2 monopoles carrying SO(3) magnetic charges. The monopole and antimonopole are indistinguishable.

It is interesting to examine the fate of these monopoles if there is a symmetry-breaking hierarchy of the form (55)

$$G = SU(3) \xrightarrow{v_1} H_1 = SO(3) \xrightarrow{v_2} H_2 = U(1), \qquad 80.$$

where $H_2 = U(1) \subset SO(3)$ is generated by

$$Q = \mathrm{diag}(\tfrac{1}{2}, -\tfrac{1}{2}, 0). \qquad 81.$$

There will, of course, be $\pi_2(H_1/H_2)$ monopoles generated by the second stage of symmetry breakdown. These are light monopoles, with core radius of order $(ev_2)^{-1}$ and mass of order v_2/e, which define topologically nontrivial loops in H_2 that can be contracted to a point in H_1.

But the light monopoles are not all the monopoles of this model; $\pi_2(G/H_2)$ is larger than $\pi_2(H_1/H_2)$, because there are topologically nontrivial loops in H_2 that cannot be contracted to a point in H_1, but are contractible in G. Thus, there are monopoles with half the magnetic charge of the minimal $\pi_1(H_1/H_2)$ monopole that are generated by the first stage of symmetry breakdown. These are heavy monopoles with a core radius of order $(ev_1)^{-1}$ and a mass of order v_1/e. They are just the Z_2 monopoles, which have been converted into Z monopoles with the Dirac magnetic charge by the physics of the second stage of symmetry breakdown. If we turn on v_2 smoothly, the Z_2 monopole, which is equivalent to its antiparticle, must choose the sign of its U(1) magnetic charge at random (55).

The heavy monopole has two cores, and most of its mass resides on its tiny inner core. But if two heavy monopoles are brought together, their inner cores can annihilate, and only the outer cores need survive. So the doubly charged light monopole can be regarded as a very tightly bound composite state of two, singly charged, heavy monopoles.

5.3 The SU(5) and SO(10) Models

The monopoles of the realistic grand unified models based on the gauge groups $G = $ SU(5) and $G = $ SO(10) have many features in common with the simpler examples considered above.

The SU(5) model (4) is the simplest gauge theory uniting the SU(3)$_c$ gauge group of the strong interactions with the [SU(2) × U(1)]$_{ew}$ gauge group of the electroweak interaction. This model undergoes symmetry breakdown at two different mass scales,

$$G = \text{SU}(5) \xrightarrow{v_1} H_1 = \{\text{SU}(3)_c \times [\text{SU}(2) \times \text{U}(1)]_{ew}\}/Z_6$$
$$\xrightarrow{v_2} H_2 = [\text{SU}(3)_c \times \text{U}(1)_{em}]/Z_3. \qquad 82.$$

Here $v_2 \sim 250$ GeV is the mass scale of the electroweak symmetry breakdown, and $v_1 \sim 10^{15}$ GeV is the mass scale of unification.

The order parameter for the symmetry breakdown at mass scale v_1 is a scalar field Φ transforming as the adjoint representation of G, which acquires the expectation value

$$\langle \Phi \rangle = \Phi_0 = v_1 \, \text{diag}(\tfrac{1}{3}, \tfrac{1}{3}, \tfrac{1}{3}, -\tfrac{1}{2}, -\tfrac{1}{2}). \qquad 83.$$

The stability group H of G is locally isomorphic to SU(3) × SU(2) × U(1), where SU(3) acts on the three degenerate eigenvectors of Φ_0/v_1 with eigenvalue $\tfrac{1}{3}$, and SU(2) acts on the two degenerate eigenvectors with eigenvalue $-\tfrac{1}{2}$. The unbroken U(1) is generated by

$$Q = \text{diag}(\tfrac{1}{3}, \tfrac{1}{3}, \tfrac{1}{3}, -\tfrac{1}{2}, -\tfrac{1}{2}), \qquad 84.$$

and, since

$$\exp(i2\pi Q) = \text{diag}[\exp(i2\pi/3), \exp(i2\pi/3), \exp(i2\pi/3), -1, -1]. \qquad 85.$$

we see that this U(1) contains the center of SU(3) × SU(2), so that the unbroken group is actually $H_1 = [\text{SU}(3) \times \text{SU}(2) \times \text{U}(1)]/Z_6$.

Equation 85 ensures that any object with trivial SU(3) triality and SU(2) duality has integer U(1) charge, in units of e. Thus, there exists a magnetic monopole in this model with the Dirac U(1) magnetic charge $g_D = 1/2e$, which also carries a Z_3 color magnetic charge and a Z_2 SU(2) magnetic charge. In an appropriate gauge, we may regard the magnetic charge carried by the monopole to be a U(1)' charge generated by

$$Q' = Q + Q_{\text{weak}} + Q_{\text{color}} = \text{diag}(0, 0, 1, 0, -1), \qquad 86.$$

where

$$Q_{\text{weak}} = \text{diag}(0, 0, 0, \tfrac{1}{2}, -\tfrac{1}{2}), \qquad 87.$$

is an SU(2) generator and

$$Q_{\text{color}} = \text{diag}(-\tfrac{1}{3}, -\tfrac{1}{3}, \tfrac{2}{3}, 0, 0) \qquad 88.$$

is an SU(3) generator. Since Q' has integer eigenvalues, a monopole with U(1)$'$ magnetic charge $g = g_D = 1/2e$ is consistent with the Dirac quantization condition.

The electroweak symmetry breakdown at mass scale v_2 leaves unbroken the U(1)$_{\text{em}}$ subgroup of [SU(2) × U(1)]$_{\text{ew}}$ generated by

$$Q_{\text{em}} = Q + Q_{\text{weak}} = \text{diag}(\tfrac{1}{3}, \tfrac{1}{3}, \tfrac{1}{3}, 0, -1). \qquad 89.$$

Since $\exp(i2\pi Q_{\text{em}})$ is a nontrivial element of the center of SU(3)$_c$, the unbroken subgroup is $H_2 = [\text{SU}(3) \times \text{U}(1)]/Z_3$, and the monopole with minimal U(1)$_{\text{em}}$ magnetic charge still carries the U(1)$'$ charge generated by Q'.

The structure of the SU(5) monopole is not much affected by the electroweak symmetry breakdown, because the magnetic charge carried by the monopole is not changed by this breakdown. There are no W and Z fields excited inside an electroweak core with a radius of order $(ev_2)^{-1}$ $\sim M_W^{-1}$, at least in the classical approximation. The true core of the monopole has a radius of order $(ev_1)^{-1} \sim 10^{-28}$ cm and the mass of the monopole is of order $(v_1/e) \sim 10^{16}$ GeV.

That the electroweak SU(2) × U(1) gauge symmetry is restored within a distance M_W^{-1} of the center of the monopole has some important consequences, though. For one thing, two monopoles with a separation much less than M_W^{-1} may orient their magnetic charges in orthogonal directions in SU(3) × SU(2) × U(1), and reduce their Coulomb repulsion to zero. For an appropriate choice of parameters, it is then possible for the attractive force between the monopoles generated by scalar exchange to cause a stable two-monopole bound state to form, with twice the minimal U(1)$_{\text{em}}$ magnetic charge (56). Also the quantum mechanical fluctuations of the W and Z fields within a distance M_W^{-1} of the center of the monopole influence the scattering of fermions by monopoles, as we see in Section 7.

The SO(10) model (57) is the next simplest realistic grand unified theory, after the SU(5) model. There are several possible choices for the symmetry-breaking hierarchy of the SO(10) model, and the properties of its monopoles depend on this choice. Rather than enumerate all the possibilities, let us focus on one particularly interesting case.

The group SO(10) is not simply connected, but has the simply connected covering group Spin(10). The center of Spin(10) is Z_2, and its 16-dimensional spinor representation is a double-valued representation of SO(10) = Spin(10)/Z_2. All representations of Spin(10) can be constructed from direct products of 16s.

Let us suppose that the order parameter for the first stage of symmetry breakdown in the SO(10) model is a scalar field Φ that transforms as the 54-dimensional representation of SO(10): Φ can be written as a traceless symmetric 10×10 matrix transforming according to

$$\Phi(x) \to \Omega(x)\Phi(x)\Omega^{\mathrm{T}}(x), \qquad\qquad 90.$$

where $\Omega(x) \in \mathrm{SO}(10)$. If Φ acquires the expectation value

$$\langle \Phi \rangle = \Phi_0 = v_1 \, \mathrm{diag}\,(2,2,2,2,2,2,-3,-3,-3,-3), \qquad\qquad 91.$$

then the unbroken subgroup H is locally isomorphic to $\mathrm{SO}(6) \times \mathrm{SO}(4)$. This group is, in turn, locally isomorphic to the direct product of $\mathrm{SU}(4)$, the covering group of $\mathrm{SO}(6)$, and $\mathrm{SU}(2) \times \mathrm{SU}(2)$, the covering group of $\mathrm{SO}(4)$.

To determine the global structure of the unbroken group, we check for nontrivial elements of $\mathrm{SU}(4) \times \mathrm{SU}(2) \times \mathrm{SU}(2)$ that act trivially in $\mathrm{Spin}(10)$. Since the fundamental spinor representation of $\mathrm{Spin}(10)$ transforms under $\mathrm{SU}(4) \times \mathrm{SU}(2) \times \mathrm{SU}(2)$ as

$$\mathbf{16} \to (4,1,2) + (\bar{4},2,1), \qquad\qquad 92.$$

we see that the element $(-\mathbb{1}_4, -\mathbb{1}_2, -\mathbb{1}_2)$ of $\mathrm{SU}(4) \times \mathrm{SU}(2) \times \mathrm{SU}(2)$ does act trivially on the spinor. Thus, the symmetry-breaking pattern is (58)

$$G = \mathrm{Spin}(10) \xrightarrow{v_1} H_1 = [\mathrm{SU}(4) \times \mathrm{SU}(2) \times \mathrm{SU}(2)]/Z_2. \qquad\qquad 93.$$

The monopoles arising from this symmetry breakdown are Z_2 monopoles carrying $\mathrm{SU}(4)$ and $\mathrm{SU}(2) \times \mathrm{SU}(2)$ magnetic charges, classified by $\pi_2(G/H_1) = \pi_1(H_1) = Z_2$.

Now suppose that, at a lower mass scale v_2, the symmetry breakdown

$$H_1 = [\mathrm{SU}(4) \times \mathrm{SU}(2) \times \mathrm{SU}(2)]/Z_2 \xrightarrow{v_2} H_2$$
$$= [\mathrm{SU}(3) \times \mathrm{SU}(2) \times \mathrm{U}(1)]/Z_6 \qquad\qquad 94.$$

occurs. [The order parameter could be a scalar field transforming as the 16-dimensional spinor representation of SO(10).] H_2 is exactly the same as the unbroken gauge group of the SU(5) model, and the monopole with the minimal U(1) magnetic charge in this SO(10) model also carries SU(3) and SU(2) magnetic charges, just like the monopole of the SU(5) model.

But, as in the example of Section 5.2, the doubly charged monopole in this model is lighter than the monopole with minimal charge (59). The minimal monopole defines a loop in H_2 that cannot be contracted to a point in H_1, but can be in G. So the core of this monopole has a radius of order $(ev_1)^{-1}$, and its mass is of order (v_1/e). The doubly charged monopole, however, has no SU(2) magnetic charge, and it defines a loop in H_2 that can be contracted to a point in H_1. It arises from the breakdown of H_1 to H_2, and has a core radius of order $(ev_2)^{-1}$ and a mass of order (v_2/e). Neither the minimal monopole nor the doubly charged monopole is much affected by the subsequent breakdown of H_2 to $H_3 = [\mathrm{SU}(3) \times \mathrm{U}(1)]/Z_3$.

In general, a grand unified theory with a complicated symmetry-breaking hierarchy may possess several stable monopoles with widely disparate masses, the monopole of minimal $\mathrm{U}(1)_{\mathrm{em}}$ charge being the heaviest. The SO(10) model described here is the simplest realistic example illustrating this possibility.

5.4 Monopoles and Strings

In Section 5.1, we saw that there are model gauge theories in which magnetic flux becomes confined to topologically stable tubes, or "strings." In some models, the strings can end on magnetic monopoles, and cause monopoles and antimonopoles to form bound pairs connected by strings. In other models, the strings cannot end; they become either infinite open strings or closed loops of string.

Our last example is a model containing both monopoles and strings (60). Although the strings in this model cannot end on monopoles, they have interesting long-range interactions with monopoles. A monopole that winds once around a string becomes an antimonopole! The model has gauge group $G = \mathrm{SO}(3)$ and a scalar order parameter Φ transforming as the 5-dimensional representation of G: Φ can be written as a traceless symmetric 3×3 matrix transforming as

$$\Phi(x) \to \Omega(x)\Phi(x)\Omega^T(x), \qquad 95.$$

where $\Omega(x) \in \mathrm{SO}(3)$. If Φ acquires the expectation value

$$\langle \Phi \rangle = \Phi_0 = v\,\mathrm{diag}(1, 1, -2), \qquad 96.$$

then the unbroken subgroup H is locally isomorphic to the SO(2) subgroup of rotations about the "z-axis," generated by

$$Q = \begin{pmatrix} 0 & -i & 0 \\ i & 0 & 0 \\ 0 & 0 & 0 \end{pmatrix}. \qquad 97.$$

But H actually has a disconnected component, because

$$\Phi_0 = \Omega_0 \Phi_0 \Omega_0^T, \qquad 98.$$

where

$$\Omega_0 = \mathrm{diag}(1, -1, -1) \qquad 99.$$

is a 180° rotation about the "x-axis." The symmetry-breaking pattern is

$$G = \mathrm{SU}(2) \xrightarrow{v} H = \mathrm{U}(1) \times Z_2,$$

and the vacuum manifold G/H is topologically equivalent to a two-dimensional sphere with antipodal points identified. [The unbroken subgroup of the SO(10) model of Section 5.3 has a similar Z_2 factor, which we did not bother to point out there (61).]

Since $\pi_2(G/H) = Z$, this model has magnetic monopoles, just like the monopoles of the SO(3) gauge theory discussed in Section 3. But these monopoles have a peculiar new feature. Since a 180° rotation about the x-axis changes the sense of a rotation about the z-axis, we have

$$\Omega_0 Q \Omega_0^T = -Q. \qquad 100.$$

Therefore, the sign of an electric or magnetic charge can be changed by a gauge transformation in H, and there is no gauge-invariant way to distinguish a monopole from an antimonopole. A "hedgehog" is not different from an "antihedgehog," because the order parameter is a "headless" vector in three-dimensional space, identified with the vector pointing in the opposite direction. We can, however, distinguish a pair of monopoles (or antimonopoles) from a monopole-antimonopole pair; the ambiguity afflicts only the sign of total charge, not the relative charge of two objects (19).

This model also contains topologically stable strings, because $\pi_1(G/H) = \pi_0(H) = Z_2$. If we perform a gauge transformation $\Omega(\varphi)$ that rotates Φ to Φ_0 at all points on the circle at spatial infinity enclosing a string, as described in Section 5.1, then this gauge transformation must have a discontinuity, at some value of φ, by an element of H in the connected component of Ω_0. The two-dimensional cross section of a string is indicated in Figure 5, where the order parameter is represented by an arrow, with the understanding that arrows pointing in opposite directions represent the same value of the order parameter.

According to Equation 100, the magnetic charge of a monopole changes sign when it crosses the discontinuity in $\Omega(\varphi)$. The location of the discontinuity is of course gauge dependent. But any monopole trajectory that winds once around the string must cross the discontinuity an odd number

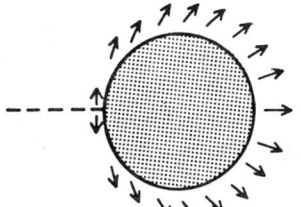

Figure 5 Cross section of a string.

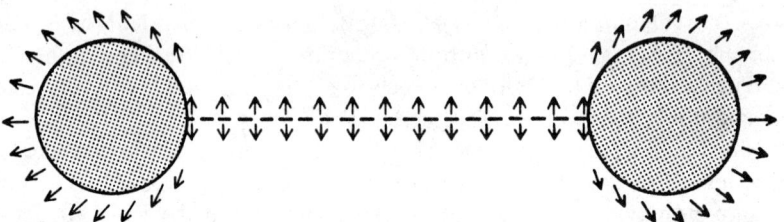

Figure 6 Cross section of a magnetically charged loop of string.

of times. We thus obtain the gauge-invariant result that a monopole that winds once around the string becomes an antimonopole (60).

There is a local criterion for distinguishing between a pair of monopoles (or antimonopoles) and a monopole-antimonopole pair; we can bring the two objects together and see whether they will annihilate or not. But this criterion is not globally well defined if strings are present. Whether they annihilate or not depends on how many times the monopoles wind around the strings before they are brought together.

Magnetic charge is conserved, so the magnetic charge lost by a monopole that winds around a string cannot disappear; it must be transferred to the string. If the string is open, the magnetic charge is transmitted to infinity along the string. But if the string is a closed loop, a finite magnetic charge density remains on the string, after it interacts with the monopole.

A cross section of a magnetically charged loop of string is sketched in Figure 6; the order parameter on a large sphere surrounding this loop is in a hedgehog configuration. The loop is a peculiar highly excited monopole, whose core has been distorted into a ring of radius R and thickness $(ev)^{-1}$. Its energy is of order $v^2 R$, plus a magnetic excitation energy of order $1/e^2 R$. The string tension of the loop causes it to oscillate with a period of order R.

The electric charge of a particle that winds around a string must also change sign, so a loop of string must be capable of supporting electric charge excitations, as well as magnetic charge excitations. The electric charge excitations of a string loop arise in much the same way as the dyonic excitations of a monopole, which are the subject of Section 6.

6. DYONS

6.1 *Semiclassical Quantization*

In Section 3, we constructed the time-independent monopole solution to the classical field equations in an SU(2) gauge theory. Now we consider the semiclassical quantization of this soliton.

The semiclassical expansion is an expansion in $\hbar$, where $\hbar^{-1}$ is a parameter multiplying the whole action. By rescaling the fields, we can write the Lagrangian (Equation 17) as

$$\mathcal{L} = \frac{1}{e^2}\left[-\frac{1}{4}F^a_{\mu\nu}F^{\mu\nu a} + \frac{1}{2}D_\mu\Phi^a D^\mu\Phi^a_a - \frac{1}{8}\left(\frac{\lambda}{e^2}\right)(\Phi^a\Phi^a - M_W^2)^2\right], \qquad 101.$$

where $F_{\mu\nu}$ and $D_\mu\Phi$ no longer depend on the gauge coupling e. We thus see that the semiclassical expansion is an expansion in e^2 with M_W and λ/e^2 fixed. In the classical limit $\hbar \to 0$, the size of the monopole remains fixed while its mass diverges like $\hbar^{-1}$.

Semiclassical quantization is carried out (in the gauge $A_0 = 0$), by expanding the Hamiltonian about the stable time-independent monopole solution. In order e, the monopole possesses a spectrum of positive-frequency vibrational excitations, which can be interpreted as meson states in the vicinity of the monopole.

Expanding about the classical solution, we also discover zero-frequency modes; these are associated with unbroken exact symmetries of the theory that act nontrivially on the solution. The time-independent monopole solutions form a degenerate set, and the zero-frequency modes are infinitesimal displacements in the manifold of degenerate solutions.

For example, the monopole solution is not translation-invariant; therefore, it has translational zero modes. The translational modes are easily quantized. To obtain an eigenstate of the Hamiltonian, we construct states that transform as irreducible unitary representations of the translation group; that is, plane wave states labeled by a momentum $\mathbf{p}$. For fixed $\mathbf{p}$, the energy of a monopole plane wave state is $O(e^2)$ in the semiclassical expansion, because the monopole mass m is $O(1/e^2)$:

$$E_\mathbf{p} = \sqrt{m^2 + p^2} = m + p^2/2m + \cdots = m + O(e^2). \qquad 102.$$

If the classical monopole solution were not rotationally invariant, it would have a moment of inertia of order $1/e^2$, and rotational excitations with energy of order e^2. But, because the monopole solution is rotationally invariant, there are no such rotational excitations.

A soliton can have zero-frequency modes associated with internal symmetries as well as space-time symmetries. In fact, the monopole solution is not invariant under a global $U(1)_{em}$ charge rotation, because the charged fields $W^\pm$ are excited in the monopole core. (Although the physical states in $A_0 = 0$ gauge are required to be invariant under time-independent local gauge transformations with compact support, they need not be invariant under global gauge transformations.) To quantize the charge rotation degree of freedom, we diagonalize the Hamiltonian by constructing irreducible representations of $U(1)_{em}$; that is, eigenstates of the electric charge Q.

Since U(1)$_{em}$ is compact, $\exp(i2\pi Q) = 1$, the eigenvalues of Q are integers. [Q is the electric charge in units of e. Half-odd-integer charge cannot occur because the monopole is invariant under the center Z_2 of SU(2).] Thus, the quantum-mechanical excitations of the fundamental monopole include *dyons*, particles that carry both magnetic and electric charge. The dyons have arisen automatically, from the semiclassical quantization of the global charge rotation degree of freedom of the monopole (62).

To determine the energies of the dyon states, we must compute the "moment of inertia" I of the monopole associated with a global charge rotation. The kinetic energy of a monopole undergoing a time-dependent charge rotation $\Omega(t) = \exp[i\vartheta(t)]$ has the form

$$L = \frac{1}{2e^2} I \dot\vartheta^2, \qquad 103.$$

where I is of order M_W^{-1}, or of order one in the semiclassical expansion. [The explicit computation of I involves some technical subtleties, as explained in (63–65).] The electric charge operator, the generator of a charge rotation, is the angular momentum conjugate to ϑ,

$$Q = \frac{\partial L}{\partial \dot\vartheta} = \frac{1}{e^2} I \dot\vartheta \qquad 104.$$

and the Hamiltonian may be written as

$$\mathcal{H} = \frac{e^2 Q^2}{2I}. \qquad 105.$$

The dyon excitations are split from the monopole ground state by an amount that is of order the electrostatic energy of a charge eQ localized on the monopole core, where Q is an integer.

The monopoles that occur in more complicated models, like those considered in Section 5, also have dyon excitations. One's naive expectation, based on the above discussion, is that these dyon states will transform as irreducible representations of the unbroken gauge group. In a realistic grand unified theory, with unbroken group SU(3)$_c$ × U(1)$_{em}$, one then expects the dyons to form color multiplets (66).

But we have already seen that this expectation is wrong. In Section 4.5 we found that global color rotations of a monopole that act nontrivially on its long-range field actually cannot be implemented. Therefore, the dyon excitations obtained by semiclassical quantization of a monopole with a color magnetic field need only have definite values of color hypercharge, the SU(3)$_c$ charge that commutes with the magnetic charge. They do not form complete color multiplets (64, 67).

The dyon excitations associated with color rotations of the monopole that act nontrivially on its long-range magnetic field fail to appear because they cannot be supported by the monopole core. These excitations are carried out to large distances by the nonabelian magnetic field, and are lost in the gluon continuum. They do appear explicitly, however, in the excitation spectrum of a widely separated monopole-antimonopole pair, with energy splittings inversely proportional to the separation of the pair (65).

6.2 The Anomalous Dyon Charge

In Section 2 we noted that the Dirac quantization condition permits dyons to have an anomalous electric charge characterized by a *CP*-violating angular parameter ϑ. We have seen that the semiclassical quantization procedure generates dyons with integer electric charge, in a *CP*-conserving theory. One wonders whether it is possible to introduce *CP* violation such that the dyons acquire anomalous charges.

In fact, it is possible (68). Let us consider adding to the Lagrange density of electrodynamics the *CP*-violating term

$$\mathscr{L}_\vartheta = \frac{\vartheta e^2}{4\pi^2} \mathbf{E} \cdot \mathbf{B}, \qquad\qquad 106.$$

where ϑ is a free parameter. In the absence of magnetic monopoles, this term is a total divergence, and has no physical consequences. But if a magnetic monopole is present, $\mathbf{B}$ is not the curl of a nonsingular vector potential, and this term has significant consequences.

Let us consider the effect of Equation 106 on the electric charge of a point monopole fixed at the origin (68, 69). In the $A_0 = 0$ gauge, the extra term modifies the momentum conjugate to A_i, and therefore also modifies the generator of an infinitesimal gauge transformation. In this gauge, physical states are invariant under finite time-independent gauge transformations that act trivially at spatial infinity. Let $\exp(i2\pi Q_\Lambda)$ be the operator that implements the gauge transformation

$$\Omega(\mathbf{r}) = \exp[i2\pi\Lambda(\mathbf{r})], \qquad\qquad 107.$$

where

$$\Lambda(\mathbf{r}=0) = 0$$
$$\Lambda(|\mathbf{r}|=\infty) = 1. \qquad\qquad 108.$$

Then, acting on physical states, we must have

$$n = Q_\Lambda = \frac{1}{2\pi}\int d^3r \frac{\partial L}{\partial \partial_0 A_i}\delta A_i = \int d^3r \left(\mathbf{E} + \frac{\vartheta e^2}{4\pi^2}\mathbf{B}\right)\cdot\left(\frac{1}{e}\nabla\Lambda\right)$$
$$= \int_{r=\infty} d^2\mathbf{S}\cdot\left(\frac{1}{e}\mathbf{E} + \frac{\vartheta e}{4\pi^2}\mathbf{B}\right) - \frac{1}{e}\int d^3r\,\Lambda\nabla\cdot\left(\mathbf{E} + \frac{e^2}{4\pi^2}\vartheta\mathbf{B}\right), \qquad 109.$$

where n is an integer, and the last equality has been obtained from an integration by parts. Since Equation 109 is satisfied by any $\Lambda(\mathbf{r})$ consistent with Equations 108, the volume integral vanishes and we obtain

$$Q = n - \frac{\vartheta}{2\pi} M, \qquad\qquad 110.$$

where Q is the electric charge of the monopole in units of e, and M is the magnetic charge in units of $g_D = 1/2e$. We have succeeded in reproducing Equation 16; now, through Equation 106, we have a dynamical interpretation of ϑ.

Since the charge spectrum (Equation 110) is unchanged when ϑ increases by 2π, one is tempted to interpret ϑ as an angular variable, and claim that the dyons parametrized by (n, ϑ) and $(n+1, \vartheta+2\pi)$, which have the same charge, are actually the same object. It is easy to see that this interpretation is correct (67, 68). Quantization of the charge rotation degree of freedom of the monopole in a theory with the term in Equation 106 is evidently equivalent to quantization in a theory without such a term, but subject to the condition

$$\exp(i2\pi Q_\Lambda) = \exp(i\vartheta), \qquad\qquad 111.$$

where $\exp(i2\pi Q_\Lambda) = \exp(i2\pi Q)$ implements a gauge transformation satisfying Equation 108. So we can think of $\exp(i\vartheta)$ as an arbitrary phase by which physical states are multiplied when acted on by a "large" gauge transformation with $\Lambda(|\mathbf{r}| = \infty) = 1$. Obviously, ϑ is an angle.

An angle ϑ can be associated with any gauge group; the dyon excitations of nonabelian monopoles, as well as abelian monopoles, may carry anomalous electric charges (67, 70). This observation seems paradoxical at first, because we know that the nontrivial commutation relations satisfied by the generators of a nonabelian gauge group require the eigenvalues of the generators to be quantized. But we have already seen how this paradox is resolved. Global gauge transformations of a nonabelian monopole cannot be defined; therefore, the dyon excitations need not form complete representations of the gauge group, and the peculiar values of the electric charge are allowed (67).

The discovery of the anomalous electric charge of the dyon has led to deep insights into the interactions of dyons and fermions, as discussed in Section 7.

6.3 Composite Dyons

The dyons we have considered so far are quantum mechanical excitations of a fundamental monopole. Another type of dyon is a composite state of a magnetic monopole and an electrically charged particle. A composite dyon has a peculiar property—it can carry half-odd-integral orbital angular momentum.

To understand this phenomenon, we consider a monopole (magnetic charge g) fixed at the origin interacting with a charged particle (electric charge e), which moves in the x–y plane. Let us imagine that the magnetic charge of the monopole turns on gradually (71). Then the z-component of the orbital angular momentum of the charged particle changes, according to Faraday's law, by

$$\Delta L_z = -(e/2\pi)\Delta\Phi_z. \qquad 112.$$

Here $\Delta\Phi_z$ is the change in the magnetic flux through a surface bounded by a circular loop in the x–y plane, centered on the monopole. But we may choose the surface bounded by the loop to pass either above or below the monopole, and the fluxes through the two surfaces differ by $4\pi g$ (not counting, of course, the flux carried by the Dirac string).

Since the choice of a surface bounding the loop is arbitrary, we must demand that the spectrum of L_z levels not depend on the choice. The spacings between L_z levels are integers; therefore, the ambiguity in L_z is undetectable if it is an integer; that is, if

$$n = (e/2\pi)(4\pi g) = 2eg. \qquad 113.$$

In yet another way, we have found that the quantum mechanics of a charged particle interacting with a magnetic monopole is consistent only if the Dirac quantization condition is satisfied.

We have also found that, if the monopole carries the Dirac magnetic charge ($n = 1$), then the L_z values are shifted up or down by half a unit. The magnetic flux through a surface bounded by a loop in the x–y plane is half the total flux emanating from the monopole. The orbital angular momentum of the charged particle is half-odd-integral. [Another way to reach this conclusion is to note that the electromagnetic field of the monopole and charged particle has an angular momentum of magnitude $eg = \frac{1}{2}$ (21).]

We conclude that the composite of an integer-spin monopole and an integer-spin charged particle can be a dyon with half-odd-integer spin (72). According to the usual connection between spin and statistics, a composite of two bosons can be a fermion!

Does the usual spin-statistics connection really hold for these objects? One might expect that the interchange of two identical composite dyons could be accomplished by merely interchanging their constituents. However, the interchange of the two dyons should in fact be performed by transporting each dyon covariantly in the gauge potential of the other (71, 73); this procedure corresponds to interchanging the electromagnetic fields of the dyons, as well as their constituents.

It is trivial to perform the covariant interchange of the dyons, if we first choose a gauge in which the vector potential vanishes. The velocity-dependent interaction of two dyons, each with magnetic charge g and electric charge e, can obviously be represented by

$$-e\mathbf{v}\cdot[\mathbf{A}(\mathbf{r})-\mathbf{A}(-\mathbf{r})], \qquad 114.$$

where $\mathbf{r}$ is the separation of the dyons, $\mathbf{v}$ the relative velocity, and $\mathbf{A}$ the monopole vector potential

$$\mathbf{A}(\mathbf{r})\cdot d\mathbf{r} = g(1-\cos\vartheta)\,d\varphi. \qquad 115.$$

The first term in the brackets in Equation 114 is due to the interaction of the electric charge of dyon 1 and the magnetic charge of dyon 2; the second term is due to the interaction of the electric charge of dyon 2 with the magnetic charge of dyon 1. Now, since

$$A_\varphi(\vartheta,\varphi)-A_\varphi(\pi-\vartheta,\varphi+\pi) = 2g = \frac{1}{ie}(\partial_\varphi\Omega)\Omega^{-1}, \qquad 116.$$

where

$$\Omega = \exp(i2eg\varphi), \qquad 117.$$

the gauge interaction, Equation 114, between the dyons can be removed by performing the gauge transformation

$$\Psi(\mathbf{r}) \to \Omega(\mathbf{r})\Psi(\mathbf{r}), \qquad 118.$$

on the two-dyon wave function Ψ.

In this gauge, the dyons may be interchanged naively, by replacing $\mathbf{r}$ by $-\mathbf{r}$. But the gauge transformation shown in Equation 118 has changed the symmetry of the wave function; when the dyons are interchanged, $\varphi \to \varphi + \pi$, Ω changes by the phase

$$\Omega \to \exp(i2\pi eg)\Omega. \qquad 119.$$

This phase is precisely what is needed to restore the usual connection between spin and statistics (73). It is $(-1)^{2l}$, where l is the orbital angular momentum of the composite dyon. It really is possible to obtain a fermion as a composite of two bosons, if $2eg$ is odd.

6.4 *Dyons in Quantum Chromodynamics*

In quantum chromodynamics, as in any gauge theory, a term of the form

$$\mathscr{L}_\vartheta = \frac{\vartheta e^2}{8\pi^2}\mathbf{E}^a\cdot\mathbf{B}^a \qquad 120.$$

can occur in the Lagrange density, where ϑ is an arbitrary parameter. In the real world, ϑ is known to be very close to zero, but it is nonetheless interesting to ask how the strong interactions would behave for different values of ϑ. For one thing, by studying the ϑ-dependence of the theory, we can gain a deeper understanding of quark confinement. For another, there may exist other "super-strong" interactions, not yet known, for which ϑ is not close to zero.

The discovery of the anomalous dyon charge, $Q = -\vartheta/2\pi$, has led to some interesting insights into the ϑ-dependence of quark confinement in quantum chromodynamics (70).

What do magnetic monopoles and dyons have to do with quark confinement? It should be possible to understand quark confinement by considering the dynamics of pure Yang-Mills theory, without quarks. If, in this theory, it is dynamically favored for color-electric flux to collapse to a tube with a characteristic width of order the hadronic size, then confinement is explained. A distantly separated quark-antiquark pair would become connected by a color-electric flux tube carrying constant energy per unit length, and the potential energy of the pair would rise linearly with the separation.

A similar phenomenon occurs in a superconductor. Magnetic flux is expelled from a superconductor (the Meissner effect), and hence collapses to a flux tube. As a result, magnetic monopoles in a superconductor would be "confined."

The Meissner effect arises because of the condensation of electrically charged Cooper pairs in the ground state of a superconductor. It is natural to suggest that quark confinement arises in quantum chromodynamics because of the condensation of color-magnetic monopoles in the vacuum state of Yang-Mills theory (70, 74). The monopole condensate would cause the Yang-Mills vacuum to expel color-electric flux and screen color-magnetic flux.

But what are the monopoles of Yang-Mills theory? We can understand their origin by choosing an appropriate gauge (70). In SU(N) Yang-Mills theory, a gauge transformation can be performed that diagonalizes, for example, the gauge field F_{12} at each point of space-time. This gauge condition generically specifies the gauge transformation up to a diagonal element of SU(N), and hence reduces the theory to an abelian $U(1)^{N-1}$ gauge theory. However, the ambiguity in the gauge transformation is a nondiagonal element of SU(N) at the isolated points in three-dimensional space where two eigenvalues of F_{12} coincide. At these isolated points, the embedding of $U(1)^{N-1}$ in SU(N) is ill defined, and magnetic monopoles appear, carrying $U(1)^{N-1}$ magnetic charges. It is the condensation of these monopoles that presumably accounts for quark confinement.

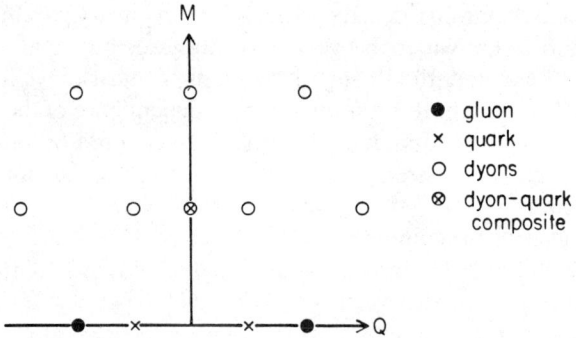

Figure 7 Electric (Q) and magnetic (M) charges in an SU(2) gauge theory with $\vartheta = \pi$.

For simplicity, consider an SU(2) gauge theory, which has only one diagonal U(1) charge and hence only one type of monopole. If $\vartheta = 0$, these monopoles carry no U(1) electric charge. Such monopoles condense, and, as a result, U(1) electric charge is confined, while U(1) magnetic charge is screened.

But, as ϑ varies from 0 to 2π, the monopole acquires an electric charge $Q = -\vartheta/2\pi$. When ϑ reaches 2π, there is again an electrically neutral monopole that condenses (Figure 3). Evidently, the object that condenses must change discontinuously for at least one value of ϑ.

In particular, for $\vartheta = \pi$, there is a monopole with twice the minimal magnetic charge (a bound state of two dyons) that is electrically neutral, and it is plausible that this object condenses, instead of the fundamental dyon (Figure 7). Thus, both elementary dyons and elementary quarks, which have electric charge $Q = \pm\frac{1}{2}$ in SU(2), are confined, but electrically neutral composites of quarks and dyons exist, which are not confined.

On the basis of this picture, one is tempted to conjecture that quarks are unconfined in SU(2) gauge theory at $\vartheta = \pi$. In fact, the liberated quark-dyon composites have orbital angular momentum 1/2, and so are bosons rather than fermions, as discussed in Section 6.3.

Similar phenomena can occur for other values of N. The discovery of the anomalous electric charge of the dyon has led us to expect a highly nontrivial dependence on the parameter ϑ in nonabelian gauge theories.

7. MONOPOLES AND FERMIONS

7.1 *Fractional Fermion Number on Monopoles*

Surprising and qualitatively new phenomena arise when we consider the quantum mechanics of electrically charged fermions interacting with magnetic monopoles.

A first indication of the subtlety of monopole-fermion interactions is obtained if we reconsider the derivation of the dyon charge spectrum (Section 6.2), including the effect of an electron coupled to the electromagnetic field. The derivation appears to go through as before, so we conclude that the allowed electric charges of the dyon are

$$Q = n - (\vartheta/2\pi)M, \qquad 121.$$

where n is an integer.

But it is known that, because of the axial anomaly (75), it is possible to rotate ϑ, the coefficient of the CP-violating $\mathbf{E} \cdot \mathbf{B}$ term in the Lagrangian, to zero, at the cost of introducing a CP-violating phase into the electron mass term (76),

$$\mathscr{L}_m = -me^{i\vartheta}\bar{\psi}_L\psi_R + \text{h.c.} \qquad 122.$$

Therefore, it should be possible to understand the origin of the anomalous dyon charge by carefully inspecting the electron vacuum polarization cloud surrounding the monopole.

Indeed, this is possible. When the Dirac equation is solved for a fermion with a complex mass in the field of a point monopole, it is found that the Dirac sea is distorted for nonzero ϑ, so that the ground state of the monopole-fermion system carries fermion number $(\vartheta/2\pi)M$ and electric charge $Q = -(\vartheta/2\pi)M$ (77). Actually, solving the Dirac equation for an electron in the field of a point monopole involves a further subtlety; it is necessary to impose a boundary condition on the electron wave function at the pole (78, 79). The above remark really holds only if the boundary condition is chosen to be CP conserving, so that the phase ϑ is the only source of CP violation in the problem. Both singular and nonsingular monopoles have the ability to carry a fractional fermion number, in a CP-nonconserving theory (80, 81).

If the mass m of the electron vanishes, its phase is not well defined, and the parameter ϑ must become unobservable. The electric charge of the monopole ground state, which is $-(\vartheta/2\pi)M$ for any nonzero m, must vanish discontinuously in the limit $m \to 0$. This behavior is not so puzzling once we realize that the anomalous charge is carried by the electron vacuum polarization cloud of the monopole. The electric charge radius of the monopole is of order m^{-1}, the electron Compton wavelength, so any observer a finite distance from the monopole center thinks that the electric charge disappears as $m \to 0$ (15, 16, 69).

In fact, all the dyonic excitations of the monopole have an electric charge radius of order m^{-1}, since it is very much preferred energetically to deposit the charge in the electron cloud, instead of on the very tiny monopole core. These excitations are split from the dyon ground state by an energy of order m, and can be regarded as (unstable) dyon-electron composites.

7.2 Monopole-Fermion Scattering

The interactions of magnetic monopoles and fermions have another, even more spectacular property. When a monopole and charged fermion collide at low energy, compared to the inverse size M_X of the monopole core, the outcome is strongly dependent on the structure of the core (15, 16). In particular, in a typical grand unified theory there are heavy gauge bosons with masses of order M_X and couplings that violate baryon-number conservation; in such a theory the cross section for baryon-number-changing scattering of a fermion by a monopole at low energy is large, and independent of M_X.

This result seems to violate a cherished principle of quantum field theory, the decoupling principle (82), which asserts that the effects of the very short-distance physics must be suppressed at low energy by a power of the short-distance scale. In this respect, monopole-fermion scattering appears to be a unique phenomenon.

To begin to understand the peculiar features of monopole-fermion scattering, recall that if a particle with electric charge e moves in the field of a point monopole with magnetic charge g, the electromagnetic field carries angular momentum

$$\mathbf{J}_{em} = -eg\hat{r} \qquad 123.$$

where $\hat{r}$ is the unit vector pointing toward the charged particle from the monopole (21, 83). If the charged particle were to pass through the monopole, this contribution to the angular momentum would change discontinuously. Therefore, conservation of angular momentum forbids the particle to pass through the pole, unless its charge or intrinsic spin can change discontinuously as it does so.

The above remark has a quantum mechanical counterpart, as is seen by solving the Dirac equation for a massless electron in the field of an abelian point monopole with $eg = \frac{1}{2}$. The wave function ψ of the electron is defined (42), as described in Section 4.5, by specifying smooth functions ψ_U and ψ_L on the upper and lower hemispheres surrounding the monopole, which satisfy a matching condition at the equator of the form shown in Equation 56. Because there is a conserved angular momentum (84)

$$\mathbf{J} = \mathbf{r} \times (-i\mathbf{\nabla} - e\mathbf{A}) - \tfrac{1}{2}\hat{r} + \tfrac{1}{2}\boldsymbol{\sigma} \qquad 124.$$

in this problem, the eigenstates of the Hamiltonian can be chosen to be eigenstates of J^2 and J_z. For the states with $J = 0$, the Dirac equation reduces to the radial equation

$$i\gamma_5 \frac{d}{dr}\chi(x) = E\chi(r), \qquad 125.$$

where

$$\psi(r, \vartheta, \varphi, t) = \frac{1}{r}\chi(r)\eta_0(\vartheta, \varphi)\exp(-iEt), \qquad 126.$$

and η_0 is the $J = 0$ "monopole harmonic" (42, 78).

The solutions to Equation 125 have an odd property; the positive-helicity ($\gamma_5 = -1$) solution is purely an outgoing wave, and the negative-helicity ($\gamma_5 = +1$) solution is purely an incoming wave. (For a positron, the helicities of the solutions are reversed.) Both solutions are singular at the origin, the location of the pole, and the Dirac equation provides no criterion for matching up the incoming and outgoing solutions. The Hamiltonian defined by the Dirac equation is therefore not self-adjoint; probability is not conserved unless the Hamiltonian is supplemented by a boundary condition at the origin (the location of the pole) relating the incoming and outgoing waves (79).

This trouble can be traced back to the unusual term $\mathbf{J}_{\text{em}} = -\frac{1}{2}\hat{r}$ in the expression for the angular momentum. An incoming (outgoing) electron must have negative (positive) helicity to be in a state with $\mathbf{J} = \mathbf{J}_{\text{em}} + \boldsymbol{\sigma} = 0$. For a positron, $\mathbf{J}_{\text{em}}$ has the opposite sign, and the helicities are reversed.

The boundary condition at the origin determines the fate of a left-handed electron, which scatters from a monopole in the $J = 0$ partial wave. But there are only two options; it becomes either a right-handed electron or a left-handed positron, because these are the only available outgoing modes with $J = 0$. The boundary condition must therefore either violate chirality (which is otherwise a good symmetry of the Hamiltonian) or require the monopole to absorb electric charge. If the charge-conserving boundary condition is chosen, then the chirality-changing $J = 0$ cross section will saturate the unitarity limit (15, 16).

The need for a boundary condition to determine the final state of an electron scattering from a point monopole is the crucial feature of monopole-fermion scattering that results in the violation of the decoupling principle. The decoupling principle leads one to expect that the amplitude for monopole-fermion scattering at energies much less than the inverse size of the monopole core does not depend on the structure of the core, except for power corrections that vanish as the size of the core goes to zero. Up to power corrections, the amplitude should be calculable in a low-energy "effective theory" in which the core is regarded as pointlike and its properties need not be specified. This expectation fails because monopole-fermion scattering is inherently ambiguous when the monopole is pointlike. Information about the core of the monopole survives in the low-energy effective theory as a boundary condition needed to specify the outcome of a scattering event; a low-energy fermion with $J = 0$ can penetrate to the core of the monopole, and be strongly influenced by its structure. In particular, the boundary condition may violate a symmetry (like baryon number) that would otherwise be a good symmetry of the low-energy effective theory.

We now see that the analysis of the scattering of a low-energy fermion by a nonsingular monopole with nonvanishing core size can be divided into two parts. First, we decide what boundary conditions must be imposed as the limit of zero core size is taken. Then the interaction of a point monopole with fermions satisfying the appropriate boundary conditions is studied. The second step is highly nontrivial. Fermion pair creation effects, which are responsible for smearing out the electric charge of the dyonic excitations of the monopole over a region with radius of order the fermion Compton wavelength, must be taken into account as fully as possible. But Rubakov (15) and Callan (16) suggested that, since only $J = 0$ fermions can penetrate to the core of the monopole, the problem can be reasonably approximated by an effective $(1+1)$-dimensional quantum field theory describing the $J = 0$ partial wave, in which the spatial coordinate is the radial coordinate r. The qualitative features of this $(1+1)$-dimensional theory are most easily glimpsed if it is converted into an equivalent "bosonized" theory (85) in which the fermions are represented by solitons. This soliton picture of monopole-fermion scattering is especially convenient when we try to understand the effects of fermion masses.

Returning to the problem of finding the appropriate boundary conditions satisfied by the fermions, let us consider, for concreteness, the case of the SU(5) grand unified model with a single generation of fermions. The magnetic charge of the SU(5) monopole is a linear combination of ordinary magnetic charge and color magnetic charge. At a distance from the monopole center much less than the characteristic hadronic size 10^{-13} cm and much greater than the radius of the core, the only fermions that interact with the monopole are those carrying Q', the corresponding combination of electric charge and color electric charge, given, in an appropriate gauge, by Equation 86. The right-handed quarks and leptons carrying nonzero Q' are

$$Q' = +1 : e_R^+ \bar{d}_{3R} u_{1R} u_{2R} \text{ (incoming)}$$
$$Q' = -1 : d_{3R} e_R^- \bar{u}_{2R} \bar{u}_{1R} \text{ (outgoing)},$$
127.

where e denotes the electron, u and d denote up and down quarks, and 1, 2, 3 are color indices. The behavior of these fermions in the field of the SU(5) monopole is identical to the behavior of an electron or positron in the field of an ordinary Dirac monopole. If fermion masses are ignored, then the right-handed (left-handed) fermions with $Q' = +1$ in the $J = 0$ partial wave are incoming (outgoing) only; for $Q' = -1$ the helicities are reversed. The new feature is that there are now four Dirac fermions interacting with the monopole, and the boundary condition at the origin causes these fermions to mix in a manner determined by the structure of the core of the monopole.

One can attempt to determine the boundary condition by solving the Dirac equation in the field of the nonsingular SU(5) monopole with finite core radius (16, 80, 86). The result is that the helicity of the incoming fermion is preserved; incoming and outgoing states are matched up as in Equations 127. We see that two units of Q' are transferred to the monopole, exciting its dyon degree of freedom.

But if we now investigate the consequences of this boundary condition, taking proper account of pair creation effects, we realize that the picture in which the incoming fermion falls to the core and deposits charge there, suggested by the solution to the Dirac equation, is not very accurate. An enormous Coulomb barrier prevents charge from being deposited on the core. It is energetically favored for the charge to be spread out over a region with a radius of order a fermion Compton wavelength. As a result, our original procedure for finding the correct boundary condition is called into question. It seems that a more appropriate boundary condition is one forbidding charge to accumulate at the origin (87).

Fortunately and remarkably, in the case of the SU(5) monopole we can obtain quite nontrivial information about the scattering process by merely demanding that none of the charges coupling to massless gauge bosons accumulate on the core (88). This constraint is especially powerful because W and Z bosons must be regarded as effectively massless at distances from the center of the monopole much less than M_Z^{-1}. Since left-handed and right-handed fermions with the same electric charge have different values of the charge $(T_3)_{\text{weak}}$, which couples to the W_3 boson, simple chirality-violating processes such as

$$e_L^- + M \to e_R^- + M \qquad\qquad 128.$$

are forbidden for massless fermions. If, for example, two $J = 0$ u quarks scatter from the monopole, there is only one possible final state of two $J = 0$ fermions; the allowed process is

$$u_{1R} u_{2R} + M \to \bar{d}_{3L} e_L^+ + M. \qquad\qquad 129.$$

The most general final fermion state consistent with conservation of all gauged charges is $\bar{d}_{3L} e^+$ accompanied by an indefinite number of the pairs $\bar{u}_R u_L$, $\bar{d}_L d_R$, and $e_L^+ e_R^-$, which carry no favor quantum numbers. So baryon-number nonconservation is forced on us, if we ignore the masses of the fermions, and the cross section for baryon-number-changing scattering is not suppressed by the small size of the monopole core.

Presumably, then, SU(5) monopoles are able to catalyze the decay of a nucleon at a characteristic strong-interaction rate (15, 16). The only property of the SU(5) model that we needed to invoke was the existence of a monopole that couples to the charge Q', at distances from the monopole center less than M_Z^{-1}, where the weak-interaction symmetries are effectively restored. In any realistic model containing an $[SU(3) \times SU(2) \times U(1)]/Z_6$ monopole (see Section 5.3) and light fermions with the standard charge assignments, the amplitude for the process in Equation 129 will be unsuppressed by the small size of the monopole core.

This process (Equation 129), with two fermions in the initial state, must occur in two steps. It is natural to inquire about the intermediate state produced when u_{1R} scatters from the monopole. What one finds (89–91) is rather subtle and bizarre; the intermediate state consists of four "semitons," each with fermion number $\frac{1}{2}$. The reaction

$$u_{1R} + M \rightarrow \tfrac{1}{2}(u_{1L}\bar{u}_{2R}\bar{d}_{3L}e_L^+) + M, \qquad\qquad 130.$$

changes baryon number and lepton number by $-\frac{1}{2}$ unit.

The semitons are destabilized by fermion mass terms or the effects of the strong color interaction. At a distance from the monopole center where these effects become important, the intermediate state in Equation 130 evolves into a final state with baryon and lepton number differing by an integer (possibly zero) from that of the initial state. One possibility is that the semitons in Equation 130 evolve into u_{1L}; chirality-violating processes like that in Equation 128 are allowed if the fermions have masses.

The evolution of semitons into "final-state" quarks and leptons is not yet understood in quantitative detail. But it is reasonable to expect that the semiton intermediate state can evolve with a probability of order one into a final state with a baryon number different from the initial state (89–91). It is also expected that adding more generations of fermions will have no qualitative effect on the baryon-number-changing processes. The main new feature in the many-generation case is that the boundary conditions and hence the scattering amplitudes depend on generalized Cabibbo-like mixing angles (92).

The above considerations strongly suggest that the baryon-number-changing cross section for a quark of energy E scattering from a monopole is of order E^{-2}, if E^{-1} is much greater than the radius of the monopole core, and much less than both the Compton wavelength of the quark and the size of a hadron. But we are really interested in the cross section for *nucleon* decay catalyzed by the monopole. There are actually two questions of experimental interest. One is, what is the cross section for capture of nuclei by the monopole? It is probably large (93), and the capture rate might conceivably control the catalysis rate in terrestrial experiments. The other is, what is the cross section for the catalysis process itself? In spite of some ambitious attempts (94), techniques do not yet exist for doing detailed quantitative calculations of the catalysis cross section. The best guess is that it is roughly geometrical, $\sigma\beta \sim 10^{-27}$ cm^2, and that the most likely final state is a positron accompanied by a pion.

8. MONOPOLES IN COSMOLOGY AND ASTROPHYSICS

8.1 *Monopoles in the Very Early Universe*

We have seen that the existence of magnetic monopoles is a very general consequence of the unification of the fundamental interactions. But to say that monopoles must exist in grand unified theories is not necessarily to say that we have a reasonable chance of observing one. If the monopole mass is really as large as 10^{16} GeV, then there is no hope of producing monopoles in accelerator experiments in the foreseeable future.

However, it is likely that the universe was once extremely hot, so hot that processes occurred that were sufficiently energetic to produce monopoles. If there are any monopoles around today, they are presumably relics of the very brief, very energetic epoch immediately following the "big bang." So it is evidently interesting to consider how many monopoles might have been produced in the very early universe (8, 9).

As the universe cooled, it is expected to have undergone a phase transition at a critical temperature T_c of order the unification mass scale M_X. When the temperature T was above T_c, the full, grand unified gauge symmetry was restored (95); the scalar field Φ, which acts as an order parameter for the breakdown of the gauge symmetry, had a vanishing expectation value. But monopoles can exist only when the gauge symmetry is spontaneously broken, so no monopoles were present when T was above T_c. When T fell below T_c, the expectation value of Φ turned on, and monopole production became possible.

Because monopoles, unlike the other superheavy particles in grand unified theories, are stable, the density of monopoles per comoving volume established in the phase transition at T of order T_c could subsequently be reduced only by annihilation of monopole-antimonopole pairs. As the universe rapidly expanded, monopoles and antimonopoles had an increasingly more difficult time finding each other, and an appreciable density of monopoles per comoving volume might have persisted.

Thus, the problem of estimating the monopole abundance may be separated into two parts. We must estimate the initial density of monopoles established during the phase transition, and we must determine to what extent the monopole density was subsequently reduced by pair annihilation.

Let us first consider the production of monopoles during the phase transition. The detailed mechanism by which monopoles were produced depends on the nature of the phase transition; in particular, on whether it was a second-order (or weakly first-order) transition, in which large fluctuations occurred, or a strongly first-order transition, in which supercooling occurred. In either case, there is no reason to believe that the monopole abundance was ever in thermal equilibrium.

In the case of a second-order (or weakly first-order) phase transition, the scalar field Φ underwent large random fluctuations when T was near T_c. As the universe expanded and cooled, the scalar field was rapidly quenched, and a large density of topological defects became frozen in; these defects are the monopoles and antimonopoles. The quenching process may be described in the following way (7): At the time when the monopoles are being produced, the scalar field Φ is uncorrelated over distances larger than some characteristic correlation length ξ. We may thus regard Φ as having a domain structure, with ξ the characteristic size of a domain. At the intersection point of several domains, each with a randomly oriented scalar field, there is some probability p that the scalar field orientation is topologically nontrivial; if so, a monopole or antimonopole must form at the intersection point. The probability p depends on the detailed structure of the monopole, but it is not very much less than one. According to this picture, the density of monopoles n established in the phase transition is

$$(n)_{\text{initial}} \sim p\xi^{-3}. \qquad 131.$$

This argument sounds suspicious, because it relies on the notion of a scalar field domain structure, even though it is always possible to make a uniform scalar field look wiggly by performing a gauge transformation. However, there is no fundamental difficulty. We can fix the gauge in an appropriate way so that the idea of a scalar field domain makes sense, and we have reached a conclusion about the density of topological defects, which is a gauge-invariant quantity.

In a second-order (or weakly first-order) phase transition, the correlation length ξ becomes large as T approaches T_c, but in the early universe causality places a limit on how much ξ can grow (96). The scalar field Φ must remain uncorrelated over distances exceeding the horizon length d_H, the largest distance any signal could have traveled since the initial singularity; thus $\xi < d_H$. In terms of the temperature T, d_H may be written as (97)

$$\xi < d_H \sim Cm_p/T^2, \qquad 132.$$

where $m_p \sim 10^{19}$ GeV is the Planck mass, which determines the expansion rate of the universe, and $C = (0.60)N^{-1/2}$, where N is the effective number of massless spin degrees of freedom in thermal equilibrium at temperature T. (In a minimal grand unified theory, $C \sim 1/20$.) Combining Equations 131 and 132, we conclude that the initial value of the dimensionless ratio n/T^3 is bounded by

$$(n/T^3)_{\text{initial}} \gtrsim p(T_c/Cm_p)^3. \qquad 133.$$

In a typical grand unified theory, with $T_c \sim 10^{15}$ GeV, $Cm_p \sim 10^{18}$ GeV, and $p \sim 1/10$, we obtain $(n/T^3)_{\text{initial}} \gtrsim 10^{-10}$.

In the case of a strongly first-order phase transition, supercooling occurs. The phase with unbroken grand unified gauge symmetry becomes thermodynamically unstable when $T < T_c$, but nonetheless persists for a while, until bubbles of the stable broken-symmetry phase eventually begin to nucleate. These bubbles expand, collide, and coalesce, filling the universe with the stable phase (98). Inside each bubble, the scalar field Φ is quite homogeneous, so that each bubble contains a negligible number of monopoles. But when the expanding bubbles collide, monopoles can be produced.

Although it is not easy to calculate in detail the initial density of monopoles produced by bubble collisions, we can obtain a lower bound on the monopole density by invoking an argument similar to the one applied above to the case of a second-order transition. Now each bubble can be regarded as a scalar field domain, and the density of monopoles produced by the collisions must exceed the probability factor p times the density of bubbles at the time they collide. Since bubbles cannot expand faster than the speed of light, each bubble must be smaller in radius than the horizon site d_H, and the density of bubbles must be greater than d_H^{-3}. We again conclude, therefore, that $n > pd_H^{-3}$, and the bound in Equation 133 still applies, except that T_c is replaced by a temperature at which bubble nucleation becomes probable.

Regardless of the nature of the phase transition, reasonably copious production of monopoles seems to be inevitable. Moreover, the monopole abundance cannot be significantly reduced by pair annihilation (8, 9, 99). The annihilation rate is determined by the monopole-antimonopole capture rate; once a bound pair forms, it quickly cascades down emitting many photons and gluons, and finally annihilates into a burst of superheavy scalar particles and X-bosons. But the capture process is relatively inefficient because the monopoles are so heavy; it fails to keep pace with the expansion of the universe if the monopole abundance n is smaller than (9)

$$(n/T^3) \sim 10^{-9}(m/10^{16} \text{ GeV}), \qquad 134.$$

where m is the mass of the monopole. (The quantity n/T^3 is convenient to consider, because it remains constant if the expansion of the universe is adiabatic, and no monopoles are created or destroyed.) Once the monopole abundance is comparable to Equation 134 or smaller, monopole-antimonopole annihilation cannot further reduce the monopole density per comoving volume.

Using the standard estimate $m \sim 10^{16}$ GeV, we see that, if the smallest possible initial monopole abundance consistent with the bound (Equation 133), $(n/T^3) \sim 10^{-10}$, is established in the phase transition, this abundance is not further reduced by annihilation at all. The only way to reduce n/T^3 further is through nonadiabatic effects that increase the entropy density, but such effects cannot dilute the monopole abundance by many orders of magnitude without at the same time diluting the baryon-number density of the universe. Neglecting generation of entropy, we conclude that the density of magnetic monopoles today is $n \sim 10^{-10} T^3$, which is comparable to the density of baryons. This conclusion is clearly absurd, if the mass of a monopole exceeds the mass of a baryon by a factor of order 10^{16}.

We have uncovered the "monopole problem," an apparently serious conflict between grand unified theories and standard big-bang cosmology. Various attempts have been made to resolve this conflict. By far the most appealing resolution of the monopole problem is offered by the inflationary universe scenario (10–12, 139).

In this scenario, a positive effective cosmological constant causes the universe to "inflate" exponentially as a function of time, after the appearance of bubbles or fluctuation regions in which the scalar field Φ is nonzero. Some monopoles are produced when bubbles or fluctuation regions first form, but they are subsequently "inflated away"; in the course of the exponential expansion, the monopole abundance is reduced to a negligible value. Eventually, after many e-foldings of expansion, the cosmological constant that drove the inflation is rapidly converted to radiation, and the universe "reheats." Its subsequent evolution is well described by the standard cosmological model.

The inflationary universe scenario is more appealing than other possible solutions to the cosmological monopole problem (22, 23, 25) because inflation solves other cosmological problems as well. It explains why the universe is nearly homogeneous and isotropic, and why the mass density of the universe today is close to the critical density required to cause it to recollapse (10). It may also explain the origin of the primordial density fluctuations that led to galaxy formation (100). As presently formulated, the inflationary-scenario is not free of flaws, but it seems likely to be essentially correct. So it is plausible that inflation is the mechanism by which the cosmological abundance of monopoles became suppressed.

Not only can inflation reduce the monopole abundance to an acceptably small level, it can easily reduce the abundance to so low a level that a monopole will never be seen (23, 101, 102). Fortunately, this last statement is not a firm prediction. Monopoles may be produced either during inflation (103) or during (104) and after (23, 101, 102) the reheating of the universe. Until the details of the scenario are better known, it will not be possible to predict accurately the monopole abundance within the context of the inflationary universe scenario.

One suggestion for suppressing the monopole abundance within the context of the standard cosmological scenario is worthy of mention. It is possible that the universe entered a superconducting phase as it cooled (105). A superconductor tries to expel magnetic flux, so monopole-antimonopole pairs would have become connected by flux tubes and annihilated rapidly (106). As the universe cooled further, it might have eventually returned to a normal nonsuperconducting phase, but only after the monopole abundance had been significantly reduced.

An interesting feature of this superconductor scenario is that a potentially interesting number of monopoles could have survived until the universe re-entered the normal phase. Although the positions of monopoles and antimonopoles are strongly correlated when monopoles are first produced, the flux tubes do not pair up monopoles and antimonopoles perfectly. Some monopoles, unable to decide which antimonopole to pair up with, may get left behind. It thus seems possible that the superconductor scenario predicts a detectable abundance of monopoles (107).

The monopole problem has exerted a healthy influence on the development of cosmology during the past few years. And if the monopole abundance is ever measured, it will severely constrain our speculations about the very early universe. But, for now, cosmology does not offer much guidance to the prospective monopole hunter; there is no definite cosmological prediction for the monopole abundance.

8.2 *Astrophysical Constraints on the Monopole Flux*

Although cosmological considerations provide no definite prediction for the monopole abundance, both the inflationary scenario and the superconductor scenario offer the possibility that the monopole abundance is both small enough to be acceptable and large enough to be detectable. Theoretical cosmology should not discourage an experimenter from looking for monopoles.

People have been looking for magnetic monopoles for a long time. But traditional monopole searches do not place significant constraints on superheavy monopoles with mass m of order 10^{16} GeV. The traditional searches have relied on the strong ionization power of a relativistic monopole (108, 109), or have sought monopoles trapped in the Earth's crust (110, 111). But a superheavy monopole would be expected to be slowly moving and very penetrating; it need not ionize heavily or stop in the earth (9, 112, 113).

How slowly moving? The monopole can be accelerated by either gravitational fields or magnetic fields; which effect is more important depends on the mass of the monopole. From gravitational fields alone, the monopole would acquire a typical galactic infall velocity of order $10^{-3}c$, regardless of its mass. To determine the effect of the magnetic field in our galaxy, recall that the field has a strength B of order 3×10^{-6} gauss and a coherence length L of order 10^{21} cm (114). A monopole with the Dirac charge g_D crossing one coherence length is accelerated to

$$v = \left(\frac{2g_D BL}{m}\right)^{1/2} \sim 10^{-3}c\left(\frac{10^{17}\text{ GeV}}{m}\right)^{1/2}. \qquad 135.$$

This magnetic acceleration is therefore more important than the gravitational acceleration for $m \lesssim 10^{17}$ GeV. The monopole does not attain a relativistic velocity for $m > 10^{11}$ GeV.

How penetrating? The stopping power of slowly moving magnetic monopoles remains a rather controversial subject, about which a little more is said in Section 9. But the energy loss in rock of a monopole with $v/c \lesssim 10^{-2}$ surely does not much exceed (115)

$$\frac{dE}{dx} \sim 100\left(\frac{v}{c}\right) \text{ GeV cm}^{-1}. \qquad 136.$$

Thus, the range in rock of a monopole with $m \sim 10^{16}$ GeV is larger than 10^{11} cm; the monopole passes through the Earth without slowing down.

Although superheavy monopoles are not easily stopped or detected, astrophysical arguments can be used to place severe limits on the flux of magnetic monopoles in cosmic rays. These limits offer valuable guidance to the prospective monopole hunter.

One stringent limit [the "Parker limit" (13)] on the monopole flux is obtained by noting that, because the magnetic field of our galaxy accelerates monopoles, the energy density $U = B^2/8\pi$ stored in the field is dissipated at the rate $dU/dt \sim \langle gn\mathbf{v} \cdot \mathbf{B} \rangle$, where n is the monopole density. By demanding that the field energy is not substantially depleted in a time τ, of order 10^8 years, required to regenerate the field, we obtain the bound

$$F = \frac{nv}{4\pi} \lesssim \frac{B}{32\pi^2 g\tau} \sim 10^{-16} \text{ cm}^{-2} \text{ s}^{-1} \text{ sr}^{-1}\left(\frac{B}{3 \times 10^{-6}\text{G}}\right)\left(\frac{10^8 \text{ y}}{\tau}\right). \qquad 137.$$

A nice feature of this flux limit is that it appears to be independent of the mass m of the monopole.

However, it is implicitly assumed in the derivation of Equation 137 that gravitational effects on the trajectory of the monopole are negligible, and we have already argued that this is not so for $m \gtrsim 10^{17}$ GeV. If a monopole enters a coherent domain of the galactic magnetic field with incident energy $\frac{1}{2}mv^2 > gBL$, then the energy ΔE it extracts from the domain is a second-order effect, due to the deflection of the monopole trajectory as it crosses the domain; on the average it is

$$\Delta E \sim (gBL)^2/\tfrac{1}{2}mv^2. \qquad 138.$$

Therefore, the rate of dissipation of magnetic field energy scales like $1/m$ for $m \gtrsim 10^{17}$ GeV, and the flux limit becomes (14)

$$F \lesssim 10^{-16} \text{ cm}^{-2} \text{ s}^{-1} \text{ sr}^{-1}(m/10^{17} \text{ GeV}), \ m \gtrsim 10^{17} \text{ GeV}. \qquad 139.$$

For $m \gtrsim 10^{20}$ GeV, a more stringent limit than Equation 139 can be obtained, which is based solely on the enormous mass of the monopole and has nothing to do with its magnetic charge (9, 14, 112, 113). The total number of monopoles in our galaxy must not exceed the mass of the galaxy. By demanding that the mass of a spherical monopole galactic halo with radius of order 30 Kpc not exceed 10^{12} solar masses, and taking the typical monopole velocity to be of order $10^{-3}c$, we obtain the flux limit (14)

$$F \lesssim 10^{-13} \text{ cm}^{-2} \text{ s}^{-1} \text{ sr}^{-1}(10^{20} \text{ GeV}/m). \qquad 140.$$

Since this limit on the flux crosses the one in Equation 139 for $m \sim 10^{20}$ GeV, we have also the mass-independent bound

$$F \lesssim 10^{-13} \text{ cm}^{-2} \text{ s}^{-1} \text{ sr}^{-1}. \qquad 141.$$

(See Figure 8.)

While monopoles are undoubtedly rare, they may play an important role in the dynamics of galaxies. The above reasoning does not exclude the possibility that monopoles make up the dark matter of galactic halos, for

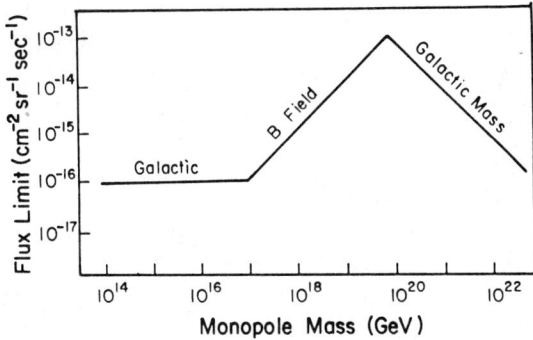

Figure 8 Astrophysical limits on the monopole flux as a function of monopole mass. "Galactic B field" labels the limit based on the energetics of the galactic magnetic field. "Galactic mass" labels the limit based on the total mass of the galaxy.

$m \gtrsim 10^{20}$ GeV. A monopole mass much larger than 10^{20} GeV seems unlikely, but a mass of order 10^{20} GeV is not that implausible; this is the typical sort of monopole mass expected in a Kaluza-Klein theory. And a monopole flux close to the limit in Equation 141 ought to be detectable; it corresponds to about one event per year in a few m^2 of detector.

The derivation of the Parker limit, Equation 139, is subject to one potentially serious criticism—the feedback of the accelerated monopoles on the galactic magnetic field has been ignored. If the monopole abundance were sufficiently *large*, then the period of magnetic plasma oscillations would be less than the time required for a monopole to cross a coherent domain of the magnetic field. The galactic magnetic field might then drive weakly damped plasma oscillations, rather than irreversible magnetic currents, and the Parker limit might be evaded by a significant margin if coherent oscillations could be maintained over many cycles (14, 116, 117). It seems likely, however, that small-scale gravitational and magnetic inhomogeneities, which are inevitably present, would destroy coherence and damp such oscillations rapidly.

In the hope of evading the Parker limit, it has also been suggested that the local flux of magnetic monopoles in the solar system greatly exceeds the ambient flux in the galaxy (118). This suggestion seems implausible on purely kinematic grounds (119).

Even more powerful limits on the monopole flux can be obtained by considering the astrophysical implications of the catalysis by monopoles of nucleon decay. The most interesting implication concerns the effects of monopoles in neutron stars (120, 121). A monopole striking a neutron star gets captured inside the star, if $m \lesssim 10^{22}$ GeV. Then, surrounded by matter at nuclear density, it catalyzes nucleon decay at a furious rate. A modest number of monopoles in a neutron star would cause the star to heat up, and emit a substantial flux of ultraviolet or x-ray photons. From observational limits on the ultraviolet and x-ray luminosity of old pulsars, it is therefore possible to derive a bound on the monopole flux F; conservatively, this bound is (120–122)

$$F \lesssim 10^{-22} \text{ cm}^{-2} \text{ s}^{-1} \text{ sr}^{-1} (\sigma\beta/10^{-27} \text{ cm}^2)^{-1}, \qquad 142.$$

where σ is the cross section for catalysis of nucleon decay by a monopole, and $\beta \sim 0.3$ is the relative velocity of the nucleon and monopole.

We infer that, if catalysis really proceeds at a strong-interaction rate, then the monopole flux must be smaller than Parker's limit (Equation 137) by at least six orders of magnitude. Monopoles must be so rare that there is little hope of observing one directly. The best way to find evidence for their existence would be by observing their effect on the luminosity distribution of neutron stars.

Of course, the limit (Equation 142) is nullified if monopoles do not catalyze nucleon decay, or do so at an insignificant rate. We expect the catalysis phenomenon to occur only if the new interactions associated with the monopole core fail to conserve baryon number, and it is surely possible to construct models for which this is not the case. An example is the SO(10) model of Section 5.3. The light monopoles associated with the $H_1 \to H_2$ breakdown in that model do not catalyze nucleon decay (59), although the heavy monopoles do, and it is easy to imagine a cosmological scenario in which these light monopoles are much more abundant than the heavy monopoles. (Note that, as is typical of monopoles that do not catalyze nucleon decay, the light monopoles carry twice the Dirac charge.) For monopoles that do not catalyze nucleon decay, the best flux limits we have are Equations 137, 139, and 140.

Still, for the experimenter who dreams of catching a monopole in the act of catalyzing nucleon decay, the bound in Equation 142 is very discouraging. An even more stringent limit can be obtained if capture of monopoles by the main sequence progenitor of the neutron star is taken into account (122). This stronger limit is more mass sensitive, however; Planck-mass monopoles, for example, would rarely be captured by main sequence stars.

Once captured by a neutron star, a monopole must be accelerated to a velocity of order c to escape. In the hope of evading the bound (Equation 142), it is worthwhile to consider whether there is any possible mechanism by which monopoles can be efficiently ejected from neutron stars at relativistic velocities (123).

It is generally believed that the core of a neutron star is a type II superconductor in which Cooper pairs of protons have condensed (124). Because the superconducting core expels magnetic flux, monopoles entering the star will eventually come to rest on the surface of the core. Typically, many magnetic flux tubes will have been trapped in the core when it went superconducting, and a monopole floating on the surface of the core will occasionally encounter the opening of a tube and drop in, penetrating the core (125).

It is conceivable that, deep within the core, there is an inner core in which charged pions condense. This pion condensate is also a type II superconductor, but its flux tubes would carry considerably higher energy per unit length than the flux tubes in the proton superconductor. Also, the flux quantum in the pion condensate would be the Dirac magnetic charge carried by the monopole, rather than half the Dirac charge as in the proton superconductor. Thus, two flux tubes in the proton superconductor would coalesce at the surface of the pion condensate, and a monopole drifting down one of them would be rapidly accelerated upon entering the pion condensate, the sizable magnetic field energy stored in the tube being

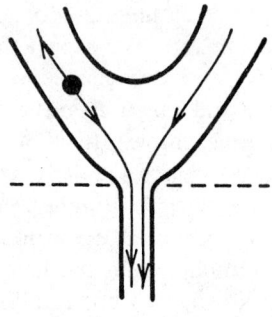 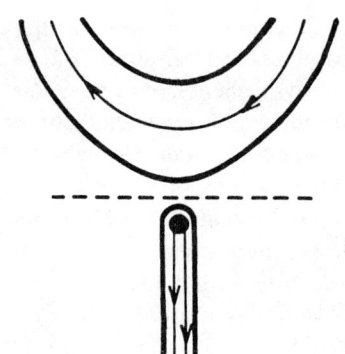

Figure 9 Magnetic monopole in a flux tube near the boundary between a proton pair condensate and a charged pion condensate. In the proton pair condensate (*left*), the magnetic flux in the tube reverses direction at the monopole, and there is no magnetic force on the monopole. In the pion condensate (*right*), the flux tube terminates on the monopole, and the monopole accelerates rapidly.

efficiently converted to monopole kinetic energy (see Figure 9). The monopole could be accelerated to a relativistic velocity, and ejected from the star! Hence, if we take the fullest advantage of our ignorance concerning the interiors of neutron stars, it is possible that the discouraging bound of Equation 142 can be evaded (123).

Even if we throw out the neutron star arguments, it may still be possible to obtain a stringent bound on the monopole flux by considering the effect of catalysis in white dwarfs; the interiors of white dwarfs are less exotic and better understood than those of neutron stars. If it is assumed that all monopoles that strike a white dwarf are captured, then, from observational limits on the luminosities of white dwarfs, we obtain a bound on the monopole flux (126) that is weaker by only about three orders of magnitude than Equation 142. Sufficiently heavy monopoles ($m \gtrsim 10^{20}$ GeV) are not captured by white dwarfs, but such heavy monopoles probably could not be ejected from a neutron star either.

All in all, the uncertaintities in the astrophysical arguments are such that it is barely possible to believe that there is an observable flux of monopoles that catalyze nucleon decay at a strong-interaction rate. Terrestrial searches for such monopoles are not completely pointless.

9. DETECTION OF MONOPOLES

9.1 *Induction Detectors*

If there are magnetic monopoles in the universe, they are surely rare, and they are probably slowly moving. Attempting to detect these monopoles is a challenging and risky experimental enterprise. But the potential rewards are so great that considerable risk is justified.

Possible techniques for detecting monopoles are briefly described here, but no attempt is made to give a complete review of recent experiments. More comprehensive reviews can be found in (127–129).

In principle, a closed loop of superconducting wire is an ideal monopole detector, because it gives an unambiguous signal whenever a monopole passes through the loop, however slowly (111, 130). To determine the effect on the loop of a monopole passing through, we may use the integrated Maxwell equation

$$\oint_\Gamma \mathbf{E} \cdot d\mathbf{r} = -\frac{d\Phi}{dt} - \frac{dQ_m}{dt}, \qquad 143.$$

which has been suitably modified to take into account the magnetic monopole current. Here Φ is the magnetic flux through a surface S_Γ bounded by the closed path Γ, and dQ_m/dt is the monopole current through the surface S_Γ. Applying Equation 143 to a path Γ entirely contained in the superconducting wire, where $\mathbf{E} = 0$, and integrating over time, we find that the change $\Delta\Phi$ in the magnetic flux linking the loop and the total magnetic charge ΔQ_m that passes through the loop are related by

$$\Delta\Phi = -\Delta Q_m. \qquad 144.$$

In particular, if a magnetic monopole with the Dirac charge g_D passes through the loop, the flux changes by two quantized flux units; the factor of two arises because the electric charge of a Cooper pair is $2e$.

The result (Equation 144) is actually obvious, because the magnetic field cannot penetrate the superconducting wire. The magnetic field lines emanating from the monopole are therefore swept back as the monopole approaches the wire, and break off, forming closed loops around the wire, as indicated in Figure 10 (131). (The field lines are allowed to break and rejoin where the magnetic field vanishes.)

We see that a monopole passing through the loop causes a sudden change in the magnetic flux linking the loop, and a corresponding shift in the dc current level in the wire, with a rise time of order the radius of the loop divided by the velocity of the monopole. For a sufficiently small loop (less than 10 cm in diameter), the shift can be easily detected by a SQUID (superconducting quantum interference device) magnetometer, provided that the loop is adequately shielded from other fluctuating magnetic disturbances.

The sizes of superconducting loop detectors are currently limited by the magnetic shielding requirement and by signal-to-noise problems, but monopole searches have been conducted with such detectors, and an

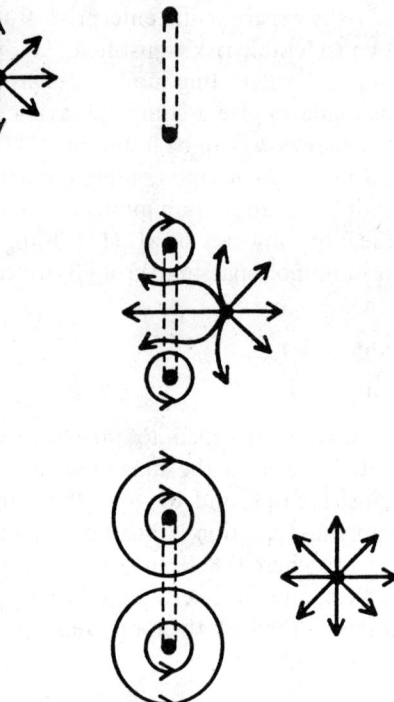

Figure 10 Bending and breaking of magnetic field lines, as a monopole passes through a loop of superconducting wire.

experimental flux limit (132) $F \lesssim 2 \times 10^{-11}$ cm^{-2} s^{-1} sr^{-1} has been obtained at the 90% confidence level. One candidate event has been seen (133), with a magnitude consistent with the Dirac magnetic charge. This event is the basis of the statement in the first paragraph of the introduction. Confirmation is still awaited.

The interpretation of this event as a magnetic monopole is not easily reconciled with the theoretical flux limits of Section 8.2. But no other completely satisfactory interpretation has yet been suggested.

9.2 Ionization Detectors

The superconducting induction detector has the significant advantage of being sensitive to monopoles of arbitrarily low velocity. But it will not be easy to construct a detector based on these principles big enough to challenge the theoretical flux limits of Section 8.2.

Larger detectors can be designed to detect the ionization loss of a monopole passing through matter. A very slowly moving ($\beta \sim 10^{-3}$)

particle with significant energy loss would be an unmistakable signal, but a negative result of a search for such events is not so easy to interpret, because the energy loss of slow monopoles is not that well understood.

The energy loss of a fast (electrically or magnetically) charged particle in matter is easily calculated, because electron encounters may be treated in the impulse approximation. But the energy loss of a slow particle depends on the details of atomic and molecular physics. One is inclined to say that the response of an atom or molecule to a very slow ($\beta \lesssim 10^{-3}$) monopole passing nearby is similar to its reponse to a magnetic field that adiabatically turns on and off; therefore, it is unlikely to become excited. But this conclusion is not necessarily correct, because the very strong magnetic field of the monopole greatly distorts the energy levels of the atom or molecule. If the ground state and an excited state closely approach each other, the adiabatic approximation may fail badly. To decide whether this occurs, detailed knowledge of the level structure in the inhomogeneous magnetic field of the monopole is needed.

As an illustration, consider a problem simple enough to be amenable to a sound theoretical analysis; the energy loss of a monopole in atomic hydrogen (134). First, imagine that a very slow monopole is incident on the nucleus of a hydrogen atom with zero impact parameter, and that the nucleus is held fixed, so that atomic recoil is neglected. If the monopole moves along the z-axis, then the z-component of angular momentum

$$J_z = [\mathbf{r} \times (\mathbf{p} - e\mathbf{A}) + \tfrac{1}{2}\boldsymbol{\sigma} - \tfrac{1}{2}\hat{n}]_z \qquad 145.$$

is conserved (84); here $\mathbf{r}$ is the electron coordinate relative to the nucleus and $\hat{n}$ is the unit vector pointing from the monopole to the electron. When the monopole is very distant from the atom, the electron Hamiltonian and angular momentum reduce to those of a simple hydrogen atom. But, because the last term in Equation 145 changes sign as the monopole moves from the far left ($\hat{n}_z = -1$) to the far right ($\hat{n}_z = +1$), we see that the passage of the monopole causes one unit of J_z to be transferred to the electron.

The low-lying levels of the hydrogen atom are sketched in Figure 11 as a function of the separation z between the nucleus and monopole. When $|z|$ is large compared to the Bohr radius a_0, the levels approach those of an unperturbed hydrogen atom; the ground state is a degenerate 1S doublet with $m = \pm\tfrac{1}{2}$, if we neglect the hyperfine splitting. Since the passage of the monopole increases m by 1, the $m = \tfrac{1}{2}$ member of the 1S doublet must evolve into the $m = \tfrac{3}{2}$ member of the 2P multiplet. If the $m = \pm\tfrac{1}{2}$ ground-state levels are occupied with equal probabilities, then there is a 50% chance that the passage of the monopole will excite the hydrogen atom.

When the positions of the nucleus and monopole coincide ($z = 0$), all three components of $\mathbf{J}$ are again conserved, and the energy levels can be computed exactly (78, 135); the ground state is an angular momentum singlet, and the first excited state is an angular momentum triplet. Thus, at $z = 0$, the levels with $J_z = -1, 0, 1$ cross. But if the impact parameter of the incident monopole is nonzero, J_z is not conserved and a level crossing cannot occur. Therefore, in the adiabatic limit, excitation of the atom occurs only for zero impact parameter.

However, if a monopole with impact parameter b moves at velocity v, excitation is likely to occur as long as the levels approach within $\hbar\omega$ of one another, where $\omega \lesssim v/b$. Calculation indicates that the hydrogen energy levels approach within a few tenths of an electron volt of one another even if b is of order a_0 (134). The excitation cross section is therefore surprisingly large; the calculated energy loss per unit density in hydrogen is (134)

$$\frac{dE}{dx} = 37(\beta/10^{-4})\text{MeV cm}^2 \text{ g}^{-1}, \qquad 146.$$

if atomic recoil is neglected. When atomic recoil is included, there is a kinematic threshold for excitation at $\beta = 1.5 \times 10^{-4}$.

This calculation illustrates that a reasonably large energy loss is possible for β as small as 10^{-4}, but also that a detailed understanding of the atomic levels in the presence of the monopole is necessary before a quantitative calculation of the energy loss can be performed. Nonetheless, a similar qualitative picture is probably applicable to more complicated materials. For example, a monopole incident on a Z-electron atom with zero impact

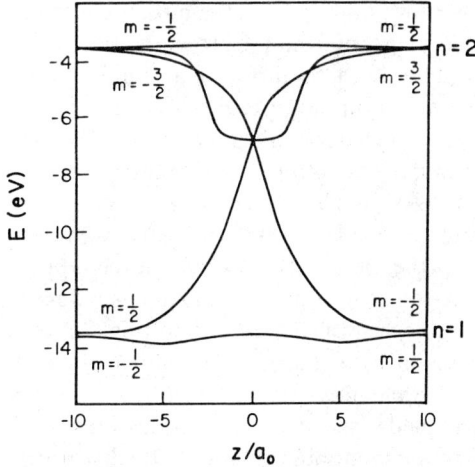

Figure 11 Electronic energy levels of a hydrogen atom in the vicinity of a Dirac monopole. The position of the monopole, relative to the atomic nucleus, is $(0, 0, z)$.

parameter will transfer Z units of angular momentum to the atom, and it seems likely that level crossings will be induced like those that occur in the hydrogen atom. The excited many-electron atom should then be able to autoionize, unlike the excited hydrogen atom, which is in a 2P state and must decay radiatively.

Other calculational schemes have also indicated that monopoles with β down to 10^{-4} have a detectable ionization loss in many materials (115). It is thus probable that existing ionization detectors are capable of detecting monopoles with $\beta \sim 10^{-4}$, and in any case we can be quite confident that monopoles with $\beta \sim 10^{-3}$ are detectable.

Since it is unreasonable to expect most of the cosmic ray monopoles incident on the Earth to have velocities much below $10^{-3}c$ (escape velocity from the galaxy is of order $10^{-3}c$), ionization detector experiments should be able to place useful limits on the cosmic ray monopole flux. The best current limit is $F \lesssim 7 \times 10^{-13}$ cm^{-2} s^{-1} sr^{-1} (136). It seems to be technically feasible to build much larger detectors that can improve this limit by several orders of magnitude.

9.3 Catalysis Detectors

Astrophysical arguments place very discouraging constraints on the flux in cosmic rays of magnetic monopoles that catalyze nucleon decay at a strong-interaction rate. It is nonetheless worthwhile to conduct experimental searches for such monopoles.

For one thing, as discussed in Section 8.2, the astrophysical arguments could be wrong. It is also conceivable that catalysis occurs at a rate small enough to be unimportant in neutron stars, but still large enough to be detectable in terrestrial experiments. The point is that the monopole may have a reasonably large cross section for capturing a nucleon; after capture it will hold onto the nucleon until it is able to catalyze its decay. Even if the nucleon must wait 10^{-6} s after capture before finally decaying, the monopole, traveling at 10^{-3} c, will have moved less than a meter. The catalysis would be no harder to observe than if it had followed 10^{-23} s after capture, as long as the detector is more than a meter long (137).

Experiments designed to search for spontaneous nucleon decay are also well suited to detect nucleon decay catalyzed by a monopole. If catalysis (or capture) occurs with a typical strong-interaction cross section, then several nucleon decay events will occur in the detector along the trajectory of the monopole. This distinctive signature allows catalyzed nucleon decay to be distinguished from spontaneous decay. Unsuccessful searches for such multiple nucleon decay events have placed the limit $F \lesssim 7 \times 10^{-15}$ cm^{-2} s^{-1} sr^{-1} on the monopole flux, assuming a catalysis cross section σ greater than 10^{-26} cm^2 (138).

The magnetic monopole continues to be strangely elusive. But just as theorists have continued to explore the wonders of the monopole with undeterred enthusiasm in spite of its elusiveness, so experimenters will continue to stalk the monopole with unfailing determination and ingenuity.

ACKNOWLEDGMENTS

I have learned about magnetic monopoles from many colleagues, especially S. Coleman, P. Nelson, and F. Wilczek. In preparing this manuscript, I have benefited from conversations with H. Sonada and N. Warner.

Literature Cited

1. Dirac, P. A. M. 1931. *Proc. R. Soc. London A* 133:60
2. Polyakov, A. M. 1974. *JETP Lett.* 20:194
3. 't Hooft, G. 1974. *Nucl. Phys. B* 79:276
4. Georgi, H., Glashow, S. L. 1974. *Phys. Rev. Lett.* 32:438
5. Pati, J. C., Salam, A. 1974. *Phys. Rev. D* 10:275
6. Georgi, H., Quinn, H., Weinberg, S. 1974. *Phys. Rev. Lett.* 33:451
7. Kibble, T. W. B. 1976. *J. Phys. A* 9-1387
8. Zeldovich, Ya. B., Khlopov, M. Y. 1978. *Phys. Lett.* 79B:239
9. Preskill, J. 1979. *Phys. Rev. Lett.* 43:1365
10. Guth, A. H. 1981. *Phys. Rev. D* 23:347
11. Linde, A. D. 1982. *Phys. Lett.* 108B:389; 114B:431
12. Albrecht, A., Steinhardt, P. J. 1982. *Phys. Rev. Lett.* 48:1220
13. Parker, E. N. 1970. *Astrophys. J.* 160:383
14. Turner, M. S., Parker, E. N., Bogdan, T. J. 1982. *Phys. Rev. D* 26:1296
15. Rubakov, V. 1981. *JETP Lett.* 33:644; 1982. *Nucl. Phys. B* 203:311; Rubakov, V. A., Serbryakov, M. S. 1983. *Nucl. Phys. B* 218:240
16. Callan, C. G. 1982. *Phys. Rev. D* 25:2141; *D* 26:2058; 1983. *Nucl. Phys. B* 212:391
17. Langacker, P. 1981. *Phys. Rep.* 72C:185
18. Ramond, P. M. 1983. *Ann. Rev. Nucl. Part. Sci.* 33:31
19. Coleman, S. 1977. In *New Phenomena in Subnuclear Physics*, ed. A. Zichichi. London: Plenum
20. Goddard, P., Olive, D. 1978. *Rep. Prog. Phys.* 41:1357
21. Coleman, S. 1983. In *The Unity of the Fundamental Interactions*, ed. A. Zichichi. London: Plenum
22. Weinberg, E. J. 1982. In *Particles and Fields—1981: Testing the Standard Model*, ed. C. A. Heusch, W. T. Kirk. New York: Am. Inst. Phys.
23. Preskill, J. 1983. In *The Very Early Universe*, ed. S. W. Hawking, G. W. Gibbons, S. Siklos. Cambridge: Cambridge Univ. Press
24. Turner, M. S. 1983. See Ref. 27, pp. 127–40
25. Lazarides, G. 1983. See Ref. 27, pp. 71–80
26. Preskill, J. 1984. See Ref. 28
27. Carrigan, R. A., Trower, W. P., eds. 1983. *Magnetic Monopoles*. New York: Plenum
28. Stone, J., ed. 1984. *Proceedings of Monopole '83*. New York: Plenum. In press
29. Aharonov, Y., Bohm, D. 1959. *Phys. Rev.* 115:485
30. 't Hooft, G. 1976. *Nucl. Phys. B* 105:538; Corrigan, E., Olive, D., Fairlie, D., Nuyts, J. 1976. *Nucl. Phys. B* 106:475
31. 't Hooft, G. 1978. *Nucl. Phys. B* 138:1
32. Strominger, A. 1983. *Commun. Nucl. Part. Phys.* 11:149; Preskill, J. 1983. See Ref. 27, pp. 111–26; Barr, S. M., Reiss, D. B., Zee, A. 1983. *Phys. Rev. Lett.* 50:317; Aoyama, S., Fujimoto, Y., Zhao, Z. 1983. *Phys. Lett.* 124B:185; Pantaleone, J. 1983. *Nucl. Phys. B* 219:367; Rubakov, V. 1983. *Phys. Lett.* 120B:191
33. Goddard, P., Nuyts, J., Olive, D. 1977. *Nucl. Phys. B* 125:1
34. Schwinger, J. 1966. *Phys. Rev.* 144:1087; Zwanziger, D. 1968. *Phys. Rev.* 176:1480, 1489
35. Glashow, S. L. 1961, *Nucl. Phys.*

22:579; Weinberg, S. 1967. *Phys. Rev. Lett.* 19:1264; Salam, A. 1968. In *Elementary Particle Physics*, ed. N. Svartholm. Stockholm: Almquist & Wiksells
36. Georgi, H., Glashow, S. L. 1972. *Phys. Rev. D* 6:2977
37. Kirkman, T. W., Zachos, C. K. 1981. *Phys. Rev. D* 24:999
38. Bogomol'nyi, E. 1976. *Sov. J. Nucl. Phys.* 24:449; Coleman, S., Parke, S., Neveu, A., Sommerfield, C. 1977. *Phys. Rev. D* 15:544
39. Prasad, M., Sommerfield, C. 1975. *Phys. Rev. Lett.* 35:760
40. Wilkinson, D., Goldhaber, A. 1977. *Phys. Rev. D* 21:1221; Weinberg, E. J. 1984. *Phys. Lett.* 136B:179
41. Goldhaber, A., Wilkinson, D. 1976. *Nucl. Phys. B* 114:317
42. Wu, T. T., Yang, C. N. 1976. *Nucl. Phys. B* 107:365; 1975. *Phys. Rev. D* 12:3845
43. Lubkin, E. 1963. *Ann. Phys.* 23:233
44. Hilton, P. J. 1953. *An Introduction to Homotopy Theory.* Cambridge: Cambridge Univ. Press; Steenrod, N. E. 1951. *The Topology of Fibre Bundles.* Princeton: Princeton Univ. Press; Nash, C., Sen, S. 1983. *Topology and Geometry for Physicists.* New York: Academic
45. Brandt, R., Neri, F. 1979. *Nucl. Phys. B* 161:253
46. Husemoller, D. 1966. *Fibre Bundles.* New York: McGraw-Hill; Steenrod, N. 1951. See Ref. 44
47. Kaluza, Th. 1921. *Sitzungsber. Preuss. Akad. Wiss. Berlin, Math. Phys. K* 1:966; Klein, O. 1926. *Z. Phys.* 37:895
48. Gross, D., Perry, M. 1983. *Nucl. Phys. B* 226:29
49. Sorkin, R. 1983. *Phys. Rev. Lett.* 51:87
50. Nelson, P., Manohar, A. 1983. *Phys. Rev. Lett.* 50:943
51. Balachandran, A. P., Marmo, G., Mukunda, N., Nilson, J. S., Sudarshan, E. C. G., Zaccaria, F. 1983. *Phys. Rev. Lett.* 50:1553
52. Gilmore, R. 1974. *Lie Groups, Lie Algebras, and Some of Their Applications.* New York: Wiley
53. Bais, F. A. 1981. *Phys. Lett.* 98B:437
54. Nielsen, H., Olesen, P. 1973. *Nucl. Phys. B* 61:45
55. Weinberg, E. J., London, D., Rosner, J. L. 1983. *Fermi Inst. Preprint, EFI 83-39.* Batavia, Ill: Fermilab
56. Gardner, C. L., Harvey, J. A. 1984. *Phys. Rev. Lett.* 52:879
57. Georgi, H. 1975. In *Particles and Fields—1974*, ed. C. E. Carlson. New York: Am. Inst. Phys.; Fritzsch, H., Minkowski, P. 1975. *Ann. Phys.* 93:193
58. Lazarides, G., Magg, M., Shafi, Q. 1980. *Phys. Lett.* 97B:87
59. Dawson, S., Schellekens, A. N. 1983. *Phys. Rev. D* 27:2119
60. Ginsparg, P., Coleman, S. 1983. Unpublished
61. Kibble, T. W. B., Lazarides, G., Shafi, Q. 1982. *Phys. Rev. D* 26:435
62. Julia, B., Zee, A. 1975. *Phys. Rev. D* 11:2227
63. Tomboulis, E., Woo, G. 1976. *Nucl. Phys. B* 107:221
64. Aboulsaood, A. 1984. *Nucl. Phys. B* 226:309; 1983. *Phys. Lett.* 125B:467
65. Coleman, S., Nelson, P. 1983. *Harvard Preprint HUTP-83/A067.* Cambridge, Mass: Harvard Univ.
66. Dokos, C., Tomaras, T. 1980. *Phys. Rev. D* 21:2940
67. Nelson, P. 1983. *Phys. Rev. Lett.* 50:939
68. Witten, E. 1979. *Phys. Lett.* 86B:293
69. Wilczek, F. 1982. *Phys. Rev. Lett.* 48:1146
70. 't Hooft, G. 1981. *Nucl. Phys. B* 109:455
71. Wilczek, F. 1982. *Phys. Rev. Lett.* 48:1144
72. Jackiw, R., Rebbi, C. 1976. *Phys. Rev. Lett.* 36:1116; 't Hooft, G., Hassenfratz, P. 1976. *Phys. Rev. Lett.* 36:1119
73. Goldhaber, A. 1976. *Phys. Rev. Lett.* 36:1122
74. Mandelstam, S. 1976. *Phys. Rep. 23C*:235
75. Adler, S. L. 1969. *Phys. Rev.* 177:2426; Bell, J. S., Jackiw, R. 1969. *Nuovo Cimento* 51:47
76. 't Hooft, G. 1976. *Phys. Rev. Lett.* 37:8; Jackiw, R., Rebbi, C. 1976. *Phys. Rev. Lett.* 37:177; Callan, C. G., Dashen, R. F., Gross, D. J. 1976. *Phys. Lett.* 63B:334
77. Grossman, B. 1983. *Phys. Rev. Lett.* 50:464; Yamagishi, H. 1983. *Phys. Rev. D* 27:2383
78. Kazama, Y., Yang, C. N., Goldhaber, A. S. 1977. *Phys. Rev. D* 15:2287
79. Goldhaber, A. S. 1977. *Phys. Rev. D* 16:1815
80. Jackiw, R., Rebbi, C. 1976. *Phys. Rev. D* 13:3398
81. Goldstone, J., Wilczek, F. 1981. *Phys. Rev. Lett.* 47:986
82. Appelquist, T., Carazzone, J. 1975. *Phys. Rev. D* 11:2856
83. Tamm, I. 1931. *Z. Phys.* 71:141
84. Fierz, M. 1944. *Helv. Phys. Acta* 17:27
85. Coleman, S. 1975. *Phys. Rev. D* 11:2088; Mandelstam, S. 1975. *Phys. Rev. D* 11:3026
86. Besson, C. 1982. PhD thesis. Princeton Univ.
87. Kazama, Y., Sen, A. 1983. *Fermilab-*

Pub-83/58-THY. Batavia, Ill: Fermilab; Yan, T.-M. 1983. Cornell Preprint CLNS-83/563. Ithaca, NY: Cornell Univ.; Balachandran, A. P., Schechter, J. 1983. Syracuse Preprint; Ezawa, Z. F., Iwazaki, A. 1983. Preprint; Yamagishi, H. 1983. Princeton preprint. Princeton Univ.
88. Grossman, B., Lazarides, G., Sanda, A. I. 1983. Phys. Rev. D 28:2109; Goldhaber, A. S. 1983. SUNY Preprint ITP-SB-83-30
89. Callan, C. G. 1983. Princeton preprint print-83-0306. Princeton Univ.
90. Sen, A. 1983. Fermilab-Pub-83/28. Batavia, Ill: Fermilab
91. Dawson, S., Schellekans, A. N. 1983. Phys. Rev. D 28:3125
92. Ellis, J., Nanopoulos, D. V., Olive, K. A. 1982. Phys. Lett. 116B:127; Bais, F. A., Ellis, J., Nanopoulos, D. V., Olive, K. A. 1983. Nucl. Phys. B 219:189
93. Fiorentini, G. 1984. See Ref. 28
94. Callan, C. G. Witten, E. 1984. Princeton preprint print-84-0054. Princeton Univ.
95. Kirzhnits, D., Linde, A. 1972. Phys. Lett. 42B:471; Weinberg, S. 1974. Phys. Rev. D 9:3357; Dolan, L., Jackiw, R. 1974. Phys. Rev. D 9:3320
96. Guth, A. H., Tye, S.-H. 1980. Phys. Rev. Lett. 44:631; Einhorn, M. B., Stein, D. L., Toussaint, D. 1980. Phys. Rev. D 21:3295
97. Peebles, P. J. E. 1971. Physical Cosmology. Princeton Univ. Press; Weinberg, S. 1972. Gravitation and Cosmology. New York: Wiley
98. Coleman, S. 1977. Phys. Rev. D 15:2929
99. Dicus, D. A., Page, D. N., Teplitz, V. L. 1982. Phys. Rev. D 26:1306
100. Guth, A. H., Pi, S.-Y. 1982. Phys. Rev. Lett. 49:1110; Hawking, S. W., Moss, I. G. 1983. Nucl. Phys. B 224:180; Bardeen, J. M., Steinhardt, P. J., Turner, M. S. 1983. Phys. Rev. D 28:679; Starobinsky, A. A. 1982. Unpublished
101. Turner, M. S. 1982. Phys. Lett. 115B:95
102. Lazarides, G., Shafi, Q., Trower, W. P. 1982. Phys. Rev. Lett. 49:1756
103. Turner, M. S. 1982. Unpublished; Goldhaber, A. S., Guth, A. H., Pi, S.-Y. 1982. Unpublished
104. Collins, W., Turner, M. S. 1983. Fermi Inst. Preprint 83-41. Chicago: Fermi Inst.
105. Langacker, P., Pi, S.-Y. 1980. Phys. Rev. Lett. 45:1
106. Vilenkin, A. 1982. Nucl. Phys. B 196:240; Bais, F. A., Langacker, P. 1982. Nucl. Phys. B 197:520
107. Weinberg, E. 1983. Phys. Lett. 126B:441
108. Fleischer, R. L., et al. 1969. Phys. Rev. 184:1393, 1398
109. Price, P. B., et al. 1978. Phys. Rev. D 18:1382
110. Kolm, H. H., et al. 1971. Phys. Rev. D 4:1285
111. Eberhard, P., et al. 1971. Phys. Rev. D 4:3260
112. Lazarides, G., Shafi, Q., Walsh, T. F. 1981. Phys. Lett. 100B:21
113. Longo, M. J. 1982. Phys. Rev. D 25:2399
114. Parker, E. N. 1979. Cosmical Magnetic Fields. Oxford: Clarendon
115. Ahlen, S. P., Kinoshita, K. 1982. Phys. Rev. D 26:2347
116. Salpeter, E. E., Shapiro, S. L., Wasserman, I. 1982. Phys. Rev. Lett. 49:1114
117. Arons, J., Blandford, R. D. 1983. Phys. Rev. Lett. 50:544
118. Dimopoulos, S., Glashow, S. L., Purcell, E. M., Wilczek, F. 1982. Nature 298:824
119. Freese, K., Turner, M. S. 1983. Phys. Lett. 123B:293
120. Kolb, E. W., Colgate, S. A., Harvey, J. A. 1982. Phys. Rev. Lett. 49:1373
121. Dimopoulos, S., Preskill, J., Wilczek, F. 1982. Phys. Lett 119B:320
122. Freese, K., Turner, M. S., Schramm, D. N. 1983. Phys. Rev. Lett. 51:1625
123. Harvey, J. A., Ruderman, M., Shaham, J. 1983. Unpublished; Harvey, J. A. 1984. See Ref. 28
124. Irvine, J. 1978. Neutron Stars. Oxford: Clarendon
125. Harvey, J. A. 1984. Nucl. Phys. B 236:255
126. Freese, K. 1983. Univ. Chicago preprint 84-0573
127. Giacomelli, G. 1983. See Ref. 27, pp. 41-70
128. Giacomelli, G. 1984. See Ref. 28
129. Barish, B. C. 1984. See Ref. 28
130. Tassie, L. J. 1965. Nuovo Cimento 38:1935
131. Cabrera, B. 1982. In Third Workshop on Grand Unification, ed. P. H. Frampton, S. L. Glashow, H. van Dam. Boston: Birkhäuser
132. Cabrera, B. 1984. See Ref. 28
133. Cabrera, B. 1982. Phys. Rev. Lett. 48:1378
134. Drell, S., Kroll, N., Mueller, M., Parke, S., Ruderman, M. 1983. Phys. Rev. Lett. 50:644
135. Malkus, W. V. R. 1951. Phys. Rev. 83:899
136. Kajino, F., et al. 1984. Phys. Rev. Lett. 52:1373
137. Goldhaber, A. S. 1983. See Ref. 27, pp. 1-17
138. Errede, S., et al. 1983. Phys. Rev. Lett. 51:245
139. Einhorn, M. B., Sato, K., 1981. Nucl. Phys. B 180:385

PION INTERACTIONS WITHIN NUCLEI

Mannque Rho

Service de Physique Théorique, CEN Saclay, 91191 Gif-sur-Yvette, France

CONTENTS

1. INTRODUCTION 531
2. STRUCTURE OF THE PION 535
3. PION AS NUCLEAR FORCE FIELD 537
   - 3.1 Paris Potential 537
   - 3.2 Pions and Spin-Isospin Modes 539
   - 3.3 Precursor Phenomenon 542
   - 3.4 Pionic Modes and Quenched g_A 545
4. "SEEING" PIONS 546
   - 4.1 Exchange Currents 546
   - 4.2 Chiral Filter: A Hypothesis 549
   - 4.3 Experimental Evidences for Soft Pions 551
5. "SEEING" $\Delta(1232)$ 557
   - 5.1 Quenching of g_A 557
   - 5.2 Δ-Hole Coupling 559
   - 5.3 Alternative to Δ-Hole Quenching 562
   - 5.4 Quark Degrees of Freedom? 564
   - 5.5 Vacuum Structure in Nuclei 565
   - 5.6 Deep Inelastic Structure Function in Nuclei 566
6. PIONS AND NUCLEON STRUCTURE 568
   - 6.1 Motivation 568
   - 6.2 Skyrme Soliton (Skyrmion) 569
   - 6.3 Leaking Baryon Charge 572
   - 6.4 From Skyrmions to Paris Potential 574
7. CONCLUSION 578

1. INTRODUCTION

One of the most exciting and challenging problems for nuclear physics is to exhibit or "see" the manifestation of the nucleon substructure, quarks and gluons, in a setting basically different from that of an isolated hadron. This is an important issue for deeper understanding of the physics of strong interactions, in particular for chiral symmetry and the question of confinement. It seems plausible that, when immersed in a baryon-rich environment as in nuclear matter, the properties of quarks change in a fundamental way: the vacuum must be readjusted, so that such fundamental properties as confinement and asymptotic freedom must be modified. In an extreme condition such as high temperature and/or high density, this seems inevitable as suggested in recent Monte Carlo simulations of quantum chromodynamics on lattice. But what is tantalizing is that such basic modification must already manifest itself at *normal* laboratory conditions. High-energy electron and heavy-ion machines that are planned for construction are expected to shed more light on this issue.

This review deals with an older problem, namely the role of pions in nuclear dynamics. This problem, although it has been with us for a long time, exposes a novel facet believed to be closely connected to the new problem mentioned above when the pion is looked at, not in terms of quark substructure per se, but as a Goldstone boson associated with the (nearly) massless up (u) and down (d) quarks. This way of describing the pion is not customary among nuclear physicists and most of the reviews available in the literature on the subject of pion-nuclear interactions have not made a full use of this aspect of pions, e.g. as a relic of chiral $SU(2) \times SU(2)$ symmetry. In this review, I follow this somewhat unorthodox approach, for, in this way, the problem of pion-nuclear interactions is seen to be closely connected to some essential properties of quantum chromodynamics.

The pion was predicted in 1935 by Yukawa (1) as a nuclear force field in analogy with electromagnetic force. It still remains, even today, as the main ingredient in the modern theory of strong interactions, at least for the longest-range part of the force. It was experimentally discovered in 1947 and since then has been most thoroughly studied. Its properties are among the best determined experimentally, including its interactions with complex nuclei. Yet the pion remains, to date, the most enigmatic and the least understood of the low-lying hadrons. The reason is that it has a dual character. On the one hand, the pion can be described as an "atomic" bound state of a quark (q) and an antiquark ($\bar{q}$) in a zero angular momentum, negative parity state. Viewed this way, the pion has no clear distinction from any other mesons, such as ρ or ω. It is an ordinary $q\bar{q}$ bag (in a bag model) or a bound $q\bar{q}$ state (in a potential model). Its mass is much lower than others, but its interactions with other hadrons can be described in the same way as other mesons are described, all with qualitatively satisfactory agreements with observation. Thus in this picture one would conclude that there is nothing special about the pion. This is the picture that emerges in both the potential model and the MIT bag model.

There is a completely different way of describing the pion; that is, the pion as a strongly collective mode. It is this, we think, that makes the pion very special when its interaction within nuclei is considered. There is compelling experimental evidence that the pion is a Goldstone boson, a relic of chiral SU(2) × SU(2) symmetry, believed to be almost exact. (The small mass of the pion, contrasted with other mesons, reflects how good the symmetry is.) Viewed as a Goldstone boson, the pion is an aristocrat, quite distinct from other mesons for several reasons. First of all, Goldstone bosons entail soft-meson theorems; thus pion-hadron interactions at low energies are reliably given by low-energy theorems. Secondly, because of its low mass, the pion dominates the long-range part of interactions between two nucleons with the vertices constrained by soft-meson theorems. And lastly—and perhaps most importantly—an effective QCD Lagrangian can be written entirely in terms of Goldstone boson fields, from which low-energy physics of mesons and baryons is to emerge. As such, the pion field has an intimate bearing on the structure of the strong interaction vacuum and excitations based upon it, in particular for the nucleon and the nucleus.

The connection between the two descriptions described in the preceding paragraphs is not known at present, although there have been many papers written on it.

The importance of the role of the pion as a Goldstone boson in nuclear physics was first noted by Brown & Green (2) in connection with three-body forces in nuclei. Current algebra and soft-pion theorems were fully developed by then, but when applied to nuclei they were nonetheless providing a new insight into phenomena hitherto obscured by complexities of the strong interaction dynamics. Chiral symmetry and its Goldstone realization were subsequently exploited in the systematization of meson exchange currents in nuclei (3), a subject initiated by Villars in 1947 (4). The advantage of this approach, as contrasted with other approaches, was that under certain circumstances that can be met in laboratories the prediction was unambiguous and model independent. There are also strong experimental evidences that this is indeed correct. What came as a complete surprise, however, is that in some cases, particularly in the electrodisintegration of deuterons at large momentum transfer and small energy transfer, the theory works way beyond its validity, a subtle hint that we are still not near a complete understanding of the apparently simple and elegant phenomena.

An important point to note here is not just that the soft-pion notion works so well in nuclei, but also what the implication might be when it does *not* apply. When an external field probes two or more nucleons at a short distance, the Goldstone mode of chiral symmetry seems no longer visible. Put differently, the pion cloud believed to surround the nucleons in nuclei appears to be inert to certain kinds of external fields. When this happens, it seems to happen precociously. An important indicator of this phenomenon is the axial-vector coupling constant $\tilde{g}_A$ in nuclear medium. To the extent that an effective coupling constant in a medium can be defined (which is open to dispute), the value $\tilde{g}_A$ over a wide range of nuclei, $10 \leq A \leq 238$, is tantalizingly close to 1, a value substantially lower than the free-space value $g_A = 1.25$. This suggests strongly that $\tilde{g}_A$ may reflect a fundamental property of nuclear medium, perhaps connected with the "vacuum" structure of the baryon-rich environment. One of our goals is to understand this.

According to a current thinking, a deeper understanding of the role of the pion in nuclei will be possible only when the structure of the nucleon itself is better understood. This is because the pion in the sense defined later plays a singularly essential role in the strong interactions of low-lying baryons. This was recognized more than two decades ago. In the early 1960s when there were no clear candidate theories for strong interactions, Skyrme (5, 6) proposed a unified model of hadrons constructed uniquely of Goldstone meson fields (sometimes referred to as chiral fields in this review) that could describe mesons and baryons at the same time. The manner in which baryons emerged from a Lagrangian containing meson fields only was highly unconventional: the spin arose from scalar fields via a topological twist. The advent of QCD, with its microscopic variables, quarks and gluons, as basic constituents of the hadrons had seemed to render Skyrme's ingenious idea only of an academic interest and banished his model to obscurity. Many physicists believed then that not only hadron structure but also nuclear dynamics must be treated explicitly in terms of quark degrees of freedom.

Several recent developments have brought profound changes to this dogma. To begin with, nobody has succeeded in calculating low-energy hadron properties purely from explicit accounts of quarks: in one way or other, one must resort to schemes that are not directly derivable from QCD, examples being the bag, string, or potential models, or to numerical simulations as in lattice gauge calculations. Furthermore there has been a growing tendency among gauge field theorists to search for effective field theories in terms of the "macroscopic" physical variables such as nucleons and mesons by integrating out the microscopic unphysical variables, such as quarks and gluons. One notable example in this direction is so-called large-N_c quantum chromodynamics (7, 8), where N_c is the number of color, believed to be three in nature.

Two parallel developments have brought Skyrme's old idea back to the forefront. Witten (9) showed that large-N_c QCD leads to an effective Lagrangian of the Skyrme type and that baryons *do* emerge as solitons from the effective theory, with the topological quantum number identified with the baryon number. In other words, the Skyrme model is to represent QCD in the long-wavelength limit, and not an alternative. Independently it has been noted (10, 11) in the context of bag models incorporating chiral symmetry (called chiral bags) that, while the color is confined, the baryon number is not confined within the bag but leaks out into the "meson cloud" surrounding the bag. This phenomenon turns out to be very natural when interpreted in terms of the topological charge identified with the baryon charge. In this respect, the "pion cloud" outside the bag plays an unusually crucial role for the structure of the nucleon inside—and perhaps also outside—of the nucleus. I believe that this aspect in its qualitative form is independent of any particular models and could provide a key to a deep understanding of nuclear structure.

2. STRUCTURE OF THE PION

In the nonrelativistic quark picture (a recent account in the context of QCD is found in 12), the pion is a bound state of a quark-antiquark pair in a potential. In the MIT bag model (13), the pion is a $q\bar{q}$ pair confined within a region, "bag." In this picture, the large splitting between the pion and other mesons (i.e. ρ) is attributed to perturbative gluon-exchange effects. In all respects, the pion is not distinct from other mesons. Simple and economical, this picture offers a description that invokes only the microscopic degrees of freedom of QCD (apart from confinement potentials or bags that are not yet derived from the theory). There is a difficulty with this picture, however, that cannot be overcome in a natural way: The low mass of the pion is an anomaly and no believable explanation exists for the multitude of pion interactions that are successfully described by the soft-pion theorems of the 1960s (reviewed in 14); one-pion exchange process between nucleons—one of the best-tested properties of the pion and, as such, a landmark in nuclear physics—poses a great problem. The cause for this defect is not hard to find. Chiral symmetry plays no obvious role in this scheme.

If the quark masses for the u quark and d quark are set equal to zero, then the QCD Hamiltonian is symmetric under the chiral SU(2) × SU(2) group. It is believed that the ground state of this theory breaks this symmetry dynamically (15–17) to isospin SU(2), the hidden symmetry manifesting itself in three Goldstone bosons (for each broken generator). But strictly zero-mass bosons do not exist in nature. This is because in reality the u- and d-quark masses are not zero, though very small on an hadronic scale $[m_u \approx 5 \text{ MeV}, m_d \approx 9 \text{ MeV}$, see Gasser & Leutwyler (18). The quark masses are believed to originate from interactions at much higher energy than the domain of QCD]. The pions (π^+, π^-, π^0) with their small masses can be identified with the would-be Goldstone bosons. For our consideration, we may ignore this small explicit symmetry breaking and assume chiral symmetry to be exact. In the light of chiral symmetry the pion appears very special: there is no obvious connection to other mesons such as ρ or ω, etc. Its small mass is natural, and soft-pion theorems (14) follow.

There has been some progress in understanding the two different features of mesons. Although not yet fully understood, QCD must undergo the dynamical symmetry breaking of Nambu-Jona-Lasinino type. Goldstone bosons appear as a zero-mass dynamical pole (as a quark-antiquark bound state) and quarks pick up masses. In this description the pion can be viewed as a collective state while other mesons are like atomic bound states (15, 19, 20). The notion of collective states vs noncollective states is quite familiar to nuclear physicists. In fact, the picture of the pion as a spin-isospin sound has recently been sharpened by means of the well-known random-phase approximation (21, 21a). One finds that as is expected of a highly collective state, the pion wave function is complicated, containing in addition to a quark-antiquark component a strong admixture of multiquark-multiantiquark components.

If the pion is a Goldstone boson, then what are the other mesons, say, $\underline{36}$ of SU(6) in the nonrelativistic quark model? What about the mesons with massive quarks?

Possible answers to these questions have been suggested. Consider, for instance, the ρ-π puzzle. It has been argued (22) that the ρ meson is a dormant "Goldstone" boson in the sense that it would be a true Goldstone boson in the static limit, becoming massive as a result of relativistic as well as symmetry-breaking [SU(6)] corrections. As for the mesons with massive quarks satisfactorily described in terms of an atomic structure, one may distinguish them as being in a different phase from that of the pion, the pair-condensed phase. The phase transition as the quark mass becomes greater than some critical mass $m_c \approx 150$ MeV has been suggested (23) and demonstrated in a simple model (20). These developments strongly suggest a subtle relation between various mesons.

Since the chiral symmetry-breaking mass scale $\bar{m} = \frac{1}{2}(m_u + m_d) \approx 7$ MeV is so small compared with the characteristic mass scale of QCD, $M \approx 500$–1000 MeV, one would expect that considering the pion as a Goldstone boson and treating it as if it were a fundamental field would be a reliable description. One can have an idea as to how good such a description might be for long-wave length processes just with these mass scales: roughly speaking the low-energy theorems of SU(2) $\times$ SU(2) should be valid within a few percent. (This rule of thumb is not always valid but the corrections are under systematic control, see Ref. 18.)

In this paper then, I adopt the picture of the pion as the Goldstone boson field; I do not attempt to connect it to the underlying $q\bar{q}$ substructure.

Replacing this aspect of QCD by an effective field theory, one is then led to a description closely approximated by the σ model (24, 25)

$$\mathscr{L} = \mathscr{L}_0 + \mathscr{L}',$$

where the SU(2) × SU(2) [or O(4) to which it is isomorphic] symmetric Lagrangian $\mathscr{L}_0$ is of the form

$$\mathscr{L}_0 = -\frac{1}{2}[(\partial_\mu \sigma)^2 + (\partial_\mu \pi)^2] - \frac{\lambda}{4}[(\sigma^2 + \pi^2) - f^2]^2$$

in terms of the scalar (σ) and pseudoscalar (π) meson fields. The symmetry-breaking term $\mathscr{L}'$, taken to be proportional to σ, accounts for the small pion mass. When $\mathscr{L}' = 0$, a nonzero vacuum expectation value $\langle \sigma \rangle$, breaking the O(4) symmetry, corresponds to a nonvanishing quark-antiquark pair condensate [$\langle \bar{\psi}\psi \rangle \neq 0$] and signals spontaneous symmetry breaking.[1] In the Lagrangian $\mathscr{L}_0$, this happens when $f^2 > 0$ as one can see from the classical potential $(\lambda/4)[(\sigma^2 + \pi^2) - f^2]^2$. By shifting $\sigma \to \sigma + \langle \sigma \rangle$, one can also see that there are Goldstone bosons π^i (with $m_\pi = 0$, if $\mathscr{L}' = 0$). The scalar meson becomes massive, which tends to ∞ as $\lambda \to \infty$. In this limit, one gets chiral symmetry realized in the nonlinear fashion, the model becoming the nonlinear σ model.

In Sections 3 and 4 the pion is treated in the framework of this σ model, with an additional chiral-invariant coupling to the nucleon field. In Section 6, we encounter the same pion field; however, nucleon fields will *not* be added by hand as we do elsewhere. The nucleon will emerge from the Goldstone boson Lagrangian as a soliton.

3. PION AS NUCLEAR FORCE FIELD

The best-established part of the nuclear force is the longest-range part mediated by a single-pion exchange (see 26 for references). In the intermediate range, exchange of two pions, both correlated and uncorrelated, dominates. At shorter ranges, the pion plays no role. Roughly, this separation leads to the structure depicted in Figure 1. Later, the regions I and III will be "derived" from an effective theory based on QCD.

3.1 *Paris Potential*

The most successful potential based on meson field theory is the Paris potential [for an updated review, see Vinh Mau (27, 28)]. The basic idea

---

[1] When $\mathscr{L}' \neq 0$, there can be two pieces: one flavor independent and one flavor dependent, the latter being nonzero even if there were no spontaneous breaking of the symmetry.

here is to separate (somewhat arbitrarily for the time being) the interaction into two regions at an internucleon distance $r \approx 0.8$ F: the long- and medium-range region for $r \gtrsim 0.8$ F and the short-range region for $r \lesssim 0.8$ F, and to treat the exterior region in terms of one-, two-, and three-pion (replaced by an ω) exchanges and the interior region as completely phenomenological. The inputs for the theoretical part—in particular for the two-pion exchange piece—are the πN phase shifts $\delta_l^{\pi N}$ and the $\pi\pi$ phase shifts $\delta_l^{\pi\pi}$ ($l = 0, 1$). Since dispersion relations are used, all the πN isobars, $\pi\pi$ resonances, and nonresonant πN and $\pi\pi$ backgrounds are automatically taken into account. The short-ranged part is parametrized (with six parameters for each isospin channel) with an energy-dependent core for the central potential, the energy dependence being required to be linear by experiments.

How well this potential describes the world data on nucleon-nucleon interactions has been amply summarized elsewhere (e.g. 28). For our purpose, the important point to emphasize is that the long- and medium-range part of the interaction is accurately given by pion exchanges alone. In a subsequent section, we see that even the short-range part may eventually be derivable from a QCD-based effective meson field theory. Thus, as envisioned by Yukawa five decades ago, the pion is established to be a bona fide nuclear force field. Thus it is natural to expect that the pion would play a preeminently important role in the nucleus. The question is, how does the pion manifest its role in nuclear processes?

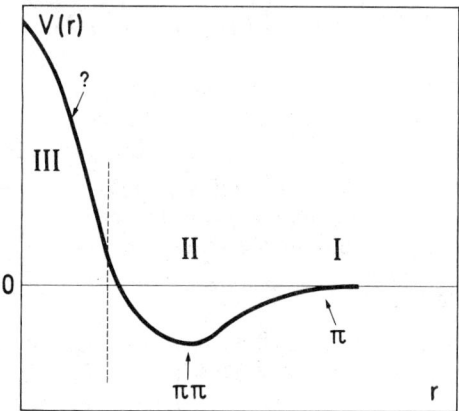

Figure 1 A schematic shape of the nucleon-nucleon potential. *I*. Region dominated by one-pion exchange; *II*. Region dominated by two-pion exchange (and three-pion exchange); *III*. Unknown region in meson theory, treated phenomenologically in the Paris potential.

3.2 Pions and Spin-Isospin Modes

The pion in nuclei interacts with nucleons through the Yukawa coupling

$$\delta \mathcal{L} = ig\bar{\psi}\gamma_5 \tau^a \psi \pi^a \quad \text{or} \quad \frac{f_\pi}{m_\pi} \bar{\psi}\gamma_5 \gamma_\mu \tau^a \psi \partial^\mu \pi^a, \qquad 1.$$

where ψ is the Fermion isodoublet field $\psi = \binom{p}{n}$, $\pi^a (a = 1, 2, 3)$ is the triplet pion field, and g and f_π are dimensionless coupling constants. To lowest order in f_π this leads to the well-known one-point exchange potential between two nucleons

$$V(\mathbf{q}, \omega) = -\frac{f_\pi^2}{m_\pi^2} \tau_1 \cdot \tau_2 \frac{\sigma_1 \cdot \mathbf{q}\, \sigma_2 \cdot \mathbf{q}}{\mathbf{q}^2 + m_\pi^2 - \omega^2}. \qquad 2.$$

An object of considerable interest is the way the pion propagates inside the nucleus [for extensive discussions, see the les Houches lectures (29); particularly Baym (30), Campbell (31), Moniz (32), Sawyer (33); also (26, 34)]. For simplicity, we consider a uniform nuclear matter but we have in mind finite nuclei as well. Then the pion propagator has the general form

$$D(\mathbf{q}, \omega) = [\omega^2 - \mathbf{q}^2 - m_\pi^2 - \Pi(\mathbf{q}, \omega) + i\varepsilon]^{-1}, \qquad 3.$$

where Π is the self-energy of the pion that contains all the dynamics of the π-nuclear interactions. The pole of the D function provides information on the pion spectrum inside the medium. The pole also signals a second-order phase transition, according to the theory of Landau & Lifshitz (35). This can be described as follows (30, 36). Let the ground-state expectation value of the pion field be denoted $\phi^a(x) = \langle \pi^a(x) \rangle$ with $a = 1, 2, 3$ the isospin index, which would vanish in the normal matter because of parity etc. The energy of the system can then be written (schematically) as a functional of ϕ:

$$E[\phi] = E_0 - \tfrac{1}{2} D^{-1} \phi^2 + \tfrac{1}{4} D^{-4} F \phi^4 + O(\phi^6), \qquad 4.$$

where E_0 is independent of ϕ, D is the connected two-point function given above, and F is the connected four-point function. We look for the minimum of the energy functional

$$\frac{\delta E[\phi]}{\delta \phi} = 0 = -D^{-1}\phi + D^{-4} F \phi^3 + O(\phi^5). \qquad 5.$$

There is a trivial solution $\phi = 0$ corresponding to the normal nuclear matter. A nontrivial solution $\phi \neq 0$ lowering the energy from that of the normal matter would lead to a pion-condensed phase. The critical point is characterized by

$$D^{-1}\phi = 0 \qquad 6.$$

as follows from the constraint $D^{-1}\phi - D^{-4}F\phi^3 + O(\phi^5) = 0$ at the minimum. This (hypothetical) pion condensation phenomenon drew considerable attention some years ago (30, 37–40).

Now let us see how D can develop a (double) pole. For symmetric nuclear matter $\omega = 0$, so the self-energy has to compensate $(\mathbf{q}^2 + m_\pi^2)$. For $\mathbf{q} = 0$, the attraction due to Equation 2 cannot contribute while the s-wave pion-nucleon coupling is repulsive. This requires that $|\mathbf{q}| \neq 0$. The question then is: How can the self-energy become sufficiently negative to compensate $(\mathbf{q}^2 + m_\pi^2)$? This boils down to asking how strongly the pion couples to nuclear excitations of the appropriate quantum number, namely $T = 1$, $J^\pi = 0^-$.

This problem has been a subject of extensive studies and we now have a rather simple, though somewhat qualitative, understanding of what is going on (41, 42). The property of the pion self-energy is determined (see Figure 2) by irreducible particle-hole interactions Γ in the pion channel. An extremely simple description of the effective particle-hole interaction can be made if one imagines Γ is generated from π and ρ exchanges (34, 42–43):

$$V_{\sigma\tau} = -\frac{f_\pi^2}{m_\pi^2}(\tau_1 \cdot \tau_2)(\sigma_1 \cdot \hat{\mathbf{q}})(\sigma_2 \cdot \hat{\mathbf{q}})\frac{q^2}{q^2 + m_\pi^2}$$

$$-\frac{f_\rho^2}{m_\rho^2}(\tau_1 \cdot \tau_2)(\sigma_1 \times \hat{\mathbf{q}}) \cdot (\sigma_2 \times \hat{\mathbf{q}})\frac{q^2}{q^2 + m_\rho^2}, \qquad 7.$$

where $q = |\mathbf{q}|$. This should be immersed in a short-range nuclear correlation, which renders the ρ exchange effective in the pion channel. (In the absence of correlation, the ρ cannot contribute in that channel, since the ρ couples transversely to the nucleon.) The precise form in which the effective potential appears depends, of course, upon the physics of short-range nucleon-nucleon interactions, which has remained ill understood up to now. (See, however, the new development in Section 6.) Roughly, one can write (for $\omega = 0$)

$$\Gamma_{\sigma\tau} = [W_l(q)\sigma_1 \cdot \hat{\mathbf{q}}\sigma_2 \cdot \hat{\mathbf{q}} + W_t(q)(\sigma_1 \times \hat{\mathbf{q}}) \cdot (\sigma_2 \times \hat{\mathbf{q}})]\tau_1 \cdot \tau_2 \qquad 8.$$

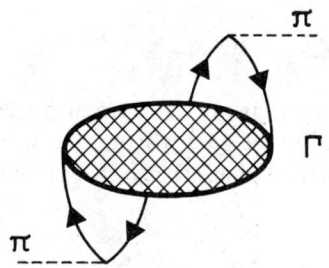

Figure 2 Irreducible particle-hole interaction Γ that appears in the pion self-energy.

with

$$W_l(q) = g'_0 - \frac{f_\pi^2}{m_\pi^2}\frac{q^2}{q^2+m_\pi^2}$$

$$W_t(q) = g'_0 - \frac{f_\rho^2}{m_\rho^2}\frac{q^2}{q^2+m_\rho^2}. \qquad 9.$$

There may be some momentum dependence in g'_0 for large q (G. Chanfray et al, 1983, unpublished) but this is not believed to be significant in the kinematic domain we are concerned with, so it can be ignored for our purpose. The g'_0 as defined this way is often referred to as the Landau-Migdal parameter. A simple estimate of g'_0 from folding the primary interaction (Equation 7) with a suitable correlation function yields (34, 42–44)

$$g'_0 \approx 0.5\text{--}0.7. \qquad 10.$$

This is also consistent with experiments (45, 46). There have been many attempts to calculate this quantity from a microscopic many-body theory, but it is hard to imagine that such calculations can be sufficiently reliable. The g'_0 represents a short-distance hadron-hadron interaction in the spin-isospin channel and we do not yet have a systematic and quantitative theory for this. (For an initial step in this direction, see Section 6.)

Excitation of the pion-like states with $T = 1$, $J^\pi = 0^-, 1^+, 2^-,\ldots$ contributes valuable information on the effective spin-isospin force. Let the initial state be $T = 0$, $J^\pi = 0^+$. Equation 8 can be rewritten to extract the relevant piece

$$(\tfrac{1}{3}W_l + \tfrac{2}{3}W_t)\boldsymbol{\sigma}_1\cdot\boldsymbol{\sigma}_2\boldsymbol{\tau}_1\cdot\boldsymbol{\tau}_2 = U(q)\boldsymbol{\sigma}_1\cdot\boldsymbol{\sigma}_2\boldsymbol{\tau}_1\cdot\boldsymbol{\tau}_2, \qquad 11.$$

where

$$U(q) = \left[g'_0 - \left(\frac{1}{3}\frac{f_\pi^2}{m_\pi^2}\frac{q^2}{q^2+m_\pi^2} + \frac{2}{3}\frac{\tilde{f}_\rho^2}{m_\rho^2}\frac{q^2}{q^2+m_\rho^2}\right)\right],$$

where $\tilde{f}_\rho$ includes effects from a short-range correlation function (44). The function $U(q)$ is plotted in Figure 3.

The repulsion at $q = 0$ [i.e. $U(0) > 0$] is rather well known from the position of collective spin-isospin modes excited at forward angle (e.g. giant Gamow-Teller resonances, see Ref. 47), and also from the quenching of the axial-vector coupling constant $\tilde{g}_A$ in nuclear medium (see the next section). There is some uncertainty as to how $U(q)$ behaves away from $q = 0$. Nonetheless, one can clearly see some qualitative features. Qualitatively one expects that, as q increases, attraction will set in in the manner suggested by Equation 11. Since g'_0 is (nearly) q independent, at some q the

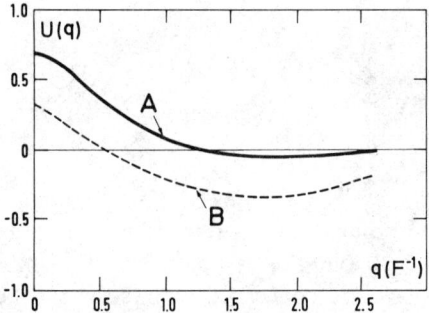

Figure 3 Irreducible particle-hole interaction in the pion channel $U(q)$ as defined in Equation 11. The curve labelled A is from (42, 46); the one labelled B is an example of a fictitious force that would give rise to a precocious pion condensation and to precursor phenomena.

attraction can win over the repulsion, leaving a net attraction. Whatever attraction is left over determines what happens in the nuclear system.

If one increases density, then the momentum distribution of the pionic mode may be made to peak at the region at which $U(q)$ is maximally negative and, given enough density, the system may subsequently make a phase transition (of course, a true phase transition can occur only in the thermodynamic limit). The critical density depends very strongly on the strength of the attraction, consequently on g'_0 on the one hand and on the momentum-dependent attraction on the other.

Unfortunately nature does not favor this intriguing phenomenon. The presently accepted values of g'_0, Equation 10, and of the (model-dependent) attractive force imply a critical density ρ_c that is much larger than the nuclear matter value ρ_0 ($\approx 0.17\,\mathrm{F}^{-3}$). Furthermore, even if it occurs at some high density, it is plausible that at the relevant density other degrees of freedom, in particular the microscopic variables of QCD (quarks and gluons), will become more relevant. If this is the case, the pion condensation phenomenon, at least in its present formulation, cannot be more than academic. (It is not excluded, however, in a different context, as discussed in Section 6.) Searches for specific signatures of this phenomenon have been unsuccessful and it still remains an open experimental question.

3.3 Precursor Phenomenon

The absence of pion condensation in nature does not, however, make the subject of pionic collective modes any less interesting. It is an important issue in its own right, as has been repeatedly emphasized by several people (48–50). In view of the intricate role the pion plays in baryon structure and consequently in nuclear structure (Section 6), it is an interesting problem—

and an important one—to map out the profile $U(q)$ (see Equation 11) as a function of q. There have been several suggestions (38, 51, 52) that a strong attraction in $U(q)$ at a momentum several times the pion mass (typically $2m_\pi \lesssim q \lesssim 4m_\pi$) could induce an anomalously large enhancement in electromagnetic, weak, or hadronic processes in some specific channels carrying the pion quantum numbers. This would be a precursor phenomenon, related to possible existence of a phase transition at some higher density.

Experimental searches for precursor phenomena have also been unsuccessful [for summary, see Weise (53) and Delorme (48)]. In some sense, this negative result was anticipated. An early argument against the possibility of precursors in a finite system was given by Barshay & Brown (54) and others (36, 55), based on what was known about pion-like states and transition amplitudes. A systematic analysis of the form of $U(q)$ by studying excitations of unnatural parity states of various angular momenta, namely $0^-, 1^+, 2^-, \ldots$, was carried out by Meyer-ter-Vehn (41) and by Speth and co-workers (42, 46). The results of the analyses were that a strongly attractive $U(q)$ in any range of q would be incompatible with the observed spectra.

The argument for this conclusion is simple and illuminating. It goes as follows. While the wave functions of some states such as the $T = 1, 1^+$ state peak at low q, those of other states such as $T = 1, 0^-, 2^-, \ldots$ and higher-spin states peak at higher q values. Therefore those states whose wave functions peak in the region where $U(q)$ is attractive will suffer downward shifts in energy, while those peaking at small q suffer upward shifts. A simple illustration was given by Meyer-ter-Vehn (41) for the $T = 1, 2^-$ state in ^{16}O. In Figure 4 is plotted the energy vs g'_0 with a schematic force of the type in Equation 11; here g'_0 controls the effective attraction or repulsion in the force. As the attraction is increased by decreasing g'_0, one notes that one particular state whose unperturbed configuration is $d_{3/2}p_{1/2}^{-1}$ sinks rapidly, crosses the lowest level (with unperturbed configuration $d_{5/2}p_{1/2}^{-1}$), and develops instability, while all the other levels undergo relatively small shifts. This behavior is a precursor phenomenon manifested in the spectrum. If one were to examine a correlation function for pion-like probes, one would then observe a dramatic enhancement corresponding to the momentum range responsible for the attraction. As the figure shows, the experimental levels are found little shifted from the unperturbed levels; in particular the would-be pionic mode is better described with $g'_0 > 0.7$.

One can also check the presence or absence of a significant attraction at $q \gtrsim 2m_\pi$ by looking at the spectra of much higher-spin states. For instance, the 12^- and 14^- states lying at 6–7 MeV in ^{208}Pb are sensitive to this region, as emphasized by Speth & Suzuki (42): the wave functions peak at

$q \approx 3m_\pi$. It was shown by a shell-model calculation (42) that these states lie close to the pure shell-model particle-hole energies, the energy shift to be accounted for being less than half an MeV. The conclusion arrived at from these calculations is that the predicted profile of $U(q)$ labelled A in Figure 3 is consistent with the available spectroscopic data and precludes any precursor phenomena in nuclei. One cannot exclude some residual attraction for $q \gtrsim m_\pi$ and it is a worthwhile effort to map this out experimentally but the chances for any precursor seem very small. The intriguing question is: Why and how has nature arranged to avoid what may have been the most natural manifestation of "pion presence" in the nuclear medium?

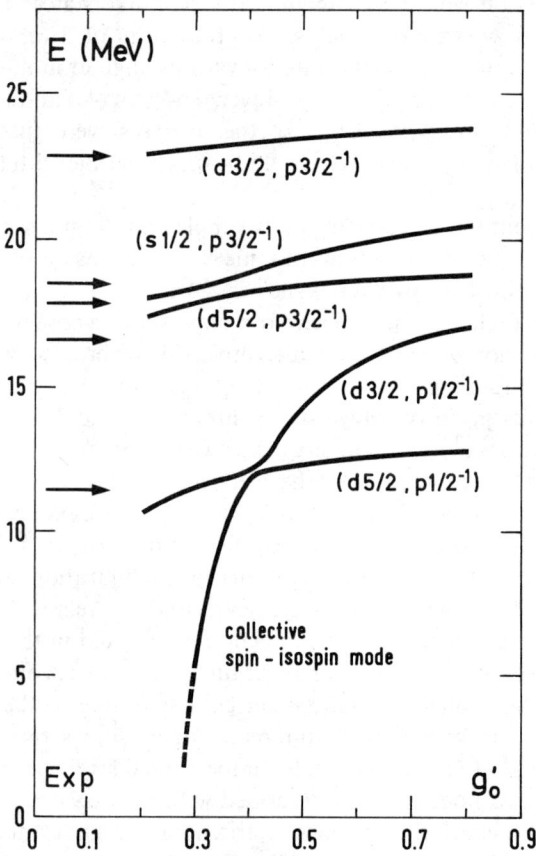

Figure 4 RPA energies for 2^- states in ^{16}O, taken from (41). The effective force used here contains only the pion contribution to the q-dependent term and hence is not fully realistic. However, the qualitative feature of the phenomenon is correct.

3.4 Pionic Modes and Quenched g_A

In the way the problem is formulated, the fate of nuclear matter depends sensitively on the numerical value of g'_0. In this regard, the question posed above immediately raises another question: Why is g'_0 found in the range 0.5–0.7 or stated otherwise why is it not 0.1 or 10?

To answer these questions, one would have to approach the problem from a more fundamental viewpoint. There is no satisfactory answer at this moment. (Some remarks are given later regarding this matter, particularly in Section 6.) Here I briefly discuss the possible connection of the quenching of the axial-vector coupling constant g_A observed in nuclear Gamov-Teller transitions to g'_0 and consequently to the absence of precursors. The quenching of g_A in a proper sense is discussed in detail in Section 5.

To see that there is a connection, let us consider the pion condensation phenomenon from the point of view of chiral symmetry. In the approach developed by Campbell, Dashen & Manassah (56), the pion-condensed phase is obtained by a "local" chiral transformation on the normal phase. Such a local transformation generates naturally the axial current A_μ. In free space, the strong interactions renormalize the axial-vector coupling of the nucleon from unity to $g_A = 1.249 \pm 0.006$, the value measured in neutron beta decay. Now pion condensation is a collective phenomenon whose interaction is dictated by the current A_μ appearing in the rotated Hamiltonian. The latter undergoes a many-body renormalization in the baryon-rich medium (the details of which are treated in Section 5). Thus the effective coupling in the medium involves, not the free-space g_A, but an effective $\tilde{g}_A$ that is empirically found to be strongly quenched. In fact in turns out to be very close to unity (57). A most appealing explanation for this quenching phenomenon is the direct participation of the isobar $\Delta(1232)$ in modifying the medium ("vacuum") structure (58, 59), through a Landau-Migdal parameter g'_0 effective in the Δ-hole channel introduced in a way analogous to that in the nucleon-hole channel. Now assume

$$(g'_0)_{NN} = (g'_0)_{N\Delta},$$

a relation referred to as a "universality" relation in the literature, and then we have a relation between the g'_0 crucial for the pionic modes and the $\tilde{g}_A$.

The $(g'_0)_{N\Delta}$ required to explain the quenching of $\tilde{g}_A$ is found to be consistent with $g'_0 \approx 0.5$–0.7. Thus it is seen that the amount of quenching in $\tilde{g}_A$ is commensurate with the inhibition of pion condensation as well as with the absence of precursors. One can understand this roughly in the following way. Whatever the basic mechanisms (some of which are discussed in Section 5), a quenched $\tilde{g}_A$ means that there is a repulsive interaction in the pionic channel. Furthermore as $\tilde{g}_A$ approaches unity, the effect of the pion cloud surrounding the nucleons in the medium disappears. This suggests an intimate connection between the observation that $\tilde{g}_A = 1$ and the absence of precursor enhancements in finite nuclei.

4. "SEEING" PIONS

While the "pion presence" within nuclei has so far eluded detection in the form of pionic collective modes, it is unambiguously seen in a different context, namely in exchange currents. The history of exchange currents is old, dating back to Villars' work in 1947. However, the experimental observation is very recent and its theoretical implication is being understood only now, thanks to the recent development in the theory of strong interactions. There is a simple regularity in the way pion degrees of freedom manifest themselves and it is believed that this holds a key to a deeper understanding of the structure of strongly interacting systems.

4.1 Exchange Currents

It may in principle be possible to formulate a fully consistent theory of many-body nuclear systems starting from a Lagrangian field theory. Indeed such an approach has been advocated by several workers, most forcibly by Walecka and his co-workers (60, 61). In this framework, there would be no need to talk about exchange currents: mesonic and nucleonic degrees of freedom could be treated on the same footing. Although there is some progress in this way of treating nuclear dynamics, it is not sophisticated enough—and hence not predictive enough—to confront the exceedingly accurate data amassed in the last decade or so. The conventional approach wherein first the nucleon degrees of freedom are treated as fully as feasible and then other degrees of freedom, e.g. mesons and nucleonic excitations, are taken into account as corrections still prove to be the most successful and informative.

Frequently, exchange currents are fraught with ambiguities, which is one reason why they have remained unobserved for so long. The ambiguities arise because exchange currents are defined according to how the nucleon-nucleon interaction is described and also how the effective configuration space that is active in the interaction is chosen; in practice, information on these is only partial. This is like the renormalization procedure in field theory. In field theory, one sets a cut-off Λ (say in momentum space), integrates out the degrees of freedom above Λ, and renormalizes coupling constants, masses, etc. in such a way that the theory is independent of how the cut-off is made. In carrying out such a program, Ward identities based on conservation laws are indispensable. In exchange currents, such a systematic approach based on Ward identities is often unavailable. (We discuss an example in connection with the axial current in Section 5.)

Fortunately there is one particular situation in which the theoretical prediction is believed to be unambiguous and free of model dependence; and that is the domain in which soft-pion theorems are applicable. Let me explain this in some detail.

Consider the scattering of two nucleons in the presence of an external current J_μ^a (μ is the Lorentz index and a the isospin or flavor index)

$$J_\mu^a(k) + \mathrm{N}(p_1) + \mathrm{N}(p_2) \to \mathrm{N}(p_1') + \mathrm{N}(p_2'). \qquad 12.$$

Here the current, which can be either the vector current V_μ^a or the axial-vector current A_μ^a, carries four-momentum $k_\mu \equiv (\mathbf{k}, k_0)$. Equation 12 applies directly to a two-nucleon process, but when a system of more than two nucleons is considered, it is a subprocess embedded in a many-body medium. To date, there is no indication for a strong multibody current other than two-body, so we assume the binary interaction suffices. To sharpen the argument as much as possible, it is illuminating to follow a reasoning based on chiral dynamics used specifically for $\pi\pi$ scattering (62, 63).

Suppose we describe the process in Equation 12 by a general chiral Lagrangian containing pion and nucleon fields only, assuming that all other degrees of freedom have been integrated out. The dynamics may then be described in terms of the number of pions exchanged and of loops in the process (Equation 12). A few examples are given in Figure 5.

According to the strategy of chiral dynamics (62, 63), we are led to construct an effective Lagrangian that is consistent with the symmetry in question, chiral SU(2) × SU(2). The Lagrangian consists of three kinds of terms, which I label $\mathscr{L}_\mathrm{M}$, $\mathscr{L}_\mathrm{F}$, $\mathscr{L}_\mathrm{I}$. The term $\mathscr{L}_\mathrm{M}$ is made of pion fields only. To be in accordance with the assumed symmetry, it must contain covariant derivatives D_μ. One can have

$$\mathscr{L}_\mathrm{M} = -\tfrac{1}{2}(D_\mu \pi)^2 + a(D_\mu \pi)^4 + b(D_\mu \pi \cdot D_\nu \pi)^2 + \text{higher derivatives} \ldots,$$

where $a, b, \ldots$ are arbitrary constants. The next term $\mathscr{L}_\mathrm{F}$ contains the nucleon field ψ and its coupling to the pion field,

$$\mathscr{L}_\mathrm{F} = -\bar{\psi}[\gamma_\mu D_\mu + m]\psi + g\bar{\psi}\gamma_\mu \gamma_5 \tau \psi \cdot D_\mu \pi.$$

These two terms are invariant under chiral transformation. Since, in nature, chiral symmetry is slightly broken as is evidenced by the pion mass $m_\pi \approx 140$ MeV, the third term $\mathscr{L}_\mathrm{I}$ is added to take into account the symmetry breaking. The net effect of this term in the context of the present application is to change the pion propagator from q_μ^{-2} to $(q_\mu^2 + m_\pi^2)^{-1}$ in all

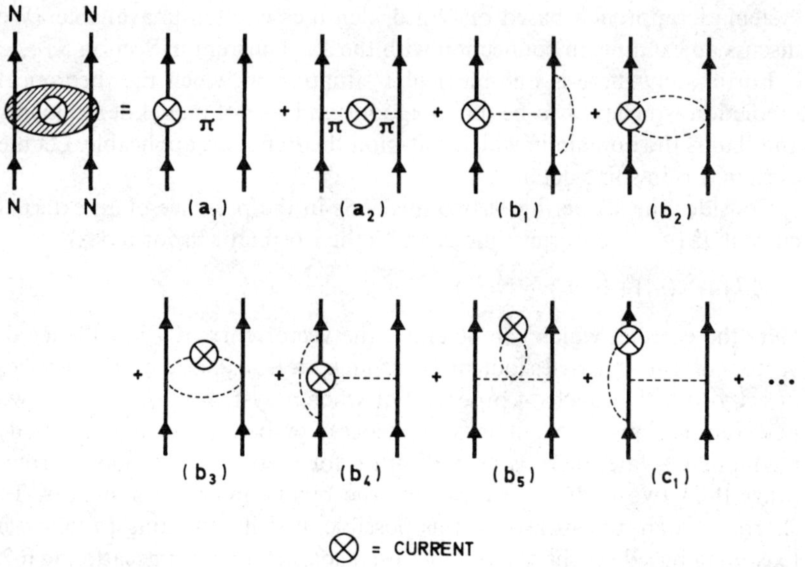

Figure 5 Chiral expansion in terms of pion exchange for the process $J_\mu^a + N + N \to N + N$. The cross denotes V_μ^a or A_μ^a. The terms (a_1) and (a_2) are referred to as "soft-pion terms"; the others are "non-soft-pion terms."

tree and loop terms to be calculated. This is the minimal prescription consistent with perturbative unitarity (62, 63). The next procedure is to calculate the graphs of the type in Figure 5 to all orders of the Lagrangian. This would then represent the full content of the theory, in particular, that of current algebra. Carrying this program to satisfaction for the process in Equation 12 is of course not practicable and there is perhaps no need for it. However, it is possible to extract very useful qualitative information by looking at a few low-order terms. This was demonstrated by Weinberg (62) for the $\pi\pi$ scattering.

The process we are considering is a lot more complex than the $\pi\pi$ scattering and no comprehensible result has yet been obtained. However, it still offers an appealing way to classify the various terms that contribute.

Let us denote by Q a typical momentum[2] carried by the exchanged pions in Equation 12. Putting the external nucleons on the mass shell, one can consider the amplitude in terms of a power series in Q

$$M = Q^D f(Q/\mu), \qquad 13.$$

[2] The effect of the pion mass is also of $O(Q)$. Therefore Q represents in an approximate way the momentum transferred or the pion mass or a combination thereof.

where f is a dimensionless function and μ a common renormalization scale on which the amplitude should not depend. In the nucleus, Q will typically range from zero to several times m_π, assumed to be small compared with typical hadronic scale ($\sim$ nucleon mass). The soft-pion theorem applies for the lowest D possible. It is easy to see on dimensional grounds that the graphs (a_1) and (a_2) in Figure 5 have the lowest D, the graphs (b_1)–(b_4) have $\Delta D = 2$ relative to (a_1) and (a_2), and so on. In the soft-pion limit, only the graphs (a_1) and (a_2) contribute. We will call these *the soft-pion terms* and the other diagrams *the non-soft-pion terms*.

The soft-pion terms are tree graphs, involving no integrations, while the non-soft-pion terms contain loops. So the ratio of the total amplitude to the soft-pion amplitude may be written (for finite Q) as

$$M/M_{\text{soft pion}} = 1 + \sum_{n \geq 1} Q^{2n} f_n(Q/\mu). \qquad 14.$$

The terms with $n \geq 1$ are the non-soft-pion corrections.

Two interesting observations can be made without detailed calculations:

1. The graphs of the type (b_1) are hadronic vertex corrections that can be phenomenologically described by form factors (when the graphs are summed to all orders). The graphs of the types (b_2) and (b_3), when summed to all orders, may be partly represented by resonances such as $\rho, \omega, \ldots$. Consequently the form-factor and vector-meson effects appear at the same level as non-soft-pion corrections. This is an important observation in understanding the experimental results.

2. The graph (c_1) with $\Delta D = 2$ can be viewed as a part of the isobar (Δ) diagram in the framework of the Chew-Low meson theory (64) (see Figure 6). Thus the Δ contribution also appears at the same level as the form-factor and vector-meson contributions.

4.2 *Chiral Filter: A Hypothesis*

The general argument given in the preceding section, supplemented with one additional hypothesis, has a surprisingly predictive power. We must

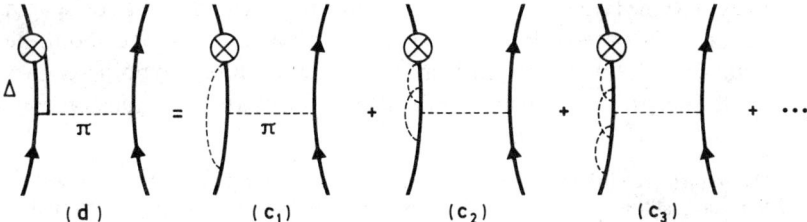

Figure 6 The Δ contribution to exchange currents as viewed in terms of pion-loop graphs in the sense of the Chew-Low theory.

assume that whenever the vertex $J_\mu + N \to N + \pi$ is enhanced relative to the current vertex without pion, then the soft-pion terms dominate over the non-soft-pion terms; otherwise the soft-pion terms are completely screened (65). This we will call a "hypothesis of chiral filter." It should be emphasized here that to date there is no rigorous justification for this notion, so it must appear ad hoc, but it seems fairly well respected by nature, a strong suggestion that it has something to do with a basic structure of the "vacuum." It remains an open question why.

Let us proceed and see what we can predict. Consider the graph (a_1) in Figure 5 where the current is either vector or axial vector.[3] The vertex V_μ^a (or A_μ^a) + N → N + π^b can be written down from current algebra or equivalently read off from the effective Lagrangian. For the vector current it is of the form

$$\varepsilon^{abc} A_\mu^c \qquad \qquad 15.$$

and for the axial-vector current it takes the form

$$\varepsilon^{abc} V_\mu^c. \qquad \qquad 16.$$

Note that because of the pion, a parity change occurs at the vertex. Since nucleons in nuclei are fairly nonrelativistic, we can look at the non-relativistic reduction of these expressions and compare them with the current vertex *without* pion emission, the latter representing the single-particle process occurring within nuclear medium that is usually referred to as impulse approximation. The results of this simple analysis are summarized in Table 1.

The prediction is deceptively[4] simple. The soft-pion term should be a predominant contribution for the space part ($\mu = 1, 2, 3$) of the meson exchange vector (isovector) current and for the time part ($\mu = 4$) of the meson exchange axial-vector current (65, 66). It came as a complete surprise that this prediction is supported by experiments beyond what was naively thought to be its range of validity. The former can be measured in magnetic dipole transitions (higher multipoles are not suitable because they involve higher momentum transfers and the transition operators are more complex) and the latter in $0^+ \leftrightarrow 0^-$ beta transitions with a change of isospin by one unit. This prediction does not say anything quantitative about the importance of the non-soft-pion terms. The "chiral filter" hypothesis goes further to suggest that whenever the soft-pion terms are large, the non-soft-

---

[3] The graph (a_2) contributes for the vector current, but not for the axial-vector current.

[4] Note the qualification *deceptively*, since I believe there is an underlying reason that is deep and goes beyond what is apparent and that is, at present, completely obscure.

Table 1 Soft-pion contribution to the exchange-current vertex compared with the impulse approximation vertex

Currents J_μ	$J_\mu + N \to N$ (impulse)		$J_\mu + N \to N + \pi$ (soft-pion)	
	$\mu = 4$	$\mu \neq 4$	$\mu = 4$	$\mu \neq 4$
V_μ	$O(1)$	$O(P/M)$	$O(P/M)^b$	$O(1)^a$
A_μ	$O(P/M)$	$O(1)$	$O(1)^a$	$O(P/M)^b$

[a] Soft-pion dominance.
[b] Soft-pion suppression.

pion terms are negligible. This is a strong assumption that need not always be valid. We have to resort to experiments to assess this point.

4.3 Experimental Evidences for Soft Pions

The most dramatic case in support of the theoretical argument given above is the electrodisintegration of the deuteron at large momentum transfer and small energy transfer. Early theoretical work on this process was done by Riska and co-workers (67), and subsequently many calculations have been repeated with various improvements (for a recent discussion, see 66, 68). It turns out, surprisingly enough, that the initial calculation remains one of the most successful compared with the recent data obtained at much higher momentum transfers.

Actually the first clear evidence for the "pion presence" in nuclei was obtained in the inverse process, the thermal neutron capture by a proton:

$$n + p \to d + \gamma. \qquad 17.$$

The 10% discrepancy persisting for many years between the measured cross section and the theory (impulse approximation) was explained for the first time by Riska & Brown (69) in terms of the soft-pion terms; this calculation has since been corroborated (70 and references therein). The deuteron electrodisintegration

$$e + d \to e' + n + p, \qquad 18.$$

by extending momentum transferred, enables the soft-pion exchange effect to show up more dramatically. This is a particularly "clean" process, since the uncertainty in the conventional treatment, that is, in terms of the nucleon degrees of freedom, is greatly minimized and consequently whatever discrepancy there is between theory and experiment can be interpreted unambiguously. The process is predominantly a magnetic dipole (M1) transition between the initial deuteron state and the final $^1S_0$ n-p state with an average energy $E_{np} \approx 1.5$ MeV. In the impulse approximation, the process is quite sensitive to the S and D contents of the deuteron wave function, especially for the squared momentum transferred around $t \equiv q^2 \approx 12$ F^{-2} at which the S and D contributions nearly cancel. This can be seen in Figure 7. While this sensitivity to the wave-function components introduces some uncertainty in the theory, it is also welcome as it greatly amplifies the nonnuclear effects that one hopes to see and that are normally suppressed. This fact was noted and elegantly exploited by Riska and collaborators (67). Their result, extended to higher momenta by Mathiot, is entirely given by the soft-pion exchange currents and is plotted in the dot-dashed line in Figure 7. When this calculation was first done, there were only a few experimental points available. Since then, data points were measured at Saclay to much larger momentum transfers (71, 72). The agreement is surprisingly good up to $t \approx 30$ F^{-2}, much beyond the domain of validity.

Unfortunately there is a fly in this ointment having to do with the

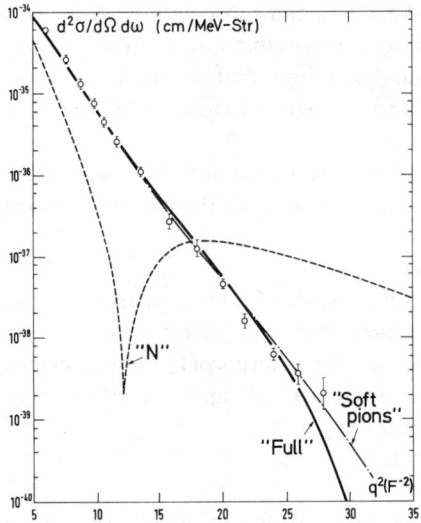

Figure 7 Electrodisintegration of the deuteron for small energy transfer ($E_{np} \approx 1.5$ MeV) and as function of momentum transfer. The data are from (71, 72), the theoretical curves from J. F. Mathiot (private communication); "*N*" corresponds to impulse approximation; "*soft-pions*" corresponds to impulse + soft-pion exchange current; and "*full*" corresponds to impulse + π and ρ exchange currents with hadronic form factors + Δ excitation current with hadronic form factors (see Figure 8).

electromagnetic form factors. The successful prediction made with the soft-pion exchange currents (Figure 5, a_1, a_2) depends upon using the vector electromagnetic form factor $F_1^V(q^2)$ and no (purely) hadronic form factors. Were one to require *a posteriori* the vector current conservation, additional terms would be needed. This is a well-known problem, for instance in the electroproduction of pions from nucleon (73). It is model dependent, however, as long as the electromagnetic form factor effects are not computed systematically with a local field theory. One possible remedy was suggested by several authors (74, 75): to replace F_1^V by the nucleon charge form factor $G_E^V(q^2)$. While restoring the current conservation, this unfortunately upsets the agreement with experiments. (The same problem occurs with the magnetic form factor of ^{3}He.)

While it constitutes one possible remedy, it is not clear that the replacement is justified otherwise. There are several reasons for this doubt. The most obvious is that the difference between F_1^V and G_E^V is of order $O(q^2/M^2)$ and corrections of this order are difficult to take systematically into account. There is no known reason to expect that higher-order corrections conspire just to give G_E^V in place of F_1^V; this is *not* what happens in the pion electroproduction (73). Furthermore the procedure is not unique; it depends sensitively on various assumptions. This is clearly an unsatisfactory situation that requires clarification.

To proceed, let us assume that one is justified in using F_1^V as in Figure 7. Now why should the soft-pion terms be relevant at such large momentum transfers at which the Q—the characteristic momentum defined in Equation 14—need not be small? The only way the result makes sense is that the non-soft-pion terms cancel among themselves. Due to complexity of the process, it is not known whether one can systematically calculate higher-order terms starting with a chiral Lagrangian and verify the cancellation. However, it is possible, with some simplifying assumptions and guidance from experiments, to estimate the non-soft-pion corrections. This has been done by several workers (76, 77).

In particular, Mathiot (77) calculated the corrections by assuming the dominance of π and ρ exchanges and the excitation of Δ, with appropriate form factors as defined in Figure 8. The result is given by the solid line in Figure 7. Note that a surprisingly large cancellation takes place among the corrective terms and persists up to momentum transfer $t \approx 25$ F^{-2}. The extent to which the cancellation occurs depends sensitively on the input parameters (coupling constants, cut-offs in the form factor, etc), but it is clear that higher-order terms, though individually large, do sum to a result considerably smaller than the soft-pion terms. Thus it is not an intrinsic smallness of the Q in Equation 14 but *suppression* of the coefficient multiplying higher powers of Q that appears to be more relevant. If this is not an accident there must then be a fundamental reason for it. We can, at the moment, offer nothing better than the "chiral filter" hypothesis.

Another process exhibiting the "pion presence" is the ^{3}He magnetic form factor (78–82). While there is an unmistakable signature for meson exchange currents, the situation here is somewhat less clear than in the deuteron electrodisintegration: complexity of the wave function obscures the validity of the various approximations made in the current operators and, more importantly, nucleons in ^{3}He are more close packed, so that soft-pion theorems may break down sooner than in the deuteron. (An evidence in support of the latter is the increasingly important role played by the isobar Δ in the quenching of magnetic dipole strength in heavy nuclei; see 83, 84.)

We now turn to a situation offering a more favorable test of the notion that chiral symmetry is "filtered" in nuclei: the time component of the axial current, $A_0(x)$. Theoretically, this operator is even simpler than the magnetic dipole (M1) operator. Because of G-parity, there is only one nonvanishing soft-pion term, namely (a_1) of Figure 5. (In the M1 case, the two soft-pion terms tend to interfere destructively, at least for low-momentum transfers, which makes the net result somewhat less important for heavy nuclei.) Unfortunately it is difficult to isolate experimentally the effects that are sensitive to the axial charge density. So far, the pion-exchange effect has been looked for in two classes of measurements: One, accurate correlation measurements of the β transition $1^+ \to 0^+$ in the famous mass-12 triad (85–87); the other, very weak $0^+ \leftrightarrow 0^-$, $\Delta T = 1$ transition rates. Measurements in different nuclei have been made

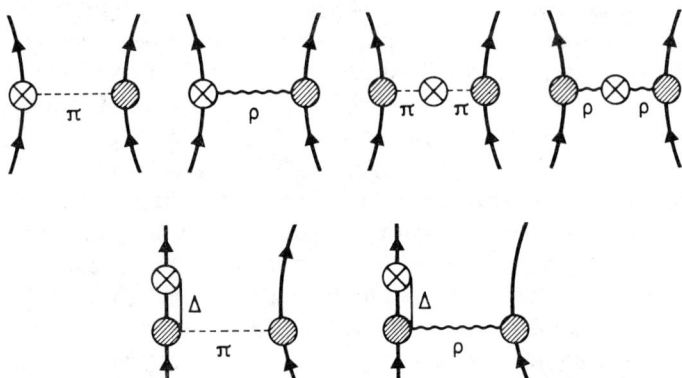

Figure 8 Phenomenological model used to take into account the non-soft-pion terms as defined in Figure 5. (See the curve labelled *"full"* in Figure 7). The hadronic form factor used is of the form: $F(k^2) = (\Lambda^2 - \mu^2)/(\Lambda^2 + k^2)$, $\Lambda_\pi = 1.25$ GeV, $\Lambda_\rho = 1.5$ GeV, $\mu_\pi = 0.14$ GeV, and $\mu_\rho = 0.77$ GeV.

in the latter class: ^{16}N(0^-) → ^{16}O(0^+) (88–90); ^{16}C(0^+) → ^{16}N(0^-) (91); ^{18}Ne(0^+) → ^{18}F(0^-) (92); ^{11}Be($1/2+$) → ^{11}B($1/2-$) (93).

One can make a simple yet fairly accurate estimate of the size of the soft-pion exchange term relative to the single-particle process. The axial charge operator is

$$A_0^\alpha(x) = A_{0I}^\alpha(x) + A_{0\pi}^\alpha(x), \qquad 19.$$

where

$$A_{0I}^\alpha(x) = -g_A \sum_i \frac{\tau_i^\alpha}{2} \frac{(\boldsymbol{\sigma}_i \cdot \mathbf{p}_i)}{M} \delta(\mathbf{x}-\mathbf{x}_i) + g_A \sum_i \frac{\tau_i^\alpha}{2} \frac{(\boldsymbol{\sigma}_i \cdot \mathbf{k})}{2M} \delta(\mathbf{x}-\mathbf{x}_i)$$

$$A_{0\pi}^\alpha(x) = \frac{m_\pi^2 g_{\pi NN}}{8\pi M F_\pi} \sum_{i<j} \frac{1}{2} (\boldsymbol{\tau}_i \times \boldsymbol{\tau}_j)^\alpha [\boldsymbol{\sigma}_i \cdot \hat{\mathbf{r}} \delta(\mathbf{x}-\mathbf{x}_i) + \boldsymbol{\sigma}_j \cdot \hat{\mathbf{r}} \delta(\mathbf{x}-\mathbf{x}_i)]$$

$$\times \frac{e^{-m_\pi r}}{m_\pi r}\left(1 + \frac{1}{m_\pi r}\right).$$

Here the subscripts I and π refer, respectively, to the single-particle and soft-pion exchange operators, $g_{\pi NN} \approx 13.7$, $g_A = 1.25$, M is the nucleon mass, m_π is the pion mass, and $r \equiv |\mathbf{r}_i - \mathbf{r}_j|$; $F_\pi \approx 93$ MeV is the pion decay constant. Evaluating the matrix elements in the Fermi gas model, one finds the ratio (48)

$$R \equiv \frac{(\text{2-body})}{(\text{1-body})} = \frac{g_{\pi NN}^2}{4\pi^2} \frac{1}{g_A^2} \frac{P_F}{M}\left[1 + \frac{m_\pi^2}{2P_F^2} - \frac{m_\pi^2}{2P_F^2}\left(1 + \frac{m_\pi^2}{4P_F^2}\right)\ln\left(1 + \frac{4P_F^2}{m_\pi^2}\right)\right],$$

where P_F is the Fermi momentum. In nuclear matter, $P_F = 1.35$ F^{-1} and $R = 0.62$. The density dependence is rather weak, the ratio reducing to $R = 0.44$ at half the nuclear matter density. More significantly, this ratio is remarkably model independent, i.e. independent of nuclear models and nuclear structure. Thus one expects the ratio to be in the range $0.4 \lesssim R \lesssim 0.6$ for any nucleus undergoing transitions via the axial charge density operator. This is a big effect, confirmed also by realistic calculations.

In all cases studied so far (96), there is no experimental evidence against this prediction. On the contrary, in some cases, experiments are definitely consistent with a large pionic contribution. For instance in the case of the mass-12 triad, careful analyses (94, 95) clearly require $R \gtrsim 0.4$ to understand the experimental correlation data. A particularly favorable case in which the effect is spectacular is the $0^+ \leftrightarrow 0^-$ transition in the mass-12 system:

$$^{16}\text{N}(0^-) \to {}^{16}\text{O}(0^+) + e^- + \bar{\nu}_e \qquad 20.$$

$$\mu^- + {}^{16}\text{O}(0^+) \to {}^{16}\text{N}(0^-) + \nu_\mu. \qquad 21.$$

An advantage of considering these two transitions is that the single-particle operators (there are two) contributing to the processes whose magnitudes are individually comparable to the soft-pion exchange term tend to cancel in the first transition, thus enhancing the soft-pion correction, while they tend to add in the second process, thus diminishing the effect. It was predicted that the soft-pion exchange current would enhance the decay rate of Equation 20 by a factor of 3 ∼ 4 relative to the rate given by the impulse approximation (65, 97). A careful calculation by Towner & Khanna (98) confirmed this prediction, which agrees with the nice measurements of Palffy et al (88) and Gagliardi et al (89). What makes this less than a triumphant confirmation of the "chiral filter" hypothesis, however, is that the calculation is obscured by corrections of the conventional nuclear structure type, namely the core polarization induced by the tensor force, that are model dependent and can be large. These corrections have to be made before a meaningful statement can be made about the role of pionic currents. They are in principle calculable but only in perturbation theory. The great uncertainty is that it is not clear whether the series converges and whether, even if it converges, a consistent, if not complete, set of graphs is calculated. Table 2, which is taken from the work of Towner & Khanna (98), illustrates this point.

The next important question to address concerns the non-soft-pion corrections. These must be small in order for the "chiral filter" hypothesis to make sense. There have been several calculations of these correction terms (99, 100). Although they are not as systematic as in the vector-current case, the result nevertheless is quite similar: there is again a strong cancellation

Table 2 The $0^+ \leftrightarrow 0^-$ β decay (Λ_β) and μ absorption (Λ_μ) in the mass-16 systems, calculated with and without soft-pion exchange and/or core polarization (C.P.) and comparison with experiments (Exp) (see 96, 98)

	Impulse			Impulse + soft-π-exchange		
	Λ_μ^a	Λ_β^b	$\Lambda_\mu/\Lambda_\beta^c$	Λ_μ^a	Λ_β^b	$\Lambda_\mu/\Lambda_\beta^c$
0th order	2.30	0.29	7.9	3.26	0.94	3.5
+ 1st order C.P.	2.30	0.18	12.7	3.11	0.64	4.8
+ 2nd order C.P.	1.48	0.12	12.2	2.00	0.42	4.8
Exp				1.56±0.11	0.41±0.05	3.8±0.5

[a] In units of 10^3 s^{-1}.
[b] In units of s^{-1}.
[c] In units of 10^3.

among higher-Q terms. In addition, unlike the M1 case, individual non-soft-pion terms are intrinsically suppressed, typically less than 10% of the soft-pion term. It would be interesting to see whether the higher-Q terms persist in being unimportant at higher momentum transfers, a relevant issue if and when nuclear axial charge form factors can be measured experimentally.

5. "SEEING" $\Delta(1232)$

We have noted in the previous section that the $O(1)$ cases in Table 1 are favorable for "seeing" pions in nuclei, largely because of a strong suppression (whose origin is unclear) of non-soft-pion contributions. We now turn to a situation in which the soft-pion term is *intrinsically* suppressed, either by particular selection rules or by trivial kinematics. Such is the case, as one can see in Table 1, for the time component of the vector current (isovector charge density) and for the space component of the axial-vector current (e.g. Gamow-Teller transitions). The former, measured in the form of nuclear charge form factors, does not for the moment lend itself to a simple theoretical interpretation, but the latter has a candidate description in terms of the nucleon excitation $\Delta(1232)$ that is extremely simple and predictive and also may have some profound implications on the vacuum structure inside the nucleus.

Let me note that since the advent of meson factories, there is nothing unusual about seeing Δ's in nuclei. The role of the Δ in nuclei has been abundantly studied in nuclear processes induced by electromagnetic probes [see, for a review, Laget (101)] and by hadronic probes [see, for a review, Moniz (102)]. The Δ is seen not only singly, but also doubly (103). I do not discuss these matters here; readers are referred to the review articles mentioned above or available elsewhere. In this section, I focus on the role of the Δ in nuclei in connection with a possible basic renormalization of a fundamental interaction caused by the baryon-rich environment, an aspect that is frequently not appreciated by nuclear physicists.

5.1 Quenching of g_A

The possibility that the axial current may be modified in nuclei was considered a long time ago by several people [e.g. Migdal (111), M. Ericson (126), Rho (1967. *Phys. Rev. Lett.* 18:671)]. To motivate the theoretical argument to be developed later, we first summarize the present status of the observations. Wilkinson (104, 105) was the first to explore systematically the possibility that, in nuclei, the axial-vector coupling constant $\tilde{g}_A$ governing Gamow-Teller transitions may be modified significantly from the free-space value $g_A \approx 1.25$. The importance of this observation is not merely that a suitably defined effective coupling constant differs, for certain transitions, from its measured free-space value, but that as in field theory there exists a global renormalization for *all* transitions that are mediated by the axial-vector current. It was noted that the effective coupling constant $\tilde{g}_A$ was systematically renormalized downward in finite nuclei (105).

Several recent developments lend strong support to the notion that the quenching phenomenon is intrinsic to the basic property of the "vacuum" defined by a baryon-rich medium. A most remarkable observation, made by Buck & Perez (57) based on a new and model-independent analysis of beta decay and magnetic moment data of the mirror nuclei, $3 \leq A \leq 39$, is that the axial coupling constant in nuclei is unity to a very good accuracy,

$$\tilde{g}_A = 1.00 \pm 0.02. \hspace{2cm} 22.$$

The fit is shown in Figure 9. An additional—and in fact more dramatic— evidence comes from the recent beautiful (p,n) data on the giant Gamow-Teller resonances (106; see, for review, 107–109). It is observed over a wide range of nuclei ($14 \leq A \leq 238$) that, while the location of the resonance is well described in terms of nucleons alone, a large portion of the strength

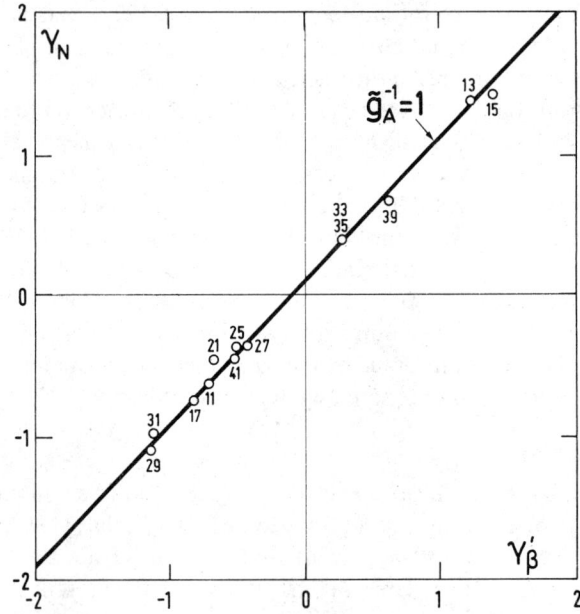

Figure 9 The Buck-Perez plot of the "gyromagnetic ratio" $\gamma_N = \mu_N/J$ vs $(-3.826) \times \gamma_\beta \equiv \gamma'_\beta$ where μ_N stands for the magnetic moments of odd-neutron members of mirror pairs and γ_β is related to the β-decay ft value for mirror pair transitions. The numerical values are taken from (57).

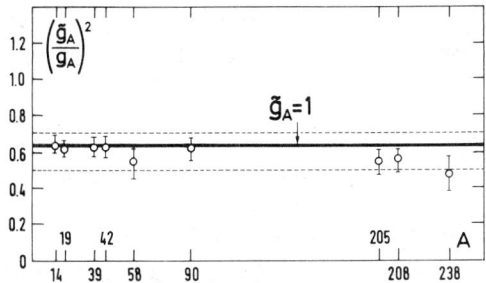

Figure 10 Fraction of Gamow-Teller sum rule strength observed $(\tilde{g}_A/g_A)^2$ in (p,n) reactions (taken from 107–109). The solid line is the Buck-Perez result $\tilde{g}_A = 1$, the area between dotted lines indicates the uncertainty due to background.

expected from the sum rule is missing from the measured strength. This paucity of strength cannot be explained (in a natural way) by the nucleon degrees of freedom alone. Some of the presently available data are plotted in Figure 10. The strength implied by the Buck-Perez constant (Equation 22) is consistent with the Gamow-Teller resonance data.

There is a bit of uncertainty on the exact amount of quenching in the Gamow-Teller strength because there is the problem of background subtraction. There is no clear-cut way of ascertaining how much constitutes uninteresting background to subtract away and hence we cannot yet be sure that all the "observed" quenching is caused by a fundamental mechanism. Within the uncertainty, however, the quenching implied by Equation 22 seems perfectly reasonable. In what follows, we will assume Equation 22 as established and discuss what we can learn from it.

5.2 Δ-Hole Coupling

We discuss in this section how the quenching can be understood. For this, a fundamental quenching is taken seriously. Later, less fundamental or more exotic mechanisms are mentioned for completeness.

The simplest and the most systematic approach is to assume that for the space part of the axial-vector current, the external field, $\mathbf{A}^a$, is completely blind to the pion field. (In what sense this is plausible is discussed below when we look at the nucleon structure proper.) This means that the renormalization of the axial current is caused by the change of environment for the β-decaying nucleon when it comes in close encounter with one or more nearby nucleons (110). It cannot occur through an influence at large distances, since at large distances it will be influenced predominantly by one-pion exchange and this, we are assuming, is not seen by the field $\mathbf{A}^a$ (see Table 1).

Written in terms of the nucleon variables only, the total axial current at zero momentum transfer is of the form

$$\mathbf{A}^a(0) = \tilde{g}_A \sum_i \frac{\tau_i^a}{2} \boldsymbol{\sigma}_i \qquad 23.$$

where the constant $\tilde{g}_A$ is so defined as to take into account the medium renormalization. Since the intrinsic operator induces a spin-isospin flip of the nucleon, the transition mediated by this operator is expected to be specific to the spin-isospin-flip component of the force. In a nucleus, such a transition corresponds, say, to a neutron in an occupied state transformed to a proton in an unoccupied state, so the relevant force would, in a particle-hole channel, be of the form (in momentum space)

$$g_0' \boldsymbol{\tau}_1 \cdot \boldsymbol{\tau}_2 \boldsymbol{\sigma}_1 \cdot \boldsymbol{\sigma}_2. \qquad 24.$$

The g_0' is, with a suitable normalization, related to the Landau-Migdal parameter (40, 111).

The static [SU(4)] quark model and equivalently the topological soliton model of Skyrme discussed below suggest that to a good approximation the spin-isospin-flip operator excites an isobar $\Delta(1232)$ as easily as it does a nucleon, with forces of the form (40)

$$\bar{g}_0' \frac{f_\pi^*}{f_\pi} (\mathbf{T}_1 \cdot \boldsymbol{\tau}_2 \mathbf{S}_1 \cdot \boldsymbol{\sigma}_2 + \boldsymbol{\tau}_1 \cdot \mathbf{T}_2 \boldsymbol{\sigma}_1 \cdot \mathbf{S}_2) \qquad 25.$$

$$\bar{\bar{g}}_0' \frac{f_\pi^{*2}}{f_\pi^2} (\mathbf{T}_1 \cdot \mathbf{T}_2 \mathbf{S}_1 \cdot \mathbf{S}_2), \qquad 26.$$

where S and T are transition spin and isospin operators, respectively, connecting N and a Δ; $f_\pi \approx 1$ is the πNN coupling constant, and $f_\pi^* \approx 2$ the πNΔ coupling constant. The interactions in Equations 24–26 are represented by the diagrams in Figure 11.

Given sufficient energy and momentum (as in a neutrino-induced

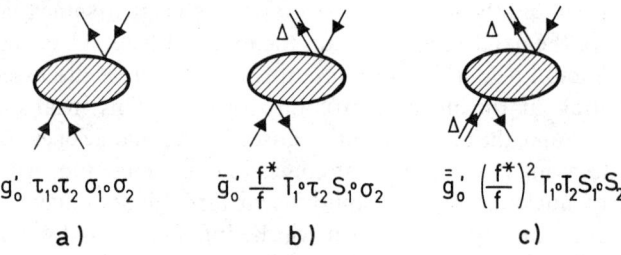

Figure 11 Definition of g_0', $\bar{g}_0'$, $\bar{\bar{g}}_0'$ in nucleon-hole and Δ-hole interactions: $f^* = \Delta$Nπ coupling constant, $f =$ NNπ coupling constant.

process), a Δ-hole could be excited directly; for zero four-momentum transfer that we are considering, Δ-hole states can still be virtually excited through the coupling in Equation 25. This coupling to virtual Δ-hole states provides a simple mechanism to account for $\delta g_A = g_A - \tilde{g}_A \neq 0$. One can make an estimate of the effect by an extremely simple model (58, 59, 112), which as confirmed by more realistic treatments (44, 113, 114) is reliable enough for our purpose. The process considered is as follows: The axial current can excite the Gamow-Teller state either directly or after a repeated excitation of virtual Δ-hole states via the forces in Equations 25 and 26. In random-phase approximation (RPA), the series can formally be summed. Let us ignore, for simplicity, all the complexity in the Δ propagator and define the Lindhard function in the form

$$L(\omega) = -\frac{(f_\pi^*)^2}{m_\pi^2}\rho\left[\frac{1}{\omega_R-\omega}+\frac{1}{\omega_R+\omega}\right], \qquad 27.$$

where ω_R is roughly the mass difference $M_\Delta - M_N \approx 2.1 m_\pi$, ρ is the average matter density, and the second term in the square bracket corresponds to a crossed Δ-N channel. With this form, the result is simple (58, 59, 115)[5]

$$\frac{\tilde{g}_A}{g_A} = [1-\tfrac{4}{9}g_0'L(0)]^{-1}. \qquad 28.$$

Using the experimental values $f_\pi^*/f_\pi \approx 2$, $f_\pi \approx 1$, and $g_0' \approx 0.6$ in pionic unit, one has

$$\frac{\tilde{g}_A}{g_A} \approx 0.67 \qquad \frac{\rho}{\rho_0} = 1$$
$$\approx 0.80 \qquad \frac{\rho}{\rho_0} = \frac{1}{2}. \qquad 29.$$

Accounting for the uncertainties in f_π^*, g_0' and for appropriate average densities, one would expect the quenching predicted in large nuclei to be between 20 and 30%.

One gets about the same answer when one performs the calculation by taking into account finite size of the nucleus. Schematically (44), one can replace $g_0'L(0)$ by $4(S/\omega_R)(A/N-Z)$, where S is the energy shift of the Gamow-Teller resonance from its unperturbed state ($S \approx 10$ MeV in ^{208}Pb) and the factor of four corresponds to $(f_\pi^*/f_\pi)^2$. The quenching comes out to be about 28%. When applied to finite nuclei, particularly for small A, there are various finite size corrections that can make some fluctuations around a uniform renormalization of fundamental nature. For instance, a

---

[5] I assume the universality relation $g_0' = \tilde{g}_0' = \bar{\tilde{g}}_0'$.

momentum-dependent Δ-hole force such as π and ρ exchange tensor forces, while vanishing in a uniform matter in the limit that momentum transfer vanishes, can contribute and should be taken into account in finite systems (55). Although the way to consistently treat this effect, which can sometimes mask the global phenomenon to a considerable degree, is more a matter of art than anything else, calculations have been done in various sophistications for beta decay (55, 96, 116) and for giant Gamow-Teller transitions (42, 53, 113).

These refined calculations support, at least qualitatively, a global renormalization of the predicted magnitude (Equations 29) and provide, in a nuclear physics approach, a mechanism for the effective constant of Buck & Perez (Equation 22). Unfortunately this mechanism does not explain why $\tilde{g}_A$ might be precisely one.

5.3 Alternative to Δ-Hole Quenching

Among the alternative explanations to the quenching so far proposed, the most serious one is that based on tensor forces. Since a tensor force couples low-lying states to very high-lying states at 200–300 MeV, its effect would be very similar to that of a simple Δ-hole coupling. It is short-ranged, therefore more or less independent of A. In fact, explicit calculations for light nuclei (for summary, see 98, 117, 118) show that the tensor-force-induced correlation *can* be substantial and may even totally account for the observed quenching (see also 119). Although the detailed structure is highly sensitive to the strength of the tensor force used (which is still controversial), this mechanism predicts that there would be some Gamow-Teller strength located above the Gamow-Teller resonance region and below the Δ peak. But to see this experimentally would require a careful separation from the background, a difficult task that has not yet been carried out.

This mechanism, in a sense trivial and less attractive than the Δ, would be more credible if one could indeed verify that the transition potential (Figure 11b) is substantially weakened compared with Figure 11a by Pauli principle, as argued by several people (98, 118, 120). It is claimed that

$$\tilde{g}'_0 \lesssim g'_0/2. \qquad 30.$$

This would hold if the Landau-Migdal parameter $\tilde{g}'_0$ were naively derived from a short-range interaction. If the interaction were very short ranged, then, because of the spin and isospin $(3/2, 3/2)$ of the Δ, the transition $NN \to N\Delta$ would be highly suppressed. (In fact, if it were of a δ function, the transition matrix element could be identically zero.) This can be stated as a cancellation between the direct and exchange terms in the $NN^{-1} \to \Delta N^{-1}$ channel. In the simple model of the π and ρ exchanges often used in the literature (98), a large cancellation does indeed occur and so $\bar{g}'_0$ does come out to be a lot smaller than g'_0.

The trouble with this argument is that a simplistic approach to a highly nonlinear quantity like $\bar{g}'_0$ (viewed from macroscopic variables such as $N, \Delta, \ldots$) can be erroneous. Individual corrections of higher-order nature can be very large and cannot be naively added. In fact, it has been shown (121, 121a) that taking into account such nonlinear effects as induced interactions in the Δ-hole interactions invalidates the naive perturbative result mentioned above.

Although the tensor correlations could certainly have a role in *some* quenching of the Gamow-Teller strength, they cannot be the *principal* agency as suggested by Arima and others. It is difficult to imagine that a fundamental quenching can be induced by conventional nuclear interactions alone. Let us see whether nature does not already rule out this alternative. For this, consider the triton β decay

$$^3H \rightarrow {}^3He + e^- + \bar{\nu}_e.$$

Since a tensor force acts more or less the same way in the trinucleon system as in heavy nuclei, one should already see its effects in this decay. The Gamow-Teller matrix element is to a good approximation written

$$M_A^{th} = \sqrt{3}(1 - \tfrac{2}{3}P_D - \tfrac{4}{3}P_{S'} + \delta), \qquad 31.$$

where P_D and $P_{S'}$ are respectively the D-state and mixed S-state probabilities and δ is the contribution from non-nucleon degrees of freedom. In this expression, the tensor-force-induced quenching corresponds to $(-2/3)P_D$. The question is: Is this term fully effective, unmodified by other agencies?

I now argue that singling out $(-2/3)P_D$ as a quenching mechanism is not physically meaningful and that only the total in the parentheses of Equation 31 calculated in a consistent way is a meaningful quantity. The crucial point here is the observation (115, 122, 123) that

$$\delta - \tfrac{2}{3}P_D \approx 0 \qquad 32.$$

fairly independently of P_D. This is seen as follows. Consider all the processes mediated by tensor operators. To second order in the tensor operator, we have the graphs given in Figure 12. The graph (a) corresponds to the term $(-2/3)P_D$, the graph (b) to the exchange currents, and the graph (c) to a contribution from the "Δ presence" in the nucleus. The last two go into δ. It is easy to see that the graph (b) has a sign opposite to that of (a) and (c), and an explicit calculation shows that, for a reasonable P_D and P_Δ, (b) cancels the sum $(a)+(c)$.

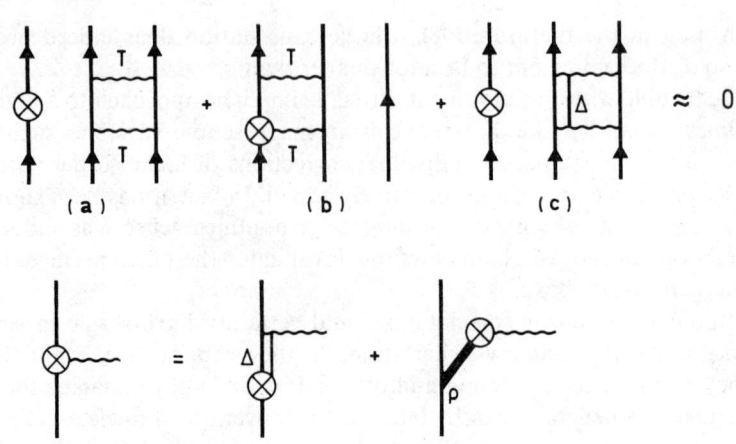

Figure 12 Cancelling diagrams in the triton Gamow-Teller matrix element. T stands for a tensor operator; the cross is a Gamow-Teller operator.

The extent of cancellation is model dependent, but nothing prevents us from assuming a total cancellation. Thus for $P_{S'} \approx 1.6\%$, the commonly accepted value, we predict

$$M_A^{th} \approx \sqrt{3}\,(0.975) \qquad \qquad 33.$$

to be compared with the experimental value (124)

$$M_A^{exp} = \sqrt{3}\,(0.970 \pm 0.008). \qquad \qquad 34.$$

Calculation with $\delta \approx 0$ and a full tensor strength would lead to $M_A^{th} \approx \sqrt{3}$ (0.92), clearly inconsistent with the experimental value. While this argument does not rule out the tensor force mechanism as a principal agent, it is very plausible from the point of view of chiral symmetry in nuclear medium. Further discussions on this point are made below.

5.4 Quark Degrees of Freedom?

The previous subsections dealt with the role of the Δ as an "elementary particle" in nuclei. The interaction was imagined to be occurring at very short distances between (two) hadrons. If we assume in the spirit of the "chiral filter" hypothesis of Section 4.2 that the Gamow-Teller operator is totally blind to pions, then, as discussed below, one can as well imagine two hadrons (bags) effectively overlapping and the spin-flip transition taking place directly on the quarks rather than on a nucleon as a whole, involving no changes other than quark spin and isospin. This is a picture, although very difficult to arrive at either from a hadronic description or from a quark-gluon description of quantum chromodynamics, predictive enough to allow us to reproduce the main results of the previous section. Such description was recently discussed in the literature (109, 114).

At first sight, the hadronic description and the quark description look very different: the former assumes the interaction between hadrons to be constant in *momentum* space and the latter takes an interaction between quarks to be constant in *coordinate* space. Subsequent discussions, however, reveal that the physics cannot be different, though the precise way that the two are related is not fully known and remains a challenging problem in nuclear physics.

Suppose modifications occur whenever two nucleons are coalesced into a single bag, say a 6-quark bag. It happens to be possible to estimate $\tilde{g}_A$ by knowing how n-quark bags respond to the axial current and the probability of an n-quark cluster in nuclei. It is not difficult to estimate the ratio $\eta^{(n)} \equiv g_A^{(n)}/g_A^{(3)}$ in bag models incorporating chiral symmetry (125)[6]

$$\eta^{(n)} \approx 1 - \frac{n-3}{9}, \quad \text{for } n \leq 12.$$

Limiting to 3-quark and 6-quark bags, we find

$$\frac{\tilde{g}_A}{g_A} \approx 1 - P_6 + \frac{1}{2}\eta^{(6)}P_6, \qquad 35.$$

where P_6 is the probability for a 6-quark cluster. In order for $\tilde{g}_A = 1$ to hold, we see that two nucleons would have to be found coalesced roughly 30% of the time. While this implies a big hadronic bag size, it is nevertheless a typical value required to explain the anomaly observed in the nucleon structure function in nuclei (EMC effect, see below). An unsatisfactory feature of this discussion, however, is as before that no special significance can be attached to the fact that $\tilde{g}_A$ may be exactly unity.

5.5 Vacuum Structure in Nuclei

It may be possible to view the phenomena discussed above as properties of a complex strong interaction vacuum as influenced by the presence of a baryon-rich environment. In this respect, the quenching of the Gamow-Teller strength may be signalling a precursor to what would happen at much higher density, at which chiral symmetry is expected to be restored. In the chiral symmetric world, $\tilde{g}_A \approx 1$ trivially. It is intriguing to ask what it means that $\tilde{g}_A$ is eactly unity in nuclei.

---

[6] This expression is not exact because of the Dirac sea polarization (10, 11) discussed in Section 6, but it is correct for $n = 3$ and $n = 12$, so we expect it to be sufficiently reliable for our purpose.

This question may be addressed in the context of an Adler-Weisberger type relation in nuclear matter (126). If one writes for mass A

$$\zeta^A \equiv \left(\frac{\tilde{g}_A}{g_A}\right)_A, \qquad \bar{\zeta} \equiv \frac{1}{g_A}$$

then Ericson's result (126) can be summarized as

$$(\zeta^A)^2 = (\bar{\zeta})^2 + [1 - (\bar{\zeta})^2]/A^t, \qquad t \approx 0.17. \qquad 36.$$

The physics behind this expression is the shadowing in π-nucleus scattering at high energy, resulting in an effective quenching of the Adler-Weisberger dispersion integral. In the limit $A \to \infty$, Equation 36 implies that $(\tilde{g}_A)_\infty = 1$. Or if the dispersion integral is completely quenched, one also gets $\tilde{g}_A \approx 1$ for appropriate mass $A \neq \infty$ (127, 128). While plausibly a manifestation of a modified vacuum structure, it is not entirely clear how this way of looking at $\tilde{g}_A$ relates to the various scenarios discussed above.

The possibility of $\tilde{g}_A$ being precisely one is particularly intriguing from the point of view of chiral symmetry. Suppose that we can proceed to describe nuclear responses using a chiral symmetric field theory. The axial current is then strictly or partially conserved. One is therefore allowed to write a Ward-Takahashi (W-T) identity, analogous to the W-T identity for the vector current (129, 129a). In the absence of nuclear matter, however, because of the pion pole (in the case of an exactly conserved axial current), g_A differs from unity (130) as contrasted with the vector current for which $g_V = 1$ from the W-T identity. This is of course because of the Goldstone-Nambu realization of chiral symmetry. We know from experiment that $g_A = 1.25$.

If $\tilde{g}_A = 1$ is not accidental, it can mean either of two possibilities. It may indicate an approach to the Wigner-Weyl mode (or chiral symmetry restored) or an intricate cancellation of the pion pole. In the first case, there would be no pion in the matter, and, in the second case, pions could be present but decoupled from the system and not seen by the Gamow-Teller operator. Whatever the case is, there must be an intricate cancellation between various (not necessarily small) terms to assure the Ward identity and $\tilde{g}_A = 1$ condition. [It seems reasonable to conjecture (129) that the above-mentioned cancellation between the core polarization, the exchange currents, etc of the type shown in Figure 12 can be a consequence of a Ward identity constraint.] It is still an unsolved problem to put these qualitative arguments on a quantitative basis, in particular by incorporating the axial Ward identity in the shell-model framework.

5.6 Deep Inelastic Structure Function in Nuclei

The recent observation that the deep inelastic lepton structure function in nuclei is different from that for a free nucleon—the so-called EMC effect

(131, 132)—may very well be related to the change of the vacuum structure discussed above. It is observed that for large momentum transfer Q^2 (such that subasymptotic effects that vanish as μ^2/Q^2 as $Q^2 \to \infty$ are suppressed), the ratio of the structure functions (iron to deuteron)

$$\lambda \equiv F_2^{Fe}/F_2^{D}$$

is enhanced ($\lambda > 1$) for small Bjorken x variable,[7] quenched for $0.3 \lesssim x \lesssim 0.8$, and then becomes enhanced again for $x > 0.8$. Several theoretical explanations have been offered: (a) Increased pion density in nuclei at low x (133–135); (b) excessive sea quarks from heavier objects, e.g. Δ(136) or nuclear clusters (137) or multiquark clusters (137–140); (c) or combinations of (a) and (b) (141).

That excess pion presence can modify quark properties in nuclear medium is plausible. Unfortunately, however the explanation relies on the elusive collective effect in the pion channel [namely an attractive $U(q)$, Figure 3] at $q \approx 2$–3 m_π, the momentum region at which the would-be precursor phenomenon is presumed to take place. What is needed is a factor-of-two enhancement in the pion response function as compared with that in free space. This is not much—hardly a precritical phenomenon—and it is not excluded that such an enhancement can come from pions in nuclei. Effects of this nature, however, should already be detectable in low-energy processes where the quantity in question is better defined theoretically. No such enhancement has been seen, as mentioned in Section 3.

What makes this mechanism less credible than the others is its extreme sensitivity to g'_0 and other parameters in $U(q)$. Unlike the axial charge transition $0^+ \leftrightarrow 0^-$, $\Delta T = 1$ discussed in Section 4.3 where the pionic enhancement (more than a factor of two) occurs precisely because the repulsive g'_0 effect is suppressed, here the repulsion (g'_0) competes with the momentum-dependent attraction as q is increased. Because g'_0 is expected to increase (if only slightly) as q is increased [this is caused by unscreening of the Δ-hole bubbles, see Brown & Rho (44)], there can be very little attraction left in any range of q.[8] It seems thus unlikely that such a mechanism can induce a basic change in the structure function to the extent seen in the experiments.

A more likely mechanism may be found in a fundamental change in the structure of quark-gluon confinement, e.g. percolation phenomenon (142) or multiquark clusters (139). In a way analogous to the quenching of g_A (Section 5.4), presence of other hadrons can modify the confinement region:

---

[7] The variable x is defined $x = Q^2/2M\nu$, where ν is the energy transfer, M the proton mass, and $Q^2 = 4EE' \sin^2 \theta/2$ the invariant square of the four-momentum transfer.

[8] It was pointed out to me by G. E. Brown (private communication) that it is even likely that $U(q)$ *never* becomes attractive at any momentum q.

While the loss of pion field in the baryon-rich environment quenches g_A, it can also change the quark momentum distribution. Thus the multiquarks confined within a region can do two things: they tend to cluster toward lower x region, thus increasing λ at $x < 0.3$, and one or several quarks can carry larger momentum than within a single nucleon, thus increasing λ at $x > 0.8$. (The depletion at $x \approx 0.6$ follows from baryon number and momentum sum rules.) Carlson & Havens (139) find that with a 30% probability for 6-quark clusters ($P_6 = 0.3$), the EMC data can be qualitatively understood. There seems to be some room, however, for additional effects at small x; these could be attributed to nuclear interactions such as the Δ or pions or a combination thereof.

6. PIONS AND NUCLEON STRUCTURE

6.1 *Motivation*

It should by now be clear to the readers that we are faced with a puzzle. On the one hand, the pion manifests itself in nuclei with an unmistakable signature; it shows up with vengeance in the electrodisintegration of the deuteron, in the magnetic form factor of ^{3}He, and in the axial charge transition. On the other hand, it precociously vanishes in other processes: the quenching of g_A could be most simply understood if pions evaporated from the system. Stated differently, the former suggests that the "bag" (or confinement size) is small and the latter that the "bag" is big.

The message is then clear. The fundamental aspect of pions in nuclei must be closely connected to the role of pions in the structure of the nucleon itself. In particular, it seems necessary to understand how chiral symmetry, whose manifestation is signalled by Goldstone pions, enters into confinement phenomena and how such inherently nonlinear dynamics is affected by the presence of other hadrons as in nuclear medium. The recent attempts to incorporate chiral symmetry into phenomenological hadron models (in particular the bag model) specifically address this problem (143–147). To date, we know of no precise and quantitative way to make the bridge between the pions in the nucleon structure and the pions within nuclear medium. This is partly because of our incomplete understanding of how pions enter into the nucleon structure (there is a controversy about this that will not be resolved in the near future) and partly because of our ignorance of how to transcribe whatever is known in the nucleon structure to many-nucleon situations.

There is, however, a startling new development that promises to offer the missing link. In this section, I discuss those aspects of the development that, I believe, will survive both theoretical and experimental tests. The story is rather old; all the essential ingredients are in fact found in a series of papers written more than two decades ago by Skyrme (5, 6). What is remarkable is that Skyrme's theory finds its justification in QCD, as beautifully shown recently by Witten (8, 9).

6.2 Skyrme Soliton (Skyrmion)

The pion field plays a particularly prominent role in Skyrme's description of the nucleon, so it is very likely to be the approach most suited to solving the problem at hand. (No claim is made that this approach supersedes such theories as the QCD on lattice or solves such fundamental problems as confinement.) In discussing this subject, I forgo the historical development. Instead I start from the modern viewpoint, QCD.

Quantum chromodynamics does not lend itself to a systematic weak coupling expansion in the color coupling g at low energy or at long distances and therefore frustrates attempts to unravel the full content of the theory. Nevertheless it is believed that there exists a hidden weak coupling: N_c^{-1}, where N_c is the number of color (7, 8). It is this $1/N_c$ expansion that we focus on in this section.

As yet, no one has succeeded in summing all the leading-order diagrams of large-N_c QCD (planar diagrams); yet studies (7, 8) show that the large-N_c expansion contains qualitative features that are consistent with empirical observations and provides answers to some of the ill-understood aspects of the strong interactions. These and other features of large-N_c QCD are reviewed extensively in the literature (see 148, 149), so we do not go into details in this paper. The main result emerging from this theory that we need for our purpose is this: For long-wavelength dynamics of the system with spontaneously broken SU(2) × SU(2) chiral symmetry, QCD in the large-N_c limit can be reduced to an effective Lagrangian field theory expressed in terms of phenomenological Goldstone bosons fields. Such a Lagrangian was already met in Section 4 in connection with the soft-pion expansion of exchange currents.

One such Lagrangian, consistent with large-N_c QCD, is the one proposed by Skyrme (5, 6),

$$\mathscr{L}^{\text{eff}} = -\frac{F_\pi^2}{4}\text{Tr}[L_\mu L_\mu] + \frac{\varepsilon^2}{4}\text{Tr}[L_\mu, L_\nu]^2 + \ldots$$

$$L_\mu = U^+\partial_\mu U$$

$$U = \frac{1}{F_\pi}[\sigma + i\boldsymbol{\tau}\cdot\boldsymbol{\pi}], \qquad U^+U = 1, \qquad 37.$$

where F_π is the pion decay constant, and ε is (as yet) an undetermined constant and ... means that higher terms in L_μ are dropped. Note that the quadratic term is unique, but the quartic term is not. In general, one can write two independent quartic terms, but this increases the number of parameters and in addition the Skyrme form is the most convenient in quantizing the theory. (It turns out that this is the same Lagrangian as in Section 4, with $a = -b$, except that there is no explicit fermion field here.)

The remarkable thing about this Lagrangian is that, being an effective theory for large-N_c QCD, it describes both low-lying mesons *and* low-lying baryons, although baryon degrees of freedom are not visible. As conjectured by Skyrme and demonstrated by several people (9, 150), miracles occur through topology: Baryons emerge from $\mathscr{L}^{\text{eff}}$ as a soliton solution to the Euler-Lagrange equation of motion. Briefly, it can be seen as follows (5, 6, 9, 150–152). The classical static solution is looked for in the "hedgehog" configuration

$$U(r) = \exp[i\tau \cdot \hat{r}\theta(r)] \qquad 38.$$

with

$$U(r) \to 1 \quad \text{as} \quad r \to \infty. \qquad 39.$$

With the spatial infinity identified as a point (Equation 39), the space is compactified to a 3-sphere, S^3 and Equation 38 represents the mapping of the space S^3 into SU(2). This mapping has a nontrivial topology or a topological knot. Mathematically, $U(r)$ of Equation 38 is an element in the third homotopy group $\pi_3[\text{SU}(2)]$, which is

$$\pi_3[\text{SU}(2)] = Z_\infty, \qquad 40.$$

a group of additive integers. Thus there exists a conserved topological current, the charge of which is characterized by an integer identified with the winding number. It has been shown that this topological charge is nothing but the baryon number B and hence, when the soliton is properly quantized, it can be identified as a baryon. Witten has shown how this happens in the SU(2) × SU(2) case as well as in the SU(3) × SU(3) case when the flavor space is extended to SU(3) by adding the strange (s) quark. The lowest-energy $B = 1$ soliton that has $K = 0$ (where $\mathbf{K} = \mathbf{J} + \mathbf{T}$) will be called a Skyrmion.

The baryon number is given by

$$B = \frac{1}{24\pi^2} \int d^3 r \varepsilon_{ijk} \text{Tr}[L_i L_j L_k]$$

$$= \frac{1}{\pi}\left\{\theta(R) - \theta(\infty) - \frac{1}{2}[\sin 2\theta(R) - \sin 2\theta(\infty)]\right\}, \qquad 41.$$

where for later convenience the lower limit of the integral is set at $r = R$. The boundary conditions are $\theta(\infty) = 0$, $\theta(0) = \pi$, so Equation 41 gives $B = 1$ for the Skyrmion. The structure of the Skyrmion is of course dictated by $\theta(r)$ and has been studied numerically (153, 154). The solution is plotted in Figure 13. When the Skyrmion is adiabatically (in the internal symmetry space) rotated, one obtains the rotational band (5, 154)

$$E_{J=T} = M_0 + \frac{J(J+1)}{2I} \qquad 42.$$

with

$$J = T = \frac{1}{2}, \frac{3}{2}, \ldots, \frac{N_c}{2}$$

and I is the moment of inertia given in terms of the Skyrmion property. Thus for $N_c = 3$, one has the usual N and Δ. The first term in Equation 42 is $O(N_c)$ and the second term $O(1/N_c)$. The rotation brings in the first quantum fluctuation to the soliton energy and corresponds to the familiar (angular momentum) projection.

Calculations including the first $1/N_c$ corrections give results on the static properties of the N and the Δ that agree with experiments within 20 to 30% (154, 155). The quality of agreement is comparable to that of nonrelativistic quark models and that of bag models. This may not be so surprising since in the limit $N_c \to \infty$, the Skyrme model, the SU(6) quark model, and the strong coupling meson theory are known to be equivalent (155a; Bardakci, K. 1983. *Lawrence Berkeley Lab. preprint*). The implication, however, is far-reaching for nuclear physics. Two extreme pictures, one a quark bag

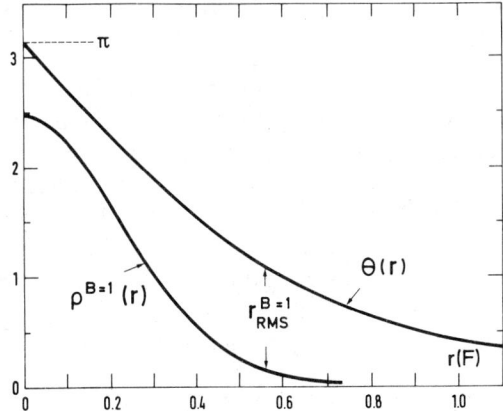

Figure 13 The Skyrmion solution $\theta(r)$ (for $B = 1$) and its baryon charge density $\rho(r)$, taken from (153).

containing three nearly zero-mass, nearly free quarks and another, containing no explicit quarks, described entirely by effective boson fields, can describe the same low-energy physics, at least for large N_c.

6.3 Leaking Baryon Charge

It is possible to go from one description to the other in a continuous manner with the help of the topological interpretation of the baryon charge.

Consider N_c massless u and d quarks confined in a spatial region that we take, for simplicity, to be a sphere with radius R. Call this a "bag." Inside the bag, the quark fields $\psi = \binom{u}{d}$ satisfy the free Dirac equation

$$\gamma \cdot \partial \psi = 0, \quad r \leq R. \qquad 43.$$

The quarks are confined by the boundary condition

$$n \cdot \gamma \psi = V(\theta)\psi, \quad r = R \qquad 44.$$

where n_μ is the unit outward normal and

$$V(\theta) = \exp[i\hat{n} \cdot \tau \gamma_5 \theta(R)].$$

For the moment, $\theta(R)$ is just a phase angle (called chiral angle), but later on $\theta(r)$ will be elevated to a dynamical variable, namely pion field. With $\theta(r)$ elevated to a dynamical variable, Equations 43 and 44 define the chiral bag model (143–147, 156).

A surprise that was recently discovered is that for $\theta(R) \neq 0$, the baryon charge $(B = 1)$ is not entirely lodged within the confinement region (10, 11). Part of the charge leaks outside of the bag, the leakage depending upon the chiral angle θ. The charge lodged inside the bag is found to be

$$B_{\text{in}} = 1 - \frac{1}{\pi}\left[\theta(R) - \frac{1}{2}\sin 2\theta(R)\right]. \qquad 45.$$

The leakage is understood in terms of polarization of the negative energy Dirac sea within the bag. For $\theta(R) \neq 0, \pi/2$, or π, the boundary condition (Equation 44) destroys the CP symmetry between the positive and negative energy quark spectra, as a consequence of which the negative energy sea depletes as θ is dialed from 0 to π. [At $\theta = \pi/2$, the spectrum is CP symmetric, but there is a zero mode and hence a depletion of exactly one half unit of the baryon charge (10, 157, 158).] This phenomenon is intimately connected to an anomaly in the baryon current due to the boundary condition with a time-varying chiral angle θ: the conservation of the baryon current breaks down.

The next remarkable thing is that the missing baryon charge can be identified with a Skyrmion living outside of the bag. Suppose that, elevated to a dynamical variable, $\theta(r)$ is governed by the Skyrme Lagrangian

(Equation 37) with Equation 38. Then we see from Equation 41 that the Skyrmion outside carries the baryon charge

$$B_{out} = \frac{1}{\pi}\left[\theta(R) - \frac{1}{2}\sin 2\theta(R)\right].\qquad 46.$$

The identification made here appears to be correct, since the sum $(B_{in} + B_{out})$ gives the correct charge $B = 1$.

The baryon charge is a topological statement, independent of details of the dynamics. Thus it does not depend upon F_π or ε in the Lagrangian that governs the dynamics of the Skyrmion. On the other hand, when the baryon charge leaks, it must take with it other charges, such as electric charge, magnetic moment, and axial charge, and this must reflect how the Skyrmion and the quarks, representing the same (long-wavelength) physics, communicate with each other. In particular, when the bag is shrunk to a point, $R = 0$, the Skyrmion must still know about the quarks even though the point bag has the vacuum baryon number. A constraint serving this purpose is found in axial-current conservation (158).

That the axial current is conserved in this hybrid bag–Skyrmion system imposes a boundary condition at the surface

$$\hat{n}\cdot\mathbf{A}_{in} = \hat{n}\cdot\mathbf{A}_{out}, \qquad r = R \qquad 47.$$

where $\mathbf{A}_{in}$ ($\mathbf{A}_{out}$) is the axial current inside (outside) the bag. For the Skyrmion governed by the Lagrangian in Equation 37, this condition reads (158)

$$F_\pi^2\left(\frac{d\theta}{dr}\right)_R\left(1 + \frac{16\varepsilon^2 \sin^2\theta(R)}{F_\pi^2 R^2}\right) = (4\pi R^3)^{-1}N_c\frac{\partial}{\partial\theta}W_{vac}(\theta)|_{\theta=\pi-\theta(R)} \qquad 48.$$

for $\theta(R) > \pi/2$.

Here W_{vac} is the energy of the Dirac sea (dependent upon θ) multiplied by R. [This expression, valid for $\theta(R) > \pi/2$, the region we are interested in for going to $R = 0$ at which $\theta(0) = \pi$, should be modified in a well-defined way for $\theta(R) < \pi/2$.] Now as $R \to 0$, both sides of Equation 48 go to a constant. The right-hand side, a property of the quark sector, is a pure number, while the left-hand side, that of a Skyrmion, is entirely given by F_π and ε. This determines ε, given F_π. One can also fix ε by the asymptotic behavior of the Skyrmion (153, 158):

$$\theta(r) \to \frac{C}{r^2} \quad \text{as} \quad r \to \infty$$

with C a constant. Through the Goldberger-Treiman relation, ε can be related to C and then to g_A. Clearly Equation 48 as $R \to 0$ must provide the same information as the asymptotic behavior. This is the required connection and provides the mechanism in which the axial charge leaks outward.

Just as the total baryon charge is a constant independent of the bag radius R, we suppose that g_A remains the same independently of R. How can this happen? It is Equation 48 that says how it happens (158). At $R \neq 0$, one more scale parameter is introduced. Since we assume no new physics is introduced by R, ε must be modified as R is changed so as to keep g_A fixed. As R is increased, the quark sector acquires more axial charge (as well as more baryon charge) and hence ε^2 must decrease in order to counterbalance it. This was verified by Brown et al (158). No other physical quantities (such as energy, magnetic moment, charge radius) are seen to vary significantly as R varies. This then gives us a qualitative explanation of why a bag can be big or small depending upon the probe. The quark sector is best suited to the processes associated with asymptotic freedom, the Skyrmion sector to those associated with nonlinear and nonperturbative phenomena. The probes seeing the former aspects will be best described by a big bag; the probes seeing the latter aspects, by a little bag.

6.4 From Skyrmions to Paris Potential

One of the most significant developments in bridging the new picture of the nucleon structure to nuclear physics is the derivation of the nucleon-nucleon potential from the Skyrmion-Skyrmion interaction (159). This subsection briefly summarizes this important piece of work.

As we saw above, low-energy physics can be described by an effective field theory of Goldstone pion fields. We will work with $R = 0$. Consider now the interaction of two Skyrmions. We assume the Born-Oppenheimer approximation can be applied and construct an adiabatic potential. When two Skyrmions are coalesced, we have an object of $B = 2$. It is well known that two solitons interact strongly in the weak coupling limit, so we expect the same thing will happen with two coalesced Skyrmions, the leading term being $O(N_c)$.

The first quantum correction leads to an $O(1/N_c)$ term. Therefore we expect that the dominant interaction between two nucleons would be given by the Skyrmion-Skyrmion interaction at short distances. Numerical calculations (153) showed that to a good approximation

$$E_B \approx \text{const } B(B+1). \qquad 49.$$

Thus the potential at zero distance is

$$V(0) \approx E_{B=2} - 2E_{B=1} \approx E_{B=1}. \qquad 50.$$

Since $E_{B=1} \gtrsim 1$ GeV numerically (153), we obtain a potential height at $r = 0$ of order of 1 GeV or greater. This is the height of the "repulsive core," here deduced entirely from a classical argument. One obtains a result similar to Equation 50 using the chiral bag model (125). In this model, a $B = 2$ object is a 6-quark bag. For a large bag, $\theta(R) \to 0$, so the $K = 0$ and $K = 1$ (recall that $\mathbf{K} = \mathbf{J} + \mathbf{T}$) levels collapse to one level, 1S, into which all six quarks can be put. However, for a small bag, the two levels split, the extent of the splitting depending upon the chiral angle $\theta(R)$. Only three (N_c) quarks can occupy the $K = 0$ level; the remaining three (N_c) quarks are elevated to the $K = 1$ level. The energy cost due to the "warping" of the quark levels leads to the repulsion shown in Equation 50, the result being roughly constant over a wide range of R. This confirms that the physics is the same in the different descriptions.

To construct the potential $V(r)$ for all r, a simplified procedure is followed by Jackson, Jackson & Pasquier (159). Instead of solving the Euler-Lagrange equation of motion for a general chiral field with $B = 2$ for a fixed separation r_{12}, they assume the product field

$$U_{B=2}(x, \mathbf{r}_1, \mathbf{r}_2) = U_{\mathbf{r}_1}(x) U_{\mathbf{r}_2}(x) \qquad 51.$$

with $U_r(x)$ given by the $B = 1$ solution (Figure 13). At large distances ($r_{12} \to \infty$), this is clearly a good approximation. It turns out to be quite reliable also for $r_{12} = 0$: the configuration of Equation 51 yields an energy only 2.5% higher than the exact $B = 2$ hedgehog energy. One expects the ansatz (Equation 51) to be least accurate in the intermediate range, roughly 0.6 to 1.2 F.

Consider now generalizing the hedgehog configuration (Equation 38) in the form

$$U(x; \alpha\beta\gamma) \equiv \exp\{i\tau_i e_{ij}[\alpha\beta\gamma]\hat{r}_j \theta(r)\}, \qquad 52.$$

where $e_{ij}[\alpha\beta\gamma]$ is a fixed orthogonal transformation describing a rotation through Euler angles $[\alpha\beta\gamma]$ of the isospin quantization axis with respect to the spatial quantization axis. For an isolated Skyrmion, the rotation does not change anything; i.e. the energy remains the same. But when the rotation is made on the Skyrmions in Equation 51 then the energy of the system will depend upon the Euler angles. Since the interaction will depend upon the relative Euler angles, it suffices to rotate only one Skyrmion, say, at $\mathbf{r}_2$. From the total energy of the system (Equation 51) evaluated for fixed r_{12} and subtracting twice the energy of a Skyrmion, the adiabatic potential $V(r_{12})$ is obtained as function of Euler angles.

At large distances, the familiar one-pion exchange potential (for zero-mass pion) is recovered (with $r \equiv r_{12}$),

$$\lim_{r \to \infty} V(r; [\alpha\beta\gamma]) = \text{const } e_{ij}[\alpha\beta\gamma] \frac{\partial^2}{\partial x_i \partial x_j}\left(\frac{1}{r}\right). \qquad 53.$$

At short distances, a universal repulsion of order of 1 GeV is obtained, with little dependence on the Euler angles. Both long- and short-distance properties seem very satisfactory. Given the Euler angle-dependent potential, it is possible to project out the channels corresponding to the nucleon-nucleon interaction and further separate out different spin-isospin dependences of the interaction. This is worked out beautifully by Jackson et al (159).

Let me summarize some of the numerical results. Write the potential in the usual form

$$V^{NN}(r) = V_c(r) + (\tau_1 \cdot \tau_2)(\sigma_1 \cdot \sigma_2)V_{\sigma\tau}(r) + (\tau_1 \cdot \tau_2)S_{12}V_{T\tau}(r) + \ldots, \qquad 54.$$

where ... stands for other components of the force not obtained in the adiabatic calculation with Skyrmions. The results are plotted and compared with the Paris potential in Figures 14, 15, and 16.

Notable features of these results are as follows:

1. The potentials at $r \gtrsim 1$ F are in satisfactory agreement with experiments (i.e. the Paris potential). On the other hand the disagreements for $r < 1$ F are not meaningful: the adiabatic potential may not be justified for comparing with experiments and, furthermore, the Paris potential, being

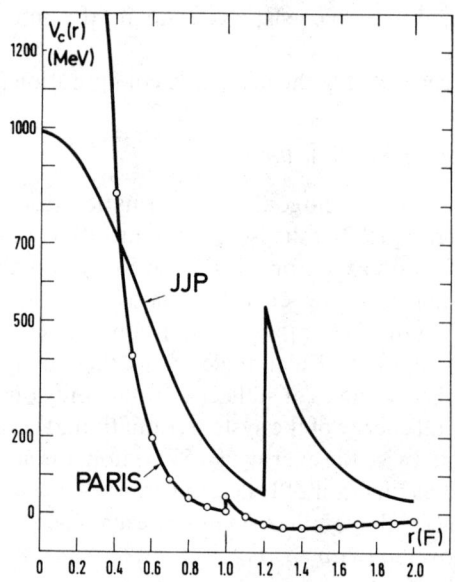

Figure 14 The 1·1 force $V_c(r)$ obtained from the Skyrmion-Skyrmion interaction (*JJP*), compared with the Paris potential (from 159). Note the scale changes at 1.0 and 1.2 F by an order of magnitude. (They do not represent discontinuities.)

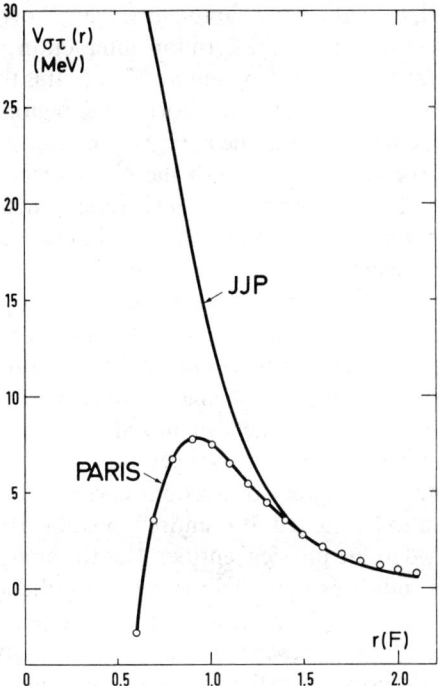

Figure 15 Same as in Figure 14 for the $\tau_1 \cdot \tau_2 \sigma_1 \cdot \sigma_2$ component $V_{\sigma\tau}(r)$.

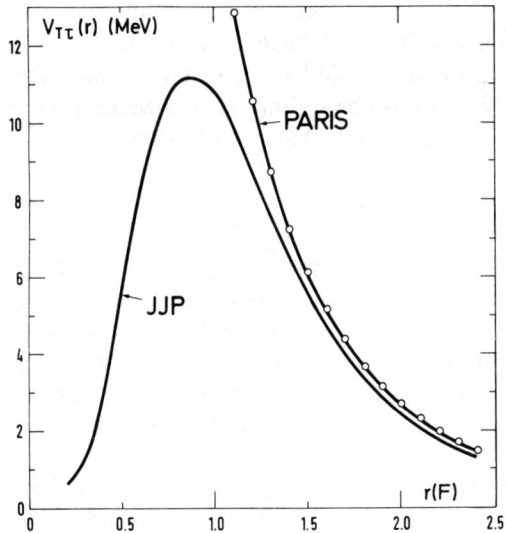

Figure 16 Same as in Figure 14 for the $\tau_1 \cdot \tau_2 S_{12}$ component $V_{T\tau}(r)$.

phenomenological, is subject to ambiguities at short distances. It is nevertheless important that the Skyrmion interaction provides descriptions, albeit qualitative, of what happens at $r < 1$ F. It is the first time that a theory has anything to say about this short-range region.

2. That the $\mathbf{1} \cdot \mathbf{1}$ component and the $\tau_1 \cdot \tau_2 \cdot \sigma_1 \cdot \sigma_2$ component appear as a strong *primary* force is consistent with the meson-exchange description. The ω and σ exchanges give rise to a strong $\mathbf{1} \cdot \mathbf{1}$ component, while the π and ρ exchanges give rise to a strong spin-isospin-dependent force. Most remarkably, the tensor force has just the right feature reflecting a cancellation between the π and ρ tensors (see 34, 40). The absence of a strong primary $\tau_1 \cdot \tau_2$ force is also consistent with the meson theory: the required component is known to come from the iteration of the pionic tensor force (160), not from the exchange of a meson.

3. While it is expected that when quantized (i.e. projected) the Skyrme Lagrangian would describe pion exchanges, it is surprising that, when the Skyrmions overlap, the degrees of freedom corresponding to the vector mesons (ρ, ω) seem to be excited. It is entirely possible that these need not correspond to any definite physical entities. But the analysis (159) in terms of Yukawa forms indicates that there is no contradiction in attributing heavy mesons to various components of the force (e.g. the tensor force illustrates the role of the ρ meson). It is noteworthy also that completely missing from the description is the intermediate-range attraction that is commonly attributed to the exchange of two pions in the σ channel. There is no meson of that quantum number in nature and so it may not be generated within the scheme; it may be necessary to second-quantize the theory and calculate the loop graphs to get the required attraction. It is intriguing that the theory seems to have a hidden symmetry of boson-fermion transformation: the Skyrmion-Skyrmion interactions seem to involve spin $-1/2$ and $-3/2$ fermions and spin -0 and -1 bosons.

7. CONCLUSION

There is no question that the pion plays a fundamental role in nuclear structure. We have seen how it manifests itself in electromagnetic and weak response functions. It is also known that deuteron properties would be extremely difficult to understand if one did not resort to the pion explicitly (161). The pion seems to take even the role that, one would have hoped, properly belongs to quarks (162): it is singularly pervasive in a baryon-rich environment.

In this review, I have argued that in nuclei the most important aspect of quantum chromodynamics is chiral symmetry and, as a relic of spontaneous breaking of the symmetry, the pion is providing us information on the structure of nuclear medium ("vacuum" in a generalized sense). The multitude of nuclear data discussed in this paper reveals that there are several facets to this. In some processes (e.g. electrodisintegration of the deuteron), nucleons and pions keep their identity to very short distances. In other cases (e.g. Gamow-Teller transitions), pions evaporate precociously. Thus one gets the paradoxical picture that the quark confinement region (or the bag) looks big and small simultaneously within a same system. In the quark picture, the way that chiral symmetry is manifested—and consequently pion interactions—depends upon the vacuum structure (i.e. quark-antiquark pair condensates in the vacuum induce chiral symmetry breaking). Thus we are naturally led to contemplate the possibility that one is seeing different facets of the vacuum in nuclei, modified in a fundamental way as a result of increased baryon density.

The key to a deeper understanding may come from further developments in the structure of the nucleon. The recently discovered topological structure of the nucleon bag—that the nucleon is a topological twist in an otherwise uniform meson background—promises to provide the required link between the nucleon and the pion and the nucleus. The leakage of the baryon charge, the axial charge, and other charges from the confinement region renders the concept of the bag size somewhat academic at least for low-energy hadron physics, and provides a qualitative resolution of the paradox referred to above. Since the bag and the Skyrmion could represent two extreme aspects of QCD, the value of a particular R represents an arbitrary separation with which physics of the two regimes is optimally taken into account: R is phenomenon dependent, not fundamental.

Although highly plausible intuitively, the connection between the pion in the Skyrmion picture and the pion "seen" in nuclei has not yet been formulated in a quantitative way. The suggestion, however, is overwhelming; all the ingredients are there. The Δ, believed to quench g_A in nuclei, appears at the same level as the nucleon. The basic degrees of freedom, assumed to be required in a meson theory, emerge when the Skyrmions come close together. The structure of the resulting potential resembles nature surprisingly closely. But the crucial link is still missing. It may require another surprise—a surprise of the sort discussed in Section 6—to find it. At the moment, we can explain things, but the explanations are still too inelegant to be entirely correct.

ACKNOWLEDGMENTS

I am indebted to many colleagues for fruitful collaborations and stimulating discussions, in particular to G. E. Brown, M. Chemtob, J. Delorme, B. Frois, A. S. Goldhaber, A. D. Jackson, K. Kubodera, F. C. Khanna, J. F.

Mathiot, E. M. Nyman, D. O. Riska, J. Speth, I. S. Towner, V. Vento, D. H. Wilkinson, and K. Yazaki—all of whom have influenced my thinking in this domain of physics. I would like to thank Jean-François Mathiot for giving me an unpublished result of his calculation (Figure 7) and for clarification on the matter and Andrew D. Jackson for the results of his Skyrmion calculations (Figures 14–16).

Literature Cited

1. Yukawa, H. 1935. *Proc. Phys. Math. Soc. Jpn.* 17:48
2. Brown, G. E., Green, A. M. 1969. *Nucl. Phys. A* 133:481
3. Chemtob, M., Rho, M. 1971. *Nucl. Phys. A* 163:1
4. Villars, F. 1947. *Helv. Phys. Acta* 20:476
5. Skyrme, T. H. R. 1961. *Proc. R. Soc. London Ser. A* 260:127
6. Skyrme, T. H. R. 1962. *Nucl. Phys.* 31:556
7. 't Hooft, G. 1974. *Nucl. Phys. B* 72:461; *Nucl. Phys. B* 75:461
8. Witten, E. 1979. *Nucl. Phys. B* 160:57
9. Witten, E. 1983. *Nucl. Phys. B* 223:422, 433
10. Rho, M., Goldhaber, A. S., Brown, G. E. 1983. *Phys. Rev. Lett.* 51:747
11. Goldstone, J., Jaffe, R. L. 1983. *Phys. Rev. Lett.* 51:1518
12. De Rùjula, A., Georgi, H., Glashow, S. L. 1975. *Phys. Rev. D* 12:147
13. Chodos, A., Jaffe, R. L., Johnson, K., Thorn, C. B. 1974. *Phys. Rev. D* 10:2599; De Grand, T., Jaffe, R. L., Johnson, K., Kiskis, J. 1975. *Phys. Rev. D* 12:2060
14. Adler, S. L., Dashen, R. F. 1968. *Current Algebra and Applications to Particle Physics.* New York: Benjamin
15. Nambu, Y., Jona-Lasinio, G. 1961. *Phys. Rev.* 122:345
16. Glashow, S., Weinberg, S. 1968. *Phys. Rev. Lett.* 20:224
17. Gell-Mann, M., Oakes, R. J., Renner, B. 1968. *Phys. Rev.* 175:2195
18. Gasser, J., Leutwyler, H. 1982. *Phys. Rep.* 87c:77
19. Finger, J., Mandula, J. E. 1982. *Nucl. Phys. B* 199:168
20. Stokar, S. 1983. *Phys. Rev. Lett.* 51:23
21. Brown, G. E. 1981. *Nucl. Phys. A* 358:39C
21a. Bernard, V., Brockman, R., Schade, M., Weise, W., Werner, E. 1984. *Nucl. Phys. A* 412:349
22. Caldi, D. G., Pagels, H. 1976. *Phys. Rev. D* 14:809
23. Pagels, H. 1979. *Phys. Rev. D* 19:3080
24. Schwinger, J. 1957. *Ann. Phys. NY* 2:407
25. Gell-Mann, M., Lévy, M. 1960. *Nuovo Cimento* 16:705
26. Rho, M., Wilkinson, D. H., eds. 1979. *Mesons in Nuclei.* Amsterdam: North-Holland
27. Vinh Mau, R. 1979. See Ref. 26, p. 151; 1982. *Nucl. Phys. A* 374:3C
28. Vinh Mau, R. 1984. *AIP Conf. Proc.* 110:187
29. Balian, R., Rho, M., Ripka, G. 1978. *Nuclear Physics with Heavy Ions and Mesons.* Amsterdam: North-Holland
30. Baym, G. 1978. See Ref. 29, p. 745
31. Campbell, D. 1978. See Ref. 29, p. 549
32. Moniz, E. J. 1978. See Ref. 29, p. 433
33. Sawyer, R. F. 1978. See Ref. 29, p. 717
34. Oset, E., Toki, H., Weise, W. 1982. *Phys. Rep.* 83:281
35. Landau, L. D., Lifshitz, E. M. 1959. *Statistical Physics*, p. 430. London/Paris: Pergamon Press
36. Pirner, H. J., Yazaki, K., Bonche, P., Rho, M. 1979. *Nucl. Phys. A* 329:49
37. Migdal, A. B. 1972. *JETP (Sov. Phys.)* 34:1184
38. Migdal, A. B. 1979. See Ref. 26, p. 941
39. Sawyer, R. F. 1972. *Phys. Rev. Lett.* 29:382
40. Weise, W., Brown, G. E. 1976. *Phys. Rep.* 27C:1
41. Meyer-ter-Vehn, J. 1981. *Phys. Rep.* 74:281
42. Speth, J., Suzuki, T. 1981. *Nucl. Phys. A* 358:139C
43. Brown, G. E., Bäckman, S. O., Osct, E., Weise, W. 1977. *Nucl. Phys. A* 286:191
44. Brown, G. E., Rho, M. 1981. *Nucl. Phys. A* 372:397
45. Ring, P., Speth, J. 1974. *Nucl. Phys. A* 235:315
46. Speth, J., Klempt, V., Wambach, J., Brown, G. E. 1980. *Nucl. Phys. A* 343:382
47. Petrovich, F., et al, eds. 1983. *Spin Excitations in Nuclei.* New York: Plenum. In press
48. Delorme, J. 1982. *Nucl. Phys. A* 374:541C

49. Ericson, M. 1984. *Prog. Part. Nucl. Phys.* 10: In press
50. Rho, M. 1981. *Nucl. Phys. A* 358:121C
51. Ericson, M. 1980. *Nucl. Phys. A* 335:309
52. Toki, H., Weise, W. 1979. *Phys. Rev. Lett.* 42:47
53. Weise, W. 1982. *Nucl. Phys. A* 374:505C
54. Barshay, S., Brown, G. E. 1973. *Phys. Lett.* 47B:107
55. Oset, E., Rho, M. 1979. *Phys. Rev. Lett.* 42:47
56. Campbell, D., Dashen, R. F., Manassah, J. 1975. *Phys. Rev. D* 12:979
57. Buck, B., Perez, S. M. 1983. *Phys. Rev. Lett.* 50:1975
58. Rho, M. 1974. *Nucl. Phys. A* 231:493
59. Ohta, K., Wakamatsu, M. 1974. *Nucl. Phys. A* 234:445
60. Walecka, J. D. 1974. *Ann. Phys. NY* 83:491
61. Serot, B. 1983. *AIP Conf. Proc.* 97:337
62. Weinberg, S. 1979. *Physica* 96A:327
63. Leutwyler, H. 1983. In *Proc. 23rd Cracow Sch. Theor. Phys. (Zakopane).* Acta Physica Polonica: In press
64. Chew, G. F., Low, F. E. 1956. *Phys. Rev.* 101:1570
65. Kubodera, K., Delorme, J., Rho, M. 1978. *Phys. Rev. Lett.* 40:755
66. Rho, M., Brown, G. E. 1981. *Commun. Nucl. Part. Phys.* 10:201
67. Hockert, J., Riska, D. O., Gari, M., Huffman, A. H. 1973. *Nucl. Phys. A* 217:14
68. Avenhövel, H. 1982. *Nucl. Phys. A* 374:521
69. Riska, D. O., Brown, G. E. 1972. *Phys. Lett.* 38B:193
70. Mathiot, J. F. 1982. *Phys. Lett.* 115B:174
71. Bernheim, M., Jaus, E., Mougey, J., Royer, D., et al. 1981. *Phys. Rev. Lett.* 46:402
72. Clemens, J.-C., et al. 1984. Unpublished
73. Fubini, S., Nambu, Y., Wataghin, V. 1958. *Phys. Rev.* 111:329
74. Fabian, W., Arenhövel, H. 1976. *Nucl. Phys. A* 298:461
75. Friar, J. L., Fallieros, S. 1976. *Phys. Rev. C* 13:2571
76. Leideman, W., Arenhövel, H. 1982. *Nucl. Phys. A* 381:365
77. Mathiot, J. F. 1984. *Nucl. Phys. A* 412:201
78. Cavedon, J. M., Frois, B., Goutte, D., Huet, M., et al. 1982. *Phys. Rev. Lett.* 49:986
79. Riska, D. O. 1980. *Nucl. Phys. A* 350:227
80. Hadjimichael, E., Goulard, B., Bornais, R. 1983. *Phys. Rev. C* 7:831
81. Strueve, W., Hajduk, C., Sauer, P. U. 1983. *Nucl. Phys. A* 405:620
82. Maize, M. A., Kim, Y. E. 1983. *Nucl. Phys. A* 407:507
83. Weise, W. 1984. *Prog. Part. Nucl. Phys.* 10: In press
84. Richter, A. 1983. *Proc. Int. Conf. on Nucl. Struct.,* p. 189. Bologna: Tipografia Comp.
85. Brandle, H., Grenacs, L., Lang, J., Roesch, L. Ph., et al. 1978. *Phys. Rev. Lett.* 40:306
86. Lebrun, P., Deschepper, Ph., Grenacs, L., Lehmann, J., et al. 1978. *Phys. Rev. Lett.* 40:302
87. Masuda, Y., Minamisono, T., Nojiri, Y., Sugimoto, K. 1978. *Phys. Rev. Lett.* 41:299
88. Palffy, L., Deutsch, J. P., Granacs, L., Lehmann, J., et al. 1975. *Phys. Rev. Lett.* 34:212
89. Gagliardi, C. A., Garvey, G. T., Wrobel, J. R., Freedman, S. J. 1982. *Phys. Rev. Lett.* 48:914; 1983. *Phys. Rev. C* 28:2423
90. Minamisono, T., Takeyama, K., Ishigai, T., Takeshima, H., et al. 1983. *Phys. Lett.* 130B:1
91. Gagliardi, C. A., Garvey, G. T., Jarmie, N., Robertson, R. G. H. 1983. *Phys. Rev. C* 27:1353
92. Adelberger, E. G., Hoyle, C. D., Swanson, H. E., Von Lintig, R. D. 1981. *Phys. Rev. Lett.* 46:695
93. Warburton, E. K., Alburger, D. E., Wilkinson, D. H. 1982. *Phys. Rev. C* 26:1186
94. Guichon, P., Samour, C. 1982. *Nucl. Phys. A* 282:461
95. Koshigiri, K., Ohtsubo, H., Morita, M. 1981. *Prog. Theor. Phys.* 66:358; 1979. *Prog. Theor. Phys.* 62:706
96. Towner, I. S. 1984. *Prog. Part. Nucl. Phys.* 10: In press
97. Guichon, P., Giffon, M., Samour, C. 1978. *Phys. Lett.* 74B:15
98. Towner, I. S., Khanna, F. C. 1981. *Nucl. Phys. A* 372:331
99. Jaeger, H. U., Kirchbach, M., Truhlik, E. 1983. *Nucl. Phys. A* 404:456
100. Kohyama, Y., Nishimura, H., Nozawa, S., Kubodera, K. 1983. *Prog. Theor. Phys.* 70:613
101. Laget, J. M. 1981. *Phys. Rep.* 69:1
102. Moniz, E. J. 1982. *Nucl. Phys. A* 374:557C
103. Schwesinger, B., Wirzba, A., Brown, G. E. 1983. *Phys. Lett.* 132B:269
104. Wilkinson, D. H. 1973. *Nucl. Phys. A* 209:470
105. Wilkinson, D. H. 1978. See Ref. 29, p. 882
106. Horen, D. J., Goodman, C. D., Foster,

C. C., Goulding, C. A., et al. 1980. *Phys. Lett.* 95B:27
107. Gaarde, C., Larsen, J. S., Rapaport, J. 1983. See Ref. 47
108. Rapaport, J. 1983. *AIP Conf. Proc.* 97:365
109. Goodman, C. 1984. *Prog. Part. Nucl. Phys.* 10: In press
110. Ericson, M., Figureau, A., Thévenet, C. 1973. *Phys. Lett.* 47B:381
111. Migdal, A. B. 1967. *Theory of Finite Fermi Systems and Applications to Atomic Nuclei.* New York: Interscience
112. Barshay, S., Brown, G. E., Rho, M. 1974. *Phys. Rev. Lett.* 32:787
113. Speth, J. 1983. See Ref. 47
114. Bohr, A., Mottelson, B. R. 1981. *Phys. Lett.* 100B:10
115. Rho, M. 1978. *Prog. Part. Nucl. Phys.* 1:105
116. Towner, I. S., Khanna, F. C. 1979. *Phys. Rev. Lett.* 42:51
117. Arima, A. 1978. *Prog. Part. Nucl. Phys.* 1:41
118. Arima, A. 1984. *Prog. Part. Nucl. Phys.* 10: In press
119. Bertsch, G. F., Hamamoto, I. 1982. *Phys. Rev. C* 26:1323
120. Arima, A., Cheon, T., Shimizu, K., Hyuga, H., Suzuki, T. 1983. *Phys. Lett.* 122B:26
121. Brown, G. E. 1984. *J. Phys.* 45(C4):479
121a. Nakayama, K., Krewald, S., Speth, J., Brown, G. E. 1984. *Phys. Rev. Lett.* 52:500
122. Green, A. M., Shucan, T. H. 1972. *Nucl. Phys. A* 188:289
123. Ichimura, M., Hyuga, H., Brown, G. E. 1972. *Nucl. Phys. A* 196:17
124. Budick, B. 1983. *Phys. Rev. Lett.* 51:1034
125. Vento, V., Rho, M. 1983. *Nucl. Phys. A* 412:413
126. Ericson, M. 1971. *Ann. Phys. NY* 63:562
127. Rho, M. 1983. See Ref. 47
128. Delorme, J., Ericson, M., Guichon, P. 1982. *Phys. Lett.* 115B:86
129. Rho, M. 1983. *AIP Conf. Proc.* 97:350
129a. Zhu, X., Wong, S. S. M. 1984. *Nucl. Phys. A* 412:391
130. Bernstein, J., Gell-Mann, M., Michel, L. 1960. *Nuovo Cimento* 16:560
131. Aubert, J. J., Bassompierre, G., Becks, K. H., Best, C., et al. 1983. *Phys. Lett.* 123B:275
132. Bodek, A., Giokaris, N., Atwood, W. B., Coward, D. H., et al. 1983. *Phys. Rev. Lett.* 50:1431
133. Lewellyn Smith, C. H. 1983. *Phys. Lett.* 128B:107
134. Ericson, M., Thomas, A. 1983. *Phys. Lett.* 128B:112
135. Friman, B. L., Pandharipande, V. R., Wiringa, R. B. 1983. *Phys. Rev. Lett.* 51:763
136. Szwed, J. 1983. *Phys. Lett.* 128B:245
137. Chemtob, M., Peschanski, R. 1984. *J. Phys. G* 10:599
138. Jaffe, R. L. 1983. *Phys. Rev. Lett.* 50:228
139. Carlson, C. E., Havens, T. J. 1983. *Phys. Rev. Lett.* 51:261
140. Pirner, H. J., Vary, J. 1983. *Heidelberg Preprint 83/02*
141. Berger, E. L., Coester, F., Wiringa, R. B. 1984. *Phys. Rev. D* 29:398
142. Baym, G. 1979. *Physica (Utrecht)* 96A:131
143. Chodos, A., Thorn, C. B. 1975. *Phys. Rev. D* 12:2733
144. Brown, G. E., Rho, M. 1979. *Phys. Lett.* 82B:177
145. Brown, G. E., Rho, M., Vento, V. 1979. *Phys. Lett.* 84B:383
146. Callan, C. G., Dashen, R. F., Gross, D. J. 1979. *Phys. Rev. D* 19:1826
147. Théberge, S., Thomas, A. W., Miller, G. A. 1980. *Phys. Rev. D* 22:2838
148. Coleman, S. 1981. *Point-like Structures Inside and Outside the Nucleon, Proc. 1979 Erice Summer Sch.*, ed. A. Zichichi. New York: Plenum
149. Witten, E. 1980. *Phys. Today*, July:38
150. Williams, J. G. 1970. *J. Math. Phys. NY* 11:2611
151. Pak, N. K., Tze, H. Ch. 1979. *Ann. Phys. NY* 117:164
152. Balachandran, A. P., Nair, V. P., Rajeev, S. G., Stern, A. 1982. *Phys. Rev. Lett.* 49:1124; 1983. *Phys. Rev. D* 27:1153
153. Jackson, A. D., Rho, M. 1983. *Phys. Rev. Lett.* 51:751
154. Adkins, G. S., Nappi, C. R., Witten, E. 1983. *Nucl. Phys. B* 228:552
155. Adkins, G. S., Nappi, C. R. 1984. *Nucl. Phys. B* 233:109
155a. Gervais, J.-L., Sakita, B. 1984. *Phys. Rev. Lett.* 52:87
156. Vento, V., Rho, M., Nyman, E. M., Jun, J. H., Brown, G. E. 1980. *Nucl. Phys. A* 345:413
157. Rho, M. 1984. *Prog. Part. Nucl. Phys.* 10: In press
158. Brown, G. E., Jackson, A. D., Rho, M., Vento, V. 1984. *Phys. Lett.* 140B:285
159. Jackson, A., Jackson, A. D., Pasquier, V. 1984. *Nucl. Phys.* In press
160. Brown, G. E., Speth, J., Wambach, J. 1981. *Phys. Rev. Lett.* 46:1057
161. Ericson, T. E. O. 1984. *Prog. Part. Nucl. Phys.* 10: In press
162. Wong, C. W. 1984. *AIP Conf. Proc.* 110:228

CUMULATIVE INDEXES

CONTRIBUTING AUTHORS VOLUMES 24–34

A

Alexander, J. M., 24:279–339
Allison, W. W. M., 30:253–98
Amaldi, U., 26:385–456
Amsel, G., 34:435–60
Andersen, J. U., 33:453–504
Appelquist, T., 28:387–499
Arad, B., 24:35–67
Arianer, J., 31:19–51
Arima, A., 31:75–105
Arnold, J. R., 33:505–37

B

Ballam, J., 27:75–138
Barnett, R. M., 28:387–499
Barschall, H. H., 28:207–37
Baym, G., 25:27–77
Bég, M. A. B., 24:379–449
Benczer-Koller, N., 30:53–84
Ben-David, G., 24:35–67
Berko, S., 30:543–81
Berry, H. G., 32:1–34
Bertrand, F. E., 26:457–509
Bienenstock, A., 28:33–113
Birkelund, J. R., 33:265–322
Blann, M., 25:123–66
Blok, H. P., 33:569–609
Bloom, E. D., 33:143–97
Boehm, F., 34:125–53
Bøggild, H., 24:451–513
Bonderup, E., 33:453–504
Brewer, J. H., 28:239–326
Bucksbaum, P. H., 30:1–52

C

Cahill, T. A., 30:211–52
Cerny, J., 27:331–51
Charpak, G., 34:285–349
Chinowsky, W., 27:393–464
Clayton, R. N., 28:501–22
Cline, D., 27:209–78
Cobb, J. H., 30:253–98
Cole, F. T., 31:295–335
Commins, E. D., 30:1–52
Cormier, T. M., 32:271–308
Crowe, K. M., 28:239–326

D

Darriulat, P., 30:159–210
de Forest, T. Jr., 25:1–26
DeTar, C. E., 33:235–64
Diamond, R. M., 30:85–157
Dieperink, A. E. L., 25:1–26
Donnelly, T. W., 25:329–405
Donoghue, J. F., 33:235–64
Dover, C. B., 26:239–317
Drees, J., 33:385–452

E

Ellis, J., 32:443–97

F

Fabjan, C. W., 32:335–89
Farley, F. J. M., 29:243–82
Feld, M. S., 29:411–54
Ferbel, T., 24:451–513
Fernow, R. C., 31:107–44
Fick, D., 31:53–74
Fisk, H. E., 32:499–573
Fleury, A., 24:279–339
Flocard, H., 28:523–96
Franzini, P., 33:1–29
Freedman, D. Z., 27:167–207
Freedman, M. S., 24:209–47
French, J. B., 32:35–64
Friar, J. L., 34:403–33
Fry, W. F., 27:209–78
Fulbright, H. W., 29:161–202

G

Gaillard, J.-M., 34:351–402
Gaillard, M. K., 32:443–97
Gaisser, T. K., 30:475–542
Geller, R., 31:19–51
Gibson, B. F., 34:403–33
Gibson, W. M., 25:465–508
Girardi, G., 32:443–97
Glashausser, C., 29:33–68
Goeke, K., 32:65–115
Goldhaber, A. S., 28:161–205
Goldhaber, G., 30:337–81
Goodman, A. L., 26:239–317

Gottschalk, A., 29:283–312
Goulding, F. S., 25:167–240
Greenberg, O. W., 28:327–86
Grunder, H. A., 27:353–92
Guinn, V. P., 24:561–91

H

Hansen, P. G., 29:69–119
Hardy, J. C., 27:333–51
Harvey, B. G., 25:167–240
Hass, M., 30:53–84; 32:1–34
Heckman, H. H., 28:161–205
Heisenberg, J., 33:569–609
Herrmann, G., 32:117–47
Hirsch, R. L., 25:79–121
Hoffman, D. C., 24:151–207
Hoffman, M. M., 24:151–207
Hughes, V. W., 33:611–44
Huizenga, J. R., 27:465–547; 33:265–322
Hung, P. Q., 31:375–438

I

Iachello, F., 31:75–105

J

Jacob, M., 26:385–456
Jackson, A. D., 33:105–41

K

Keefe, D., 32:391–441
Keller, O. L. Jr., 27:139–66
Kim, Y. E., 24:69–100
Kirsten, F. A., 25:509–54
Kleinknecht, K., 26:1–50
Kohaupt, R. D., 33:67–104
Kolb, E. W., 33:645–96
Kota, V. K. B., 32:35–64
Krisch, A. D., 31:107–44
Kuo, T. T. S., 24:101–50
Kuti, J., 33:611–44

L

Lach, J., 29:203–42
Lal, D., 33:505–37

583

Lande, K., 29:395–410
Lane, K., 28:387–499
Lanford, W. A., 34:435–60
Lattimer, J. M., 31:337–74
Lawson, J. D., 34:99–123
Lee-Franzini, J., 33:1–29
Lingenfelter, R. E., 32:235–69
Litherland, A. E., 30:437–73
Ludlam, T., 32:335–89

M

Mahaux, C., 29:1–31
Mark, J. C., 26:51–87
Matthiae, G., 26:385–456
Maurette, M., 26:319–50
McGrory, J. B., 30:383–436
Measday, D. F., 29:121–60
Meyerhof, W. E., 27:279–331
Miller, G. A., 29:121–60
Mills, F. E., 31:295–335
Montgomery, H. E., 33:383–452
Moretto, L. G., 34:189–245
Müller, B., 26:351–83
Murnick, D. E., 29:411–54
Myers, W. D., 32:309–34

N

Nagamiya, 34:155–87
Neumann, R. D., 29:283–312

O

Oeschger, H., 25:423–63

P

Palmer, R., 34:247–84
Pantell, R. H., 33:453–504
Payne, G. L., 34:403–33
Peck, C. W., 33:143–97
Pendleton, H. N., 30:543–81
Perkins, D. H., 34:1–52
Perl, M. L., 30:299–335
Pethick, C., 25:27–77
Picasso, E., 29:243–82
Pigford, T. H., 24:515–59
Pondrom, L., 29:203–42
Povh, B., 28:1–32
Preskill, J., 34:461–530

Primakoff, H., 31:145–92
Protopopescu, S. D., 29:339–93

Q

Quentin, P., 28:523–96

R

Ramaty, R., 32:235–69
Ramond, P., 33:31–66
Ramsey, N. F., 32:211–33
Randrup, J., 34:155–87
Reay, N. W., 33:539–68
Reedy, R. C., 33:505–37
Renton, P., 31:193–230
Rho, M., 34:531–82
Rolfs, C., 28:115–59
Rosen, S. P., 31:145–92
Rosenfeld, A. H., 25:555–98

S

Sak, J., 30:53–84
Sakurai, J. J., 31:375–438
Samios, N. P., 29:339–93
Sanford, J. R., 26:151–98
Saudinos, J., 24:341–77
Sauli, F., 34:285–349
Sauvage, G., 34:351–402
Scharff-Goldhaber, G., 26:239–317
Schramm, D. N., 27:37–74; 27:167–207
Schröder, W. U., 27:465–547
Schwitters, R. F., 26:89–149
Sciulli, F., 32:499–573
Seaborg, G. T., 27:139–66
Segrè, E., 31:1–18
Seki, R., 25:241–81
Selph, F. B., 27:353–92
Shifman, M. A., 33:199–233
Sidwell, R. A., 33:539–68
Silbar, R. R., 24:249–77
Simpson, J. A., 33:323–81
Sirlin, A., 24:379–449
Sivers, D., 32:149–75
Söding, P., 31:231–93
Sorba, P., 32:443–97
Speth, J., 32:65–115
Stanton, N. R., 33:539–68

Steigman, G., 29:313–37
Stephens, F. S., 30:85–157
Sternheim, M. M., 24:249–77
Strauch, K., 26:89–149
Swiatecki, W. J., 32:309–34
Symons, T. J. M., 34:155–187

T

Taulbjerg, K., 27:279–331
Taylor, T. B., 25:406–21
Tigner, M., 34:99–123
Tollestrup, A. V., 34:247–84
Trautmann, N., 32:117–47
Trautvetter, H. P., 28:115–59
Truran, J. W., 34:53–97
Tubbs, D. L., 27:167–207
Tubis, A., 24:69–100
Turner, M. S., 33:645–96

V

Vandenbosch, R., 27:1–35
Vogel, P., 34:125–53
Voss, G.-A., 33:67–104

W

Wagoner, R. V., 27:37–74
Wahlen, M., 25:423–63
Walecka, J. D., 25:329–405
Watt, R. D., 27:75–138
Weidenmüller, H. A., 29:1–31
Wetherill, G. W., 25:283–328
Wiegand, C. E., 25:241–81
Wilczek, F., 32:177–209
Wildenthal, B. H., 30:383–436
Wilkin, C., 24:341–77
Williams, W. S. C., 31:193–230
Winick, H., 28:33–113
Wiss, J. E., 30:337–81
Wolf, G., 31:231–93
Wozniak, G. J., 34:189–245

Y

Yan, T.-M., 26:199–238
Yodh, G. B., 30:475–542
Yoshida, S., 24:1–33

CHAPTER TITLES, VOLUMES 24–34

PREFATORY CHAPTER

Fifty Years Up and Down a Strenuous and Scenic Trail	E. Segrè	31:1–18

ACCELERATORS

The Fermi National Accelerator Laboratory	J. R. Sanford	26:151–98
Heavy-Ion Accelerators	H. A. Grunder, F. B. Selph	27:353–92
Synchrotron Radiation Research	H. Winick, A. Bienenstock	28:33–113
Ultrasensitive Mass Spectrometry with Accelerators	A. E. Litherland	30:437–73
The Advanced Positive Heavy Ion Sources	J. Arianer, R. Geller	31:19–51
Increasing the Phase-Space Density of High Energy Particle Beams	F. T. Cole, F. E. Mills	31:295–335
Inertial Confinement Fusion	D. Keefe	32:391–441
Progress and Problems in Performance of e^+e^- Storage Rings	R. D. Kohaupt, G.-A. Voss	33:67–104
The Physics of Particle Accelerators	J. D. Lawson, M. Tigner	34:99–123
Superconducting Magnet Technology for Accelerators	R. Palmer, A. V. Tollestrup	34:247–84

ASTROPHYSICS

Neutron Stars	G. Baym, C. Pethick	25:27–77
Element Production in the Early Universe	D. N. Schramm, R. V. Wagoner	27:37–74
The Weak Neutral Current and Its Effects in Stellar Collapse	D. Z. Freedman, D. N. Schramm, D. L. Tubbs	27:167–207
Experimental Nuclear Astrophysics	C. Rolfs, H. P. Trautvetter	28:115–59
Isotopic Anomalies in the Early Solar System	R. N. Clayton	28:501–22
Cosmology Confronts Particle Physics	G. Steigman	29:313–37
Experimental Neutrino Astrophysics	K. Lande	29:395–410
Baryon Number and Lepton Number Conservation Laws	H. Primakoff, S. P. Rosen	31:145–92
The Equation of State of Hot Dense Matter and Supernovae	J. M. Lattimer	31:337–74
Gamma-Ray Astronomy	R. Ramaty, R. E. Lingenfelter	32:235–69
Elemental and Isotopic Composition of the Galactic Cosmic Rays	J. A. Simpson	33:323–81
Cosmic-Ray Record in Solar System Matter	R. C. Reedy, J. R. Arnold, D. Lal	33:505–37
Grand Unified Theories and the Origin of the Baryon Asymmetry	E. W. Kolb, M. S. Turner	33:645–96
Nucleosynthesis	J. W. Truran	34:53–97
Magnetic Monopoles	J. Preskill	34:461–530

ATOMIC, MOLECULAR, AND SOLID STATE PHYSICS

Atomic Structure Effects in Nuclear Events	M. S. Freedman	24:209–47
Kaonic and Other Exotic Atoms	R. Seki, C. E. Wiegand	25:241–81
K-Shell Ionization in Heavy-Ion Collisions	W. E. Meyerhof, K. Taulbjerg	27:279–331
Synchrotron Radiation Research	H. Winick, A. Bienenstock	28:33–113
Advances in Muon Spin Rotation	J. H. Brewer, K. M. Crowe	28:239–326
The Parity Non-Conserving Electron-Nucleon Interaction	E. D. Commins, P. H. Bucksbaum	30:1–52
Proton Microprobes and Particle-Induced X-Ray Analytical Systems	T. A. Cahill	30:211–52

Positronium	S. Berko, H. N. Pendleton	30:543–81
Channeling Radiation	J. U. Andersen, E. Bonderup, R. H. Pantell	33:453–504
Superconducting Magnet Technology for Accelerators	R. Palmer, A. V. Tollestrup	34:247–84
Nuclear Reaction Techniques in Materials Analysis	G. Amsel, W. A. Lanford	34:435–60

BIOLOGY AND MEDICINE

Environmental Aspects of Nuclear Energy Production	T. H. Pigford	24:515–59
Synchrotron Radiation Research	H. Winick, A. Bienenstock	28:33–113
Intense Sources of Fast Neutrons	H. H. Barschall	28:207–37
Diagnostic Techniques in Nuclear Medicine	R. D. Neumann, A. Gottschalk	29:283–313

CHEMISTRY

Chemistry of the Transactinide Elements	O. L. Keller Jr., G. T. Seaborg	27:139–66
Advances in Muon Spin Rotation	J. H. Brewer, K. M. Crowe	28:239–326
Rapid Chemical Methods for Identification and Study of Short-Lived Nuclides	G. Herrmann, N. Trautmann	32:117–47

EARTH AND SPACE SCIENCES

Radiometric Chronology of the Early Solar System	G. W. Wetherill	25:283–328
Fossil Nuclear Reactors	M. Maurette	26:319–50
Isotopic Anomalies in the Early Solar System	R. N. Clayton	28:501–22
Cosmology Confronts Particle Physics	G. Steigman	29:313–37
Experimental Neutrino Astrophysics	K. Lande	29:395–410
Particle Collisions Above 10 TeV As Seen By Cosmic Rays	T. K. Gaisser, G. B. Yodh	30:475–542
Elemental and Isotopic Composition of the Galactic Cosmic Rays	J. A. Simpson	33:323–81
Cosmic-Ray Record in Solar System Matter	R. C. Reedy, J. R. Arnold, D. Lal	33:505–37

FISSION AND FUSION ENERGY

Environmental Aspects of Nuclear Energy Production	T. H. Pigford	24:515–59
Nuclear Safeguards	T. B. Taylor	24:407–21
Fossil Nuclear Reactors	M. Maurette	26:319–50
Inertial Confinement Fusion	D. Keefe	32:391–441

INSTRUMENTATION AND TECHNIQUES

Identification of Nuclear Particles	F. S. Goulding, B. G. Harvey	25:167–240
Low Level Counting Techniques	H. Oeschger, M. Wahlen	25:423–63
Blocking Measurements of Nuclear Decay Times	W. M. Gibson	25:465–508
Computer Interfacing for High-Energy Physics Experiments	F. A. Kirsten	25:509–54
The Particle Data Group: Growth and Operations—Eighteen Years of Particle Physics	A. H. Rosenfield	25:555–98
The Fermi National Accelerator Laboratory	J. R. Sanford	26:151–98
Hybrid Bubble-Chamber Systems	J. Ballam, R. D. Watt	27:75–138
Synchrotron Radiation Research	H. Winick, A. Bienenstock	28:33–113
Intense Sources of Fast Neutrons	H. H. Barschall	28:207–37
Advances in Muon Spin Rotation	J. H. Brewer, K. M. Crowe	28:239–326
Nuclei Far Away from the Line of Beta Stability: Studies by On-Line Mass Separation	P. G. Hansen	29:69–119
Hyperon Beams and Physics	J. Lach, L. Pondrom	29:203–42
The Muon $(g-2)$ Experiments	F. J. M. Farley, E. Picasso	29:243–82
Diagnostic Techniques in Nuclear Medicine	R. D. Neumann, A. Gottschalk	29:283–312
Applications of Lasers to Nuclear Physics	D. E. Murnick, M. S. Feld	29:411–54

Transient Magnetic Fields at Swift Ions Traversing Ferromagnetic Media and Applications to Measurements of Nuclear Moments	N. Benczer-Koller, M. Hass, J. Sak	30:53–84
Proton Microprobes and Particle-Induced X-Ray Analytical Systems	T. A. Cahill	30:211–52
Relativistic Charged Particle Identification by Energy Loss	W. W. M. Allison, J. H. Cobb	30:253–98
Ultrasensitive Mass Spectrometry with Accelerators	A. E. Litherland	30:437–73
The Advanced Positive Heavy Ion Sources	J. Arianer, R. Geller	31:19–51
Increasing the Phase-Space Density of High Energy Particle Beams	F. T. Cole, F. E. Mills	31:295–335
Beam Foil Spectroscopy	H. G. Berry, M. Hass	32:1–34
Rapid Chemical Methods for Identification and Study of Short-Lived Nuclides	G. Herrmann, N. Trautmann	32:117–47
Electric-Dipole Moments of Particles	N. F. Ramsey	32:211–33
Calorimetry in High-Energy Physics	C. W. Fabjan, T. Ludlam	32:335–89
Inertial Confinement Fusion	D. Keefe	32:391–441
Progress and Problems in Performance of e^+e^- Storage Rings	R. D. Kohaupt, G.-A. Voss	33:67–104
Physics with the Crystal Ball Detector	E. D. Bloom, C. W. Peck	33:143–97
Channeling Radiation	J. U. Andersen, E. Bonderup, R. H. Pantell	33:453–504
Superconducting Magnet Technology for Accelerators	R. Palmer, A. V. Tollestrup	34:247–84
High-Resolution Electronic Particle Detectors	G. Charpak, F. Sauli	34:285–349
Nuclear Reaction Techniques in Materials Analysis	G. Amsel, W. A. Lanford	34:435–60

NUCLEAR APPLICATIONS

Environmental Aspects of Nuclear Energy Production	T. H. Pigford	24:515–59
Applications of Nuclear Science in Crime Investigation	V. P. Guinn	24:561–91
Status and Future Directions of the World Program in Fusion Research and Development	R. L. Hirsch	25:79–121
Nuclear Safeguards	T. B. Taylor	25:407–21
Global Consequences of Nuclear Weaponry	J. C. Mark	26:51–87
Intense Sources of Fast Neutrons	H. H. Barschall	28:207–37
Diagnostic Techniques in Nuclear Medicine	R. D. Neumann, A. Gottschalk	29:283–312
Proton Microprobes and Particle-Induced X-Ray Analytical Systems	T. A. Cahill	30:211–52
Ultrasensitive Mass Spectrometry with Accelerators	A. E. Litherland	30:437–73
The Equation of State of Hot Dense Matter and Supernovae	J. M. Lattimer	31:337–74
Nucleosynthesis	J. W. Truran	34:53–97
Nuclear Reaction Techniques in Materials Analysis	G. Amsel, W. A. Lanford	34:435–60

NUCLEAR REACTION MECHANISMS—HEAVY PARTICLES

Reactions Between Medium and Heavy Nuclei and Heavy Ions of Less than 15 MeV/amu	A. Fleury, J. M. Alexander	24:279–339
Positron Creation in Superheavy Quasi-Molecules	B. Müller	26:351–83
Heavy-Ion Accelerators	H. A. Grunder, F. B. Selph	27:353–92
Damped Heavy-Ion Collisions	W. U. Schröder, J. R. Huizenga	27:465–547
High Energy Interactions of Nuclei	A. S. Goldhaber, H. H. Heckman	28:161–205
Nuclei at High Angular Momentum	R. M. Diamond, F. S. Stephens	30:85–157
Polarization in Heavy Ion Reactions	D. Fick	31:53–74
Resonances in Heavy-Ion Nuclear Reactions	T. M. Cormier	32:271–308

Fusion Reactions Between Heavy Nuclei	J. R. Birkelund, J. R. Huizenga	33:265–322
Nuclear Collisions at High Energies	S. Nagamiya, J. Randrup, T. J. M. Symons	34:155–87
The Role of Rotational Degrees of Freedom in Heavy-Ion Collisions	L. G. Moretto, G. J. Wozniak	34:189–245

NUCLEAR REACTION MECHANISMS—LIGHT PARTICLES

Meson-Nucleus Scattering at Medium Energies	M. M. Sternheim, R. R. Silbar	24:249–77
Proton-Nucleus Scattering at Medium Energies	J. Saudinos, C. Wilkin	24:341–77
Knock-Out Processes and Removal Energies	A. E. L. Dieperink, T. De Forest Jr.	25:1–26
Preequilibrium Decay	M. Blann	25:123–66
Excitation of Giant Multipole Resonances through Inelastic Scattering	F. E. Bertrand	26:457–509
Element Production in the Early Universe	D. N. Schramm, R. V. Wagoner	27:37–74
Experimental Nuclear Astrophysics	C. Rolfs, H. P. Trautvetter	28:115–59
High Energy Interactions of Nuclei	A. S. Goldhaber, H. H. Heckman	28:161–205
Intense Sources of Fast Neutrons	H. H. Barschall	28:207–37
Nuclear Physics with Polarized Beams	C. Glashausser	29:33–68
Hopes and Realities for the (p, π) Reaction	D. F. Measday, G. A. Miller	29:121–60
Alpha Transfer Reactions in Light Nuclei	H. W. Fulbright	29:161–202
Theory of Giant Resonances	K. Goeke, J. Speth	32:65–115
Rapid Chemical Methods for Identification and Study of Short-Lived Nuclides	G. Herrmann, N. Trautmann	32:117–47
Inelastic Electron Scattering from Nuclei	J. Heisenberg, H. P. Blok	33:569–609
Nucleosynthesis	J. W. Truran	34:53–97
Pion Interactions Within Nuclei	M. Rho	34:531–82

NUCLEAR STRUCTURE

Resonance Fluorescence of Excited Nuclear Levels in the Energy Range 5–11 MeV	B. Arad, G. Ben-David	24:35–67
Shell-Model Effective Interactions and the Free Nucleon-Nucleon Interaction	T. T. S. Kuo	24:107–50
Post-Fission Phenomena	D. C. Hoffman, M. M. Hoffman	24:151–207
Knock-Out Processes and Removal Energies	A. E. L. Dieperink, T. De Forest Jr.	25:1–26
Preequilibrium Decay	M. Blann	25:123–66
Electron Scattering and Nuclear Structure	T. W. Donnelly, J. D. Walecka	25:329–405
Blocking Measurements of Nuclear Decay Times	W. M. Gibson	25:465–508
The Variable Moment of Inertia (VMI) Model and Theories of Nuclear Collective Motion	G. Scharff-Goldhaber, C. B. Dover, A. L. Goodman	26:239–317
Excitation of Giant Multipole Resonances through Inelastic Scattering	F. E. Bertrand	26:457–509
Spontaneously Fissioning Isomers	R. Vandenbosch	27:1–35
Delayed Proton Radioactivities	J. Cerny, J. C. Hardy	27:333–51
Hypernuclei	B. Povh	28:1–32
Self-Consistent Calculations of Nuclear Properties with Phenomenological Effective Forces	P. Quentin, H. Flocard	28:523–96
Nuclear Physics with Polarized Beams	G. Glashausser	29:33–68
Nuclei Far Away from the Line of Beta Stability: Studies by On-Line Mass Separation	P. G. Hansen	29:69–119
Hopes and Realities for the (p, π) Reaction	D. F. Measday, G. A. Miller	29:121–60
Alpha Transfer Reactions in Light Nuclei	H. W. Fulbright	29:161–202
Applications of Lasers to Nuclear Physics	D. E. Murnick, M. S. Feld	29:411–54
Transient Magnetic Fields at Swift Ions Traversing Ferromagnetic Media and Applications to Measurements of Nuclear Moments	N. Benczer-Koller, M. Haas, J. Sak	30:53–84
Nuclei at High Angular Momentum	R. M. Diamond, F. S. Stephens	30:85–157
Large-Scale Shell-Model Calculations	J. B. McGrory, B. H. Wildenthal	30:383–436
Polarization in Heavy Ion Reactions	D. Fick	31:53–74

The Interacting Boson Model	A. Arima, F. Iachello	31:75–105
Beam Foil Spectroscopy	H. G. Berry, M. Hass	32:1–34
Statistical Spectroscopy	J. B. French, V. K. B. Kota	32:35–64
Theory of Giant Resonances	K. Goeke, J. Speth	32:65–115
Resonances in Heavy-Ion Nuclear Reactions	T. M. Cormier	32:271–308
The Macroscopic Approach to Nuclear Masses and Deformations	W. D. Myers, W. J. Swiatecki	32:309–34
Fusion Reactions Between Heavy Nuclei	J. R. Birkelund, J. R. Huizenga	33:265–322
Inelastic Electron Scattering from Nuclei	J. Heisenberg, H. P. Blok	33:569–609
The Role of Rotational Degrees of Freedom in Heavy-Ion Collisions	L. G. Moretto, G. J. Wozniak	34:189–245
Recent Progress in Understanding Trinucleon Properties	J. L. Friar, B. F. Gibson, G. L. Payne	34:403–33
Pion Interactions Within Nuclei	M. Rho	34:531–82

NUCLEAR THEORY

Time Description of Nuclear Reactions	S. Yoshida	24:1–33
The Theory of Three-Nucleon Systems	Y. E. Kim, A. Tubis	24:69–100
Shell-Model Effective Interactions and the Free Nucleon-Nucleon Interaction	T. T. S. Kuo	24:101–50
Post-Fission Phenomena	D. C. Hoffman, M. M. Hoffman	24:151–207
Meson-Nucleus Scattering at Medium Energies	M. M. Sternheim, R. R. Silbar	24:249–77
Proton-Nucleus Scattering at Medium Energies	J. Saudinos, C. Wilkin	24:341–77
Neutron Stars	G. Baym, C. Pethick	25:27–77
Electron Scattering and Nuclear Structure	T. W. Donnelly, J. D. Walecka	25:329–405
The Variable Moment of Inertia (VMI) Model and Theories of Nuclear Collective Motion	G. Scharff-Goldhaber, C. B. Dover, A. L. Goodman	26:239–317
Element Production in the Early Universe	D. N. Schramm, R. V. Wagoner	27:37–74
Self-Consistent Calculations of Nuclear Properties with Phenomenological Effective Forces	P. Quentin, H. Flocard	28:523–96
Recent Developments in Compound-Nucleus Theory	C. Mahaux, H. A. Weidenmüller	29:1–31
Large-Scale Shell-Model Calculations	J. B. McGrory, B. H. Wildenthal	30:383–436
The Interacting Boson Model	A. Arima, F. Iachello	31:75–105
The Equation of State of Hot Dense Matter and Supernovae	J. M. Lattimer	31:337–74
Statistical Spectroscopy	J. B. French, V. K. B. Kota	32:35–64
Theory of Giant Resonances	K. Goeke, J. Speth	32:65–115
The Macroscopic Approach to Nuclear Masses and Deformations	W. D. Myers, W. J. Swiatecki	32:309–34
Nuclear Matter Theory: A Status Report	A. D. Jackson	33:105–41
Grand Unified Theories and the Origin of the Baryon Asymmetry	E. W. Kolb, M. S. Turner	33:645–96
Nucleosynthesis	J. W. Truran	34:53–97
Recent Progress in Understanding Trinucleon Properties	J. L. Friar, B. F. Gibson, G. L. Payne	34:403–33
Pion Interactions Within Nuclei	M. Rho	34:531–82

PARTICLE INTERACTIONS AT HIGH ENERGIES

Meson-Nucleus Scattering at Medium Energies	M. M. Sternheim, R. R. Silbar	24:249–77
Proton-Nucleus Scattering at Medium Energies	J. Saudinos, C. Wilkin	24:341–77
Inclusive Reactions	H. Bøggild, T. Ferbel	24:451–513
The Physics of e^+e^- Collisions	R. F. Schwitters, K. Strauch	26:89–149
The Fermi National Accelerator Laboratory	J. R. Sanford	26:151–98
The Parton Model	T.-M. Yan	26:199–238
Diffraction of Hadronic Waves	U. Amaldi, M. Jacob, G. Matthiae	26:385–456
Neutrino Scattering and New-Particle Production	D. Cline, W. F. Fry	27:209–78
Psionic Matter	W. Chinowsky	27:393–464

Charm and Beyond	T. Appelquist, R. M. Barnett, K. Lane	28:387–499
Hopes and Realities for the (p, π) Reaction	D. F. Measday, G. A. Miller	29:121–60
Hyperon Beams and Physics	J. Lach, L. Pondrom	29:203–42
Large Transverse Momentum Hadronic Processes	P. Darriulat	30:159–210
The Tau Lepton	M. L. Perl	30:299–335
Charmed Mesons Produced in e^+e^- Annihilation	G. Goldhaber, J. E. Wiss	30:337–81
Particle Collisions Above 10 TeV As Seen By Cosmic Rays	T. K. Gaisser, G. B. Yodh	30:475–542
High Energy Physics with Polarized Proton Beams	R. C. Fernow, A. D. Krisch	31:104–44
Hadron Production in Lepton-Nucleon Scattering	P. Renton, W. S. C. Williams	31:193–230
Experimental Evidence on QCD	P. Söding, G. Wolf	31:231–93
The Structure of Neutral Currents	P. Q. Hung, J. J. Sakurai	31:375–438
What Can We Count On? A Discussion of Constituent-Counting Rules for Interactions Involving Composite Systems	D. Sivers	32:149–75
Calorimetry in High-Energy Physics	C. W. Fabjan, T. Ludlam	32:335–89
Charged-Current Neutrinonteractions	H. E. Fisk, F. Sciulli	32:499–573
Upsilon Resonances	P. Franzini, J. Lee-Franzini	33:1–29
Progress and Problems in Performance of e^+e^- Storage Rings	R. D. Kohaupt, G.-A. Voss	33:67–104
Physics with the Crystal Ball Detector	E. D. Bloom, C. W. Peck	33:143–97
Sum Rule Approach to Heavy Quark Spectroscopy	M. A. Shifman	33:199–233
Muon Scattering	J. Drees, H. E. Montgomery	33:383–452
Cosmic-Ray Record in Solar System	R. C. Reedy, J. R. Arnold, D. Lal	33:505–37
Measurement of Charmed Particle Lifetimes	R. A. Sidwell, N. W. Reay, N. R. Stanton	33:539–68
Internal Spin Structure of the Nucleon	V. W. Hughes, J. Kuti	33:611–44
Grand Unified Theories and the Origin of the Baryon Asymmetry	E. W. Kolb, M. S. Turner	33:645–96
Nuclear Collisions at High Energies	S. Nagamiya, J. Randrup, T. J. M. Symons	34:155–87
High-Resolution Electronic Particle Detectors		34:285–349
Hyperon Beta Decays	J.-M. Gaillard, G. Sauvage	34:351–402
Magnetic Monopoles	J. Preskill	34:461–530

PARTICLE SPECTROSCOPY

Kaonic and Other Exotic Atoms	R. Seki, C. E. Wiegand	25:241–81
The Particle Data Group: Growth and Operations—Eighteen Years of Particle Physics	A. H. Rosenfeld	25:555–98
The Physics of e^+e^- Collisions	R. F. Schwitters, K. Strauch	26:89–149
Psionic Matter	W. Chinowsky	27:393–464
Quarks	O. W. Greenberg	28:327–86
Charm and Beyond	T. Appelquist, R. M. Barnett, K. Lane	28:387–499
Light Hadronic Spectroscopy: Experimental and Quark Model Interpretations	S. D. Protopopescu, N. P. Samios	29:339–93
The Tau Lepton	M. L. Perl	30:299–335
Charmed Mesons Produced in e^+e^- Annihilation	G. Goldhaber, J. E. Wiss	30:337–81
Experimental Evidence on QCD	P. Söding, G. Wolf	31:231–93
Upsilon Resonances	P. Franzini, J. Lee-Franzini	33:1–29
Physics with the Crystal Ball Detector	E. D. Bloom, C. W. Peck	33:143–97
Sum Rule Approach to Heavy Quark Spectroscopy	M. A. Shifman	33:199–233
Bag Models of Hadrons	C. E. DeTar, J. F. Donoghue	33:235–64

Measurement of Charmed Particle Lifetimes	R. A. Sidwell, N. W. Reay, N. R. Stanton	33:539–68

PARTICLE THEORY

Proton-Nucleus Scattering at Medium Energies	J. Saudinos, C. Wilkin	24:341–77
Gauge Theories of Weak Interactions (Circa 1973–74 C. E.)	M. A. B. Bég, A. Sirlin	24:379–449
The Parton Model	T.-M. Yan	26:199–238
Diffraction of Hadronic Waves	U. Amaldi, M. Jacob, G. Matthiae	26:385–456
Quarks	O. W. Greenberg	28:327–86
Charm and Beyond	T. Applequist, R. M. Barnett, K. Lane	28:387–499
Light Hadronic Spectroscopy: Experimental and Quark Model Interpretations	S. D. Protopopescu, N. P. Samios	29:339–93
Baryon Number and Lepton Number Conservation Laws	H. Primakoff, S. P. Rosen	31:145–92
Experimental Evidence on QCD	P. Söding, G. Wolf	31:231–93
The Structure of Neutral Currents	P. Q. Hung, J. J. Sakurai	31:375–438
What Can We Count On? A Discussion of Constituent-Counting Rules for Interactions Involving Composite Systems	D. Sivers	32:149–75
Quantum Chromodynamics: The Modern Theory of the Strong Interaction	F. Wilczek	32:177–209
Physics of Intermediate Vector Bosons	J. Ellis, M. K. Gaillard, G. Girardi, P. Sorba	32:443–97
Gauge Theories and Their Unification	P. Ramond	33:31–66
Sum Rule Approach to Heavy Quark Spectroscopy	M. A. Shifman	33:199–233
Bag Models of Hadrons	C. E. DeTar, J. F. Donoghue	33:235–64
Internal Spin Structure of the Nucleon	V. W. Hughes, J. Kuti	33:611–44
Grand Unified Theories and the Origin of the Baryon Asymmetry	E. W. Kolb, M. S. Turner	33:645–96
Magnetic Monopoles	J. Preskill	34:461–530

RADIATION EFFECTS

K-Shell Ionization in Heavy-Ion Collisions	W. E. Meyerhof, K. Taulbjerg	27:279–331
Synchrotron Radiation Research	H. Winick, A. Bienenstock	28:33–133
Calorimetry in High-Energy Physics	C. W. Fabjan, T. Ludlam	32:335–89
Channeling Radiation	J. U. Andersen, E. Bonderup, R. H. Pantell	33:453–504

WEAK AND ELECTROMAGNETIC INTERACTIONS

Atomic Structure Effects in Nuclear Events	M. S. Freedman	24:209–47
Gauge Theories of Weak Interactions (Circa 1973–74 C. E.)	M. A. B. Bég, A. Sirlin	24:379–449
CP Violation and K^0 Decays	K. Kleinknecht	26:1–50
The Weak Neutral Current and Its Effects in Stellar Collapse	D. Z. Freedman, D. N. Schramm, D. L. Tubbs	27:167–207
Neutrino Scattering and New-Particle Production	D. Cline, W. F. Fry	27:209–78
Hypernuclei	B. Povh	28:1–32
Charm and Beyond	T. Applequist, R. M. Barnett, K. Lane	28:387–499
Hyperon Beams and Physics	J. Lach, L. Pondrom	29:203–42
The Muon $(g-2)$ Experiments	F. J. M. Farley, E. Picasso	29:243–82
Cosmology Confronts Particle Physics	G. Steigman	29:313–37
Experimental Neutrino Astrophysics	K. Lande	29:395–410
The Parity Non-Conserving Electron-Nucleon Interaction	E. D. Commins, P. H. Bucksbaum	30:1–52
The Tau Lepton	M. L. Perl	30:299–335
Charmed Mesons Produced in e^+e^- Annihilation	G. Goldhaber, J. E. Wiss	30:337–81

Positronium	S. Berko, H. N. Pendleton	30:543–81
Baryon Number and Lepton Number Conservation Laws	H. Primakoff, S. P. Rosen	31:145–92
The Structure of Neutral Currents	P. Q. Hung, J. J. Sakurai	31:375–438
Electric-Dipole Moments of Particles	N. F. Ramsey	32:211–33
Physics of Intermediate Vector Bosons	J. Ellis, M. K. Gaillard, G. Girardi, P. Sorba	32:443–97
Charged-Current Neutrino Interactions	H. E. Fisk, F. Sciulli	32:499–573
Muon Scattering	J. Drees, H. E. Montgomery	33:383–452
Measurement of Charmed Particle Lifetimes	R. A. Sidwell, N. W. Reay, N. R. Stanton	33:539–68
Inelastic Electron Scattering from Nuclei	J. Heisenberg, H. P. Blok	33:569–609
Internal Spin Structure of the Nucleon	V. W. Hughes, J. Kuti	33:611–44
Proton Decay Experiments	D. H. Perkins	34:1–52
Low-Energy Neutrino Physics and Neutrino Mass	F. Boehm, P. Vogel	34:125–53
Hyperon Beta Decays	J.-M. Gaillard, G. Sauvage	34:351–402
Magnetic Monoples	J. Preskill	34:461–530